Lecture Notes in Production Engineering

Lecture Notes in Production Engineering (LNPE) is book series that reports the latest research and developments in Production Engineering, comprising:

- Biomanufacturing
- Control and Management of Processes
- Cutting and Forming
- Design
- Life Cycle Engineering
- Machines and Systems
- Optimization
- Precision Engineering and Metrology
- Surfaces

LNPE publishes authored conference proceedings, contributed volumes and authored monographs that present cutting-edge research information as well as new perspectives on classical fields, while maintaining Springer's high standards of excellence. Also considered for publication are lecture notes and other related material of exceptionally high quality and interest. The subject matter should be original and timely, reporting the latest research and developments in all areas of production engineering. The target audience of LNPE consists of advanced level students, researchers, as well as industry professionals working at the forefront of their fields. Much like Springer's other Lecture Notes series, LNPE will be distributed through Springer's print and electronic publishing channels.

Indexed in Scopus *Indexed in EI Compendex* To submit a proposal or request further information please contact Anthony Doyle, Executive Editor, Springer (anthony.doyle@springer.com).

Konrad Wegener · Markus Bambach
Editors

4th International Conference on Thermal Issues in Machine Tools (ICTIMT2025)

 Springer

Editors
Konrad Wegener
Department of Machine Tools
and Manufacturing, Inspire AG
ETH Zürich
Zurich, Switzerland

Markus Bambach
Department of Mechanical and Process
Engineering
ETH Zürich
Zurich, Switzerland

ISSN 2194-0525   ISSN 2194-0533 (electronic)
Lecture Notes in Production Engineering
ISBN 978-3-032-01193-0   ISBN 978-3-032-01194-7 (eBook)
https://doi.org/10.1007/978-3-032-01194-7

This work was supported by Prof. Markus Bambach.

This Springer imprint is published by the registered company Springer Nature Switzerland AG
The registered company address is: Gewerbestrasse 11, 6330 Cham, Switzerland

If disposing of this product, please recycle the paper.

Preface

Machining is energy conversion. All energy supplied to the machine tool ends in heat, also the intentionally provided energy to the process zone. The only exception is the tiny amount incorporated in the state change of the material as workpiece and chips. Understanding the heat conversion and flow in the process zone is of crucial importance for the correct setup of the respective process, for the wear of the tool, the process signature onto the material and for the operation of the machine tool. Machining processes therefore cannot be discussed, the process result not predicted without taking the energy conversion and the development of temperature as the most important state variable into account. Temperature and the temperature time profile on the material decide on the state and microstructure of the material and the properties and quality of the manufactured part. A significant effort in machining is therefore dedicated to the attempt to control the temperature in the process zone by cooling but still more efficient by correct process setup, which requires besides experience a good deal of simulation technology.

From the energy release in the process zone but also from necessary drives and the environment, a manufacturing machine is influenced.

Thermal distortion and thermal errors in machine tools are the most frequent reasons for defective parts. When the knowledge- and technology transfer organisation inspire AG started to grow in 2002 and IWF of ETH Zürich restarted, the most urgent whish of Swiss machine tool manufacturers according to an industry survey was to develop means against thermal distortions of machines. CIRP colleague Jim Brian stated "whatever you design, it becomes a thermometer". The next milestone in this topic was the CIRP-keynote paper of 2012 on "thermal issues in machine tools" prepared by a large scientific group since 2009, collecting the relevant state of the art at that time and initiated a lot of research throughout the world. The International Conference on Thermal Issues in Machine Tools, ICTIMT, a conference series having taken place in Dresden, Prague, Dresden, and now in Zurich was launched out of the SFB TR 96 funded by the German research foundation (DFG) by the involved scientists and as a forum to discuss findings and developed methods both for the heat release in the process zone and the thermal influence on machine tools.

As the physics of heat transfer and thermal elongation is invincible, three different approaches remain: thermo-energetically optimized design, tempering of shop floors and machine tools and thermal compensation using the axes of the machine and as required auxiliary compensation axes with small strokes. The first is generally recommended, whatever comes afterwards. Tempering is straightforward and therefore the most frequently applied method, but also the least sustainable one, as it consumes energy and/or requires startup time partly even with scrap production to reach the thermal equilibrium due to the inbuilt heat sources. Compensation especially while using the machine axes is the most elegant way for better accuracy.

Thermal compensation replaces resource expenditure by a knowledge-based approach, which was the general guideline for the EU Horizon2020 program. Thermal compensation is one of the rare cases, where sustainability and increase of accuracy coincide. General outline of compensation methods is based on three different functions and can thus be described with a morphological box: information supply by sensors and the CNC-program, control system based on a machine model and actuation like machine axes or auxiliary heat sources and sinks. The measurement of thermal displacements is a challenge as such, as they are superposed with large strokes of the machine tool. But also the isolation and identification of individual errors are a task to be solved. An important topic today is the influence of fluidic media and thus of hollow spaces especially but not only the working area with the influences of process and process cooling. Modelling for compensation requires sufficient precise models that are capable to deliver correction data in real time. This then is the domain of meta models, which today are frequently based on artificial intelligence (AI) in terms of self-learning compensation models. Physics-supported AI in most cases provides the best tradeoff between effort and success.

For both topical blocks machine and process CFD simulation becomes important and still suffers from the bad predictability of heat transfer coefficients to the machine body for the modelling of the machine and to tool, workpiece and fluid for the simulation of the process behavior. Careful measurements are crucial technologies and in most cases need to be supported by modelling and simulation because the accessibility of the process zone or the area of interaction is limited.

With all efforts in compensation of manufacturing machines only deterministic errors can be compensated. This emphasizes that a design for compensation must generate machine tools with excellent repeatability as primary goal.

Increasing the repeatability, increasing the predictability by enhancing models and simulation approaches and enhancing the measurability of thermal deviations as well as tracing them back to their origin are the tasks for future progress in the field to achieve still higher accuracy of machining results.

A vital role plays the cooperation between industry and academia in this topic. Provide understanding and explore possible approaches is the domain of academia, implementation, integration of sensors and providing powerful interfaces to the control as well as long term experience under real industry conditions the one of the industry. To promote and deepen this collaboration are one of the most important goals of the ICTIMT 2025 conference

We would like to sincerely thank all scientists and industrial partners, the CIRP conformance committee, the scientific committee, having all contributed to the ICTIMT2025 in Zurich.

June 2025

Markus Bambach
Advanced Manufacturing Lab Zurich
(AMLZ)
Department of Mechanical and Process
Engineering, ETH Zürich
Zurich, Switzerland

Josef Mayr
Zurich, Switzerland

Konrad Wegener
Department of Machine Tools
and Manufacturing
Inspire AG
Zurich, Switzerland

Presentation of CIRP

CIRP was founded in 1951 with the aim to address scientifically, through international co-operation, issues related to modern production science and technology. The International Academy for Production Engineering takes its abbreviated name from the French acronym of College International pour la Recherche en Productique (CIRP) and includes some 500 members from 50 countries. The number of members is intentionally kept limited, so as to facilitate informal scientific information exchange and personal contacts.

CIRP has some 180 Fellows and Honorary Fellows who are internationally recognized scientists elected to be CIRP members for life. Due to the limited number of CIRP Fellows, the election of a Fellow is a lengthy, rigorous process ensuring the highest possible academic standards.

CIRP includes some 150 Associate members well known scientists, with high potential, elected typically for a period of three years with the possibility of renewal. A number of Associate members eventually become Fellows. Some Associate members may also belong to fields related to Manufacturing.

CIRP, although an academic organization, encourages the participation of industry in its activities. There are some 150 Corporate members who follow the research work of the academic members of CIRP and very often contribute to the information exchange within CIRP by presenting their views on industrial needs and perspectives.

Young Research Affiliates and some Invited members, particularly from countries not yet involved in CIRP, are also included in the CIRP community.

In a recent development, there is work under way to establish a CIRP Network of young scientists active in manufacturing research.

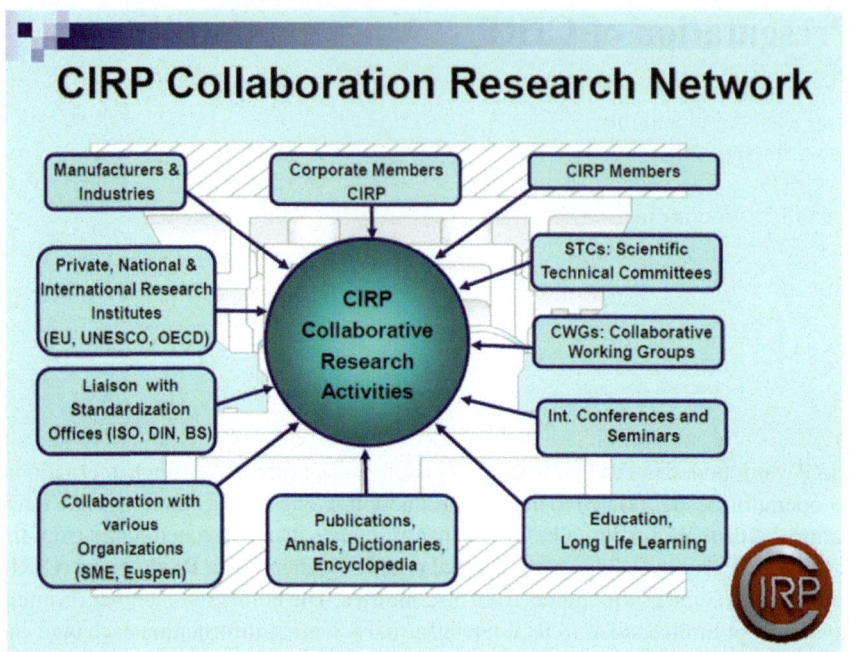

CIRP is organized along the lines of a number of Scientific Technical Committees (STCs) and Working Groups (WGs), covering many areas. CIRP organizes annually a General Assembly and the so-called Winter Meetings. In the General Assembly (GA), which lasts for a week, there is an intensive technical program with over 140 technical paper presentations from different fields of manufacturing, a number of keynote papers, at the opening of the conference, as well as technical work within the STCs.

In parallel, there is a social program, aiming at making the culture of the General Assembly site known to the members and also at creating an informal environment for information exchange among the members. The winter meetings are always organized in Paris and last three days. Moreover CIRP organizes, through its membership, a number of conferences, notably the Manufacturing Systems Seminar and a number of other conferences with relevant topics. CIRP members also organize a variety of conferences, under the sponsorship of CIRP.

The main publications of CIRP are the CIRP Annals under ISI standards with two volumes; Volume I, with refereed papers presented in the GA by Fellows, Associate, Corporate and Invited members and Volume II with refereed keynote papers. Then there is the Journal of Manufacturing Science and Technology, with four volumes. There are also proceedings of CIRP conferences available online on Procedia-CIRP, tri-lingual dictionaries of production engineering and their Encyclopedia of Production Engineering, etc. A Newsletter is published twice a year. Currently the *CIRP Annals* and the *Journal of Manufacturing Science and Technology* are published by

Elsevier, while Springer Verlag publishes the Dictionaries of Production Engineering and the Encyclopedia of Production Engineering.

The CIRP organization includes besides the President, who is elected annually, the Council and a number of other committees ensuring a continuous improvement and reflecting the changing needs of manufacturing science and technology.

CIRP has its headquarters in Paris, staffed by permanent personnel and welcomes potential corporate members and interested parties in CIRP publication and activities in general.

CIRP Scientific Technical Committees

Ten Scientific and Technical Committees (STCs) that are responsible for their subject areas of interest and for coordinating CIRP run collaborative research projects.

Additionally, working groups are created to investigate the special problems of education and training associated with new manufacturing technology.

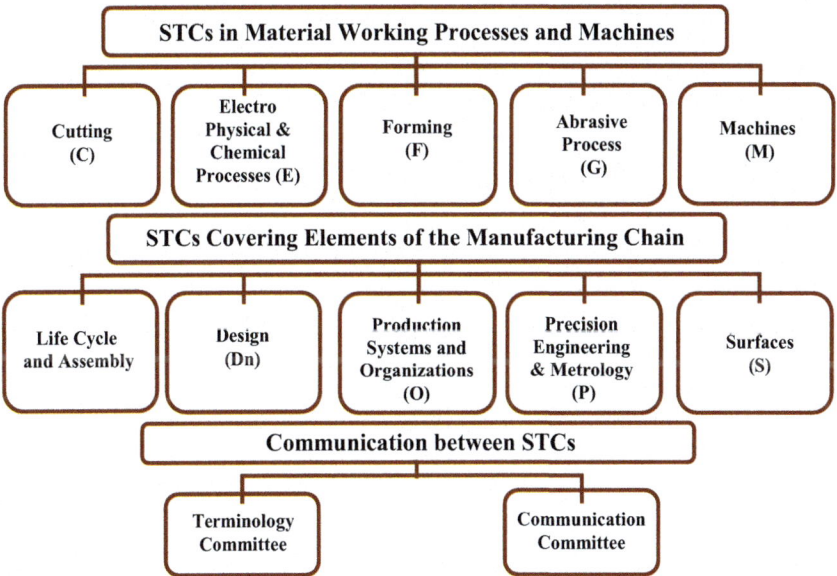

Editorial Committee ICTIMT2025

International Scientific Committee
Dr. Gorka Aguirre, Spain
Prof. Dr. Andreas Archenti, Sweden
Prof. Dr.-Ing. Markus Bambach, Switzerland
Prof. Dr. Friedrich Bleicher, Austria
Prof. Dr.-Ing. Christian Brecher, Germany
Prof. Dr. Erhan Budak, Turkey
Berend Denkena, Germany
Prof. Dr.-Ing. Steffen Ihlenfeldt, Germany
Prof. Dr.-Ing. Martin Dix, Germany
Prof. Dr. Alkan Donmez, United States of America
Dr. Otakar Horejs, Czech Republic
Prof. Dr. Soichi Ibaraki, Japan
Prof. Dr. Jerzy Jedrzejewski, Poland
Prof. Dr. Yasuhiro Kakinuma, Japan
Dr. Petr Kolar, Czech Republic
Prof. Dr. Daisuke Kono, Japan
Dr. Sebastian Lang, Switzerland
Dr. Martin Mares, Czech Republic
Prof. Dr. Atsushi Matsubara, Japan
Prof. Dr. Réné Mayer, Canada
Dr. Josef Mayr, Switzerland
Prof. Dr.-Ing. Hans-Christian Möhring, Germany
Dr.-Ing. Lars Penter, Germany
Dr. Mathieu Ritou, France
Prof. Dr.-Ing Masakazu Soshi, United States of America
Prof. Dr. Matej Sulitka, Czech Republic
Prof. Dr.-Ing. Konrad Wegener, Switzerland
Prof. Dr.-Ing. Matthias Weigold, Germany
Prof. Dr.-Ing. Michael Zäh, Germany

Contents

Process-Integrated Thermal Behavior in Manufacturing

Estimation of Thermal Boundary Conditions in Rotating Bearing by Data Assimilation

Ayato Ishigaki and Daisuke Kono

Abstract As machine tools become faster and more precise, thermal simulation is becoming increasingly important in spindle design and in understanding the operating conditions. Thermal simulation requires appropriate thermal boundary conditions. To set boundary conditions, the relationship between the spindle temperature distribution and parameters of bearing such as thermal contact conductance, heat generation and preload is particularly important. However, it is difficult to directly measure those boundary conditions during spindle operation. Although there have been studies on thermal boundary conditions of bearings, there have been few studies on those in real operating conditions. In this study, the thermal boundary conditions of a rotating bearing were estimated by the particle filter which is one of data assimilation methods. The boundary conditions were estimated by iterative finite element thermal simulations and comparison of the simulated results with the experimental temperature distribution measured by a thermal camera. The maximum error in temperature simulation with the obtained boundary conditions was 0.5 K. The obtained thermal boundary conditions also indicated the heat transfer and heat generation contribution of the inner and outer races of the bearing.

Keywords Machine tools · Thermal errors · Vision-based measurement · Camera calibration

1 Introduction

During machine tool operation, heat generated from heat sources such as spindle bearings and motors is transferred throughout the machine and changes the temperature distribution. The change in temperature distribution causes deformation of the machine tool and thermal error. In the spindle, thermal expansion causes changes in

A. Ishigaki · D. Kono (✉)
Department of Micro-Engineering, Kyoto University, Kyoto, Japan
e-mail: kouno.daisuke.8w@kyoto-u.ac.jp

© The Author(s) 2026
K. Wegener and M. Bambach (eds.), *4th International Conference on Thermal Issues in Machine Tools (ICTIMT2025)*, Lecture Notes in Production Engineering,
https://doi.org/10.1007/978-3-032-01194-7_1

the bearing preload, contact angle and stiffness, which affect the dynamic characteristics of the spindle. Therefore, accurate prediction of the temperature distribution and thermal displacement distribution of the machine tool is important in machine tool design and control.

To predict the temperature distribution using a thermal model of spindles, appropriate thermal boundary condition setting is necessary. In particular, the heat generated in the front bearing of the spindle is significant and has a large impact on the temperature distribution of the spindle. The thermal boundary conditions of the bearing are complex because the boundary conditions of the bearing, such as heat generation and heat transfer, vary with temperature and preload [1]. The heat generation is modelled based on the friction torque of the bearing [2–5]. The heat transfer in the bearing is modeled considering the contact thermal resistance [3–5]. However, the estimation error of the temperature using such models can reach 5–10 K and is not small [3]. Thus, further studies including the experimental analysis of the complex thermal boundary conditions are still important. Although the direct measurement of these boundary conditions has been tried [6], it is not easy for a practical spindle during rotation because the temperature measurement of the rotating elements such as balls and inner rings is difficult.

Higher accuracy in the parameter estimation can be achieved by data assimilation that enables parameter optimization with the combination of the experiment and simulation. The data assimilation has been used to estimate the condition dependent boundary conditions in feed drives and spindles during operation [7, 8]. By similar methodologies, precise models for high-speed spindles have also been developed considering the interaction between the thermal boundary conditions and mechanical conditions such as preload and clearance in bearings [9, 10]. Although these studies showed that the data assimilation is effective to estimate the complex thermal boundary conditions, the details of bearing model are not clearly described in their works. For a more detailed analysis, the heat generation in bearings and the thermal contact conductance between the race and shaft/housing were estimated by data assimilation [11]. However, the model in [11] still lacks details because it ignores the heat generation difference between the inner and outer rings and the heat transfer through the bearing.

In this study, the thermal boundary conditions of a rotating bearing were estimated by the particle filter which is one of data assimilation methods. The heat generation between the inner ring and ball, the heat generation between the outer ring and ball, and the thermal contact conductance between the inner and outer rings and ball were separately estimated. Those boundary conditions were estimated by iterative finite element thermal simulations and comparison of the simulated results with the experimental temperature distribution measured by a thermal camera.

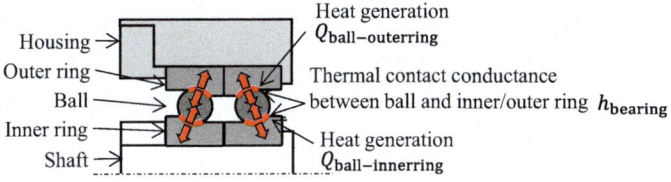

Fig. 1 Schematic of spindle bearings and objective thermal boundary conditions

2   Estimation Method of Thermal Boundary Conditions

2.1   Objective Thermal Boundary Conditions of Bearings

In this study, we focused on angular contact ball bearings with a back-to-back config-uration, which are commonly used in the front bearings of machine tool spindles. Figure 1 shows the schematic of the cross-section of the bearing and the thermal boundary conditions to be estimated. As shown in the figure, the heat generation $Q_{\text{ball-innerring}}$ between the inner ring and ball, the heat generation $Q_{\text{ball-outerring}}$ between the outer ring and ball, and the thermal contact conductance h_{bearing} between the inner and outer rings and ball are to be estimated. Because the thermal boundary conditions of the bearing to be calculated depend on preload and rotational speed, but not on the structure of the shaft and housing, the thermal boundary conditions can be applied even when the same bearing is used in different structures, by calculating the thermal boundary conditions for multiple preloads and rotational speeds in advance.

2.2   Estimation Method

Among the thermal boundary conditions of the bearing, the heat generation of the inner ring $Q_{\text{ball-innerring}}$ and outer ring $Q_{\text{ball-outerring}}$, and the thermal contact conduc-tance h_{bearing} cannot be estimated simultaneously. This is because the apparent heat flux from the bearing to the shaft side and housing side can be obtained from the temperature of the shaft and housing. This apparent heat flux includes heat transfer by $Q_{\text{ball-innerring}}$, $Q_{\text{ball-outerring}}$ and h_{bearing}, but the contribution of these cannot be sepa-rated. Therefore, it is necessary to estimate the heat generation and thermal contact conductance separately.

As shown in Fig. 2a, let the apparent heat flux to the shaft side be Q_{shaft}, the apparent heat flux to the housing side be Q_{housing}, the inner ring temperature be $T_{\text{innerring}}$, the outer ring temperature be $T_{\text{outerring}}$, and the ball temperature be T_{ball}. Also, let the heat flux due to $Q_{\text{ball-innerring}}$ that is transmitted to the inner ring be $Q_{\text{innerring}}$, and the heat flux due to $Q_{\text{ball-outerring}}$ that is transmitted to the outer ring be $Q_{\text{outerring}}$. Here, $Q_{\text{innerring}}$ and $Q_{\text{outerring}}$ are assumed to be proportional to the total heat generation of the bearing. Because the apparent heat flux includes the heat flux from

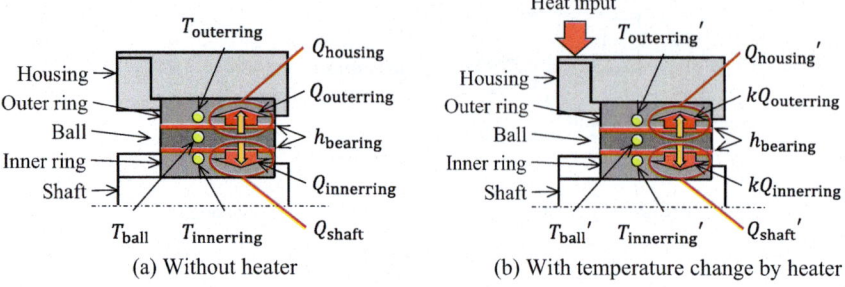

Fig. 2 Thermal boundary conditions of bearing with and without heater

the heat source and heat transfer through the bearing, $Q_{\text{innerring}}$ and $Q_{\text{outerring}}$ can be expressed as follows.

$$Q_{\text{shaft}} = Q_{\text{innerring}} + h_{\text{bearing}}\left(T_{\text{ball}} - T_{\text{innerring}}\right) \tag{1}$$

$$Q_{\text{housing}} = Q_{\text{outerring}} + h_{\text{bearing}}\left(T_{\text{ball}} - T_{\text{outerring}}\right) \tag{2}$$

In order to obtain h_{bearing} from Eqs. (1) and (2), it is necessary to eliminate the unknown variables $Q_{\text{innerring}}$ and $Q_{\text{outerring}}$. Therefore, as shown in Fig. 2b, a new heat source is installed on the housing side, and the heat flux and temperature distribution are changed from the state shown in Fig. 2a. At this time, we assume that when the heat generated by the entire bearing changes by a factor of k due to the temperature change, $Q_{\text{innerring}}$ and $Q_{\text{outerring}}$ also change by the factor of k in proportion to the total heat as follow:

$$k = \frac{Q'_{\text{innerring}}}{Q_{\text{innerring}}} = \frac{Q'_{\text{outerring}}}{Q_{\text{outerring}}} \tag{3}$$

where $Q'_{\text{innerring}}$ and $Q'_{\text{outerring}}$ are the heat flux with the temperature change corresponding to $Q_{\text{innerring}}$ and $Q_{\text{outerring}}$, respectively.

Then, the apparent heat flux to the shaft Q'_{shaft} and the apparent heat flux to the housing Q'_{housing} with the temperature change are expressed by the following equations:

$$Q'_{\text{shaft}} = kQ_{\text{innerring}} + h_{\text{bearing}}\left(T'_{\text{ball}} - T'_{\text{innerring}}\right) \tag{4}$$

$$Q'_{\text{housing}} = kQ_{\text{outerring}} + h_{\text{bearing}}\left(T'_{\text{ball}} - T'_{\text{outerring}}\right) \tag{5}$$

where $T'_{\text{innerring}}$ is the inner ring temperature with the temperature change, $T'_{\text{outerring}}$ is the outer ring temperature with the temperature change, and T'_{ball} is the ball temperature with the temperature change. Combining Eqs. (1), (2), (4) and (5), h_{bearing} can be obtained as follows:

$$h_{\text{bearing}} = -\frac{\left(Q'_{\text{shaft}} - kQ_{\text{shaft}}\right) - \left(Q'_{\text{housing}} - kQ_{\text{housing}}\right)}{\left(T'_{\text{innerring}} - kT_{\text{innerring}}\right) - \left(T'_{\text{outerring}} - kT_{\text{outerring}}\right)} \quad (6)$$

The heat flux and temperature required to calculate Eq. (6) are obtained using data assimilation with particle filter. The data assimilation enables obtaining physical quantities that are difficult to measure, such as inner ring temperature and heat flux. Using the obtained h_{bearing}, the heat generation in the bearing is estimated by performing data assimilation again. The specific method is described in Sect. 3.

3   Experiment for Estimation of Thermal Boundary Conditions

3.1   Experimental Setup and Method

In order to obtain the measured temperature for the data assimilation, experiments were conducted. The schematic of the experimental setup is shown in Fig. 3. The rotating shaft was supported by bearings, and the rotation of the motor was transmitted to the shaft via a coupling and torque meter. Angular contact bearings (7010C, JIS B 1513) were used in back-to-back configuration. A brushless motor (CPH50, Coreless Motor Co., Ltd.) was used to rotate the shaft. A torque meter (UTMII-5Nm) was attached between the motor and the shaft to measure the shaft's rotational torque. A 30W heater was attached because temperature measurements are necessary to estimate h_{bearing} using Eq. (6).

A thermal camera and K-type thermocouples were used to measure the temperature. The temperature measurement points are shown in Fig. 3c. Seven points S_{1-7} were set on the shaft and ten points H_{1-10} were set on the housing. Because the emissivity of the surface of the experimental device varies depending on the material and surface treatment, blackbody tape with an emissivity of 0.95 was attached to all measurement points. The ambient temperature, outer ring temperature of the bearing, and housing temperature at the point H_2 were measured using thermocouples. The specifications of the experimental apparatus are shown in Table 1.

In the experiment, temperature measurements were taken continuously with the heater off and with the heater on. The motor rotation speed was set to 6650 min^{-1}. In both cases, measurements were continued until a steady state was reached based on the thermocouple temperature.

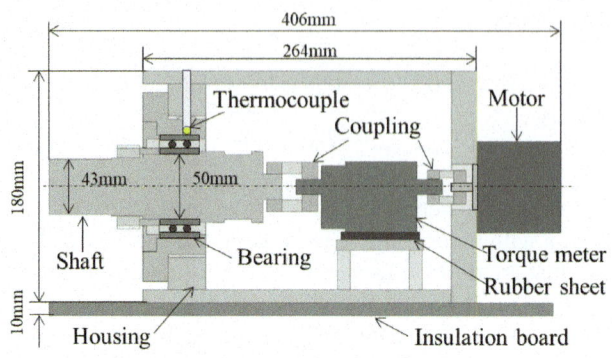

(a) Schematic drawing

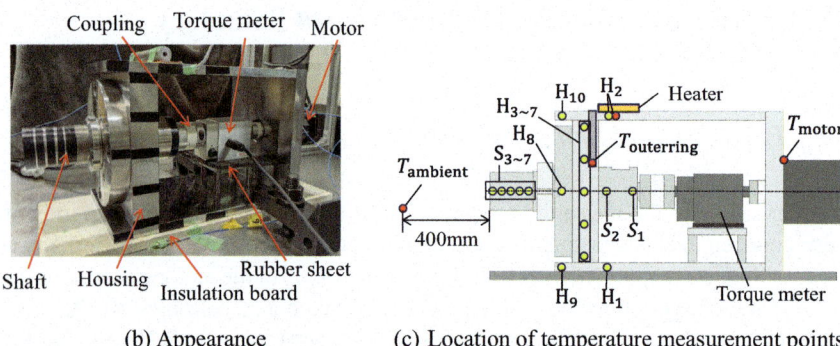

(b) Appearance   (c) Location of temperature measurement points

Fig. 3 Schematic and appearance of experimental setup

Table 1 Specifications of experimental setup

Bearing	Inner diameter (mm)	50
	Outer diameter (mm)	80
	Contact angle (°)	15
	Preload (N)	152
Motor	Voltage (V)	24
	Torque (Nm)	0.2
	Rotation speed (min^{-1})	6650
	Power (W)	140
Thermal camera	Image size (pixel)	640 × 480
	Measurement range (°C)	− 40 ~ 120
	Minimum detection limit (K)	0.08
	Measurement accuracy (K)	± 2
	Frame rate (fps)	30

3.2   Data Assimilation Method

Using a particle filter, we performed data assimilation for the steady-state heat transfer analysis using the finite element method (FEM) and the experiment described in Sect. 3.1. In order to reduce the number of unknown variables in a single data assimilation, we decided to perform data assimilation separately for the shaft side and the housing side in the estimation of $h_{bearing}$.

Figure 4a shows a cross-sectional view of the simulation model used to estimate the thermal boundary conditions on the shaft side. The thermal boundary conditions to be estimated are the apparent heat flux Q_{shaft} to the shaft side, the thermal contact conductance h_{shaft} between the shaft surface and the air, and the heat flux $Q_{coupling}$ from the coupling. Temperature values measured at points S_{1-7} and ambient temperature were used for data assimilation.

Figure 4b shows a cross-sectional view of the simulation model used to estimate the thermal boundary conditions on the housing side. The thermal boundary conditions to be estimated are the apparent heat flux $Q_{housing}$ to the housing side, the thermal contact conductance $h_{housing}$ between the housing surface and the air, and the heat flux Q_{heater} from the heater to the housing. The influence of the heat from the motor and ground are represented by T_{Motor} and T_{ground}, respectively. Temperature values measured at points H_{1-10} and ambient temperature were used for data assimilation.

The likelihood w_k^i for ith particle to evaluate the accuracy of the temperature estimation was obtained by the square root of the temperature deviation as follows:

$$w_k^i = \frac{1}{\left(\sqrt{2\pi\sigma^2}\right)^M} \exp\left(-\frac{\sum_{j=1}^{M}\left(t_{kj}^i - y_j\right)^2}{2\sigma^2 M}\right) \tag{7}$$

where t_{kj}^i is the temperature estimated by simulation, y_j is the temperature measured in experiment, M is the number of temperature measurement points, σ is the observation noise corresponding to the tolerance of the measurement error, j is the temperature

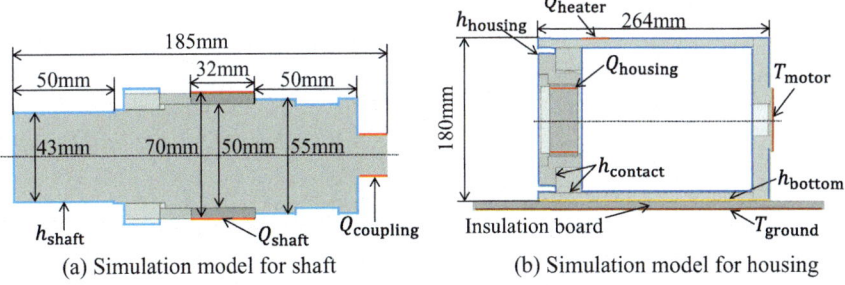

(a) Simulation model for shaft   (b) Simulation model for housing

Fig. 4 Simulation model used in data assimilation to estimate thermal contact conductance $h_{bearing}$ of bearing

measurement point number, and k is the iteration number in the data assimilation. Table 2 summarizes the parameters used in the simulation. For both simulations, the number of particles was set to 100. The observation noise was set to 0.2.

Figure 5 shows a cross-sectional view of the simulation model used for data assimilation to estimate the heat generation of the bearing. In this simulation, the bearing was simplified as laminated cylinders. Because the heat generation and heat transfer of the torque meter were unknown, the torque meter was omitted from the model, and instead the heat flux transmitted from the torque meter via the coupling was set at the right end of the shaft.

The major materials used in the experimental apparatus are S45C and S50C carbon steels for mechanical structures, SUJ2 high-carbon chromium bearing steel for bearings, and glass fiber for the heat insulating plate. COMSOL Multiphysics Version 6.2, a commercial FEM software, was used for modeling and analysis.

Table 2 Initial conditions in data assimilation

		Without heater	With heater
Simulation for shaft side	Q_{shaft} (W/m^2)[a]	3000	3000
	h_{shaft} (W/m^2 K)[a]	70	70
	$Q_{coupling}$ (W/m^2)[a]	1000	1000
	$T_{ambient}$ (°C)	18.6	18.9
	$h_{contact}$ (W/m^2 K)	2000	2000
Simulation for housing side	$Q_{housing}$ (W/m^2)[a]	2000	1000
	$h_{housing}$ (W/m^2 K) [a]	10	10
	Q_{heater} (W/m^2)[a]	0	7000
	$T_{ambient}$ (°C)	18.6	18.9
	T_{motor} (°C)	38.9	41.6
	$h_{contact}$ (W/m^2 K)	2000	2000
	h_{bottom} (W/m^2 K)	500	500

[a] Parameters highlighted in gray were updated in data assimilation. The initial values for those parameters are listed in the table

Fig. 5 Simulation model used in data assimilation to estimate heat generation $Q_{ball-innerring}$ and $Q_{ball-outerring}$ in inner and outer rings

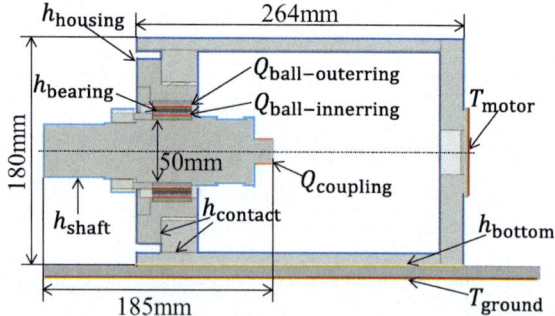

3.3 Data Assimilation Results for Estimating Thermal Contact Conductance of Bearing

Figure 6 shows a steady-state thermal image taken by the thermal camera. When the heater was used to apply heat flux to the housing, the temperature increased by 1–17 °C at each measurement point compared to the case where no heater was used. The temperature values at the measurement points were obtained from each thermal image and used as reference temperature in data assimilation.

Figure 7 shows the update history of Q_{shaft} and $Q_{housing}$ in the data assimilation. In Fig. 7a, Q_{shaft} with the heater is slightly larger than Q_{shaft} without the heater. In contrast, in Fig. 7b, $Q_{housing}$ with the heater is smaller than $Q_{housing}$ without the heater. This result is consistent with the estimation that the heat given by the heater transferred to the shaft through the housing and bearing. This result also indicates that the apparent heat such as Q_{shaft} and $Q_{housing}$ is influenced by the other boundary conditions of the setup.

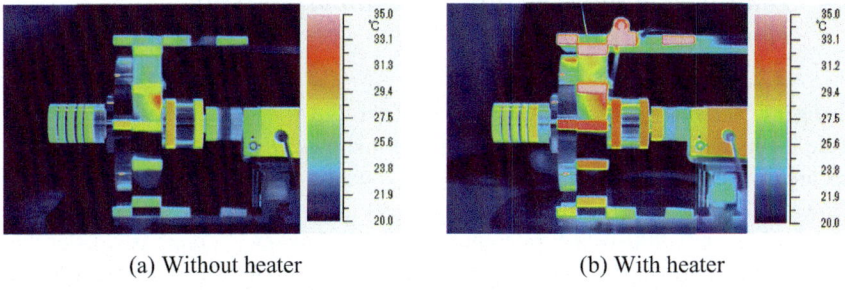

(a) Without heater   (b) With heater

Fig. 6 Measured temperature distribution with and without heater

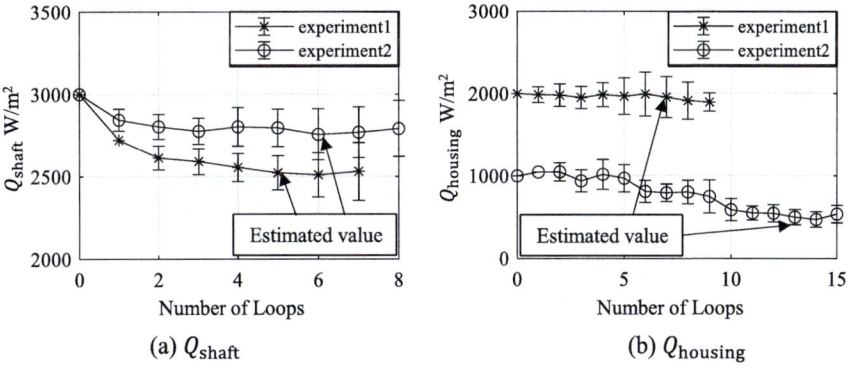

(a) Q_{shaft} (b) $Q_{housing}$

Fig. 7 Update history of Q_{shaft} and $Q_{housing}$ in the data assimilation. The error bar represents the standard deviation. The estimated value in the second iteration from the last iteration was employed as the final estimation result

Table 3 Thermal conditions
estimated by data assimilation

	Without heater	With heater
Q_{shaft} (W/m^2)	2523	2757
$Q_{housing}$ (W/m^2)	1957	500
$T_{innerring}$ (°C)	32.6	33.9
$T_{outerring}$ (°C)	32.8	36.0
$Q_{bearing}$ (W)	28.6	20.1

In Fig. 7, when the likelihood did not increase in the last three iterations, the estimated value in the second iteration from the last iteration was employed as the final estimation result. Those thermal conditions estimated in the data assimilation were summarized in Table 3. The factor k was obtained as 0.705 by Eq. (3). Then, $h_{bearing}$ was obtained as 973 W/m^2 K by Eq. (6).

3.4 Data Assimilation Results for Estimating Heat Generation of Bearing

Using the thermal conditions obtained in Sect. 3.3, $Q_{ball\text{-}innerring}$ and $Q_{ball\text{-}outerring}$ were estimated by data assimilation. Figure 8 shows $Q_{ball\text{-}innerring}$ and $Q_{ball\text{-}outerring}$ updated in the data assimilation. $Q_{ball\text{-}innerring}$ was estimated to be 2710 W/m^2 as shown in Fig. 8a. The estimated $Q_{ball\text{-}innerring}$ is larger than Q_{shaft} shown in Table 3 by approximately 200 W/m^2. In contrast, the estimated $Q_{ball\text{-}outerring}$ is smaller than $Q_{housing}$ by almost similar amount. This result indicates that the heat generated at the inner ring transferred to the housing through the bearing.

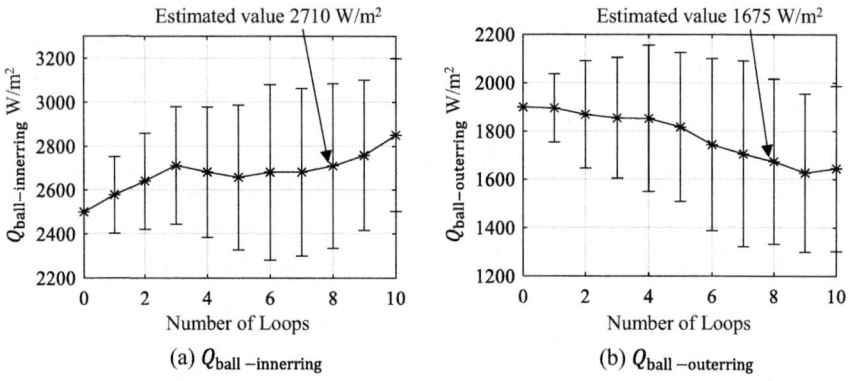

(a) $Q_{ball-innerring}$ (b) $Q_{ball-outerring}$

Fig. 8 Update history of $Q_{ball-innerring}$ and $Q_{ball-outerring}$ in the data assimilation. The error bar represents the standard deviation

Fig. 9 Error of temperature estimated using the obtained boundary conditions

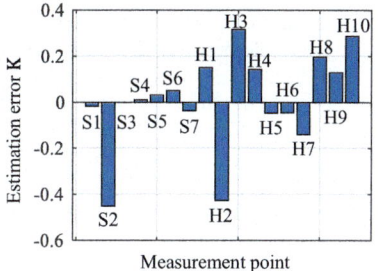

Figure 9 shows the error of the temperature estimated using the obtained boundary conditions. The results shows that the estimation accuracy is good because the maximum error is 0.5 K. However, the uncertainty of the obtained boundary conditions is large because the standard deviation reached 20% of the estimated value at the maximum as shown in Fig. 8. The accuracy improvement of the temperature measurement is effective to reduce the uncertainty. The FEM model in the data assimilation may also need modification because the maximum error exceeds the used observation noise. These improvements are considered in our future work.

4 Conclusions

The thermal boundary conditions of a rotating bearing were estimated by data assimilation. The methodology was proposed to estimate the heat generation at the inner and outer rings and the thermal contact conductance between the inner and outer rings and ball. Those boundary conditions were estimated by the particle filter that uses iterative finite element thermal simulations and comparison of the simulated results with the experimental temperature distribution measured by a thermal camera. The maximum error in temperature simulation with the obtained boundary conditions was 0.5 K. The obtained boundary conditions indicated that the apparent heat transfer to the housing included the heat generated at the outer ring and heat transferred from the inner ring though the bearing. However, the large uncertainty of the obtained boundary conditions is a remained issue because the standard deviation reached 20% of the estimated value at the maximum. The accuracy improvements of the temperature measurement and thermal modelling to reduce the estimation uncertainty are considered in future works.

References

1. Mayr, J., Jedrzejewski, J., Uhlmann, E., Donmez, M.A., Knapp, W., Härtig, F., Wendt, K., Moriwaki, T., Shore, P., Schmitt, R. Brecher, C., Würz, T., Wegener, K.: Thermal issues in machine tools. CIRP Ann. **61**, 771–791 (2021)

2. Zhang, Y., Li, X., Hong, J., Yan, K., Li, S.: Uneven heat generation and thermal performance of spindle bearings. Tribol. Int. **26**, 324–335 (2018). https://doi.org/10.1016/j.triboint.2018.04.035
3. Bossmanns, B., Tu, J.: A thermal model for high speed motorized spindles. Int. J. Mach. Tool Manuf. **39**, 1345–1366 (1999). https://doi.org/10.1016/S0890-6955(99)00005-X
4. Than, V.T., Huang, J.H.: Nonlinear thermal effects on high-speed spindle bearings subjected to preload. Tribol. Int. **96**, 361–372 (2016). https://doi.org/10.1016/j.triboint.2015.12.029
5. Takabi, J., Khonsari, M.M.: Experimental testing and thermal analysis of ball bearings. Tribol. Int. **60**, 93–103 (2013). https://doi.org/10.1016/j.triboint.2012.10.009
6. Takeuchi, Y.R., Dickey, J.T., Frantz, P.P.: A methodology for measuring thermal properties of bearings in motion (2004)
7. Thiem, X., Rudolph, H., Krahn, R., Ihlenfeldt, S., Fetzer, C., Müller, J.: Adaptive thermal model for structure model based correction. In: International Conference on Thermal Issues in Machine Tools, pp. 67–82 Springer (2023)
8. Thiem, X., Riedel, M., Kauschinger, B., Müller, J.: Principle and verification of a structure model based correction approach. Procedia CIRP **46**, 111–114 (2016)
9. Jedrzejewski, J., Kowal, Z., Kwaśny, W., Modrzycki, W.: Hybrid model of high speed machining centre headstock. CIRP Ann. Manuf. Technol. **53**, 285–288 (2004)
10. Jędrzejewski, J., Kowal, Z., Kwaśny, W., Modrzycki, W.: High-speed precise machine tools spindle units improving. J. Mater. Process. Technol. **162–163**, 615–621 (2005)
11. Tan, F., Chen, H., Peng, J., Deng, C.: Efficient boundary conditions identification in thermal simulation of the spindle system with reduced order model and differential evolution algorithm. Case Stud. Thermal Eng. **59**, 104526 (2024). https://doi.org/10.1016/j.csite.2024.104526

Surface Integrity Parameters in the Milling of Hardened 1.6580 Steel Under Different Cooling Conditions

Amirhossein Ranjbar, Mohammad Amirshirzad, Behzad Jabbaripour, and M. K. Gupta

Abstract Steel 1.6580 is one of the most useful types of alloyed steel in the machinery and mold-making industries. The significant attention has been paid to it due to its favorable mechanical properties and affordable price. Additionally, wear resistance of this steel is capable of enhancing by both through and case hardening treatment. Since hardened steel machining process poses numerous challenges and can potentially affect the subsurface characterization and mechanical properties of machined parts, it is essential to have deep understanding of the machining parameters on the surface tribological and metallurgical aspects of surface integrity such as microhardness variation. In this study, milling process effects with single edge carbide tool (APKT1003PDSR) has been investigated under Dry and Minimum Quantity of Lubrication (MQL) environments. The analysis on surface topography and microhardness beneath the machined surface clearly demonstrate the effectiveness of MQL on sub-surface quality of machined parts. This enhancement attributed to the adhered particles reduction. Moreover, the MQL process proficiently regulated the metallurgical alterations induced by the heat produced in the cutting zone, thus maintaining the surface hardness of the machined component. In addition, the findings of this work not only offer evidence for a better understanding of metallurgical changes in the machined surface, but also provide valuable insights for industry professionals. These insights can assist in selecting the most appropriate machining parameters based on the functional requirements of the component.

A. Ranjbar
Faculty of Mechanical Engineering, Opole University of Technology, Opole, Poland

A. Ranjbar · M. Amirshirzad · B. Jabbaripour
Department of Mechanical Engineering, Central Tehran Branch, Islamic Azad University, Tehran, Iran

M. K. Gupta (✉)
Faculty of Mechanical Engineering, Opole University of Technology, 76 Proszkowska St, Opole 45-758, Poland
e-mail: munishguptanit@gmail.com

Faculty of Mechanical Engineering, Department of Machining, Assembly and Engineering Metrology, VSB Technical University of Ostrava, 17. Listopadu, Ostrava 2172/15708 00, Poruba, Czech Republic

© The Author(s) 2026
K. Wegener and M. Bambach (eds.), *4th International Conference on Thermal Issues in Machine Tools (ICTIMT2025)*, Lecture Notes in Production Engineering,
https://doi.org/10.1007/978-3-032-01194-7_2

Keywords Steel · Surface integrity · Machining · Cutting temperature

1 Introduction

Alloy steels are extensively used materials in industries. Due to the significant role of machining in the manufacturing process, researchers have paid considerable emphasis on examining how machining impacts the characteristics of these alloys [1]. Microscopic properties of the surface created from machining operations have an influential impact on performance and economics of the machining process and manufactured components produced by alloy steels [2]. This is particularly critical with manufactured parts exposed by wear and cyclic forces. Further, the different nature of each machining process affect the tribological properties [3, 4]. Temperature generation is also one of the parameters that crucially evaluated during machining process because its directly affect the mechanical and metallurgical aspects of manufactured parts [4]. To prevent the heat concentration at the cutting zone, cutting fluids are used at the tool and workpiece. Previously, applying oil–water emulsion also called flood cooling, was the only method of cooling but nowadays, the green and environmentally friendly cooling methods play a vital role in the machining industry. For instance, minimum quantity lubrication (utilizing oil particles with compressed air) or cryogenic fluids (Liquid CO_2 and N_2), among the most prominent modern cooling methods.

The investigation of various cooling techniques and their effects on the machined component properties has attracted significant attention from researchers in recent years. For instance, Mia et al. [5] conducted a statistical analysis to investigate the surface integrity. The results of their study not only identified the most impactful parameter on surface integrity components but also demonstrated that in MQL methods, increasing the volume of misted oil, and consequently reducing the temperature, improves surface properties. Tunc et al. [6] have recommended the use of MQL in the milling of stainless steels, as it reduces the localized temperature at the cutting zone. The application of this method has been shown to improve surface roughness parameters and subsurface residual stress values. Additionally, the efficiency of MQL is increased by some nano particles like Al_2O_3 or SiO_2 and the authors claim that this additive can improve the thermal conductivity of oil. Bai et al. [7] studied the effects of different nano particles on machining variation in MQL environment. Tazehkandi et al. [8] investigated the tool wear and surface topography under optimized MQL method. The results indicated that the optimized MQL method reduces the size of lubricant droplets and increases the surface covered by cooling. Nogoc et al. [9] introduces a novel approach for applying hybrid nanofluids in MQL hard machining. The results indicate that hybrid nano cutting oils in MQL outperform mono nanofluids, achieving lower cutting temperatures, more stable surface roughness (Ra), back force (Fp), and cutting force (Fc), due to enhanced cooling and lubrication.

Due to the widespread industrial application of 1.6580 steel in the machinery and component manufacturing sectors particularly in its hardened state (It is worth noting that the alloy investigated in this study particularly in its hardened state accounts for over 20% of the alloy steel consumption in countries such as Iran.) there is a noticeable gap in the availability of evidence regarding the influence of machining parameters which results in machining characteristics. This issue is especially critical in industrial applications that often facing with the lack of advanced inspection equipment. The findings of this study aim to assist practitioners in selecting optimal machining conditions especially through the use of single-edge cutting tools, which are commonly employed to reduce operational costs, so as to achieve the best possible surface properties, such as reduced friction coefficient and enhanced hardness, while simultaneously preserving tool integrity. From an academic standpoint, this research addresses the significant voids, as most previous studies have focused on this alloy in its non-hardened condition, with limited data available on the metallurgical behavior of the machined surface under various machining environments. Therefore, the present study seeks to bridge the gap between industrial practice and academic research by providing insight into these aspects and it is the most innovative practical aspect of this work. As it was mentioned before, the primary objective of this study is to bridge the existing gap in both academic and industrial domains regarding the investigation of subsurface properties in the machining of hardened 1.6580 alloy using a single-edge cutting tool.

2 Materials and Methods

The facilities and equipment used during the experiments are shown in Fig. 1. The workpiece with the size of 125 * 25 * 20 mm has been used for the milling experiments. The work piece was hardened before machining by quenching process. The specimens were subjected to a quench and temper heat treatment process. Initially, the hardened component was held at a temperature of 870 °C for 40 min. A vacuum furnace was employed for this purpose. Subsequently, the part was quenched using mineral quenching oil. To reduce excessive hardness and to achieve uniform metallurgical properties, tempering was conducted at 230 °C for 90 min, followed by slow cooling to ambient temperature. Hardness measurements indicated an average hardness of approximately 52 HRC, with a variation of ± 3 HRC across different locations on the component. The used tool was a single edge carbide cutting insert (APKT1003PDSR with Al_2O_3 coating). The holder used is EM-AP10-D12-C12-L120-Z01, produced by the Iranian X-Hold company. Additionally, the experiments were conducted using a VCM CNC machine DMG Mori Dura 635, equipped with a Siemens controller and a maximum spindle speed of 12,000 RPM. A total of 8 tests were conducted with a constant cutting depth of 1.25 mm, cutting speeds of 50 and 100 m/min, and feed rates of 0.09 and 0.18 mm/rev, under both dry and minimum quantity lubrication (MQL) conditions. The adjusted parameters of MQL were determined through preliminary experiments and the details are given in Table 1. The MQL system employed in the tests was custom-designed and fabricated, as depicted in Fig. 2. For lubrication fluids in this MQL system, Behran Tarash 55 cutting oil was utilized for lubrication and cooling at the cutting zone.

An optical microscope (OLYMPUS BX51M) was used to measure the tool wear values and the machined surfaces. The scanning electron microscope (SEM) was to used to analyze tool wear mechanism. The subsurface microhardness measurements were performed with Vickers index SHIMADZU microhardness tester which was located in Amirkabir university of technology and a 200-g load was applied for 10 s. The hardness measurements were continued until the measured value approached the bulk material hardness or three consecutive identical readings were obtained.

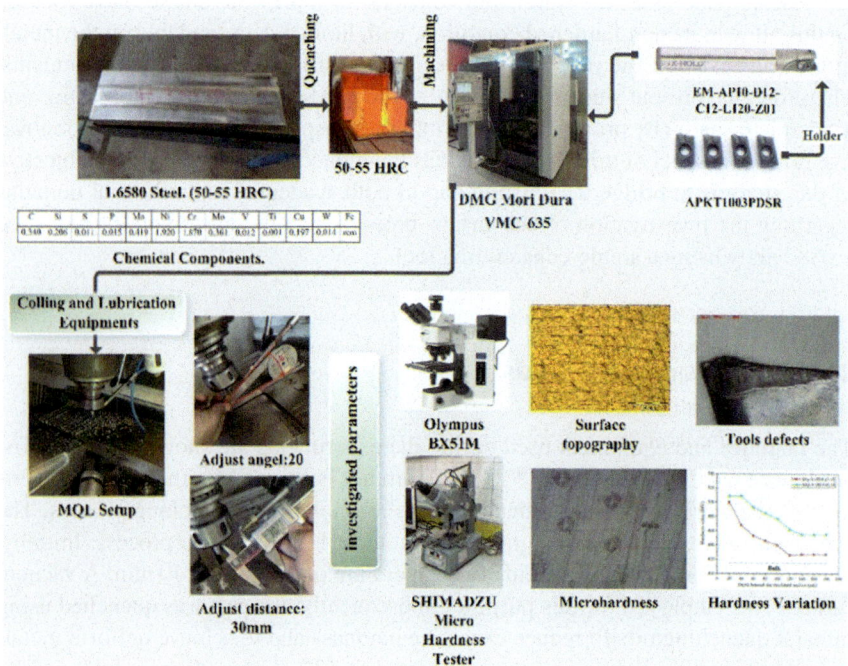

Fig. 1 Experimental setup in the current experiments

Table 1 MQL parameters used in this study		
Air pressure	8 bar
Oil flow rate	70 ml/h
Target face	Rake face
Distance to rake face	30 mm
Adjustment angle	20°
Injector nozzle diameter	1 mm

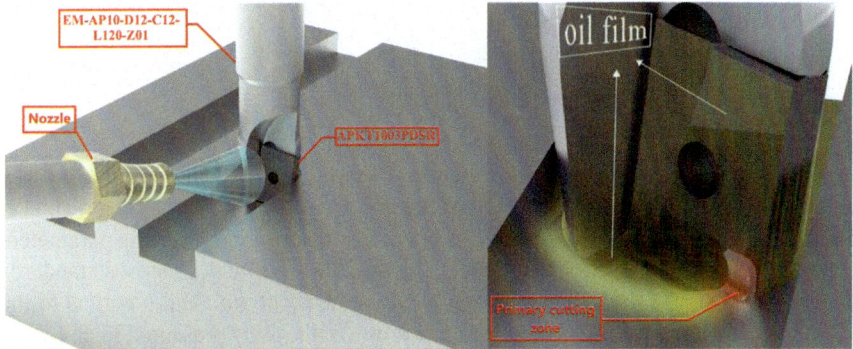

Fig. 2 Mechanism of MQL system

3 Thermal Aspects of Minumum Quantity of Lubrication (MQL)

In MQL, the 60–100 ml/hour volume of oil is delivered to the cutting zone via a high-pressure air stream and oil mist covers the tool faces and the workpiece with a very thin layer (5–10 μm). The covered film layer mitigates the friction's coefficient by tool and workpiece direct contact prevention. Further, oil penetration specially on harsh cutting conditions, causes heat reduction by evaporation. On the other hand, the high-pressure air flow, dissipate the heat from cutting zone also [10]. MQL substantially influences the development of built-up edge and modifies the chip morphology. These modifications minimize the contact area between the tool face and the chip, hence reducing machining forces. The drop in forces directly results in less energy dissipation as heat, therefore leading to a reduction in cutting temperature. The cooling mechanism in MQL is achieved through a constant convective rate of heat transfer. In machining, certain critical regions are more affected by temperature, including the primary cutting zone and the rake face, which is in direct contact with the separated chip. Although a significant portion of the generated heat is dissipated from the machining zone via chip flow, part of the heat penetrates the workpiece and can impact tool performance, as shown in Fig. 2. To dissipate heat from this region, the use of lubrication and cooling sprays is highly effective due to their high heat transfer coefficient [11].

4 Results and Discussion

4.1 Cutting Temperature Analysis

In this study, an empirical formula, which was derived from a semi-experimental and mathematical model, was used for initial cutting zone calculation. Baralic et al. [12] has focused on initial heat in cutting zone prediction by examining cutting parameters during the milling of 30CrNiMo8 steel. This steel, known under various standards as 1.6580, AISI 4340, and UNS 43400, is fully consistent with and identical to the alloy used in this study. The chemical composition of the alloy presented in this study matches that of our research. The researchers developed a mathematical model based on the Artificial Neural Network (ANN) technique, which enabled them to propose a model to calculate the initial heat in the cutting zone. They have also conducted experiments to demonstrate the effectiveness of their formula, in which the accuracy of the formula's performance was estimated with an approximate deviation of 1 °C. The established mathematical model is illustrated in (1) Where T is the temperature at starting point of machining, S_Z is feed per tooth in mm/tooth, a is axial depth of cut in mm, and V is the cutting speed in m/min.

$$
\begin{aligned}
T = {}& 305.1491 \times V^{2.5238} \times Sz^{-0.20344} \times a^{-0.64328} \\
& \times \exp(-0.42316(lnV)^2 - 0.1933(lnSz)^2 + 0.37903(lna)^2 \\
& + (0.42383lnv * lna) - (0.44884lnSz * lna))
\end{aligned}
\tag{1}
$$

Figure 3 illustrates the cutting temperature obtained during dry and MQL machining. To calculate the cutting heat for MQL methods, the coefficients provided in study [13] were utilized. The graph indicates that the highest heat occurs in dry machining, followed by MQL. Although the outcomes represent the initial cutting zone temperature, it is evident that as machining time increases, these values will also rise. Overall, the temperature in the MQL environment is approximately 35% lower than in the dry machining environment. The mechanism of heat reduction is already presented in Sect. 3.

4.2 Surface Topography

In this work, surface topography is examined because the microscopic effects formed on the machined surface are one of the critical parameters that determining the part's performance. Although these defects are invisible to naked eyes, their presence significantly reduces the components efficiency. Moreover, investigating these defects are crucial, as the presence of certain defects, such as micro-cracks, may lead to the sudden failure of the component. Figure 4 depicts the surface topography

Fig. 3 Cutting temperature under dry and MQL condition

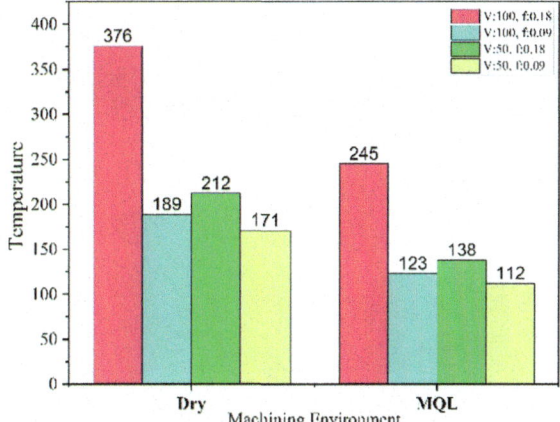

alternations at different cutting parameters and cooling environments in 50X magnification. In dry machining where heat transfer only occurs through air convection due to the absence of cooling fluids. the cutting zone exposed to significant higher heat concentration compared to the MQL environment [14]. This elevated heat can lead to adhesion of chip particles to the workpiece surface. Subsequent detachment of these particles may result in cavities on the machined surface. Then these particles detaching in subsequent tool pass, may result in cavities on the machined surface [15]. Smaller particles adhering to the workpiece surface might also become trapped beneath the cutting tool during successive tool passes. Consequently, the rotation of the tool can rub these particles against the surface and leading to the formation of grooves between the primary feed marks [16]. The increased presence of grooves contributes to higher surface friction and roughness [17]. A significant fewer surface defects are observed on machined surfaces, when the MQL mist has been applied to the cutting zone. This method by moderate reduction of temperature, leads to decrease the numbers of cavities in term of size and number in comparison with dry machining.

In MQL condition, significantly fewer surface defects are observed. This method, by moderate reduction of temperature, leads to decrease the numbers of cavities in term of size and number in comparison with dry machining. Moreover, the formation of a lubrication film on the workpiece surface prevents direct contact between chip particles and the surface, and causes reducing the friction coefficient between the tool and the workpiece. Additionally, the oil evaporation contributes to further temperature reduction at the cutting zone [18, 19] By reducing the feed rate to 0.09 mm/tooth, when the cutting parameters approach the finishing process, as illustrated in Fig. 4, the cutting zone temperature also decreases, and the surface defects density is significantly reduced. the defects on the surface are primarily limited to small adhered particles and minor cavities in dry machining. Similar findings have been reported in the study [20].

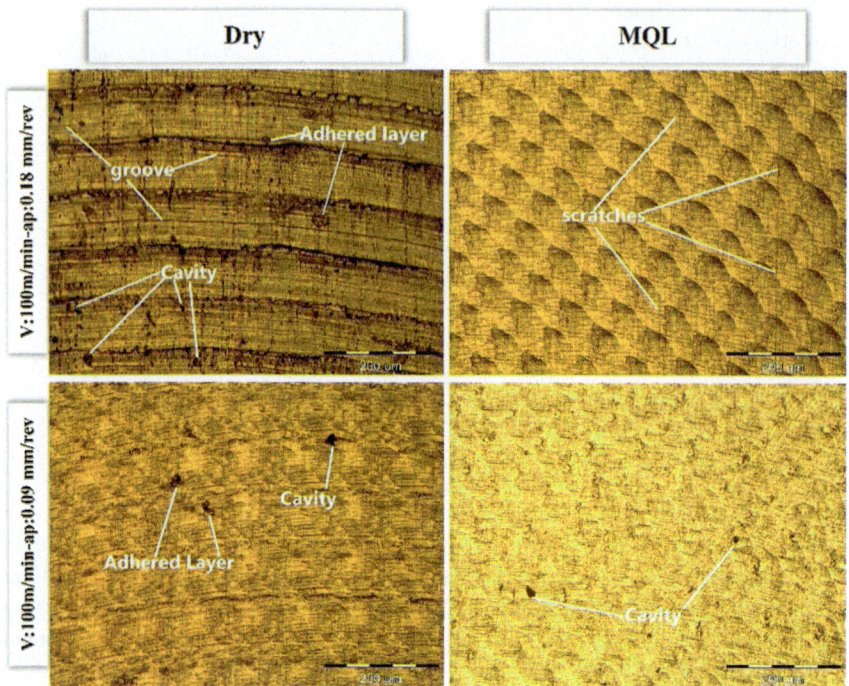

Fig. 4 Machined surface topography under dry and MQL condition at 50X magnification

4.3   Tool Wear

Evaluating tool wear is crucial because it have direct impact on the economic efficiency of machining operations. Additionally, all surface integrity parameters, including roughness, geometric and dimensional accuracy, forces, microhardness, microstructure, are directly influenced by the tool [21]. The SEM images of the tool's primary flank face at a cutting speed of 100 m/min, a feed rate of 0.18 mm/rev, and a depth of cut of 1.25 mm in both MQL and Dry environment are illustrated in Fig. 5. In these experiments, flank wear is the most observed abrasion mechanism in carbide tools. In this case, the direct contact between the chips and the primary flank face led to an increase in the friction coefficient, heat generation, and sequential formation and disintegration of chips along the tool's faces. This repeated process leads to the tool coating degradation and detachment of material from the tool surface [22]. As shown in Fig. 5, in the absence of cooling and lubrication fluids, the tool's cutting edge undergoes plastic deformation. In other words, with the continuous chip flow and the associated rise in temperature, the tool loses its mechanical properties, and its geometric integrity is compromised under machining forces. Prolonging machining under these conditions may result in tool failure [23]. The presence of burn marks on tools indicates a chemical reaction between the tool material and the oxygen in

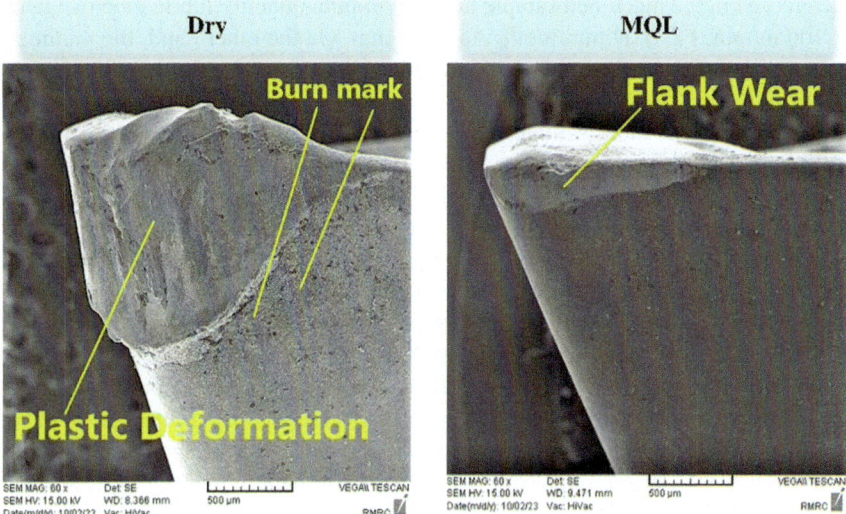

Fig. 5 Tools wear under dry and MQL condition at V: 100 m/min, F: 0.18 mm/rev, a_p: 1.25 mm

the air, caused by the increased temperature in the cutting zone [24]. This defect has occurred only under dry machining conditions. The presence of this defect abatements the functional properties of the tool coatings. Additionally, Carbide tools are particularly sensitive to temperature, and the cobalt phase of tools loses its mechanical properties entirely at approximately 800 °C. Furthermore, even at temperatures as low as half this value, the mechanical strength of the tool is reduced by 50% [18]. Consequently, during dry machining, the tool's hardness decreases due to the elevated temperatures, resulting in accelerated wear beyond expected levels. In summary, the significant increase in tool wear under dry conditions is primarily attributed to the higher temperatures caused by the absence of lubricating fluids.

4.4 Microhardness

Another critical parameter influencing the surface integrity of machined components is the evaluation of microhardness variations beneath the machined surface and within the layers adjacent to the machined surface [18]. The microhardness variation for all samples is presented in Fig. 6. As shown, the Bulk area exhibits a hardness value of 415HV, and an increase in hardness relative to the bulk is observed beneath the surface of all machined samples in both dry and MQL conditions. Microhardness variation plots also demonstrate that, as the distance from the machined surface increases, the hardness decreases and gradually approaches the bulk hardness value. The chart indicates that the maximum hardness is approximately 518 HV, which has been record in the machined sample under minimum quantity lubrication (MQL) at V: 100 m/min, Fz: 0.09 mm/tooth, Ap: 1.25 mm. On the other hand, the minimum hardness, HV420, is observed in the machined sample under dry conditions at V: 100 m/min, F_z: 0.18 mm/tooth, and A_p: 1.25 mm. The chart further indicates that hardness variations in all samples decrease sharply up to a depth of 140 μm, after which they stabilize within a consistent range. In all experiments, except for those with v: 50 m/min and F_z: 0.09 mm/rev (the lowest cutting parameters) the subsurface hardness in the dry machining method was lower than that obtained under MQL. Moreover, it is clear that in dry method, the hardness reduction trend is sharper than MQL. In other words, this indicate that the hardness achieved in dry machining is lower than MQL, and the reduction in hardness trend is also more notable.

Generally, subsurface hardness variations in machined surfaces can be attributed to two primary factors. Firstly, strain hardening and work hardening, caused by severe plastic deformation during the cutting process, leads to grain refinement and densification beneath the surface. And secondly, is the heat generated penetration from rapid plastic deformation, which diffuses into the underlying layers [25, 26]. The machining experiments that have been conducted under MQL environment, archived dramatically higher microhardness values than the dry ones. This higher hardness value beneath the machined surface proves the work hardening happened as a result of plastic deformation. The stress generated in plastic deformation during chip-removal process compresses the subsurface grain and swifts it in feed direction alignment. thereby enhancing strain hardening in the subsurface layers, which manifests as an increase in hardness. Therefore, augmenting strain hardening in the subsurface layers, resulting in a boost in hardness [27].

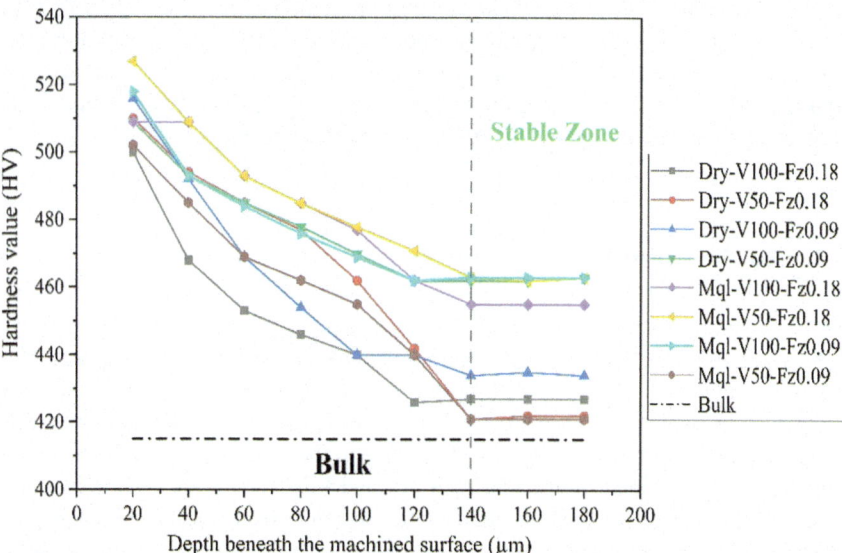

Fig. 6 Hardness variation in all experiments

One of the key factors which govern this critical variation is cutting speed and feed rate, as they directly affect the plastic deformation rate. The results clearly demonstrate that the effect of cutting speed on subsurface hardness surpasses that of cutting feed. higher cutting speeds lead to elevated subsurface hardness due to the intensified strain hardening rate. Similar findings have been reported in studies [28, 29] supported by comparable evidence. Additionally higher cutting speed is the main reason for more flank wear in the tools. Machining with higher flank wear is required grater cutting force for machining operation and contributing to the observed increase in hardness in the machined samples [18].

In result of through hardening by quenching process before machining experiments, it would be necessary to properly understand the concept of Continuous-Cooling-Transformation (CCT) graph in 1.6580 alloy steel as shown in Fig. 7. This graph has been improved and replotted based on the findings of study [30]. This graph describes the metallurgical phase change in non-equilibrium cooling. Examining this diagram reveals that non-equilibrium cooling (rapid temperature drops) prevents the diffusion of carbon atoms into the BCC structure of the pearlite phase. Instead, carbon becomes trapped in a needle form within the crystalline lattice, resulting in the formation of martensite with a BCT crystalline structure in the steel. Moreover, the diagram confirms that although the austenite phase fully occurs within the temperature range of 713–775 °C, an increase in temperature beyond 330 °C will lead to the partial transformation of this phase into austenite and pearlite. The proportion of this transformation increases progressively with rising temperature [28, 30]. Additionally, a higher cooling rate not only results in the formation of softer phases than martensite but also leads to a reduction in hardness. This is analogous to the occurrence in the dry method, where the component is cooled in the ambient environment. However, under minimum lubrication conditions, the heat is dissipated through oil spraying at a significantly higher rate.

In dry machining, the cutting parameters generate heat in the cutting zone that exceeds the phase transformation threshold. As the machined part gradually cools to ambient temperature under equilibrium cooling conditions, needle-shaped martensitic phase particles have sufficient time to diffuse into the crystalline lattice. This diffusion results in material softening in the heat-affected region, approximately 140 μm thick. Essentially, the annealing phenomenon reduces the hardness of the machined samples in the absence of cooling fluids, emphasizing the critical importance of thermal management in machining processes. In this case, due to the higher thermal load, the subsurface grains in the machined area are likely to grow larger, which generally results in lower hardness compared to finer grains [31]. However, the results for v: 50 m/min and f_z: 0.09 mm/tooth differ significantly. This series of experiments involved the lowest cutting speed and feed per tooth. Under these conditions, the highest subsurface hardness was observed during dry machining. This suggests that, when machining conditions are not severe and the thermal effects from plastic deformation are negligible, the most influential factor on hardness variations is the interaction between the tool and the surface. In the other word, this manner in which the tool moves and rubs against the surface impacts the rate of plastic deformation and the forces exerted on the workpiece during machining. In

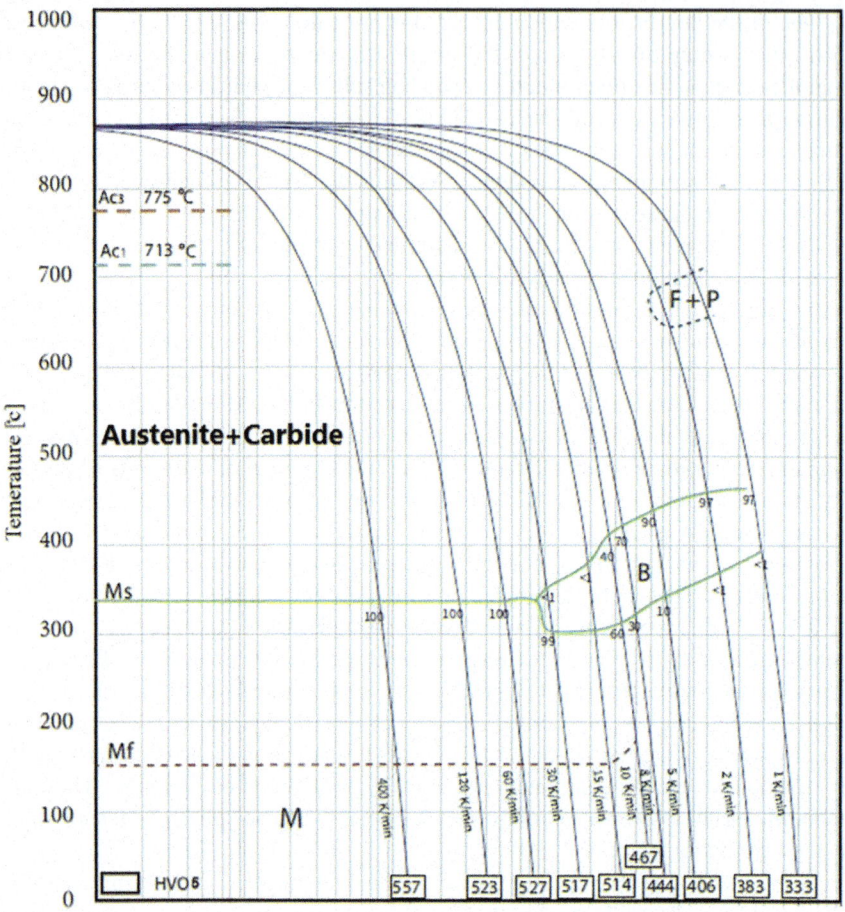

Fig. 7 Continuous-cooling-transformation of 1.6580 steel. Modified from [30]

MQL, the presence of an oil film between the tool and the workpiece not only reduces friction but also prevents direct contact. This helps preserve the surface properties and maintain greater surface integrity.

5 Conclusions

1. Despite the inherent brittleness of hardened alloy steel, detached particles are capable of adhering to the machined surface and create defects such as adhered layer and grooves. This phenomenon is common in harsh cutting parameters, when the heat generation is considerable. However, employing the MQL mist in

the cutting area protect the machined topography due to its cooling and direct contact prevention.

2. In machining process of hardened steel, high cutting parameters beside a dry environment led to irreparable damage to the tool like nose plastic deformation. Nevertheless, when machining in an MQL environment is conducted, the tool's defects is only summarized to flank wear.

3. The subsurface hardness of the machined workpiece is influenced by two parameters: the intensity of plastic deformation and the heat penetration into the subsurface. As long as the penetrated heat does not reach the phase transformation range of hardened steel from martensite to austenite, the impact of the plastic deformation rate on hardness will be more significant. Therefore, to maintain the hardening properties during the machining of hardened alloy steels, the use of cutting fluids appears to be essential.

4. The study confirms that MQL machining contributes to higher subsurface microhardness values due to reduced heat impact and enhanced strain hardening effects. In contrast, dry machining results in lower hardness and a steeper decline trend in subsurface microhardness due to thermal softening effects.

References

1. Wagri, N.K., Petare, A., Agrawal, A., Rai, R., Malviya, R., Dohare, S., Kishore, K.: An overview of the machinability of alloy steel. Mater. Today: Proc. **62**, 3771–3781 (2022)
2. Jabbaripour, B., Sadeghi, M.H., Shabgard, M.R., Faraji, H.: Investigating surface roughness, material removal rate and corrosion resistance in PMEDM of γ-TiAl intermetallic. J. Manuf. Process. **15**(1), 56–68 (2013)
3. Jabbaripour, B., Ansari, M.J., Taheri, A.: Investigating delamination, uncut fibers, tolerances and surface topography in high speed drilling of polypropylene composite reinforced with carbon fibers and calcium carbonate nanoparticles. Int. J. Mater. Form. **14**(6), 1417–1437 (2021)
4. Sales, W.F., Schoop, J., da Silva, L.R., Machado, Á.R., Jawahir, I.S.: A review of surface integrity in machining of hardened steels. J. Manuf. Process. **58**, 136–162 (2020)
5. Mia, M., Bashir, M.A., Khan, M.A., Dhar, N.R.: Optimization of MQL flow rate for minimum cutting force and surface roughness in end milling of hardened steel (HRC 40). Int. J. Adv. Manuf. Technol. **89**, 675–690 (2017)
6. Tunc, L.T., Gu, Y., Burke, M.G.: Effects of minimal quantity lubrication (MQL) on surface integrity in robotic milling of austenitic stainless steel. Procedia CIRP **45**, 215–218 (2016)
7. Bai, X., Li, C., Dong, L., Yin, Q.: Experimental evaluation of the lubrication performances of different nanofluids for minimum quantity lubrication (MQL) in milling Ti–6Al–4V. Int. J. Adv. Manuf. Technol. **101**, 2621–2632 (2019)
8. Tazehkandi A.H., Shabgard M., Tutunchi A.: Improving the performance of machining parameters in the turning process of Inconel 686 by using Cryo-MQL method. Lubrication Science (2024)
9. Ngoc, T.B., Duc, T.M., Tuan, N.M., Hoang, V.L., Long, T.T.: Machinability assessment of hybrid nano cutting oil for minimum quantity lubrication (MQL) in hard turning of 90CrSi steel. Lubricants **11**(2), 54 (2023)
10. Yıldırım, Ç.V., Kıvak, T., Sarıkaya, M., Şirin, Ş: Evaluation of tool wear, surface roughness/topography and chip morphology when machining of Ni-based alloy 625 under MQL, cryogenic cooling and CryoMQL. J. Market. Res. **9**(2), 2079–2092 (2020)

11. Jamil, M., Khan, A.M., Gupta, M.K., Mia, M., He, N., Li, L., Sivalingam, V.: Influence of CO_2-snow and subzero MQL on thermal aspects in the machining of Ti–6Al–4V. Appl. Therm. Eng. **177**, 115480 (2020)
12. Baralic, J., Mitrovic, A., Petrovic, S.S., Djurovic, S., Nedic, B.: Neural network for enhancement of end milling processes through accurate prediction of temperature in the cutting zone. J. Braz. Soc. Mech. Sci. Eng. **46**(6), 1–9 (2024)
13. Baldin, V., da Silva, L.R., Davis, R., Jackson, M.J., Amorim, F.L., Houck, C.F., Machado, Á.R.: Dry and MQL milling of AISI 1045 steel with vegetable and mineral-based fluids. Lubricants **11**(4), 175 (2023)
14. Ozdemir E., Gullu A.: Investigation of the surface quality and machinability of AISI 316LVM. Proc. Inst. Mech. Eng. Part E: J. Process Mecha. Eng. (2023)
15. Bordin, A., Sartori, S., Bruschi, S., Ghiotti, A.: Experimental investigation on the feasibility of dry and cryogenic machining as sustainable strategies when turning Ti–6Al–4V produced by additive manufacturing. J. Clean. Prod. **142**, 4142–4151 (2017)
16. Biček, M., Dumont, F., Courbon, C., Pušavec, F., Rech, J., Kopač, J.: Cryogenic machining as an alternative turning process of normalized and hardened AISI 52100 bearing steel. J. Mater. Process. Technol. **212**(12), 2609–2618 (2012)
17. Arslan, A., Masjuki, H.H., Kalam, M.A., Varman, M., Mufti, R.A., Mosarof, M.H., Khuong, L.S., Quazi, M.M.: Surface texture manufacturing techniques and tribological effect of surface texturing on cutting tool performance: a review. Crit. Rev. Solid State Mater. Sci. **41**(6), 447–481 (2016)
18. Khanna, N., Shah, P.: Comparative analysis of dry, flood, MQL and cryogenic CO_2 techniques during the machining of 15-5-PH SS alloy. Tribol. Int. **146**, 106196 (2020)
19. Thakur, D.G., Ramamoorthy, B., Vijayaraghavan, L.: Influence of minimum quantity lubrication on the high speed turning of aerospace material superalloy Inconel 718. Int. J. Mach. Mach. Mater. **13**(2–3), 203–214 (2013)
20. Eynian M. Chatter stability of turning and milling with process damping. Doctoral dissertation, University of British Columbia (2013)
21. Guo, Y.B., Liu, C.R.: Mechanical properties of hardened AISI 52100 steel in hard machining processes. J. Manuf. Sci. Eng. **124**(1), 1–9 (2002)
22. Machado, Á.R., Diniz, A.E.: Tool wear analysis in the machining of hardened steels. Int. J. Adv. Manuf. Technol. **92**, 4095–4109 (2017)
23. Nordgren, A., Samani, B.Z., Saoubi, R.M.: Experimental study and modelling of plastic deformation of cemented carbide tools in turning. Procedia CIRP **14**, 599–604 (2014)
24. Shokrani, A., Al-Samarrai, I., Newman, S.T.: Hybrid cryogenic MQL for improving tool life in machining of Ti–6Al–4V titanium alloy. J. Manuf. Process. **43**, 229–243 (2019)
25. Warren, A.W., Guo, Y.B.: On the clarification of surface hardening by hard turning and grinding. Trans. North Am. Manuf. Res. Inst. SME **34**, 309–316 (2006)
26. Zhang, S., Li, W., Guo, Y.B.: Process design space for optimal surface integrity in finish hard milling of tool steel. Prod. Eng. Res. Devel. **6**, 355–365 (2012)
27. Leadebal, W.V., Jr., de Melo, A.C., de Oliveira, A.J., Castro, N.A.: Effects of cryogenic cooling on the surface integrity in hard turning of AISI D6 steel. J. Braz. Soc. Mech. Sci. Eng. **40**, 1–4 (2018)
28. Gorni, A.A.: Steel forming and heat treating handbook. São Vicente, Brazil (2011)
29. Wika, K.K., Litwa, P., Hitchens, C.: Impact of supercritical carbon dioxide cooling with minimum quantity lubrication on tool wear and surface integrity in the milling of AISI 304L stainless steel. Wear **426**, 1691–1701 (2019)
30. Močnik T. (unpublished) Optimizacija toplotne obdelave jekla 30CRNIMo8 za doseganje ustreznih mehanskih lastnosti po preseku. Doctoral dissertation. Univerza v Ljubljani, Naravoslovnotehniška fakulteta
31. Gupta, M.K., Niesłony, P., Sarikaya, M., Korkmaz, M.E., Kuntoğlu, M., Królczyk, G.M.: Studies on geometrical features of tool wear and other important machining characteristics in sustainable turning of aluminium alloys. Int. J. Precis. Eng. Manuf. Green Technol. **10**(4), 943–957 (2023)

Empirical and Simulation-Based Analysis of Heat Partition in Orthogonal Milling

Hui Liu, Markus Meurer, and Thomas Bergs

Abstract Accurate prediction of heat partition during milling is essential for optimizing tool life, surface integrity, and process efficiency. However, conventional methods either require high computational effort or provide limited local temperature data, especially under complex milling conditions. This study presents a combined experimentalâŁ"analytical framework for thermal analysis in milling. A custom-designed test bench enables quasi-two-dimensional milling with orthogonal cuts and allows high-speed thermographic imaging of the tool and workpiece. An analytical model employing a parabolic heat source is used to simulate the temperature distribution in both components. The model incorporates material-specific thermophysical properties and is calibrated using inverse analysis based on experimental temperature measurements. Validation results show close agreement between simulated and measured thermal fields, confirming the reliability of the method. The findings demonstrate that cutting speed significantly affects workpiece temperature by increasing thermal power input, while the feed rate mainly influences heat carried away by the chip. The proposed framework enables efficient and accurate estimation of heat partitioning with minimal computational cost, offering a practical tool for thermal process optimization in milling. The method is particularly suitable for rapid evaluation of cutting conditions and could serve as a foundation for adaptive control strategies in smart manufacturing.

Keywords Transient thermal modeling · Milling process · Analytical heat transfer model · Cutting edge temperature · Heat partitioning

H. Liu (✉) · M. Meurer · T. Bergs
Manufacturing Technology Institute (MTI), RWTH Aachen University, Aachen, Germany
e-mail: h.liu@mti.rwth-aachen.de

M. Meurer
e-mail: m.meurer@mti.rwth-aachen.de

T. Bergs
e-mail: t.bergs@mti.rwth-aachen.de

K. Wegener and M. Bambach (eds.), *4th International Conference on Thermal Issues in Machine Tools (ICTIMT2025)*, Lecture Notes in Production Engineering,
https://doi.org/10.1007/978-3-032-01194-7_3

1 Introduction

During machining, over 85 % of the mechanical work performed by the cutting tool is converted into heat [1], resulting in a rapid temperature increase in the cutting zone. In dry cutting, a part of the generated heat is removed by the chips, while the remaining heat is transferred to the cutting tool and the workpiece. The heat conducted into the workpiece raises the surface temperature significantly. In combination with high mechanical loads, this temperature increase can alter the surface microstructure. As a result, microcracks and tensile residual stresses may form, degrading the mechanical properties of the workpiece and surface integrity [2]. Heat transferred to the cutting tool increases tool temperature, accelerates thermally induced wear, and causes thermal deformation, which may displace the tool center point and compromise dimensional accuracy [3]. Therefore, a detailed understanding of heat generation and partitioning in machining is essential for effective thermal management, improved surface quality, and extended tool life.

As thermal energy is directly correlated with temperature, experimental studies on heat generation in cutting processes primarily rely on temperature measurements. However, intermittent cutting operations, especially milling, pose significant challenges for accurate thermal data acquisition from both the tool and the workpiece. Complex tool and workpiece geometries hinder sensor placement, while the constantly changing tool engagement leads to rapid temperature fluctuations. These conditions demand high-speed, responsive sensors capable of capturing transient thermal behavior with precision. To address these challenges, Sato et al. [4] embedded optical fibers into an indexable insert milling cutter and connected them to a two-color pyrometer via an integrated optical coupling in the tool holder, enabling real-time tool temperature measurement. Similarly, Akhtar et al. [5] placed multiple thermocouples within the workpiece to monitor temperature variations during side milling. While such point-based methods offer valuable localized data, they are limited by complex sensor integration and provide no insight into the overall temperature distribution, making them inadequate for comprehensive thermal analysis.

Thermal imaging is a more effective technique for global temperature measurement during milling. Lazoglu et al. [6] used infrared thermography to monitor tool temperature during face milling and combined the results with simulations to identify peak thermal loads. Langenhorst et al. [7] applied similar techniques to monitor workpiece temperature and explore heat input under varying process parameters. Despite their advantages, thermal imaging methods also face limitations. The non-perpendicular alignment between measurement surfaces and the infrared camera due to complex geometries introduces angular errors and reduces accuracy. In addition, the cutting zone is often obscured by chips and the tool body, making direct observation difficult. Thermal imaging also captures only the surface temperatures of exposed areas on the chip, tool, or workpiece, which may not accurately reflect the steep temperature gradients and true thermal conditions inside the cutting zone.

An alternative approach was introduced by Augspurger [8], who developed an experimental setup using specially designed tools without a helix angle, combined with a linear machine table feed, to enable milling with orthogonal cuts. This configuration enables cutting in the Z-direction under constant conditions, effectively simplifying the process to a two-dimensional operation. In the following, the term orthogonal milling refers to this simplified cutting method. The setup provides unobstructed optical access to the machining zone and allows for more accurate analysis of temperature distribution. Despite its advantages, the specialized tools required for this setup were costly and not widely available. The present study builds upon Augspurger's concept by modifying the tool design to reduce complexity while preserving suitability for thermal investigations in milling.

The heat transferred to the tool and the workpiece can be determined indirectly through inverse analysis based on measured temperature data. The most direct approach for this purpose is numerical simulation. Commercial software packages such as ABAQUS, ANSYS, and COMSOL provide dedicated modules for temperature simulation. However, due to the complexity of model setup and the high computational cost, numerical simulations are primarily used in research contexts. In contrast, analytical models offer mathematically simpler formulations and do not require commercial software for implementation, making them particularly valuable for industrial applications. Most analytical models for workpiece temperature are derived from solutions involving moving heat sources, as presented by Carslaw and Jaeger [9]. Building on this foundation, Richardson et al. [10] developed an analytical model for temperature distribution in the workpiece using a modified Jaeger solution for a flat moving heat source in a simulated infinite solid. This model is widely applied for calculating workpiece temperatures in two-dimensional milling.

In the field of tool temperature modeling, foundational work was carried out by Stephenson and Ali [11], who used Green's function to solve differential equations describing the temperature distribution in tools under interrupted cutting conditions. Wu et al. [12] later enhanced Stephenson's model by extending the energy term to account for heat losses during the non-cutting phase. Oliveira et al. [13] introduced a further advancement by incorporating the thermal effects of tool coatings into a transient temperature model. Most recently, the authors of this paper proposed a transient tool temperature model that includes the effects of convective heat transfer, thereby improving the accuracy of temperature prediction under realistic cutting conditions [14].

This study analyzes the heat transferred to the cutting tool and workpiece during milling under various cutting speeds and feed conditions, using a combination of experimental methods and analytical modeling. The following sections first describe the experimental setup, including the cutting tools and workpiece materials used, along with their relevant properties. Section 3 provides a brief overview of the analytical model employed in this study and explains how the model is used to inversely determine the heat partition. Section 4 presents a discussion of the experimental and simulation results, including the validation of the simulation models and an analysis of the relationship between heat partition and cutting parameters. The final section summarizes the key findings and outlines directions for future research.

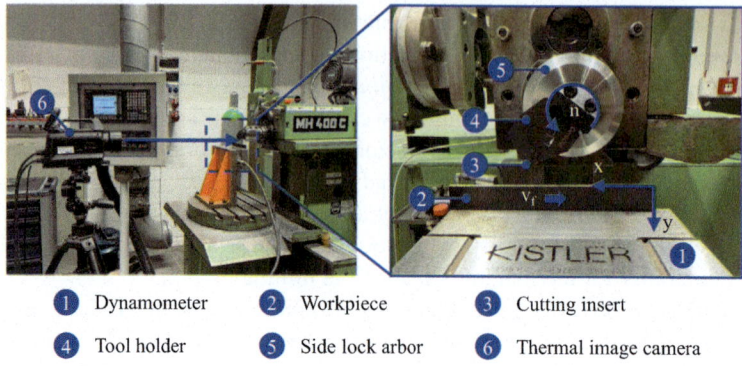

① Dynamometer   ② Workpiece   ③ Cutting insert

④ Tool holder   ⑤ Side lock arbor   ⑥ Thermal image camera

Fig. 1 Experimental setup of temperature measurement in orthgonal milling

2 Experimental Setup

The orthogonal milling test bench was set up on a MAHO MH 400 C horizontal milling machine, as shown in Fig. 1. The feed motion was limited to the x-axis and was realized through the horizontal movement of the machine tool's table, while the main spindle remained stationary without any feed movement. This setup ensured that the machining zone remained fixed in the machine tool's coordinate system, allowing the thermal imaging camera to capture high-quality images of the cutting area from a fixed position.

A FLIR SC7500 thermal camera was positioned at the same height as the spindle axis and aligned perpendicularly to both the workpiece and the side face of the cutting tool. Before the experiment, the camera was calibrated and configured to measure temperatures in the range of 0 to 300 °C. To ensure consistent emissivity close to 1, the tool and workpiece surfaces were coated with matte black paint.

The cutting tool was a custom-designed grooving insert made by Sandvik from H13A cemented carbide. It featured a $\gamma = 12$ ° rake angle and a $\alpha = 3$ ° clearance angle, without any coating or chip breaker. The insert was mounted on a Sandvik tool holder 570-32R123H23B with a 570-2C 32 218 shank, and combined with an SK 40 steep taper holder with side-locking mandrel form ADB from Garant.

The workpiece material was AISI 1045 steel, machined into rectangular plates with dimensions of 100 mm × 3 mm × 50 mm. The workpiece was mounted on a Kistler multi-component dynamometer type 9129AA and positioned perpendicular to the machine table. Process forces were measured along the x-axis (feed direction) and y-axis (vertical direction), in accordance with the coordinate system illustrated in Fig. 1. The cutting force F_c was calculated using the following equation:

$$F_c = F_x \cdot \cos \omega - F_y \cdot \sin \omega \tag{1}$$

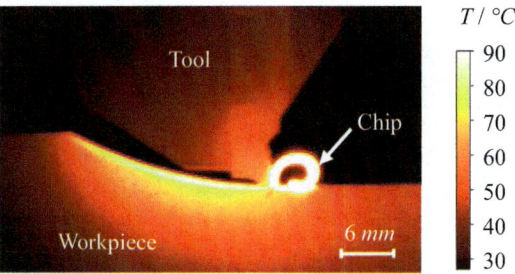

Fig. 2 Example of a recorded temperature distribution in the cutting zone after 10 *s* of machining (cutting speed v_c = 150 m/min, feed per tooth f_z = 0.15 mm)

Table 1 Investigated cutting parameters

Cutting speed v_c / m/min	Axial depth of cut a_p / mm	Radial depth of cut a_e / mm	Feed per tooth f_z / mm
90, 120, 150	3	6	0.15, 0.25

where ω is the tool rotation angle with respect to the vertical direction. F_x and F_y are the measured forces in the x- and y-directions, respectively.

The thermal camera recorded temperature data at a frame rate of 800 Hz. After 10 s of cutting, which corresponds to approximately 50 cuts at a cutting speed of v_c = 90 m/min and about 80 cuts at v_c = 150 m/min, the system reached a quasi-steady thermal state. For model validation and evaluation, the temperature field captured at the moment the tool disengaged from the workpiece was used as the reference. An example of the recorded thermal image is shown in Fig. 2.

All milling operations were performed in down-milling mode. The cutting parameters investigated are summarized in Table 1. Each test was performed twice using a new tool to ensure repeatability and measurement reliability.

3 Analytical Modeling of Cutting Temperatures

The present applies the analytical model developed in the authors' previous work [14] to calculate the heat transfer into the cutting tool and the workpiece. As the investigation focuses solely on dry cutting, the convection term from the original model is excluded. The cutting tool is modeled as a three-dimensional orthogonal body. For simplification, a new tool coordinate system is introduced, which differs from the previously used cutting coordinate system. A square heat source is applied to the rake face, as shown in Fig. 3. The contact area was identified by microscopic analysis after machining. The side surfaces of the tool are assumed to be adiabatic. In the tool length direction (X-axis), the body is treated as semi-infinite due to its distance from

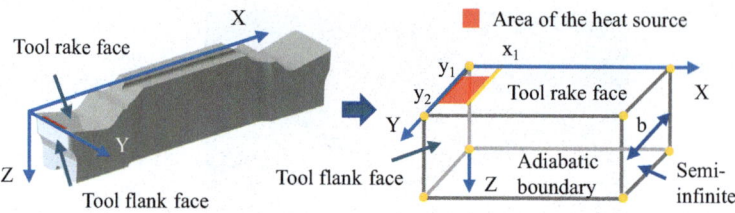

Fig. 3 Schematic of the tool model with heat input at the tool–chip interface

the cutting zone. The bottom surface in contact with the tool holder is simplified as an adiabatic boundary, accounting for thermal contact resistance. The resulting temperature distribution in the tool, T_{tool}, is given by the following equation:

$$T_{tool}(x, y, z, t) = \frac{\alpha_t}{\kappa_t} \int_0^t \dot{q}_t \cdot \left\{ \frac{2}{\sqrt{4\pi\alpha_t(t-\tau)}} \exp\left[-\frac{z^2}{4\alpha_t(t-\tau)}\right] \right\} \cdot$$

$$\frac{1}{2} \cdot \left\{ -\mathrm{erf}\left(\frac{y-y_2}{\sqrt{4\alpha_t(t-\tau)}}\right) + \mathrm{erf}\left(\frac{y+y_2}{\sqrt{4\alpha_t(t-\tau)}}\right) + \mathrm{erf}\left(\frac{y+y_2-2b}{\sqrt{4\alpha_t(t-\tau)}}\right) + \right.$$

$$\left. \mathrm{erf}\left(\frac{y-y_1}{\sqrt{4\alpha_t(t-\tau)}}\right) - \mathrm{erf}\left(\frac{y+y_1}{\sqrt{4\alpha_t(t-\tau)}}\right) - \mathrm{erf}\left(\frac{y+y_1-2b}{\sqrt{4\alpha_t(t-\tau)}}\right) \right\} \cdot$$

$$\frac{1}{2} \cdot \left\{ -\mathrm{erf}\left(\frac{x-x_1}{\sqrt{4\alpha_t(t-\tau)}}\right) + \mathrm{erf}\left(\frac{x+x_1}{\sqrt{4\alpha_t(t-\tau)}}\right) \right\} d\tau + T_r$$

$$(2)$$

where α_t is the thermal diffusivity and κ_t is the thermal conductivity of the tool material. The heat flux $\dot{q}_t$ is modeled as a periodic rectangular pulse function, where the frequency corresponds to the rotational speed of the tool and the pulse width represents the cutting engagement time per revolution. The function erf refers to the error function.

For the workpiece, the milling temperature model developed by Richardson et al. [10] is used. The temperature distribution in the workpiece T_{wp} is calculated by combining Jaeger's moving band heat source solution with a sinusoidal heat flux distribution. The integration is carried out over the angular contact arc between the rotating tool and the workpiece surface. The model setup is illustrated in Fig. 4. The model assumes that the workpiece extends infinitely in the z-direction and in the negative y-direction. The heat flux is considered proportional to the instantaneous chip thickness, which varies with the angular position along the contact arc. In the z-direction, the heat source is also treated as infinite. The surface of the workpiece is modeled as an adiabatic boundary. The resulting equation is given as:

Fig. 4 Analytical model of temperature distribution in the workpiece based on Ref. [10]

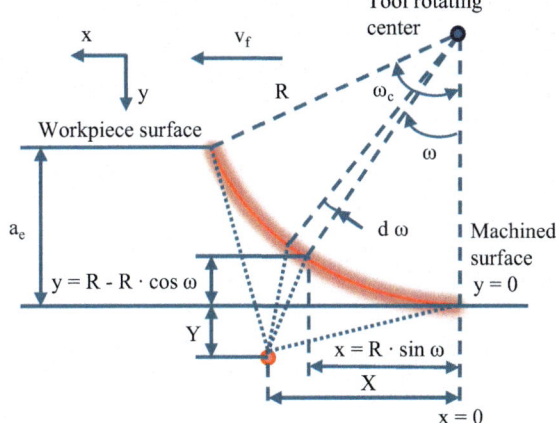

$$T_{wp}(x,z) = \frac{1}{\pi \kappa_{wp}} \int_0^{\omega_c} \dot{q}_m \cdot r_{tool} \cdot \cos \omega \cdot \frac{\sin \omega}{\sin \omega_c} \cdot e^{-v_f(x - r_{tool} \sin \omega)/2\alpha_{wp}} \cdot$$

$$K_0 \left\{ \frac{v_f}{2\alpha_{wp}} \sqrt{(x - r_{tool} \sin \omega)^2 + [y + r_{tool}(1 - \cos \omega)]^2} \right\} d\omega + T_r \tag{3}$$

where α_{wp} is the thermal diffusivity and κ_{wp} is the thermal conductivity of the workpiece material. The angle ω_c represents the angle between the tool and the vertical direction at the moment it first enters the workpiece. K_0 is the hyperbolic Bessel function of the second kind. The parameters r_{tool} and v_f refer to the tool radius and feed rate, respectively. The maximum heat flux $\dot{q}_m$ can be derived from the average heat input $\dot{q}_{avg}$ using the following relation:

$$\dot{q}_m = \dot{q}_{avg} \cdot \left(\frac{\omega_c \sin \omega_c}{1 - \cos \omega_c} \right) \tag{4}$$

To determine the effective heat input into the tool and the workpiece, Equations (2) and (3) are used to simulate their respective temperature fields. The input heat flux is iteratively adjusted until the simulated peak temperature matches the experimental measurement. The total mechanical power is calculated by integrating the measured torque. Assuming that 90 % of this power is converted into heat, the heat partition into the workpiece can be derived accordingly.

The thermophysical properties used in the simulation are listed in Table 2. Tool properties were determined experimentally, while workpiece data were taken from the material datasheet. Since the temperature dependence of material properties is not considered in the tool and workpiece models, all parameters were taken at room temperature. The model accuracy was validated by comparing the simulated temperature distributions with experimental results.

Table 2 Thermophysical properties of the tool and workpiece materials

Property	Unit	Value
Thermal conductivity of the tool κ_t	W/ (m K)	116
Thermal diffusivity of the tool α_t	mm²/s	39.2
Thermal conductivity of the workpiece κ_{wp}	W/ (m K)	48.03
Thermal diffusivity of the workpiece α_{wp}	mm²/s	13.95

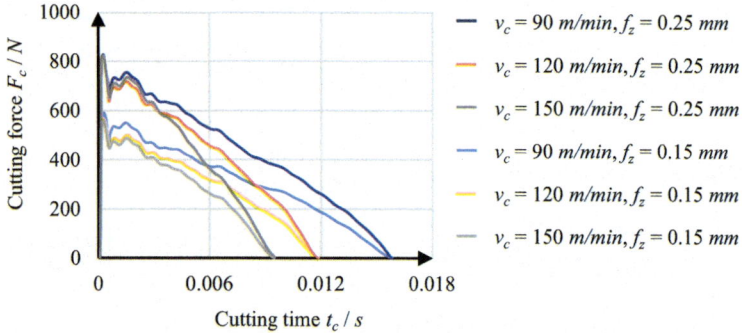

Fig. 5 Cutting force under investigated conditions

4   Results and Discussion

This section presents the experimental and simulation results, focusing on the thermal behavior during the milling process. The cutting force was analyzed first due to its direct influence on mechanical work and heat generation. Figure 5 shows the cutting force calculated using Eq. 1. The curve was smoothed using the weighted average method.

A peak load was observed when the tool engaged the workpiece, causing slight initial system oscillations. As the undeformed chip thickness decreased, the cutting force gradually declined. The feed per tooth significantly affected the maximum cutting force, with higher feed values resulting in greater forces. In contrast, cutting speed had negligible influence. Mechanical work per revolution was determined by integrating the cutting force over time.

Figure 6 presents the mechanical work and corresponding cutting power per tool revolution. At constant feed per tooth, the mechanical work per revolution remained nearly unchanged across different cutting speeds. A higher feed per tooth resulted in increased mechanical work. Cutting speed had minimal effect on the cutting force. Although lower speeds extended the time per revolution, the total energy input per

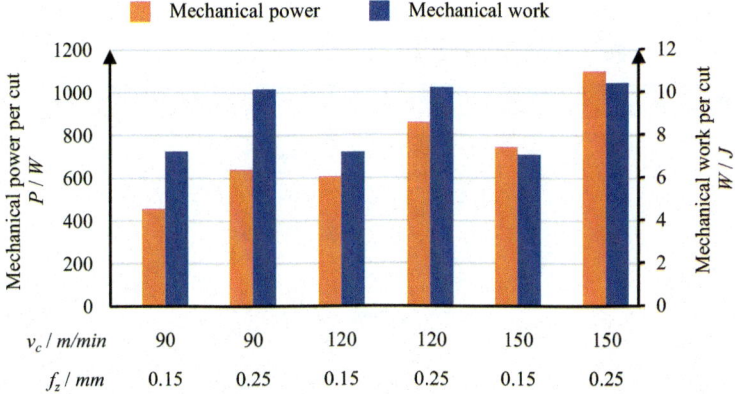

Fig. 6 Mechanical work and average power per tool revolution

revolution was similar. In contrast, higher cutting speeds raised the material removal rate, which led to an increase in process power.

The temperature distribution in the workpiece was simulated using Eq. 3. Figure 7 shows the temperature field at the moment the tool disengaged from the workpiece, based on both simulation and thermal imaging data. The continuous heat map represents the experimental result, while the dashed lines show the simulation output. The simulated peak temperature matched the experimental value. A perfect match would make the dashed contours no longer visible, as they would blend with the color of the thermal image.

The experimental results showed that the workpiece temperature increased significantly with rising cutting speed. This behavior is attributed to the higher cutting power, which results in greater thermal power input. Although an increase in feed per tooth also led to higher cutting forces and more heat generation, most of the heat was carried away by the chips. As a result, the corresponding rise in workpiece temperature was relatively limited. Therefore, it can be concluded that cutting speed is the dominant factor affecting the thermal load on the workpiece.

A comparison between the experimental and simulation results revealed good agreement during the initial engagement of the tool with the workpiece. However, at the moment when the tool exited the workpiece, the simulation noticeably underestimated the temperature. This discrepancy is primarily due to the down-milling process, where the undeformed chip thickness gradually decreases. Near the tool exit, the undeformed chip thickness falls below the cutting edge radius, and the material removal mechanism transitions from shearing to ploughing. During this phase, the increased friction between the tool and workpiece leads to significant heat generation, resulting in a rise in surface temperature [15]. This ploughing-induced thermal effect was not accounted for in the current simulation model.

Similar to the workpiece, the tool temperature was influenced by both cutting speed and feed per tooth. Figure 8 shows the thermal image of the tool alongside

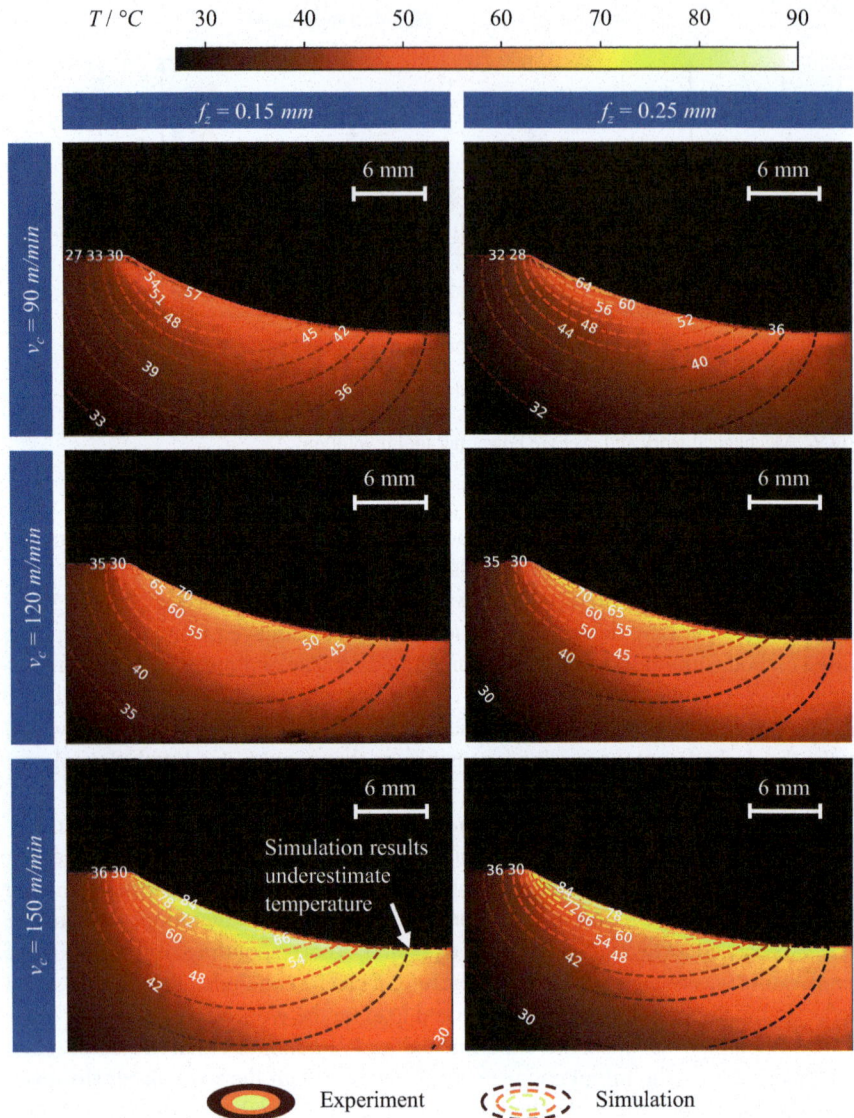

Fig. 7 Comparison of experimental and simulated workpiece temperature distribution

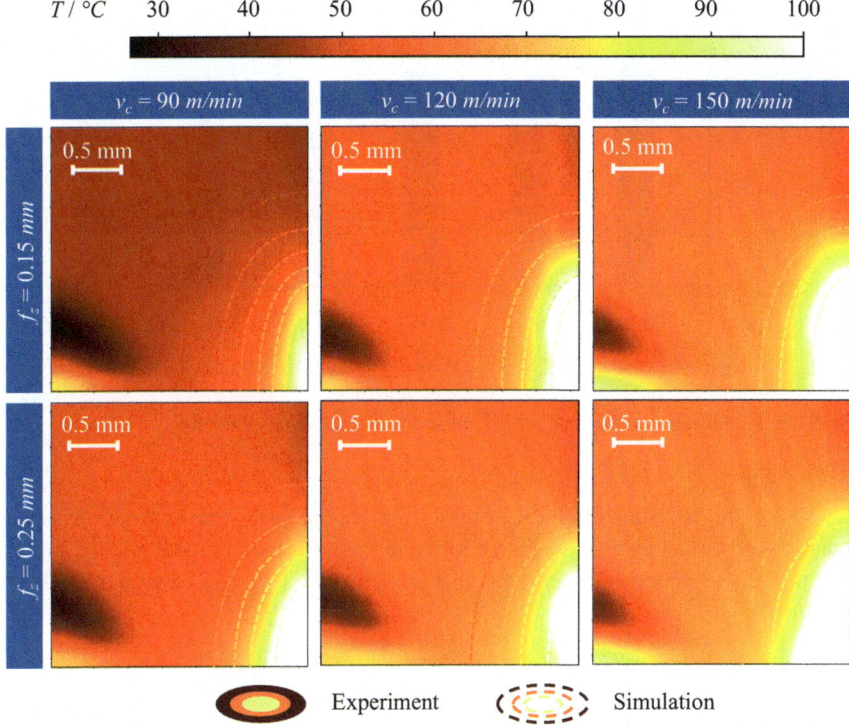

Fig. 8 Comparison of experimental and simulated tool temperature distribution

the temperature distribution simulated using Eq. 2. The thermal imaging results indicated that the high-temperature region was concentrated primarily in the toolâĹ"chip contact area. Outside this region, the temperature dropped sharply. As cutting speed increased, the overall tool temperature rose significantly. The feed per tooth had a relatively minor effect on the overall tool temperature but noticeably affected the extent of the high-temperature zone. Overall, good agreement was observed between the simulation and experimental results, demonstrating that the tool temperature model accurately captured the thermal load on the tool. However, the resolution of the measured temperature field was limited, particularly near the tool edge. This reduced resolution restricts the accuracy of the comparison between simulation and experiment.

The temperature simulation provided estimates of the heat conducted into the tool and the workpiece. By comparing these values with the total cutting energy derived from the cutting force, the heat partition between the tool and the workpiece was determined. Figure 9 summarizes the heat partition under the investigated conditions. The results showed that increasing cutting speed or feed per tooth led to a slightly higher proportion of heat entering the workpiece, primarily due to the increased thermal power in the cutting zone. In contrast, the heat conducted into the tool

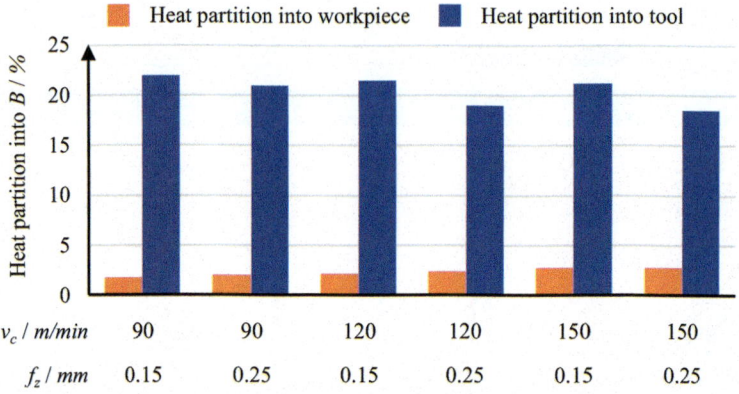

Fig. 9 Heat partition in orthogonal milling under the investigated conditions

decreases. This is because higher cutting speeds lead to faster chip evacuation, which reduces the contact time between the tool and the heat source. Additionally, higher feed per tooth generates larger chips that absorb more heat and carry it away from the cutting zone.

5 Conclusion and Outlook

This study presented a combined experimental and analytical approach to investigate heat partition in orthogonal milling. A specially designed test bench enabled precise thermal measurements using high-speed thermography, while analytical models were employed to estimate the temperature distribution in both the tool and workpiece. The simulation results showed good agreement with experimental data, validating the effectiveness of the model.

The findings revealed that cutting speed is the dominant factor influencing workpiece temperature due to its impact on thermal power input. Although increased feed per tooth raised cutting force and total heat generation, most of the heat was carried away by the chips, resulting in only a moderate temperature rise in the workpiece. In contrast, higher cutting speeds and feed rates reduced the proportion of heat conducted into the tool, primarily due to more efficient heat removal by faster and larger chips.

The proposed method offers an efficient and accurate way to estimate heat distribution in milling without relying on computationally intensive simulations. However, the results also revealed limitations in the analytical model for workpiece temperature. The predicted temperature field deviated from the actual distribution, particularly in cases with small undeformed chip thickness, where ploughing effects become significant. These effects were not considered in the current model. Future work should focus on extending the analytical framework to include ploughing-induced thermal effects, especially near the tool exit, and incorporate temperature-dependent material properties to improve simulation accuracy. In addition, the relatively low resolution of the thermal images captured for the cutting tool limited the reliability of model validation. Future experiments should apply higher-resolution thermal imaging to enhance the accuracy of temperature field measurements on the tool.

Acknowledgements The authors gratefully acknowledge the German Research Foundation (DFG) for funding the SFB TRR 96 Transfer Project T14 (Project No. 174223256).

References

1. Rittel, D., Zhang, L., Osovski, S.: The dependence of the Taylor-Quinney coefficient on the dynamic loading mode. J. Mech. Phys. Solids. **107**, 96–114 (2017)
2. Jiang, X., Wei, Y., Zhou, J., Zhan, K., Ding, Z., Liang, S.: Residual stress generation and evaluation in milling: a review. Int. J. Adv. Manuf. Technology. **126**, 3783–3812 (2023)
3. Putz, M., Schmidt, G., Semmler, U., Oppermann, C., Bräunig, M., Karagüzel, U.: Modeling of heat fluxes during machining and their effects on thermal deformation of the cutting tool. Procedia CIRP **46**, 611–614 (2016)
4. Sato, M., Tamura, N. & Tanaka, H.: Temperature variation in the cutting tool in end milling. J. Manuf. Sci. Eng. **133** (2011)
5. Akhtar, W., Lazoglu, I.: A novel analytical algorithm for prediction of workpiece temperature in end milling. CIRP Ann. **71**, 57–60 (2022)
6. Lazoglu, I., Bugdayci, B.: Thermal modelling of end milling. CIRP Ann. **63**, 113–116 (2014)
7. Langenhorst, L., Sölter, J., Heinzel, C.: Partitioning of primary shear zone heat in face milling. CIRP Ann. **71**, 53–56 (2022)
8. Augspurger, T.: Thermal analysis of the milling process. Rheinisch-Westfälische Technische Hochschule Aachen (2018)
9. Carslaw, H., Jaeger, J.: Conduction of heat in solids. Clarendon Pr (1980)
10. Richardson, D., Keavey, M., Dailami, F.: Modelling of cutting induced workpiece temperatures for dry milling. Int. J. Mach. Tools Manuf. **46**, 1139–1145 (2006)
11. Stephenson, D., Ali, A.: Tool temperatures in interrupted metal cutting. J. Eng. Ind. **114**, 127–136 (1992)
12. Wu, B., Cui, D., He, X., Zhang, D., Tang, K.: Cutting tool temperature prediction method using analytical model for end milling. Chin. J. Aeronaut. **29**, 1788–1794 (2016)
13. Oliveira, G., Ribeiro, S., Guimarães, G.: An inverse procedure to estimate the heat flux at coated tool-chip interface: a 3D transient thermal model. Int. J. Adv. Manuf. Technol. **112**, 3327–3341 (2021)
14. Liu, H., Rodrigues, L., Meurer, M., Bergs, T.: A three-dimensional analytical model for transient tool temperature in cutting processes considering convection. CIRP J. Manuf. Sci. Technol. **43**, 1–14 (2023)
15. Basuray, P., Misra, B., Lal, G.: Transition from ploughing to cutting during machining with blunt tools. Wear **43**, 341–349 (1977)

Liquid Cooling System for Tool Steel Coating on Gears by Directed Energy Deposition

Takumi Nagashima, Masaya Yokota, Shiho Takemura, Takanori Mori, Yoko Hirono, Teppei Maki, and Yasuhiro Kakinuma

Abstract An improvement in the performance of gears in the drive train is required due to the recent shift to electric vehicles. In this study, directed energy deposition (DED), a type of additive manufacturing, was utilized as a coating technology to get higher-performance gears. However, the mechanical properties of coating layers and heat-affected zone (HAZ) depend on the thermal histories. Therefore, a liquid cooling system consisting of a heatsink and a chiller which can adjust water temperature and flow rate was developed to improve the hardness of gear teeth by coating with DED. The performance of the liquid cooling system was evaluated by the temperature transition during the coating process and by measuring the hardness of the coating layer. The combination of the developed liquid cooling system and thermal conduction grease reduced the maximum temperature and exhibited a higher cooling rate during the process. As a result, the hardness of the gear teeth was successfully increased.

Keywords Directed energy deposition · Coating · Thermal compensation

1 Introduction

Over the last years, the automotive sector has been increasingly moving toward the production of electric vehicles to address both current and the future energy and climate crises [1]. Since the gears of electric vehicles are used under severe conditions such as high torque and high-speed rotation, they are required to have high tooth surface strength [2]. At present, carburizing heat treatment is used for

T. Nagashima · M. Yokota · S. Takemura · Y. Kakinuma (✉)
Department of System Design Engineering, Keio University, Yokohama, Japan
e-mail: kakinuma@sd.keio.ac.jp

T. Mori · Y. Hirono
DMG Mori Co., Ltd, Tokyo, Japan

T. Maki
Nissan Motor Co., Ltd, Yokohama, Japan

© The Author(s) 2026   45
K. Wegener and M. Bambach (eds.), *4th International Conference on Thermal Issues in Machine Tools (ICTIMT2025)*, Lecture Notes in Production Engineering,
https://doi.org/10.1007/978-3-032-01194-7_4

surface hardening of gear teeth to achieve this performance. However, since setup changes are typically required in heat treatment processes, it results in longer lead times.

In recent manufacturing technologies, metal additive manufacturing (AM) has greatly impacted across various industries, such as automotive, aerospace, medical, and oil and gas industries [3, 4]. The AM process is distinguished by extremely localized melting and solidification, and it enables the formation of hierarchical microstructures with enhanced mechanical properties [5]. In the process of Directed Energy Deposition (DED), a type of metal AM, metal powders are supplied to a melt pool which is formed by irradiating a laser or electron beam for deposition as shown in Fig. 1. One major advantage of DED is that it eliminates the need for vacuum chambers. As a result, it is particularly suitable for metal coatings to enhance the surface properties of tools [6]. Since the DED coating process does not require setup changes and allows process integration, it leads to shorter lead times. Additionally, the DED process enables the individual treatment of components, allowing functional properties to be customized for each specific part. Moreover, by coating expensive materials onto low-cost substrates, it is possible to reduce the overall use of costly materials while achieving the desired functionality.

Therefore, this study applied DED for tool steel coating on gears to get high performance. In the DED process, a typical cooling rate ranges from 10^3 to 10^5 [7]. From this characteristic of rapid heating and cooling, many studies on the influence of process parameters, thermal cycles, and cooling rate on the mechanical properties or the microstructures in the DED process have been conducted. Ostlaza et al. [8] studied the relationship between the mechanical properties of MMC coating and process parameters. Omar et al. [9] studied the effects of scanning speed and powder feed rate on the hardness of AISI D2, and showed that higher cooling rates lead to increased hardness. Wang et al. [10] studied the influences of laser power and the scanning speed on the microstructure and ductility of Inconel738. Guan et al. [11] clarified that the increase in scanning speed leads to a higher cooling rate, which results in finer cell structures and denser dislocations in a ternary equiatomic alloy. Patil et al. [12] studied the influence of cooling rate on microstructure and hardness of laser

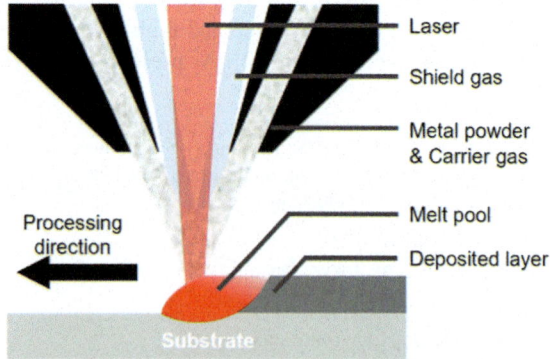

Fig. 1 Scheme of directed energy deposition

Laser

Shield gas

Metal powder & Carrier gas

Melt pool

Deposited layer

Processing direction

Substrate

DED Inconel 718 by changing the laser power and the scanning speed. In this study, higher cooling rate caused by higher scanning speed made the grainsizes smaller and improved microhardness. However, most of the research focuses on cooling rate by changing process parameters such as laser power or scanning speed, and methods of increasing cooling rate independent on process parameters are limited. Moreover, in the DED coating process, the temperature of the coated part rises during laser irradiation due to heat accumulation, then it gradually cools down. Since higher cooling rate hardens steels due to martensite forming and a long holding time during heat treatment causes grain growth, it is important for the strength of coating layers and substrates to manage the cooling rate after the coating. For these reasons, a liquid cooling system for coating on gears, consisting of a heatsink and a chiller which can adjust water temperature and flow rate, was developed to harden the gear teeth. The system was attached to the gear while and after coating. The performance of the system and the effect of water temperature (as one of the parameters) on cooling rate and hardness were experimentally evaluated.

2 Experiment

2.1 Materials

A tool steel PLASweld Ferro55 (Voestalpine Co.) (size range 45–125 μm) was utilized as material powder. Table 1 shows the chemical composition of Ferro55. The substrate of the gear was SCr420H (ISO 20Cr4H), chromium steel. Table 2 shows the chemical composition of SCr420H.

Thermal conduction grease (type G776-100, Shin-Etsu Chemical co., ltd) was utilized to enhance the cooling ability of the developed cooling system. The thermal conductivity was 1.3 W/(m K) [13].

Table 1 Chemical composition of Ferro55 (wt%) [14]

C	Si	Mn	Cr	Mo	Fe
0.35	0.3	1.1	7.0	2.2	Bal

Table 2 Chemical composition of SCr420H [15]

C	Si	Mn	Cr	Mo
0.17–0.23	0.15–0.35	0.55–0.95	0.85–1.25	–
Ni	P	S	Cu	Fe
– 0.25	– 0.03	– 0.03	– 0.3	Bal

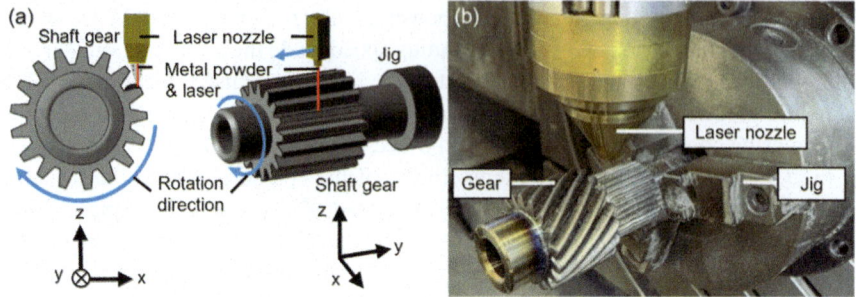

Fig. 2 **a** Gear coating method by DED **b** Experimental setup

2.2   Method of Gear Coating by DED

The coatings on the gears were conducted using a 5-axis DED machine (LASERTEC
65 DED hybrid, DMG Mori co., ltd). The gear coating method by DED and experimental setup are shown in Fig. 2. To coat gear teeth, a gear was rotating at high
speed while laser scanning along the rotational axis of the workpiece. The position
of the laser nozzle was not set vertically above the gear but shifted along the x-axis
from the center of the gear, so that melt pool could be formed on the gear flank.
In addition, although the focus of the powder supply was adjusted to align with the
outer diameter of the gear teeth, the powder particles were blown into the melt pool
formed on the gear flank.

There were two experiments, laser irradiation experiment without powder supply
and the coating experiment. The laser irradiation experiment was conducted to investigate the performance of the cooling system with and without thermal conduction
grease. Then, the coating experiment was conducted to investigate the effect of the
system on the hardness of gear teeth. The experiment conditions of DED are shown in
Table 3, and the experimental conditions of the liquid cooling system in each experiment are shown in Table 4. The experimental condition of the system of the coating
experiment with liquid cooling was the maximum cooling ability, the combination
of the lowest water temperature, the highest flow rate, and with grease.

2.3   Liquid Cooling System

A developed heatsink made of copper is shown in Fig. 3a. To remove heat efficiently,
a heatsink was designed to be inserted into the gear, and water could flow inside the
heatsink as shown in Fig. 3b. It was connected to a chiller (RKS401J-MV-00000,
Orion Machinery co., ltd) which can adjust water temperature (5–40 °C) and flow rate
(3.5–7.2 l/min). Thus, the maximum cooling ability of the chiller was the combination
of 5 °C and 7.2 l/min. Since the heatsink inserted into the gear and the components of
water ports were connected by a rotary joint, the component did not rotate during the

Table 3 Experimental conditions of directed energy deposition

Parameter	Unit	Value	
		Laser irradiation	Coating
Laser power (Preheating)	W	1600	1200
Laser power (Coating)	W	–	1600
Rotating speed	min^{-1}	34.6	
Feed rate at outer diameter of the gear	mm/min	6000	
Scan speed of laser nozzle	mm/min (mm/rev)	20 (0.58)	
Scan length	mm	40	
Powder supply	g/min	0	9.0
Carrier gas supply	l/min	6.0	
Shield gas supply	l/min	8.0	

Table 4 Experimental conditions of liquid cooling system

Experiment		Water temperature [°C]	Flow rate [l/min]	Use of Grease
Laser irradiation	Without cooling	–	–	–
	Cooling 20 °C without grease	20	3.8	–
	Cooling 20 °C with grease	20	3.8	○
	Cooling 5 °C with grease	5	3.8	○
Coating	Without cooling	–	–	–
	With liquid cooling	5	7.2	○

rotational coating. A configuration of the developed liquid cooling system is shown in Fig. 4a.

2.4 Evaluation

A flow meter (FD-H22F, KEYENCE Co.) and water-temperature meters (FI-T25, KEYENCE Co.) were attached to the inlet and outlet of the chiller to monitor the water temperature and the flow rate as shown in Fig. 4b. The temperature at the tooth surface of the gear during and after the coating was measured using an infrared thermal camera (InfRec R500 Pro, Nippon Avionics Co.). The camera was set inside the DED machine as shown in Fig. 4c. The analysis point of the temperature is indicated in Fig. 5a. Since this point was fixed even during rotation, the average temperature

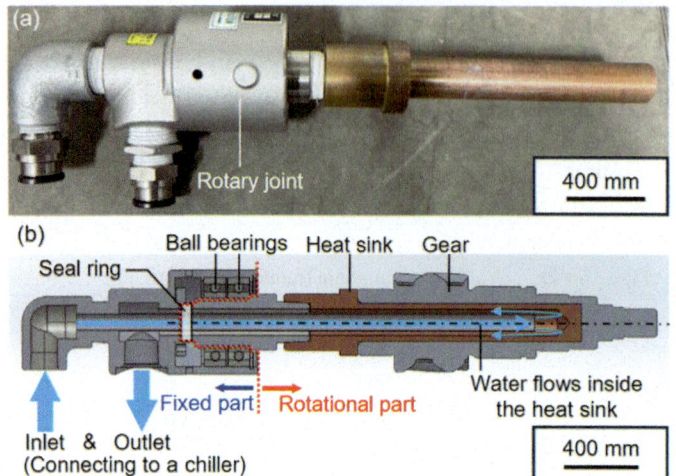

Fig. 3 **a** Appearance of developed heatsink **b** Cross-section of the gear with the heatsink

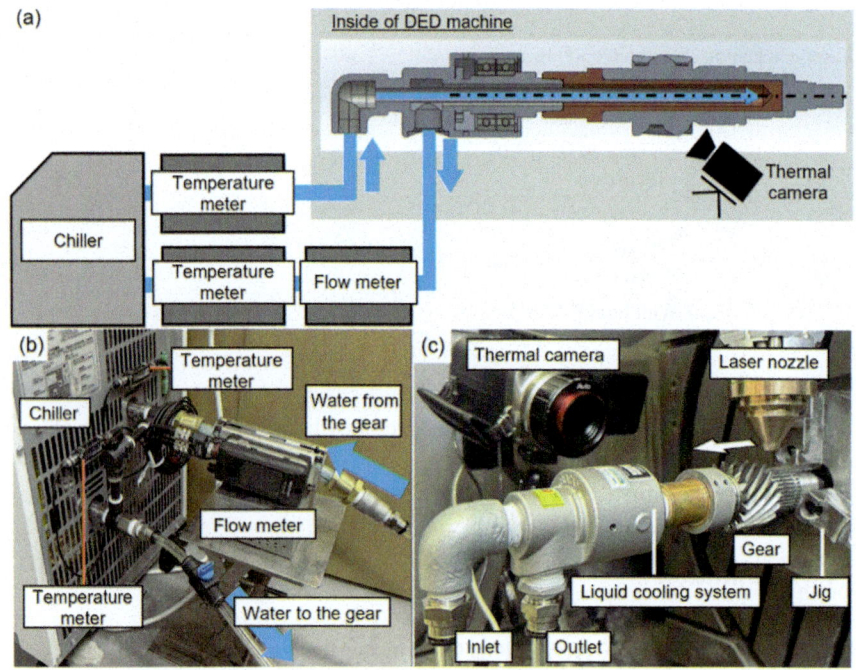

Fig. 4 **a** Configuration of developed liquid cooling system **b** Appearance of the system around a chiller **c** Appearance of the system inside of DED machine

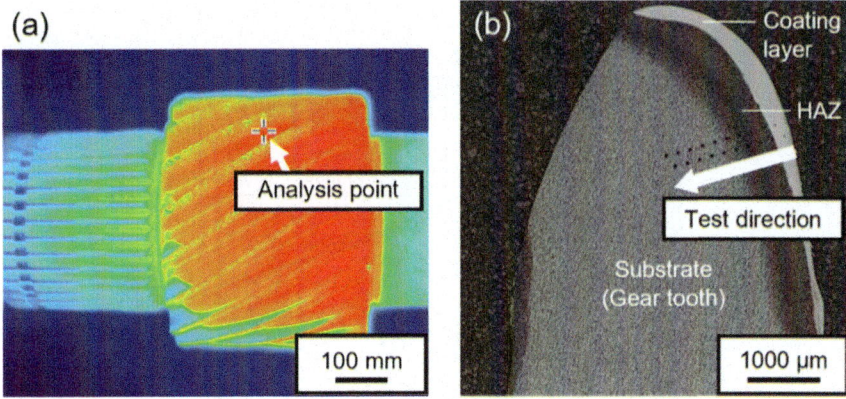

Fig. 5 Methods of evaluation **a** Temperature analysis point **b** Hardness test direction

profile on the same cross-section could be obtained. After the coating experiment, the coated gear teeth were cut, resin-embedded and polished. The hardness of the cross section of the gear teeth was measured using a micro-Vickers hardness tester (Q30A+, QATM). The hardness test was conducted from around the pitch point on the gear flank toward the interior as shown in Fig. 5b. To clarify the relationship between thermal history and hardness, the temperature analysis point was located at the same position as the sectioned area. The cross-sections were observed with a digital microscope (VHX-5000 series, KEYENCE Co.).

3 Results

3.1 Laser Irradiation Experiment

Figure 6 shows the temperature of the gear teeth surface during the laser irradiation experiment. Comparing the temperatures during laser irradiation, the maximum temperatures with liquid cooling were lower than those without cooling. After laser irradiation, each temperature gradually decreased. The temperatures with liquid cooling dropped close to room temperature at approximately 240 s, 300 s, and 500 s for the conditions of cooling at 5 °C with grease, cooling at 20 °C with grease, and cooling at 20 °C without grease, respectively. In contrast, the temperature without cooling remained at 365 °C even after 600 s.

Moreover, the cooling rates in the temperature range of 500–400 °C, as indicated by the gray background, increased not only due to liquid cooling but also with the application of grease and a reduction in water temperature.

The developed liquid cooling system effectively suppressed the temperature rise during the process by increasing heat capacity through the heatsink and the internal

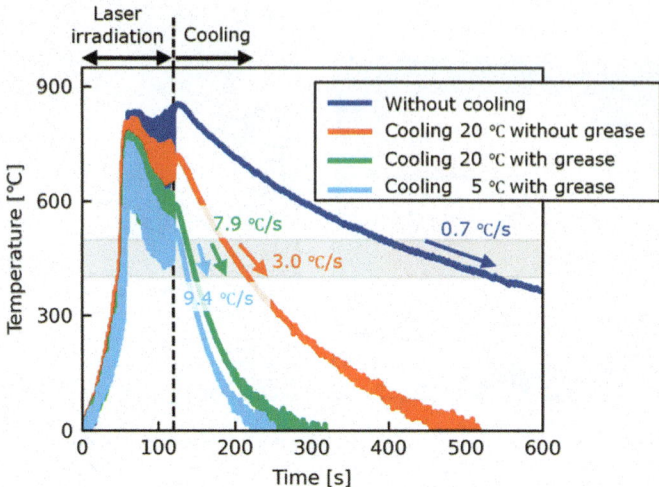

Fig. 6 The temperatures of gear teeth of the laser irradiation experiment. The maximum and minimum temperatures in the calculation range (500–400 °C) correspond to the lowest temperature at the end of laser irradiation and around the highest temperature at the end of temperature measurement (600 s) among the four conditions within the same temperature range. The cooling rates in the temperature range of 500–400 °C, highlighted in gray, were 0.7 °C/s, 3.0 °C/s, 7.9 °C/s, and 9.4 °C/s under the conditions: without cooling, cooling at 20 °C without grease, cooling at 20 °C with grease, and cooling at 5 °C with grease, respectively

water. The peak temperature during the process can be controlled by adjusting water temperature or the use of grease. Additionally, thermal conduction grease and lower water temperatures further enhanced the cooling rates. The performance of the system with and without grease was evaluated in this experiment.

3.2   Coating Experiment

3.2.1   Temperature of Gear Teeth

Figure 7 shows the temperatures of the gear teeth surface during the coating experiment. Some temperature data could not be measured due to the limited storage capacity of the thermal camera. The missing data, represented by dotted line, were interpolated. The range (500–300 °C) in which cooling rates were calculated is indicated with a gray background.

The maximum temperatures recorded during the coating process were 631 °C with liquid cooling and 1248 °C without cooling. The cooling rates for the temperature range of 500–300 °C were 0.5 °C/s without cooling and 7.3 °C/s with liquid cooling.

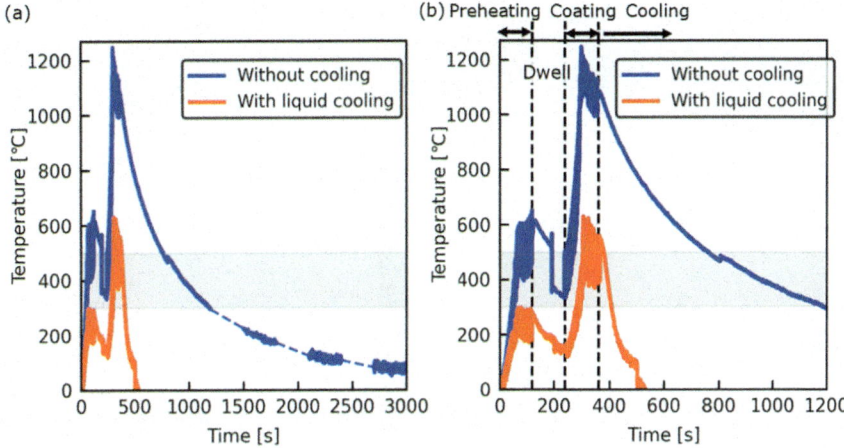

Fig. 7 Temperatures of gear teeth of the coating experiment **a** 0–3000 s **b** 0–1200 s. The maximum and minimum temperatures in the calculation range (500–300 °C) correspond to the lowest temperature at the end of coating and the highest temperature at the end of temperature measurement (1200 s) in both conditions for comparison within the same temperature range

Similar to the laser irradiation experiment, the developed liquid cooling system suppressed the temperature rise during the process and significantly enhanced the cooling rate.

3.2.2   Hardness of Coated Gear Tooth

The optical microscope images of gear tooth cross-section are shown in Fig. 8 and the results of the hardness test on the cross-section of the coated gear teeth are shown in Fig. 9. The hardness of both the coating layer and the heat-affected zone (HAZ) increased with the application of the developed liquid cooling system. The average hardness of the coating layer without liquid cooling (from the surface to 0.3 mm depth) was 533 HV, whereas with liquid cooling, it increased to 650 HV (from the surface to 0.2 mm depth).

Similarly, the average hardness of the HAZ (0.4–1.6 mm from the surface) was 250 HV without liquid cooling and increased to 318 HV (0.3–0.9 mm from the surface) with liquid cooling.

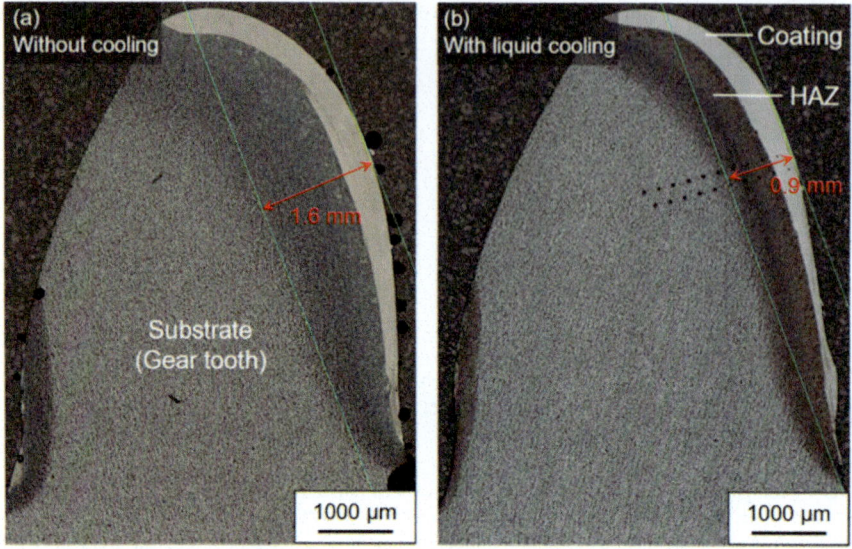

Fig. 8 Gear tooth cross-section **a** Without cooling **b** With liquid cooling. The HAZ at the hardness testing point reached a depth of 1.6 mm from the surface without cooling, and 0.9 mm with cooling

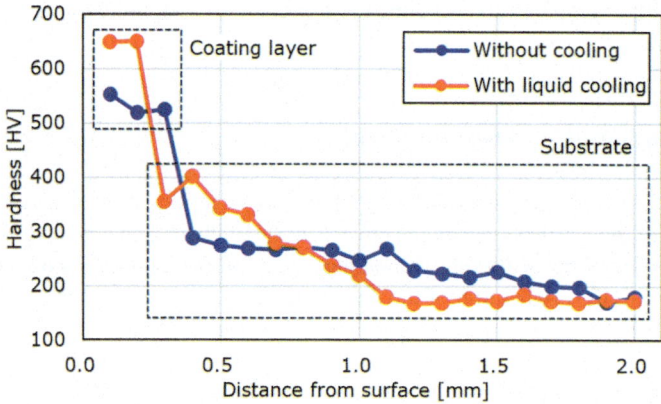

Fig. 9 Hardness test result of gear tooth

4   Discussion

4.1   *Thermal Contact Resistance*

Heat flux $q\left[\text{W/m}^2\right]$ is expressed as follows:

$$q = -\lambda \frac{\partial T}{\partial x} \tag{1}$$

Fig. 10 Scheme of thermal contact resistance generation [16]

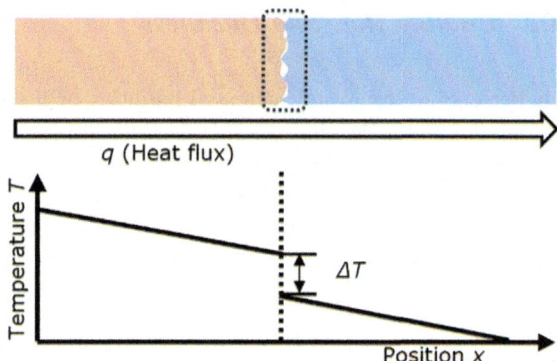

where λ is the thermal conductivity, $\frac{\partial T}{\partial x}$ is the temperature gradient. A large temperature gradient was observed due to the inserted heatsink and internal water. Additionally, a lower water temperature resulted in a greater temperature gradient.

When two solids come into contact, the actual contact area is much smaller than the apparent one due to surface roughness and shape irregularities. Additionally, heat transfer is hindered because air, which has low thermal conductivity, fills the gaps in the non-contact areas. As shown in Fig. 10 [16], a rapid temperature drop $\Delta T [\mathrm{K}]$ occurs within a very narrow region of the contact surface. This temperature drop is caused by thermal contact resistance $R [\mathrm{K} \cdot \mathrm{mm}^2/\mathrm{W}]$ which is given as follows [17]:

$$R = \frac{\Delta T}{q} \tag{2}$$

The use of thermal conduction grease reduced the thermal contact resistance by filling the gaps in the non-contact areas, thereby increasing the heat flux.

Consequently, the heat flux was enhanced by liquid cooling, which created a larger temperature gradient, and by thermal conduction grease, which reduced the thermal contact resistance, ultimately leading to increased cooling rates.

The chiller can lower the water temperature to 5 °C. However, if the water temperature is lower than room temperature, the heatsink may shrink, increasing the gaps between the heatsink and the gear, which in turn raises the thermal contact resistance. Therefore, the use of thermal grease is highly beneficial for the developed liquid cooling system, as it helps efficiently remove heat while allowing the system to operate even with lower water temperatures.

On the other hand, the use of grease may interfere with the automatic exchange of workpieces in industrial production lines. Thus, alternative methods to grease may need to be explored for practical implementation.

4.2 Cooling Rate and Hardness

Higher cooling rates and increased hardness of the coating layer and HAZ were achieved with the developed liquid cooling system.

The hardness of steel is strongly influenced by its martensitic phase fraction, while the presence of Fe_3C may also play a role. The fraction of martensite in the coating layer and HAZ could have been increased by a higher cooling rate using liquid cooling. Martensite has a fine microstructure and the strength improvement due to grain refinement is well known and is described by the Hall–Petch equation [18]:

$$\sigma = \sigma_0 + kd^{-1/2} \tag{3}$$

where σ, σ_0, k, and d are the yield strength, the friction stress, a constant, and the grain size, respectively. This equation indicates that smaller grain sizes lead to increased strength. In addition, martensite grain size is influenced by prior austenite grain size. Prior austenite grain size during quenching is determined by the maximum heating temperature, with a tendency to increase as the temperature rises [19]. This suggests that the reduction of the maximum temperature during the coating, achieved by the cooling system, has contributed to smaller prior austenite grain size.

Thus, it can be inferred that the higher cooling rate and the suppression of the maximum temperature by the developed liquid cooling system enhanced the hardness of the coating layer and HAZ by promoting martensite formation and grain refinement.

5 Conclusion

In this study, the performance of the developed liquid cooling system, consisting of a heatsink and a chiller, for the tool steel coatings on gears using the DED process was evaluated by measuring the temperatures of gear teeth during the processing and the hardness of the coated gear teeth.

The developed liquid cooling system effectively reduced the peak temperatures during the process and significantly increased cooling rates. This effect was particularly pronounced due to the increase in heat flux, which resulted from the enhanced temperature gradient caused by lower water temperature and the reduced thermal contact resistance achieved by using thermal conduction grease. Furthermore, the system improved the hardness of the coating layer and HAZ by increasing cooling rates.

References

1. Ramos, M.R., Ruiz-Gálvez, M.E.: The transformation of the automotive industry toward electrification and its impact on global value chains: inter-plant competition, employment, and supply chains. Eur. Res. Manag. Bus. Econ. **30**, 100242 (2024)
2. Watanabe, Y.: Materials and surface modification technologies for sliding parts such as gears with functions demanded for next generation automobiles. J. Jpn. Soc. Tribol. **66**(3), 175–181 (2021). (in Japanese)
3. Armstrong, M., Mehrabi, H., Naveed, N.: An overview of modern metal additive manufacturing technology. J. Manuf. Process. **84**, 1001–1029 (2022)
4. Patel, S., Liu, Y., Siddique, Z., Ghamarian, I.: Metal additive manufacturing: principles and applications. J. Manuf. Process. **131**, 1179–1201 (2024)
5. Wang, Y.M., et al.: Additively manufactured hierarchical stainless steels with high strength and ductility. Nat. Mater. **17**, 63–71 (2018)
6. Park, G.-W., et al.: Effect of residual stress on pore formation in multi-materials deposited via directed energy deposition. Addit. Manuf. **81**, 104016 (2024)
7. Ahn, D.-G.: Directed energy deposition (DED) process: state of the art. Int. J. Precis. Eng. Manuf.-Green Technol. **8**, 703–742 (2021)
8. Ostlaza, M., Arrizubieta, J.I., Lamikiz, A., Ukar, E.: Study on the flexural behaviour and bonding strength of WC-Co metal matrix composite coatings produced by laser directed energy deposition. Surf. Coat. Technol. **463**, 129538 (2023)
9. Omar, S.M.T., Plucknett, K.P.: The influence of DED process parameters and heat-treatment cycle on the microstructure and hardness of AISI D2 tool steel. J. Manuf. Process. **81**, 655–671 (2022)
10. Wang, G., Yang, L., Dai, G., Chen, C., Qin, Y., Yang, S.: Influence of processing parameters on the thermal history, microstructure and tensile properties of Inconel 738 produced by laser directed energy deposition. Opt. Laser Technol. **177**, 111081 (2024)
11. Guan, S., et al.: Revealing thermal behavior, cracking behavior, phase and microstructure formation of a ternary equiatomic alloy additively manufactured using directed energy deposition. Addit. Manuf. **78**, 103897 (2023)
12. Patil, M.A., Ghara, T., Das, B., Kulkarni, D.M.: Influence of cooling rate on microstructure, dislocation density, and associated hardness of laser direct energy deposited Inconel 718. Surf. Coat. Technol. **495**, 131575 (2025)
13. G776-100 Thermally Conductive Grease (Low Viscosity Type) Shin-Etsu Chemical, MISUMI. Accessed: 18 Apr. [Online]. Available: https://jp.misumi-ec.com/vona2/detail/223004970 849/?ProductCode=G776-100
14. Metal Powders—voestalpine Böhler Welding. Accessed: 31 Jan 2025. [Online]. Available: https://www.voestalpine.com/welding/global-en/products/product-search/metal-powders/
15. Japan Industrial Standards, JIS G 4052 Structural steels with specified hardenability bands, (2016) (in Japanese)
16. Xie, M.: Experimental study on contact thermal resistance of common thermal interface materials. In: Proceedings of the Eighth Asia International Symposium on Mechatronics, pp. 1250–1260 (2022)
17. Shimizu, S., Kikumori, K., Sakamoto, H., Tamaoki, K.: Quantitative measuring method of thermal contact resistance considered real contact area. J. Jpn. Soc. Precis. Eng. **71**(8), 1026–1030 (2005) (in Japanese)
18. Kato, M.: Hall-Petch relationship and dislocation model for deformation of ultrafine-grained and nanocrystalline metals. Mater. Trans. **55**(1), 19–24 (2014)
19. Hayashi, T., Kurosawa, N., Yamada, K.: The method of prior austenite grain refining using induction hardening. JFE Tech. Rep. **23**, 4–9 (2009). (in Japanese)

Investigation of the Influence of Temporally and Spatially Resolved Heat Fluxes at the Cutting Edge on the Maximum Tool Temperature During Side Milling

Simon Winter, Tim Göttlich, Hui Liu, Thomas Bergs, and Reinhold Kneer

Abstract The maximum machining speed in milling processes is significantly limited by the upper temperature limit of the materials of the tools used. During the cutting process, large heat fluxes occur at the cutting edge of the tool, resulting in short term temperature peaks. The precise measurement of these temperature peaks is very challenging, due to their highly transient nature and limited spatial accessibility. Analytical and numerical simulations are therefore a suitable tool to determine the thermal load on the milling cutter. Current FEM modeling approaches for machining processes often require substantial computational resources, resulting in significant CPU time and memory usage. Therefore, the aim of this work is to develop a fast and computationally efficient simulation approach to predict transient tool temperatures, with a focus on the accurate identification of temperature peaks. In a previous study, a coupled approach was used to determine the heat fluxes at the cutting edge caused by the milling cutter's engagement in the workpiece. For this purpose, an FEM machining simulation of the milling process was combined with a simpler heat conduction simulation of the cutter. The heat fluxes then served as boundary conditions in the heat conduction simulation to calculate the resulting temperature field. In this approach, a spatially and temporally constant geometry engagement and thus a spatially and temporally constant heat flux into the cutting edge has been taken into account up to now. In the present work, the temporally and spatially varying nature of the cutting edge heat flux is considered and its influence on the resulting temperature field and especially the maximum temperature of the tool is investigated. The results of the extended thermal simulation model are presented and compared with the results of the previous work.

Keywords Transient thermal modeling · Side milling · Finite-element-method · Cutting edge heat transfer · 3D heat conduction simulation

S. Winter (✉) · T. Göttlich · R. Kneer
Institute of Heat and Mass Transfer, RWTH Aachen University, Aachen, Germany
e-mail: winter@wsa.rwth-aachen.de

H. Liu · T. Bergs
Manufacturing Technology Institute, RWTH Aachen University, Aachen, Germany

© The Author(s) 2026
K. Wegener and M. Bambach (eds.), *4th International Conference on Thermal Issues in Machine Tools (ICTIMT2025)*, Lecture Notes in Production Engineering,
https://doi.org/10.1007/978-3-032-01194-7_5

1 Introduction

During the milling process, approximately 90 % of the mechanical cutting power is dissipated into heat [1]. Around 15 % of the heat generated flows from the cutting zone into the tool, the rest is divided between the chip and the work piece. As a result of this heat flow, large temperature gradients develop, leading to increased temperatures within the tool. As a result the milling tool is exposed to locally varying thermal expansion inside the milling cutter. This induces thermal stresses that increase wear and thus reduce durability. In addition, inhomogeneous expansion can lead to distortion of the center of the tool, making the entire milling process less accurate. Thus, precise knowledge of the tool temperature field is of great importance. Therefore, some experimental work was carried out to determine the cutter temperature during the milling process. The easiest way to do this is to use thermocouples in the milling cutter, although measurement errors can occur due to the thermal inertia of the thermocouples [2]. To avoid this, optical measuring systems such as two-color pyrometers can be used to determine a local cutter temperature [3]. An infrared camera can be used to resolve the entire temperature field of the milling cutter instead of one measuring point [4]. However, optical measurement of the milling cutter temperature requires continuous optical access, which can be restricted by rotation, detaching chips or the cooling lubricant used. This makes it difficult to take detailed temperature measurements of the cutter during the cutting process. Therefore, using modeling approaches to estimate the cutter temperature during the machining process is a reasonable alternative. Analytical approaches such as the Greens function are often only valid for simple tool geometries and uniform heat flux distributions, such as orthogonal cutting [5]. To map the complex geometry and the transient heat flux distributions along the tool cutting edge, numerical modeling is better suited to determine the three-dimensional temperature field. For this purpose, a numerical modeling approach was developed in a previous work by Helmig et al. with which the transient temperature field of the milling cutter during the machining process can be determined [6]. For this purpose, the resulting heat flux into the tool cutting edge is first calculated using a computationally intensive three-dimensional chip formation simulation, which requires a considerable amount of CPU time and memory. These are then provided as boundary conditions for a much less computationally intensive three-dimensional thermal conduction simulation of the milling cutter in order to determine the resulting temperature field. This approach reduces the overall computational effort, leading to shorter simulation times. In the present work, the influence of a spatially resolved heat flux at the tool cutting edge compared to average heat fluxes on the temperature field and especially on the maximum temperature of the milling cutter is to be investigated.

2 Setup of the Modeling Approach

The milling cutter examined is a solid carbide end mill from Sandvik Coromant, type 2P342-1600-CMA 1740 with a diameter of $d = 16$ mm. It consists of four cutting edges and a flute helix angle of $FHA = 38°$. The geometry of the cutter is shown in Fig. 1 on the left. The cutter is not considered in its entirety, but only at the cutting edges over a length of $l = 25$ mm, which represent the region of interest.

First, three-dimensional chip formation simulations are carried out with the milling cutter using the commercial software *ABAQUS 2021* to determine the resulting spatially resolved heat flux into the respective tool cutting edge. A coupled Euler-Lagrangian (CEL) approach is used to model the interaction between the milling cutter and the workpiece, in which the milling cutter is defined as a rigid body in the Lagrangian domain. The workpiece is defined in the Eulerian domain using a Johnson-Cook material model for AISI 1045 steel for the thermo-viscoplastic behavior of the material. In this study, a side milling process using the rolling surface with a milling depth of $a_p = 6$ mm and a cutting velocity of $v_c = 150 \frac{m}{min}$ is examined as a representative process in the simulation. The detailled cutting parameters of the test case are as follows (Table 1).

The heat input into the milling cutter results from the dissipation of the cutting power. The heat flux at the respective engaging cutting edge is subsequently determined as a function of time and location with a chip formation simulation.

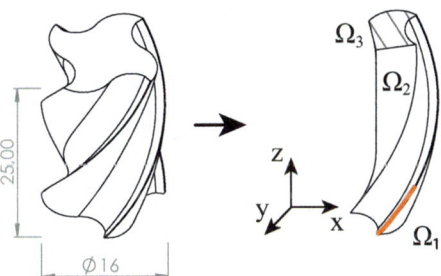

Fig. 1 Left: Geometry of the milling cutter. Right: Computational domain of the quarter milling cutter with boundary conditions. For boundary condition Ω_1, a heat flux boundary condition representing the machining process is applied. The outer sides Ω_2 of the tool and top of the tool Ω_3 are adiabatic

Table 1 Cutting parameters of the test case

Cutting speed v_c	Feed f	Cutting width a_e	Cutting depth a_p
$150 \frac{m}{min}$	0.2 mm	3 mm	6 mm

Table 2 Thermophysical properties of the milling cutter

Density ρ	Heat capacity c_p	Heat conductivity k
$114800 \frac{kg}{m^3}$	$243 \frac{J}{kgK}$	$81.78 \frac{W}{mK}$

In the next step, this is then transferred into a boundary condition for the thermal analysis of the milling cutter. This is carried out in *MATLAB 2023* in the *Partial Differential Equation Toolbox*, in which the three-dimensional transient heat conduction equation is solved in order to determine the resulting temperature field T:

$$\rho c_p \frac{\partial T}{\partial t} = k \frac{\partial^2 T}{\partial x^2} + k \frac{\partial^2 T}{\partial y^2} + k \frac{\partial^2 T}{\partial z^2} \tag{1}$$

The thermophysical properties density ρ, heat capacity c_p and thermal conductivity k are assumed to be constant. Table 2 shows the corresponding values:

The geometry of the cutter is divided into a numerical calculation grid and subsequently the equation is discretized using the finite volume method. For reasons of computational efficiency, only one cutting edge and only one quarter of the cutter are modeled by using the symmetry condition. The shape of the computational domain with the respective boundary conditions is shown in Fig. 1 on the right.

To solve the heat conduction equation, a homogeneous temperature field with a uniform temperature of 30 °C is selected as the starting condition for the milling cutter. For the required boundary conditions, the outer surfaces of the cutter are selected as adiabatic for simplification (Ω_2 in Fig. 1). The boundary condition on the milling cutter shank is also selected as adiabatic (Ω_3 in Fig. 1). The heat flows from the tool into the machine structure and the environment are not yet considered here, but will be the focus of future work. Due to these selected adiabatic boundary conditions, the tool temperature will tend to be slightly overestimated. The only boundary condition that can still influence the cutter temperature is the boundary conditions on the cutting edges, which is the focus of this work.

In the first step, the temporal and local heat flux density $\dot{q}''(x, y, z, t)$ from the chip formation simulation is applied as a boundary condition to the respective cutting edge (Ω_1 in Fig. 1) in order to map the machining process. Due to the cutter geometry and the circumferential face milling with the rolling surface, only one cutting edge is in contact with the workpiece at any one time, where the cutting power is dissipated. The twisted shape of the cutting edge leads to the spatial and temporal dependence of the heat flux at the cutting edges shown in Fig. 2 at the top. Within one revolution of the milling cutter, lasting $\tilde{T}$=20.1 ms with these cutting parameters, each of the four cutting edges is engaged for a quarter of the time of one period, thus 5 ms. For the remaining $\frac{3}{4}$ of the rotation, the milling cutter is not in contact with the workpiece during side milling and the cutting edge heat flux drops to $0 \frac{W}{mm^2}$. The local heat flux at each point of the cutting edge has a similar progression over time. When it first

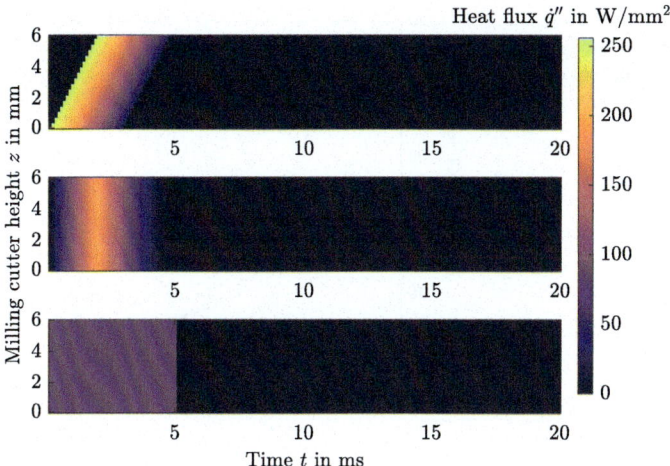

Fig. 2 Top: Case 1—Locally and temporally dependent heat flux at the cutting edge. Middle: Case 2—Temporally dependent heat flux at the cutting edge. Bottom: Case 3—Temporally and locally averaged heat flux at the cutting edge

enters, it reaches its maximum of 256 $\frac{\text{W}}{\text{mm}^2}$ and then decreases until it drops again to 0 $\frac{\text{W}}{\text{mm}^2}$ when it exits.

In order to further reduce the complexity and computational effort of the thermal simulation and make it as efficient as possible, the temperature field is calculated for two additional configurations of the heat flux boundary condition.

On the one hand, a time-dependent heat flux is selected here, in which an average heat flux is selected for the engaging part of the cutting edge, which can be seen in Fig. 2 in the middle. The periodic behavior can also be seen there, but the maximum of the heat flux only reaches its peak value at 191 $\frac{\text{W}}{\text{mm}^2}$.

On the other hand, the largest simplification is selected, in which the engaging part of the cutting edge receives a temporally and spatially constant heat flux during the engagement, as was already selected in the previous work by Helmig et al. [6]. The resulting heat flux can be seen in Fig. 2 at the bottom. The average heat flux per cutting edge in the engagement is 87 $\frac{\text{W}}{\text{mm}^2}$. The heat flux resulting form the temporal and local averaging is equivalent to the local averaging of the heat fluxes over the engagement period. However, it is referred to in the following as local and temporal averaged, as this corresponds to that of Helmig et al. in the previous work.

The resulting temperature field, in particular the maximum temperature of the cutter, is subsequently calculated as a function of these heat flux boundary conditions.

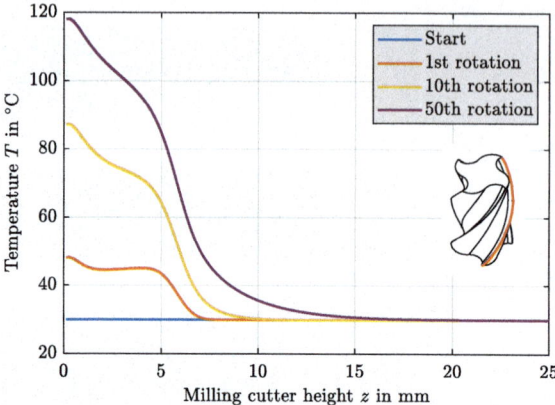

Fig. 3 Local temperature profile along the cutting edge of the cutter at the end of the respective revolution

3 Results

First, the local temperature profile along the cutting edge at the end of different numbers of revolutions for case 1 is shown in Fig. 3. Initially, only the temperature field of the area of the cutting edge where the heat flux is introduced by the machining is affected. Here, the highest temperature is at the tip of the cutting edge, as the geometry of the cutting edge restricts the dissipation of heat into the solid body of the milling cutter. With increasing time, the heat flows along the cutting edge and also leads to an increase in temperature in the areas that are not engaged. The maximum temperature occurs at the tip of the cutting edge at $z = 0$ mm. Therefore, the temperature profile over time at this critical point is analyzed in the following.

Figure 4 shows the temperature curve of all three cases at the tip of the cutting edge of the milling cutter $z = 0$ mm during the first three revolutions. The period duration $\tilde{T}$ of one revolution is 20 ms. During the first $\frac{1}{4}$ revolution, the areas on

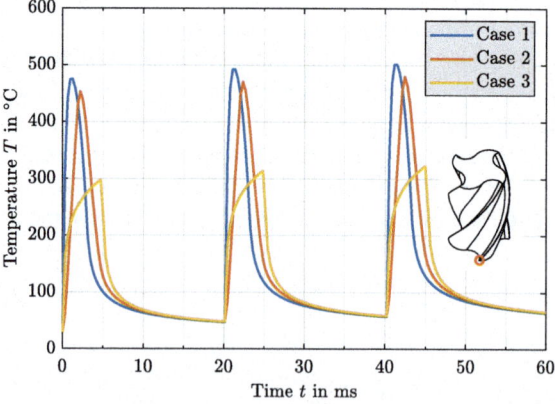

Fig. 4 Temperature progression over time at the cutting edge tip at $z = 0$ mm for the first three revolutions for the three cases

the cutting edge of the milling cutter engage with the workpiece. In all three cases, the temperature profiles exhibit oscillating behavior over the three revolutions, characterized by steep temperature gradients that gradually attenuate as the maximum is approached. After reaching this maximum, the temperature decreases sharply at first, then more gradually as it nears the minimum by the end of the revolution.

The thermal energy is either transferred directly to the observed point at the cutting edge by the dissipated heat from the machining process by the cutting engagement or by heat conduction from surrounding hotter areas of the cutter. At the same time, heat flows towards the areas in the cutter body with lower temperatures, such as in the center of the cutter or at the areas of the cutting edge that are not engaged. During the engagement, the incoming heat flow from the machining process dominates the energy balance and the temperature rises. After a balance of incoming and outgoing heat flow is reached at the time of the temperature maximum, the outgoing heat flow dominates the energy balance for the duration of the remaining $\frac{3}{4}$ revolution, as the temperature is higher than in the rest of the body and the temperature drops. This behavior is repeated in every revolution with the period duration $\widetilde{T}$.

Comparing the three curves, it is noticeable that the maximum temperature in case 1 is reached first with temporal and spatial resolution of the engagement after 1 ms and is the highest among the 3 cases. For case 2 and case 3, the maxima are later in time at 2 ms and 5 ms and also 20 K or 270 K lower. At the beginning of each revolution, case 1 and case 3 exhibit the steepest temperature gradients, whereas the gradient in case 2 is slightly more moderate. In case 1 and case 2, the gradients remain largely unattenuated until near their respective temperature maxima, while in case 3 a significant flattening occurs well before the maximum is reached. However, it can be seen that the area under the peaks is the same in each case, as the same amount of heat is introduced in each case.

In case 1 the point at the tip of the cutting edge initially experiences a high heat flux that decreases over time. As the temperature at this point rises, heat is simultaneously conducted into the cooler regions inside the milling tool body. Although the direct heat input at the considered point diminishes over the duration $\widetilde{T}$ of the revolution, adjacent points on the cutting edge above the observed point receive a time delayed heat flux from the milling process. Consequently, after the initial direct heat flux inflow, additional heat is conducted from neighboring regions onto the cutting edge. Thus, the temperature at the observed point increases rapidly due to the high initial heat flux and continues to rise by conduction from hotter areas even after the highest value of the heat flux from cutting engagement has passed the observed point. Only if the current contact surface is far enough away the conductive heat inflow to the observed point falls below the heat flow flowing from this point to cooler areas of the milling cutter body. At that stage, the temperature at the observed point begins to decline. The pronounced local and time dependent effects in case 1 result in a faster and higher temperature rise, followed by a more rapid cooling compared to case 2 and case 3.

In case 2, all points along the cutting edge experience simultaneously an identical, time varying heat flux. Initially, the heat flux at the tip of the cutting edge is lower compared to case 1 and case 3, before increasing to a maximum and finally declining toward the end of the engagement period. This temporal behavior results in an overall lower temperature gradient compared to case 1. Due to the lack of spatial resolution of the heat source, the peak heat flux at the observed point is lower than in case 1, which is why the maximum temperature is lower than in case 1.

In case 3 all points along the cutting edge experience the same spatially and temporally constant heat flux during the $\frac{1}{4}\widetilde{T}$ engagement period. Locally, at the beginning of the cutting process, this heat flux exceeds that of case 2 for all points on the cutting edge, which leads to a comparatively steeper temperature gradient. As the temperature rises rapidly at the point of the cutting edge, the heat flow to the colder areas in the cutter increases accordingly. Since the constant heat flux is lower than the maximum fluxes observed in case 1 and case 2, but still remaining above their minima, the resulting temperature maximum is lower and shifted to the right, occurring at $t = \frac{1}{4}\widetilde{T} = 5\,\mathrm{ms}$. The total energy input during the engagement period is equivalent to that case 1 and case 2.

Figure 5 shows the maximum temperature of the milling cutter for each rotation for the three cases, so that a longer temperature trend can be examined. It can be clearly seen that all three curves show a logarithmically increasing maximum temperature, as the internal energy of the cutter continues to increase over time. The curve of case 1 shows the highest temperatures, followed by the curves of case 2 and case 3. After 50 revolutions the maximum temperature of case 1 is 560 °C, while case 2 is 540 °C and case 3 is 380 °C. Due to the averaging of the applied heat flux from the machining process, the maximum temperature in case 2 and case 3 is underestimated by 20 °C and 180 °C, respectively, compared to case 1, in which a spatially and temporally resolved heat flux is taken into account. It is evident that the maximum temperatures vary depending on the calculation method. However, for each rotation, the temperature difference corresponds to a constant offset.

However, it is interesting to note that the offset between the maximum temperatures of the different methods remains constant over the revolutions. This is due to the fact that the energy input into the milling cutter is the same for all three cases.

Fig. 5 Maximum temperature of the milling cutter for each revolution of the three cases

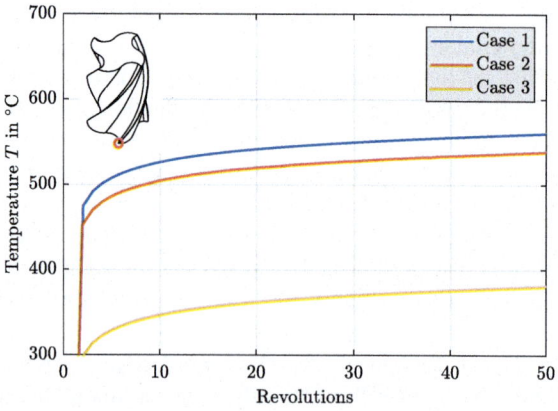

In the future, this could mean that if this offset is known, the time-dependent maximum temperature of the cutter could also be determined using an averaged heat flux and then corrected using the offset. The comparison of the computing times indicates that case 1 requires the most computing time under the same computational grid conditions, while case 2 requires 60% less computing time compared to case 1. Case 3 requires 92% less computing time compared to case 1 and is therefore the most computationally efficient solution.

4 Conclusion

A novel approach for estimating the transient temperature field of a milling cutter during machining, which was first presented in Helmig et al. [6], was further investigated with regard to the influence of spatially and temporally differently resolved heat sources on the resulting temperature field. The resulting spatially and time-resolved heat flows into the cutting edge of the milling cutter were determinded using a three-dimensional chip formation simulation. Subsequently, the averaged heat fluxes were transferred to the thermal simulation of the milling cutter as a boundary condition in order to determine the resulting temperature field and the maximum temperatures.

The determined temperature fields for the different resolutions of the applied heat fluxes show considerable differences in their maximum values. It was shown that the use of spatial and temporal averaging considerably reduces the calculation time. However, this averaging also leads to a considerable underestimation of the maximum temperatures at the cutting edge, which manifests itself in a constant offset between the different simulation variants. In future analyses, this systematic behavior of this offset will be further investigated in order to identify simulation methods that achieve a lower computational effort with consistently accurate temperature estimates. Although the simulation shows reasonable results, future work needs to validate the simulated temperature data with experimental data to confirm the accuracy of the calculated cutter maximum temperature and the reliability of the modeling approach.

References

1. Taylor, G.I., Quinney, H.: The latent energy remaining in a metal after cold working. Proc. R. Soc. Lond. Ser. A Contain. Pap. Math. Phys. Character **143**(849), 307–326 (1934). https://doi.org/10.1098/rspa.1934.0004. ISSN 0950-1207
2. Wang, K.K., Wu, S.M., Iwata, K.: Temperature responses and experimental errors for multitooth milling cutters. J. Eng. Ind. **90**(2), 353–359 (1968). https://doi.org/10.1115/1.3604640. ISSN 0022-0817
3. Sato, M., Tamura, N., Tanaka, H.: Temperature variation in the cutting tool in end milling. J. Manuf. Sci. Eng. **133**(2) (2011). https://doi.org/10.1115/1.4003615. ISSN 1087-1357

4. Augspurger, T., Meurer, M., Liu, H., Mattfeld, P., Bergs, T.: Experimental study of the connection between process parameters, thermo-mechanical loads and surface integrity in machining Inconel 718. Procedia CIRP **87**, 59–64 (2020). https://doi.org/10.1016/j.procir.2020.02.081. ISSN 2212-8271
5. Karaguzel, U., Bakkal, M., Budak, E.: Modeling and measurement of cutting temperatures in milling. Procedia CIRP **46**, 173–176 (2016). https://doi.org/10.1016/j.procir.2016.03.182. ISSN 2212-8271
6. Helmig, T., Liu, H., Winter, S., Bergs, T., Kneer, R.: Development of a tool temperature simulation during side milling. In: Ihlenfeldt, S. (Hrsg.) 3rd International Conference on Thermal Issues in Machine Tools (ICTIMT2023). Lecture Notes in Production Engineering, pp. 308–317. Springer International Publishing, Cham (2023). https://doi.org/10.1007/978-3-031-34486-2_22. ISBN 978-3-031-34485-5

Experimental and Simulative Investigation of Cutting Edge Geometry Effects on Surface Integrity in Turning of Inconel 718

Anna Kibireva, Hui Liu⬤, Markus Meurer⬤, and Thomas Bergs

Abstract The performance and lifespan of safety-critical components, such as turbine discs made from Inconel 718, are significantly influenced by residual stresses and surface microstructural characteristics resulting from thermomechanical effects during machining. Elevated temperatures generated in the cutting zone play a critical role, with most heat being removed by chips, while some residual heat transfers to the workpiece and tool. This thermal effect not only accelerates tool wear but also promotes undesirable tensile residual stresses, adversely affecting fatigue strength and overall mechanical performance. Optimizing machining processes through improved tool design can significantly mitigate tensile residual stresses, enhancing fatigue life. This study investigates the effect of cutting tool edge geometry on residual stress and microstructural changes during the turning of Inconel 718. Various tool geometries were analyzed to determine their specific influences on thermomechanical loads in the cutting zone. Additionally, a chip formation simulation validated by experimental data was developed to provide deeper insights into the underlying physical mechanisms. The findings enhance the understanding of how cutting tool geometry affects surface integrity and form a foundation for future function-oriented tool design strategies.

A. Kibireva (✉) · H. Liu · M. Meurer · T. Bergs
Manufacturing Technology Institute (MTI), RWTH Aachen University, Aachen, Germany
e-mail: a.kibireva@mti.rwth-aachen.de

H. Liu
e-mail: h.liu@mti.rwth-aachen.de

M. Meurer
e-mail: m.meurer@mti.rwth-aachen.de

T. Bergs
e-mail: t.bergs@mti.rwth-aachen.de

K. Wegener and M. Bambach (eds.), *4th International Conference on Thermal Issues in Machine Tools (ICTIMT2025)*, Lecture Notes in Production Engineering,
https://doi.org/10.1007/978-3-032-01194-7_6

1   Introduction

The performance and lifespan of safety-critical components, such as turbine discs made from Inconel 718, depend on their reliability under cyclic loading [1, 2]. One of the key factors influencing the reliability and fatigue behavior of these components is surface integrity [3]. A crucial aspect of surface integrity is the presence of tensile stresses, which are recognized as being one of the factors that can lead to crack propagation. During the cutting process, residual stresses occur in workpieces due to plastic deformations and thermal impacts. In addition to high deformation, thermal energy is also introduced into the workpiece edge zone. The mechanical stress, in which the material is loaded beyond its yield point, leads to strong plastic deformations in the workpiece edge zone. The thermal stress can cause microstructural changes, phase transformations and hardness changes [4]. Due to the high loads on the tool, low thermal conductivity of Inconel 718, increased tool wear, and excessive heat generation resulting from machining, the selection of the appropriate tool remains a significant challenge in the machining process [1].

The level and depth of the material influence and the resulting stress state in the workpiece edge zone depend on a large number of parameters. In addition to the workpiece-cutting material combination, this primarily includes the macro and micro geometry of the cutting tool. Experimental investigations have shown that the components of cutting force increase with both the cutting edge radius and the chamfer angle [5]. Additionally, the cutting force is primarily influenced by the cutting edge form factor K, which indicates the ratio of the cutting edge segments on rake and flank faces [6]. Moreover, the use of a rounded cutting edge leads to a larger shear zone with plastic material flow under the newly created surface [7].

However, identifying the influence of the cutting edge geometry on residual stresses remains challenging. The experimental investigations in [8–10], showed that an increase in the cutting edge radius led to higher compressive residual stresses in workpiece surface layer. The possible reason for this is that when machining with rounded cutting edges material is subjected to high normal stress on the surface with increasing cutting edge radius. This results in higher induced plastic deformations. In contrast, some studies [2, 11, 12] observed an increase in residual tensile stresses with the cutting edge radius during turning. This was attributed to the fact that cutting temperatures rise with the cutting edge radius due to increased friction. Thus, the influence of cutting tool geometry on the surface integrity remains controversial. Although it is well known that thermomechanical load has a significant influence on residual stresses, it is difficult to determine the stresses on the workpiece directly using experimental methods. One of the methods to broaden the knowledge about the mechanisms loads in the cutting zone is modeling. Compared to empirical experiments, finite element modeling (FEM) approaches allow to quantify the thermomechanical loads in the cutting zone.

This paper presents an empirical investigation into the influence of cutting edge microgeometry on the thermomechanical loads during the face turning process of Inconel 718 workpieces. The study further examines the resulting surface residual stresses and microstructural changes. Additionally, 3D-FEM simulations are conducted to support and extend the experimental findings by providing deeper insight into the underlying mechanisms.

2 Methodology

2.1 Workpiece material and cutting tools

For the experiments, disc-shaped workpieces made of Inconel 718 with a diameter of $D = 184$ mm and a width of $b = 12$ mm were used. The microstructure of the material, as well as chemical and mechanical properties are shown in Fig. 1. The workpiece microstructure shows γ-matrix, recognizable as a white phase in the microstructure and the dark colored δ-phase, present both at the grain boundaries and in the matrix. The material shows a homogeneous grain size distribution.

Indexable inserts with soldered, uncoated cubic boron nitride (CBN) tips from LMT TOOLS were used for all experiments. The DNGA150412 geometry, featuring various cutting edge preparations, was employed in conjunction with a DDHNL2525M15-type toolholder. The cutting edge microgeometry, measured using an LMI TECHNOLOGIES MICRO CAD optical system, is presented in Table 1. In addition to different cutting edge radii, chamfered tools were also examined, since

Inconel 718	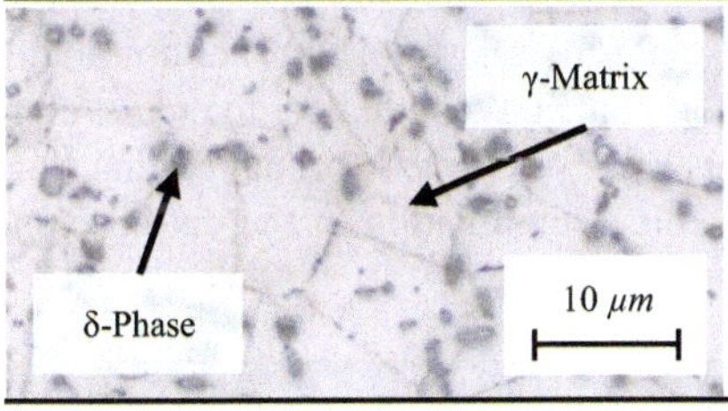	Physical Properties		
	γ-Matrix	Young's Modulus	$E\ [GPa]$	217
		Tensil Strength	$R_m\ [MPa]$	1365
	10 µm	Yield Strength	$R_{p0,2}\ [MPa]$	1121
δ-Phase		Hardness	$[HV]$	420
Thermal Conductivity	Specific Heat Capacity	Thermal Expansion	Density	Poisson's Ratio
$\lambda\ [W{\cdot}K/m]$	$c\ [J/kg{\cdot}K]$	$\alpha\ [10^{-6}/K]$	$\rho\ [kg/m^3]$	$v\ [-]$
11.9	441	12.2	8220	0.3
Chemical Composition				
Element	Ni	Cr	Fe	Nb	Mo
wt [%]	54.28	18.25	17.05	5.04	2.98
Element	Co	Si	Mn	C	N
wt [%]	0.3	0.249	0.048	0.034	0.501

Fig. 1 Microstructure, chemical and mechanical properties of Inconel 718

Table 1 Microgeometry of cutting tools

Parameter	Type A	Type B	Type C	Type D
Geometry	DNGA150412			
Effective rake angle γ [deg]	−6	−6	−6	−6
Cutting edge radius r_β [μm]	4	19	15	15
Chamfer length l_{ch} [μm]	–	–	150	150
Chamfer angle γ_{ch} [deg]	–	–	10	25

chamfers can enhance cutting edge stability and tool life [13, 14]. A new cutting insert was used for each experiment. Comparing cutting tool geometries A, B, C, and D aims to assess the effects of cutting edge radius and chamfer geometry on the residual stresses and microstructure of Inconel 718.

2.2 Experimental Setup

The experimental investigation of the face turning of Inconel 718 was performed on the DMG MORI NEF 600 CNC Turning and Milling Center. Figure 2 shows the experimental setup.

The workpieces were clamped externally in a three-jaw chuck. The tool holder was mounted on a Kistler 91299AA three-component dynamometer to measure the force components. Although the experiments were performed without cutting fluid supply, compressed air was used to avoid chip accumulation in the cutting zone

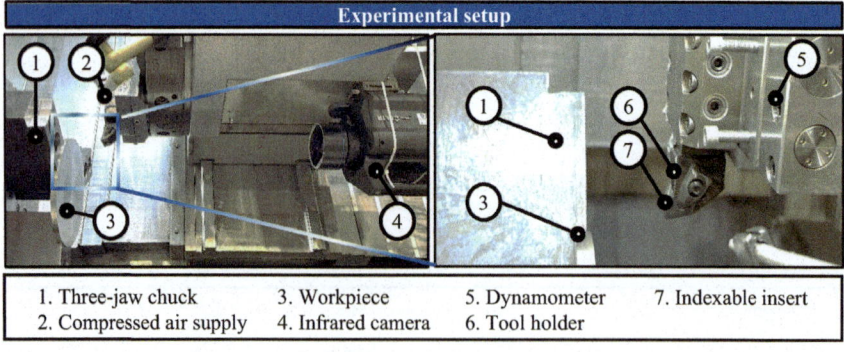

1. Three-jaw chuck	3. Workpiece	5. Dynamometer	7. Indexable insert
2. Compressed air supply	4. Infrared camera	6. Tool holder	

Fig. 2 Experimental setup

Table 2 Cutting parameters

Cutting speed v_c (m/min)	Depth of cut a_p (mm)	Feed f (mm)
150	0.25	0.1

and re-cutting of the chip. The cutting experiments were performed for face turning process, and the cutting parameters are shown in Table 2. The experiments were repeated twice to ensure statistical reliability.

In addition to cutting force measurements, a FLIR X6580 SC thermographic camera, with a frame rate of 355 Hz and resolution of 640 x 512, calibrated for a temperature range of 0 to 300°C, was used to record the temperature field on the machined workpiece surface. An example of temperature evaluation is shown in Fig. 3. The maximum temperature on the workpiece was measured in the evaluation area and then compared to the simulation results. To ensure accuracy and minimize variability, the mean values of the maximum surface temperatures were calculated from 10 consecutive frames captured by the thermal camera.

Residual stresses were measured after the cutting process using the X-Ray diffraction method on an Xstress G2R X-ray diffractometer from Stresstech GmbH, following the $\sin^2 \Psi$ method. While factors such as grain size, surface roughness, and inhomogeneous deformation can sometimes result in scatter exceeding the uncertainty of individual measurements, X-ray diffractometry remains one of the most precise techniques for residual stress analysis [15]. Measurements were taken on the surface in the radial direction σ_r (aligned with the feed direction) and the tangential direction σ_τ (aligned with the cutting velocity). The residual stresses were specifically evaluated on the face surface of the workpieces at the largest diameter, with three measurement spots of $\varnothing$ 2.0 mm each. Additionally, microstructural analysis of the workpiece rim zone was conducted for each sample, and the depth of deformation lines h was measured across the surface layer.

Fig. 3 Temperature field evaluation

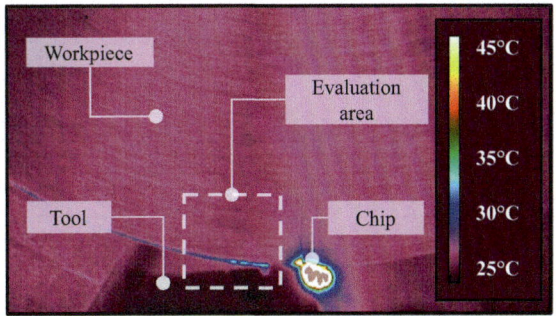

Fig. 4 Simulation setup

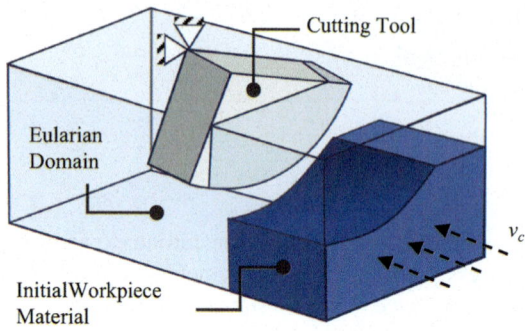

2.3 *Simulation model setup*

To theoretically explain the influence of cutting edge geometry on residual stresses, a 3D chip formation simulation was performed. The simulation model was developed using the CEL approach and implemented with the commercial FEM software ABAQUS 2021. Figure 4 illustrates the simulation setup. The workpiece was discretized within the Eulerian domain using a hexahedral mesh with a minimum element size of 5 μm, established through a previous mesh sensitivity analysis [21]. The tool was modeled as a Lagrangian rigid body, with wear effects neglected. This assumption was justified by the use of a new cutting tool for each test. The tool was discretized with a tetrahedral mesh having a minimum element size of 5 μm, and the simulations were conducted for a cutting process duration of 0.05 seconds. To approximate realistic conditions, the model was extended to a sufficient length to allow partial cooling and stress relaxation, bringing the simulated workpiece closer to room temperature. However, this approach does not fully account for the relaxation process. Consequently, the model is intended for qualitative analysis and explanation rather than precise quantitative predictions.

For the simulation of elasto-plastic material behavior, the Johnson-Cook (JC) material model [17] (Eq. 1) was selected, with its parameters detailed in Table 3, where A represents yield stress, B and n are strain parameters, C is strain rate parameter, m is thermal softening parameter.

Table 3 Johnson-Cook parameters for Inconel 718 [20] and friction model parameters [21]

A (MPa)	B (MPa)	n (−)	C (−)	m (−)	T_{ref} (°C)	T_m (°C)	$\dot{\varepsilon}_0$ (s^{-1})	D_1 (−)	D_2 (−)
1200	1284	0.54	0.006	1.2	25	1800	0.001	0.04	1.2
D_3 (−)	D_4 (−)	D_5 (−)	$u_{pl,f}$ (−)	p (−)	β (−)	μ_0 (−)	T_0 (°C)	m_r (−)	
(−)1.45	0.04	0.89	0.0165	0.7	3	0.36	660	0.55	

$$\sigma = (A + B\varepsilon^n)\left[1 + C \ln\left(\frac{\dot{\varepsilon}}{\dot{\varepsilon}_0}\right)\right]\left[1 - \left(\frac{T - T_{ref}}{T_{melt} - T_{ref}}\right)^m\right] \tag{1}$$

The damage initiation model [18] was applied to simulate serrated chip formation, as shown in Eq. 2:

$$\varepsilon_f = \left(D_1 + D_2 \cdot \varepsilon^{D_3 \cdot \eta_s}\right)\left[1 + D_4 \ln\left(\frac{\dot{\varepsilon}}{\dot{\varepsilon}_0}\right)\right]\left[1 + D_5 \cdot \left(\frac{T - T_{ref}}{T_{melt} - T_{ref}}\right)\right] \tag{2}$$

where ε_f is failure strain, η_s is defined as ratio of the normal stresses and von Mises equivalent stress, and the parameters D_1-D_5 are damage initiation parameters. The damage evolution model (Eq. 3), based on effective plastic displacement upl was also implemented in the scope of previous work [16]. In this model, D_e represents damage variable, and β and p are damage parameters.

$$D_e = \frac{1 - e^{-\beta \cdot \frac{u_{pl}}{u_{pl.f}}}}{1 - e^{-\beta}} \cdot p \tag{3}$$

Additionally, a temperature-dependent friction model, as proposed by Puls [19], was employed in this work and shown in Eq. 4. In this model, μ represents the apparent friction coefficient, m_r is the thermal parameter, and T_0 denotes the critical temperature. The corresponding parameters can be found in Table 3.

$$\mu = \mu_0 \cdot \left(1 - \left(\frac{T - T_0}{T_{melt} - T_0}\right)^{m_r}\right) \tag{4}$$

The cutting force components in feed direction F_f and in the direction of the cutting velocity F_c were compared with the experimental results. Additionally, the simulated temperatures of the machined surfaces, measured 4.5 mm behind the cutting area, were compared with thermographic images.

3   Results

Figure 5 presents an example of the FEM simulation results for the cutting process. The simulation accuracy is validated by comparing forces and temperature. The simulation provides additional information about the stress and strain as well as thermal loads in the cutting zone.

The measured and simulated cutting force components in the cutting direction F_c, the feed direction F_f and thrust direction F_p are presented in Fig. 6. As it can be seen, the cutting force components rise with a higher cutting edge radius. Varying the cutting edge radius from $r_\beta = 4\,\mu m$ (Type A) to $r_\beta = 19\,\mu m$ (Type B) led to an increase

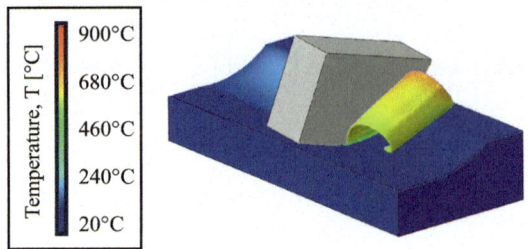

Fig. 5 Simulation results example: $r_\beta = 4\ \mu$m, $v_c = 150\ m/min$, $a_p = 0.25$ mm, $f = 0.1$ mm

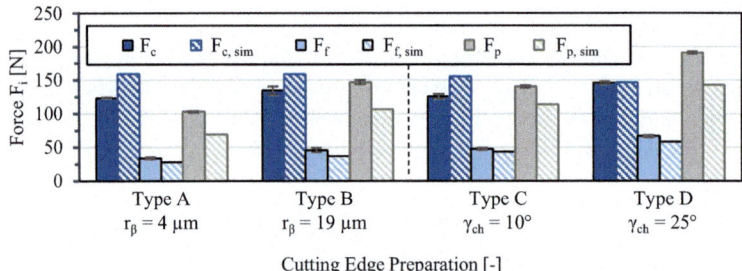

Cutting Edge Preparation [-]

Fig. 6 Experimental and simulative cutting force components

in cutting force F_c by 9 %. Moreover, the F_f and F_p components exhibited a more pronounced increase of 39–52 %. The increase in feed forces is consisted with the previous findings in the literature [11]. When using chamfered tools, the cutting force components show a strong correlation with an increased chamfer angle. A 16 % rise in F_c and a 53 % increase in F_f were observed when transitioning from Tool C to Tool D. This increase in cutting forces can be attributed to the compression of material beneath the cutting edge, leading to intense elastic and plastic deformations [22].

The simulation results for the cutting force component F_c show deviations for the cutting edge radius $r_\beta = 4\ \mu$m. One possible explanation is that the material model calibration used in previous studies ($r_\beta = 25\ \mu$m) did not account for the micro-tool configuration applied in this research. It should be noted that the simulation is suitable for qualitative rather than quantitative comparisons.

Figure 7 illustrates the results of the workpiece surface temperature measurements. The recorded temperatures are influenced by the emissivity coefficient, which varies depending on surface geometry, roughness, and color. During face turning, the removal of the surface layer creates a freshly machined surface with an emissivity coefficient lower than that of a black body ($\epsilon < 1$). Although the machine room is shielded from ambient radiation, the thermal camera, calibrated for surfaces with an emissivity of 1, may still underestimate the actual surface temperature. Consequently, the thermal images were not used for quantitative analysis. However, qualitative comparisons of the temperature data remain valid.

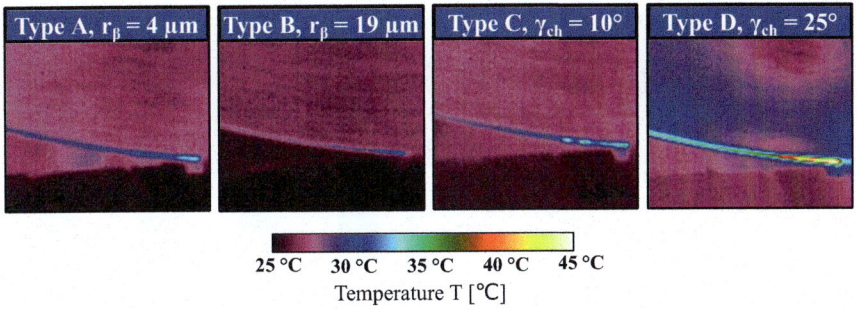

Fig. 7   IR-camera temperature measurement

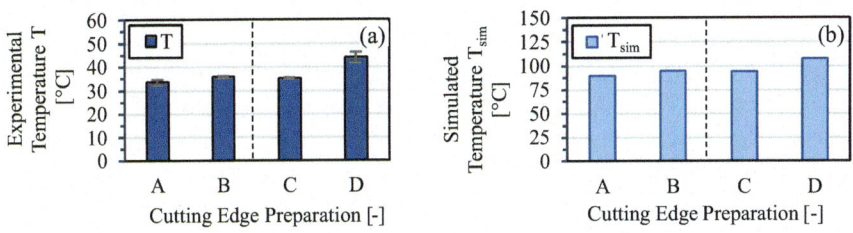

Fig. 8   Temperature results: experimental temperature (**a**), simulated temperature (**b**)

As it can be seen from both simulated and experimental results (Fig. 8a and b), the temperature trends correlate. Using the cutting inserts with $r_\beta = 19~\mu$m (Type B) leads to a slight increase in maximum temperature, up to approximately 5–8 %. Applying a chamfered geometry with a chamfer angle of 10 ° (Type B) had no significant influence on temperature changes. However, increasing the chamfer angle from 10 ° to 25 ° (Type D) resulted in a rise from 9 K to 13 K for both the experimental and simulated results, respectively. The chamfer creates a larger negative rake angle in the contact zone, leading to increased friction and heat generation.

The results of the residual stress measurements are presented in Fig. 9. Tensile residual stresses were observed on the surfaces of all workpieces, measured along the cutting velocity direction (σ_τ) In the tangential direction, the tensile residual

Fig. 9 Experimental residual stresses

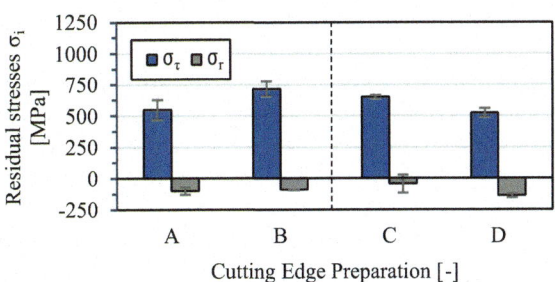

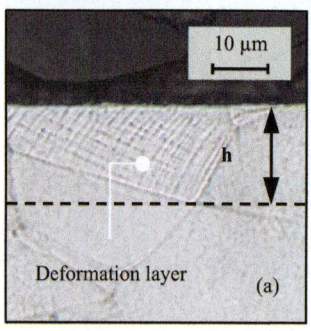
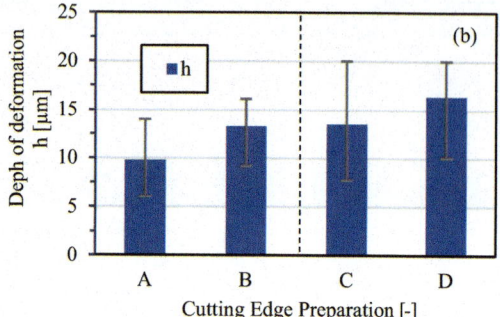

Fig. 10 Measurement of deformation layer: microstructure (**a**), depth of deformation layer (**b**)

stresses σ_τ exhibited an increase of up to 30 % with change in the cutting edge radius from $r_\beta = 4$ μm (Type A) to $r_\beta = 19$ μm (Type B). Among the analyzed cutting edge preparations, the type B exibited the most pronounced tensile residual stresses, reaching σ_τ of 713 MPa. Although elevated temperatures during the cutting process are generally associated with increased tensile residual stresses, this phenomenon cannot be attributed to thermal effects alone. The observed residual stress levels result from a complex interaction between thermal and mechanical loads, with temperature contributing to the stress state but not being the sole determining factor, thereby intensifying the thermal impact on the surface and enhancing residual stress levels. These findings are consistent with previous studies reported in [23] and [24].

The use of tools with a chamfer angle γ_{ch} of 10 ° for type C resulted in an 8 % reduction in tensile residual stresses compared to set B. In contrast to the results observed with changes in cutting edge radius, further increasing the chamfer angle γ_{ch} to 25 ° (Type D) led to a noticeable decrease in tensile stresses by 20 %. A possible explanation for this phenomenon is that mechanical effects, such as increased strain and strain rates, were more pronounced under these conditions. This interpretation is also supported by the simulation results shown in Figure 12, which reveal higher stress levels for chamfered tools, confirming the significant mechanical influence on residual stress formation. Higher mechanical loads contributed to the development of compressive stresses, thereby reducing tensile residual stresses [22].

Unlike the residual stresses in the cutting velocity direction σ_τ, the residual stresses in the feed direction σ_r exhibited compressive stresses for all cutting tool geometries. Variations in the cutting edge radius showed no significant influence on the residual stresses in the radial direction. Although Type D exhibits elevated temperatures, the compressive residual stresses are still dominant due to the stronger mechanical impact from plastic deformation, which outweighs the thermal effects. Similar effects were observed in [22]. Furthermore, higher mechanical loads also caused greater deformation in the rim zone, as evidenced by the deformation layer measurements shown in Fig. 10a. Microscopic images were taken in the feed direction, and the thickness of the deformation layer h was measured for five samples in each test. The results were summarised in Fig. 10b. The findings confirmed the mechanical effect,

Fig. 11 Distribution of Mises stress after cutting time of 0.05 seconds.

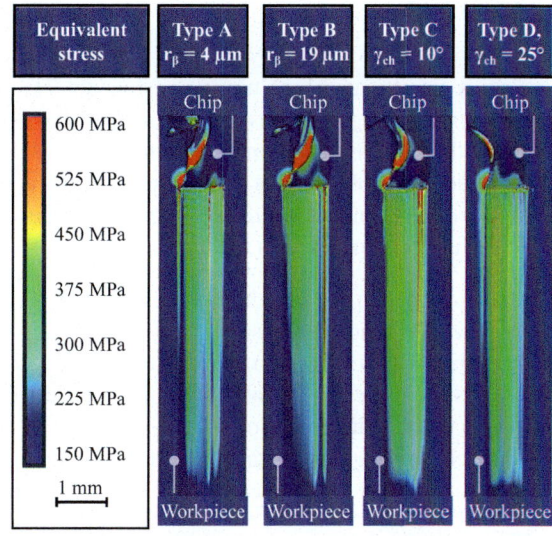

Fig. 12 Simulated temperatures in cutting zone

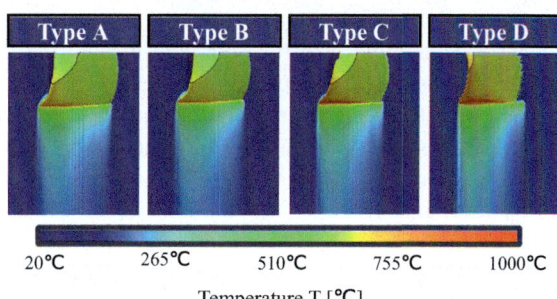

as both tools Type B and D exhibited greater depths of deformation h compared to other configurations.

Based on the temperature, residual stress and deformation results, it can be deduced that the chamfered geometry amplifies its influence on the surface layer. This effect is directly linked to the increased thermo-mechanical load on the workpiece within the cutting zone. This can also be confirmed in the chip formation simulation, as illustrated in Fig. 11. The stresses were higher for chamfered tools, both within the cutting zone and on the machined surface, aligning well with the experimental results.

In Fig. 12, a comparison of the temperatures in the cutting zone is presented. As it can be seen from the simulation results, the chamfered geometry has a greater influence on the temperature in the cutting zone compared to the cutting edge radius. The highest temperatures were observed for Type C and Type D. This is also supported by experimental observation.

4   Conclusion

This study investigated face turning of Inconel 718 using CBN indexable inserts with varying microgeometries, complemented by the development of a 3D-FEM model of the process. The numerical simulation validated the experimental results and provided valuable insights into the temperature and stress distribution within the workpiece. Based on the experimental and numerical findings reported in this paper, the following conclusions can be made:

- Increasing the cutting edge radius from $r_\beta = 4$ μm to 19 μm resulted in a 30 % rise in tensile residual stresses.
- The application of a chamfered cutting edge geometry reduced tensile residual stresses in the tangential direction σ_τ while enhancing compressive residual stresses in the radial direction σ_r.
- As the cutting edge radius and chamfer angle increased, the depth of the affected deformation layer h also increased.

This study lays the groundwork for future research on the fatigue life of components and the influence of tool micro- and macro-geometrical features on surface integrity. Upcoming investigations will expand the range of cutting edge geometries and consider the effects of tool wear to gain a comprehensive understanding of their effects on microstructural changes and residual stress formation. Furthermore, future studies will focus on advanced simulation-based models that incorporate relaxation and cooling processes, enabling more accurate prediction of residual stress development.

Acknowledgements The authors would like to thank the Deutsche Forschungsgemeinschaft (German Research Foundation) for funding the project 503700894.

References

1. La Monaca, A., Murray, J., Liao, Z., Speidel, A., Robles-Linares, J., Axinte, D., Hardy, M., Clare, A.: Surface integrity in metal machining–Part II: Functional performance. Int. J. Mach. Tools Manufact. **164**, 103718 (2021)
2. Outeiro, J., Pina, J., M'Saoubi, R., Pusavec, F., Jawahir, I.: Analysis of residual stresses induced by dry turning of difficult-to-machine materials. CIRP Ann. **57**, 77–80 (2008)
3. Jawahir, I., Brinksmeier, E., M'Saoubi, R., Aspinwall, D., Outeiro, J., Meyer, D., Umbrello, D., Jayal, A.: Surface integrity in material removal processes: Recent advances. CIRP Ann. **60**, 603–626 (2011)
4. Liao, Z., La M. A., Murray, J., Speidel, A., Ushmaev, D., Clare, A., Axinte, D., M'Saoubi, R.: Surface integrity in metal machining–Part I: Fundamentals of surface characteristics and formation mechanisms. Int. J. Mach. Tools Manufact. **162**, 103687 (2021)
5. Denkena, B., Biermann, D.: Cutting edge geometries. CIRP Ann. **63**, 631–653 (2014)
6. Denkena, B., Lucas, A., Bassett, E.: Effects of the cutting edge microgeometry on tool wear and its thermo-mechanical load. CIRP Ann. **60**, 73–76 (2011)

7. Shaw, M.: Metal Cutting Principles. Oxford University Press (2005)
8. Hua, J., Umbrello, D., Shivpuri, R.: Investigation of cutting conditions and cutting edge preparations for enhanced compressive subsurface residual stress in the hard turning of bearing steel. J. Mater. Process. Technol. **171**, 180–187 (2006)
9. Shen, Q., Liu, Z., Hua, Y., Zhao, J., Lv, W., Mohsan, A.: Effects of cutting edge microgeometry on residual stress in orthogonal cutting of inconel 718 by FEM. Mater. (Basel, Switzerland)
10. Thiele, J., Melkote, S., Peascoe, R., Watkins, T.: Effect of cutting-edge geometry and workpiece hardness on surface residual stresses in finish hard turning of AISI 52100 steel. J. Manufact. Sci. Eng. **122**, 642–649 (2000)
11. Nasr, M., Ng, E., Elbestawi, M.: Modelling the effects of tool-edge radius on residual stresses when orthogonal cutting AISI 316L. Int. J. Mach. Tools Manuf. **47**, 401–411 (2007)
12. Outeiro, J., Dias, A., Jawahir, I.: On the effects of residual stresses induced by coated and uncoated cutting tools with finite edge radii in turning operations. CIRP Ann. **55**, 111–116 (2006)
13. Coelho, R., Silva, L., Braghini, A., Bezerra, A.: Some effects of cutting edge preparation and geometric modifications when turning INCONEL 718^{TM} at high cutting speeds. J. Mater. Process. Technol. **148**, 147–153 (2004)
14. Sugihara, T., Enomoto, T.: High speed machining of inconel 718 focusing on tool surface topography of CBN tool. Procedia Manufact. **1**, 675–682 (2015)
15. Fitzpatrick, M., Fry, A., Holdway, P., Kandil, F., Shackleton, J., Suominen, L.: Determination of Residual Stresses by X-ray Diffraction (2005)
16. Peng, B.: Multiscale modeling of thermomechanical loads in the broaching of direct aged inconel 718. (Rheinisch-Westfälische Technische Hochschule Aachen)
17. Johnson, G., Cook, W.A.: Constitutive model and data for metals subjected to large strains, high strain rates and high temperatures. In: Proceedings of the 7th International Symposium On Ballistics, The Hague, Netherlands (1983)
18. Johnson, G., Cook, W.: Fracture characteristics of three metals subjected to various strains, strain rates, temperatures and pressures. Eng. Fract. Mech. **21**, 31–48 (1985)
19. Puls, H., Klocke, F., Lung, D.: Experimental investigation on friction under metal cutting conditions. Wear **310**, 63–71 (2014)
20. Seimann, M., Peng, B., Fischersworring-Bunk, A., Rauch, S., Klocke, F., Döbbeler, B.: Model-based analysis in finish broaching of inconel 718. Int. J. Adv. Manufact. Technol. **97**, 3751–3760 (2018)
21. Peng, B., Bergs, T., Klocke, F., Döbbeler, B.: An advanced FE-modeling approach to improve the prediction in machining difficult-to-cut material. Int. J. Adv. Manufact. Technol. **103**, 2183–2196 (2019)
22. Hua, J., Shivpuri, R., Cheng, X., Bedekar, V., Matsumoto, Y., Hashimoto, F., Watkins, T.: Effect of feed rate, workpiece hardness and cutting edge on subsurface residual stress in the hard turning of bearing steel using chamfer+hone cutting edge geometry. Mater. Sci. Eng. A. **394**, 238–248 (2005)
23. Liu, Y., Xu, D., Agmell, M., Saoubi, R., Ahadi, A., Stahl, J., Zhou, J.: Numerical and experimental investigation of tool geometry effect on residual stresses in orthogonal machining of Inconel 718. Simul. Modell. Pract. Theory. **106**, 102187 (2021)
24. Elsheikh, A., Shanmugan, S., Muthuramalingam, T., Thakur, A., Essa, F., Ibrahim, A., Mosleh, A.: A comprehensive review on residual stresses in turning. Adv. Manuf. **10**, 287–312 (2022)

Measurement and Analysis of the Thermal Load During BTA Deep Hole Drilling

L. Brause, S. Michel, R. Schmidt, and D. Biermann

Abstract Boring and Trepanning Association (BTA) deep hole drilling is often used to machine bores with large diameters (D > 40 mm) and a length-to-diameter ratio greater than ten (l/D > 10). Typically, the tool cutting edges on the BTA drill head are arranged asymmetrically. This results in radial forces that are supported by guide pads on the bore wall. This leads to forming processes on the bore surface and feed grooves are smoothed. In addition to improving the roughness, this also results in changes in the subsurface zone of the bore with regard to the microstructure, the hardness and the residual stress. In experimental investigations, the thermal load on the outer cutting insert and the guide pads is determined. The available sampling rate of a fiber-optic two-color pyrometer is used to achieve a high local resolution of the temperature measurements. Experiments were performed with the optical measurement fiber located inside the workpiece to determine the temperature at the cutting edge and the guide pads.

Keywords BTA (Boring and trepanning association) · STS deep hole drilling · Temperature · Fiber-optic two-color pyrometer

1 Introduction

Boring and Trepanning Association (BTA) or Single Tube System (STS) deep hole drilled components are often large, expensive and exposed to intense loads in their application. The components are often also safety relevant. The BTA deep hole drilling process is designed for large diameters and length-to-diameter ratios of $l/D \geq 10$. The process is used industrially for diameters $D = 6 \ldots 1500$ mm. For example, typical components manufactured using the BTA deep hole drilling process are turbine shafts or hydraulic cylinders. Other conventional and common deep hole drilling methods are single-lip deep hole drilling and ejector deep hole drilling. A

L. Brause (✉) · S. Michel · R. Schmidt · D. Biermann
Institute of Machining Technology (ISF), TU Dortmund University, Dortmund, Germany
e-mail: lucas.brause@tu-dortmund.de

© The Author(s) 2026
K. Wegener and M. Bambach (eds.), *4th International Conference on Thermal Issues in Machine Tools (ICTIMT2025)*, Lecture Notes in Production Engineering,
https://doi.org/10.1007/978-3-032-01194-7_7

83

special feature of BTA deep hole drilling is the evacuation of chips through the inside of the drill head and boring bar, ensuring that the bore surface is not affected by the contact with chips [1, 2].

Figure 1 shows an example of a type 12 BTA drill head with a diameter of $D = 60$ mm. The asymmetrical tool design leads to high radial forces, which are supported by the guide pads on the bore wall. The drill head thus guides itself in the process, which improves the straightness of the bore. The BTA deep hole drilling process is performed using deep hole drilling oil. This is necessary for chip evacuation in the process and improves process reliability. A typical volume flow rate in a BTA deep hole drilling process is $\dot{V} = 300$ l/min with a pressure of $p \approx 10$ bar. The mixture of oil and chips is evacuated through the chip hole in the drill head and the boring bar, on which the drill head is mounted [1, 2].

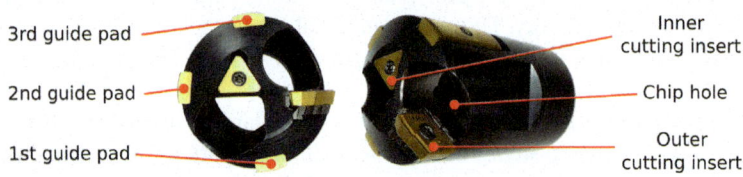

Fig. 1 BTA drill head type 12

Mechanical and thermal loads applied to the bore during the deep hole drilling process affect the surface integrity, which includes the shape of the bore surface and material characteristics of the bore subsurface layer. The targeted conditioning of the surface integrity due to the contact of the outer cutting insert and guide pads is important for the durability and reliability of the machined component [3, 4].

Figure 2 schematically shows the interaction of the outer cutting insert and the guide pads with the workpiece. The outer cutting insert forms the bore, which is finally leveled, burnished and affected by the guide pads. These mechanisms create a thermomechanical load on the bore subsurface and lead to significant changes in surface integrity.

Investigations have already been performed with regard to the axial force and torque occurring during BTA deep hole drilling. The normal force on the first guide pad was also examined. Further research results were obtained on the average temperature in the bore subsurface. However, the detailed distribution of the temperatures on the outer cutting insert and in particular on the guide pads was not investigated [6–10].

Temperature was measured underneath the first guide pad using thermocouples. Temperatures of $\vartheta \approx 800\,°C$ were measured with workpiece materials that differ from the here shown investigations [10]. An infrared camera was also used for temperature measurement to observe the workpiece from outside the bore during the BTA deep hole drilling [5].

Studies on the temperature in the bore subsurface with thermocouples showed that increasing the cutting speed and increasing the feed results in higher temperatures when drilling AISI 304 L. BTA deep hole drilling in AISI 4140+QT showed an increased temperature in the subsurface with higher feeds. A higher cutting speed results in an increased temperature, which could indicate thermal hardening of the material. The potential and challenges of temperature measurement with the fiber-optic two-color pyrometer were also demonstrated [9].

In addition, the BTA deep drilling process has been simulated. For this, numerical models were used. On the one hand, the residual stresses induced in the bore subsurface layer were modeled with corresponding theorems. On the other hand, the thermomechanical stress was modeled using the Coupled Eulerian-Langrarian (CEL) method [11, 12].

Further calculations to determine the temperature of the outer cutting insert were simulated by Shi et al. Using thermocouples at a greater distance of $a = 1.5$ mm from the bore surface, thermal process characteristics were recorded during the deep hole drilling process in Inconel 718 and used for the simulation model. It was observed that the thermal load on the outer cutting insert is at its maximum in the region behind the chip breaker [13].

2 Experimental Framework

2.1 Materials and Workpiece Preparation

Two different alloys were analyzed in this study. One of the materials is the chromium-nickel steel AISI 304 L. The other alloy investigated is the quenched and tempered AISI 4140+QT. The mechanical characteristics of the alloys are listed in Table 1.

For the experiments, the cylindrical workpieces were sawn to a length of $l = 500$ mm. Afterward, they were turned to an outer diameter of $D = 75$ mm and an $\alpha = 30°$ chamfer was created for clamping the workpiece in the BTA deep hole drilling machine. In addition, bores with a diameter $D = 1$ mm were drilled on a turn

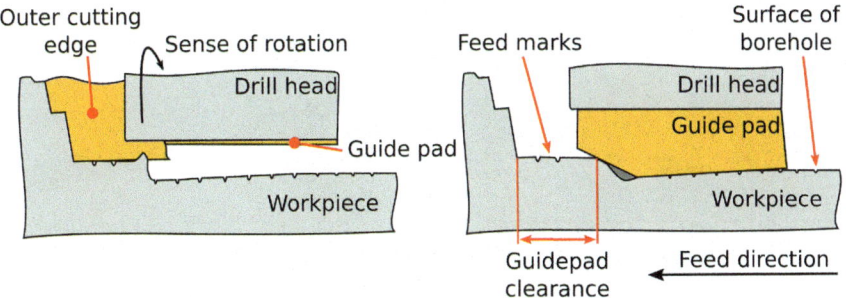

Fig. 2 Schematic interaction between outer cutting insert, guide pad and bore wall according to [5]

Table 1 Mechanical properties of the workpiece materials

Material	AISI 304 L	AISI 4140+QT
Tensile strength R_m	599 MPa	978 MPa
Yield strength R_p	260 MPa	873 MPa
Hardness HB	161	194

mill center through the side of the workpiece to insert the fiber-optic to measure the temperature with the fiber-optic two-color pyrometer.

2.2 Experimental Setup

The experiments were carried out on a *Giana GBB 560* BTA deep hole drilling machine. The machine is shown in Fig. 3 (left). Due to the positioning of the fiber-optic for temperature measurement, the BTA deep hole drilling process was executed with a stationary workpiece. The oil flow rate during the experiments was $\dot{V} = 300$ l/min. The oil is added to the process through the oil pressure head. The BTA deep hole drilling machine has a maximum drive power at the tool spindle of $P = 50$ kW and a maximum speed of $n = 1600$ 1/min. Bores with a maximum length of $l = 3000$ mm can be drilled on the shown BTA deep hole drilling machine.

For the experiments a type 12 drill head from *botek Praezisionsbohrtechnik GmbH* was used. The drill head with indexable cutting inserts and guide pads is shown in Fig. 1 and has a diameter of $D = 60$ mm. The cutting inserts and guide pads with TiN coating were used for drilling in AISI 4140+QT. For the experiments in the alloy AISI 304 L, uncoated cutting inserts and guide pads as well as the coating *Botek BX* (TiAlN) were used. The cutting inserts and guide pads are made of P20 carbide.

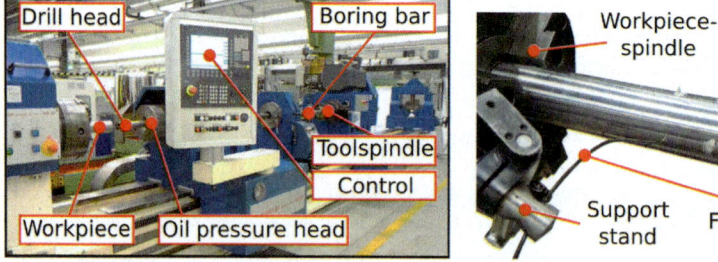

Fig. 3 Giana GBB 560 deep hole drilling machine (left) and measurement setup at the workpiece (right)

2.3 Measurement Setup

The *FIRE-3* fiber-optic two-color pyrometer from *en2Aix* was used to measure the temperature of the outer cutting insert and the guide pads. A fiber-optic was placed in the workpiece to measure the temperature. The exact temperature calculation is explained below. Figure 3 (right) shows the workpiece with an attached fiber-optic.

The fiber-optic two-color ratio pyrometer was developed to measure the temperature on surfaces with varying emissivity and difficult accessibility. The emissivity does not need to be known for measurements with the fiber-optic two-color pyrometer. The temperature is calculated from the spectral radiation of two wavelengths of light using a damping polynomial of the fiber-optic [14]. The *Pigtail WF 300* fiber-optic was used in the experimental setup. The fiber-optic has a diameter of $DF = 330$ μm. The sampling rate of the pyrometer was set to $\nu = 100$ kHz during the experiments. The fiber-optic is inserted into the side of the workpiece in the experimental setup. For this purpose, bores with a diameter of $D = 1$ mm were drilled during the workpiece preparation. These bores are tapered with a brass tube. This allows the fiber-optic to be inserted into the workpiece safely and without fracturing. In the experimental setup, the fiber-optic is bonded with a cold hardening adhesive.

3 Data Processing

The fiber-optic two-color ratio pyrometer uses the relation of two light wavelengths. The intensity of the wavelengths is recorded as electrical voltages U_i for those two measuring channels. Figure 4 shows an example of a section of a measuring channel.

The system noise can be recognized in the zoomed-in diagram on the right side. The amplitudes of the measurements are recorded when the outer cutting insert or the guide pads pass by. For data processing, the system noise in both measurement channels is filtered as first step. The measured voltages U_i are then amplified by the set gain factor on the measuring device and are normalized. To calculate the temperature ϑ, the damping polynomial of the fiber-optic from Eq. 1 is used with the normalized voltages $U_{n,i}$.

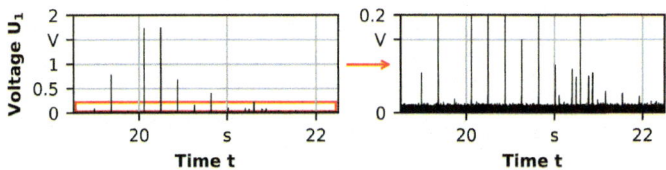

Fig. 4 Voltage measurement in FIRE-3 fiber-optic two color pyrometer

$$\vartheta = \sum_{j=0}^{5} = p_j \cdot \left(\frac{U_{n,1}}{U_{n,2}}\right)^j \tag{1}$$

The filtered measured points shown in Fig. 5 result from the given relationship between time and distance due to the process parameters feed and cutting speed. The first measurements show the contact between the cutting edge of the outer cutting insert and the workpiece.

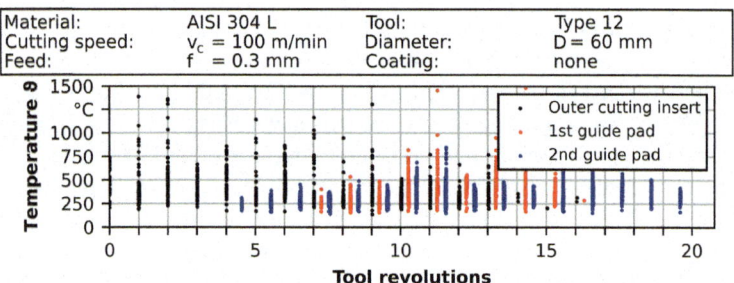

Fig. 5 Temperatures according to tool revolutions

Due to the relationship regarding the design of the drill head, the outer cutting insert is measured after each full revolution of the tool system. All measuring points between two full revolutions represent the temperatures at the guide pads. The first guide pad is shifted $\Psi_1 \approx 90°$ on the drill head to the outer cutting insert. The second guide pad is shifted by $\Psi_2 \approx 180°$ to the outer cutting insert. The measured locations can be assigned to the guide pads.

Measured temperatures on the first guide pad are shown in red in the diagram. The temperatures measured on the second guide pad are blue. Compared to the temperatures on the outer cutting insert shown in black, it can be determined that the temperatures of the guide pads are initially recorded after a few revolutions of the drill head. This results from the axial distance between the cutting edge corner of the outer cutting insert and the position of the guide pads. The stages of temperature measurement with the fiber optic are shown in Fig. 6. The detailed illustration shows the initial contact of the fiber-optic with the indexable outer cutting insert in the drill head. In phase 1, the fiber-optic is cut by the cutting edge. Phase 2 shows the cut fiber optic used to measure the temperatures of the indexable cutting insert and guide pads.

Due to the demanding application and the fragility of the fiber-optic, unreliable results were sometimes obtained. Despite the repetition of experiments, the measurement proves to be challenging in some cases due to occurring process dynamics and the fragility of the fiber optic.

As a novel contribution the measurement position in the axial direction can be calculated by multiplying the revolutions of the drill head with the feed. The exact

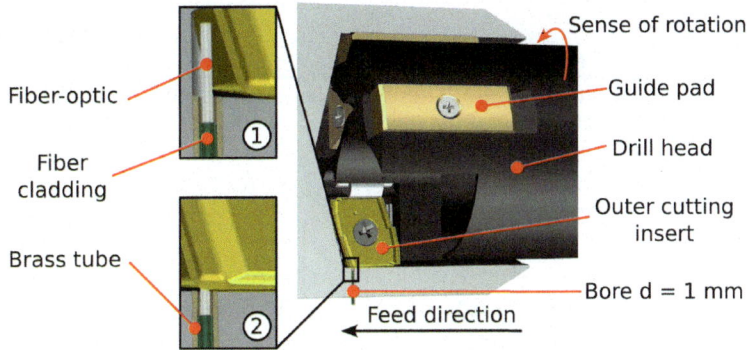

Fig. 6 Stages of fiber-optic two-color measurement at the outer cutting insert according to [9]

position in the radial direction can be calculated using the first measured location, the drill head diameter and the cutting speed.

The maximum measured temperature per full drill head revolution is used to estimate the temperature on the outer cutting insert. The temperature of the outer cutting insert is visualized in axial direction by approximating the measured maxima using a polynomial function to reduce the effect of outliers. The temperatures of the guide pads are determined from the calculated spots in axial direction and radial direction. This is used to identify the exact location of the detected temperature on the guide pad.

4 Thermal Load on the Outer Cutting Edge

In the following diagrams, the temperature at the outer cutting insert is analyzed. For this purpose, the temperature is shown in a diagram with the polynomial approximation already described. A section of the cutting insert with the corresponding measurement spots is shown above the diagrams. The color represents the temperature at the specific measured location.

Figure 7 (left) illustrates the temperatures at the outer cutting insert during BTA deep hole drilling in AISI 304 L with a feed of $f = 0.1$ mm. Higher temperatures are reached with a cutting speed of $v_c = 80$ m/min compared to a cutting speed of $v_c = 60$ m/min. With a cutting speed of $v_c = 60$ m/min, temperatures $\vartheta < 500$ °C are measured at the outer cutting insert. With a cutting speed of $v_c = 80$ m/min the maximum approximated temperature is $\vartheta = 891$ °C. It is noticeable that the highest temperature occurs in the area of the chip breaker.

Figure 7 (right) also shows the temperatures on the outer cutting insert with a feed of $f = 0.1$ mm and at a cutting speed of $v_c = 80$ m/min and $v_c = 100$ m/min for outer cutting inserts coated with Botek BX. The BTA deep hole drilled material is AISI 304 L. With the cutting speed of $v_c = 100$ m/min results a maximum

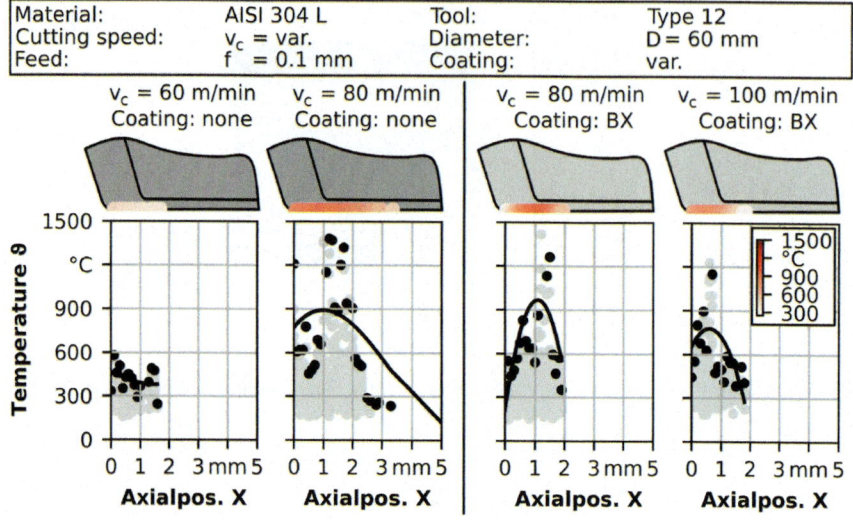

Fig. 7 Temperatures on the outer cutting insert in AISI 304 L

approximated temperature of $\vartheta = 776\,°C$. This behavior may be attributable to thermal softening of the alloy. Again, the maximum temperature of the cutting insert is indicated near the chip breaker.

Figure 8 (left) shows temperatures resulting from the experiments with a feed of $f = 0.3\,mm$ and TiN coating on the cutting insert. The BTA deep hole drilled alloy is AISI 4140+QT. The cutting speeds shown in this diagram are $v_c = 60\,m/min$ and $v_c = 100\,m/min$.

It is evident that the temperature with a cutting speed of $v_c = 100\,m/min$ is significantly higher than with a cutting speed of $v_c = 60\,m/min$. The approximated maximum temperature at the lower cutting speed is $\vartheta = 739\,°C$. At the higher investigated speed, the approximated temperature is $\vartheta = 913\,°C$. With a cutting speed of $v_c = 60\,m/min$, it is evident that no measured temperature can be analyzed for a temporary period of one revolution in the area of the chip breaker. It is possible that the bore in which the fiber-optic is located has become clogged. At the observed cutting speed of $v_c = 100\,m/min$, the maximum approximated temperature in the area of the chip breaker can be recognized.

Figure 8 (right) shows the influence of the feed on the temperature on the outer cutting insert. The feeds $f = 0.15\,mm$ and $f = 0.3\,mm$ with a cutting speed of $v_c = 100\,m/min$ are analyzed in the diagrams. The BTA deep hole drilled alloy was AISI 4140+QT with TiN-coated cutting insert.

These diagrams also show that the temperature in the area of the chip breaker is at its maximum. In addition to the higher local measurement resolution with a lower feed, it is evident that the temperature $\vartheta = 813\,°C$ in the area of the chip breaker with a reduced feed is lower than with a feed of $f = 0.3\,mm$.

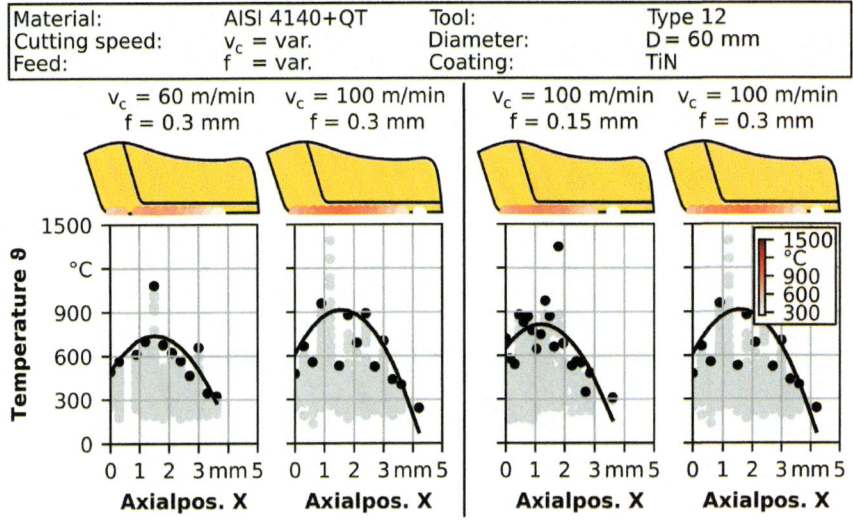

Fig. 8 Temperatures on outer cutting insert in AISI 4140+QT

5   Thermal Load on the Guide Pads

The following figures show the temperatures determined on the first guide pad. Figure 9 (left) shows the temperatures on the guide pad with a feed of $f = 0.3$ mm and a cutting speed of $v_c = 100$ m/min. The uncoated guide pad was used for BTA deep hole drilling in AISI 304 L.

It is evident that no temperatures are measured in the area of the chamfer of the guide pad due to the increased distance to the fiber-optic and the deep hole drilling oil in between. Successful measurements are obtained in the contact area of the guide pad. The temperature in the center of the guide pad is higher than in the peripheral area. Here, the leveling effect occurs, which is reflected in the contact zone of the guide pad [6].

Figure 9 (right) shows the temperatures on the first guide pad during the BTA deep hole drilling process with a feed of $f = 0.2$ mm and a cutting speed of $v_c = 80$ m/min. The Botek BX coating is applied to the guide pad. Compared to the uncoated guide pad, it can be noted that the measured spots were recorded with fewer interruptions. The maximum measured temperature on the guide pad with a feed of $f = 0.2$ mm and a cutting speed of $v_c = 80$ m/min is $\vartheta = 1242$ °C.

Material:	AISI 304 L	Tool:	Type 12
Cutting speed:	v_c = var.	Diameter:	D = 60 mm
Feed:	f = var.	Coating:	var.

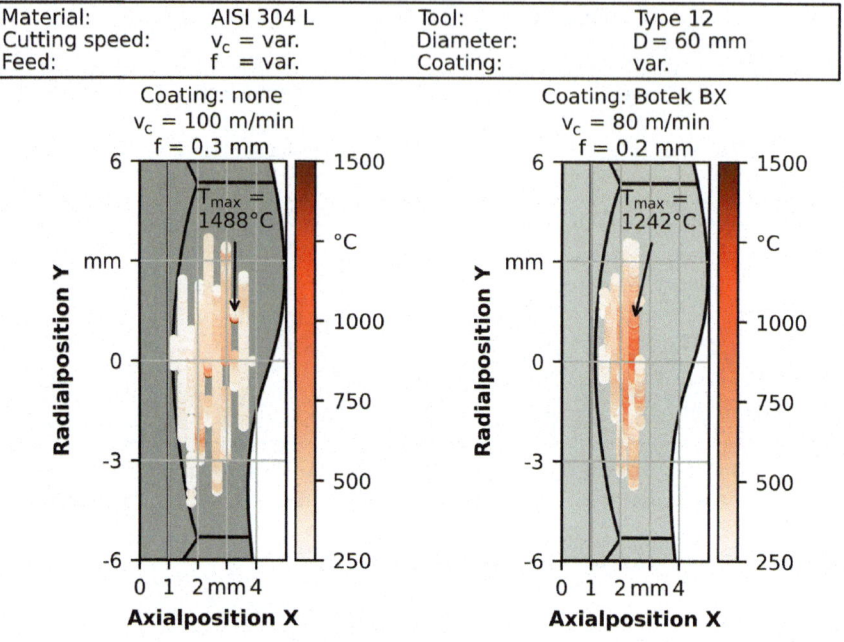

Fig. 9 Temperatures on first guide pad

6 Conclusion and Outlook

After an introduction with information on BTA deep hole drilling and the presentation of the experimental framework, which includes the investigated materials and the deep hole drilling machine, the data processing is discussed. As a novel contribution, the section shows how the temperatures at the guide pads and the outer cutting edge are calculated using the correlation of feed and cutting speed in combination with the sampling rate of the two-color pyrometer.

This measurement method and extensive evaluation of the measurement signal shows that an exact local resolution of the temperatures measurement is possible. It can be concluded that an increased feed leads to higher temperature at the outer cutting insert when drilling AISI 304 L. It can also be noted that the maximum temperature at the outer cutting insert occurs in the area of the chip breaker.

This article shows a temperature measurement with high resolution on the guide pad. Temperatures up to $\vartheta \approx 1500\,°C$ are reached with a high temperature gradient on the surface.

In addition, the limits of the measuring system have also been demonstrated. Continuous measurement of the guide pad temperature is only possible to a limited level. This can be solved, for example, by developing a sensor system integrated into the guide pad. The shown temperatures already provide information for development. In addition, continuous in-process measurements would be possible with an integrated data transmission system. These measurements provide the opportunity to adapt a process control with information about the influence on surface integrity or wear on the guide pad.

Temperature measurements for other processes like single-lip deep hole drilling, reaming, tapping or roller burnishing are possible with the shown method.

Acknowledgements The project is funded by the Deutsche Forschungsgemeinschaft (DFG, German Research Foundation)—Project number 500498267 and the Fraunhofer Gesellschaft. The authors would also like to thank the project partners Fraunhofer Institute for Surface Engineering and Thin Films (IST), BGTB GmbH, TIBO Tiefbohrtechnik GmbH and botek Präzisionsbohrtechnik GmbH.

References

1. Biermann, D., Bleicher, F., Heisel, U., Klocke, F., Möhring, H.-C., Shih, A.: Deep hole drilling. CIRP Ann. **67**(2), 673–694 (2018)
2. VDI-Richtlinie: Deep-hole drilling, VDI-3210 Part 1. Beuth Verlag (2006)
3. Strodick, S., Berteld, K., Schmidt, R., Biermann, D., Zabel, A, Walther, F.: Influence of cutting parameters on the formation of white etching layers in BTA deep hole drilling. tm – Technisches Messen **87**(11), 674–682 (2020)
4. Strodick, S., Walther, F., Schmidt, R., Brause, L., Zabel, A., Biermann, D.: Mikromagnetische Charakterisierung der Bohrungsintegrität. wt Werkstattstechnik online **112**(11/12), 756–761 (2022)
5. Abrahams, H.: Untersuchungen zum Führungsleistenverschleiß und zur Prozessdynamik beim BTA-TiefÂbohren austenitischer Stähle. Dissertation, TU Dortmund University (2016)
6. Fuß, H.: Aspekte zur Beeinflussung der Qualität von BTA-Tiefbohrungen. Dissertation, TU Dortmund University (1986)
7. Schmidt, R., Strodick, S., Walther, F., Biermann, D., Zabel, A.: Influence of the process parameters and forces on the bore sub-surface zone in BTA deep-hole drilling of AISI4140 and AISI 304 L. Procedia CIRP **87**, 41–46 (2020)
8. Schmidt, R., Strodick, S., Walther, F., Biermann, D., Zabel, A.: Analysis of the functional properties in the bore sub-surface zone during BTA deep-hole drilling. Procedia CIRP **88**, 318–323 (2020)
9. Schmidt, R., Brause, L., Strodick, S., Walther, F., Biermann, D., Zabel, A.: Measurement and analysis of the thermal load in the bore subsurface zone during BTA deep hole drilling. Procedia CIRP **107**, 375–380 (2022)
10. Schmidt, R.: Untersuchungen zum Einfluss des thermomechanischen Belastungskollektivs auf die Oberflächenkonditionierung beim BTA-Tiefbohrprozess, Dissertation, TU Dortmund University (2024)
11. Huang, X., Schmidt, R., Strodick, S., Walther, F., Biermann, D., Zabel, A.: Simulation and modelling of the residual stress state in the sub-surface zone of BTA deep-hole drilled specimens with eigenstrain theory. Procedia CIRP **102**, 150–155 (2021)
12. Biermann, D., Schmidt, R., Strodick, S., Walther, F., Zabel, A.: Numerical modelling of the BTA deep hole drilling process. Procedia CIRP **123**, 470–475 (2024)
13. Shi, Y., et al.: BTA deep hole vibration drilling for nickel-based alloys: cooling patterns and cutter tooth wear mechanisms. Materials **15**(22), 8178 (2022)
14. Müller, B., Renz, U.: Development of a fast fiber-optic two-color pyrometer for the temperature measurement of surfaces with varying emissivities. Rev. Sci. Instrum. **72**, 3366–3374

Design Measures for Enhancing Heat Conduction in CFRP Motor Spindle Components

Leonie Kilian[D], **Patrick Fehn**[D], and **Matthias Weigold**[D]

Abstract In motor spindles, effective cooling of the bearings and consequently, efficient heat transfer from the bearings to the cooling channels, is essential for achieving high rotational speeds and extending service life. While CFRP offers advantages such as high specific stiffness and low thermal expansion, its low through-thickness thermal conductivity presents a challenge. In filament-wound cylindrical components, the laminate thickness aligns with the radial direction. In motor spindles, this radial heat conduction is critical for transferring heat from the bearings to the cooling channels. This article presents a systematic literature review on measures to enhance thermal conductivity, examining their applicability in motor spindle applications. Building on this, an approach for the development of components optimized for thermal conductivity for use in motor spindles is discussed. This serves as the foundation for a systematic method to improve heat flow through the thickness direction of CFRP motor spindle components.

1 Introduction

The usage of carbon fibre-reinforced polymers (CFRP) increases in several applications. CFRP has a low density, high specific stiffness and high damping properties. With its anisotropic behaviour depending on the fibre direction and amount of fibres in relation to the polymer matrix the material parameters can be designed according to the operational requirements. This applies for the mechanical behaviour as well as for the thermal behaviour. Since the fibres posses a negative coefficient of coefficient of thermal expansion (CTE), CFRP laminates can be designed to a wide range of thermal expansion coefficients. This study focuses on composites with epoxy resin as matrix. Regarding the thermal conductivity, polyacrylonitrile (PAN)-based fibres have a conductivity that is 20–400 times higher, using pitch-based fibres even 5.000 times higher, than the conductivity of the epoxy matrix. The epoxy matrix has a low

L. Kilian (✉) · P. Fehn · M. Weigold
Institute for Production Management, Technology and Machine Tools, Technical University of Darmstadt, Darmstadt, Germany
e-mail: l.kilian@ptw.tu-darmstadt.de

© The Author(s) 2026

K. Wegener and M. Bambach (eds.), *4th International Conference on Thermal Issues in Machine Tools (ICTIMT2025)*, Lecture Notes in Production Engineering,
https://doi.org/10.1007/978-3-032-01194-7_8

thermal conductivity of $\lambda_{ep} = 0.21 \frac{W}{mK}$ [1]. In composite materials, fibres are aligned along one plane's extension, with no fibres oriented in the thickness direction. So, in the thickness direction the properties of the resin dominate the behaviour of the composite. As a result, due to the poor thermal properties of the polymeric resin, CFRP exhibits low thermal conductivity in the laminate thickness (out-of-plane) direction.

Cylindrical parts are commonly produced using a filament winding process, where fibres are oriented around a core at specific angles to the longitudinal axis. This alignment allows heat flow through the fibres in both the axial and tangential directions. However, the radial direction, which corresponds to the laminate thickness direction, presents the same thermal conductivity challenges as the out-of-plane direction in flat composites. The mainly cylindrical parts of motor spindle systems (shaft, housing, spacer sleeves) are also manufactured by a filament winding process. In motor spindle systems the radial direction plays a key role in determining heat conduction from the inside to the cooling channels. Besides power losses of the motor and the friction in the tool clamping device the friction within the bearings is the main source of heat in motor spindles [2]. The bearing temperature determines the service life of the bearings [3], so heat conduction from the bearings to the cooling channels through housing or bearing caps is essential. Over the past two decades, numerous studies have investigated conductivity-enhancement in CFRP. This article provides an overview of the results of a systematic literature review. The evaluation of these studies forms the basis for a methodical approach to investigate and improve the heat flow through the thickness direction of CFRP motor spindle components. The use of composites in motor spindle systems requires excellent mechanical behaviour, especially a high stiffness. In full carbon motor spindles the CTE is designed to be very small (and negative) in the axial direction. In hybrid systems, where steel (e.g., housing) and CFRP parts (e.g., shaft) are combined, the CTE of the composite parts is adjusted to be similar to the CTE of the steel parts, to prevent stresses resulting from different thermal expansions. Thus, the effects of the conductivity-enhancing measure on both the mechanical behaviour and the CTE must also be considered to ensure that the advantages of CFRP are preserved. An additional constraint is that the measure must be compatible with the filament winding process in accordance with industrial standards, while minimizing any additional cost to the already expensive CFRP manufacturing process.

2 Previous Studies on Methods to Enhance Heat Conduction in CFRP

In the past several measures to increase the through-thickness conductivity of CFRP have been investigated. Table 1 shows the results of a systematic literature review. The relevant studies either use highly thermally conductive fillers mixed with the matrix or additional elements like fibres, wires or rods in the thickness (z-) direction. In some studies [4–7] pure epoxy samples were investigated, so the interaction between fibres

Table 1 Overview of reviewed studies on enhancing the through-thickness conductivity of fibre-reinforced composites and epoxy resin. The evaluations and data in the sources differ, leading to variations in the listed information. The enhanced conductivity λ_2, along with the percentage deviation relative to the initial value (calculated as $\frac{\lambda_2 - \lambda_1}{\lambda_1} \cdot 100$) is provided if stated in the source. For the CTE the percentage increase relative to the initial value is given. A space (–) means that either no data was provided in the source or it is deemed not relevant. The reviewed studies are evaluated for their applicability in motor spindle components manufactured via filament winding. The last column indicates whether a method is deemed significant for further investigation ($\oplus$) or not ($\ominus$).

References	Description of method (Loading in vol% or wt%)	λ_2 in W/(m K) (percentage deviation)	CTE in %	Mechanical properties	
[17]	Weaving Copper wires ($\varnothing = 0.25$ mm) as z-fibers, vacuum-assisted resin transfer molding technique	(+12.99 %)	–	–	$\ominus$
[23]	Creating PVDF/BN/CF composites via hot pressing, BN (15 wt% $\varnothing = 1$ μm), CF (15 wt%, $\varnothing = 11$μm, length = 200 μm)	1.89	–	–	$\ominus$
[24]	MWCNTs@Fe$_3$O$_4$ nanofluid (MFNF) as sizing agent in UD laminate	(+ 112.23%)	–	ILSS + 21.69% bending strength + 39.51%	$\ominus$
[9]	Network of MWCNT (0.1–0.5 Vol%) and hBN filler (10–30 Vol%)	3.36 (+ 140%)	–	–	$\ominus$
[6]	Weaving highly conductive films (highly oriented graphite film (1.97%) and copper film (1.74%)) in UD laminate				$\oplus$
	CU	21.45 4775%	4753	ILSS –13–17%	
	HOGF	49.19 11080%	11029	ILSS –19–21%	

(continued)

Table 1 (continued)

References	Description of method (Loading in vol% or wt%)	λ_2 in W/(m K) (percentage deviation)	CTE in %	Mechanical properties	
[25]	Al_2O_3 filler ($\varnothing = 0$ nm)	–	–	–	①
[26]	GO@Fe_3O_4 nanofluid sizing agent (2.5 wt%), Sizing treatment	3.099	129	–	①
[18]	Pins of PAN-based carbon fibre M55J and pitch-based carbon XN90–60S, pin distances: 3.0, 6.0 and 8.4 mm	8.8	1367	–	①
[16]	Hybrid nanofiller: introducing SiC nanowires in CF by electrophoretic deposition & grafting graphene on CF in weave laminate		–	–	①
[5]	Nano-alumina (nAl_2O_3) filler (0.5–2.0 wt%, $\varnothing = 13$ nm) in woven carbon fabric	–	–	–	①
[13]	Nanosized graphene & microsized Cu, metaparticles (0.26 wt%), Ultrasonication & spraying	6.5	520	–	①
[27]	Surface modification and copper plating CF	–	–	–	①
[28]	Morphological effect of CF	–	–	–	①
[29]	Copper Coating of CF	0.9 (+ 50%)	–	–	①
	C Nickel Coating of CF	2.9 (+ 383.3%)			
[30]	Graphene-coated CF	(+ 16.67%)		IISS 37.04%	①

(continued)

Table 1 (continued)

References	Description of method (Loading in vol% or wt%)	λ_2 in W/(m K) (percentage deviation)	CTE in %	Mechanical properties	
[31]	Chopped pitched based CF, graphite flake in UD and cross-ply laminate (2.5–7.5 Vol%)	2.21 (+ 166.3%)	–	–	⊕
[7]	Layer-by-layer spray coating of UD laminate (Prepreg) with inorganic crystal filler			Flexural strength in percent of pure CFRP	⊕
	aluminium (0.01 wt%)	1.350 (+ 32%)	–	–4.76%	
	aluminium (0.1 wt%)	1.462 (+ 60%)	–	–25.04%	
	copper (0.01 wt%)	1.650 (+ 81%)	–	–4.76%	
	copper (0.1 wt%)	1.634 (+ 79%)	–	+ 0.14%	
	magnesium (0.01 wt%)	1.710 (+ 87%)	–	–8.99%	
[32]	Graphene oxide filler ($\varnothing$ = 500 nm) in UD laminates (vacuum assisted resin infusion moulding) (6.3 Vol%)	0.81–0.85 (+ 7.8%)	–	IISS + 37%	⊕
[10]	Micro-nano BN fillers in UD prepreg laminate	(+ 188.24%)	–	–	⊖
[33]	Copper and hBN particles, coating CF tapes with electrophoretic deposition, modulate direction of hBN via electric field (vacuum assisted resin infusion molding)	2.16 (+ 217%)	166	Decrease in tensile strength	⊖

(continued)

Table 1 (continued)

References	Description of method (Loading in vol% or wt%)	λ_2 in W/(m K) (percentage deviation)	CTE in %	Mechanical properties	
[15]	Spray-coating carbon nanotubes (CNTs) and graphene nanoplatelets (GNPs) on fibre	1500 (+ 650%)	–	–	⊕
[34]	Carbon nanotube/copper/carbon fibre hierarchical composites by electrophoretic deposition	2.025 (+ 215%)	–	IISS 60.72MPa (+ 39.5%)	⊘
[35]	Modified BN (with KH550 surface treatment) particles in mesophase pitch-based carbon fibers (20 vol%)	7.9 (+ 79.5%)	–	–	⊘
[36]	Electroplated copper particles in fabric composites	0.69 (+ 68.4%)	–	–	⊘
[37]	MWCNTs (1–3 wt%, $\varnothing$ = 10–15 nm) and graphite nanoplatelets (GNPs)(1–15 wt%, 20–25 nm) fillers in UD laminate GNPs 15 wt%	0.96 (+ 48%)	–	–	⊕
[38]	Diamond powder $\varnothing$ = 0.5–1 μm (0–12 Vol%) in woven fabric T300 fibre (PAN based) YS90A fibre (pitch based)	1.8 + 125% 2.69 + 177%	– –	– –	⊕

(continued)

Table 1 (continued)

References	Description of method (Loading in vol% or wt%)	λ_2 in W/(m K) (percentage deviation)	CTE in %	Mechanical properties	
[19]	Al foils + Cu spheres Ø = 2...6 mm, distance: 2 mm Cu spheres Ø = 2 mm	7.6 + 1166.7%	+ 11.8	Tensile strength 450 MPA −46.4%	⊘
[39]	Carbon/carbon composites, SiC nanofibres on fibre (sizing)	–	–	–	⊘
[40]	Al$_2$O$_3$ encapsulated by graphene oxide, functionalization of Al$_2$O$_3$, glass fibre/epoxy composite	–	–	–	⊘
[4]	Al$_2$O$_3$ (Ø = 80−100 μm) (2.5−25 Vol%)	2.1 (+190,5%)	–	–	⊕
[41]	Diamond powder (5−50 wt%)	4.5 (+650%)	–	–	⊕
[14]	Ag flakes Ø = 20 μm (5 Vol%), mixing in epoxy, PAN-& pitch-based fabric	1.86 (+ 70%)	–	Tensile strength −17%	⊕
[42]	Electroplating and rapid thermal annealing of Cu shells	–	–	–	

(continued)

Table 1 (continued)

References	Description of method (Loading in vol% or wt%)	λ_2 in W/(m K) (percentage deviation)	CTE in %	Mechanical properties	
[20]	ANSYS evaluation, Cu rod & conductive coating Cu rod without coating Cu rod with coating	+ 500% + 9000%	–	–	⊕
[43]	Diamond particles (0.1–1 wt%)		–	Increase in Vickers hardness	⊕
[44]	Carbon foam, Sandwich construction	19.21	1243	–	⊕
[45]	WCNT@SiO$_2$ (0.5–1 wt%)	0.213 + 47.9%	–	–	⊕
[46]	Networked α-alumina fibre (42–52 Vol%)	4.2	–	–	⊕
[47]	Nanostructuring the interlaminar interface with carbon black ($\varnothing = 0.1\,\mu$m) (0.8 wt%)	210%	–	–	⊕
[21]	3D-weaving (5.5 Vol% pitch-based z-yarn) Cu fibres (4.3 Vol%) and pitch-based fill-fibre	21.93 (+ 2863.5%) 14.53 (+ 1863.5%)	–	–	⊕
[22]	Three-Dimensional Woven Fiber Architectures	–	–	–	⊕
[48]	MWCNT, carbon fiber/phenolic composite (7 wt%)	396	–	–	⊕

and the matrix with filler and the influence on the mechanical properties were not taken into account. Nevertheless, these studies can serve as a foundation for measures to be developed.

Reference [8] gives a recent overview on the impact of fillers in CFRP. For enhancement of thermal conductivity Aluminium Oxide (Al_2O_3) and Boron Nitride (BN) or hexagonal BN (hBN) as filler material are named. hBN filler is part of a variety of studies listed in Table 1. For an effective use of hBN filler complex manufacturing processes are required to form directed nano-networks [9, 10] or to create nano-composites [11, 12]. Additionally, many studies use multi-walled carbon nanotubes (MWCNT), graphite and copper (Cu) particles as filler material. These measures that involve spraying particles onto the composite during the laminating process or mechanically mixing filler particles into the epoxy resin demonstrated positive results. The spraying measure was investigated with different filler materials by [7, 13–15].

Measures that use electrophoretic deposition [10, 16] of particles are excluded from further considerations since this process is not practical for a filament winding process. Measures that change the sizing of fibres or the molecular structure of the polymers [9] are excluded for now as well, because of the complexity of these processes. Since these polymers or fibres are currently not available on the market the measures are not suitable for industrial production of motor spindle components.

Additional elements like fibres, wires or rods in the laminate thickness (z) direction [17–20] or 3D-weavings [21, 22] are often complex in manufacturing and often influence the mechanical properties negatively since they disrupt the fibre orientation.

2.1   Findings and Conclusions from the Literature Review

Investigating the methods listed in Table 1 the most common measure is to enhance the conductivity of the matrix through filler particles. For most fillers this can be practical for the filament winding process. For the impregnation process the fibres are guided through a resin bath. If the particles are added to the resin bath, depending on the areal weight of the particles, the settling of particles at the bottom of the resin bath must be considered. Additionally, to ensure that a sufficient number of particles adhere to the fibres, the adhesion of the particles on the impregnated fibres until deposition on the winding core cannot be overlooked. If epoxy with filler particles is suitable for the filament winding process depends on the filler material and has to be investigated in detail. Most methods in Table 1 use the vacuum assisted resin infusion moulding (VARIM) technique due to the high areal weight of the filler particles to prevent particle settling.

Adding particles to the composite by spraying a suspension on the composite during the manufacturing process seems a valid option as well. It involves an additional step during the manufacturing process but should be able to be automated or at least can be done without interrupting the winding process. Also, a localized deposition in specific segments of the components seams possible.

The integration of additional elements or fibres in the thickness direction requires additional efforts and has the described limitations regarding the mechanical properties. The integration into the winding process must be part of further research. Considering that in motor spindles a localized heat flow to the cooling channels is desired, additional elements seem to be a valid option and should be investigated.

The review highlights the need for further investigations to meet the specific requirements of the filament winding process. No study has examined the integration within the winding process, so the measures can only serve as a guideline, and direct transfer is not possible. To include elements in the radial direction has to be investigated in further studies. Additionally, the complex thermo-mechanical behaviour of the motor spindle system, which operates under extremely tight tolerance constraints, must be considered. Many studies leave certain aspects unaddressed, often providing insufficient details on sample preparation. Additionally, CTE and mechanical properties are often not examined. Future research must consider these specific requirements of motor spindle components.

3 Modelling of Heat Flow and Heat Conduction in CFRP

To assess the impact of enhancement methods, a fundamental understanding of heat conduction is essential. These principles provide the foundation for an analytical model, which serves as an efficient tool for representing thermal behaviour and evaluating the effects of various parameters. Heat transfer models have been developed since the nineteenth century, based on Fourier's (1822, Théorie analytique de la chaleur) fundamentals, Maxwell and Lord Rayleigh published their work on heat transfer. In the 20th century Bruggeman (1935) modelled thermal conductivity through disperse systems. Early models, such as those by Thornburg and Pears (1965), attempted to describe longitudinal and transverse heat transfer in unidirectional composites, but often lacked accuracy due to neglected interface conditions. Springer and Tsai (1967) improved upon this by incorporating shear properties, while Clayton introduced an iterative approach. Hashin later established upper and lower bounds for through-thickness thermal conductivity in transversely isotropic materials (1983). Subsequent refinements included models considering fiber coatings (Hatta and Taya 1986), interfacial thermal resistance (Hasselman and Johnson 1987), and orthotropic carbon fibres (Hasselman 1993). Further improvements came from void-inclusive models (Krach and Advani 1996) and models for woven-fabric composites, introduced by Knappe and Martinez-Freire (1965) and later refined for balanced (Dasgupta and Agarwal 1992) and unbalanced fabrics (Ning and Chou 1995). More recently, Thomann et al. (2004) developed a model for commingled yarn preforms, and Turias et al. proposed artificial neural networks to predict thermal conductivity (2004). Despite these advancements, most models remain limited to in-plane heat transfer. Out-of-plane thermal conductivity in fibre-reinforced plastics typically reaches only about four times the matrix conductivity. For accurate modelling multiple models must be combined [21].

3.1 Fundamentals of Heat Conduction

The transfer of heat through solid material is called heat conduction and is described by the heat flow $\dot{Q}$. The heat flux $\dot{q}$ is defined as the heat flow $\dot{Q}$ through a surface element dA. The thermodynamic energy balance for a closed volume dV where temperature gradients exist but no mechanical power is present can therefore be written as

$$\frac{dE}{d\tau} = \dot{Q} = -\int \dot{q} \cdot n \, dA \qquad (1)$$

with the normal vector n.

From the first law of thermodynamics we know, that between two surfaces of different temperature a steady heat flow will occur. Fourier's law of heat conduction states that the heat flux is directly proportional to the temperature gradient. With the thermal conductivity $\lambda = \lambda(T, p)$ the heat flow and the heat flux can be written as

$$d\dot{Q} = -\lambda(\text{grad } T) \, n \, dA = -\lambda \frac{\partial T}{\partial n} dA \, ; \; \dot{q} = -\lambda(\text{grad } T). \qquad (2)$$

3.2 Steady, One-Dimensional Conduction of Heat

In the case of steady heat transfer the heat flux and the temperatures at each location are constant over time. With time-independent T, $\frac{\partial T}{\partial \tau} = 0$. By considering a body with thickness δ and relatively large extensions in the other directions the assumption that the heat flows only in one direction is valid.

At the plate's surface with coordinate z_1, the temperature is set to T_1, and at the surface with coordinate z_2, the temperature is set to T_2. Due to the large dimensions in the x-y-plane, the temperature field depends only on the z-coordinate. As a result, the heat flux density is proportional to the thermal conductivity λ of the material, the applied temperature difference $T_2 - T_1$, and inversely proportional to the body thickness δ [49]

$$\dot{q} = \frac{\dot{Q}}{A} = -\lambda \cdot \frac{T_2 - T_1}{\delta} = -\lambda \cdot \frac{T_2 - T_1}{z_2 - z_1}. \qquad (3)$$

When a heat flow $\dot{Q}$ is conducted through a body of finite thickness δ, the body acts as a thermal resistance. This thermal conduction resistance R_λ is expressed in terms of the temperature difference ΔT as [50]

$$R_\lambda = \frac{\Delta T}{\dot{q} A}. \qquad (4)$$

For a body in the form of a flat plate, where the dimensions in the x-y-plane are large compared to the thickness δ in the z- direction, it follows from (2) and (3) [51]

$$R_{\lambda P} = \frac{\Delta T}{\dot{q} A} = \frac{\delta}{\lambda A}. \tag{5}$$

3.3 Heat Conduction of Heat Through Multi-layer Walls

For a wall composed of multiple layers of materials with different thermal conductivities λ_j arranged in series, the overall thermal resistance is given by

$$R_{\lambda,\text{sc}} = \sum_j R_{\lambda,j}. \tag{6}$$

The overall thermal conductivity or effective thermal conductivity of a multilayered wall of total thickness δ follows with the individual thermal conductivities λ_j and the thickness of the individual layers δ_j to

$$\frac{1}{\lambda_{\text{eff,sc}}} = \frac{1}{\delta} \sum_j \frac{\delta_j}{\lambda_j} \tag{7}$$

For walls with layers in parallel connection the thermal resistance is given by

$$R_{\lambda,\text{pc}} = \frac{1}{\sum_j \frac{1}{R_{\lambda,j}}}. \tag{8}$$

The overall thermal flow is the addition of the individual thermal flows through each layer. The overall thermal conductivity (λ_{eff}) of parallel layered walls is calculated by a weighted average of the individual thermal conductivities. With the volume fraction φ_j and the thermal conductivity λ_j of the respective material follows

$$\lambda_{\text{eff,pc}} = \sum_j \varphi_j \lambda_j. \tag{9}$$

Considering a plate of thickness δ with surface area A_1 and λ_1 with a bolt of A_B and λ_B the heat flows through both materials in parallel and the effective thermal resistance follows with (5) and (8) to

$$R_{\text{eff}} = \frac{\delta}{\lambda_1 A_1 + \lambda_B A_B}. \tag{10}$$

With $A_B = L_B s_B$, $A_1 = Ls - A_B$ and $A_{total} = Ls$ after rearranging it follows

$$\lambda_{eff} = \frac{\lambda_1 (Ls - L_B s_B) + \lambda_B L_B s_B}{Ls}. \tag{11}$$

Replacing $\beta = \frac{L_B s_B}{Ls}$ the effective thermal conductivity can be written as

$$\lambda_{eff} = \lambda_1 (1 - \beta) + \lambda_B \beta. \tag{12}$$

In fibre-reinforced composites the material properties of the composite are a combination of the fibres and the matrix properties λ_m. The thermal conductivity of the fibre varies along the fibre ($\lambda_{f\parallel}$) and in the fibre thickness direction ($\lambda_{f\perp}$). In fibre direction of an unidirectional (UD) layer fibres and matrix form a parallel connection and the thermal conductivity $\lambda_\parallel$ can be determined by adapting (9). Analogous to (12), where the area ratio β of the wall and the bolt is considered, $\lambda_\parallel$ is calculated using the fibre volume fraction φ.

$$\lambda_\parallel = \lambda_{f\parallel} \cdot \varphi + \lambda_m \cdot (1 - \varphi). \tag{13}$$

The thermal conductivity of an UD layer perpendicular to the fibre is given by [1]

$$\lambda_\perp = \frac{\lambda_{f\perp} + \lambda_m + (\lambda_{f\perp} - \lambda_m) \cdot \varphi}{\lambda_{f\perp} + \lambda_m - (\lambda_{f\perp} - \lambda_m) \cdot \varphi} \cdot \lambda_m. \tag{14}$$

By looking at the enhancement measures (Sect. 2) it is obvious that not only fibres and epoxy matrix have to be taken into account but also filler particles. Numerous studies in the literature have attempted to develop models for estimating the effective thermal conductivity of heterogeneous materials. To calculate the conductivity of a matrix with filler particles (λ_{mf}) the comparatively simple models are the series model and the parallel model. With the parallel model λ_{mf} is calculated analogous to (13) replacing $\lambda_{f\parallel}$ with the conductivity of the filler particles λ_{fp} and φ with the filler volume fraction v_{fp}. The series model can be written as [37]

$$\frac{1}{\lambda_{mf}} = \frac{1 - v_{fp}}{\lambda_m} + \frac{v_{fp}}{\lambda_{fp}} \tag{15}$$

and the geometric mean model is given in [37] as

$$\lambda_{mf} = \lambda_{fp}{}^{v_{fp}} \cdot \lambda_m{}^{1 - v_{fp}}. \tag{16}$$

3.4   Analytic Models of Composite with Filler Particles

To analytically describe the conductivity of a CFRP composite in the thickness direction with filler particles in the matrix the formulas to calculate $\lambda_\perp$ (14) and the selected model for λ_{mf} have to be combined. Different analytic models were reviewed by [52, 53]. Common models are the series, the parallel and the geometric mean model, the Maxwell-Eucken model and the Hamilton-Crosser model. They differ in their consideration of particle size and geometry, particle interaction, particle packing in the matrix or interfacial boundaries. Reference [37] compared the results of different analytical models with experimental results. All models underestimated the thermal conductivity of the nano-reinforced polymers. The calculated conductivities differed by up to 115% depending on the weight fraction of the filler.

In this study, a first comparison of the series, the parallel and the geometric mean model is conducted. With the series model (15) and (14) $\lambda_\perp$ becomes

$$
\lambda_{\perp,s} = \frac{\lambda_{f\perp} + \left(\left(\frac{1-v_{fp}}{\lambda_m} + \frac{v_{fp}}{\lambda_{fp}}\right)^{-1}\right) + \left(\lambda_{f\perp} - \left(\left(\frac{1-v_{fp}}{\lambda_m} + \frac{v_{fp}}{\lambda_{fp}}\right)^{-1}\right)\right) \cdot \varphi}{\lambda_{f\perp} + \left(\left(\frac{1-v_{fp}}{\lambda_m} + \frac{v_{fp}}{\lambda_{fp}}\right)^{-1}\right) - \left(\lambda_{f\perp} - \left(\left(\frac{1-v_{fp}}{\lambda_m} + \frac{v_{fp}}{\lambda_{fp}}\right)^{-1}\right)\right) \cdot \varphi}
$$
$$
\cdot \left(\left(\frac{1-v_{fp}}{\lambda_m} + \frac{v_{fp}}{\lambda_{fp}}\right)^{-1}\right). \tag{17}
$$

Using the parallel model (13) it follows

$$
\lambda_{\perp,p} = \frac{\lambda_{f\perp} + (\lambda_{fp} \cdot v_{fp} + \lambda_m \cdot (1 - v_{fp})) + (\lambda_{f\perp} - (\lambda_{fp} \cdot v_{fp} + \lambda_m \cdot (1 - v_{fp}))) \cdot \varphi}{\lambda_{f\perp} + (\lambda_{fp} \cdot v_{fp} + \lambda_m \cdot (1 - v_{fp})) - (\lambda_{f\perp} - (\lambda_{fp} \cdot v_{fp} + \lambda_m \cdot (1 - v_{fp}))) \cdot \varphi}
$$
$$
\cdot (\lambda_{fp} \cdot v_{fp} + \lambda_m \cdot (1 - v_{fp})). \tag{18}
$$

Using the geometric mean model (16) it follows

$$
\lambda_{\perp,g} = \frac{\lambda_{f\perp} + (\lambda_{fp}{}^{v_{fp}} \cdot \lambda_m{}^{1-v_{fp}}) + (\lambda_{f\perp} - (\lambda_{fp}{}^{v_{fp}} \cdot \lambda_m{}^{1-v_{fp}})) \cdot \varphi}{\lambda_{f\perp} + (\lambda_{fp}{}^{v_{fp}} \cdot \lambda_m{}^{1-v_{fp}}) - (\lambda_{f\perp} - (\lambda_{fp}{}^{v_{fp}} \cdot \lambda_m{}^{1-v_{fp}})) \cdot \varphi} \cdot (\lambda_{fp}{}^{v_{fp}} \cdot \lambda_m{}^{1-v_{fp}}). \tag{19}
$$

$\lambda_{f\perp}$ is normally not given and can hardly be measured. Reference [1] gives a $\lambda_{f\perp}$ of 1.7 W/(m K) and [37] of 0.915 W/(m K) for TORAY T300. Alternatively, if $\lambda_\perp$ is known for pure CFRP a comparison of pure CFRP and CFRP with filler particles can be used. At first the formula for $\lambda_\perp$ (14) has to be rearranged for $\lambda_{f\perp}$. The resulting $\lambda_{f\perp}$ can then be used to calculate the $\lambda_\perp$ of the CFRP with filler particles.

Calculating $\lambda_\perp$ with the values listed in Table 2 we get the following results:

- without filler with (14): $\lambda_\perp = 0.507$
- with filler with (17): $\lambda_{\perp,s} = 0.548$

Table 2 Example values used in the calculation of thermal conductivity

Parameter	Value
$\lambda_{f\perp}$ (T300)	1.7 W/(m K)
λ_m (epoxy)	0.21 W/(m K)
λ_{fp} (copper)	390 W/(m K)
φ	0.55
v_{fp}	0.10

- with filler with (18): $\lambda_{\perp,p} = 12.912$
- with filler with (19): $\lambda_{\perp,g} = 0.846$.

This example shows significant differences between the models. Analogous to the calculation of nano-reinforced polymers (just matrix with filler particles) [37] $\lambda_{\perp,s}$ is very small and seems to underestimate the real conductivity. This example shows the need for further investigation and careful consideration when using analytical models for estimating the effect of conductivity-enhancing measures.

4 Approach for Investigating and Developing Conductivity-Enhancement Measures for CFRP Motor Spindle Components

Sections 2 and 3 highlight the need for further research, particularly for complex applications such as motor spindles. Therefore, this article presents an approach for the development and investigation of measures to enhance thermal conductivity.

The approach can be divided into a sequential experimental path and an analytic path (Fig. 1). Starting with flat plates, the thermal conductivity of different laminates and conductivity-enhancing design measures are evaluated. A setup is designed to generate one-dimensional heat flow through the plate thickness. Additionally, the influence of these design measures on mechanical behaviour can be easily assessed through tensile tests on the flat plates. The most promising results are then transferred to cylindrical components and evaluated in a static, non-rotational test setup under ambient conditions. Along with evaluating the thermal conductivity of the cylindrical parts, the impact of the manufacturing process (filament winding) and surrounding conditions, as well as the thermal strain and displacement, are also assessed. To evaluate the thermal behaviour under operational conditions, a motor spindle test bench is used. This setup demonstrates the combined effects of mechanical and thermal factors on CFRP motor spindle components and their impact on operation. Additionally, comparing the heat conditions in the static setup with those in the motor spindle test bench serves as validation of the replication of conditions in the static setup. For an efficient way to evaluate different parameters and their influence on

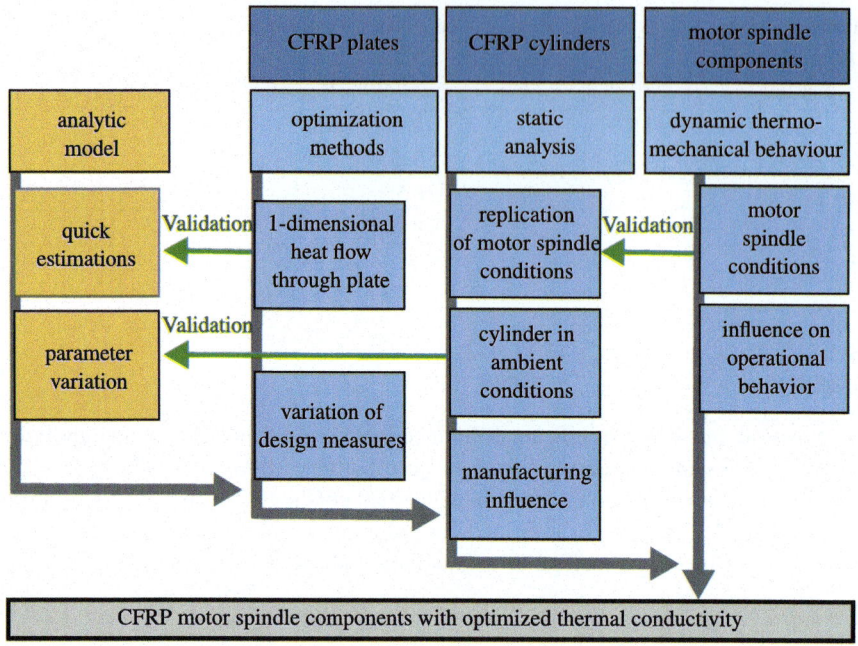

Fig. 1 Experimental and analytical approach to design CFRP motor spindle components with enhanced thermal conductivity

the thermal conductivity analytic models are used. They represent one-dimensional heat flow through plates and cylinder walls. The analytic models are validated and optimised through the experimental results.

5 Summary

This study provides a review of conductivity enhancement measures through the use of fillers or additional elements in CFRP. To apply conductivity enhancement measures in motor spindles, further investigations are required to assess their effects on mechanical properties and their integration into the filament winding process. For this purpose, an approach is proposed. Different aspects and influencing factors that need to be considered are discussed. This forms the foundation for a systematic approach to improve heat flow through the thickness direction of CFRP in motor spindle components.

Acknowledgements This work was supported by the Federal Ministry for Economic Affairs and Climate Action (BMWK) in the research projects SPOTLIGHT (03LB2061) and DynOLight (KK5031616 KU4).
Ethics Approval Not relevant

References

1. Schürmann, H.: Konstruieren mit Faser-Kunststoff-Verbunden (SpringerLink Bücher). Springer, Berlin (2005). https://doi.org/10.1007/b137636
2. Abele, E., Altintas, Y., Brecher, C.: Machine tool spindle units. CIRP Ann. **59**(2), 781–802 (2010). ISSN: 00078506. https://doi.org/10.1016/j.cirp.2010.05.002
3. Denkena, B., Bergmann, B., Klemme, H.: Cooling of motor spindles—a review (2020). https://doi.org/10.15488/10698
4. Agrawal, A., Satapathy, A.: Experimental investigation of micro-sized aluminium oxide reinforced epoxy composites for microelectronic applications. Procedia Mater. Sci. **5**, 517–526 (2014). https://doi.org/10.1016/j.mspro.2014.07.295
5. Yadhav, B.L., Chandan, C., Raghavendra, T., Suresha, B., Govinda Raju, H.K.: Role of alumina filler on thermal properties of carbon-epoxy nanocomposites. Materi. Today Proc. **59**(1), 166–170 (2022). https://doi.org/10.1016/j.matpr.2021.10.387
6. Ouyang, Z., Rao, Q., Peng, X.: Significantly improving thermal conductivity of carbon fiber polymer composite by weaving highly conductive films. Compos. Part A Appl. Sci. Manuf. **163**(9), 107–183 (2022). https://doi.org/10.1016/j.compositesa.2022.107183
7. Lee, E., Cho, C.H., Hwang, S.H., et al.: Improving the vertical thermal conductivity of carbon fiber-reinforced epoxy composites by forming layer-by-layer contact of inorganic crystals. Materials (Basel, Switzerland) **12**(19), 1944–1996. https://doi.org/10.3390/ma12193092
8. Kennedy, S.M., Vasanthanathan, A., Jeen Robert, R.B., Amudhan, K., Nagendran, M.: Impact of fillers in enhancing the properties of cfrp composites–a comprehensive exploration. Next Res. **2**(1), 100–117 (2025). https://doi.org/10.1016/j.nexres.2024.100117
9. Wang, Z., Wu, Z., Al Masoud, N., et al.: Effective three-dimensional thermal conductivity networks in polystyrene/multi-walled carbon nanotubes/aluminum oxide@hexagonal boron nitride composites based on synergistic effects and isolated structures. Adv. Compos. Hybrid Mater. **6**(3), 20648 (2023). https://doi.org/10.1007/s42114-023-00702-8
10. Zheng, X., Kim, S., Park, C.W.: Enhancement of thermal conductivity of carbon fiber-reinforced polymer composite with copper and boron nitride particles. Compos. Part A: Appl. Sci. Manuf. **121**(5), 449–456 (2019). https://doi.org/10.1016/j.compositesa.2019.03.030
11. Dong, J., Cao, L., Li, Y., Wu, Z., Teng, C.: Largely improved thermal conductivity of pi/bnns nanocomposites obtained by constructing a 3d bnns network and filling it with agnw as the thermally conductive bridges. Compos. Sci. Technol. **196**, 108–242 (2020). https://doi.org/10.1016/j.compscitech.2020.108242
12. Yao, Y., Sun, J., Zeng, X., Sun, R., Xu, J.-B., Wong, C.-P.: Construction of 3d skeleton for polymer composites achieving a high thermal conductivity. Small **14**(13), e1704044 (2018). https://doi.org/10.1002/smll.201704044
13. Lee, E., Son, I., Lee, J.H.: Starfish surface-inspired graphene-copper metaparticles for ultrahigh vertical thermal conductivity of carbon fiber composite. Compos. Sci. Technol. **199**, 108–385 (2020). https://doi.org/10.1016/j.compscitech.2020.108385
14. Wang, S., Haldane, D., Gallagher, P., Liu, T., Liang, R., Koo, J.H.: Heterogeneously structured conductive carbon fiber composites by using multi-scale silver particles. Compos. Part B Eng. **61**(9), 172–180 (2014). https://doi.org/10.1016/j.compositesb.2014.01.049
15. Li, Y., Zhang, H., Liu, Y.: et al.: Synergistic effects of spray-coated hybrid carbon nanoparticles for enhanced electrical and thermal surface conductivity of cfrp laminates. Compos. Part A: Appl. Sci. Manuf. **105**(1), 9–18 (2018). https://doi.org/10.1016/j.compositesa.2017.10.032
16. Wang, T., Song, Q., Zhang, S., et al.: Simultaneous enhancement of mechanical and electrical/thermal properties of carbon fiber/polymer composites via sic nanowires/graphene hybrid nanofillers. Compos. Part A Appl. Sci. Manuf. **145**(12), 106–404 (2021). https://doi.org/10.1016/j.compositesa.2021.106404
17. Lu, N., Sun, X., Wang, H., et al.: Synergistic effect of woven copper wires with graphene foams for high thermal conductivity of carbon fiber/epoxy composites. Adv. Compos. Hybrid Mater. **7**(1), 1 (2024). https://doi.org/10.1007/s42114-024-00840-7

18. Li, M., Fang, Z., Wang, S., Gu, Y., Zhang, W.: Thermal conductivity enhancement and syner-gistic heat transfer of z-pin reinforced graphite sheet and carbon fiber hybrid composite. Int. J. Heat Mass Transfer. **171**(1), 121093 (2021). https://doi.org/10.1016/j.ijheatmasstransfer.2021. 121093
19. Yu, G.-C., Wu, L.-Z., Feng, L.-J., Yang, W.: Thermal and mechanical properties of carbon fiber polymer-matrix composites with a 3d thermal conductive pathway. Compos. Struct. **149**(12), 213–219 (2016). https://doi.org/10.1016/j.compstruct.2016.04.010
20. Yu, H., Nonn, A., Schneiders, S., Heider, D., Advani, S.G.: An approach to enhance through-thickness thermal conductivity of polymeric fiber composites. Int. J. Heat Mass Transf. **59**(7), 20–28 (2013). https://doi.org/10.1016/j.ijheatmasstransfer.2012.11.055
21. Schuster, J., Heider, D., Sharp, K., Glowania, M.: Thermal conductivities of three-dimensionally woven fabric composites. Compos. Sci. Technol. **68**(9), 2085–2091 (2008). https://doi.org/10.1016/j.compscitech.2008.03.024
22. Sharp, K., Bogdanovich, A.E., Tang, W., Heider, D., Advani, S., Glowiana, M.: High through-thickness thermal conductivity composites based on three-dimensional woven fiber architec-tures. AIAA J. **46**(11), 2944–2954 (2008). https://doi.org/10.2514/1.38108
23. Yi, G., Li, J., Henderson, L.C., et al.: Carbon fiber/boron nitride fillers for enhancing through-plane thermal conductivity of poly(vinylidene fluoride): Synergistic effect and mechanism. Compos. Commun. **51**(3), 102090 (2024). https://doi.org/10.1016/J.COCO.2024.102090
24. Li, J., Jiang, N., Cheng, C., et al.: Preparation of magnetic solvent-free carbon nanotube/fe3o4 nanofluid sizing agent to enhance thermal conductivity and interfacial properties of carbon fiber composites. Compos. Sci. Technol. **236**(1), 109980 (2023). https://doi.org/10.1016/j. compscitech.2023.109980
25. Ouyang, Y., Bai, L., Tian, H., Li, X., Yuan, F.: Recent progress of thermal conductive ploymer composites: Al2o3 fillers, properties and applications. Compos. Part A Appl. Sci. Manuf. **152**(3), 106685 (2022). https://doi.org/10.1016/j.compositesa.2021.106685
26. Cheng, H., Xing, L., Zuo, Y., et al.: Constructing nickel chain/mxene networks in melamine foam towards phase change materials for thermal energy management and absorption-dominated electromagnetic interference shielding. Adv. Compos. Hybrid Mater. **5**(2), 755–765 (2022). https://doi.org/10.1007/s42114-022-00487-2
27. Mahmood, R.Y., Alobaedy, A.: Enhancement of thermal stability and wettability for epoxy/cu coated carbon fiber composites. Iraqi J. Phys. **18**(47), 55–61 (2020). https://doi.org/10.30723/ ijp.v18i47.611
28. Wang, G., Gao, M., Yang, B., Chen, Q.: The morphological effect of carbon fibers on the thermal conductive composites. Int. J. Heat Mass Transfer **152**(5), 119477 (2020). https://doi. org/10.1016/j.ijheatmasstransfer.2020.119477
29. Bard, S., Schönl, F., Demleitner, M., Altstädt, V.: Copper and nickel coating of carbon fiber for thermally and electrically conductive fiber reinforced composites. Polymers **11**(5) (2019). https://doi.org/10.3390/polym11050823
30. Cheng, X., Zhang, J., Wang, H., Wu, L., Sun, Q.: Improving the interlaminar shear strength and thermal conductivity of carbon fiber/epoxy laminates by utilizing the graphene–coated carbon fiber. J. Appl. Polymer Sci. **136**(7), 3 (2019). https://doi.org/10.1002/app.47061
31. Fang, Z., Li, M., Wang, S., Gu, Y., Li, Y., Zhang, Z.: Through-thickness thermal conductivity enhancement of carbon fiber composite laminate by filler network. Int. J. Heat Mass Transfer **137**(15), 1103–1111 (2019). https://doi.org/10.1016/j.ijheatmasstransfer.2019.04.007
32. Senis, E.C., Golosnoy, I.O., Dulieu-Barton, J.M., Thomsen, O.T.: Enhancement of the electrical and thermal properties of unidirectional carbon fibre/epoxy laminates through the addition of graphene oxide. J. Mater. Scie. **54**(12), 8955–8970 (2019). https://doi.org/10.1007/s10853-019-03522-8
33. Zhang, S., Gao, L., Han, J., et al.: Through-thickness thermal conductivity enhancement and tensile response of carbon fiber-reinforced polymer composites. Compos. Part B Eng. **165**, 183–192 (2019). https://doi.org/10.1016/j.compositesb.2018.11.114
34. Yan, F., Liu, L., Li, M., Zhang, M., Xiao, L., Ao, Y.: Preparation of carbon nan-otube/copper/carbon fiber hierarchical composites by electrophoretic deposition for enhanced

thermal conductivity and interfacial properties. J. Mater. Sci. **53**(11), 8108–8119 (2018). https://doi.org/10.1007/s10853-018-2115-9

35. Fan, B., Liu, Y., He, D., Bai, J.: Enhanced thermal conductivity for mesophase pitch-based carbon fiber/modified boron nitride/epoxy composites. Polymer **122**(25), 71–76 (2017). https://doi.org/10.1016/j.polymer.2017.06.060

36. Yu, S., Park, K., Lee, J.-W.: et al.: Enhanced thermal conductivity of epoxy/cu-plated carbon fiber fabric composites. Macromolecular Res. **25**(6), 559–564 (2017). https://doi.org/10.1007/s13233-017-5114-9

37. Kostagiannakopoulou, C., Fiamegkou, E., Sotiriadis, G., Kostopoulos, V.: Thermal conductivity of carbon nanoreinforced epoxy composites. J. Nanomater. **2016**(6), 1–12 (2016). https://doi.org/10.1155/2016/1847325

38. Srinivasan, M., Maettig, P., Glitza, K.W., et al.: Out of plane thermal conductivity of carbon fiber reinforced composite filled with diamond powder. Open J. Compos. Mater. **06**(02), 41–57 (2016). https://doi.org/10.4236/ojcm.2016.62005

39. Chen, J., Xiao, P., Xiong, X.: The mechanical properties and thermal conductivity of carbon/carbon composites with the fiber/matrix interface modified by silicon carbide nanofibers. Mater. Design **84**(1), 285–290 (2015). https://doi.org/10.1016/j.matdes.2015.06.085

40. Yao, Y., Zeng, X., Guo, K., Sun, R., Xu, J.-B.: The effect of interfacial state on the thermal conductivity of functionalized Al_2O_3 filled glass fibers reinforced polymer composites. Compos. Part A Appl. Sci. Manuf. **69**(14), 49–55 (2015). https://doi.org/10.1016/j.compositesa.2014.10.027

41. Srinivasan, M., Maettig, P., Glitza, K.W., Sanny, B., Schumacher, A.: Multiscale calculation for increasing the thermal conductivity of carbon fiber composite with diamond powder (2014). https://doi.org/10.13140/RG.2.1.5035.8169

42. Yu, S., Park, B.-I., Park, C., Hong, S.M., Han, T.H., Koo, C.M.: Rta-treated carbon fiber/copper core/shell hybrid for thermally conductive composites. ACS Appl. Mater. Interfaces. **6**(10), 7498–7503 (2014). https://doi.org/10.1021/am500871b

43. Rakha, S.A., Khan, R.R., Khurram, A.A., Fayyaz, A., Zakaullah, M., Munir, A.: Mechanical properties of epoxy composites with low content of diamond particles. J. Appl. Polymer Sci. **127**(5), 4079–4085 (2013). https://doi.org/10.1002/app.38029

44. Sihn, S., Ganguli, S., Anderson, D.P., Roy, A.K.: Enhancement of through-thickness thermal conductivity of sandwich construction using carbon foam. Compos. Sci. Technol. **72**(7), 767–773 (2012). https://doi.org/10.1016/j.compscitech.2012.02.003

45. Cui, W., Du, F., Zhao, J., et al.: Improving thermal conductivity while retaining high electrical resistivity of epoxy composites by incorporating silica-coated multi-walled carbon nanotubes. Carbon **49**(2), 495–500 (2011). https://doi.org/10.1016/j.carbon.2010.09.047

46. Hojo, F., Kagawa, H., Takezawa, Y.: Synthesis of a polymer composite with networked .alpha.-alumina fiber and evaluation of its thermal conductivity. J. Ceramic Soc. Japan **119**(1391), 601–604 (2011). https://doi.org/10.2109/jcersj2.119.601

47. Han, S., Lin, J.T., Yamada, Y., Chung, D.: Enhancing the thermal conductivity and compressive modulus of carbon fiber polymer-matrix composites in the through-thickness direction by nanostructuring the interlaminar interface with carbon black. Carbon **46**(7), 1060–1071 (2008). https://doi.org/10.1016/j.carbon.2008.03.023

48. Kim, Y.A., Kamio, S., Tajiri, T., et al.: Enhanced thermal conductivity of carbon fiber/phenolic resin composites by the introduction of carbon nanotubes. Appl. Phys. Lett. **90**(9), 2767 (2007). https://doi.org/10.1063/1.2710778

49. Hannoschöck, N.: Wärmeleitung und -transport: Grundlagen der Wärme- und Stoffübertragung, 1st edn. Springer, Berlin (2018). https://doi.org/10.1007/978-3-662-57572-7. [Online]. Available: http://nbn-resolving.org/urn:nbn:de:bsz:31-epflicht-1591369%7D

50. Baehr, H.D., Stephan, K.: Heat and Mass Transfer, 3rd edn. Springer, Berlin (2011).https://doi.org/10.1007/978-3-642-20021-2. [Online]. Available: http://nbn-resolving.org/urn:nbn:de:bsz:31-epflicht-1606852%7D

51. Herwig, H., Moschallski, A.: W ä rme ü bertragung: Physikalische Grundlagen und ausf ü hrliche Anleitung zum L ö sen von Aufgaben (Springer eBooks Computer Science and Engineering), 4. überarbeitete und erweiterte Auflage. Springer Vieweg, Wiesbaden (2019). ISBN 978-3-658-26400-0. https://doi.org/10.1007/978-3-658-26401-7
52. Pietrak, K., Wisniewski, T. S.: A review of models for ective thermal conductivity of composite materials. J. Power Technol. **95**(1) (2014)
53. Yang, X., Liang, C., Ma, T., et al.: A review on thermally conductive polymeric composites: Classification, measurement, model and equations, mechanism and fabrication methods. Adv. Compos. Hybrid Mater. **1**(2), 207–230 (2018). https://doi.org/10.1007/s42114-018-0031-8

Influence of Thermoelectrically Controlled Temperature of Spindle Front Bearings on the Warm-Up Time After a Spindle Stop

Eckart Uhlmann, Mitchel Polte, Florian Triebel, Thomas Pache, and Roland Binninger

Abstract In various industrial sectors, including optics, automotive engineering, and information technology, there is an increasing demand for the production of components with high level of geometrical accuracy. In high-precision machining, a significant proportion of geometric inaccuracies can be attributed to thermally induced displacements of the tool center point (TCP). In particular, variations in the thermal load, such as changes in internally induced heat flux and the inherent thermal inertia of the machine tool, lead to prolonged periods required to reach thermal steady state. These delays adversely affect machining accuracy, particularly since the TCP is coupled to the condition of the motorized spindle's front bearings, which are affected by internal heat sources and external heat sinks. To enhance the thermomechanical stability, a motorized milling spindle was equipped with a thermoelectric temperature control system, enabling precise control of the temperatures at the outer races of the front and rear bearings, as well as at the stator of the motor. This experimental setup was employed to investigate the effect of thermoelectric temperature control on the time required to reach steady state of axial tool displacement following a spindle stop. To simulate changes in the tool or workpiece, the spindle was brought to a stop from rotational speeds ranging from 10,000 to 40,000 1/min for a predetermined period. The influence of temperature control during this downtime on the subsequent time required to reach steady state was analyzed. The results show that thermoelectrical induction of heat during downtime can significantly reduce the time required to reach steady state conditions. The findings of this study demonstrate that, compared

E. Uhlmann · M. Polte · F. Triebel (✉) · T. Pache
Institute for Machine Tools and Factory Management IWF, Technische Universität Berlin, Berlin, Germany
e-mail: triebel@iwf.tu-berlin.de

E. Uhlmann · M. Polte
Fraunhofer Institute for Production Systems and Design Technology IPK, Berlin, Germany

R. Binninger
Fraunhofer Institute for Physical Measurement Techniques IPM, Freiburg, Germany

© The Author(s) 2026

K. Wegener and M. Bambach (eds.), *4th International Conference on Thermal Issues in Machine Tools (ICTIMT2025)*, Lecture Notes in Production Engineering, https://doi.org/10.1007/978-3-032-01194-7_9

to a constant temperature, an adaptively controlled increase in temperature during downtime results in a reduction of the steady state time by a minimum of 62%.

Keywords Thermoelectric temperature control · Peltier modules · Spindle warm-up time · Thermally induced displacement

1 Introduction

A growing demand for components with high levels of geometrical accuracy is evident in various industrial sectors, including optics, automotive engineering, and information technology [1]. The manufacturing of these components is constrained by the machining accuracy of the machine tools utilized. In the field of high- and ultra-precision manufacturing, the machining accuracy of machine tools is considerably limited by their thermal behavior. Alterations in heat flow rates $\dot{Q}_{ind}$ within the machine tool and its components, attributable to electrical and mechanical power losses, lead to varying temperature fields, consequently resulting in thermally induced deformations δ_{th}. The thermally induced deformations δ_{th} of machine tool components can affect the positional correlation between tool and workpiece, thereby adversely affecting the machining accuracy [2]. Mayr et al. [2] stated, that up to 75% of the geometrical errors of machined workpieces can be attributed to the thermal behavior.

Motorized spindles are considered as a primary heat source in machine tools [3]. Consequently, they play a crucial role when it comes to optimization of the thermal behavior of machine tools. According to Denkena et al. [4], heat flow rates $\dot{Q}_{ind}$ in motorized spindles are induced through the conversion from electrical to mechanical energy E in the rotor and stator as well as through friction within bearings. The induced heat flow rates $\dot{Q}_{ind}$ are varying depending on the actual rotational speed n and load torque M_l. The direct connection between the shaft of motorized spindles and the tool or the workpiece exerts considerable influence on the achievable machining accuracy. Denkena et al. [4] conclude that passive measures for cooling motorized spindles have advantages regarding the operational cost and resource efficiency, while active cooling systems offer a substantially higher cooling potential which is necessary to achieve the full potential of modern machine tools. Consequently, active fluid cooling systems are widely used in high-performance spindles [4]. However, state-of-the-art fluid cooling systems are not reactive to variations in the induced heat flow rates $\dot{Q}_{ind}$ [5].

In addressing the issue of dynamic imbalance between induced heat flow rates $\dot{Q}_{ind}$ and dissipated heat flow rates $\dot{Q}_{dis}$ in motorized spindles, Liu et al. [6] proposed a cooling strategy based on power adjustment. The researchers employed analytical modeling and prediction strategies to ensure enhanced cooling performance that is adapted to variable energy dissipation.

The implementation of localized cooling mechanisms at locations with increased induced heat flow rates $\dot{Q}_{ind}$, such as bearing seats, in combination with a global cooling system, has the potential to further enhance the motorized spindle's thermal behavior. Jonath et al. [7] and Liang et al. [8] investigated cooling systems based on heat pipes. It was demonstrated that the implementation of such locally acting cooling systems results in less axial shaft displacement Δl_s compared to an identical system that uses water cooling.

An alternative approach for realizing localized cooling mechanisms is the utilization of Peltier modules. Peltier modules utilize the inverse Seebeck effect, the Peltier effect, to transport heat through the typically prismatic shaped modules. The Peltier effect, first discovered by Jean Peltier [9], is observed when an electric current I passes through two conductors composed of different materials. In addition to the induced Joule heat flow rate $\dot{Q}_J$, the Peltier heat flow rate $\dot{Q}_P$ is induced at one junction and absorbed at the other. The magnitude of the Peltier heat flow rate $\dot{Q}_P$ is proportional to the electrical current I and dependent on the characteristic Peltier coefficient Π_{XY} for combination of material X and Y [10]. A Peltier module is composed of multiple semiconductor legs arranged in an alternating pattern of n- and p-doped elements, which are electrically connected in series. The semiconductors are enclosed in a thermal parallel arrangement by two ceramic plates, thereby ensuring electrical isolation. By applying an electric current I and controlling its direction, a heat flow rate $\dot{Q}$ can be transferred from one side to the other. The applicability of Peltier modules in the context of rapid control of temperatures ϑ in motorized spindles has been investigated by Ngo et al. [11], Fan et al. [12], and Uhlmann et al. [13, 14], among others. Experiments and simulations have shown that, compared to conventional cooling, steady state time t_{st} and axial shaft displacement Δl_s of motorized spindles can be significantly reduced [13].

In order to meet high requirements in machining accuracy, it is necessary to reach and hold a thermal steady state after changes in the induced heat flow rate $\dot{Q}_{ind}$. The reduction of steady state time t_{st} following a spindle stop has the potential to reduce the non-productive time following a change of tool or workpiece. This in turn can enhance productivity and reduce energy consumption per machined part. Figure 1 shows the typical behavior regarding the axial shaft displacement Δl_s of a high-frequency motorized spindle with state-of-the-art water cooling. In this example a motorized milling spindle of type Z62 from Nakanishi Jaeger GmbH, Ober-Mörlen, Germany, was operated at a rotational speed n = 55,000 1/min, followed by a stop for a downtime of t_d = 300 s. Thereafter, the spindle was set to the previous operational state with rotational speed of n = 55,000 1/min.

Figure 1 shows the axial shaft displacement Δl_s measured at a reference plane of the tool holder. The tool holder utilized is described by Uhlmann et al. [14]. Starting in steady state, immediately after spindle stop the axial shaft displacement abruptly falls to $\Delta l_s = -7$ μm. During downtime t_d the spindle slowly contracts further. After restart of the spindle, first the axial shaft displacement Δl_s increases rapidly, evolving to a slower rising and reaching a steady state. The high gradient of the axial shaft displacement Δl_s after stopping and restarting the spindle can be attributed to the centrifugal forces F_C acting on the elements of the angular contact ball bearings typically used within high-frequence spindles and the gyroscopic moment M_g, e.g., described by Jedrzejewski and Kwasny [15]. The difference in the gradient of the axial

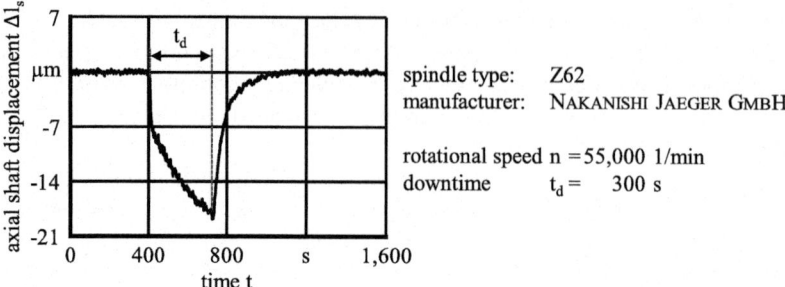

Fig. 1 Axial shaft displacement Δl_s of a motorized milling spindle of type Z62 from Nakanishi Jaeger GmbH operated at rotational speed n = 55,000 1/min and stopped for a downtime t_d = 300 s

shaft displacement Δl_s after stopping and restarting the spindle can be explained by a different absolute value of the acceleration a for starting and stopping the spindle. It has been determined that the time t_{start} required to reach a specific rotational speed n is approximately 70% higher than the time t_{stop} required to stop the spindle. The slower changes in axial shaft displacement Δl_s can be attributed to the thermal behavior [15]. The objective of the work presented in this paper is to investigate the influence of thermoelectrically controlled temperatures ϑ inside the spindle on induced axial shaft displacement Δl_s and the time t_{st} required for axial shaft displacement Δl_s to reach steady state conditions.

2 Experimental Setup and Methodology

With the aim to improve the thermomechanical behavior of machine tools, a prototype of a motorized spindle with a thermoelectric temperature control system based on Peltier modules has been developed and manufactured [13, 14, 16]. The test bench for investigating this thermoelectric temperature control system is composed of the spindle itself, a cooling unit and various measurement and control equipment.

The design of the spindle is based on a commercially available motorized milling spindle of type Z62 from the Nakanishi Jaeger GmbH. The spindle shaft is supported by a pair of single row angular contact ball bearings at both the front and rear ends of the spindle. The housing contains the bearing seats and the stator of the synchronous motor. The motor is controlled by a frequency inverter of type SD2T made by SIEB & MEYER AG, Lüneburg, Germany, and reaches rotational speeds of n ≤ 60,000 1/min.

The square-based Peltier modules are located between the housing and the water cooler. Each six Peltier modules with an edge length l = 23 mm are positioned at the outer circumference of the front and rear bearings, while twelve Peltier modules with an edge length l = 30 mm are placed around the stator, likewise. The prismatic shape of the modules results in a hexagonal outer surface of the housing and inner surface of the water cooler. The Peltier modules allow for control of heat flow rates $\dot{Q}$ inside the spindle and enable an independent temperature control by means of a PID controller, which regulates the respective heat flow $\dot{Q}$ between heat sources and sinks. Resistance thermometers of type Pt1000 made by Therma Thermofühler GmbH, Lindlar, Germany, are implemented to measure temperatures ϑ inside the spindle. Axial shaft displacement Δl_s is measured by an eddy current sensor of type ES-S04-C-CAx connected to a measuring system of type DT3071 both made by Micro-Epsilon Messtechnik GmbH & Co. KG, Ortenburg, Germany. The eddy current sensor is mounted on a measuring adapter made of FeNi36 and measures axial shaft displacement Δl_s at the reference plane of the tool holder, described by Uhlmann et al. [14]. The measurements and controls are integrated into a modular real-time system of type Adwin-Pro II made by Jäger Computer Gesteuerte Messtechnik GmbH, Lorsch, Germany. Measurement and control on the Adwin-Pro II run at a frequency f = 4 kHz.

For the experimental investigation, three measurement phases are defined. Initially in phase 1, the spindle is run at constant rotational speed n with controlled temperatures at the front bearings ϑ_{fb}, the stator ϑ_{st}, and the rear bearings ϑ_{rb} at a setpoint $\vartheta_{fb} = \vartheta_{st} = \vartheta_{rb} = 25$ °C. After reaching steady state conditions, regarding axial shaft displacement Δl_s, in phase 2 the spindle is stopped for a downtime of $t_d = 300$ s. During this downtime t_d the setpoint of the temperature at the front bearings is changed to 25 °C $\leq \vartheta_{fb} \leq 50$ °C. This variation in setpoint is henceforth referred to as downtime temperature ϑ_d. Subsequent to downtime t_d, in phase 3 the spindle is set back to the original operating conditions of phase 1.

The time required for the axial shaft displacement Δl_s to reach a steady state is determined applying a threshold-based method on the measured data. This time period is henceforth referred to as steady state time t_{st}. First, the mean value of axial shaft displacement $\overline{\Delta l_s}$ at steady state is determined using the measured data for the time 1500 s $\leq$ t $\leq$ 1800 s after re-starting the spindle. Then a tolerance band is defined around the mean axial shaft displacement $\overline{\Delta l_s}$. The steady state is defined as the point at which the measured data is first bounded by the specified tolerance band and subsequently remains within its boundaries. This study examines two different tolerance band widths $w_1 = 2$ μm and $w_2 = 1$ μm to address different accuracy requirements. It is imperative to note that a tolerance band width of $w_2 = 1$ μm signifies that the boundary of the tolerance band is located at $\pm w_2/2 = \pm 0.5$ μm around the mean axial shaft displacement $\overline{\Delta l_s}$.

The experiments described above are performed for rotational speeds n $\in$ { 10,000; 20,000; 30,000; 40,000} 1/min. For each rotational speed n downtime temperature is set to $\vartheta_d \in$ {25; 30; 35; 40; 45; 50} °C. Figure 2 shows an exemplary measurement with a visualization of both tolerance bands to determine steady state time t_{st}.

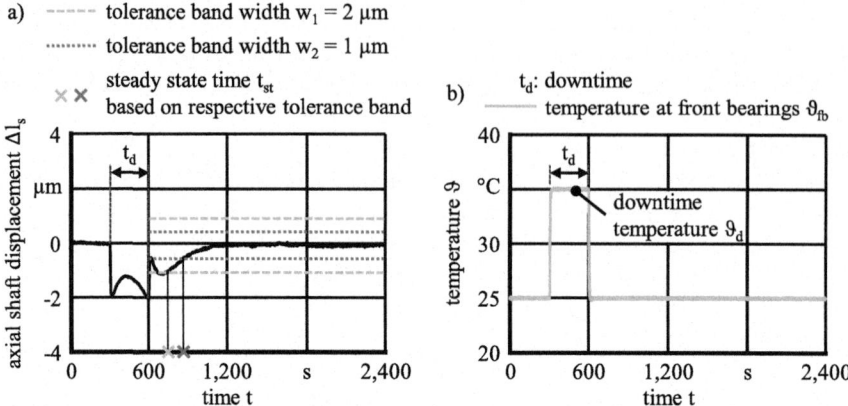

Fig. 2 Exemplary measurement at rotational speed n = 30,000 1/min with downtime temperature ϑ_d = 35 °C; **a** axial shaft displacement Δl_s with tolerance bands to determine steady state time t_{st}; **b** temperature at front bearings ϑ_{fb}

3 Results and Discussion

The repeatability of the method for determining steady state time t_{st}, as outlined in Sect. 2, was determined through the execution of a total number of N = 5 repetitions of a measurement each conducted under identical parameters. Figure 3 depicts axial shaft displacement Δl_s of the spindle prototype operated at rotational speed n = 30,000 1/min and downtime temperature of ϑ_d = 25 °C. Apparently, in phase 1 and phase 3 axial shaft displacement Δl_s aligns throughout all five measurements. In phase 2 a predominantly vertical shift can be observed. It is hypothesised that this shift can be attributed to the occurrence of nonlinear effects within the rear bearings. The outer races of the rear bearings are pretensioned by springs, allowing a small degree of axial movement within the bearing seat in the housing to ensure a constant preload on the angular contact ball bearings. Due to friction, the axial position of the outer races of the rear bearings varies after a spindle stop, influencing axial shaft displacement Δl_s. Subsequent to the restart of the spindle, the parallel vertical shift immediately disappears and the measurements of the axial shaft displacement Δl_s realign.

In Table 1 steady state times t_{st} regarding defined tolerance band widths w_1 = 2 μm and w_2 = 1 μm are presented. For a total number of N = 5 measurements, each conducted under identical parameters, as shown in Fig. 3, the method results in a mean steady state time of $\overline{t_{st}}$ = 197.5 s with a standard deviation of s_t = 4.9 s regarding the tolerance band width w_1 and a mean steady state time of $\overline{t_{st}}$ = 271.0 s with a standard deviation of s_t = 7.3 s regarding the tolerance band width w_2. These findings indicate that the proposed method for determining the steady state time t_{st} is suitable of producing repeatable results.

Figure 4 depicts the results of the experiments, described in Sect. 2, with rotational speeds n ∈ {10,000; 20,000; 30,000; 40,000} 1/min and downtime temperature ϑ_d

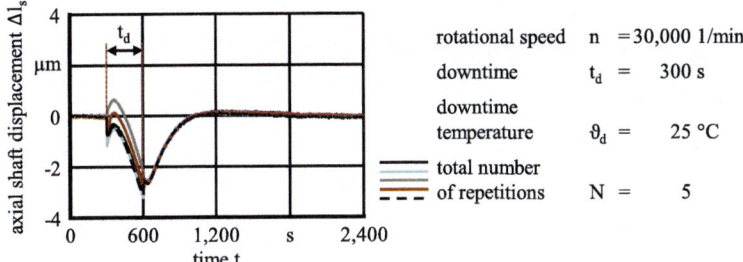

Fig. 3 Total number of N = 5 measurements at rotational speed n = 30.000 1/min with downtime temperature of $\vartheta_d = 25\ °C$

Table 1 Steady state time t_{st} according to the method outlined in Sect. 2 for a total number of N = 5 measurements under identical conditions, see Fig. 3

Tolerance band width w_i	steady state time t_{st} for experiment number					Mean steady state time $\overline{t_{st}}$	Standard deviation s_t
	1	2	3	4	5		
2 μm	204.0 s	200.7 s	193.2 s	197.3 s	192.3 s	197.5 s	4.9 s
1 μm	279.4 s	274.3 s	267.1 s	273.6 s	260.5 s	271.0 s	7.3 s

$\in \{25;\ 30;\ 35;\ 40;\ 45;\ 50\}$ °C. The measurements demonstrate that variation of temperature at the front bearings ϑ_{fb} during downtime t_d exerts an influence on the axial shaft displacement Δl_s in phase 2 and phase 3 for all rotational speeds n considered.

During phase 2, the downtime temperature ϑ_d exerts an influence on the gradient of the axial shaft displacement Δl_s. Maintaining the temperature at the front bearings at $\vartheta_d = \vartheta_{fb} = 25\ °C$ throughout the whole experiment, axial shaft displacement Δl_s exhibits a reduction during downtime t_d. After restart of the spindle, in phase 3, axial shaft displacement Δl_s rises without overshooting until steady state is reached.

Higher downtime temperatures ϑ_d first lead to a decline in the rate of reduction and subsequently lead to an increase of axial shaft displacement Δl_s during phase 2. At the beginning of phase 2, an increase in downtime temperature ϑ_d results in a greater reduction of the axial shaft displacement Δl_s. In phase 3, following the restart of the spindle, an increase in downtime temperature ϑ_d results in a rising peak of axial shaft displacement Δl_s. The phenomenon leads to an overshoot before axial shaft displacement Δl_s reaches steady state.

The rising peaks of the axial shaft displacement Δl_s observed at higher downtime temperatures ϑ_d can be attributed to the thermal expansion of the angular contact ball bearings. Following an abrupt change in the setpoint of the temperature at the front bearings ϑ_{fb}, the temperature of the outer races is initially affected. Subsequently, a time delay occurs before the temperature of the inner races is affected.

The results of the determination of steady state time t_{st} regarding a tolerance band width of $w_1 = 2\ \mu m$ and $w_2 = 1\ \mu m$ are presented in Table 2. For further evaluation, steady state time t_{st} of the measurement with unchanged temperature at the front

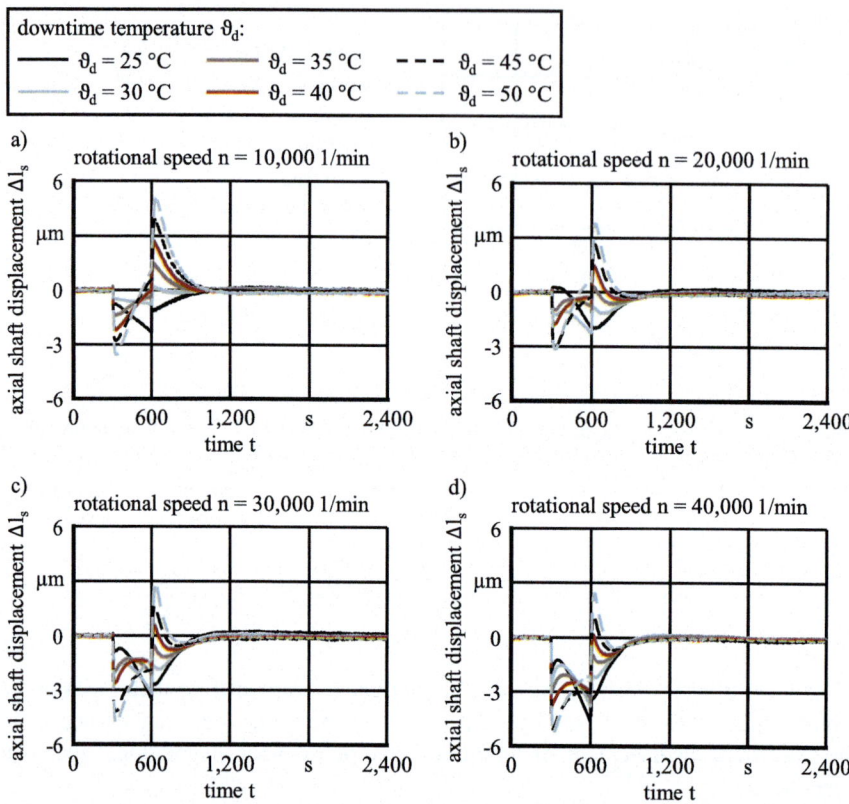

Fig. 4 Measurement of the axial shaft displacement Δl_s with variation of the temperature at the front bearings $\vartheta_{fb} \in \{25; 30; 35; 40; 45; 50\}$ °C during downtime of $t_d = 300$ s; **a** rotational speed n = 10,000 1/min; **b** rotational speed n = 20,000 1/min; **c** rotational speed n = 30,000 1/min; **d** rotational speed n = 40,000 1/min

bearings of $\vartheta_{fb} = 25$ °C throughout the whole experiment is defined as reference sequence. It is evident, that for every rotational speed n under consideration and each defined tolerance band width w_i there is a downtime temperature ϑ_d leading to a shorter steady state time t_{st} compared to the reference sequence.

The implemented Peltier modules provide a thermal decoupling of the internal structure of the spindle from the fluid cooling system. Enabling a temperature at the front bearings $\vartheta_{fb} \geq 25$ °C during the downtime t_d can mitigate the temperature reduction and subsequent thermal contraction of the internal spindle structure. As a result, the thermal state of the spindle is closer to steady state when heat flow rates $\dot{Q}_{ind}$ are induced by restarting the spindle.

At rotational speed n = 10,000 1/min a minimum steady state time $t_{st,min}$ is achieved with a downtime temperature of $\vartheta_d = 30$ °C regarding both tolerance band widths w_i. However, in this case the tolerance band width $w_1 = 2$ µm is never

Table 2 Steady state time t_{st} for variation of rotational speed $n \in \{10{,}000; 20{,}000; 30{,}000; 40{,}000\}$ 1/min and downtime temperature $\vartheta_d \in \{25; 30; 35; 40; 45; 50\}$ °C

Tolerance band width w_i	Steady state time t_{st} with downtime temperature ϑ_d					
	$\vartheta_d = 25$ °C	$\vartheta_d = 30$ °C	$\vartheta_d = 35$ °C	$\vartheta_d = 40$ °C	$\vartheta_d = 45$ °C	$\vartheta_d = 50$ °C
at rotational speed n = 10,000 1/min						
2 μm	59.1 s	0.0 s	67.2 s	152.8 s	200.1 s	242.8 s
1 μm	213.9 s	1.0 s	140.7 s	234.2 s	280.4 s	320.1 s
at rotational speed n = 20,000 1/min						
2 μm	197.2 s	156.8 s	0.0 s	54.6 s	98.0 s	129.5 s
1 μm	294.6 s	278.6 s	209.4 s	83.9 s	129.0 s	166.7 s
at rotational speed n = 30,000 1/min						
2 μm	211.6 s	184.5 s	136.6 s	3.7 s	57.9 s	90.6 s
1 μm	282.9 s	275.4 s	263.1 s	259.3 s	78.9 s	113.3 s
at rotational speed n = 40,000 1/min						
2 μm	183.4 s	168.7 s	169.9 s	6.9 s	39.7 s	71.1 s
1 μm	239.2 s	230.6 s	248.7 s	235.9 s	196.2 s	90.3 s

exceeded, resulting in a steady state time of $t_{st} = 0.0$ s. An increase in the rotational speed n necessitates an increase in the downtime temperature ϑ_d, leading to the minimum steady state time $t_{st,min}$. This phenomenon can be attributed to the increased heat flow rates $\dot{Q}_{ind}$ which are induced at higher rotational speeds n. Maintaining the temperature level within the internal spindle structure during downtime t_d necessitates a higher downtime temperature ϑ_d. At rotational speed n = 40,000 1/min the minimum steady state time $t_{st,min}$ is achieved with a downtime temperature of $\vartheta_d = 40$ °C in the tolerance band width w_1 and a downtime temperature of $\vartheta_d = 50$ °C in the tolerance band width w_2.

It can be observed, that for every tolerance band width w_i under consideration the minimum steady state time $t_{st,min}$ correlates positively with rotational speed n. This phenomenon can be attributed to an increasing influence of a spindle stop on the axial shaft displacement Δl_s at higher rotational speeds n. The lowest difference between the reference steady state time $t_{st,ref}$ and the minimum steady state time $t_{st,min}$ can be observed at rotational speed n = 40,000 1/min regarding the tolerance band width w_2. Here, the minimum steady state time of $t_{st,min} = 90.3$ s, leads to a reduction of 62% compared to the reference.

It can be concluded, that the method of increasing the temperature at the front bearings ϑ_{fb} during downtime t_d has a positive effect on the steady state time t_{st}. In order to achieve optimal results, it seems crucial to apply an appropriate temperature at the front bearings ϑ_{fb} during downtime t_d, taking into account the prevailing conditions.

4 Conclusion and Outlook

The thermal behavior of motorized spindles has a significant influence on machining accuracy of machine tools when it comes to high- and ultra-precision manufacturing. A novel temperature control system for motorized spindles based on Peltier modules was used to control the temperature at the front bearings ϑ_{fb} of a prototypically realized motorized milling spindle. It has been demonstrated, that the set point of the temperature at the front bearings ϑ_{fb} during a spindle stop with a downtime of $t_d = 300$ s has a relevant influence on the achievable steady state time t_{st} in the subsequent restart phase.

A series of experiments were conducted to determine the influence on steady state time t_{st} varying rotational speed $n \in \{10,000; 20,000; 30,000; 40,000\}$ 1/min and downtime temperature $\vartheta_d \in \{25; 30; 35; 40; 45; 50\}$ °C. The determination of steady state was executed through the implementation of a threshold-based method. The increase in the set point of the temperature at the front bearings ϑ_{fb} during a spindle stop resulted in a reduction of the time to reach steady state t_{st} by a minimum of 62%. The main finding of this research is that systematic control of the temperature of spindle components during downtimes t_d, taking into account the prevailing conditions, has evident influence on the achievable time to steady state t_{st} after spindle restart.

Using high performance motorized spindles equipped with thermoelectric temperature control systems as an alternative to conventional fluid cooling systems in the domain of high- and ultra-precision manufacturing, offers the potential of enhancing productivity. This enhancement is achieved by reducing the time to reach steady state t_{st} after every spindle downtime t_d. In most automated processes, the spindle downtimes t_d can be determined in advance. In order to minimize steady state times t_{st} following spindle downtimes t_d in manufacturing processes, it should be a future strategy to implement targeted downtime temperatures ϑ_d for each downtime t_d. Consequently, the impact of variable downtimes t_d will be thoroughly examined in subsequent research work. Additionally, future work may consider first and higher order variations of downtime temperature ϑ_d and an expanded array of rotational speeds n.

Acknowledgements This work was funded within the Fraunhofer-DFG transfer program by Fraunhofer Society and Deutsche Forschungsgemeinschaft (DFG, German Research Foundation) - 529738427.

References

1. Uhlmann, E., Mullany, B., Biermann, D., Rajurkar, K.P., Hausotte, T., Brinksmeier, E.: Process chains for high-precision components with micro-scale features. CIRP Ann. **65**(2), 549–572 (2016)

2. Mayr, J., Jedrzejewski, J., Uhlmann, E., Alkan Donmez, M., Knapp, W., Härtig, F., Wendt, K., Moriwaki, T., Shore, P., Schmitt, R., Brecher, C., Würz, T., Wegener, K.: Thermal issues in machine tools. CIRP Ann. **61**(2), 771–791 (2012)
3. Brecher, C., Wissmann, A.: Compensation of thermo-dependent machine tool deformations due to spindle load: investigation of the optimal transfer function in consideration of rough machining. Prod. Eng. Res. Devel. **5**(5), 565–574 (2011)
4. Denkena, B., Bergmann, B., Klemme, H.: Cooling of motor spindles—a review. Int. J. Adv. Manuf. Technol. **110**(11–12), 3273–3294 (2020)
5. Shabi, L., Weber, J.: Analysis of the energy consumption of fluidic systems in machine tools. Proc. CIRP **63**, 573–579 (2017)
6. Liu, T., Gao, W., Tian, Y., Zhang, D., Zhang, Y., Chang, W.: Power matching based dissipation strategy onto spindle heat generations. Appl. Therm. Eng. **113**, 499–507 (2017)
7. Jonath, L., Luderich, J., Brezina, J., Gonzalez Degetau, A. M., Karaoglu, S.: Improving the thermal behavior of high-speed spindles through the use of an active controlled heat pipe system. In: 3rd International Conference on Thermal Issues in Machine Tools (ICTIMT2023), Lecture Notes in Production Engineering. Springer International Publishing, Cham (2023)
8. Liang, F., Gao, J., Xu, L.: Investigation on a grinding motorized spindle with miniature-revolving-heat-pipes central cooling structure. Int. Commun. Heat Mass **112**, 104502 (2020)
9. Peltier, J.C.A.: Nouvelles expériences sur la caloricité des courants électrique: new experiments on the heat effects of electric currents. Ann. Chim. Phys. **56**, 371–386 (1834)
10. Heywang, W.: Amorphe und polykristalline Halbleiter. Springer, Berlin, Heidelberg (1984)
11. Ngo, T.-T., Wang, C.-C., Chen, Y.-T., Than, V.-T.: Developing a thermoelectric cooling module for control temperature and thermal displacement of small built-in spindle. Therm. Sci. Eng. Prog. **25**, 100958 (2021)
12. Fan, K., Xiao, J., Wang, R., Gao, R.: Thermoelectric-based cooling system for high-speed motorized spindle I: design and control mechanism. Int. J. Adv. Manuf. Technol. **121**, 3787–3800 (2022)
13. Uhlmann, E.: Recent advances in precision, sustainability and safety of machine tools. J. Mach. Eng. **23**, 5–25 (2023)
14. Uhlmann, E., Polte, M., Triebel, F., Salein, S., Temme, P., Hartung, D., Perschewski, S.: Reduction of warm-up period after machine downtime by means of a thermoelectric tempered motorized milling spindle. In: euspen Special Interest Group: Thermal Issues, Zurich (2022)
15. Jedrzejewski, J., Kwasny, W.: Modelling of angular contact ball bearings and axial displacements for high-speed spindles. CIRP Ann. **59**(1), 377–382 (2010)
16. Uhlmann, E., Polte, J., Salein, S., Iden, N., Temme, P., Hartung, D., Perschewski, S.: Entwicklung einer thermoelektrisch temperierten Motorspindel. wt Werkstattstechnik online **110**(5), 299–305 (2020)

Thermal Error Compensation in Machine Tools

Advancing Sustainable Manufacturing: Multidimensional Optimization of the Thermal Behavior of Machine Tools

Christian Brecher, Stephan Neus, Ralph Klimaschka, Janis Schäfer, and Alexander Steinert

Abstract To achieve the sustainable operation of machine tools, it is essential to address the optimization problem consisting of productivity, quality, and energy use. The thermal behavior significantly influences these parameters, thereby directly affecting sustainable machine operation. The Collaborative Research Center CRC/TRR96, funded by the German Research Foundation DFG, has proposed approaches to systematically tackle this conflict. Building on the foundational knowledge and transferring it into practical applications, control systems can effectively be developed and implemented. One such application is the model predictive control of active cooling systems in machine tools for compensation of thermoelastic displacements during operation, thus enhancing the machine accuracy. However, to optimally select the operating point of the active cooling system from the holistic sustainability perspective, it is crucial to consider the effects on all dimensions of the optimization problem. The design of a multivariable model predictive control is demonstrated using the example of the spindle cooling. The approach involves extending the system modeling and developing of a sustainability index.

Keywords Sustainability · Manufacturing · Machine tool · Thermal behavior · Multidimensional optimization · Active cooling system · Model predictive control

1 Sustainable Manufacturing

Profound global change processes have intensified the demand for sustainable action in recent years. The manufacturing industry is confronted with the accompanying sustainability transformation. The efficient production of industrial goods has been one of the fundamental pillars of prosperity in Germany. To maintain the standard of

C. Brecher (✉) · S. Neus · R. Klimaschka · J. Schäfer · A. Steinert
Laboratory for Machine Tools and Production Engineering (WZL), RWTH Aachen University, Aachen, Germany
e-mail: c.brecher@wzl.rwth-aachen.de

K. Wegener and M. Bambach (eds.), *4th International Conference on Thermal Issues in Machine Tools (ICTIMT2025)*, Lecture Notes in Production Engineering,
https://doi.org/10.1007/978-3-032-01194-7_10

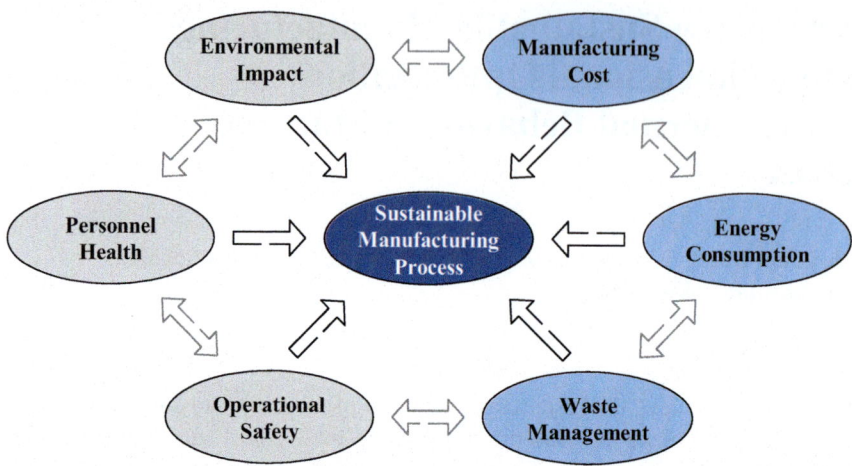

Fig. 1 Six major elements affecting the sustainability of manufacturing processes [1]

living in the future, it is therefore essential not only to focus on sustainable products but also to ensure sustainable production processes.

Sustainable production comes along with several facets. In [1], six major elements significantly affecting the sustainability of manufacturing processes are identified (see Fig. 1). Thus, the design of sustainable manufacturing processes requires solving a multidimensional optimization problem. Manufacturing costs, energy consumption, and waste management can be considered measurable elements, whereas environmental impact, personnel health, and operational safety cannot be easily or uniquely determined by the parameters of the manufacturing system.

In this context, the machine tool plays a pivotal role. To meet the growing demand for investment and consumer goods, machine tools of varying degrees of automation are of critical importance. In Germany, the production of machine tools represents the highest-revenue sector in mechanical and plant engineering [2]. Whereas in the past, the demands placed on machine tools have largely been defined by productivity, quality, and availability, future operations must also be aligned with ecological evaluation criteria. While productivity-enhancing innovations through increasing power density aim at a reduction in production costs, they inevitably lead to higher energy input. As power density increases, higher power losses are incurred. Higher power losses generally lead to increasing heat input in the machine, resulting in more pronounced temperature gradients and displacement fields, which can contribute to quality deviations and thus the production of waste.

As a result, the machine tool faces a complementary requirement profile. The interplay of productivity, quality, and energy use is the center stage of optimizing manufacturing costs, waste management, and energy consumption. To achieve sustainable operation of machine tools, it is essential to address this optimization problem. The thermal behavior has a significant impact on the highlighted parameters, thereby directly impacting the sustainable operation.

Already twelve years ago, the Collaborative Research Center CRC/TRR96, funded by the German Research Foundation DFG, proposed a visionary approach to systematically tackle this conflict between productivity, quality, and energy use (P-Q-E) for machining trailblazing to meet today's demand for sustainable manufacturing. The research focus was the thermoenergetic optimization of machine tools.

2 Foundation: CRC/TRR96 as an Enabler

2.1 Profile of the CRC/TRR96

The CRC/TRR96 was initiated to contribute to solving the trilemma of productivity, accuracy, and energy use in machining (see Fig. 2). As a result of increasing cutting performance due to further developments in cutting materials and machine elements and thus higher energy input, power losses increased, meaning that the economic efficiency of production was no longer guaranteed due to thermally induced errors and the effort required to control them. Solving the trilemma requires in-depth fundamental knowledge in subdomains, as well as cross-location and interdisciplinary collaboration for its integration. Consequently, a collaborative effort was established between the research sites in Aachen, Chemnitz, and Dresden, bringing together various fields of expertise, including mechanical engineering, electrical engineering, mathematics, and computer science.

The pursued solution approach is based on measures that enable the control of the process despite increasing power losses and without the need for additional energy measures. Moreover, thermally unsteady environmental conditions and the operational conditions typical of small and medium-sized batch production are considered.

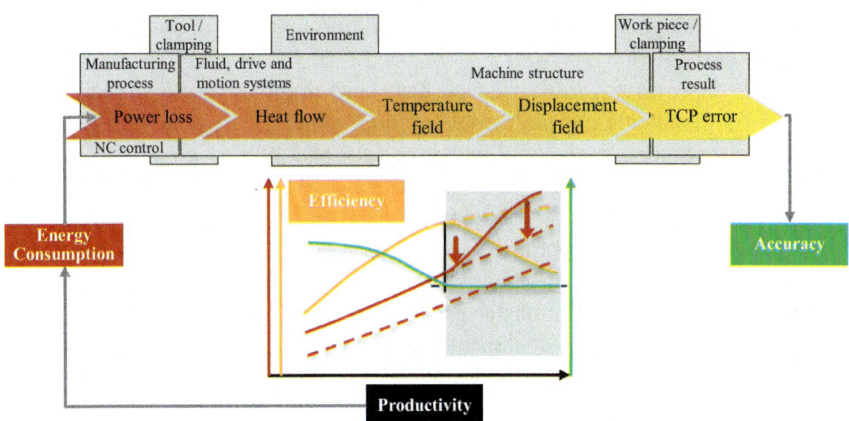

Fig. 2 Solving the conflict of productivity, accuracy and energy use (from [3], licensed under CC-BY 4.0)

Over the course of its duration, the CRC/TRR96 investigated and developed effective corrective and compensation solutions concerning the thermoelastic machine behavior. These solutions are designed to enable precise machining in the context of future energy-efficient production conditions. For ruling and solving the optimization problem between productivity, quality, and energy, the overall problem was divided into successive subproblems in three phases.

Phase 1: Fundamental Phase

In this phase, the foundational modeling and parameterization studies were conducted at the component level. The objective was to explore the underlying principles, phenomena, influencing factors, and solution approaches for controlling thermal behavior. These research efforts were complemented by experiments conducted on subproject-specific test benches, leading to the successful development of submodels for components and assemblies, as well as measurement and modeling methods. Initial solutions were tested in real machine environments.

Phase 2: Integration Phase

The second phase aimed to increase the complexity of the analysis objects. The focus of the research shifted to assemblies, with representative complex machine tool assemblies selected as integration objects: the motor spindle, machine frame, and feed axis. The scientific work of the subprojects was concentrated on these integration objects, with the goal of investigating the interactions of individual effects.

Phase 3: Demonstration Phase

Next, the research was applied to the entire machine. Building on the investigations on submodels and overall models, corrective and compensatory solutions were extended to the application for the entire machine under real operating conditions. This phase introduced new scientific challenges due to the numerous uncertainties and parameter fluctuations that must be considered. The models needed to be extended, and new solutions for managing variable operating conditions, such as online identification, were developed.

Along the three phases, three channels have been employed to transfer knowledge and drive developments into industrial solutions: transfer projects with renowned companies as partners, lump-sum funded projects, and industry conferences.

In total, over 140 researchers from 16 institutes and chairs collaborated on developing a comprehensive set of solutions tackling the thermoenergetic behavior of machine tools. The whole spectrum of topics was covered by 23 subprojects along the thermoenergetic chain, five of which were transfer projects with renowned companies as partners. Six further transfer projects are currently ongoing. The research results have been thoroughly documented and published through a variety of channels. A total of over 480 publications have been generated, of which over 300 have undergone peer review. The doctoral theses are particularly notable, with around 40 completed dissertations. All contributions culminate in the publication of a comprehensive final book, offering a solution-oriented description of all findings and results. In addition, a software-based documentation and assistance system, DOCASYS, was developed for users in industry. It serves as documentation of the developed solution methods for reducing the thermal error and suggests customized solutions for the user's application.

2.2 Exemplary Subprojects of the CRC/TRR96

In the CRC, fundamental research was initially conducted across nearly all areas relevant to the topic of thermal behavior. It was only through the acquired understanding and newly gained model knowledge that integration into assemblies and the machine could subsequently take place. Exemplary subprojects are presented below.

Example 1: Thermal Motor Model

Electric drives can generate several kilowatts of power loss. Therefore, the calculation of the thermal behavior of electric drives in machine tools is essential. Creating the model requires knowledge of the internal structure of the motors. However, the geometry data is not available in most cases. Within the CRC, approaches for the approximate determination of motor parameters were developed.

The motor model is built using the method of thermal source networks. For the model parameterization of asynchronous and synchronous motors, the relevant motor dimensions are initially required. These are the main dimensions of the rotor and stator as well as the nut geometry. The essential geometric data can be approximately determined through fixed ratios, which have been derived from a large number of examined motors. Another crucial task is the parametrization of thermal conductivity, specific heat capacity, and density. An electric motor is composed of various materials. In addition to homogeneous materials such as the copper short-circuit cage or the magnets, whose properties can be obtained from the literature, the material mixtures in windings and laminations are particularly significant. The parameters of these mixtures are calculated for both radial and axial directions, considering the direction of the heat flux. In addition to the heat capacities and heat transfer, it is decisive to determine the nodes within the motor model at which the various loss components are introduced. These include current heat losses, magnetization losses, additional losses due to harmonics in the air gap, and gas friction losses within the air gap. Each of these loss components is assigned to specific nodes, such as the stator tooth or winding head, to accurately simulate the motor's thermal behavior [4].

Example 2: Friction Behavior of Feed Axes

The feed axis is one of a machine tool's key assemblies, integral within the flux of force. As power density demands increase, feed axis components like ball screws and linear guides are exposed to higher loads, leading to increased friction in the respective rolling contacts and, consequently, more heat generation and power losses. Accurate models for these components' heat flow become pivotal for assessing and mitigating its adverse impacts on both machining quality and energy efficiency.

As part of phase 2 of the CRC, extensive investigations into the tribological behavior of feed axis components were conducted to address this task through metrological analyses on test benches and the development of mathematical models. The focus was on studying the influence of different boundary conditions such as rolling element geometries, component sizes, preload classes, wipers, and lubricants regarding individual friction contribution. The test benches also allowed for applying variable load collectives over extended periods to create a realistic set of operation conditions. The proposed model, shown in Fig. 3, utilizes the component's motion profile along with given boundary conditions to compute a friction force, which in turn generates a local heat input. This heat input influences a coupled thermoelastic finite element (FE) simulation model at defined nodes, enabling the quantification of the resulting interactions between machine components and structural elements, such as thermally induced forces. Within the model, a load-independent friction curve (Stribeck curve) is analytically approximated and extended by a load factor curve to accurately determine the friction forces acting on the component. Furthermore, it was observed that components with identical specifications but from different manufacturers can exhibit varying frictional behavior due to dissimilarities in manufacturing tolerances and variations in preload settings [5].

Example 3: Simulation Runtime Reduction

Predictive thermoelastic simulation models as part of a digital twin yield optimization potential for both the design phase and the operation of a machine tool. These models often incorporate an FE model, which can require significant computational time and memory resources, depending on their dimensionality and the numerical solvers used. Various methods were analyzed and further refined within the CRC to reduce runtime and thereby increase simulation efficiency.

Firstly, to address the dimensionality challenge, model order reduction (MOR) techniques were investigated. In its basic concept, MOR identifies essential equations, removes redundancies in the system matrices, and by this, substantially reduces matrix dimensions whilst maintaining sufficient accuracy. The dimension reduction can scale in the range from millions of degrees of freedom in the input model to a few hundred in the reduced output model. For enhancing practical utilization, CRC researchers built a MOR tool in the MATLAB environment that is embedded in

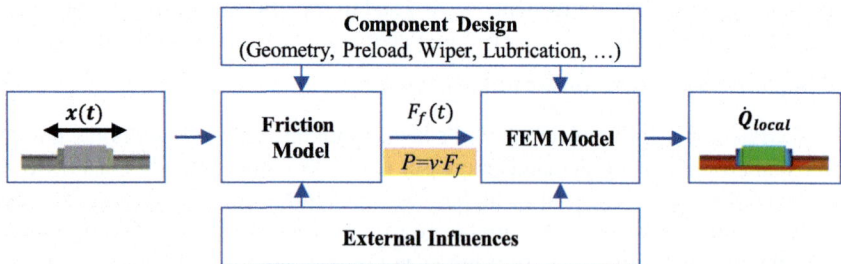

Fig. 3 Model for the friction behavior of feed axis (from [5], licensed under CC-BY 4.0)

an extension of the finite element analysis (FEA) software ANSYS. However, the matrices for the underlying state base model can also be extracted from other FE software. The tool integrates different MOR techniques such as balanced truncation (BT), Padé approximation, and Krylov algorithm (IRKA) with their specific application depending on error threshold and processing time requirements. Furthermore, the CRC addressed practical considerations for applying MOR in thermoelastic simulations, such as the discussions on when to deploy overall machine models (less input/output links) versus assembly-specific (flexible design changes) or even hybrid models (trade-off between these two). Additionally, a two-step reduction strategy proved advantageous for large-scale models, along with the separation of local time-dependent nonlinearities, which can occur due to thermally induced stiffness variations in guides and time-independent parts [6].

Secondly, to further reduce computational efforts, enhanced numerical methods to solve the underlying equations were developed as part of the CRC. The geometric coupling of machine components in conjunction with the physical coupling of heat conduction and linear elasticity requires tailored numerical methods for simulating temperature and deformation fields. Assemblies with relative motion impose additional computational complexity due to the time- and location-dependency of the heat transfer. While spatial parallelization using domain decomposition methods is well established, parallel time discretization methods remain an active area of research due to their sequential nature. The CRC researchers proposed a parallel time integration method (PARAeXP) to enhance the efficiency of solving transient thermoelastic simulations by decomposing the heat conduction equations – a system of partial differential equations (PDE) – into the homogeneous and partial part of the PDE. These parts are then distributed across different CPU cores and logically linked. Runtime experiments showed a fourfold speedup for models with constant heat conduction coefficients when using the PARAeXP approach compared to conventional methods [7].

Transfer to Industry

Throughout the CRC, continuous exchange with industry stakeholders was maintained through transfer projects, lump-sum funded projects, and industry conferences. An exemplary transfer project was the thermoenergetic optimized cooling of machine tool structures. Geometric optimization of the cooling circuits in a concrete machine frame, combined with individual cooling strategies for each circuit, significantly reduced power consumption [3]. In a lump-sum funded project, an AI-based spindle correction was realized based on a hybrid database of empirical and simulative data [8]. As part of an industry conference, a successfully implemented data-driven approach for modeling the thermoelastic TCP error using the encoder difference and artificial neural networks was presented [9].

3 Application: Model Predictive Control of Active Cooling Systems of Machine Tools

Building on the foundational knowledge established by the CRC and transferring it into practical applications, real control systems can be implemented. One such example is the model predictive control of active cooling systems in machine tools to compensate for thermally induced errors at the TCP. The trilemma of productivity, quality, and energy use becomes particularly clear in this case.

In recent decades, the active temperature control of machine components, especially electrical machines, has formed the basis for a continuous increase in the power density of these components. This in turn favored and supported productivity gains in machining. Furthermore, active cooling systems have recently also been increasingly used for the temperature control of structural components, such as machine beds or frame structures. As this compensating measure reduces thermal deformations, the machining accuracy and quality are targeted by this approach. However, studies have shown that active cooling systems in machine tools significantly contribute to the overall electrical energy consumption due to a comparatively simple design and inefficient operating conditions [10, 11]. Optimizing the design and operation of active cooling systems can therefore address the areas P, Q, and E contributing to sustainability in machine tools.

This requires a demand-oriented and target-driven adjustment of the cooling conditions, variable in terms of time and location. In concrete terms, the constant cooling that is often prevalent today must be replaced by an intelligently operating demand-based control system.

A model predictive control of active cooling systems of machine tools for compensation of thermoelastic displacements was developed. The approach presented below achieves a reduction in thermally induced TCP displacements through controlled operation of the hardware-modified active cooling system. The structure of the approach can be derived from the following four hypotheses:

Hypothesis 1

By controlling and homogenizing the structural temperatures of a machine tool, thermoelastic TCP displacements can be reduced.

Hypothesis 2

The inlet temperatures of individually controllable cooling circuits are suitable control variables for influencing structural temperatures.

Hypothesis 3

Controlling the structural temperatures by adjusting the flow temperatures requires a model-based future forecast due to large time constants.

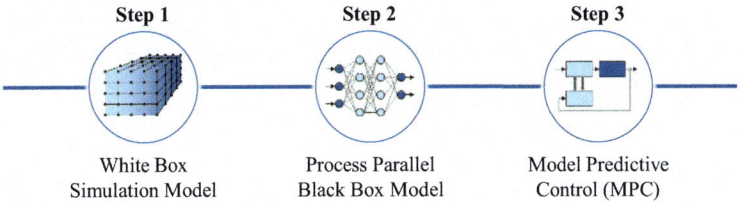

Fig. 4 Relevant steps for controlled operations of active cooling systems in machine tools [11]

Hypothesis 4

Black box surrogate models trained using complex white box simulation models are suitable for a process-parallel forecast of the structural temperatures of machine tools.

For testing the research hypotheses, a bottom-up approach breaks down the modeling process into three steps (see Fig. 4).

Step 1: White Box Simulation Model

The system model forms the heart of a model predictive control, enabling it to include forecasts and predictions into the control loop. Due to process-parallel control operations, the system model must meet the requirement for extremely short calculation times in addition to the inherent requirement for high prediction accuracy. Unfortunately, the thermal behavior of machine tools is affected by a wide range of boundary conditions and shows complex dependencies and interactions between these. Historically, time-consuming and cost-intensive metrological investigations were found to be able to accurately capture this complex behavior, as simulation models require abstraction and simplification measures limiting the usability of these models. Simulation models, however, allow a resource-efficient investigation of the thermal behavior, for which a precise modeling of thermal boundary conditions is essential.

Within the CRC, as mentioned above, a wide spectrum of modeling techniques for thermal boundary conditions, including active cooling systems, electric machines, machine roller elements, roller-based linear motion components, or machine spindle units, has been developed. Furthermore, mathematical model order reduction algorithms have been applied to the thermoelastic simulation of machine tools. Examples for the integration of these modeling approaches can be found in [11–13].

By integrating these modeling techniques into a white box simulation model of a 5-axis CNC machine tool, the thermal behavior can be simulated accordingly. The exemplary results in Fig. 5 compare the measured and simulated average structural temperatures of X, Y, and Z slide and the bed. During the test, varying inlet temperatures of the active cooling system consisting of four individual circuits (see Fig. 7) were applied. As the results show, the mean absolute errors (MAE) reach a maximum of 0.2 °C. It can be summarized that the CRC/TRR 96 white box simulation approach allows predicting the thermal behavior of machine tools with sufficient accuracy. By using this simulation-based virtual image, the thermal machine behavior can be derived for basically any set of boundary conditions, avoiding complex and expensive metrological machine investigation (Fig. 6).

Step 2: Black Box Surrogate Model

The white box simulation model and the mentioned advantages of a low-cost and fast data generation come with a significant disadvantage, as it cannot be embedded into a process parallel control environment to actively adjust the active cooling systems.

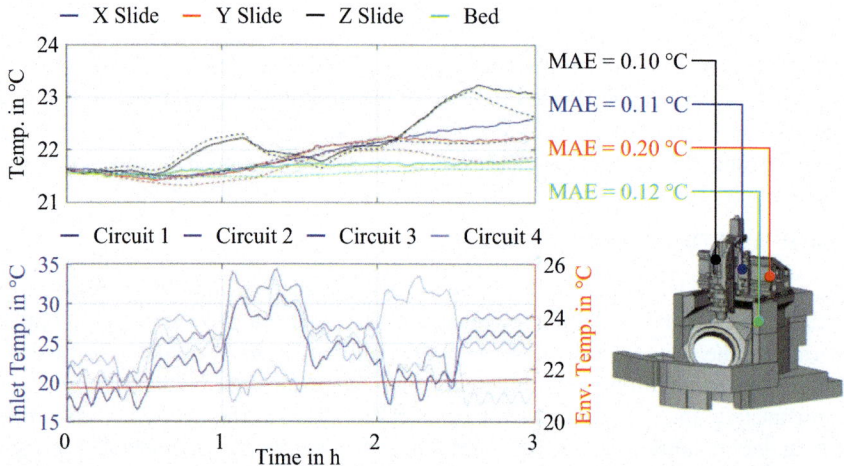

Fig. 5 Exemplary simulation result (measured data in solid lines, simulated data in dashed lines) [11]

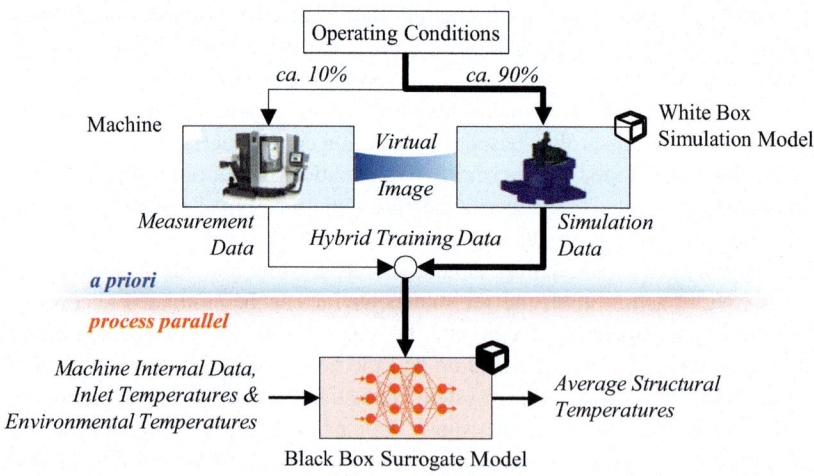

Fig. 6 Training of the black box surrogate model [11]

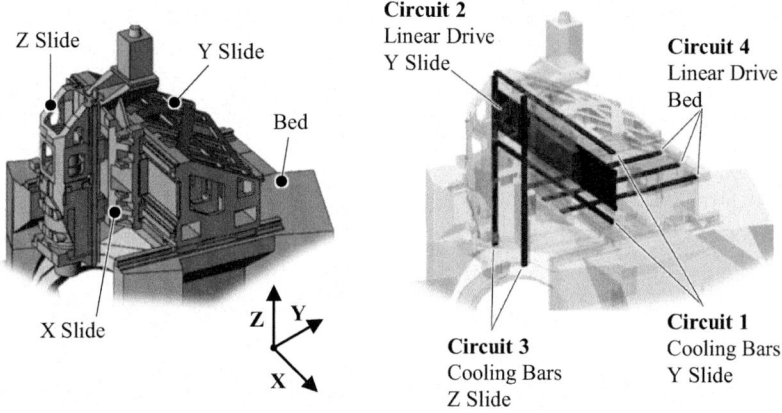

Fig. 7 Cooling circuits of the active cooling system of the exemplary 5-axis machine tool [11]

This integration requires the iterative solution of a multidimensional optimization problem for the ongoing recalculation of optimal target values: How shall the individual inlet temperatures of the four cooling circuits based on the current boundary conditions be adjusted to homogenize the average structural temperatures?

According to the research hypothesis 4, virtually simulated datasets describing the thermal machine behavior (see Step 1) are used to train a black box surrogate model. The training data set is expanded by a small amount of measurement data to incorporate phenomena not represented in the white box simulation model (see Fig. 6).

As a black box model, a nonlinear autoregressive artificial neural network with external input with the following structure was found to work best for the thermal behavior of machine tools (Table 1).

The black box surrogate model predicts the training data sets provided by the white box simulation model with a maximum MAE of 0.17 °C and a 50,000-times higher performance considering the calculation times. It is therefore suitable to function as an embedded system model as the core of the model predictive control.

Step 3: Model Predictive Control (MPC)

Implementation of the MPC structure (see [11]) requires a modification of the active cooling system. The analyzed 5-axis machine tool is equipped with cooling structures

Table 1 Artificial neural network hyperparameters [11]

Hidden layer: 10 neurons	Output layer: 4 neurons
Input delay: 0.25	Feedback delay: 1.00
Transfer function: Satlins	Data structure: Step progression
Training function: Bayesian regularization	

Centrifugal pumps, 3-way ball valves with
controllable rotary actuators   Flow and temperature sensors

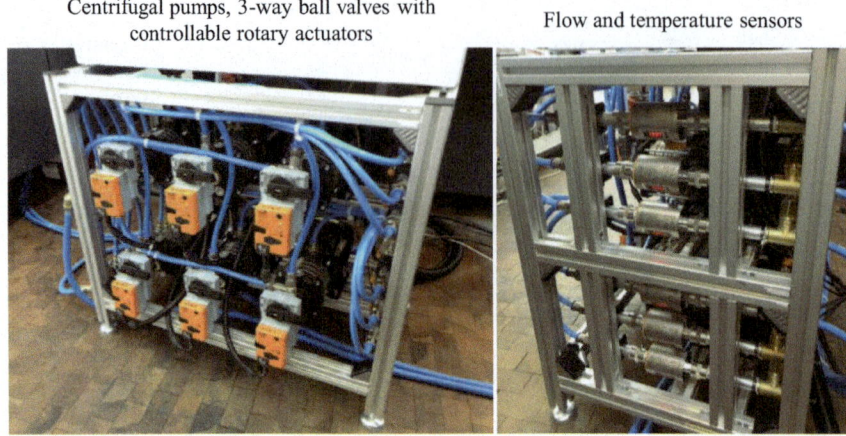

Fig. 8 Cooling infrastructure to provide individual inlet temperatures to each cooling circuit [11]

for electric machines and structural machine components, closely placed to roller-based linear motion components such as ball screw nuts or linear guides. For the purpose of this research, four individually controlled circuits are used (see Fig. 7). With these circuits, large parts of the structural machine components' temperatures can be actively influenced.

Each cooling circuit is equipped with a circular pump and a volume flow sensor to provide a controlled volume flow (see Fig. 8). Furthermore, 3-way ball valves with controllable actuators allow mixing cold and hot water, the latter provided by an energy-saving stratified storage system, to allow setting inlet temperatures. While the volume flow control is almost without time delay, a PIC controller can be applied. In the case of the temperature control, however, a large time delay requires a metrologically parameterized Smith predictor to stabilize the control measures [11].

In metrological investigations, favorable average structural temperatures from a thermoelastic point of view were identified and form the target values w of the control. Within the control cycle of one minute, the cost function $f(u, x)$ is calculated using the black box surrogate model to derive the actual average structural temperatures $y(u, x)$ using the inputs u (inlet temperatures) and x (machine internal data, such as feed velocities, axis positions, and motor currents).

$$e = w - y(u, x) := f(u, x) \rightarrow \min f(u, x)_2^2$$

In the following, a mathematical optimization is performed: To minimize the control deviation, new inlet temperatures x are found to ensure that the average structural temperatures aim for the target values. To calculate the control deviation, a forecast horizon of 15 min was found to be effective.

Although the approach focuses on controlling the average structural temperatures, the thermoelastic displacements of the TPC are considered to verify the effectiveness

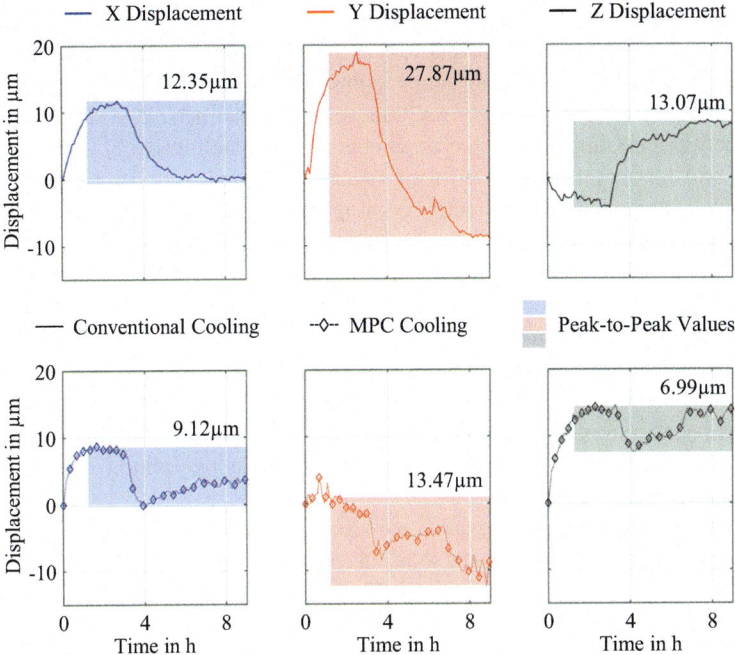

Fig. 9 Exemplary results of the MPC cooling [11]

of the approach. Overall, a significant reduction in the overall displacement behavior can be achieved by the MPC cooling, as shown in the direct comparison in Fig. 9. The displacement reduction in three-dimensional space reaches 47% considering the peak-to-peak values after the warm-up phase. In the exemplary test, the thermal machine behavior was varied by axis movements.

4 Vision: Multivariable Control for Sustainable Manufacturing

The presented MPC of the active cooling system can effectively increase the machine accuracy. The inlet temperatures of the cooling circuits prove to be suitable control variables for influencing structural temperatures. By controlling and homogenizing these temperatures, thermoelastic displacements of the TPC can be reduced.

However, to optimally select the operating point of the active cooling system from a sustainability perspective, the effects of the cooling parameters on all dimensions of the optimization problem must be accounted for. Only with the help of this knowledge can a multivariable MPC be set up that – despite variable operating conditions – adjusts the cooling parameters in real-time during operation to ensure that an optimum operating point of the cooling system is continuously maintained.

The cooling system generally supplies several consumers. A distinction can be made between cooling circuits for cooling electrical machines, e.g., spindle or feed drives, and for temperature control of structural components. In a first step, this approach is illustrated using the active cooling system of main spindles for enhancing the sustainability of their operation.

The main spindle forms the core of a cutting machine tool, as it provides the power required for the process. As a highly stressed component, whose performance defines productivity and whose load-displacement behavior significantly influences the achievable workpiece quality, it has a central role in sustainable manufacturing. In milling applications, modern main spindles are usually cooled motor spindles. The aim has been to maximize power density to achieve high machining performance with compact, lightweight spindle designs. Therefore, active spindle cooling plays a critical role.

In commercial motor spindles, jacket cooling near the stator is typically used, with shaft cooling being a less common solution to reduce thermal spindle growth. When the machine tool is activated, an uncontrolled central pump is engaged, which continuously circulates cooling fluid under constant conditions. The supply temperature is maintained within a defined temperature window using a recirculating chiller in two-point control. However, there is no linkage between the cooling power and the actual cooling demand, which, on one hand, results in unnecessarily high energy consumption and, on the other hand, leaves untapped potential for improving machining accuracy through compensation of thermoelastic deformations. Long, productivity-limiting warm-up phases can also be reduced through targeted temperature control using the active cooling system, thereby enhancing productivity. Furthermore, active cooling systems are used to limit the maximum motor or winding temperature and can have a significant impact on preload forces and maximum bearing pressures. Thus, the active cooling system of a spindle significantly influences resource usage during operation, the service life of key components, as well as the achievable manufacturing quality and productivity, which in turn has substantial implications for sustainability.

The central research question arises from the summarized problem statement: **How should active cooling systems in machine tool spindles be designed to achieve intelligent, demand-based, and thus sustainable temperature regulation?**

The following two steps are essential to address the research question:

Step 1: Extension of the System Model

First, the system behavior under different constant cooling parameters is investigated, both simulatively and empirically. The measured variables include those relevant for the sustainable operation of a motor spindle: axial spindle growth, overall electrical power consumption, motor temperature, and warm-up time. In addition to the electrical power consumption of the spindle itself, the power consumption of the cooling unit is to be recorded during the process to later evaluate the impact of potential cooling strategies on the overall energy consumption. Next to the system variables that can be measured, the stiffness and fatigue behavior of a spindle are key consequences of the prevailing pressures in the spindle bearings. Since direct measurement of this critical parameter is not feasible, the spindle system is thermomechanically modeled. Based on the model, internal bearing parameters can thus be provided. Subsequently, a digital thermoenergetic spindle twin is created. For this, the generated data set is transferred into a system model to efficiently describe the system behavior. The model uses input data that describe the thermal boundary conditions. These include machine-internal data, such as motor current or rotational speed, as well as the cooling parameters applied to the cooling system. The output variables to be determined are the ones introduced earlier:

- axial spindle growth
- bearing pressure
- motor temperature
- electrical power consumption
- warm-up time.

In Fig. 10, input and output variables are shown.

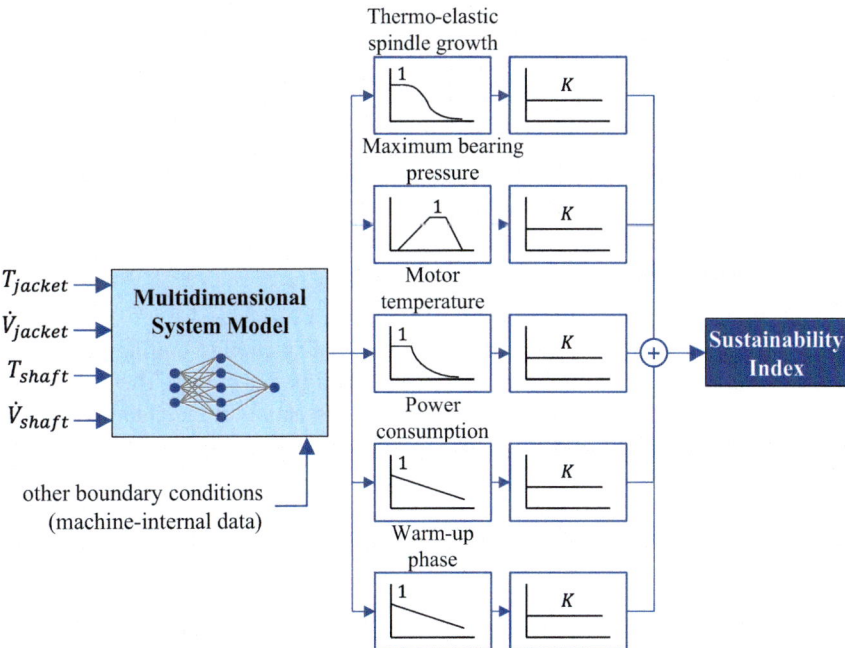

Fig. 10 Formation of a sustainability index for spindle operations

Step 2: Formation of a sustainability index

It is necessary to derive a sustainability index from these variables, based on which the current state can be assessed. The system model, which provides the five considered variables, is extended to translate these into a comprehensive sustainability index (see Fig. 10). However, it is important to consider that the variables are to be treated differently. While a reduction in power consumption directly correlates with a contribution to sustainability, a motor temperature falling below a limit value does not provide any additional benefit. Furthermore, the five target variables that the sustainability index should consider can be incorporated with different weightings. The weighting ultimately depends on the optimization objectives of the user. In the case of tightly toleranced components to be manufactured, thermoelastic spindle growth may assume greater importance than in a scenario without critical tolerances that must be maintained.

Finally, the system model, extended by the sustainability model, is integrated into a control architecture. The core of the architecture is a model predictive control system, as described in Chap. 3. The sustainability index serves as the control variable, while the cooling parameters act as the manipulated variables. In line with the model predictive approach, the controller can perform a thermal look-ahead based on current load data and determine the expected course of the control variable. Subsequently, an optimizer provides new cooling parameters, minimizing the control deviation.

The proposed approach demonstrates that the multivariable control of the active cooling system has the potential to significantly improve the sustainability of spindle operation. To further increase the sustainable operation of the entire machine, it is essential to extend the multivariable control approach beyond spindle cooling and apply it to the overall machine system. This requires:

1. Extension of the White Box Simulation Model
2. Development of a Process Parallel Black Box Model
3. Sustainability Modeling
4. Implementation of a Multivariable MPC.

This involves the definition of appropriate control variables. Next to the cooling parameters, such as the inlet temperatures of the cooling circuits, cutting parameters or parameters of the cooling lubricant supply could be integrated. These variables directly affect the thermal behavior, making them essential for maintaining optimal operating conditions. By adjusting these parameters in real-time during the process, the machine can continuously operate at an optimal operating point from a sustainability perspective, thereby enhancing the overall sustainability of the machine tool operation.

5   Summary and Outlook

Sustainable manufacturing is crucial but achievable. The complexity of future challenges will require effective collaboration and linking of different technical domains. Thermic behavior will play a central role in ensuring sustainable operation of machine tools. The interplay between vertical research depth and horizontal collaboration has been demonstrated as a key factor, as shown by the Collaborative Research Center (CRC). Looking ahead, the next steps will involve a holistic consideration of all sustainability dimensions, integrating the machine tool into the factory context and extending the factory's scope to include global supply chain networks that are affected by volatile political, economical, and ecological circumstances.

Both short-term and long-term challenges demand a flexible, user-centric approach to operations. For instance, rising electricity prices evoke the need for reduced energy consumption, while customer demands may require higher productivity. Supporting machine operators, production planners, and managers through intelligent and connected co-pilot systems, powered by data-driven models, will be pivotal. These systems must accommodate individual goals and criteria, as different focus might be placed on the respective sustainability factors. Additionally, the weighting of these factors may need to be adjusted dynamically by the co-pilot systems based on changing circumstances, offering resilience in rapidly evolving manufacturing environments.

Acknowledgements   The authors would like to thank the Deutsche Forschungsgemeinschaft (DFG, German Research Foundation) for funding the Collaborative Research Center CRC/TRR 96.

References

1. Kutz, M. (ed.): Mechanical Engineers' Handbook. Materials and Mechanical Design. Wiley, Hoboken, NJ (2006)
2. Statista: Umsatz im deutschen Maschinenbau nach Sektoren 2024 I Statista. Stand: 02.04.2025. https://de.statista.com/statistik/daten/studie/173637/umfrage/branchenumsatz-des-maschinen baus-in-deutschland-nach-sektoren/. Zugriff am 02.04.2025
3. Brecher, C.: Thermo-energetische Gestaltung von Werkzeugmaschinen. Springer Fachmedien Wiesbaden, Wiesbaden (2025)
4. Winkler, S., Werner, R.: Der Elektroantrieb als thermo-energetische Blackbox. In: Brecher, C. (Hrsg.) Thermo-energetische Gestaltung von Werkzeugmaschinen. Wiesbaden: Springer Fachmedien Wiesbaden 2025, S. 105–115
5. Neus, S., Steinert, A., Kneer, F., et al.: Untersuchung von Maschinenkomponenten. In: Brecher, C. (Hrsg.): Thermo-energetische Gestaltung von Werkzeugmaschinen. Wiesbaden: Springer Fachmedien Wiesbaden 2025, S. 61–70
6. Vettermann, J., Aumann, Q., Saak, J., et al.: Rechenzeitsparende Modellierung. In: Brecher, C. (Hrsg.): Thermo-energetische Gestaltung von Werkzeugmaschinen. Wiesbaden: Springer Fachmedien Wiesbaden 2025, S. 199–212
7. Naumann, A.: Effiziente transiente thermo-elastische Simulation von Werkzeugmaschinen. In: Brecher, C. (Hrsg.): Thermo-energetische Gestaltung von Werkzeugmaschinen. Wiesbaden: Springer Fachmedien Wiesbaden 2025, S. 235–244

8. Brecher, C., Lohrmann, V., Steinert, A., et al.: Via KI thermische Spindelverlagerungen korrigieren. https://www.maschinenmarkt.vogel.de/via-ki-thermische-spindelverlagerungen-korrigieren-a-856c947e31206100b5c85044a243df8e/
9. Brecher, C., Dehn, M., Neus, S.: A data-based model of the thermo-elastic TCP error using the encoder difference and neural networks. In: International Conference on Thermal Issues in Machine Tools, pp. 119–131 (2023)
10. Brecher, C., Bäumler, S., Jasper, D., et al.: Energy efficient cooling systems for machine tools. In: Dornfeld, D.A. (ed.) Leveraging Technology for a Sustainable World. Proceedings of the 19th CIRP Conference on Life Cycle Engineering, University of California at Berkeley, Berkeley, USA, May 23–25, 2012, pp. 239–244. Springer, Berlin, Heidelberg
11. Steinert, A.: Modellprädiktive Regelung aktiver Kühlsysteme von Werkzeugmaschinen zur Kompensation thermo-elastischer Verlagerungen. Dissertation, Rheinisch-Westfälische Technische Hochschule Aachen; Apprimus Verlag (2024)
12. Brecher, C., Ihlenfeldt, S., Neus, S., et al.: Thermal condition monitoring of a motorized milling spindle. Prod. Eng. Res. Devel. **13**(5), 539–546 (2019)
13. Vettermann, J., Steinert, A., Brecher, C., et al.: Compact thermo-mechanical models for the fast simulation of machine tools with nonlinear component behavior. Automatisierungstechnik **70**(8), 692–704 (2022)

Thermal Aspects Towards the Fully Compensated Machine Tool

Josef Mayr, Konrad Wegener, and Petr Kaftan

Abstract The thermal behaviour of machine tools is the greatest challenge of precision in manufacturing and counteracting it with a controlled environment swallows an immense amount of effort and energy. As thermal errors result from physics and physics is invincible, the solution must be generated from another principle as a workaround. The vision of a fully compensated machine tool intends to remove the thermal errors for machine tools that operate under normal shop floor conditions with the help of numerical control. Research in the field of machine tools at inspire AG and the Institute of Machine Tools and Manufacturing (IWF) at ETH Zurich is following the vision of the fully compensated machine tool. This keynote paper provides an overview of past and current work on how this vision is being pursued in the field of thermal issues in machine tools and related manufacturing systems. The paper concludes with a summary and assessment of the state of the art and provides an outlook on the challenges in thermal error research on the way towards the fully compensated machine tool by illustrating that up to now, not all thermal axis errors of machine tools can be measured in a sufficiently short time and be compensated by data driven models, and the physical models for predicting the thermal behaviour often do not achieve the required accuracy.

Keywords Machine tool, Thermal error, Test-Pieces, Modelling, Compensation

J. Mayr (✉) · K. Wegener · P. Kaftan
inspire AG, Zurich, Switzerland
e-mail: josef.mayr@inspire.ch

K. Wegener
e-mail: wegener@iwf.mavt.ethz.ch

P. Kaftan
e-mail: petr.kaftan@inspire.ch

© The Author(s) 2026   147
K. Wegener and M. Bambach (eds.), *4th International Conference on Thermal Issues in Machine Tools (ICTIMT2025)*, Lecture Notes in Production Engineering,
https://doi.org/10.1007/978-3-032-01194-7_11

1   Introduction

The accuracy of machine tools is one of the major properties of a machine tool and is decisive in high-end manufacturing. Nevertheless, there is a growing demand for high-precision machined workpieces. Examples can be found in the medical industry, in microelectronics, in aerospace and space applications, and in the energy sector. Up to now, it has been common practice to take measures associated with high effort and energy consumption, such as the air conditioning of shop floors and the excessive cooling of machine components. Due to the restriction of the precision of rotary axes, often still, the linear axes of machine tools are used for realizing the tool movement during machining of high-precision parts, e.g., in milling. This restriction limits the design of workpieces that can be machined and is contrary to the growing demand for high-precision, multi-axis machined workpieces. Contrary to this requirement, machine tools are increasingly exposed to price competition, and significantly higher demands regarding manufacturing productivity are made. For reasons of resource efficiency, the climate impact of machine tools in production is increasingly being taken into consideration. Wegener et al. [1] show that compensation is an outstanding, cheap technology that transforms the machine tool design from originally resource-based technologies into knowledge-based technologies. A numerical compensation approach is as good as the underlying models are capable of mapping the respective machine error to be compensated. Modelling techniques and their quality are, therefore, the key competence and the basis for higher production efficiency [2]. Increasing automated compensation for machine tools requires that the machines also work increasingly autonomously and, therefore, become more intelligent. As Wegener et al. [3] show, it is important to ensure that the user is not bored by the machine's autonomy but that the machine communicates with him in a meaningful way. An intelligent approach, proposed in Refs. [4, 5] and shown in Fig. 1, integrates the capabilities of the user and the machine's intelligence in a way that the machine can learn from the user and vice versa, which will be crucial on the way to a fully compensated machine tool.

What generally applies to the machine tool for compensation also applies in particular to the consideration of the thermal issues of machine tools. A deep knowledge of the processes regarding thermal issues in the machine tool is crucial in order to ultimately create the models for compensation. Two main approaches can be chosen, or as usual, a combination of both, for getting to understand and become able to describe the effects. The measurement technologies for investigating thermal errors and their simulations. The 2012 CIRP keynote paper by Mayr et al. [6] is a milestone in the field and initiating quite some research. Therefore, it addresses three areas of research activity: measurement technology, computations, and the compensation or reduction of thermal effects. The influence of fluids is further considered in a separate chapter in [6]. Fluids perform a wide range of tasks in machine tools, as shown in [7], and have a decisive influence on the thermal behaviour of the machine and the resulting machining errors. Conversely, the temperature of the fluids also influences the behaviour of the machine in terms of dynamics and geometric accuracy, as

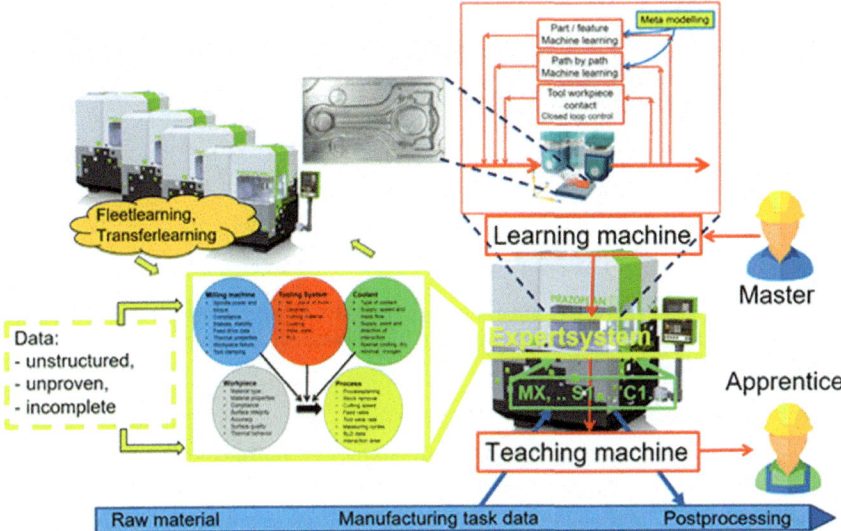

Fig. 1 Concept of an intelligent machine tool integrating the operator's capabilities for increased autonomy [5]

Stoop et al. [8] demonstrate using the example of a motor spindle. Therefore, when designing spindles, it is crucial not only to optimize them for dynamic behaviour but also to consider the kinematic motion characteristics and thermal behaviour in the design phase [9, 10].

2 Investigation of Thermal Behaviour by Measurement Techniques

The most important fluidic element influencing the thermal behaviour of the machine tool, but also the machine users' or skilled workers' behaviour, is the atmosphere, particularly the hall environment. The climatic values, air temperature, air velocity, air humidity, heat radiation, and the chemical composition of the air are influenced by many different factors. Taking all of them into account is almost impossible. Mayr et al. [11] developed a shop floor environment monitoring system taking into account the opening of the hall door and the outside temperature next to the hall door, as shown in Fig. 2. With the measurement setup used as input for a grey box compensation model, the authors show that the thermal peak-to-peak errors caused by the environmental fluctuations of a precise 5-axis machine tool can be reduced from 4.0 to 2.2 μm.

Metal-working fluids (MWF) also have a significant influence on the thermal behaviour of the machine. Mayr et al. [12] show that both the resulting errors and

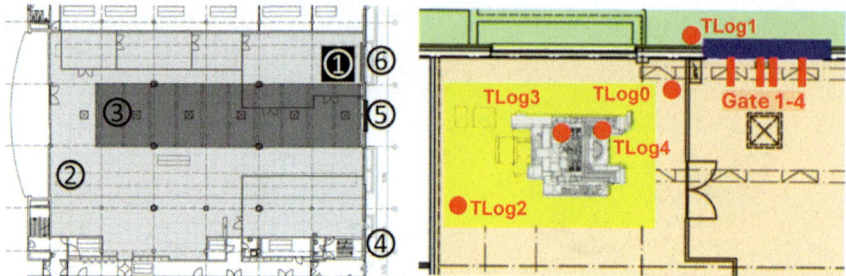

Fig. 2 Left: Shop floor plan and location of the machine tool; 1: machine tool, 2: shop floor layout, 3: skylights, 4: outer wall, 5: hall door, 6: windows, Right: Location of temperature (TLog0–TLog4) and inductive (Gate1–Gate4) sensors [11]

the transient behaviour of the machine change when the machine operates in dry or wet machining conditions. The authors further demonstrate that the behaviour can be significantly influenced by the fluid temperature control. Hernández-Becerro et al. [13] investigate the influence of different cooling concepts on a 5-axis bed-type milling machine. They illustrate that the thermal behaviour of the particular tool centre point (TCP)-displacements in the machine under investigation cannot be influenced by the cooling concept but are largely influenced by the environment. Pavlicek et al. [14] present a measurement setup that monitors the behaviour of a tool grinding machine in parallel with the process. The authors demonstrate the protective mechanism that protects the sensors during the process and simultaneously show that switching from oil as a coolant to CO_2 has a significant impact on the thermal behaviour of the tool grinding machine. Figure 3 shows the measurement setup and the different behaviour depending on the MWF used on the machine.

For thermal compensation, on-machine measurement is of crucial importance. Mayr et al. [15] and Mayr [16] show that the thermal errors of machine tools influence

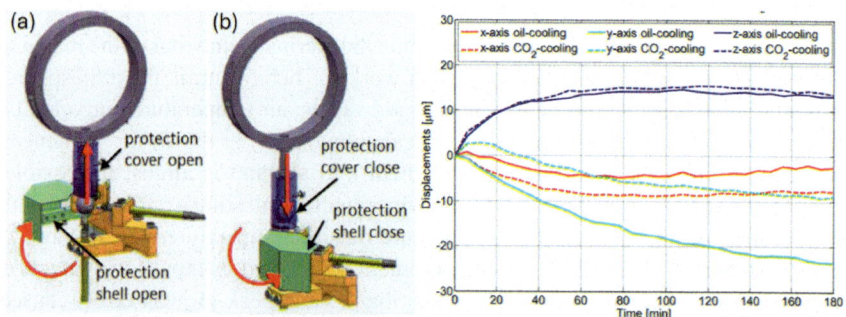

Fig. 3 Left: **a** Measurement setup with protection cover in measurement configuration and **b** in process configuration during the use of oil and CO_2, Right: measurement of the TCP displacements as a function of the cooling medium oil or CO_2 [14]

the geometrical errors of the machine and that these errors can also be considered as time-dependent position, orientation, and component errors. Gebhard [17] shows that this approach must be extended by taking into account the displacements of functional surfaces. However, measuring the transient behaviour of all geometrical errors of a 5-axis machine tool with sufficient speed and precision is not possible with the measurement setups available today on the machines. Therefore, it is often limited to the investigation of individual thermal load cases in a few axis positions of the machine tool's workspace or to the measurement of individual axis errors in a few degrees of freedom, as shown by Blaser et al. [18] for the example of a gantry-type machine tool. The measurement results of the axis errors are used to model them with an Auto Regressive model with eXogenous input (ARX) approach. Zimmermann et al. [19] extend the discrete R-test measurement cycle used by Blaser [20] and Gebhard [17] to achieve a separation of spindle- and table-side thermal errors in the z-direction of a milling machine, which is particularly important for machine tools with swivel tables. A combination of the discrete R-Test measurement with an artifact that can be inserted into a notch in the machine table, for additional measuring linear axis errors, is shown by Zimmermann et al. [21]. The authors show that this artifact allows both position and orientation errors, and partial component errors of the machine tool can be measured in a sufficiently short time and taken into account in the compensation.

Most milling machines today have a touch probe to be automatically mounted from the tool changer (ATC) into the spindle. With the touch-probe, an artifact of known geometry on the workpiece side can be probed to measure thermal machine tool displacements and, e.g., compare them with the thermal machine tool errors computed from a thermal compensation model of the machine. On machines with a rigid tool magazine, as is the case with Swiss-type automatic lathes, for example, touch probes are not available. Kaftan et al. [22] developed the torque limit skip (TLS) approach to measure the thermal displacements by recording motor current. A comparison of the measurement results between TLS and touch probe measurement shows that the method is, in principle, accurate enough to record the thermal displacements, shown in Fig. 4. In an extension of the approach, as shown in [23], the method is used to determine the end positions of the ball screw drive of a linear axis. Linear interpolation can thus be used to estimate the thermal positioning errors in the axial direction.

3   Investigation of Thermal Behaviour by Numerical Simulations

As shown, the information about the thermal behaviour of the machine tool is still challenging to measure and often has limited significance. In addition, measurements for investigations often take days and weeks to investigate the thermal behaviour of machine tools. For this reason, numerical simulation technology based on physical models, such as the Finite Element Method (FEM), is a suitable technology for simulating the behaviour of the machines. These physical models can be used to efficiently simulate individual load cases, such as the behaviour of linear axes [25] or entire axis systems [26]. The simulation results often open up the possibility of virtually testing design measures, as shown by Mayr et al. [27] using the example of cooling concepts for ball screw systems. Mayr et al. [28] pointed out that the simulation results often provide the scope for design measures that can be taken to improve the thermal behaviour of the machine tools. In physical modelling, the representation of boundary conditions is crucial. Using the example of linear axes, Ess et al. [29] use a multibody dynamics model (MBS), derived from the FEM model of the machine tool, illustrated in Fig. 5, to model the dynamic forces on the coupling points, which in turn represent the boundary condition for the thermal model of the FEM simulation, illustrated in Fig. 5 right.

Figure 6 displays the entire thermal chain of causes, given in [1], which is the basis for the numerical physical-based computations. The proper formulation of the boundary conditions is of particular importance. For an example of a cooling system on a machine tool, Ess et al. [30] describe energy losses that locally influence the temperature field of the machine tool.

Züst et al. [31] describe the energetic behaviour of the machine tool by describing individual subsystems and can thus predict the thermal state of the fluid cooling circuits. From this, Mohammadi et al. [32] and Züst [33] developed a holistic energy model with which the transient energetic behaviour of machine tools can be predicted. Züst et al. [34] use the method to precisely predict the energetic behaviour of a 5-axis milling machine and to visualize the energetic flows in the machine. The example of the visualization of the energetic flows in a tool grinding machine is shown in Fig. 7. Rhiner et al. [35] further show that the energy measurement system and the modelling approach can be used to illustrate the benefits of thermal compensation methods regarding the energy efficiency of machine tools.

In addition to the energy flows in machine tools, the environment has a significant influence on the modelling of thermal behaviour, as already shown in Sect. 2 [37]. In particular, modelling the fluidic behaviour in enclosures is still a major challenge. Pavlicek et al. [38] developed a method with which an empirical metamodel can

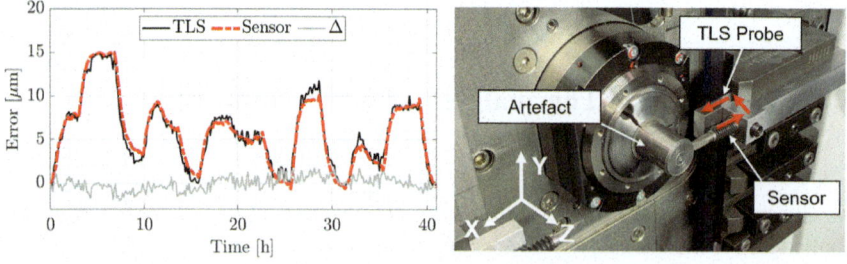

Fig. 4 Comparison between torque limit skip (TLS) measurements and touch probe (Sensor) measurements when investigating the thermal errors of a Swiss-type automatic lathe [24]

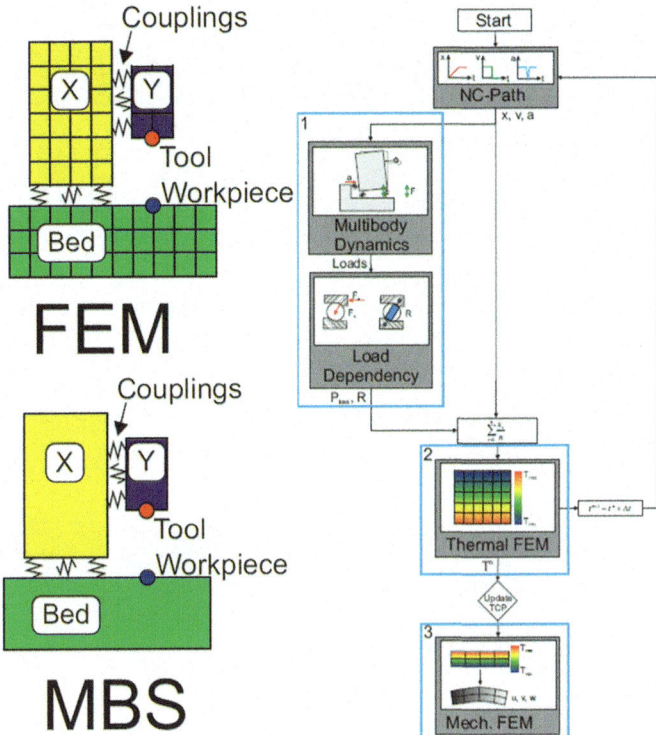

Fig. 5 Conversion from a FEM model to a multibody dynamics model and flowchart of the simulation approach [29]

be derived from CFD simulation to describe these boundary conditions. Pavlicek et al. [39] and Hernandes-Becerro et al. [40] model the influence of enclosures on machine tools as boundary conditions in FEM simulations. Another method is to derive the metamodels from reference bodies, as shown by Mayr et al. [36] and Pavlicek [41]. The FEM simulation models offer the possibility to evaluate the thermal behaviour of machine tools in the frequency domain. Mayr et al. [42] compute the thermal resonance frequency for a machine tool and show that its consideration offers great advantages in design measures to optimize the thermal behaviour of machine tools. Using the example of the thermal influence of covering the machine components with material of low thermal conductivity, it is shown how the thermally induced displacements on the TCP can be reduced in a frequency-specific manner in the range of the thermal resonance frequency.

In order to increase the efficiency of numerical simulations, Model Order Reduction (MOR) methods have become established in FEM simulation technology. Hernandez-Becerro et al. [43] and Hernandez-Becerro [44] use the MORe tool, developed at inspire and IWF, to evaluate the thermal behaviour of a 5-axis machine tool in the frequency and time domain. With the reduced thermo-mechanical model and

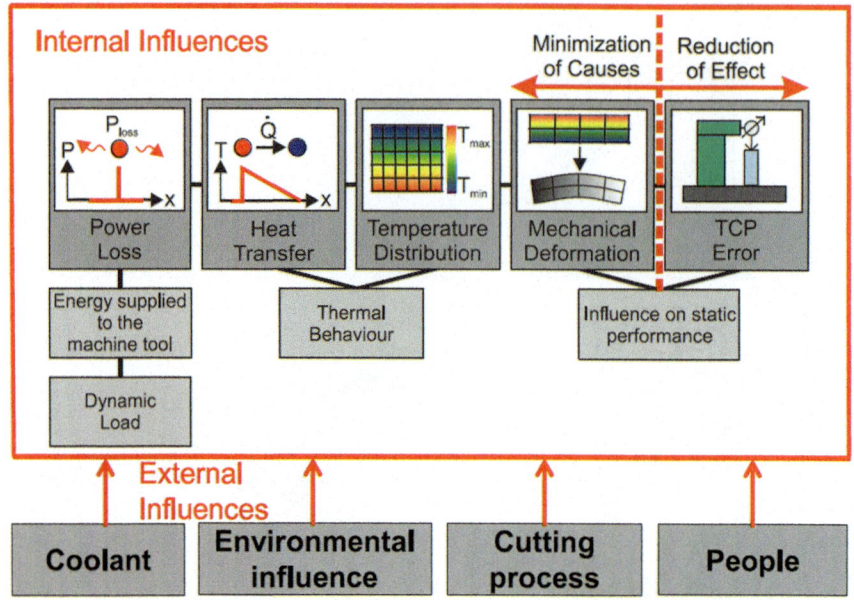

Fig. 6 Thermal chain of causes for TCP-errors [1]

a Monte-Carlo simulation placed on top of it, Hernandez-Becerro et al. [45] are able
to simulate the influence of different boundary conditions on the thermal behaviour,
especially the displacements at the TCP of a machine tool. By incorporating real measurement data, which primarily describes the fluidic behaviour of the environment
and the machine tool, the simulation approach of Hernandez-Becerro et al. [46] can
be used to simulate precisely the thermal behaviour of a 5-axis milling machine. The
implementation of varying boundary conditions in MOR simulation models is due to
their mathematical characteristics usually makes it impossible. Hernandez-Becerro
et al. [47] developed a workaround by implementing different poses of the axes in
the reduced model to fulfil this task. Using the same approach, Hernandez-Becerro
et al. [48] successfully optimized the MWF circuit of a 5-axis column-type milling
machine in a way that its influences on the displacements at the TCP are minimized.
Although physical thermo-mechanical models are very efficient for compensation,
as shown by Mayr et al. [49] and Ess [50] for a tool grinding machine, a sufficient
reduction of the thermal errors on the TCP could only be achieved with limitations.
Nevertheless, Blaser et al. [51] demonstrate that the MOR simulation models can
very efficiently simulate the transient behaviour of the machine tool and are therefore suitable for comparing compensation approaches and models in order to find a
suitable compensation model and structure with relatively optimized measurement
effort.

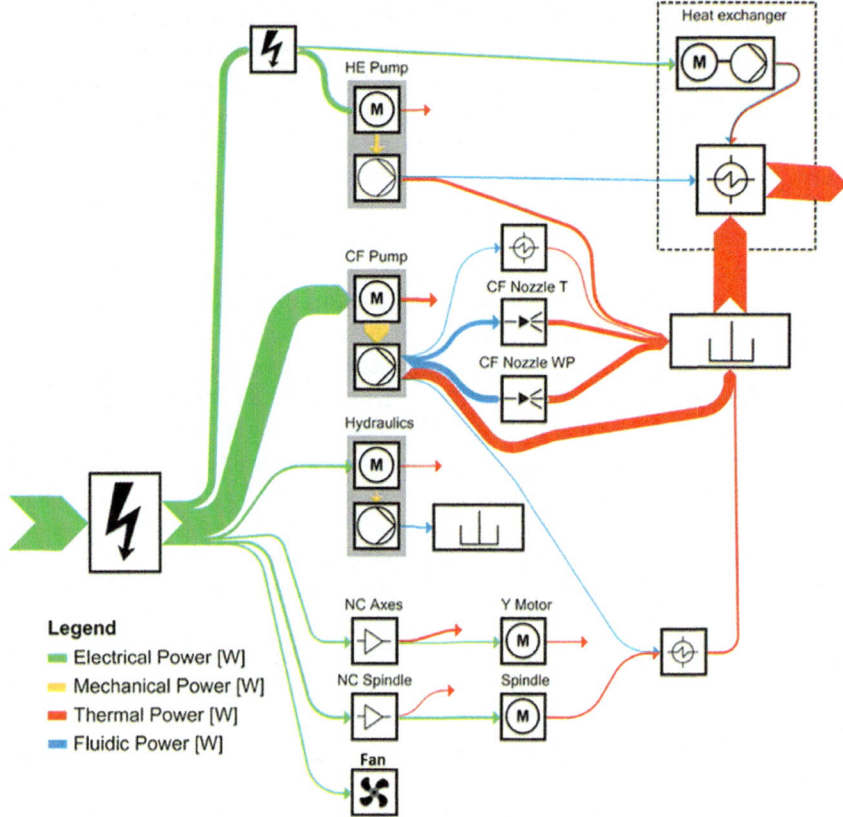

Fig. 7 Visualization of the energy flow of a tool grinding machine during processing [36]

4 Thermal Error Reduction by Compensation

Although much progress has been made in recent years in the field of research on thermal compensation of machine tools, implementation on the machine tool's control is still often an open topic or only possible to a limited extent. One method applied in industry is that after machining, the workpiece is measured using a coordinate measuring machine. Any deviations from the target values are recorded and transferred to specialized software. This software computes basic correction values, typically focusing on parameters such as tool length. These correction values are subsequently sent to the machine control system, where adjustments are made, for example, through tool length correction by applying "virtual wear." Due to the sequential nature of this process, there is typically a minimum time delay of two workpieces before the corrections take effect. This method is particularly well-suited for high-volume production. Additionally, various commercial software solutions are available to support and automate this workflow.

Another possibility for the implementation is an additional industrial PC (IPC). The communication between a machine tool's control and the IPC can be through standard communication protocols available. In this process, the machine sends various model inputs, such as measurement data, axis positions, and power consumption, to the IPC. The IPC uses a thermal compensation model, computing the correction values, which are sent back to the machine's control system. The compensation is implemented by adjusting the machine's axis position, for example, by shifting the coordinate system. However, it is important to note that in many cases, access to the system is restricted, and there are no commercially available solutions to address this issue. Furthermore, the correction values can often only be applied after an NC (Numerical Control) block or NC routine, which can delay the compensation process.

A hardware solution for self-optimizing thermal error compensation typically involves communication between the IPC and the CNC (Computer Numerical Control) system, often facilitated by PLC (Programmable Logic Controller) variables. This communication enables real-time adjustments and ensures that thermal errors are compensated dynamically during the machine tool's operation. The implementation of this solution is highly dependent on the specific machine tool's control manufacturer. Without their cooperation or support, it is not feasible to integrate the system effectively. Moreover, there are currently no standardized interfaces or commercial solutions available that can provide a universal approach to thermal error compensation across different machine models and control systems. With the right setup, online compensation of thermal errors of machine tools is possible, allowing for continuous real-time correction during machine operation. This capability ensures improved precision and reduces the impact of thermal errors on machining accuracy.

When implementing thermal error compensation on machine tools, it is important to define a maximum compensation value that is not exceeded. The maximum compensation value should be higher than the expected thermal errors to ensure that all expected thermal errors can be compensated. However, it should be small enough to prevent damage to the machine, workpiece, or tool. Incorrect sensor data or over-compensation can lead to costly consequences, including wear and tear on the equipment.

Thermal error compensation significantly enhances the precision of the machine tool by compensating inaccuracies caused by temperature influences. Researchers and industrial engineers should be aware that such systems increase the precision of the machine and are subject to export regulations, which may restrict the distribution or the use of the technology.

Although Wegener et al. [52] and Gebhard [17] show that simplified physical models are suitable for reducing the thermally induced errors of a rotary-swivel-axis unit of a 5-axis milling machine by up to 80%, the models can only be used to a limited extent for entire machine tools. There are many reasons for this. On the one hand, the computing power required for very large models is still immense, and this computing power is not available for the real-time computations of thermally induced errors of the machine tool on its control system. In addition, there are uncertainties in the description of the material models and the changing boundary conditions, such as convection, which can only be achieved with additional effort, particularly in the case of order-reduced models, and the thermo-mechanical models of the machine tools must be coupled with corresponding electro-mechanical or, e.g., multi-body simulations to describe the load-specific heat generation. For this reason, phenomenological models that describe the behaviour of the machines with reduced physical information, grey-box models, or without physical information, black-box models, have been increasingly used in recent years. Gebhard et al. [53] and Wegener et al. [54] could reduce the thermal errors of a 5-axis machine with a data-driven grey-box model derived from measured data by more than 70%. Figure 8 illustrates the measured errors of the machine with and without compensation. The error band was significantly reduced with thermal error compensation.

As can be clearly seen in Fig. 8, thermal error compensation can reduce both the peak values and the range of errors. Zimmermann [55] developed a visualization method to display the distribution of thermal errors in a bar chart. Figure 9 shows the statistical distribution of volumetric errors for a 5-axis machine tool, uncompensated, with compensation for the three main thermal errors and compensation for at least 15 individual thermal errors. Such evaluations are helpful to derive information, e.g., the machine capability index, for the planning of statistical process control.

Mayr et al. [56] were able to master thermal compensation even with the application of MWF in the workspace and reduced the thermal errors significantly up to more than 90% with phenomenological models integrated in the machine tools' control. Mayr et al. [57] extend this approach to automatic model generation, using on-machine measurements and a software routine to generate the model. This approach, successfully implemented at the industrial partner site, is crucial to be able to adapt models for a large number of machines. In order to take into account different influences that affect the thermal behaviour of the machine, Mayr et al. [58] developed a staggered approach based on phenomenological models. This approach examines

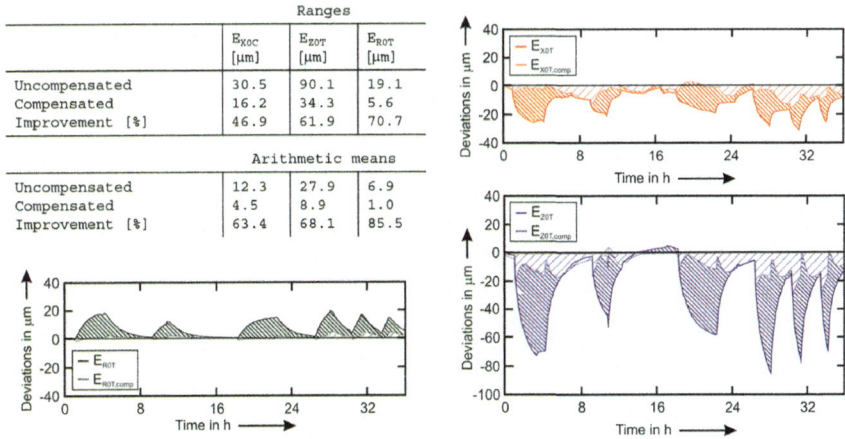

	Ranges		
	E_{XOC} [µm]	E_{Z0T} [µm]	E_{R0T} [µm]
Uncompensated	30.5	90.1	19.1
Compensated	16.2	34.3	5.6
Improvement [%]	46.9	61.9	70.7
	Arithmetic means		
Uncompensated	12.3	27.9	6.9
Compensated	4.5	8.9	1.0
Improvement [%]	63.4	68.1	85.5

Fig. 8 Measured location errors EX0T, EZ0T and ER0T with and without compensation [53]

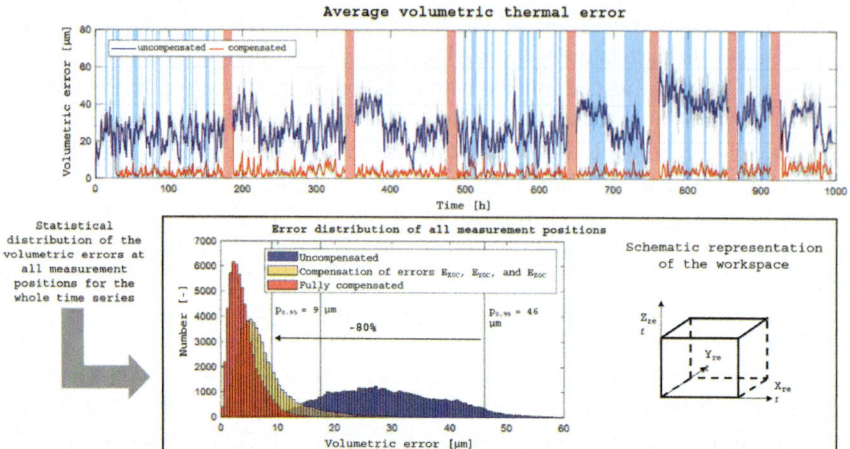

Fig. 9 Statistical evaluation of the volumetric thermal machine tool error for an uncompensated, partially compensated, three main errors, and fully compensated, 15 individual thermal errors, machine tool [55]

separating the thermal load cases and models axis-specifically thermal errors. A separate model is created for each error and each load case in order to map a large number of thermal influences. The errors of the individual models are superimposed to compensate for the machine tool with a holistic compensation model. By relating the energetic behaviour of the machine and the thermal errors at the TCP, Mayr et al. [59] show that the cooling capacity of the internal machine cooling of a 5-axis milling machine can be used as an input variable for a phenomenological model to reduce the thermally induced errors of a rotary and swivelling axis unit of a machine tool by up to 80%.

A major challenge when using data-driven models is long-term stability. Usually, the models are trained for one state and one environmental condition, season, etc., and therefore have a limited range of validity. This was solved by Blaser et al. [60] with an automated geometrical probing cycle on the machine tool, where it becomes possible to significantly reduce the thermal errors of 5-axis machine tools for any boundary conditions and machine states. However, the measuring time interval must be short enough in order to exclude the transient behaviour of the machine tool during measuring, which is why Blaser et al. [61] and Blaser [18] use the Adaptive Learning Control (ALC) approach, in which a data-driven phenomenological self-learning compensation model is trained by measurements on the machine. If the specified tolerance band is exceeded, measured on the machine, the model is automatically re-trained and re-parameterised.

Blaser et al. [62] developed also an extension of the approach in which the phenomenological model is replaced by an AutoRegression model with eXogenous input (ARX), and named it Thermal Adaptive Learning Control (TALC). The advantage of this approach, illustrated in Fig. 10, is that the ARX models exhibit significantly

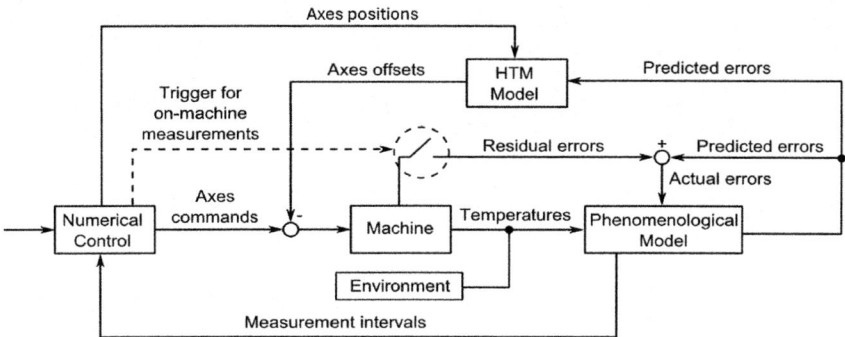

Fig. 10 Schematic diagram of thermal compensation procedure with TALC for numerically controlled machine tools [62]

improved long-term stability behaviour but still require automated re-training of the model. However, it turns out that the models with time series require a relatively long re-training period if the tolerance limits are exceeded. Mayr et al. [63] combine the TALC approach with a weighted least squares approach in order to be able to take into account the production history of the machine during retraining by simulating the missing measurement data. As Blaser et al. [64] show, the TALC approach can be used to take into account and successfully compensate for machine errors caused by short-term effects, such as the addition of MWF, if the measurement intervals are sufficiently finely resolved.

Since, as shown, the states on the machine change over time, in [55, 65] an extension of the TALC approach, including adaptive input selection, is introduced. Figure 11 shows the iterative process for input selection and model recalibration. Figure 12 shows an example of the error curve for the machine tool error EX0C and the optimum sensor set for each time interval. With adaptive input selection, the number of recalibrations could be reduced from seven, with a static sensor set, to two for a 108 h measurement cycle. Zimmermann et al. [66] mention as a limitation that the method, presented in [65], involves a separate input selection and modelling step, which leads to an iterative process. By extending the method with a Group-LASSO approach in combination with particle swarm optimization, the process of input selection and the modelling step can be parallelized. The combination of TALC and the newly developed method for optimal input selection achieves precise and robust compensation results with adaptive model inputs, which shows a very good long-term behaviour when applied to test series with a high variance of thermal load cases.

Another optimization potential of the TALC is the reduction of the necessary control measurements. Zimmermann et al. [67] present a method for recognizing an unknown thermal state of the machine on the basis of a one-class support vector machine in order to avoid periodic measurements in the machine. The procedure is illustrated in Fig. 13. With the method, the number of control measurements can be reduced from 160 to 35 in a series of measurements over 200 h with only tiny

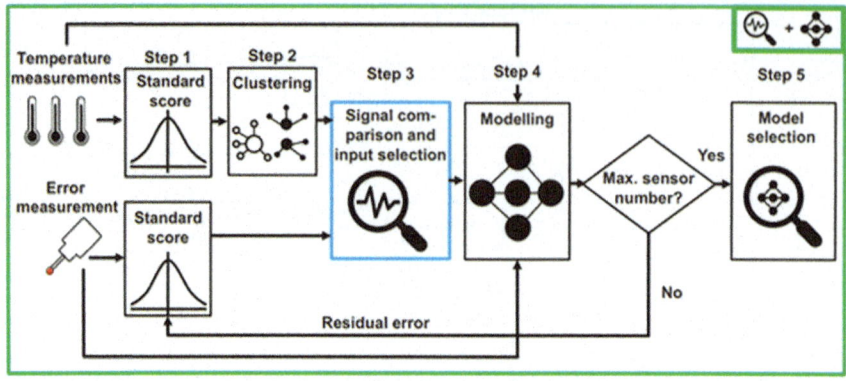

Fig. 11 Iterative procedure of the input selection and modelling [65]

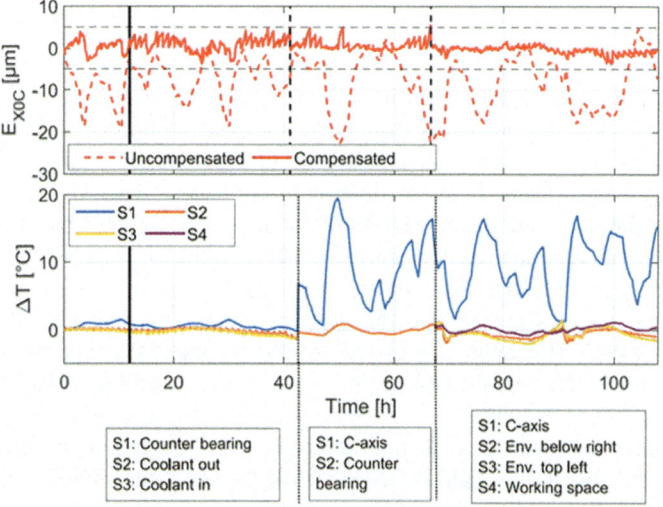

Fig. 12 Uncompensated and compensated thermal error EX0C over a 108 h stochastic load cycle. The listed sensors (S) for each time interval are defined as the optimal model inputs, selected by the approach from [65]

influence on the compensation results. In order to verify the method of state detection, in particular the ability to find a global minimum, Zimmermann et al. [68] perform a statistical analysis of the results, which verifies the capabilities of the approach.

Kaftan et al. [69] apply the TALC method to a Swiss-type automatic lathe. Although the modelling result with ARX is sufficiently accurate, due to the special nature of the machine, a non-observable state occurs when the machine door is opened. The information about the thermal state is the last measurement before the process interruption and the first measurement after the process interruption [70]. By implementing a Random Forest Regression (RFR) model, Kaftan et al. [71] could

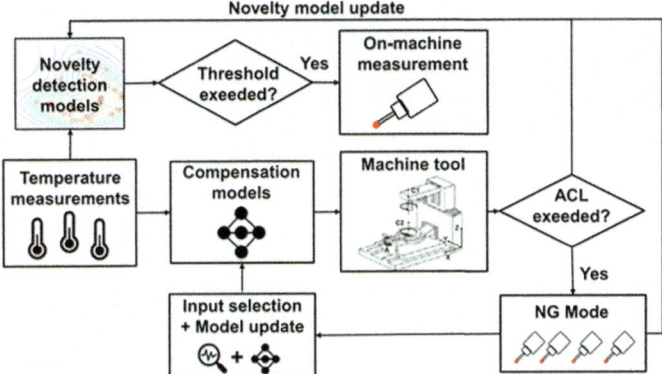

Fig. 13 Integration of the novelty detection model for on-demand triggered ST updates into the TALC [67]

cope with that effect with sufficient accuracy. Figure 14 shows the ARX-RFR model schematic for thermal error compensation. Figure 15 illustrates the modelling and compensation results using the ARX model and the ARX-RFR model.

Transferring the data-driven compensation models from one machine to another has not been investigated in detail and is part of ongoing research work. Stoop et al. [72] developed a federated learning approach with which knowledge can be transferred between the machines via a database in the cloud. Each machine works autonomously. The models are exchanged via the cloud and updated in the cloud with the information obtained from the connected machines. The connection of two similar machines, which are of the same type using different C-axis drive systems, is shown schematically in Fig. 16. Both machines are 5-axis milling machines of the

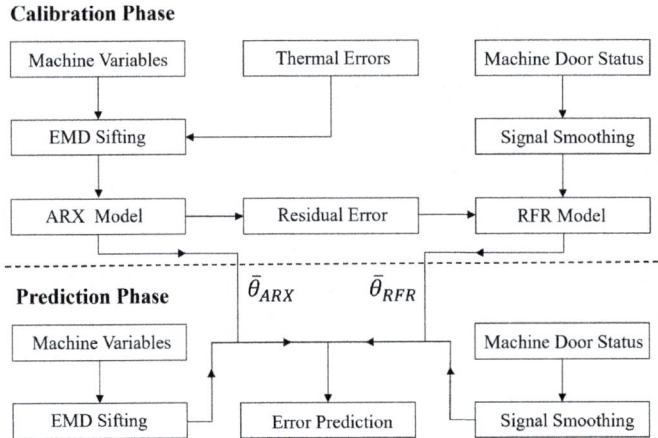

Fig. 14 The ARX-RFR model schematic for thermal error compensation [71]

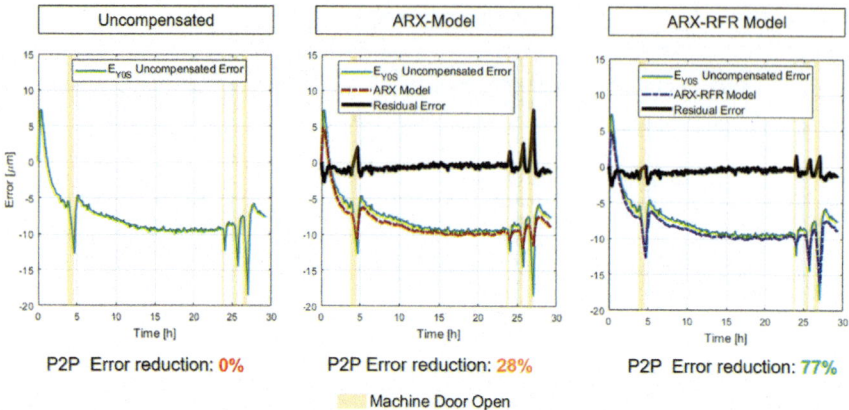

P2P Error reduction: **0%**   P2P Error reduction: **28%**   P2P Error reduction: **77%**

Machine Door Open

Fig. 15 Thermal compensation results of ARX (top) and ARX-RFR (bottom) for a lathe. Red stripes indicate times when the lathe's workspace door is opened [70]

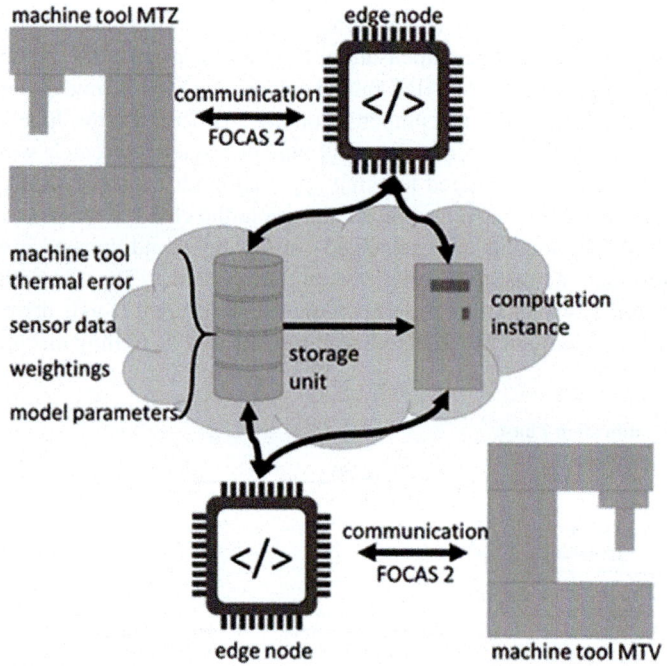

Fig. 16 The communication and computation setup between 2 machine tools of the same type [73]

same type NMV5000DCG. The MTV machine, with a worm drive rotary table and one axis cooling circuit for the linear axes, the main spindle, and the rotary-swivel axes unit, is located in a temperature-controlled environment. The MTZ machine, with a direct-drive rotary table and two separate cooling circuits, one for the linear axes and the main spindle and a second for the rotary-swivel axis unit, is located together with some other machines and devices in a workshop with temperature fluctuations between 16 and 40°C. The exchange of the learned models between them required retraining, but the model of the MTZ required less retraining on the MTV than vice versa, showing that the MTZ has a better learning success due to the poor environment [73]. The advantage of the federated learning approach is that each machine retains its autonomy, and models and data can still be exchanged. It is not necessary to transfer the pure measurement data, which can be problematic in terms of data protection, but only pre-processed data is exchanged.

Data-driven thermal error compensation can use a wide range of suitable input sets and models. Lang et al. [74] use both power and temperature measurements for internal and external influence variables and are thus able to reduce the root mean squared error (RMSE) of the volumetric error of a 5-axis milling machine by 72%. The authors further show that models such as ARX models are more suitable than LSTM and feedforward ANN models, which can also achieve acceptable accuracy but require more parameters and thus training effort. The prediction accuracy of LSTM neural networks is investigated by Rhiner et al. [35]. The Monte Carlo Dropout (MC-Dropout) method is used to determine the uncertainty of LSTM predictions based on the given database. Figure 17 demonstrates the Monte Carlo Dropout procedure for the estimation of the uncertainty, together with simulation results.

In particular, more complex data-driven models are often subject to a very large database, which is usually not available when investigating the thermal behaviour of machine tools. Lang et al. [75] present a study in which different approaches for preprocessing and potentially extending the input data by linear interpolation, cubic spline interpolation, and zero-order lag interpolation, all of them further extended by Gaussian noise, for compensation models are investigated. The authors show that using data augmentation represents an increase in volumetric error modelling from 48.2% to 65.8% without the need for additional measurement effort when using linear interpolation with Gaussian noise. As shown by Zimmermann et al. [65], the

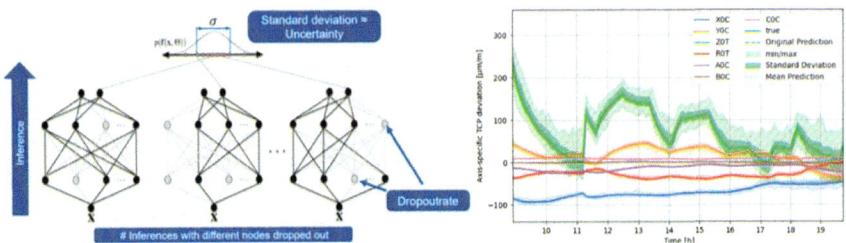

Fig. 17 Left: Illustration of Monte Carlo Dropout in neural networks to estimate uncertainty; Right: Illustration of the prediction of each axis-specific error, including the uncertainty estimate resulting from MC-Dropout [35]

choice of sensors used as model input is crucial for the quality of the compensation model. Another important factor is the correct positioning of the sensors on the machine. Lang et al. [74] develop a method to derive optimal sensor positions from a digital twin of the machine by intelligently clustering the nodes on the surface. In Fig. 18, the integration of the thermo-mechanical model into the digital twin of the machine is shown. Figure 19 illustrates the clustering on the machine tool's surface nodes, derived from a systematic load of axes movements of the machine tool in an uncontrolled environment.

Because training the data-driven compensation models still takes a lot of time, the effort to further reduce this unproductive measurement is obvious. Lang et al. [76] estimate the temperature vector of an order-reduced machine tool's finite element simulation model online using a Kalman filter and a small amount of temperature measurement data, illustrated in Fig. 20. The real-time capability allows the model to be used as a model for thermal error compensation, where the method can reduce up to 60% of the thermally induced errors at the TCP without the need for TCP-displacement measurements. In Fig. 20, the concept of the use of the Kalman filter for TCP displacement estimations is shown.

Another possibility to reduce measurement time during machine production is to measure the thermal error only when there is some indication that the compensation model has run out of tolerance and needs to be recalibrated. This can be achieved with probabilistic models, such as the Gaussian Process model, which provides not only

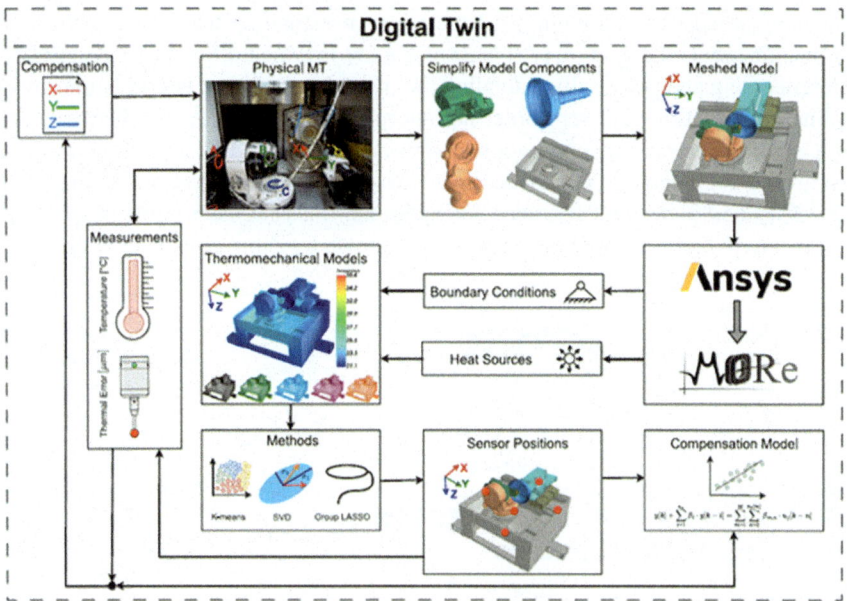

Fig. 18 Overview of the development of the thermomechanical model and its integration in the digital twin of the MT [75]

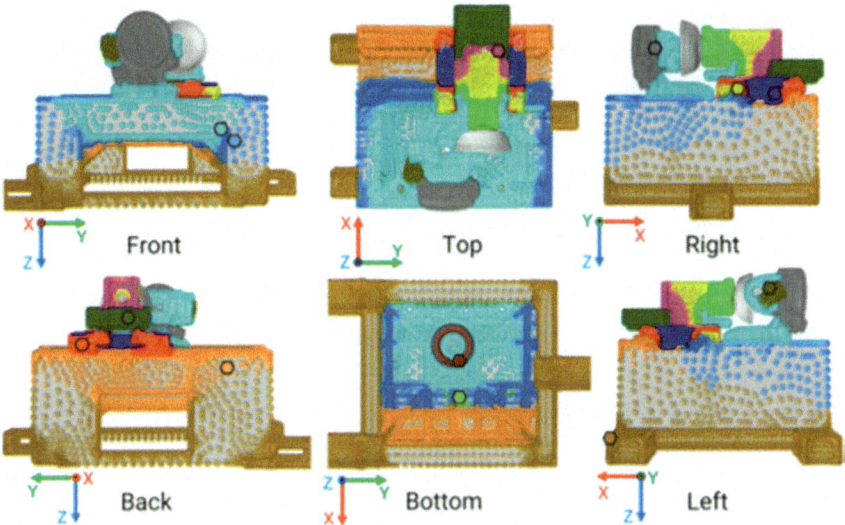

Fig. 19 K-means clustering on the machine tool's surface nodes with a total of 13 clusters evaluated. The clustered regions are illustrated by the different nodal colours [75]

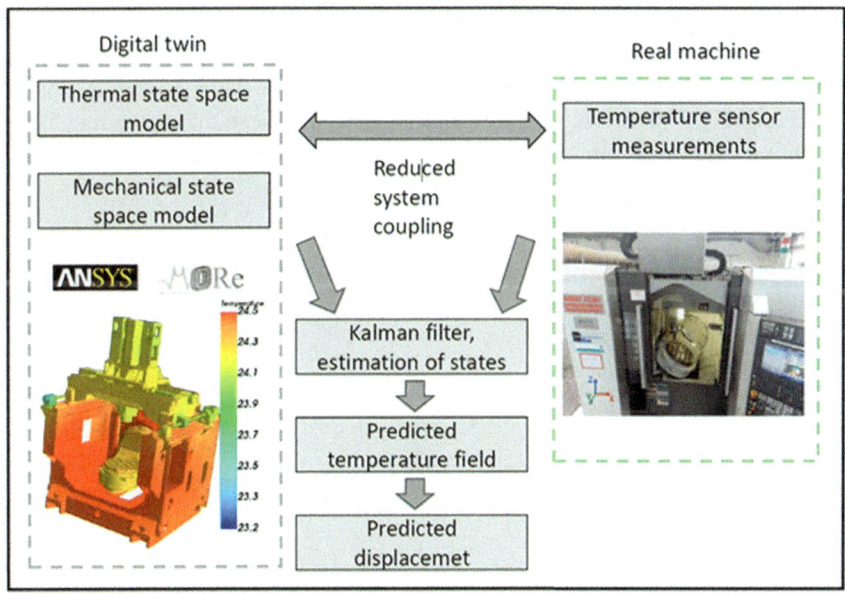

Fig. 20 Concept of the use of the Kalman filter together with the FEM model of the machine tool for thermal error prediction [76]

an estimate of the thermal error but also an uncertainty associated with the prediction. A certain threshold for a maximum allowable uncertainty can then be used to trigger a control measurement, as shown in Fig. 21. Kaftan [24] therefore developed the Adaptive Gaussian Process model which ensures that control measurements only take place in regions of uncertainty and do not interrupt the prediction when the errors are compensated well.

5 Validation of the Compensation Approach with a Multi-axis Machined Workpiece

Despite all the theoretical considerations of systematic evaluation on manufactured test pieces, the compensation success must be found on machined parts, which also introduce a number of additional difficulties that must be overcome. The thermal behaviour of the part plays a decisive role here. The temperature of the raw part is not necessarily the same as that of the machine tool. Then the heating of the part depends on the contact between the clamping device and the blank and, of course, on the heat given off to the part in the process zone, both of which must be included in the model. A first approach is the test specimen of an impeller with artificially extended production times presented in [77]. The repeating geometry of the impeller is always manufactured at the same axis position. In this order, the blank has to be clamped centrally on the rotary table. The evaluation of the geometry deviations on the impeller is shown in Fig. 22. It can be obviously seen that under real 5-axis machining conditions, the geometrical errors of the impeller can be significantly reduced by thermal compensation means.

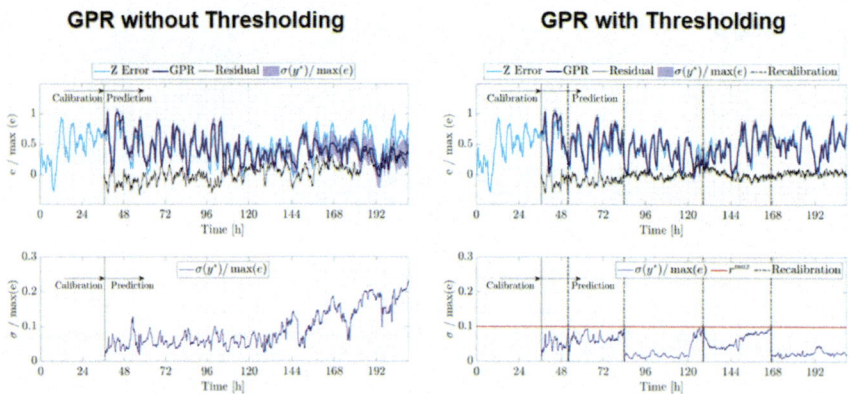

Fig. 21 Compensation schematic with the GPR model

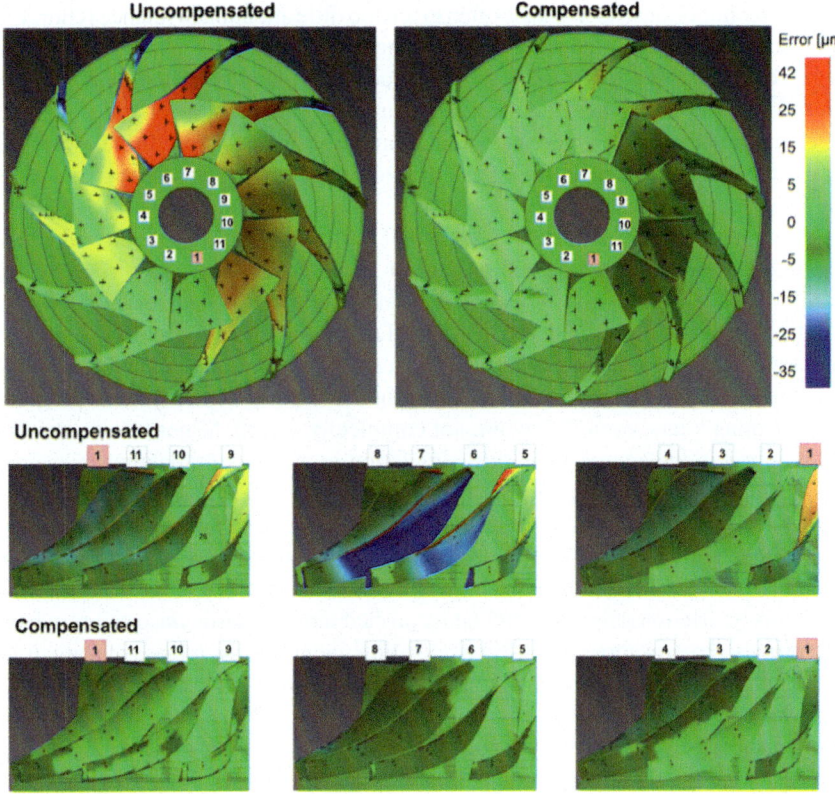

Fig. 22 3D visualization of the uncompensated and the compensated machined impeller [77]

6 Conclusion and Outlook

Research in the field of machine tools at inspire AG and the Institute of Machine Tools and Manufacturing (IWF) at ETH Zurich is following the vision of the fully compensated machine tool. This paper highlights the thermal aspects of the research work of the past years that paved the way to the fully compensated machine tool. The paper highlights the need for action, which is manifold, ranging from measurement tasks to numerical simulation methods and compensation approaches. As shown, it is thematically impossible to separate the areas of responsibility and would significantly slow down the development process.

In particular, research in the field of self-learning data-driven compensation strategies, such as the TALC approach developed at IWF and inspire, has made significant progress in recent years. Aspects that are considered here are the acquisition of measurement data on the machine, the modelling of errors, the increase in productivity, and the reduction of the climate impact of the machine tools in production.

Another important aspect that is often given too little attention in science is how the knowledge gained can be communicated. Zimmermann et al. [67] use the example of a series of laboratory courses to show how the complex interconnection can be taught. Zimmermann et al. [68] divide the subject area into the sub-aspects of evaluating the energetic behaviour, numerical simulation, and compensation. In this way, the complex subject matter can be quickly taught to course participants through practical exercises.

Although, as shown, a very broad range of topics must be considered and major steps have been taken in research, it is clear that there is still a way to go before machine tools can ultimately work autonomously and be fully compensated. Important developments on this path are in the field of measurement technology to record all thermal errors in the workspace, which can also be carried out online and in parallel to the process on the machine. The solutions for this will certainly make progress in the coming years, even if they are not completely solved. Major efforts are also still required in the area of numerical simulation in order to ultimately achieve the required accuracy for mapping the thermal effects. As shown, great progress has been made, but the achievable compensation results are far behind those of the data-driven models. Finally, as the work of Kaftan [70] in particular shows, the precision of the thermal compensation of machine tools will find its limits in mechanics. Most deterministic friction effects are still being pushed into stochastic uncertainty, which increasingly requires more precise consideration, including the release of static friction in bolted joints due to thermally induced stresses. Nevertheless, if the mechanics improve, as is already the state of the art for high-precision machine tools, the measuring accuracy of the available measuring technology will inevitably be a limiting factor, and the modelling precision will have to be significantly increased.

Future developments for high-precision compensation of machine tools still require a great effort in research work, as many effects have only been rudimentarily investigated. On the one hand, the possibilities of accessibility to the machine control system for compensation. Furthermore, physical simulation methods such as MORe, developed by Daniel Spescha [78], already offer the possibility of computing thermal errors online, albeit with severe limitations in terms of variability. The first possible applications have already been identified, and research in physical modelling and simulation technology will enable a wider range of applications in the near future.

Data-driven, self-learning, and retrained models are also expected to undergo development steps in the near future, which will enable them to be used more widely. The models will be able to cover more load cases without significantly losing accuracy. In particular, the handling of production disturbances and production planning variations have so far in research only been considered to a limited extent. Furthermore, research is needed into the transfer of models in order to allow machines to learn from machines. To date, research work has often only included the machine, which is not least due to the universal usability of the machines.

Nevertheless, measurements with the thermal test workpieces show that the temperature change of the workpiece significantly influences the thermal defects on the workpiece. This also includes the influences of the MWF, workspace extraction, and machine tool temperature control. Furthermore, the influence of process parameters, tool wear, and the influence of chips on machining accuracy concerning thermal effects have been considered poorly in research. Finally, all future developments must also take into account their influence on the climatic effect, and machine tools must be able to reliably maintain their precision in, e.g., non-air-conditioned workspaces.

References

1. Wegener, K., Weikert, S., Mayr, J.: Age of compensation-challenge and chance for machine tool industry. Int. J. Autom. Technol. **10**(4), 609–623 (2016)
2. Wegener, K., Weikert, S., Rüttimann, N., Ess, M., Mayr, J., Burkhalter, J., Lengwiler, A.: Modeling as a basis for efficient production. In: Sustainable Production for Resource Efficiency and Ecomobility, pp. 115–134 (2010)
3. Wegener, K., Weiss, L., Mayr, J., Gittler, T.: Konnektivität von Maschinen, Lernen und Intelligenz. In: 4. Innovation Forum Zerspanungstechnologie (2022)
4. Wegener, K., Weikert, S., Mayr, J., Maier, M., Ali Akbari, V.O., Postel, M.: Operator integrated–concept for manufacturing intelligence. J. Mach. Eng. **21**, 5–28 (2021)
5. Wegener, K., Maier, M., Ostad Ali Akbari, V., Mayr, J.: AI R&D for real factory. In: The 20th International Machine Tool Engineer's Conference (IMEC2024), pp. 36–63. Japanese Machine Tool Builders' Association (2024)
6. Mayr, J., Jedrzejewski, J., Uhlmann, E., Donmez, M.A., Knapp, W., Härtig, F., Wendt, K., Moriwaki, T., Shore, P., Schmitt, R., Brecher, C., Würz, T., Wegener, K.: Thermal issues in machine tools. CIRP Ann. **61**(2), 771–791 (2012)
7. Wegener, K., Mayr, J., Merklein, M., Behrens, B.-A., Aoyama, T., Sulitka, M., Fleischer, J., Groche, P., Kaftanoglu, B., Jochum, N., Möhring, H.-C.: Fluid elements in machine tools. CIRP Ann. **66**(2), 611–634 (2017)
8. Stoop, F., Mayr, J., Wegener, K.: Aerostatic stiffness and damping analysis for high-speed air bearings in ultra-precision machine tools. In: Euspen's 22nd International Conference & Exhibition, pp. 157–160 (2022)
9. Stoop, F., Meier, S., Civelli, P., Mayr, J., Wegener, K.: Multi-variable rotor dynamics optimization of an aerostatic spindle. CIRP J. Manuf. Sci. Technol. **42**, 12–23 (2023)
10. Stoop, F.: Luftgelagerte Spindel mit aktivem magnetischem Aktuator. PhD Thesis, Diss ETH Zürich (2024)
11. Mayr, J., Weikert, S., Egeter, M., Wegener, K.: Shop floor conditions versus precision on 5-axis machine tools. Lamdmap **11**, 299–308 (2015)
12. Mayr, J., Gebhardt, M., Massow, B.B., Weikert, S., Wegener, K.: Cutting fluid influence on thermal behavior of 5-axis machine tools. Procedia CIRP **14**, 395–400 (2014)
13. Hernández-Becerro, P., Blaser, P., Mayr, J., Weikert, S., Wegener, K.: Measurement of the effect of the cutting fluid on the thermal response of a five-axis machine tool. In: Laser Metrology and Machine Performance XII (2017)
14. Pavliček, F., Beer, Y., Weikert, S., Mayr, J., Wegener, K.: Design of a measurement setup and first experiments on the influence of CO_2-cooling on the thermal displacements on a machine tool. Procedia CIRP **46**, 23–26 (2016)
15. Mayr, J., Ess, M., Weikert, S., Wegener, K.: Calculating thermal location and component errors on machine tools. In: ASPE Proceedings 24 Annual Meeting, pp. 128–131 (2009)
16. Mayr, J.: Beurteilung und Kompensation des Temperaturgangs von Werkzeugmaschinen. PhD Thesis, Diss ETH Zürich (2009)
17. Gebhard, M.: Thermal Behaviour and Compensation of Rotary Axes in 5-Axis Machine Tool. PhD Thesis, Diss ETH Zürich (2014)

18. Blaser, P., Hauschel, C., Rüttimann, R., Hernández-Becerro, P., Mayr, J., Wegener, K.: Thermal characterization and modelling of a gantry-type machine tool linear axis. In: Euspen's 19th International Conference & Exhibition, pp. 166–169 (2019)

19. Zimmermann, N., Mayr, J., Wegener, K.: Extended discrete R-test as on-machine measurement cycle to separate the thermal errors in Z-direction. In: Proceedings of the Euspen's Special Interest Group: Thermal Issues, pp. 21–24 (2022)

20. Blaser, P.: Adaptive Learning Control for Thermal Error Compensation. PhD Thesis, Diss ETH Zürich (2020)

21. Zimmermann, N., Kartenbender, J.-M., Mayr, J., Wegener, K.: On-machine measurement cycle for the adaptive thermal error compensation of linear axes. In: Euspen's 22nd International Conference & Exhibition, pp. 149–152 (2022)

22. Kaftan, P., Porquez, F., Mayr, J., Pomodoro, K., Keel, M., Trombert, D., Wegener, K.: Thermal error measurement and compensation with torque limit skip in Swiss-type lathe manufacturing. Precis. Eng. **88**, 315–323 (2024)

23. Kaftan, P., Porquez, F., Mayr, J., Pomodoro, K., Trombert, D., Wegener, K.: Measurement of ball screw thermal elongation with torque limit skip. In: Proceedings of the Euspen's Special Interest Group: Thermal Issues (2023)

24. Kaftan, P.: Thermal Error Compensation of High Precision Swiss-type Lathes. PhD Thesis, Diss ETH Zürich (2025)

25. Mayr, J., Ess, M., Weikert, S., Wegener, K.: Simulation and prediction of the thermally induced deformations on machine tools caused by moving linear axis using the FDEM simulation approach. In: ASPE Proceedings 23 Annual Meeting, pp. 168–171 (2008)

26. Ess, M., Mayr, J., Weikert, S., Wegener, K.: Thermal model of machine tool feed drivetrains. In: ASPE Proceedings Annual Meeting, pp. 456–461 (2011)

27. Mayr, J., Ess, M., Weikert, S., Wegener, K.: Comparing different cooling concepts for ball screw systems. In: ASPE Proceedings Annual Meeting, pp. 978–981 (2010)

28. Mayr, J., Ess, M., Weikert, S., Wegener, K.: Thermal behaviour improvement of linear axis. In: 11th Euspen International Conference (2011)

29. Ess, M., Weikert, S., Wegener, K., Mayr, J.: Dynamic Loads and Thermal Errors on Machine Tools, pp. 1–5. ETH Zürich (2012)

30. Ess, M., Mayr, J., Weikert, S., Wegener, K.: An energy model for the calculation of losses and their effects on machining accuracy. In: Euspen's 12th International Conference & Exhibition, pp. 515–518 (2012)

31. Züst, S., Gontarz, A., Pavlíček, F., Mayr, J., Wegener, K.: Model-based prediction approach for internal machine tool heat sources on the level of subsystems. Procedia CIRP **28**, 28–33 (2015)

32. Mohammadi, A., Züst, S., Mayr, J., Blaser, P., Sonne, M.R., Hattel, J.H., Wegener, K.: A methodology for online visualization of the energy flow in a machine tool. CIRP J. Manuf. Sci. Technol. **19**, 138–146 (2017)

33. Züst, S.: Model-Based Optimization of Internal Heat Sources in Machine Tools. PhD thesis, Diss ETH Zürich (2017)

34. Züst, S.D., Blaser, P., Mayr, J., Knapp, W., Wegener, K.: Thermo-energetic issues in 5-axis machine tools. In: Proceedings of the MTTRF 2017 Annual Meeting (2017)

35. Rhiner, L., Lang, S., Mayr, J., Bambach, M.: Investigating the energy efficiency of thermal error compensation in machine tools. In: 3rd Special Interest Conference on Precision Engineering for Sustainability (2024)

36. Mayr, J., Pavlicek, F., Züst, S., Blaser, P., Hernandez-Becerro, P., Weikert, S., Wegener, K.: Thermal error research, an overview. Lamdamap **12**, 10–31 (2017)

37. Mayr, J., Weikert, S., Wegener, K.: Comparing the thermo-mechanical behavior of machine tool frame designs using a FDM-FEA simulation approach. In: ASPE Proceedings 22 Annual Meeting, pp. 17–20 (2007)

38. Pavlíček, F., Dietz, F., Blaser, P., Züst, S., Mayr, J., Weikert, S., Wegener, K.: An approach for developing meta models out of fluid simulations in enclosures of precision machine tools. In: ASPE Proceedings Annual Meeting, pp. 456–461 (2016)

39. Pavlíček, F., Pamies, D.P., Mayr, J., Züst, S., Blaser, P., Hernández Becerro, P., Wegener, K.: Using meta models for enclosures in machine tools. In: Conference on Thermal Issues in Machine Tools, pp. 159–168 (2018)

40. Hernández Becerro, P., Zimmermann, N., Pavlíček, F., Blaser, P., Knapp, W., Mayr, J., Wegener, K.: Learning efficient modeling and compensation for thermal behavior of machine tools. In: Proceedings of the 2018 MTTRF Annual Meeting, pp 269–274 (2018)

41. Pavlicek, F.: Meta Models for the Calculation of Heat Transfer Coefficients on Enclosures. PhD Thesis, Diss TU Chemnitz (2019)

42. Mayr, J., Ess, M., Pavlíček, F., Weikert, S., Spescha, D., Knapp, W.: Simulation and measurement of environmental influences on machines in frequency domain. CIRP Ann. **64**(1), 479–482 (2015)

43. Hernández-Becerro, P., Mayr, J., Wegener, K.: Efficient thermo-mechanical model of a precision 5-axis machine tool. In: Proceedings of the Euspen's Special Interest Group: Thermal Issues (2016)

44. Hernandez-Becerro, P.: Efficient Thermal Error Models of Machine Tools. PhD Thesis, Diss ETH Zürich (2020)

45. Hernández-Becerro, P., Purtschert, J., Konvicka, J., Buesser, C., Schranz, D., Mayr, J., Wegener, K.: Reduced-order model of the environmental variation error of a precision five-axis machine tool. J. Manuf. Sci. Eng. **143**, 021005 (2021)

46. Hernandez-Becerro, P., Mayr, J., Wegener, K.: Reduced thermo-mechanical model of a rotary table of a 5-axis precision machine tool. In: ASPE Spring Topical Meeting Design and Control of Precision Mechatronic Systems (2020)

47. Hernández Becerro, P., Mayr, J., Blaser, P., Pavlíček, F., Wegener, K.: Model order reduction of thermal models of machine tools with varying boundary conditions. In: Conference on Thermal Issues in Machine Tools, pp. 169–178 (2018)

48. Hernández Becerro, P., Blaser, P., Mayr, J., Wegener, K.: Design improvement of the cutting fluid supply of a large 5-axis machine tool. In: ASPE Proceedings 22 Annual Meeting (2018)

49. Mayr, J., Ess, M., Weikert, S., Wegener, K.: Compensation of thermal effects on machine tools using a FDEM simulation approach. In Proceedings Lamdampa 9 (2009)

50. Ess, M.: Simulation and Compensation of Thermal Errors of Machine Tools. PhD Thesis, Diss ETH Zürich (2012)

51. Blaser, P., Mayr, J., Wegener, K.: Simulation based comparison of thermal error modelling methods for machine tools. In: Proceedings of the Euspen's Special Interest Group: Thermal Issues, pp. 116–119 (2020)

52. Wegener, K., Gebhardt, M., Mayr, J., Knapp, W., Blaser, P.: Thermal compensation for 5-axis machine tools with physical model. In: Euspen's 14th International Conference & Exhibition, pp. 323–326 (2014)

53. Gebhardt, M., Mayr, J., Furrer, N., Widmer, T., Weikert, S., Knapp, W.: High precision grey-box model for compensation of thermal errors on five-axis machines. CIRP Ann. **63**(1), 509–512 (2014)

54. Wegener, K., Gebhardt, M., Mayr, J., Knapp, W.: Thermal issues in 5-axis machine tools. In: Proceedings MTTRF 2013 (2014)

55. Zimmermann, N.: Selbstlernende thermische Fehlerkompensation für Werkzeugmaschinen. PhD Thesis, Diss ETH Zürich (2022)

56. Mayr, J., Blaser, P., Knapp, W., Wegener, K.: Compensation of cutting fluid influences on five axis machine tools. In: MTTRF 2016 Annual Meeting, pp. 101–118 (2016)

57. Mayr, J., Müller, M., Weikert, S.: Automated thermal main spindle & B-axis error compensation of 5-axis machine tools. CIRP Ann. **65**(1), 479–482 (2016)

58. Mayr, J., Tiberini, T., Blaser, P., Wegener, K.: Thermal error compensation of 5-axis machine tools using a staggered modelling approach. In: Conference on Thermal Issues in Machine Tools, pp. 421–430 (2018)

59. Mayr, J., Egeter, M., Weikert, S., Wegener, K.: Thermal error compensation of rotary axes and main spindles using cooling power as input parameter. J. Manuf. Syst. **37**, 542–549 (2015)

60. Blaser, P., Gebhardt, M., Mayr, J., Knapp, W., Wegener, K.: Automatic compensation of thermally induced errors on five-axis machine tools. In: Euspen-Special Interest Group Meeting: Thermal Issues, pp. 302–309 (2014)

61. Blaser, P., Pavliček, F., Mori, K., Mayr, J., Weikert, S., Wegener, K.: Adaptive learning control for thermal error compensation of 5-axis machine tools. J. Manuf. Syst. **44**, 302–309 (2017)

62. Blaser, P., Mayr, J., Wegener, K.: Long-term thermal compensation of 5-axis machine tools due to thermal adaptive learning control. MM Sci. J. (2019)

63. Mayr, J., Blaser, P., Ryser, A., Hernandez-Becerro, P.: An adaptive self-learning compensation approach for thermal errors on 5-axis machine tools handling an arbitrary set of sample rates. CIRP Ann. **67**(1), 551–554 (2018)

64. Blaser, P., Mayr, J., Pavliček, F., Hernández-Becerro, P., Wegener, K.: Adaptive learning control for thermal error compensation on 5-axis machine tools with sudden boundary condition changes. In: Conference on Thermal Issues in Machine Tools, pp. 339–348 (2018)

65. Zimmermann, N., Lang, S., Blaser, P., Mayr, J.: Adaptive input selection for thermal error compensation models. CIRP Ann. **69**(1), 485–488 (2020)

66. Zimmermann, N., Büchi, T., Mayr, J., Wegener, K.: Self-optimizing thermal error compensation models with adaptive inputs using Group-LASSO for ARX-models. J. Manuf. Syst. **64**, 615–625 (2022)

67. Zimmermann, N., Breu, M., Mayr, J., Wegener, K.: Autonomously triggered model updates for self-learning thermal error compensation. CIRP Ann. **70**(1), 431–434 (2022)

68. Zimmermann, N., Mayr, J., Wegener, K.: Statistical analysis of self-optimizing thermal error compensation models for machine tools. In: Proceedings of the Euspen's Special Interest Group: Thermal Issues, pp. 22–23 (2022)

69. Kaftan, P., Mayr, J., Wegener, K.: Thermal model of a Swiss-type Lathe. In: Proceedings of the Euspen's Special Interest Group: Thermal Issues, pp. 21–24 (2022)

70. Kaftan, P., Mayr, J., Wegener, K.: Thermal compensation of sudden working space condition changes in Swiss-type Lathe machining. In: International Conference on Thermal Issues in Machine Tools, pp. 15–27 (2023)

71. Wegener, K., Mayr, J., Zimmermann, N., Kaftan, P.: Intelligent Thermal Compensation of Machine Tools, pp. 61–71 (2022)

72. Stoop, F., Mayr, J., Sulz, C., Bleicher, F., Wegener, K.: Fleet learning of thermal error compensation in machine tools. In: 26th IEEE ETFA, pp. 1–4 (2021)

73. Stoop, F., Mayr, J., Sulz, C., Kaftan, P., Bleicher, F., Yamazaki, K., Wegener, K.: Cloud-based thermal error compensation with a federated learning approach. Precis. Eng. **79**, 135–145 (2023)

74. Lang, S., Zimmermann, N., Mayr, J., Wegener, K., Bambach, M.: Thermal error compensation models utilizing the power consumption of machine tools. In: International Conference on Thermal Issues in Machine Tools, pp. 41–53 (2023)

75. Lang, S., Zorzini, M., Scholze, S., Mayr, J., Bambach, M.: Sensor placement utilizing a digital twin for thermal error compensation of machine tools. J. Manuf. Syst. **80**, 243–250 (2025)

76. Lang, S., Lampert, N., Mayr, J., Wegener, K., Bambach, M.: Training efficient and compensating fast: data augmentation for thermal error compensation models of machine tools. In: Proceedings of the Euspen's Special Interest Group: Thermal Issues (2024)

77. Zimmermann, N., Mayr, J., Wegener, K.: An action-oriented teaching approach for intelligent and energy efficient precision manufacturing. NAMRC **50**(33), 961–969 (2023)

78. Spescha, D., Weikert, S., Wegener, K.: Simulation in the design of machine tools. In: Reinventing Mechatronics, pp. 163–177. Springer (2020)

Practical Implementation of Complex Thermal Models on NC Controls

Tim Boye and Florian Sellmann

Abstract Thermally induced displacements in machine tools cause defective parts and limit five-axis machining accuracy. Compensation methods can reduce these displacements with minimal hardware effort, but due to their complexity these methods are rarely used. ARX models show promise for adaptive compensation of displacements within the kinematic chain of a machine tool. Experiments show significant reduction in displacements across the five-axis workspace. Implementing ARX models on NC controls in production faces constraints like low acceptance of production interruptions. ARX models are complex black-box models, making verification difficult. This paper presents a modified model structure that simplifies implementation, improves prediction quality, and reduces sensor noise. It also addresses the problem of interruption of production by necessary measurements. The model is implemented on a TNC 640 control with a five-axis machine tool, with training and testing conducted in a climate chamber.

1 Introduction

Modern machine tools are capable of milling entire parts from solid material in progressively shorter times. The enhanced performance of these tools, coupled with growing automation that enables the complete finishing of all six sides of a workpiece, results in higher material removal rates. This increased volume of material removal is accompanied by a higher power output in the machine tools, which subsequently leads to greater heat input into the machine. Consequently, structural temperature changes ultimately cause a displacement between the TCP (Tool Center Point) and the table.

T. Boye (✉) · F. Sellmann
Dr. Johannes Heidenhain GmbH, Traunreut, Germany
e-mail: boye@heidenhain.de

F. Sellmann
e-mail: sellmann@heidenhain.de

© The Author(s) 2026

K. Wegener and M. Bambach (eds.), *4th International Conference on Thermal Issues in Machine Tools (ICTIMT2025)*, Lecture Notes in Production Engineering,
https://doi.org/10.1007/978-3-032-01194-7_12

175

Basically, there are two ways to counteract this displacement:

1. Control the temperature and cool or temper the corresponding components.
2. Compensate for the resulting displacement through an appropriate thermal model on the machine control.

The first option is generally preferable, as it addresses the error at its root. Ideally, the machine would no longer experience thermal changes, fundamentally solving the problem. Consequently, many machine tool manufacturers choose to temper their machines either partially or completely.

However, fully thermally stabilizing a machine tool proves to be highly complex and costly. It requires numerous cooling channels and sophisticated temperature management systems, which demand significant power. Despite these higher costs and increased energy consumption, not all heat sources can be effectively eliminated. Consequently, hybrid systems are often developed, incorporating both cooled structural components and control-based compensation.

The compensation technique for partially cooled systems generally needs to address minor thermal errors. However, due to the cooling or temperature control, it is significantly more complex in terms of modeling compared to uncooled systems. The integrated cooling systems must be considered as additional heat sources and sinks.

2   Thermal Compensation

A common method for building thermal models is the phenomenological approach. In this approach, observed thermal displacements are described using a certain model structure based on temperatures or power flows.

For example, Zimmermann [1] presents an adaptive ARX model structure. It is based on an adaptive LARS regression, which adjusts the model parameters to the identified displacements. The performance of this model was demonstrated on a real manufactured impeller part.

In Zimmermann [1], 11 other model-based thermal compensation methods [2–12] are listed. All models have different numbers of parameters and thermal load cases. Due to the listed high measurement times of up to 132 h, only models that consider the entire kinematic chain, as presented in Zimmermann [1] or [15], seem advantageous. However, it remains unclear for all methods to what extent the chosen model can represent the thermally measured error. Although significant reductions in error are shown, the reason for the remaining error is not mentioned. Therefore, we propose an approach that evaluates the quality and performance of the selected model based on the residual error and its variation.

3 Determination of the Kinematic Chain and Evaluation of Model Choice

To determine the kinematically relevant errors, the method described in [13] is used. Many different methods for identifying kinematic parameters are described. All of them are using a touch probe and different artifacts. We limit ourselves to the method that can determine all essential kinematic errors of a five-axis machine in a short time (5 min). This is made possible by an eccentrically placed precision sphere on the worktable. The position of this sphere is measured in different swivel positions using the touch probe.

In terms of modeling, we choose a model that includes all translational and rotational location errors of rotary and linear axes as well as the position of the sphere on the table. In [14], it is highlighted that this model has the best ratio of generality to model simplicity. This model consists of 14 parameters:

1. Three orthogonalities between the three linear axes.
2. Four positional errors each for two rotary axes.
3. X, Y, Z position of the precision sphere.

This model has proven to be very universal and can generally be identified on any five-axis table-machine with a well conditioned identification-matrix. A key prerequisite is that the linear workspace can be largely scanned by probing the sphere. This is usually possible for machines with table-side rotary axes. Usually large portal machines cannot be adequately characterized with this model and this type of measurement method.

Under these conditions, the method has many advantages over other measurement methods:

- Inexpensive measuring device configuration, which is implemented in almost every machine tool.
- Wireless and automatically exchangeable from tool magazine.
- Long battery life of several days.
- Process-safe against coolant and chips.
- Modular, space-saving design: The sphere position can be adapted to the conditions of the machines and their processes.
- Measurement accuracy $< 1\,\mu$m: Since the determination of the sphere position is based only on neighborhood accuracy, high measurement accuracy can be achieved.
- All relevant positional errors for all five axes can be determined in a short time.

The combination of all these advantages is essential for efficiently capturing the kinematics of a machine with short measurement times for a single sphere position and high reliability for long term measurements.

Despite all the advantages of this method, it must be mentioned that it is an identifying method. Kinematic parameters are determined from the measured errors using a least-squares algorithm. This approach has been known for a long time [16]. In our

example, all 11 location errors of all axes are identified. Due to a significantly overdetermined measurement, a residual error remains. This error cannot be described by
the model. In time series measurements, this residual error must remain almost constant. If this residual error changes over time, the chosen model cannot adequately
describe the structural changes. As a result of this the model must be readjusted.

Figure 1 shows an sample measurement in a AC table machine. The kinematic
structure of this machine is: TOOL-Z-Y-BASE-X-A-C-Table. The black box in the
upper left represents the workspace of the linear axes. The measurement points fill the
workspace of the linear axes well. From these measurements, the depicted kinematic
parameters can be identified, and a residual error of 5 μm remains. Measurement and
residual errors are displayed in the bottom right right figure as absolute values at the
corresponding measuring positions. Figure 2 shows a time series measurement that
is mapped to the corresponding parameters of the chosen model and has a residual
error. This is on average 3 μm and a maximum of 7 μm. This residual error limits the
achievable accuracy of the kinematic model in the five-axis workspace. The change
in the residual error over time is on average only 0.7 μm and a maximum of 2.5 μm.
With an assumed measurement accuracy of 1 μm, this change in model deviation
is acceptable. If we change the model to only 4 translational location errors of the
rotary axes and sphere position, the value on average nearly doubles to 1.3 μm.
The maximum change in the residual error more than doubles to 6 μm. This simple
model is therefore not as suitable for description of the thermal behavior of this
certain machine. Although our chosen model cannot reproduce all the errors of the

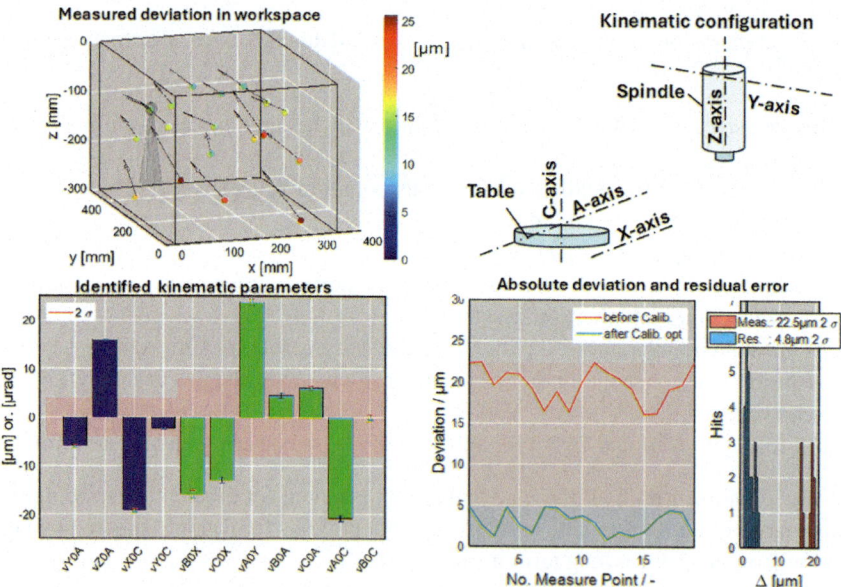

Fig. 1 Single example measurement of AC table machine

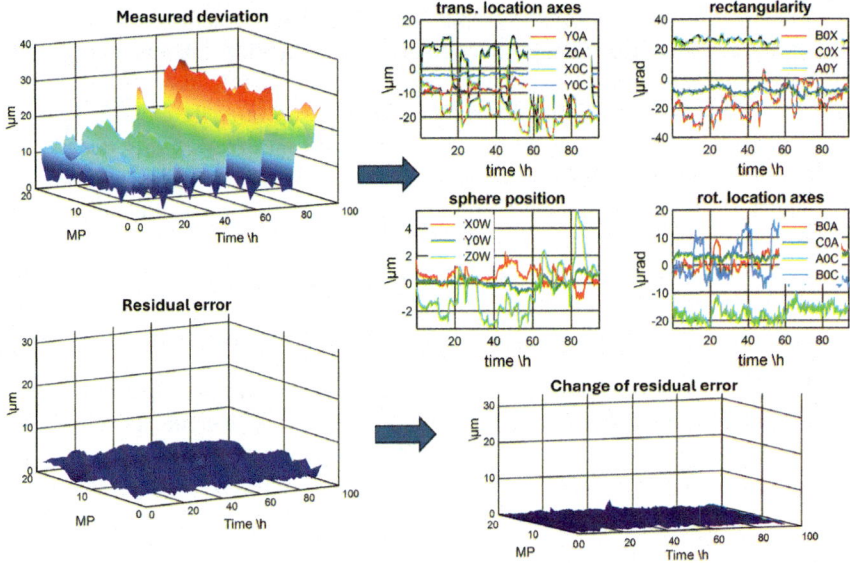

Fig. 2 Continuous determined kinematic parameters and remaining residual deviation and change of residual error

real machine, it seems to be able to represent the main thermal changes within the measurement accuracy. In this case, the change in the residual error is within the desired limits. It should be noted that this change is related to the applied thermal loads (see Chap. 6). These should generally represent the maximum loads. If other thermal loads are applied, the change in the residual error must be re-examined. This allows the validity of the selected 14-parameter model to be proven.

4 Thermal Compensation Model

The parameter trends depicted in Fig. 2 are the foundation of a thermal model, along with the corresponding synchronous temperature trends (not shown).

In [14], for example, an ARX model is identified based on the measured temperature trends. The identification of the model is based on a "grouped LASSO" regression, through which both the number of parameters and the dimension of the model are adjusted. It is shown that the model is very powerful and reduces the maximum error from 78 to 16 μm over a period of approximately 800 h. During this time, various load cases were considered.

The aim of this work is to implement the performance of such compensation on an NC control. The following boundary conditions arise:

1. Adapting the model through several hours of re-measurements is undesirable in a productive environment. Production processes generally cannot be interrupted by relearning cycles.
2. The ARX structure is similar to that of an IIR filter, meaning the identified parameters must be fed back into the model with several time delays. Consequently, the model requires parameter sets from the past within the specified time frame and model dimension. When the machine is restarted, these values are not available and have to be determined. This would only be possible by a measurement with a touch probe and precision sphere and the associated identification run. This approach is not desirable in a production environment where measurements need to be taken for approximately one hour after a machine is restarted.

Therefore, we confine ourselves to a static, fixed model that optimally represents the thermal environment. To prevent the undesired measurements before each machine start, we change the model to an FIR structure. Although this increases the number of parameters, feedback of the model's output values is no longer necessary.

5   Identification of the FIR Model

The procedure for creating a model for the kinematic parameters from temperature data is divided into three steps:

1. Sorting out redundant temperature data.
2. Multitask LASSO regression for each individual kinematic parameter.
3. Weighted ridge regression with extended FIR structure of the temperature data identified in step 2.

Eliminating redundant temperature data enhances the robustness of the compensation and improves the conditions for subsequent multitask LASSO regression. It is generally advantageous for temperature data to cover a wide range to achieve maximum resolution. Therefore, redundant temperature data with the largest signal range are selected.

The multitask LASSO regression is a linear regression with mixed L1/L2 norm:

$$\left(\frac{1}{2 \cdot n_{samples}} \right) \cdot ||Y - XW||^{2_{Fro}} + \alpha \cdot ||W||_{21} \tag{1}$$

with

$$||W||_{21} = \sum_i \sqrt{\sum_j W\{i, j\}^2} \tag{2}$$

In this expression Y corresponds to the kinematic parameters and the features X to the temperature data. The factor α can regulate the weighting matrix and thus reduce the number of selected temperature data. The strength of the weighting is based on the achieved prediction results and the number of desired maximum temperature data.

The regression result created in step 2 provides the necessary output data for a subsequent weighted ridge regression. The regression criterion is as follows:

$$\min||Y - XW||_2^2 + \alpha \cdot ||W|| \tag{3}$$

Compared to a standard ridge regression, α is changed to a vector that continuously increases with the age of the features X. In this way, the regression creates a typical settling behavior of an FIR structure. From a physical point of view, this delayed settling behavior represents the inadequacy of the sensor position and the transient behavior of heat conduction.

Figure 3 shows the comparison between multitask LASSO regression and the extended FIR model exemplified for two parameters. The orthogonality B0X and the displacement Z0A have a significant influence on the kinematic behavior in our example. If the statistical spread of the prediction error is only slightly reduced by the FIR structure, maximum values and outliers can be significantly decreased. For thermal compensation, maintaining a corridor is much more important than minimizing the error precisely. The goal of the compensation is to keep the error

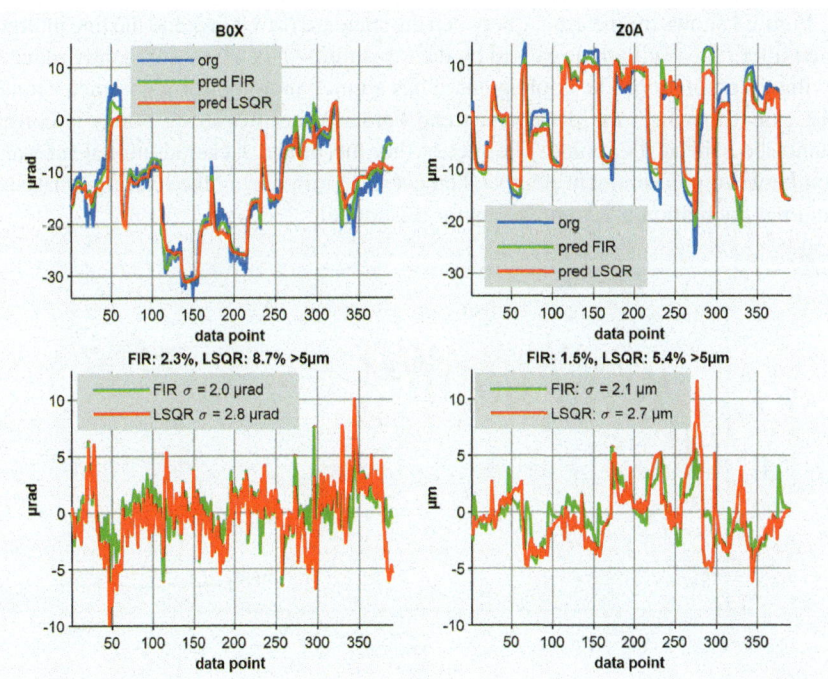

Fig. 3 Model prediction of FIR and LSQR model

within a desired corridor for various thermal boundary conditions. The FIR structure also has the advantage over the ARX structure that the noise of the temperature measurements can be significantly reduced through fine interpolation of the filter parameters.

For temperature measurement, a high frequency data acquisition is possible compared to determining the kinematic parameters. While the kinematic parameters are only available every 15 min due to the measurement method, temperatures can be determined every second easily. Minimizing the noise of the kinematic parameters is particularly important because kinematic parameters should change as uniformly as possible during machine operation in compensation mode. In our case, we identify a filter length of 15. This means that the filter has a temporal length of $15 \cdot 15$ min = 225 min. For fine interpolation, we use a factor of 3, resulting in 45 filter parameters per temperature X. Experiments showed that further fine interpolation is no longer advantageous. The benefit of the finally interpolated model is very evident in the parameter Z0A. During thermal tests with a spindle speed, the load is always interrupted by the 5-min sphere probing measurement. This results in a strong temperature drop in the spindle sensor. This high-resolution temperature trend is not reflected in the model's learning data. The model only relies on data in 15-min intervals for the learning cycle. However, temperature data continuously flows in during compensation operation. This leads, for example, to the parameter trend shown in Fig. 4.

Figure 4 shows the difference between the standard (raw) filter and the fine interpolated filter. The oscillations caused by the measurement can be significantly reduced by fine interpolation. The cooling curve has a much more uniform parameter trend. The extent to which the parameter trend between the measuring points is correct cannot be said at this point. The fact is that the model lacks additional information between measurement points taken every 15 min during the learning phase, so minimal fluctuation is advantageous for this model.

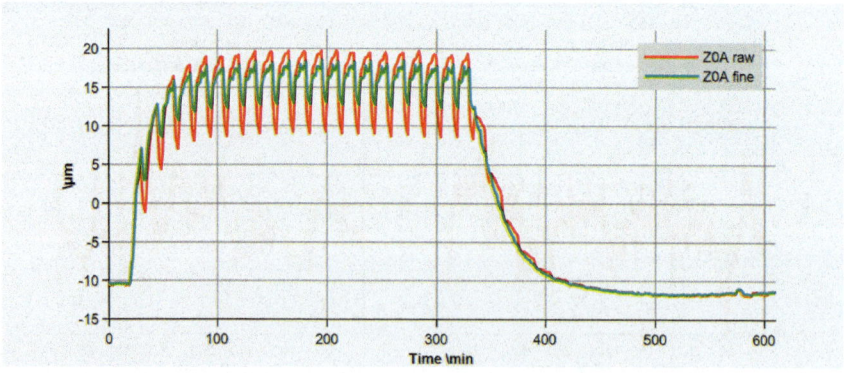

Fig. 4 Comparison between standard and fine interpolated FIR model

6 Experimental Results with the FIR Model

For the experiments, the AC table machine with a travel path of $300 \times 400 \times 300$ mm of the linear axes was used. A precision sphere is clamped eccentrically on the table, so the working volume is well covered as shown in Fig. 1. The machine is controlled with a TNC 640 from HEIDENHAIN. The machine has structural cooling of all linear drives and the spindle. The compensation model is implemented in the PLC.

The machine was placed in a climate chamber for all experiments, where the ambient temperature can be varied. Temperature values and the installation position on the machine structure can be seen in Fig. 5. For the 100-h learning phase of the model, temperature trends similar to a day-night cycle and sudden changes in temperature from 22 °C to 30 °C were applied. After a a temperature step, the temperature was held for 24 h. During the test phase with and without active compensation, the climate chamber was operated in a 60-h day-night cycle with an amplitude increased by 20%. To stress the structure with internal heat sources, all axes and the spindle were moved at different times during both the learning and test phases. The axes were operated with a feed rate between 0 and 13,000 mm/min. Single-axis movements, three-axis movements, and five-axis movements were performed. The spindle was operated at different speeds from 0 to 18,000 rpm.

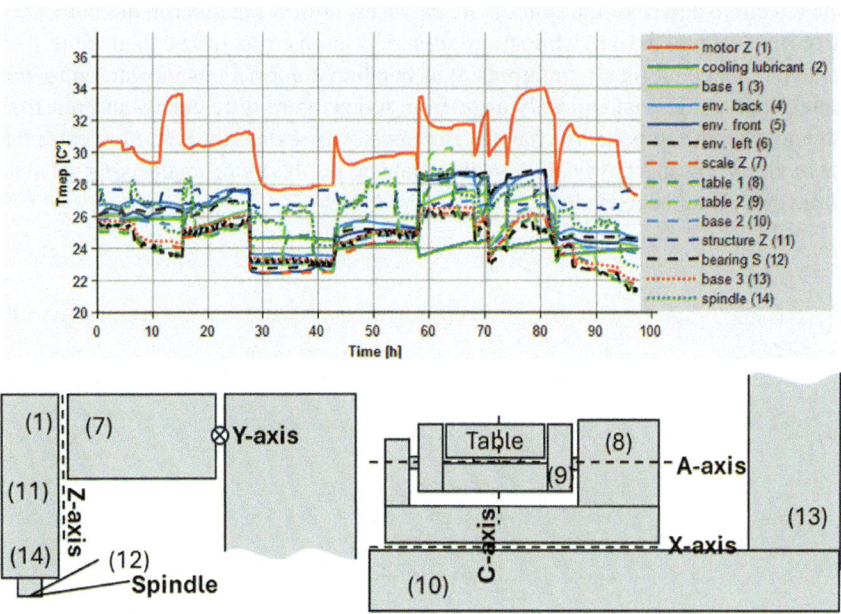

Fig. 5 Measured temperatures during testrun and placement of sensors

These heating cycles were cyclically interrupted every 10 min to perform a sphere probing measurement of 5 min. The cycle time of the determined parameters is therefore 15 min. The determined sphere positions and the data acquisition of the temperature sensors are implemented on the NC control, so no additional PC was needed. For testing purposes, it is generally recommended to place sensors close to all internal heat sources, key structural components, the surroundings, and coolants. Unnecessary sensors are then eliminated by the LASSO algorithm. A total of 20 temperature sensors are installed on the control. Four sensors were already installed by the manufacturer, and all others were additionally applied to the machine specifically for the tests. To minimize the noise of the temperature measurement data, the sensors are read every second, and averaged over one minute. These filtered measurements are then stored in an overall database and secondly in a ring buffer for the compensation model. The model itself is then updated every minute. The results of the model's learning phase are already shown in Fig. 2. During the learning phase, deviations in the sphere position of up to 35 μm occur. The maximum deviations can be seen between 60 and 100 h. In this range there are also the step responses caused by the ambient temperature. Through the multitask LASSO regression, 14 of the 20 available sensors were selected as relevant. For each of the 8 kinematic parameters, a maximum of 7 temperature sensors are needed. The identified model can be represented in the form of the individual step responses of each sensor. Figure 6 shows the step responses of the multitask LASSO regression and the FIR model for the 8 most changing kinematic parameters. It is evident that both models show similar behavior. The red curve describes the sum of all responses. It turns out that rotational location errors tend to zero in sum, while translational location errors have a final value. This behavior can be physically interpreted, as bending deformations in a structure with homogeneous material can only arise from temperature differences, and not from a homogeneous temperature change. The transient response therefore indicates that the model works in a physically sensible manner. Furthermore, an unstable behavior of the model can be excluded.

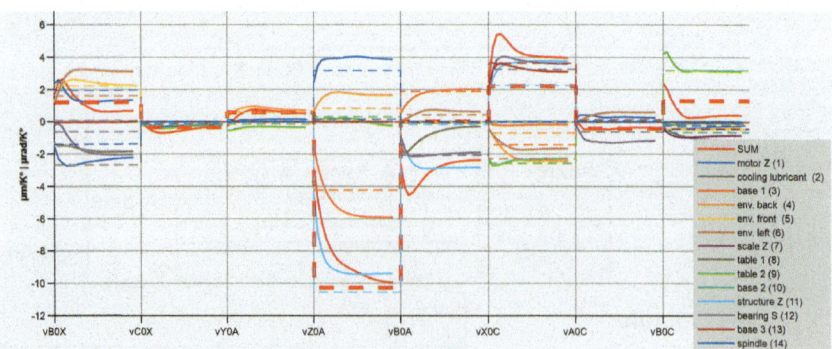

Fig. 6 Step response of temperatures for kinematic error. LASSO regression: –, FIR model: —

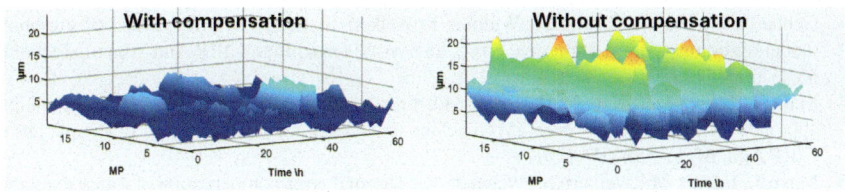

Fig. 7 Measured results during test phase with and without compensation

The results of the test phase are shown in Fig. 7. While the test phase without compensation is above 10 μm in 50% of all measurements, the measured error with the FIR model can be kept below 10μm in 98% of all measurements. The maximum deviation is 23 μm and 13 μm, respectively, and the mean value is 10.5 μm and 4.3 μm, respectively. An improvement of 50% initially seems small. It should be noted that the deviations not compensated for were minimized by a medium static calibration. This minimizes the percentage improvement. The initial state is therefore less interesting than the target accuracy of less than 10 μm.

7 Conclusion

In this paper we discuss the practical use of complex thermal compensation models in milling machine tools. The literature shows many promising approaches, for example the ARX Model. By using any thermal model, we propose a method to evaluate this model for a thermal compensation. This is done by analyzing the residual error of the model and, in particular, its changes over time. We also describe an approach to implement a compensation model on an NC control by addressing some practical issues from manufacturing environments. Re-measurements during production are in general undesirable. Therefore, we stay with a static model. We present an FIR structure that reduces the noise of temperature sensors and enables compensation without recalibration after a reboot. A straightforward method to identify this model is presented and results comparing measurements with and without thermal compensation on a test-data set are shown.

References

1. Zimmermann, N., Müller, E., Lang, S., Mayr, J., Wegener, K.: Thermally compensated 5-axis machine tools evaluated with impeller machining tests. CIRP J. Manuf. Sci. Technol. **46**, 19–35 (2023)
2. Guo, S., Mei, X., Jiang, G.: Geometric accuracy enhancement of five-axis machine tool based on error analysis. Int. J. Adv. Manuf. Technol. **105**(1–4), 137–153 (2019)

3. Gebhardt, M., Mayr, J., Furrer, N., Widmer, T., Weikert, S., Knapp, W.: High precision grey-box model for compensation of thermal errors on five-axis machines. CIRP Ann. Manuf. Technol. **63**(1), 509–512 (2014)

4. Mayr, J., Blaser, P., Ryser, A., Hernandez-Becerro, P.: An adaptive self-learning compensation approach for thermal errors on 5-axis machine tools handling an arbitrary set of sample rates. CIRP Ann. **67**(1), 551–554 (2018)

5. Mayr, J., Egeter, M.; Weikert, S.; Wegener, K.: Thermal error compensation of rotary axes and main spindles using cooling power as input parameter. J. Manuf. Syst. **37**, 542–549 (2015)

6. Blaser, P., Mayr, J., Wegener, K.: Long-term thermal compensation of 5-axis machine tools due to thermal adaptive learning control. MM Sci. J. 3164–3171 (2019)

7. Zimmermann, N., Büchi, T., Mayr, J., Wegener, K.: Self-optimizing thermal error compensation models with adaptive inputs using Group-LASSO for ARX-models. J. Manuf. Syst. **64**, 615–625 (2022)

8. Zimmermann, N., Kartenbender, J-M., Wegener, K.: On-machine measurement cycle for the adaptive thermal error compensation of linear axes. In: Euspen's 22nd International Conference & Exhibition. Geneva, Switzerland (2022)

9. Liu, J., Ma, C., Wang, S.: Data-driven thermally-induced error compensation method of high-speed and precision five-axis machine tools. Mech. Syst. Signal Process. **138**, 106538 (2020)

10. Bitar-Nehme, E., Mayer, JRR.: Modelling and compensation of dominant thermally induced geometric errors using rotary axes' power consumption. CIRP Ann. 2–5 (2018)

11. Vu Ngoc, H., Mayer, JRR., Bitar-Nehme, E.: Deep learning LSTM for predicting thermally induced geometric errors using rotary axes' powers as input parameters. CIRP J. Manuf. Sci. Technol. **37**, 70–80 (2022)

12. Zimmermann, N., Lang, S., Mayr, J., Wegener K.: The impact of self-learning thermal error compensation models on the accuracy of 4-axis thermal test pieces. In: ICPE, pp. 4–6 (2022)

13. Rahman, M.: Calibration of machine tools using on machine probing of an indigenous artefact. Ph.D. Thesis, École Polytechnique de Montréal (2016)

14. Brecher, C., Behrens, J., Lee, TH., Charlier, S.: Calibration of five-axis machine tool using r-test procedure. In: Laser Metrology and Machine Performance XII (2017)

15. Zimmermann, N.: Selbstlernende thermische Fehlerkompensation für Werkzeugmaschinen. Ph.D. Thesis, ETH-Zürich (2022)

16. Gautier, M.: Numerical calculation of the base inertial parameters of robots. IEEE International Conference on Robotics and Automation, vol. 2, pp. 1020–1025 (1990)

Reducing Thermally Induced Position and Orientation Errors of a Precision Machine Tool's Rotary Axis due to Rotation by Using Hydrostatic Bearings

Fabian A. Tripkewitz, Matthias Fritz, and Matthias Weigold

Abstract Most rotary axes of precision five-axis machines use highly preloaded precision roller bearings. These enable a low error motion and maximum stiffness. As a drawback, the high preload increases frictional torque, causing significant frictional losses under rotation, resulting in thermally induced position and orientation errors. Due to the inherent kinematic limitation of five-axis machines, some resulting errors are not compensable. Therefore, hardware optimization is necessary to obtain the required stiffness but reduce thermally induced position and orientation errors due to axis rotation. A promising approach is the application of shallow recess hydrostatic bearings, where a thin fluid film separates the stationary and rotating component, improving heat transfer based on heat conduction via a thin film. This paper experimentally investigates two rotary axes integrated in two identical machines, one with precision roller bearings and one with shallow recess hydrostatic bearings. The application of the hydrostatic bearing shows a significant reduction of 78–95% in position and orientation errors relative to the roller bearing. Experiments further show the effectiveness of an additional active shaft cooling for a hydrostatic bearing, reducing errors by an additional 10–38%.

1 Introduction

The application of manufacturing processes requiring a rotating workpiece on precision five-axis machine tools challenges the bearing technology of the rotary axes. Typically, such precision machine tools use highly preloaded precision roller bearings. These bearings enable low error motion and high static stiffness to reduce load-induced position and orientation errors. However, the high preload increases

F. A. Tripkewitz (✉) · M. Fritz
Kern Microtechnik GmbH, Eschenlohe, Germany
e-mail: fabian.tripkewitz@ametek.com

F. A. Tripkewitz · M. Weigold
Institute of Production Management, Technology and Machine Tools (PTW), Technical University Darmstadt, Darmstadt, Germany

© The Author(s) 2026

K. Wegener and M. Bambach (eds.), *4th International Conference on Thermal Issues in Machine Tools (ICTIMT2025)*, Lecture Notes in Production Engineering,
https://doi.org/10.1007/978-3-032-01194-7_13

frictional power loss during rotation [1], heating the components and resulting in thermally induced position and orientation errors. Furthermore, roller bearings are characterized by a limited heat transfer between the rotating and stationary component due to the small contact surface and finite number of rollers. The stationary component is often actively cooled, which, combined with the limited heat transfer through the rollers, results in a large temperature gradient between the rotating and stationary component. In addition, the increased shaft, respectively machine table temperature, will heat the clamped component, leading to further deviations [2]. An economical solution to reduce the error is the application of compensation techniques using different methods and input variables [3]. Multiple researchers have shown the possibility of modeling the thermally induced errors of rotary axes, enabling a significant reduction in most errors [4, 5]. However, there are two major limitations: the repeatability of the errors [6] and the kinematics of the five-axis machines, which cannot compensate for all calculated errors [7], as the third angular degree of freedom (DOF) cannot be manipulated [8].

Because of this inherent kinematic limitation, hardware optimization is necessary to reduce thermally induced position and orientation errors due to axis rotation and increase the reproducibility of the remaining errors. Based on the advantageous stiffness, damping, minimum runout, and reproducibility [1], precision machines often use hydrostatic bearings [9], ideally with reduced film height [10]. Here, the heat transfer between the bearing components mainly results from heat conduction via the thin film and convection [11]. A special type of hydrostatic bearing is the shallow recess bearing, which forms a large contact area with a small film thickness, improving conductive heat transfer between the stationary and rotating component.

This article presents an experimental comparison of a conventional high-precision roller bearing and a hydrostatic shallow recess bearing in an industrial application regarding thermally induced position and orientation errors due to rotation. As the measurements highlight the significant influence of the temperature increase of the shaft, experiments also investigate the influence of an active shaft cooling with temperature-controlled hydrostatic fluid at low pressure.

2 Theory

2.1 Nomenclature of the Position and Orientation Errors

Figure 1 depicts the reference coordinate system and the nomenclature of the errors used in this article. The origin of the reference coordinate system matches the intersection of the rotary and swivel axis, respectively, the kinematic center of the machine. The axial reference surfaces of the zero-point clamping system (ZPCS), or in general terms the table of the machine, is at z_T. The nomenclature of the thermally induced position and orientation errors of the axis is according to ISO 230-7 [12]. In addition, the position of the axial reference surface is also relevant for an explicit description

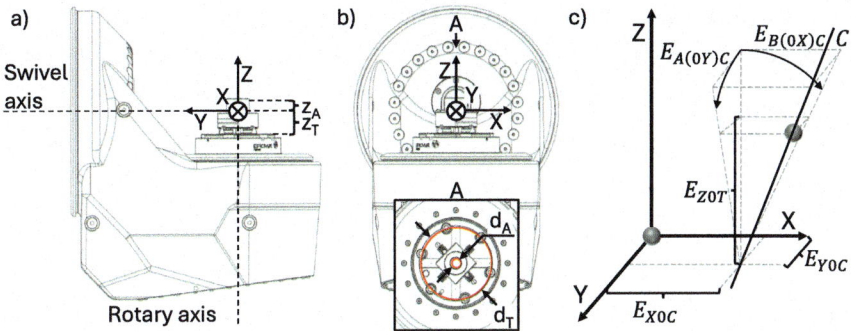

Fig. 1 Reference coordinate system relative to the hydrostatic rotary axis in side (**a**) and front view (**b**) and the nomenclature for the position and orientation errors based on ISO 230-7 and Gebhardt (**c**) [2, 12]. View A depicts a top view of the machine table

of the position of a workpiece. According to Gebhardt, the position error of the axial growth of the machine table E_{Z0T} is introduced for this purpose [2]. The zero position error of the C-axis E_{C0C} is of minor significance for continuous turning operations and neglected to reduce measurement complexity. The covered errors are summarized in Table 1 and schematically depicted in Fig. 1c.

2.2 Determination of the Position and Orientation Errors

Position and orientation errors are determined indirectly using eight touch-trigger probe (TTP) measurements. For probing, the machine table holds an artifact with a circular hole of the diameter d_A at the height z_A concentric to the axis of rotation. The size of the artifact allows probing the reference surfaces on the machine table at the diameter d_T, adding no thermal length in the Z-direction. Table 2 summarizes the nominal probing positions, probing direction, and the resulting values. Assuming a rotationally symmetrical thermoelastic error of the machine table and artifact that is

Table 1 Relevant position and orientation errors to describe the thermoelastic errors of a rotary axis according to ISO 230-7 and Gebhardt [2, 12]

Symbol	Error
E_{X0C}	Error of the position of C in X-axis direction
E_{Y0C}	Error of the position of C in Y-axis direction
E_{Z0T}	Error of the axial growth of the machine table
$E_{A(0Y)C}$	Error of the orientation of C in A-Axis direction; squareness of C to Y
$E_{B(0X)C}$	Error of the orientation of C in B-Axis direction; squareness of C to X

Table 2 Sequence of measuring points to calculate the position and orientation errors

Nominal position			Probing direction	Measured value
x	y	z		
$d_A/2$	0	z_A	X+	X, Y
0	$d_A/2$	z_A	Y+	
$-d_A/2$	0	z_A	X−	
0	$-d_A/2$	z_A	Y−	
$d_T/2$	0	z_T	Z−	Z_1
0	$d_T/2$	z_T	Z−	Z_2
$-d_T/2$	0	z_T	Z−	Z_3
0	$-d_T/2$	z_T	Z−	Z_4

The first four measuring points are part of the HEIDENHAIN probing cycle 421, with the circle's center coordinates as an output

constant throughout the probing cycle, Eqs. 1–5 describe the mathematical relations between the measured values and the position and orientation errors.

$$E_{A(0Y)C} = \frac{Z_2 - Z_4}{d_T} \tag{1}$$

$$E_{B(0X)C} = \frac{Z_3 - Z_1}{d_T} \tag{2}$$

$$E_{X0C} = X - z_A \cdot E_{B(0X)C} \tag{3}$$

$$E_{Y0C} = Y + z_A \cdot E_{A(0Y)C} \tag{4}$$

$$E_{Z0T} = \frac{Z_1 + Z_2 + Z_3 + Z_4}{4} \tag{5}$$

3 Methodology

For comparison of the two bearing principles, we use two machines of the Micro-Platform with ISO 10791-2 V [w C' B' b X Y Z (C) t] kinematics from KERN

MICROTECHNIK [13]. Both use identical base machines with hydrostatic bearings and direct drives on the linear axes but different swivel and rotary axes. The MicroHD possesses lifetime greased high-precision cylindrical roller bearings (axial and radial runout <0.5 µm) in the swivel and rotary axes, whereas the MicroHD+ uses oil hydrostatic shallow recess bearings (axial and radial runout <0.1 µm). Regarding mechanical properties, the maximum speed of both rotary axes is identical, but the hydrostatic version possesses an increased static stiffness. Furthermore, both the outer race of the bearing and the stator of the direct drive of both axes are water-cooled with an identical total volume flow. The active shaft cooling of the hydrostatic axis is initially deactivated to enable an isolated comparison of the bearing principles. The only difference is the geometry of the casting that connects the rotary axis to the swivel axis, which differs locally due to topology optimization of the hydrostatic version regarding static stiffness.

The chain of cause and effect for thermal errors starts with the electrical power consumption of the direct drive to rotate the axis. The direct drive and the bearing convert the electrical power to thermal power, which is then distributed into the structure by heat transfer, changing the initial temperature distribution. Due to the thermal expansion of the components, that difference in temperature distribution leads to thermoelastic deformations, resulting in the position and orientation errors of the axis [14].

Therefore, the presented methodology evaluates the drive power, temperature changes, and the resulting position and orientation errors for both bearing types.

3.1 Test Setup

Figure 2a shows an overview of the test setup. The test setup uses a MicoHD or a MicroHD+ from KERN MICROTECHNIK, depending on the investigated bearing principle. Otherwise, the test setup is identical for both bearing principles. Each machine uses a TC52 LF TTP from BLUM-NOVOTEST and a ZPCS of type Power-Chuck P from EROWA, holding an artifact of type ER-008617, also from EROWA. Three independent DAQ systems record the drive power, temperatures, and geometric errors during the test procedure. The first DAQ system is the machine, which probes the artifact and provides information on the drive's electrical power consumption. While the machine saves the data of the TTP in a text file, a measuring computer with the software TNCscope of the machine controls manufacturer, HEIDENHAIN, records the drive power. The second DAQ system is an ALMEMO 2690-8A from AHLBORN, which records the temperature of the casting at the tree

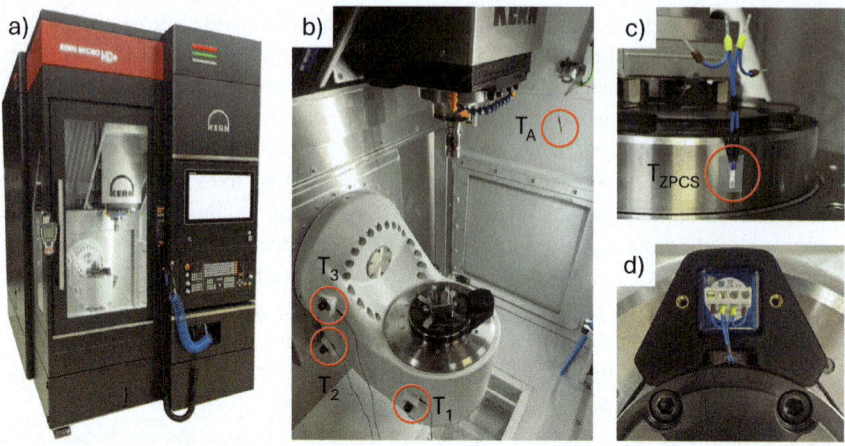

Fig. 2 Test setup for the hydrostatic version with the linear axis in waiting position in overview (**a**) and detailed view (**b**). Three temperature sensors measure the temperature of the casting ($T_1 - T_3$) while one sensor measures the ambient temperature T_A close to the waiting position of the touch-trigger probe. (**c**) shows the placement of the rotating temperature sensor measuring the zero-point clamping system's temperature T_{ZPCS} and (**d**) depicts the temperature sensor connected to the miniature data logger with the removed lid

positions $T_1 - T_3$ and the ambient temperature T_A with negative temperature coefficient (NTC) thermistors of type FNA 611 from AHLBORN. Double-sided thermal tape (heat conductivity: 1.3 W/mK) connects the temperature sensors to the casting to improve the heat transfer. In addition, a foam piece preloads the temperature sensor onto the casting, as depicted in Fig. 2b. The position of the three sensors in the XY-plane is identical for both bearing types but differs slightly in the X-direction for T_2.

The third DAQ system is a battery-powered miniature data logger from ESYS, type blueDAN Pt100, that records the temperature of the rotating ZPCS T_{ZPCS} via a PT-100 thermistor of type M 1020 from YAGEO NEXENSOS. To improve thermal coupling, double-sided thermal tape holds the sensor to the circumference of the ZPCS, as shown in Fig. 2c. A 3D-printed holder made of polyoxymethylene (POM) is strapped to the ZPCS and secures the miniature data logger relative to it. A compressed piece of foam between the holder and the sensor, depicted in Fig. 2d, ensures contact of the temperature sensor with the ZPCS under rotation and encapsulates the temperature sensor from the ambient air. As thermal effects are not time-critical, the recorded data of the three DAQ systems is synchronized by time afterward.

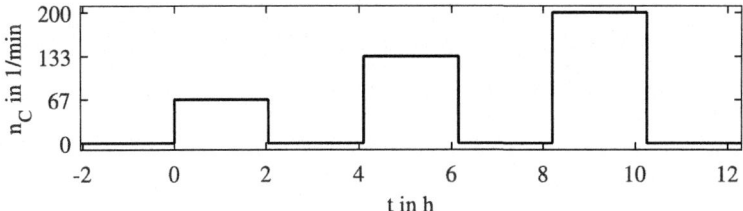

Fig. 3 Nominal rotational speed of the rotary axis n_C during the test procedure as a function of time t

3.2 Test Procedure

The test procedure investigates three different rotational speeds n_C to examine the relationship between rotational speed and thermally induced errors. Each rotational speed includes a warm-up and a cool-down phase. The rotational speed of the warm-up phases is constant at 67, 133, and 200 rpm, whereas the rotational speed of the cool-down phases is zero rpm. A duration of 2 h was previously determined to achieve thermal equilibrium at the end of each phase. The test procedure starts with a cool-down phase for equal initial conditions for each rotational speed. Figure 3 shows the corresponding nominal rotation speed of the test procedure as a function of time t.

As described in Sect. 2.2, measuring the position and orientation errors requires a stationary axis. Therefore, the axis rotation stops during each warm-up phase, the axis orients for measurement, the machine probes the axis, and the axis accelerates back to its nominal speed. The measuring cycle, respectively the time interval with deviation from the nominal speed, takes approximately 30 s. To reduce the influence of the measurement, the measuring cycle is performed after every three minutes of axis rotation with the nominal rotational speed, resulting in a total cycle time of 3.5 min. A phase consists of 35 cycles, resulting in a total duration of 2.05 h per phase. If not probing, the TTP rests at the waiting position, as shown in Fig. 2a. During the test procedure, the DAQ systems record the drive power at approximately 10 Hz and the temperatures at 0.2 Hz. Seven measurements are carried out for each bearing principle for statistic evaluation. In addition, seven measurements at maximum rotational speed are carried out with the hydrostatic bearing and activated shaft cooling for comparison.

4   Results and Discussion

4.1 Electrical Power Consumption

The root cause of the thermal errors is the electrical power consumption P_{el} of the system under investigation. As the average power consumption is of interest and to reduce the effect of noise based on the low sampling rate, 50 samples are averaged,

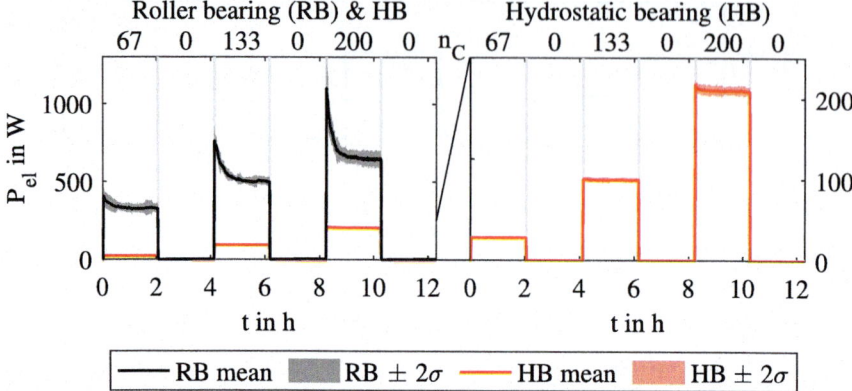

Fig. 4 Electrical power consumption P_{el} of the axis drive over the time t of the roller and hydrostatic bearing (left side) and a zoomed-in section of the hydrostatic bearing (right side). The figure does not depict the power consumption during each measuring cycle for readability. In addition, the rotational speed n_C at each sequence is added. The graphs show the mean and the standard deviation σ with coverage factor k = 2 of seven measurements

resulting in a reduced sampling rate of 0.2 Hz. Based on the reduced sampling rate, Fig. 4 depicts the two bearing variants' mean electrical power consumption, each supplemented by the expanded standard deviation. The figure does not show the power consumption during the measuring cycles to improve readability.

Three significant differences exist in the power consumption of the two bearing principles: the magnitude, variation during one phase, and their standard deviation. At 67 rpm, the mean power consumption of the hydrostatic version is lower by a factor of 11–14 compared to the roller bearing. This difference reduces with rotational speed, resulting in a reduced mean power consumption of the hydrostatic version by a factor of three to five at 200 rpm. Based on the viscous friction of the hydrostatic bearing, the power consumption is approximately constant for 67 and 133 rpm. Only at 200 rpm a minor relative reduction in mean power consumption of 4.4% is noticeable, with a maximum expanded relative standard deviation of 3.6%. In contrast, the roller bearing possesses a reduction in mean power consumption independent of the rotational speed and increased standard deviation. For comparison, the roller bearing possesses a relative reduction in mean power consumption at 200 rpm of 42.1% with a maximum expanded relative standard deviation of 15.0%. The significant reduction in power consumption and the increased standard deviation may be the result of thermally induced changes in preload and lubrication.

4.2 Temperature Change

As a consequence of the electrical power input, the temperature of the system changes. Figure 5 shows the temperatures T_1 of the structure and T_{ZPCS} of the ZPCS. Due to

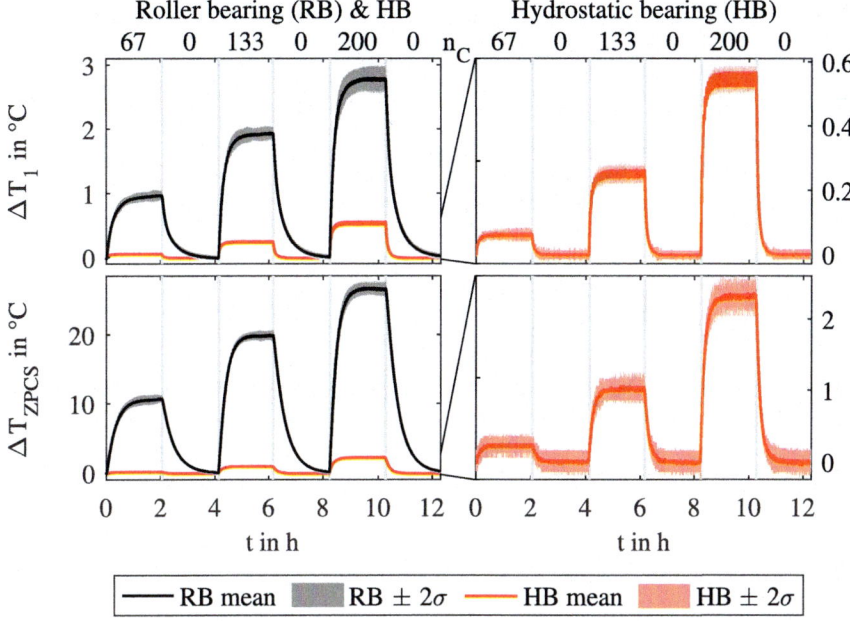

Fig. 5 Temperature of the structural component T_1 and the zero-point clamping system T_{ZPCS} over the time t of the roller and hydrostatic bearing (left side) and a zoomed-in section of the hydrostatic bearing (right side). In addition, the rotational speed n_C at each sequence is added. The graphs show the mean and the standard deviation σ with coverage factor k = 2 of seven measurements

the deviation of the nominal rotational speed for probing, the temperatures converge to a semi-saturated state with cyclic temperature change with the frequency of the probing cycle. Using a hydrostatic bearing compared to a roller bearing reduces the mean temperature of the structure T_1 at 67 rpm by a factor of fifteen and at 200 rpm by a factor of five. The reduction in temperature change is even more dominant for the temperature of the ZPCS T_{ZPCS}, resulting in a factor of 43 at 67 rpm and a factor of eleven at 200 rpm. In addition, applying a hydrostatic instead of a roller bearing reduces the mean time constant at the structure by 56–71% and at the ZPCS by 37–57%.

4.3 Position and Orientation Errors

The resulting position and orientation errors are measured and calculated according to Sect. 2.2. For an isolated evaluation of the short-term errors due to the axis rotation, a straight line is fitted to the last five measurements of each of the four cool-down phases with zero rpm (t ≈ 0, 4, 8, and 12 h) and subtracted from the measured values. Figure 6 summarizes the result.

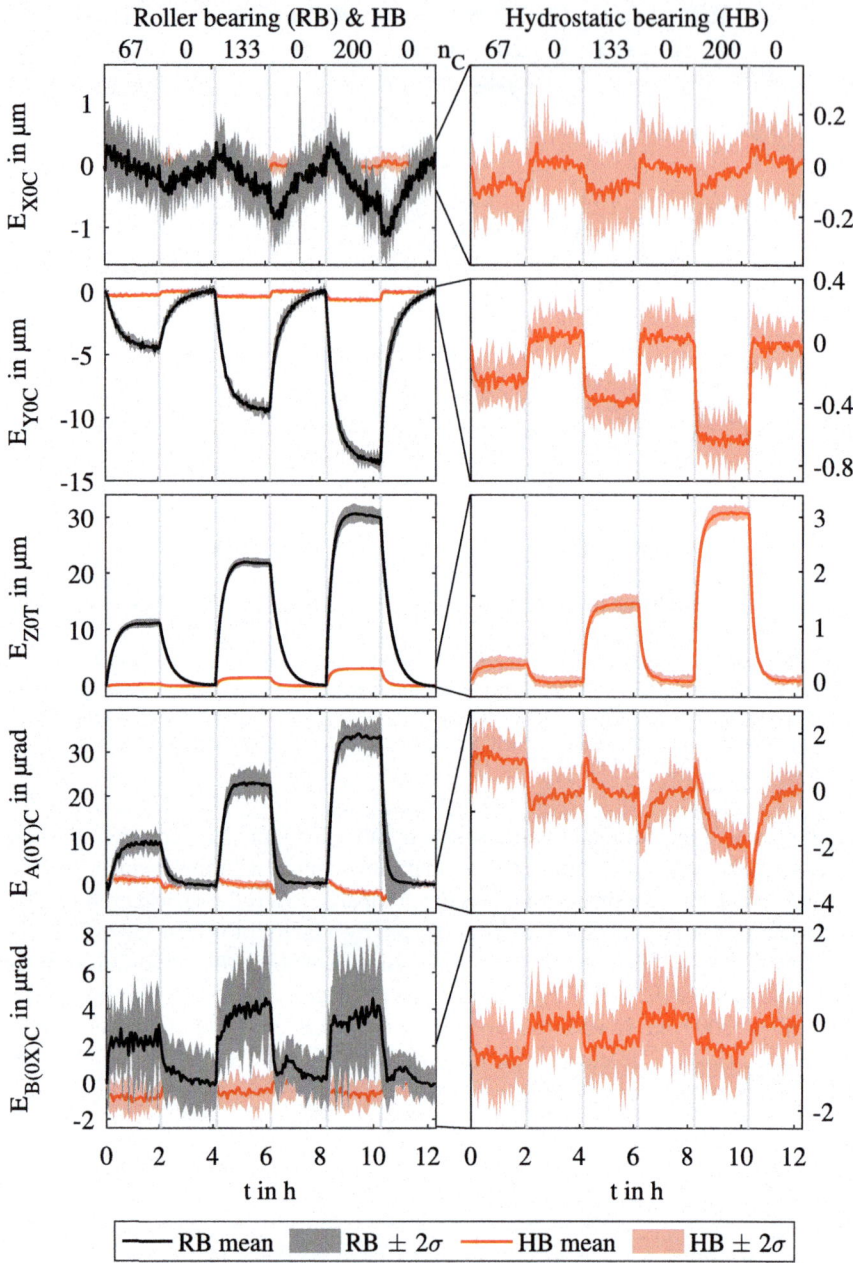

Fig. 6 Thermally induced position and orientation errors over the time t of the roller and hydrostatic bearing (left side) and a zoomed-in section of the hydrostatic bearing (right side). The nomenclature is according to ISO 230-7 and Gebhardt [2, 12]. In addition, the rotational speed n_C at each sequence is added. The graphs show the mean and the standard deviation σ with coverage factor k = 2 of seven measurements

Due to the existing thermo-symmetry along the X-axis, E_{X0C} is the most minor position error. For the roller bearing version, there is no asymptotic but a linear behavior increasing with rotational speed, reaching a maximum error of $-1.1\,\mu$m. In contrast, the hydrostatic version shows a constant offset of approximately $-0.14\,\mu$m, corresponding to a reduction of 88%. For E_{Y0C}, the magnitude of both bearing principles is proportional to the rotational speed but possesses a different behavior. The roller version possesses a PT1 behavior with a maximum magnitude of $-13.7\,\mu$m at 200 rpm. In contrast, the hydrostatic version shows again a step-shaped error with a maximum magnitude of $-0.7\,\mu$m at maximum rpm, corresponding to a reduction of 95%. The position error E_{Z0T} is the only one with a PT1 behavior for both bearing principles. The error of the roller bearing version increases linearly with rotational speed, resulting in a maximum error of 30.7 μm and a time constant of 11.4 min at 200 rpm. In contrast, the error of the hydrostatic version increases progressively with rotational speed with a maximum magnitude of 3.1 μm and a time constant of 8.6 min at 200 rpm, corresponding to a reduction in magnitude by 90% and in time constant by 25%.

The orientation error $E_{A(0Y)C}$ of the roller version shows a PT1 behavior with a magnitude proportional to the rotational speed, resulting in a maximum error of 34.4 μrad at 200 rpm. In contrast, the hydrostatic version shows no uniform error profile and possesses a maximum error of $-3.4\,\mu$rad at 200 rpm, corresponding to a reduction of 90%. A systematic error may cause this maximum error for the hydrostatic version due to the calculation method combined with the reduced time constant in the Z-direction. The orientation error $E_{B(0X)C}$ is the only error with a step-shape for both bearing principles without specific dependence on the rotational speed. For the roller version, the maximal error is 4.6 μrad, and for the hydrostatic version $-1\,\mu$rad, resulting in a reduction of 78%.

4.4 Heat Transfer Between Bearing Components

Due to the significant difference in power consumption, the tested rotational speeds allow no investigation based on equal power consumption. Therefore, the power consumption of the hydrostatic bearing at 200 rpm of (212 ± 6) W served as a reference, and the rotational speed of the roller bearing was varied to match it. This approach resulted in a power consumption of (214 ± 12) W at a rotational speed of 40 rpm. Based on this finding, seven additional warm-up and cool-down phases were measured for the roller bearing at a rotational speed of 40 rpm.

Applying equal electrical power input, the maximum increase in mean temperature of the structure T_1 is $(0.55 \pm 0.02)\,°$C for the hydrostatic bearing and $(0.53 \pm 0.07)\,°$C for the roller bearing. Based on the difference within the expanded standard deviation, the stationary cooling can be assumed identical for both bearing principles. In contrast, the mean temperature of the ZPCS T_{ZPCS} increases $(2.3 \pm 0.2)\,°$C for the hydrostatic bearing and $(6.0 \pm 0.6)\,°$C for the roller bearing, corresponding to a relative increase in temperature of the roller bearing to the hydrostatic bearing by

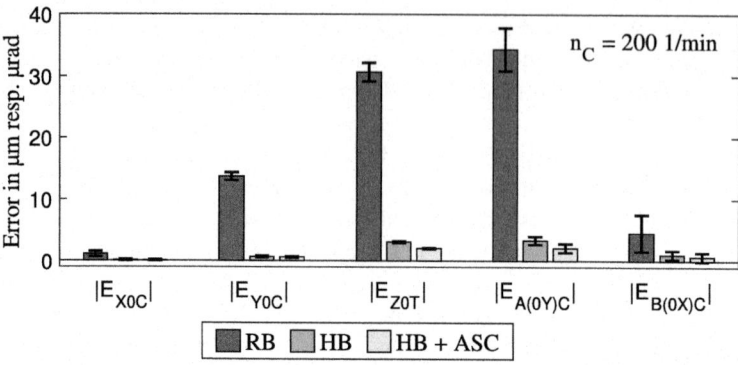

Fig. 7 Absolute values of the position and orientation errors for the rotary axis with a roller bearing (RB), hydrostatic bearing (HB) and hydrostatic bearing with active shaft cooling (HB + ASC) at the maximum tested rotation speed of $n_C = 200$ rpm at thermal equilibrium. The error bars represent the expanded standard deviation σ with a coverage factor k = 2 based on seven measurements

160%. The reduced temperature gradient of the hydrostatic bearing demonstrates the superior heat transfer between its components relative to a roller bearing and enables a reduction in position and orientation errors at equal electrical power input.

4.5 Active Shaft Cooling (ASC)

An established measure to reduce thermally induced errors in motor spindles is the application of active shaft cooling (ASC) [15]. Therefore, the hydrostatic bearing's rotor provides twelve radially symmetrical arranged cooling circuits, supplied with oil at low pressure (<0.1 MPa) with a total volume flow of three liters per minute at a temperature of (20 ± 0.1) °C. The mean temperature of the structure T_1 shows no significant difference with activated ASC, whereas the mean temperature of the ZPCS T_{ZPCS} reduces by 30%. The reduction in temperature change is also reflected in the resulting position and orientation errors. Figure 7 shows the resulting absolute errors of the hydrostatic bearing with ASC in thermal equilibrium at a speed of 200 rpm compared to the roller bearing and the hydrostatic bearing without ASC. The position error E_{Y0C} reduces by 10%, whereas all other errors decrease by 33–38% relative to the hydrostatic bearing without active shaft cooling. Based on the measurements, the ASC further reduces the temperature change of the rotating component, reducing position and orientation errors from axis rotation.

5 Conclusion

This article investigates two different bearing principles for rotary axes in terms of their capability to reduce thermoelastic errors from axis rotation. For this, two identical five-axis machine tools were compared experimentally, one using the industry standard high-precision roller bearings, and the other using shallow recess hydrostatic bearings in the rotary axis. The measurements confirm the superiority of the hydrostatic bearing, showing an error reduction of 78–95% relative to the roller bearing. Key factors for the improved thermal properties of the hydrostatic bearing compared to the roller bearing are the reduced driving power due to reduced friction and the improved heat transfer based on the oil film. In addition, the repeatability of hydrostatic bearings reduces the standard deviation of its errors, enabling further error reduction by compensation. Furthermore, experiments show the effectiveness of an additional ASC for the hydrostatic rotary axis, enabling an additional error reduction by 10–38%.

Competing Interests

Fabian A. Tripkewitz and Matthias Fritz are employees of Kern Microtechnik GmbH.

References

1. Abele, E., Altintas, Y., Brecher, C.: Machine tool spindle units. CIRP Ann. (2010). https://doi.org/10.1016/j.cirp.2010.05.002
2. Gebhardt, M., Schneeberger, A., Weikert, S., Wegener, K: Thermally caused location errors of rotary axes of 5-axis machine tools. Int. J. Autom. Technol. (2014). https://doi.org/10.20965/ijat.2014.p0511
3. Zimmermann, N., Büchi, T., Mayr, J., Wegener, K.: Self-optimizing thermal error compensation models with adaptive inputs using group-LASSO for ARX-models. J. Manuf. Syst. (2022). https://doi.org/10.1016/j.jmsy.2022.04.015
4. Gebhardt, M., Mayr, J., Furrer, N., Weikert, S., Knapp, W.: High precision grey-box model for compensation of thermal errors on five-axis machines. CIRP Ann. (2014). https://doi.org/10.1016/j.cirp.2014.03.029
5. Mayr, J., Egeter, M., Weikert, S., Wegener, K.: Thermal error compensation of rotary axes and main spindles using cooling power as input parameter. J. Manuf. Syst. (2015). https://doi.org/10.1016/j.jmsy.2015.04.003
6. Gao, W., Ibaraki, S., Donmez, A.: Machine tool calibration: measurement, modeling, and compensation of machine tool errors. Int. J. Mach. Tools Manuf. (2023). https://doi.org/10.1016/j.ijmachtools.2023.104017
7. Gebhardt, M., Ess, M., Weikert, S., Wegener, K.: Phenomenological compensation of thermally caused position and orientation errors of rotary axes. J. Manuf. Process. (2013). https://doi.org/10.1016/j.jmapro.2013.05.007
8. Wegener, K., Weikert, S., Mayr, J.: Age of compensation: challenge and chance for machine tool industry. Int. J. Autom. Technol. (2016). https://doi.org/10.20965/ijat.2016.p0609
9. Schellekens, P., Rosielle, N., Vermeulen, H., Vermeulen, M., Wetzels, S., Pril, W.: Design for precision: current status and trends. CIRP Ann. (1998). https://doi.org/10.1016/S0007-8506(07)63243-0

10. Wegener, K., Mayr, J., Merklein, M., Behrens, B-A., Aoyama, T., Sulitka, M., Fleischer, J., Groche, P., Kaftanoglu, B., Jochum, N., Möhring, H.C.: Fluid elements in machine tools. CIRP Ann. (2017). https://doi.org/10.1016/j.cirp.2017.05.008
11. Chen, D., Bonis, M., Zhang, F., Dong, S.: Thermal error of a hydrostatic spindle. Precis. Eng. (2011). https://doi.org/10.1016/j.precisioneng.2011.02.005
12. ISO 230-7:2015-05: Test Code for Machine Tools—Part 7: Geometric Accuracy of Axes of Rotation. International Organization for Standardization, Geneva (2015)
13. ISO 10791-2:2023-04: Test Conditions for Machining Centres—Part 2: Geometric Tests for Machines with Vertical Spindle (Vertical Z-Axis). International Organization for Standardization, Geneva (2023)
14. Ess, M.: Simulation and Compensation of Thermal Errors of Machine Tools. ETH Zürich (2012)
15. Denkena, B., Bergmann, B., Klemme, H.: Cooling of motor spindles: a review. Int. J. Adv. Manuf. Technol. (2020). https://doi.org/10.1007/s00170-020-06069-0

Geometric Thermal Error Compensation Using Subassembly Models Enhanced by Adaptive Learning Control

Christian Naumann⬤, Philipp Klimant⬤, Sebastian Lang⬤, Josef Mayr,
Markus Bambach⬤, Konrad Wegener⬤, Christoph Habersohn⬤,
and Friedrich Bleicher⬤

Abstract Thermal errors remain the most difficult type of manufacturing inaccuracy of cutting machine tools due to the challenges in predicting or avoiding them. Thermal compensation using joined submodels for all relevant machine subassemblies reduces the complexity for all submodels because they handle smaller, simpler geometries and less heat sources/sinks. Model training can be done using simulations, where the thermal errors of each subassembly can be computed. With enough training data, simple regression models suffice to predict the thermal error at the subassembly

C. Naumann (✉) · P. Klimant
Fraunhofer Institute for Machine Tools and Forming Technology IWU, Chemnitz, Germany
e-mail: christian.naumann@iwu.fraunhofer.de

P. Klimant
e-mail: philipp.klimant@iwu.fraunhofer.de

P. Klimant
Fraunhofer IWU, Professorship Virtual Technologies, Hochschule Mittweida–University of
Applied Sciences, Mittweida, Germany

S. Lang · J. Mayr · K. Wegener
Inspire AG, Zurich, Switzerland
e-mail: sebastian.lang@inspire.ch

J. Mayr
e-mail: josef.mayr@inspire.ch

K. Wegener
e-mail: konrad.wegener@inspire.ch

M. Bambach
ETH Zurich, AMLZ, Zurich, Switzerland
e-mail: mbambach@ethz.ch

C. Habersohn · F. Bleicher
TU Vienna, IFT, Vienna, Austria
e-mail: habersohn@ift.at

F. Bleicher
e-mail: bleicher@ift.at

201

K. Wegener and M. Bambach (eds.), *4th International Conference on Thermal Issues in
Machine Tools (ICTIMT2025)*, Lecture Notes in Production Engineering,
https://doi.org/10.1007/978-3-032-01194-7_14

level. This method is demonstrated on a machine tool and validated using measurement data. The main drawbacks of the geometric compensation method are that it is difficult to train and optimize the subassembly models from measurement data of a specific machine. To solve this issue, the residual error is predicted with adaptive learning control using ARX models, which are trained from thermal measurements and enable the overall model to overcome differences between simulation and real machine. They also allow the model to adapt to changing thermal conditions and untrained thermal load cases, thereby increasing the overall accuracy and robustness significantly. This showed a reduction of the volumetric root mean square error from 44 to 7 μm. One final issue is the integration of thermal compensation models into the machine tool control. The paper describes different methods of realizing the control integration and challenges of obtaining real-time thermal position offsets.

Keywords Thermal error · Error compensation · Machine tool precision · Regression analysis · Artificial neural network · Thermal adaptive learning control

1   State of the Art in Thermal Error Compensation

Thermal errors in cutting machine tools refer to relative displacements between cutting tool (tool center point—TCP) and workpiece as a result of thermal influences. Thermal influences may, e.g. be waste heat from the machine or the cutting process (electric losses, friction, etc.) or ambient temperature changes. Even after decades of research, thermal errors continue to reduce the machining precision and can lead to reduced productivity or even scrap, as was indicated by an industrial survey performed by Bräunig et al. in 2018 [1].

1.1   Methods for Thermal Error Reduction

An effective method for reducing thermal errors during machining is thermal error compensation, which involves the measurement or model-based prediction of the thermal error with a subsequent application of a corresponding positioning offset in the machine tool control.

One recent example of correlative compensation is a model based on Gaussian process regression developed by Wei et al. [2], which expands on work of Zimmermann et al. from 2020 [3]. It proved to be superior to ridge regression, principle component regression (PCR), thermal autoregressive with exogenous input (ARX) and long short-term memory (LSTM) neural network models, when tested under different working conditions in an investigation on a Victor Vcenter-55 three-axis machine tool. In another work by Chiu et al. [4], a standard neural network with hyperparameter optimization and correlation-based input variable selection to minimize collinearity was successfully employed to reduce the thermal error of an AWEA VP-2012 gantry machine tool by between 30 and 90%.

One of the early examples of a phenomenological model is the use of first and second order time-delay transfer functions. Brecher et al. [5] have used such a model to deal with both internal heat sources from spindle and three kinematic axes and from the ambient temperature via measurement-based model training and thereby achieved an error reduction of more than 80%.

The adaptive learning control using ARX models (TALC) developed by Blaser [6] also falls into this category. It uses a specially developed on-machine measurement cycle to achieve self-adaptability of the compensation model parameters. Mares et al. [7] used a similar ARX model as a basis and updated it for untrained thermal load cases by adding transfer functions. In their investigation, this improved the prediction accuracy measured by root mean square error (RMSE) from 56 to 92% without needing to change the original model parameters.

Online-capable simulation models, which are most often model order reduced (MOR) FEM simulations, can likewise be suitable for thermal error compensation. This requires the accurate modelling of the geometry, the prevailing boundary conditions along with a good parametrization of the model, as is described by Ess [8]. Thiem et al. [9] use such models in their structure model-based correction and have developed a method for updating the model parameters to account for changing ambient conditions or machine tool wear. Lang et al. [10] introduced a Kalman filter to estimate the thermal state of a reduced order model directly, which allowed the physics-based compensation of the thermal error with only a few temperature sensors.

Baum et al. [11] use integrated deformation sensors (IDS), essentially long CFRP bars with a 1D displacement sensor, to measure the lengthening and bending of large machine assemblies combined with a geometric-kinematic model to compute the corresponding TCP displacement. In an investigation on a three-axis vertical machining center, this enabled them to reduce the thermal error by over 85% in the most critical direction. The IDS concept used by Baum et al. is derived from a measurement system introduced by Biral et al. in 2006 [12].

Overall, there are numerous highly effective compensation methods for reducing the thermal error of machine tools under variable working and ambient conditions. Most of them are, however, still limited in their prediction accuracy when several more complex scenarios overlap. Correlative models, e.g., usually have trouble extrapolating beyond the trained thermal load cases. Phenomenological models have difficulties when internal heat sources/sinks with changing loads overlap with changing ambient conditions. Methods such as TALC can then use measurements to recalibrate the model, but this is still limited and it interrupts the production process. Simulation-based methods require a great effort to correctly parametrize them and many heat transfer and convection parameters can only be estimated roughly, which limits the overall accuracy of these models. Using IDS is a very robust method but not all machine assemblies allow for their usage. Aside from these general considerations, there are many different types of machine tools in terms of heat sources and sinks (types and locations), kinematics, process types, etc. which make some compensation strategies more effective than others.

1.2 Strategy and Structure of the Paper

A strategy, which has proven to be effective in many fields, is to break down complex problems into smaller, more manageable components. In terms of thermal error compensation, this can be done by calculating the errors of smaller subassemblies individually and then adding up the resulting estimations. Each subassembly has far fewer relevant heat sources and sinks and therefore an overall simpler thermo-elastic behavior. While the idea is simple, its implementation has a number of challenges. The first question is how to split the machine tool into subassemblies and how many partitions are sensible. Then, the perhaps largest problem is in obtaining data on the thermo-elastic behavior of individual assemblies. Finally, most models require some kind of measurement-based optimization or periodic recalibration in order to respond to changes in ambient conditions, from machine tool wear or simply due to untrained load scenarios. This raises the question of how to achieve this optimization while using such a composite compensation model.

The following chapters will answer these questions and demonstrate the effectiveness of the suggested approach, which was first introduced as a concept in 2024 [13]. This current work expands the initial paper by a more detailed description of the methodology, a thorough validation, a secondary model optimization step and some concepts for control integration.

Chapter 2 describes the geometric error compensation using assembly submodels and validates this method first on simulation data and later on machine tool measurements. Chapter 3 describes the thermal adaptive learning control, which uses machine tool measurements to first train and then intermittently check and recalibrate an autoregressive model with exogenous input. This method will be used to significantly reduce the residual thermal error and compare the stacked approach to a direct application of a TALC model. Chapter 4 investigates different methods for the control integration of thermal error compensation, specifically for machine tools with a Siemens CNC control. Finally, Chapter 5 concludes with a critical assessment of the new compensation method and an outlook on the next steps in its further development and validation.

2   Geometric Thermal Error Compensation Using Subassembly Models

Modern machine tools have a large number of heat sources and heat sinks. For this paper, a DMU 80 eVo from DMG MORI, shown in Fig. 1, is used to demonstrate the suggested methodology. The DMU 80 eVo has five axes, three linear axes and a swivel rotary table to enable full 5D milling. This means, that there are (at least) five motors, six linear guides with twelve carriages, three ball screw drives (BSDs), each with nuts and bearings, gear box and bearings for the swivel rotary table, cooling systems for motors, guides and some of the bearings and numerous auxiliary systems for tool changing, chip removal, cutting fluid supply, electricals systems and more. All of these components have a thermal footprint with different heat flows and different activation times. Moving further along the thermo-elastic effect chain, the resulting temperatures within the machine tool are shaped by the geometry of each component, its thermal material properties and the connection between adjacent components. This demonstrates the complexity of modelling the entire thermo-elastic behavior of a machine tool.

Much simpler would be to deal with only a single part or assembly of a machine tool, such as the machine base or one of the linear slides. This is the idea behind the suggested geometric compensation method and is explained and tested in the subsequent sections. While this does not eliminate the need to deal with all of the mentioned heat sources and sinks and their interaction, it simplifies the task of model parametrization by focusing on small, mostly contained subsystems. This complexity reduction comes from adding the kinematic structure to the otherwise black-box regression model and also includes some modelling error as a trade-off for the model

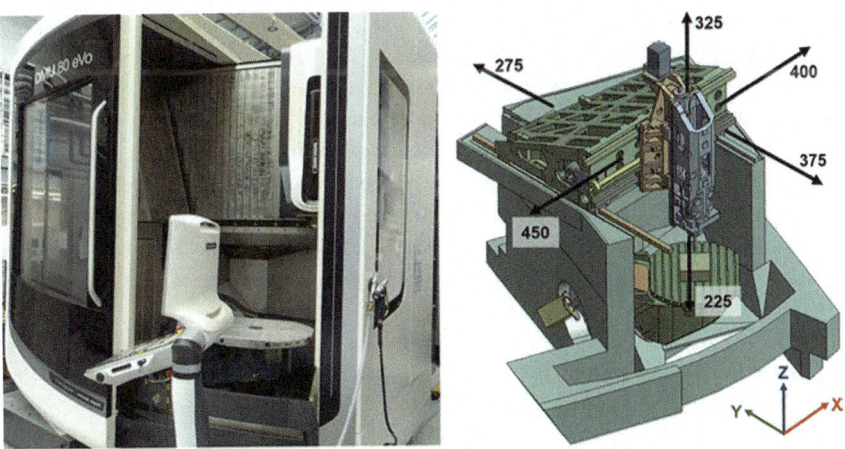

Fig. 1 DMU 80 eVo; left: photo machine tool, right: CAD model without housing, with workspace dimensions in [mm] (from [14], licensed under CC-BY 4.0)

simplifications. Another benefit is that this approach reduces the number of training simulations required for model parametrization.

2.1 Model Structure and Training Data Generation

As mentioned, the goal is to break down the thermal error to the subassembly level. First, the type of model to be used for these components must be determined. The introduction has outlined the most common types of compensation models. Simulation-based compensation has no need for submodels. Phenomenological models work well, but they would need to be trained from simulation data instead of the more common method of experimental parametrization. Compensation with integrated deformation sensors (IDS) can be highly effective, if enough sensors are installed. Due to the required cost and space for their installation, there are very few commercial machine tools equipped with IDS. The focus is therefore on data-driven methods such as regression models and neural networks. These require sensor data input, like temperatures or control data on motor currents and cooling systems.

The next problem is how to divide the machine tool and how many subdivisions are needed. Here, the general rule should be to divide only as few times as necessary. The reasons are that more submodels require more effort for training data generation and model training and that in the end, all residual errors of the subassembly error prediction add up, which may either cancel out some of those errors or it may lead to a much larger total prediction error. Beyond this, the subdivision depends on the machine tool kinematics and the location and number of heat sources and sinks. The best strategy is therefore to try the coarsest subdivision and further subdivide large assemblies if no reliable compensation model can be found. For the DMU 80 eVo, the first attempt could be made with a submodel for the swivel rotary table and one for the column with the three linear slides. For the table, a further subdivision into swivel console and rotary unit could be considered. For the column, the subdivision into base and the three linear slides would be the next plausible step.

Obtaining training data, particularly the subassembly deformations, whether for the suggested regression-based models or, e.g., phenomenological models is difficult and costly to obtain from measurements. If there is just one subdivision into table and column, this may be possible and worth considering, especially for other machine tools without a housing, where direct measurement of component displacements is possible.

For the DMU 80 eVo, this is, however, too impractical. Instead, model training from simulation data is the most effective method. The basis for the following is a functioning, parametrized and validated FEM simulation model capable of computing temperature and deformation fields of the machine tool for different thermal load cases. The necessary steps for creating such a model for the DMU 80 eVo have been described by Geist et al. [15]. Using this ANSYS model, additional deformation probes can be added and subsequently exported from the thermo-elastic

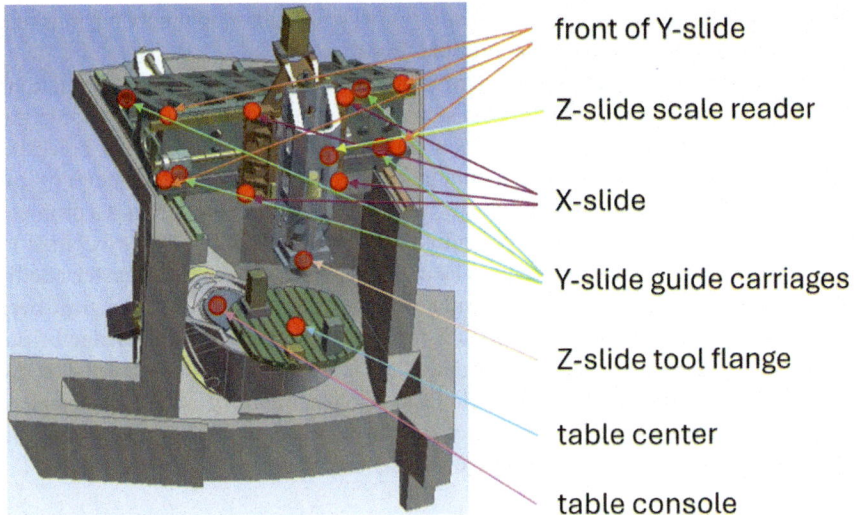

front of Y-slide

Z-slide scale reader

X-slide

Y-slide guide carriages

Z-slide tool flange

table center

table console

Fig. 2 Locations for deformation probes on the DMU 80 eVo for Z-error calculation; dark spheres show visible locations and opaque spheres are hidden behind solids

simulation. These probes can be used to calculate the displacement at interface locations between the defined machine tool assemblies. Different sets of probes may be necessary for different error components. Particularly for the linear axes, the location and effect of linear scales must also be taken into account. Figure 2 shows the probes used for the Z-displacement of the DMU 80 eVo.

Some error components may require linear interpolation of the displacement values from several probes, which can also enable the component error calculation for different axis positions. To give an example for this, consider the Z-error for the Y-slide. This error component describes the vertical deformation of this subassembly from the linear guide carriages of the Y-axis to the guide carriages of the X-axis. If the Y-slide warms up more strongly on the left than on the right, e.g. due to the X-motor being located on the left or due to many machine tool movements on the left side of the X-axis, then the left side of the Y-slide also expands more than the right side. Therefore, the Z-error of the Y-slide can depend on the X-axis position. To account for this, the Z-error of the Y-slide is calculated from four displacement probes located near the four end points of the X-axis linear guides, see Fig. 2. The Z-error should be linearly interpolated between the average Z-error of the left two probes and the average Z-error of the right two probes for the current X-axis position. Adding two more probes in the middle of the X-axis would allow to account for bending of the Y-slide also.

It was mentioned above, that the division into subassemblies should be started from a very coarse (or no) splitting, tested and then refined iteratively as needed. While this is sensible, the maximum reasonable number of subdivisions should be devised from the very start, so that all the simulation data, which is needed for model computation, already includes all necessary deformation probes for subsequent subdivisions and simulations do not need to be run multiple times.

The introduction mentioned the complexities of using thermo-elastic simulations for error compensation, so it may seem strange that these same simulations are being used here as the foundation of this new compensation approach. There is, however, a large difference between using a simulation offline to train a phenomenological model and using the simulation online by itself. Correct parametrization is a problem both need to deal with, although the online simulation is much more susceptible to parametrization errors like inaccurate heat loads or convective heat transfer coefficients. More importantly, in online simulations, the prediction errors often grow over time, as small inaccuracies in the parameters and an incomplete knowledge of the thermal machine tool history shift the simulated temperature field ever further away from the actual current temperature field inside the machine tool. When simply using a simulation to train a phenomenological model, it is acceptable to define a specific self-contained load case with a specific starting condition.

2.2 Model Training and Validation on Simulation Data

In order to test the feasibility of the composite compensation approach, the multi-section thermal load case shown in Table 1 was simulated. The table gives the axis and spindle speeds in percentages of their maximum values during constant cyclic movements. Standby also contains small motor losses in position-control mode. The resulting machine tool temperatures for this three-part load profile, as simulated in ANSYS Mechanical, are shown in Fig. 3.

The simulated thermal error for the load profile is shown in Fig. 4, where the error has already been separated into the component thermal error contributions from the machine bed, the linear slides, the linear scale for the Z-axis and the swivel rotary table. From these subassembly errors, the total relative error is computed

Table 1 Simulated load case description

N°	Axis movement	Spindle	Cutting fluid
1a	Y 75%	100%	OFF
1b	Y 75%	100%	ON
1c	Standby	Standby	OFF
2a	Z 75%	100%	OFF
2b	Z 75%	100%	ON
2c	Standby	Standby	OFF
3a	X 30% Y 30% Z 30% B 30% C 30%	100%	OFF
3b	X 30% Y 30% Z 30% B 30% C 30%	100%	ON
3c	Standby	Standby	OFF

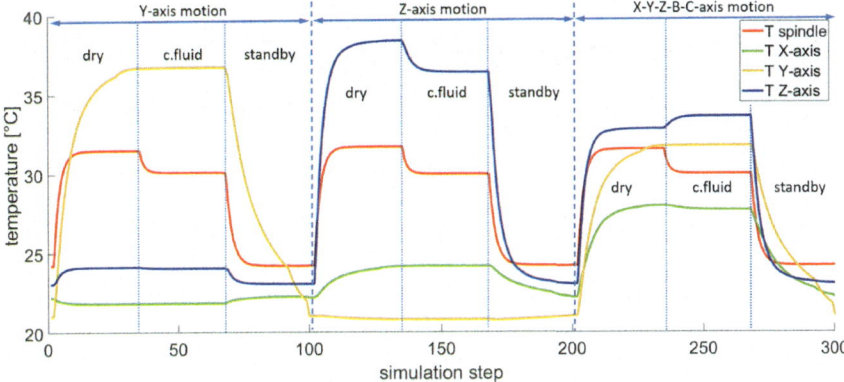

Fig. 3 Simulated temperatures for specified load case

by subtracting the table error from the sum of all the other (column) component errors. Note, that the linear scale is omitted from this. While the subassembly errors themselves may need to be adjusted for different axis positions, once they have been computed, simple addition suffices to obtain the total error.

Most of the error components have relatively simple correlations between temperature and deformation, though in some cases there are delays between temperature change and corresponding thermal deformation change. Therefore, a hybrid model was used, which combines first order time delay functions for the temperature sensors and piecewise multi-linear characteristic diagrams for mapping the time-shifted temperatures onto the deformations. The optimal time delay coefficients were determined by iteratively testing values via the interval bisection method and comparing the resulting covariance between time-delayed temperature and deformation (component thermal error). Since there are only eight temperature sensors in

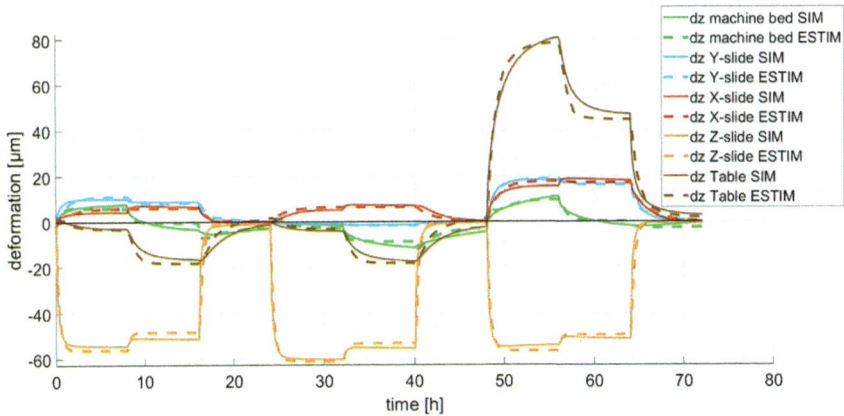

Fig. 4 Simulated subassembly z-error contributions for specified load profile

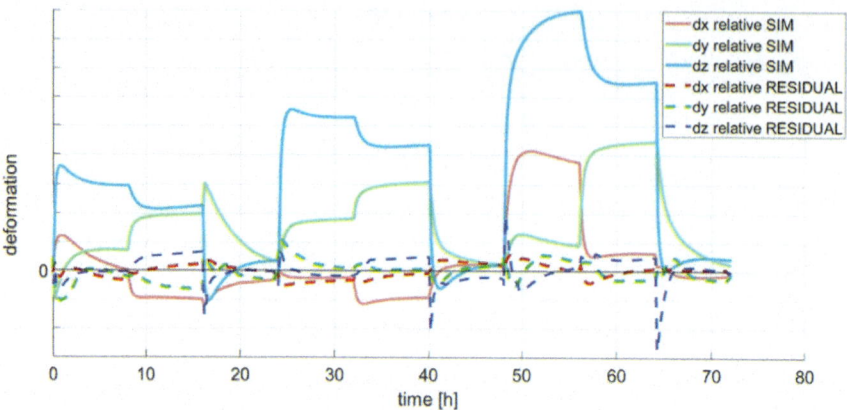

Fig. 5 Relative total simulated errors (solid) versus residual errors after compensation (dashed) for specified load profile

total and a manageable set of training data, the input variable selection for the characteristic diagram was done by testing all combinations. Figure 5 shows the total relative TCP error and the residual error after compensation, showing a significant RMSE reduction of over 80%.

2.3 Model Validation on Measurement Data

In order to test the effectiveness of the new compensation method for a non-trained, production-relevant load case, a 15 h measurement with changing axis and spindle loads was performed on the DMU 80 eVo. The load case comprises nine sections, with eight simulated single-part production load sections and a final cool-down phase, see Fig. 6.

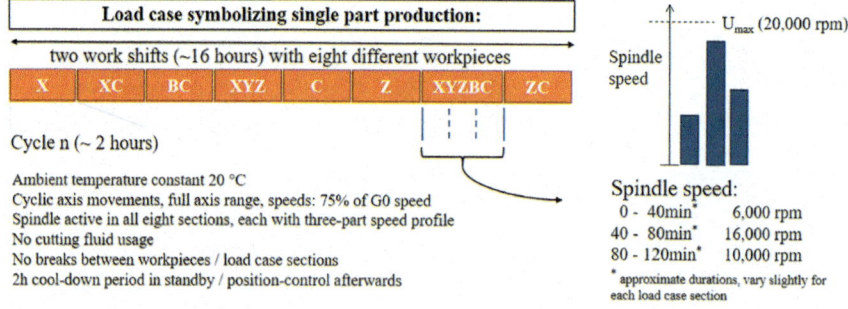

Fig. 6 Load case description for the test load case

Figure 7 shows the recorded temperature profile for the test load case and Fig. 8 the corresponding sensor locations as well as the position for the displacement measurement (measurement point MP2.1).

The measured relative TCP displacement for the test load case is shown in Fig. 9.

Applying the developed compensation model on this dataset in an offline test has provided the error reduction shown in Fig. 10. The relative TCP error shown in Fig. 10 was reduced, as measured by the RMSE, by 50%, 39% and 76% for the X-, Y- and Z-direction, respectively. Further improvements are possible by improving the simulation parameters (e.g. heat source loads and convection coefficients) used

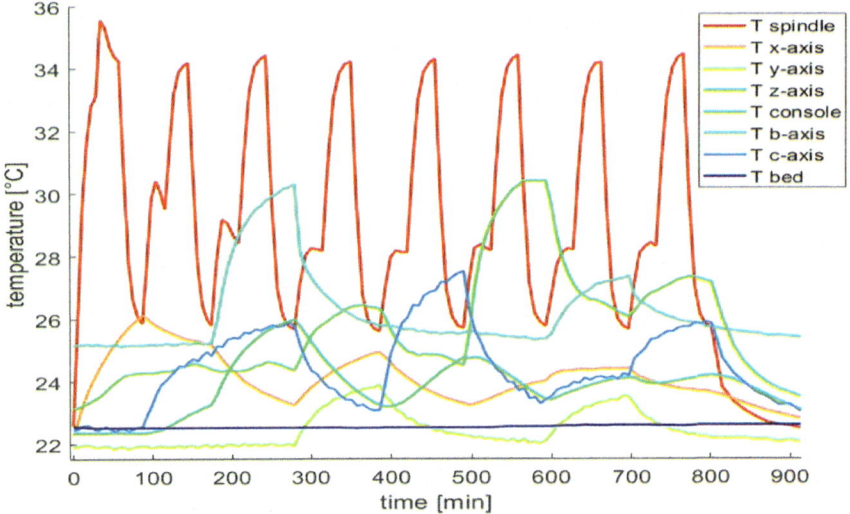

Fig. 7 Measured temperature profile for the test load case

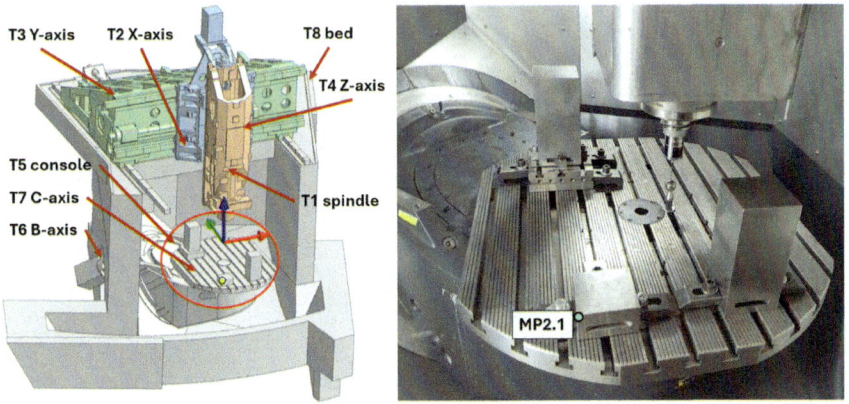

Fig. 8 Temperature sensor positions on the DMU 80 eVo and measurement point MP2.1

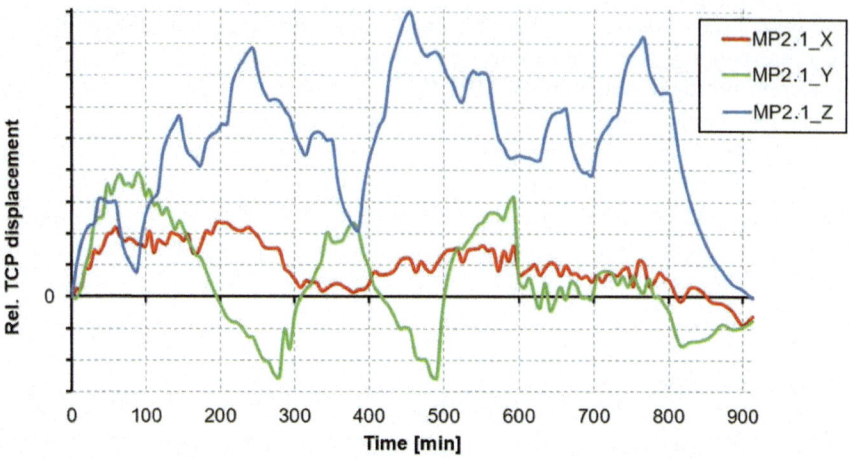

Fig. 9 Measured thermal error for the test load case at MP2.1

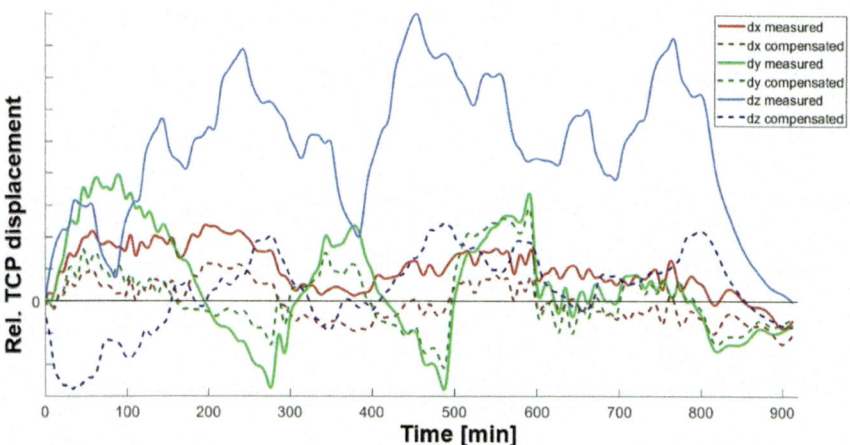

Fig. 10 Measured versus compensated thermal error using the composite model at MP2.1

to train the compensation model and also by adding more training simulations for an expanded training database.

3 Thermal Adaptive Learning Control

This section introduces the data-driven TALC model architecture [6] that is used to first compensate for the measured thermal error directly and then alternatively in a stacked approach to compensate for the residual error from the assembly models.

3.1 TALC Compensation Principle and Model Structure

To predict thermal errors in machine tools using data-driven models, ARX models are very suitable and thus applied here. Since thermal errors can be described by a linear time-invariant dynamic system, they are represented as a weighted sum of current and past outputs, as well as current and past inputs. This allows for a simple consideration of thermal hysteresis within the model. Formally, the model is given by

$$y[k] + \sum_{i=1}^{n_a} a_i y[k-i] = \sum_{m=1}^{M} \sum_{j=1}^{n_b} b_{j,m} u_m[k-j]$$

where $y[k]$ is the output at time step k, such as one specific thermal error for example in X-direction. This means, that there are N of those multiple-input–single-output (MISO) models, where N is the number of thermal errors that can be measured and compensated. $u_m[k]$ represents an input, in this case a temperature sensor measurement at time step k, of which there are in total M sensors selected for this model. The model parameters a_i and $b_{j,m}$ are fitted using weighted least squares on the training data and subsequent recalibrations. The order $n_{a,max}$ defines how many past outputs are considered, while $n_{b,max}$ specifies how many past inputs could influence the current time step. These maximum values are set as hyperparameters, while the specific values within the defined maximum are chosen for each input m specifically using the Akaike Information Criterion (AIC) [16]. The selection of model inputs involves K-means clustering (with the number of clusters defined by the Calinski-Harabasz Criterion [17]) followed by cross-correlation analysis to prevent collinear model inputs.

Figure 11 shows how such an ARX model can be integrated in the TALC structure. Periodic on-machine measurements check if the residual is within the defined action control limit (ACL), set at 10 μm. 10 μm is an arbitrary value, but has proved to be a good compromise between the frequency of necessary model updates and the achievable precision and is also an acceptable limit for this type of machine tool. If the residual exceeds its ACL, a not-good (NG) mode is activated. In this NG mode, measurements are gathered for half an hour and used to update (retrain) the compensation model.

3.2 Compensation of the Residual Thermal Error

The introduced ARX model can either be applied directly to the measured thermal error or alternatively to the residual thermal error of the subassembly models (introduced in Sect. 2) in a stacked approach. Figure 12 shows the results of the application of the ARX model generated for this stacked approach for the Y-error and also includes the model inputs used. I.e., the measured Y-error is the compensated

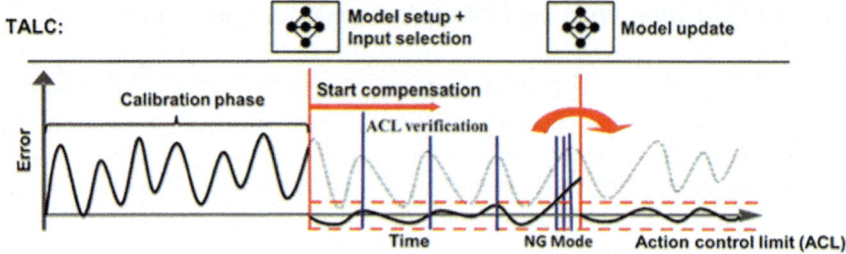

Fig. 11 TALC approach of model fitting with retraining when the residual error exceeds the ACL during verification measurements [18]

error after the subassembly model was already applied (see Fig. 10) and is used to train the ARX model for the first 10 hours. The Y-error is shown as an example and was chosen, because it is relatively large compared to the uncompensated error. It can be seen in Fig. 12, that the error can be reduced further than the residual of the subassembly model. In total, there are two NG modes, where the residual of one of the errors exceeded the ACL of 10 μm.

The training error (data from 0 to 10 h) is very low, indicating that the inputs explain most of the error well. The 10-h training period, however, is rather short, as unseen behavior still occurs afterwards, requiring a model retraining.

Figure 13 shows the volumetric error for all tested compensation models as well as the measured uncompensated error. The volumetric error is the Euclidean distance of the X-, Y- and Z-error allowing for one metric to judge the accuracy of the predictions and the machine tool in general.

Table 2 shows the RMSE for all tested models by axis and volumetric error. For the TALC model, only the validation data is used, which is indicated in Table 2 by a *. Thus, no data from the initial training phase or during the NG modes is considered in the RMSE calculation. It can be seen that the machine is exposed to significant thermal effects leading to an average volumetric RMSE of 44.3 μm, of which the Z-error dominates.

The subassembly model already achieves a significant error reduction of 69%. The TALC models, however, achieve even better results. The volumetric RMSE reduction of the direct TALC model of 81% is higher than that of the subassembly model. The stacked TALC reduces the error of the subassembly model by a further 56% leading to an overall error reduction of the stacked model by 85%. This is achieved by one less NG mode compared to the direct TALC, which is why the reductions cannot be compared one to one, as the direct TALC model uses slightly more training data while performing slightly worse, indicating the benefit of the stacked subassembly and TALC approach leading to the highest total thermal error reduction.

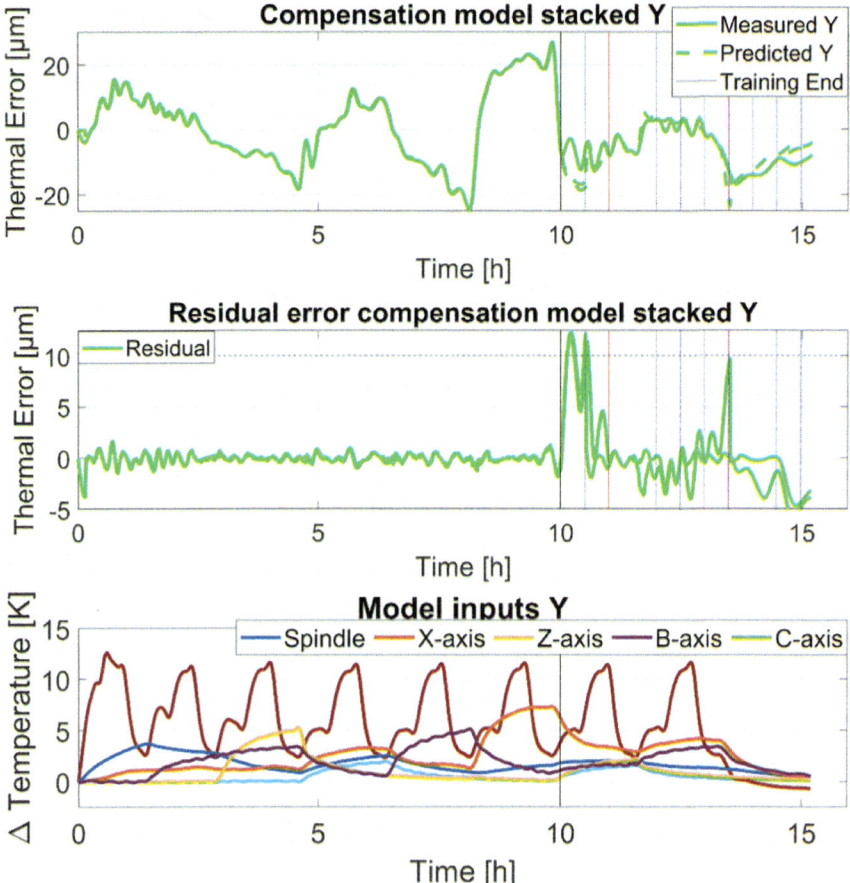

Fig. 12 Thermal error in Y-direction compensated by the stacked approach (subassembly model and subsequent TALC) and the TALC model inputs in the bottom graph

4 Methods for Control Integration of Thermal Error Offsets

One main task of the CNC system of a machine tool is the trajectory planning of the cutting tool relative to the workpiece. Therefore, an interpolator calculates the tool path in advance depending on motion parameters like feed, acceleration or jerk limitations, as well as geometrical parameters like tool length, diameter and orientation. Beside these input requirements, modern NC systems have implemented algorithms to compensate for geometric errors of the mechanical structure. In current systems, these compensation algorithms work with static values, which are parametrized during commissioning or maintenance operations and are not adapted to the present machine tool condition, such as its current thermal state.

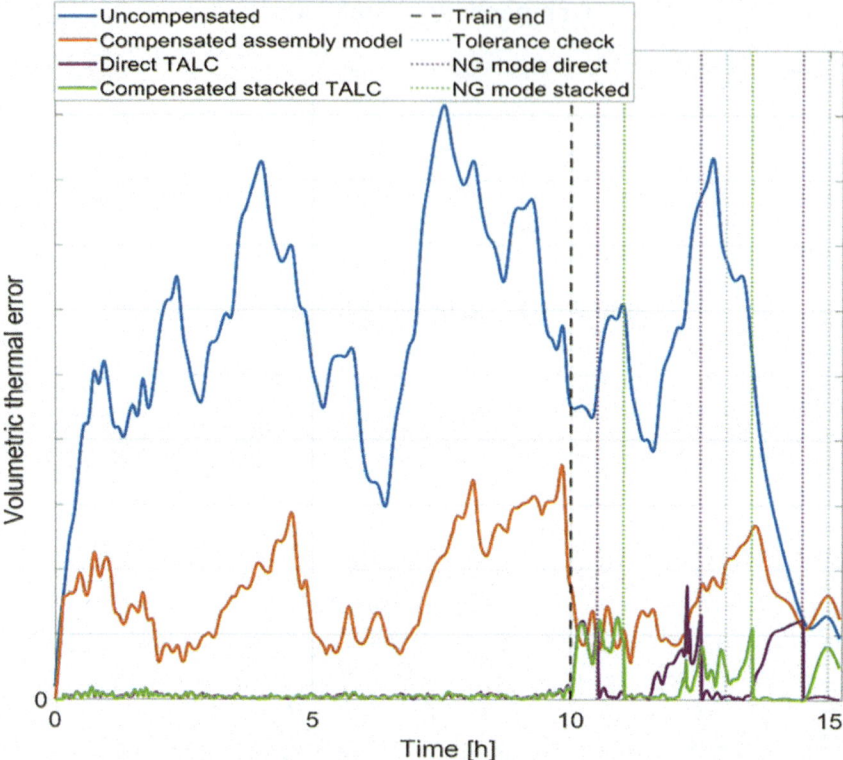

Fig. 13 Volumetric errors for the uncompensated case, after compensation with the subassembly model, after direct TALC compensation and after compensation with the stacked model

Table 2 Reduction of RMSE by different models

Model	X [μm]	Y [μm]	Z [μm]	Volumetric [μm]
Uncompensated	12.3	17.0	52.9	56.9
Subassembly model	6.1	10.4	12.7	17.5
TALC	4.0*	4.7*	6.3*	8.5*
Assembly model + TALC	5.2*	4.9*	1.6*	6.8*

The main task of an adaptive geometric compensation system based on the thermal condition of a machine tool is to provide the path generator with proper compensation offsets. Due to the look-ahead function of the interpolator, which calculates the tool path in advance, and the latency of the field bus systems, a real-time calculation of the thermal deviation of the current position and playback of the compensation by an external device is not possible. Also, a more complex calculation of the momentary compensation within the NC kernel is limited due to the calculation power and system integration possibilities.

One solution to overcome these limitations is to provide the NC control with a complete compensation matrix, which provides the compensation of the whole workspace and not just the momentary TCP position. The NC kernel then selects and interpolates the compensation values depending on the actual TCP location from the position encoders. Additionally, the compensation matrix can be updated in suitable time intervals from an external device, which can perform more complex calculations than the NC system and has less restrictive real-time requirements. The machine transmits, e.g. via the logic control (PLC), the measured temperatures to the external devices and after compensation matrix calculation, the updated matrix entries back to the NC. The slower matrix updates are possible due to the slow temperature field changes (high thermal inertia) of the mechanical structure.

The advantage of this kind of compensation is that several CNC systems have already implemented compensation functionality based on a matrix solution. They are mainly focused on eliminating component and assembly tolerances, but may also be repurposed for thermal error compensation. Nevertheless, it is irrelevant to these functions, for which reason a geometric deformation of a machine tool structure occurs. Figure 14 shows, how thermal error compensation using compensation matrices can be achieved using existing CNC functionality.

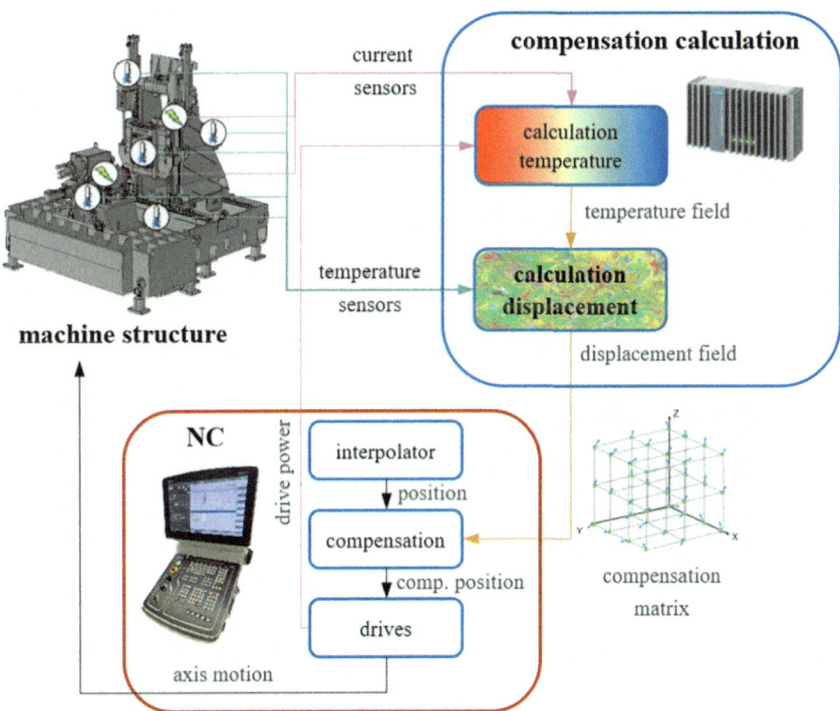

Fig. 14 Control integration concept and workflow for online thermal error compensation

The following two sections present two possible implementations for a Sinumerik One CNC using different internal compensation systems of the control unit.

4.1 MEC—Multi-error Compensation

This matrix-based compensation function represents the latest development and is still being continuously improved by Siemens. It is based on a multidimensional, axis-aligned grid for the support points, accommodating up to five base axes that can compensate up to three machine axes. Additionally, it is possible to activate and weight/scale more than one compensation table to switch between or combine different error effects.

Since Sinumerik One CNC software version Vx.25, it is also possible to update the values of the compensation tables via an NC command at runtime and therefore change the behavior of the machine tool during the running production. For the unit (e.g. external edge device or industrial PC) responsible for determining the temperature-induced deformation, this means that compensation values must be determined at all support points of the compensation matrix. This can be achieved through multiple simulation runs using identical temperature and power data at varying axis positions.

4.2 VCS—Volumetric Compensation System

This compile cycle is based on the 21-component error model according to ISO 230. By this, the real orientation of the axis system and the orientation error and deviation of each carrier/slide are defined and the total deviation of the TCP is calculated along the kinematic chain. The main advantage of this compensation is the inclusion of the kinematic chain, which also takes the tool dimension and orientation into account.

Similar to the MEC, an update of the compensation table values can be done at runtime by the NC code. For the unit responsible for determining the temperature-induced deformation, this means that rather than calculating a compensation matrix for a large set of workspace support points, axis errors (e.g. 6 per linear axis) can be calculated in accordance with the ISO 230 standard. This approach can reduce the required simulation effort necessary for training the compensation models.

4.3 Updating the Compensation Table at Runtime

The update of the compensation values can be done according to the control structure seen in Fig. 15. The right side of the schematic shows an independent system, which calculates the mechanical deviation of the machine tool structure based on the momentary temperature state and power consumption. This system can run directly on the NC or PLC, if the computations are not too complex, but it may also be an edge device, an industrial PC or even a cloud computer. The results, i.e. the computed updated compensation table, will be provided periodically to the CNC system in the form of an NC program file, which is transferred to the memory of the control unit via FTP or a mounted network drive.

In the CNC system, an NC program-dependent update of the compensation algorithm is implemented. For example, the update routine can be implemented in the tool change cycle. During tool changes, it is ensured that the tool is not in contact with a machined part and sudden micro-movements introduced by a change of the geometrical compensation have no effect on the production/workpiece. This way, the machine tool producer can implement the compensation independent of the machining program and ensure periodic updates of the compensation values. Similarly, updates could typically be safely performed after any G0 command. Additionally, a command-based update can be provided to the CAM programmer enabling enhanced machining accuracy also during long-lasting cutting processes without tool change.

The advantage of the above-introduced procedure is the use of the built-in compensation algorithm of the CNC system, which ensures performance and reliability. Also, a failure or a real-time violation of the system responsible for the prediction of the thermal error does not affect the production capability of the machine tool. Should such a failure occur, then only the update of the compensation table values would be halted and production could still continue safely, if possibly with less accuracy. A possible response to such a situation could be informing the user via a warning message and offering a manual compensation table update outside of active cutting operations.

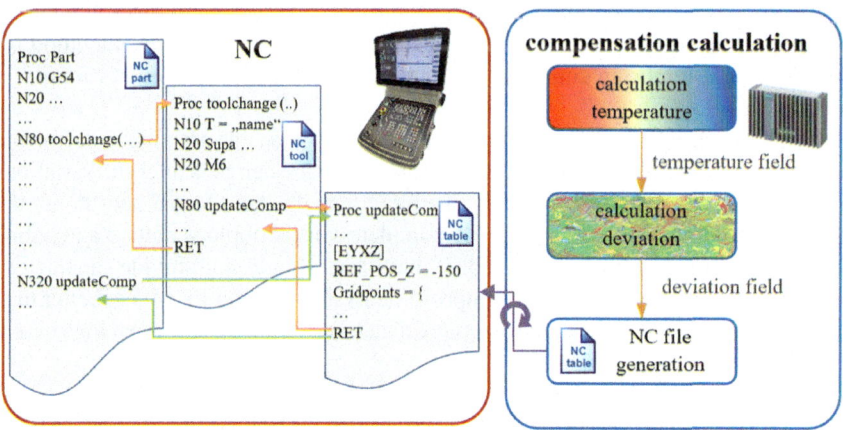

Fig. 15 Schematic for compensation table update during runtime

5   Conclusion and Outlook

Complex machine tools with a large number of heat sources and sinks still pose a significant problem for thermal error compensation methods. Building on a concept suggested in 2024, this paper improves and validates the new method for compensation using subassembly models. The paper starts by partitioning the machine tool into a minimal set of subassemblies, which is demonstrated for the DMU 80 eVo. Subsequently, training data for all subassemblies is obtained from a set of FEM simulations. The first validation used a simulation of a long, composite load case, where the compensation model achieved an error reduction of around 80%. A subsequent validation using measured test data with changing axis loads then showed a thermal error reduction of between 39 and 76%. The lower accuracy is due to imprecise simulation parameters and the untrained load cases.

In order to achieve a further reduction of this residual thermal error, the thermal adaptive learning control (TALC) was applied. It uses short sequences of measurement data to train an ARX model and recalibrates itself as needed, i.e. when the specified ACL is violated. To determine, when updates of the model are needed, periodic test measurements are performed. The TALC managed to further reduce the thermal error of the measured load case to below 20%. The TALC alone, without a previous compensation, was also highly effective with an only 4 percent higher volumetric error than the stacked method. Nevertheless, since the TALC requires production interruptions, which cost both time and energy, the use of the apriori trained subassembly model is still a good way to achieve a high basic accuracy. The TALC based on the subassembly residual error also performed slightly better and required one less model update, indicating that the stacked approach outperforms a purely data-driven approach.

The final part of the paper investigates different methods for control integration of thermal error compensation routines. Multi-error compensation and volumetric compensation, two useful approaches are described in detail. Finally, a method for control integration using compensation tables, which can be updated at runtime, is presented.

Future work will build on this method by improving the training data for the assembly compensation model, so that higher accuracy can be achieved. Variations in the ambient temperatures will also be more heavily investigated and the model tested against them. Once this is done, a complete control implementation and online test of the method will be performed. If these tests should not provide satisfactory error reductions, then the standard improvements must be employed, i.e. better data (simulation parameter optimization), more data (simulation of additional load cases) or other/additional model inputs.

Acknowledgements This research was created in part during the Cornet project GeoComp 322 EBR funded by Germany's Federal Ministry for Economic Affairs and Climate Action (BMWK) and the Österreichische Forschungsförderungsgesellschaft mbH (FFG). This work was co-financed by Innosuisse–Swiss Innovation Agency, Project no. 27835.1 IP-ENG.

References

1. Bräunig, M., Regel, J., Richter, C., Putz, M.: Industrial relevance and causes of thermal issues in machine tools. In: Conference proceedings, 1st conference on thermal issues in machine tools. Wissenschaftliche Scripten, Auerbach/Vogtland, pp. 127–139 (2018)

2. Wei, X., Ye, H., Miao, E., Pan, Q.: Thermal error modelling and compensation based on Gaussian process regression for CNC machine tools. Precis. Eng. **77**, 65–76 (2022). https://doi.org/10.1016/j.precisioneng.2022.05.008

3. Zimmermann, N., Lang, S., Blaser, P., Mayr, J.: Adaptive input selection for thermal error compensation models. CIRP Ann. **69**(1), 485–488 (2020). https://doi.org/10.1016/j.cirp.2020.03.017

4. Chiu, H.-S., Chang, C.-H., Huang, Y.-C., Lai, Y.-C., Yang, C.-J., Chen, Y.-B.: An efficient thermal error prediction model using neural networks and key temperature points for gantry machining centers. J. Mech. **39**, 529–539 (2023). https://doi.org/10.1093/jom/ufad042

5. Brecher, C., Hirsch, P., Weck, M.: Compensation of thermoelastic machine tool deformation based on control internal data. CIRP Ann. **53**(1), 299–304 (2004). https://doi.org/10.1016/S0007-8506(07)60702-1

6. Blaser, P.: Adaptive learning control for thermal error compensation. Dissertation ETH Zurich (2020)

7. Mares, M., Horejs, O., Straka, M., Sveda, J., Kozlok, T.: An update of thermal error compensation model via on-machine measurement. MM Sci. J. **2022**, 6275–6282 (2022). https://doi.org/10.17973/MMSJ.2022_12_2022150

8. Ess, M.: Simulation and compensation of thermal errors of machine tools. Dissertation ETH Zurich (2012)

9. Thiem, X., Rudolph, H., Krahn, R., Ihlenfeldt, S., Fetzer, C., Müller, J.: Adaptive thermal model for structure model based correction. In: 3rd ICTIMT 2023, LNPE Springer, pp. 67–82 (2023). https://doi.org/10.1007/978-3-031-34486-2_6

10. Lang, S., Talleri, S., Mayr, J., Wegener, K., Bambach, M.: Kalman filter-driven state observer for thermal error compensation in machine tool digital twins. Manufactur. Lett. **41**, 208–218 (2024). https://doi.org/10.1016/j.mfglet.2024.09.025

11. Baum, C., Brecher, C., Klatte, M., Lee, T.H., Tzanetos, F.: Thermally induced volumetric error compensation by means of integral deformation sensors. Procedia CIRP **72**, 1148–1153 (2018). https://doi.org/10.1016/j.procir.2018.03.045

12. Biral, F., Bosetti, P., Oboe, R., Tondini, F.: A new direct deformation sensor for active compensation of positioning errors in large milling machines. Adv. Motion Control '06 vol. 1, IEEE, pp. 126–131 (2006)

13. Naumann, C., Klimant, P., Lang, S., Mayr, J., Oborin, E., Einspieler, C., Bleicher, F., Wegener, K.: Hybrid thermal error compensation model using mixed targeted assembly error models. Euspen SIG meeting thermal issues, Eindhoven 2024 (2024). https://www.euspen.eu/resource/hybrid-thermal-error-compensation-model-using-mixed-targeted-assembly-error-models-2/

14. Naumann, C., Geist, A., Kumar, T.S., Weber, J., Steiert, C., Voigt, I., Plum, F., Thiem, X., Bertaggia, N., Glänzel, J., Weber, J., Zontar, D., Brecher, C., Ihlenfeldt, S.: Anwendungsbeispiel DMU 80 eVo. In: Thermo-energetische Gestaltung von Werkzeugmaschinen–Praxishandbuch, ed. Brecher, C., Springer, pp. 489–522 (2025). https://doi.org/10.1007/978-3-658-45180-6_29

15. Geist, A., Naumann, C., Glänzel, J., Putz, M.: Methods for determining thermal errors in machine tools by thermo-elastic simulation in connection with thermal measurement in a climate chamber. MM Sci. J. pp. 6575–6581 (2023). https://doi.org/10.17973/MMSJ.2023_06_2023049

16. Akaike, H.: A new look at the statistical model identification. IEEE Trans. Autom. Control **19**(6), 716–723 (1974). https://doi.org/10.1109/TAC.1974.1100705

17. Calinski, T., Harabasz, J.: A dendrite method for cluster analysis. Commun. Stat. **3**(1), 1–27 (1974)

18. Blaser, P., Pavlíček, F., Mori, K., Mayr, J., Weikert, S., Wegener, K.: Adaptive learning control for thermal error compensation of 5-axis machine tools. J. Manuf. Syst. **44**(2), 302–309 (2017). https://doi.org/10.1016/j.jmsy.2017.04.011

Development and Integration of Temperature Measurement on a Machining Center for Precision Machining Used for a Data Driven Thermal Distortion Compensation Considering Real Production Conditions

Christoph Habersohn, Christoph Einspieler, Nikita Nobel, and Friedrich Bleicher

Abstract Temperature-induced geometric errors in manufacturing processes are a critical challenge for ensuring high precision of machined parts. These errors are largely influenced by the thermal behavior of machine components during operation, such as machine beds, travelling columns, fixtures and spindle units. This study presents the development and integration of an in-process temperature measurement system designed to monitor the thermal conditions of a machining center during the manufacturing process operated in a real production environment. The system incorporates temperature sensors placed in key areas of the machine tool structure, capturing the temperature field that impacts the machine tool components directly. The focus is put on the system's design, including sensor placement, wiring, data acquisition, and the integration of performance data gathered from the CNC such as spindle speed, feed rates and axis drive power consumptions. The data acquisition is supported by a post-process measurement of a geometrical feature on the machined workpieces. In a first step, the collected temperature and machine performance data are analyzed to identify patterns that contribute to the predication of thermal distortions, resulting in a geometric error. These results show a significant enhancement of machining accuracy and offer a promising methodology in precision machining for improving spatial thermal process stability by enabling adaptive control based on real-time thermal compensation feedback.

Keywords Thermal error · In-process measurement · Compensation · Machine tool

C. Habersohn (✉) · C. Einspieler · N. Nobel · F. Bleicher
Institute of Production Engineering and Photonic Technologies, TU Wien, Vienna, Austria
e-mail: habersohn@ift.at

© The Author(s) 2026
K. Wegener and M. Bambach (eds.), *4th International Conference on Thermal Issues in Machine Tools (ICTIMT2025)*, Lecture Notes in Production Engineering,
https://doi.org/10.1007/978-3-032-01194-7_15

1 Introduction and Motivation

The precision of manufacturing processes strongly depends on stable geometric accuracy conditions of machine tools. One of the most significant challenges in this context is the thermal deformation of machine structures, which leads to geometric errors within kinematic chains. These thermal effects directly impact the dimensional accuracy and surface quality of machined workpieces, compromising process stability.

To assess geometric deviations, standardized evaluation methods, such as the ISO 230 for linear and angular errors in motions of machine tool components, are commonly applied. However, these assessments are typically conducted under no-load conditions, failing to capture the thermal effects that occur in real production environments.

Currently available CNC options for thermal error compensation and volumetric error correction also face limitations when dealing with transient geometric deviations caused by temporarily changing thermal influences. In general, these systems rely on mathematical models that describe the positional distance between two points in the workspace or adjust the position and orientation of coordinate systems [1]. As a result, compensation is typically limited to the machine's state at the time of delivery, without considering long-term thermal behavior. Moreover, few systems exist that continuously record temperature data over extended periods, further hindering accurate compensation under real operating conditions.

Zimmermann et al. [2] propose a self-learning thermal error compensation strategy for 5-axis machine tools, aiming to replace conventional energy-intensive cooling methods with intelligent, data-driven approaches. By automatically characterizing machine behavior and adapting compensation models, the method significantly reduces thermal errors across various load cases. Demonstrated by the machining of impellers, the approach enhances machining accuracy under fluctuating temperatures, advancing sustainable precision manufacturing.

Due to the high energy consumption of spindle cooling systems, Mori et al. [3] developed an on–off cooling method in order to achieve more efficient cooling. The method was successfully tested on two machining centers with minimal thermal fluctuations.

The influence of cutting fluid on machine tools has been studied by Mayr et al. [4], investigating the differences between dry and wet cutting. The results show a larger thermal error when cutting fluid is used during the machining process.

Wiessener et al. developed a thermal test piece for 5-axis machine tools equipped with a rotary table. This enables the visualization of thermally induced errors in the three linear axes and two angular errors at the tool center point [5].

Since the thermal influences of especially the rotary axes of 5-axis machine tools have not been studied extensively in the past, Gebhardt [6] focuses on characterizing the thermal properties of the rotary and swivel axes. The author achieved an error reduction of up to 85%.

Volumetric error modeling is a commonly used method for characterizing errors in machine tools. However, thermally induced errors are difficult to incorporate by compensation algorithms because of their time-variant and stochastic nature. Baum et al. [7] propose a new method of including thermal errors into the volumetric model of a machine tool by directly measuring structural deformations. While classical temperature measurements can only gather a limited amount of thermal influences, the structural deformations contain the results of all thermal sources. Results showed a correction of thermally induced errors of a vertical machining center from $\pm 200\,\mu m$ down to $\pm 25\,\mu m$ in the most critical axis direction.

A common method for analyzing the thermal distortion properties and errors of machine tools is the Finite Element Method. For the calculation, the heat transfer coefficient is necessary as an input. Especially for housings and enclosures, it is usually obtained by computation-intensive and therefore time-consuming CFD simulations. Pavliček [8] developed a novel metamodel allowing for a fast calculation of the heat transfer coefficients of different housings. Results show that the new method can indeed keep up with traditional CFD simulations while requiring significantly less computational efforts.

Brecher et al. [9] present a simulation model that combines the precision of geometric measurements with the dynamic adaptability of transient thermal simulations. The methodology details the processing and integration of geometric measurement data into the simulation framework to improve the predictive accuracy of thermal error compensation under real-time thermo-elastic conditions.

A major drawback of existing thermal compensation systems is the lack of long-term real production environment data, as most measurements are performed under controlled conditions in laboratory scale that do not reflect actual manufacturing environments. Conducting measurements during production often interrupts the manufacturing process, altering the thermal conditions and leading to non-representative results. To achieve an accurate compensation methodology, it is essential to correlate thermal measurement data with post-process evaluations of the machined workpieces. Furthermore, thermal influences vary significantly over extended periods due to factors such as day-night cycles, seasonal changes between summer and winter, and the gradual wear of machine components. Without continuous long-term monitoring, existing systems cannot account for these dynamic conditions, resulting in limited effectiveness for real production applications.

2 Experimental Setup–Machine Tool and Process

For the implementation of the experimental setup, a machining center FlexModul from Krause-Mauser was selected for a high precision fine boring operation equipped with a hydro-static spindle and a radial position control of the fine boring tool. It was equipped with a CNC system Sinumerik ONE and upgraded with an industrial edge device for data acquisition. The machining center features a three-axis inverse kinematics configuration where the workpiece is moved relatively to stationary spindle

 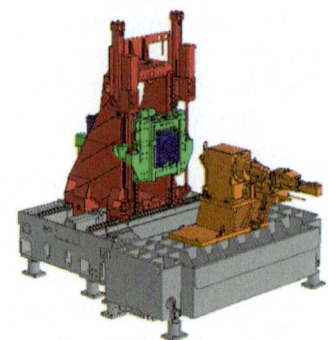

Fig. 1 FlexModul and the three axes mechanical structure

units. The workpiece setup by the used fixture is oriented vertically, i.e. in Y-axis direction. In this setup, spatial investigations are also possible by changing the fixture orientation. In total, two spindles are implemented, the hydrostatic spindle for precision boring and a conventional motor spindle for a honing operation, allowing for enhanced flexibility and precision in machining operations. This kind of machine tool is designed for mass production of automotive parts with a focus on the manufacturing of connecting rods in the presented use case.

As depicted in Fig. 1, the machine tool structure features a moving column configuration, where the workpiece is clamped on. This type of machining center offers a robust and rigid design to ensure stability during the machining process. Additionally, it is equipped with standard auxiliary systems, such as a cooling system and a chip conveyor to support efficient operation and maintain process reliability. The machining sequence is the fine boring of high-precision connecting rods for high-performance engines like race cars. Therefore, the shape and position of the piston pin bore and the crankshaft pin bore on the connecting rod are manufactured within a tolerance range of just a few micrometers.

To ensure the quality of the machined parts, a 100% inspection using an automated tactile measurement system is performed, where the critical geometrical parameters are measured and documented independent of the influence of a worker. Based on the metrological investigation, the process is manually adapted by an operator changing settings at the machine tool and parameters in the CNC-program (see Fig. 2). As a result, process adjustments take effect with a delay of at least one or a certain number of workpieces, and it is not possible to respond to changes in the current process.

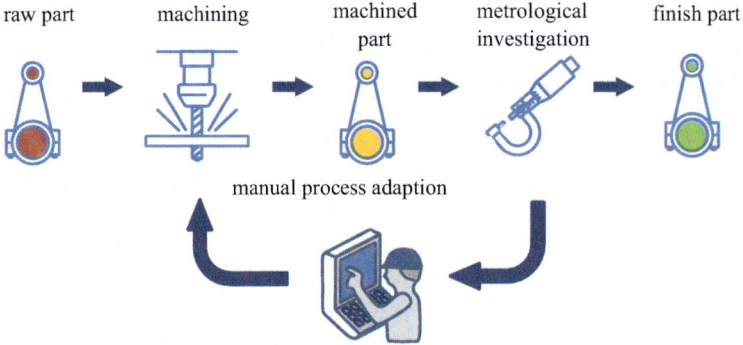

Fig. 2 Manual closed loop control strategy of connection rod production

3 Data Acquisition

The data acquisition system is designed to gather both temperature measurements and real-time data directly derived from the numerical control system, enabling a thorough understanding of the machine's operational conditions. A total of 24 PT100 temperature sensors were selected for their accuracy and reliability, and they are strategically placed at critical points within the machine tool structure. These sensor placements are chosen to capture the most relevant thermal data as depicted in Fig. 3. The sensors are positioned in friction-intensive areas, such as the guide-ways and the ball screw nut, where wear and heat generation are most prominent. Additionally, sensors are placed at significant heat sources and sinks, including the coolant temperature, which plays a vital role in temperature regulation and overall system stability.

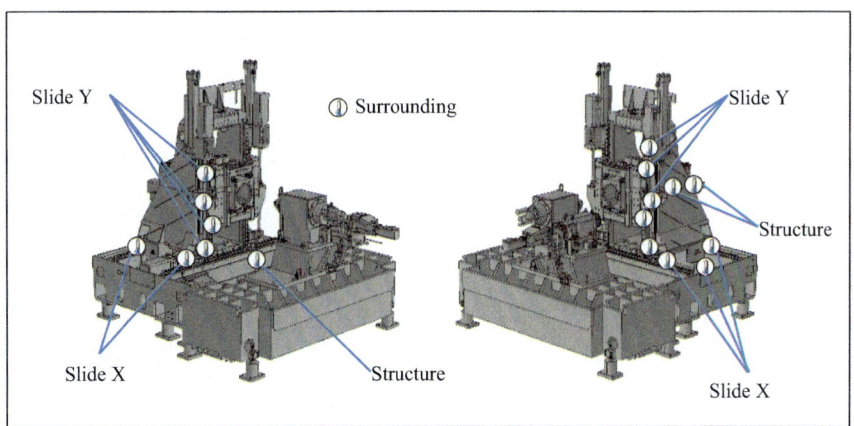

Fig. 3 Sensor positions at the machine tool kinematics

To ensure a comprehensive thermal profile of the machine, temperature monitoring is not limited to the mechanical components alone. The system also tracks ambient temperature as well as temperature changes in selected structural elements of the machine, such as the frame and supporting components. This holistic approach of temperature measurement allows for an in-depth analysis of the thermal behavior throughout the entire system, helping to identify potential thermal imbalances, heat accumulations, or cooling inefficiencies. The data from these sensors is continuously monitored and analyzed, providing valuable insights into the machine's thermal dynamics and enabling an in-process optimization strategy.

The temperature sensors are integrated into the machines PLC bus system via a Siemens ET200 bus participant (Fig. 4), which allows the seamless transmission of sensor data across the network. This data is then collected over the NC-system using a Siemens Industrial Edge device, which provides the capability to acquire low- as well as high-frequency data such as machine position and power consumption at the interpolator cycle-time (IPO time rate). The software solution employed is Siemens' software AnalyzeMyWorkpiece (Fig. 5) for edge devices, which includes a planner module for job definition and management. Within this framework, a specific job is configured to collect XYZ positional data and power measurements from seven machine axes at the interpolation cycle time including the spindles and the radial position control of the fine boring tool. This data is subsequently stored in a JSON file for further analysis and processing. Additionally, a total of 32 temperature sensor readings and R-parameters are monitored every second, providing precise in-process tracking of the system's thermal behavior. These readings are also stored in the same file, ensuring consistency and traceability.

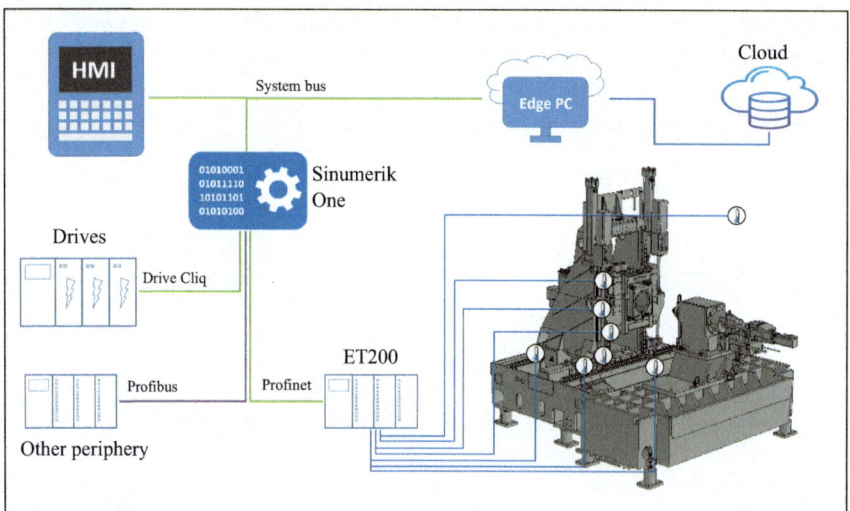

Fig. 4 Communication scheme

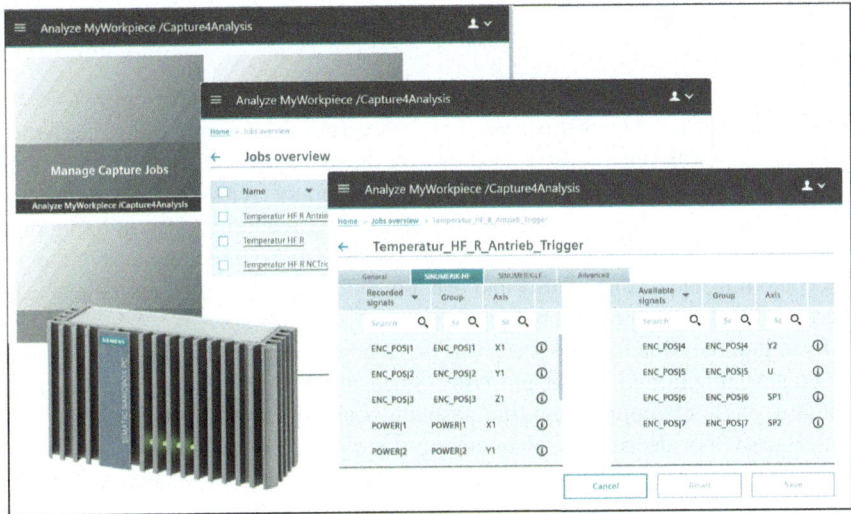

Fig. 5 Configuration AnalyseMyWorkpiece

To ensure continuous and uninterrupted data recording, the system is designed to start data acquisition at the power up of the CNC. This is achieved through a start trigger that is linked to the release of the axis controllers' power, which occurs when the machine tool is first powered up and the axes are set for operation. This mechanism guarantees that the data recording process starts automatically and continuously remains active throughout the machine's operational cycle, capturing data without any time gaps or interruptions.

In addition to the in-process machine data recording, a 100% inspection of the manufactured components is carried out to document the quality of the machined workpieces. During this automated inspection process, critical dimensions such as the cylindrical diameter of the boreholes and the distances between specific features are precisely measured in one clamping setup. These measurements are automatically recorded and stored in a centralized database. By fusing this inspection data with the machine data, a comprehensive quality control system is established. The collected inspection data is synchronized with the corresponding machine data by a time stamp, allowing for a direct correlation between the manufacturing process and the component's dimensional characteristics. This synchronization enables real-time monitoring of the production quality and facilitates the detection of any deviations from the desired specifications, ensuring that each part meets the required standards for functionality and performance in the use phase. Furthermore, the collected data can be used for long-term traceability purposes, providing a complete and detailed record of each component's production history, which is essential for both the quality assurance and continuous process optimization.

4   First Results

The advantage of these recordings lies in the use of data derived from real-world production over an extended period. The data was collected over a time period of more than a year for this study. This allows for the inclusion of effects and variations that typically do not manifest in a controlled laboratory environment, or are not apparent due to the limited duration of standard test recordings. By utilizing data from an actual manufacturing process, it is possible to capture and analyze factors that could impact product quality and performance in a way that is not replicable in short-term and controlled test setups.

In an initial evaluation, the focus was placed on examining the distance between the crankshaft pin and piston pin bore. Since these boreholes are manufactured within a single clamping setup, the absolute positioning within the machine tool workspace is not decisive for maintaining the distance between them. Thus, it is solely determined by the axis movements and the structural deformations.

Since the NC program is continuously adjusted based on the measurements in order to ensure that the machined distance remains within specified tolerances, the first step was to determine the theoretical distance defined by the NC-program. To achieve this, the position of the centers of the two bore holes were extracted from the high-frequency X/Y encoder data, and the Euclidean distance between these two points was cyclically calculated (Fig. 6). This procedure includes all applied corrections made by the operator, whether they are issued by tool corrections or adjustments based on parameters within the NC program itself. By incorporating these corrections into the analysis, the evaluation reflects the real machining deviation in the measurement data.

For the analysis, neither analytical nor simulation-based deformation models are utilized. Instead, a purely data-driven approach is applied. This methodology avoids the complexity of simulations thus lacking capabilities for in-process-control, particularly those involving contact and bearing behavior. Moreover, effects that are typically neglected in simulations are inherently accounted for within the measurement data. Consequently, the solution space remains unrestricted.

To find a possible relation between temperature and geometrical measures, the canonical correlation approach is applied [10 pp.256–269] [11] The canonical correlation approach tries to find linear combinations, such that the correlation is maximized,

$$\arg \max_{a,b} Cor\left(a^T X, b^T Y\right)$$

where X is the matrix of the temperature sensors and Y is the matrix of the geometric characteristics of the machined workpiece. The positive and negative weights are of interest indicating the significance of parameters in the correlation. Parameters at weights near zero are less important. In general, such dimension reduction methods have a drawback in interpretation. Some authors remark that such an approach works if it leads to simplification and interpretability [12, 13].

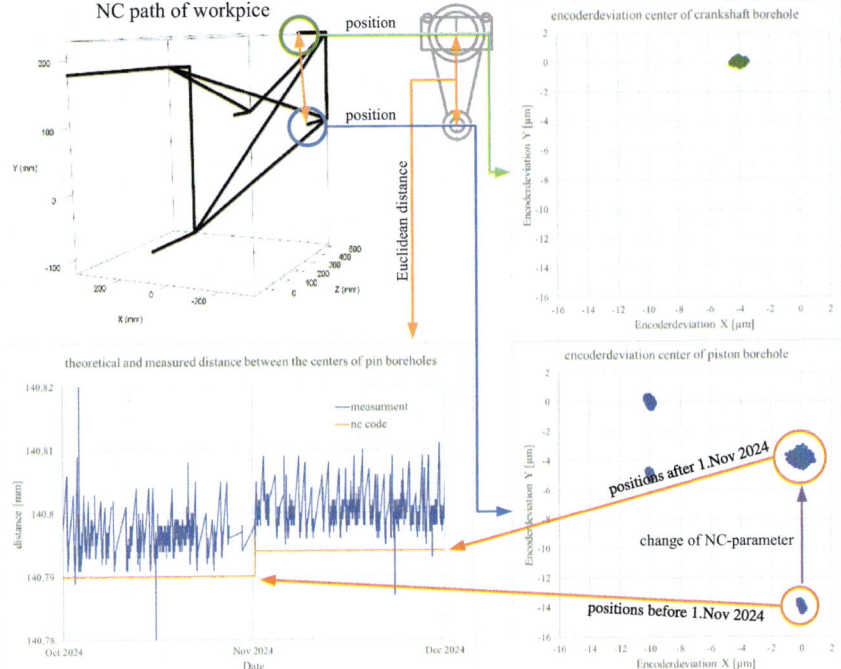

Fig. 6 NC-path determining the theoretical distance between the two borehole centers

Table 1 shows the weights, where the correlation is maximized. Two clusters of sensors are important. These are the spindle cluster and the linear guide cluster (green color). The combinations of the temperatures from these clusters correlate with the essential effects for thermal distortions. For example, the linear guide cluster has positive and negative weights. Such a combination of weights indicates that the temperature differences within this cluster have an impact on the geometrical characteristics of the connection rod.

To derive an appropriate model-based prediction of thermal distortions, this effect was considered in several derived models from these two temperature clusters. If the residuals from these models satisfy the assumptions of the homoscedasticity, non-autocorrelation and normal distribution, then the corresponding model can be used in application. The essential relation indicating the axis distance of the connecting rod borehole centers primarily involves the temperature sensors located at the spindle bearings of the Y-slide on both the right and left side. Additionally, significant correlations are observed with the temperature sensors positioned at the linear guides of the X-slide, both on the front- (towards the workspace) and at the backside (Fig. 7). These observations suggest a bending of the traveling column. Such a bending can affect the resulting accuracy of the distance between the crankshaft pin and piston pin bore. However, the influence of additional but unknown effects should be considered.

Table 1 Weights with maximized correlation

	X slide guide backside right	X slide guide frontside right	X slide guide backside left	X slide guide frontside left	X slide ball screw left	Moving column frontside	Moving column backside	Y slide guide upper right	Y slide guide upper left	Y slide guide lower right	Y slide guide lower left	Base structure upper	Y slide spindle bearing lower right	Y slide spindle bearing upper left	Y slide spindle bearing lower left	Y slide ball screw right	Y slide ball screw left	Base structure lower	flow cooling system	return cooling system	surrounding
	ϑ_1	ϑ_2	ϑ_3	ϑ_4	ϑ_5	ϑ_6	ϑ_7	ϑ_8	ϑ_9	ϑ_{10}	ϑ_{11}	ϑ_{12}	ϑ_{13}	ϑ_{14}	ϑ_{15}	ϑ_{16}	ϑ_{17}	ϑ_{18}	ϑ_{19}	ϑ_{20}	ϑ_{21}
	-0.31	0.15	*0.77*	*-0.69*	0	*-0.74*	0.16	**-0.35**	0.02	**0.23**	**0.26**	0	0	0.15	*0.64*	NA	-0.16	-0.03	0	-0.02	-0.03

Note: underlaying light green is denoted in bold and underlaying green is denoted in italic

Y slide spindle bearing lower right ϑ_3 Y slide spindle bearing lower left ϑ_{15}

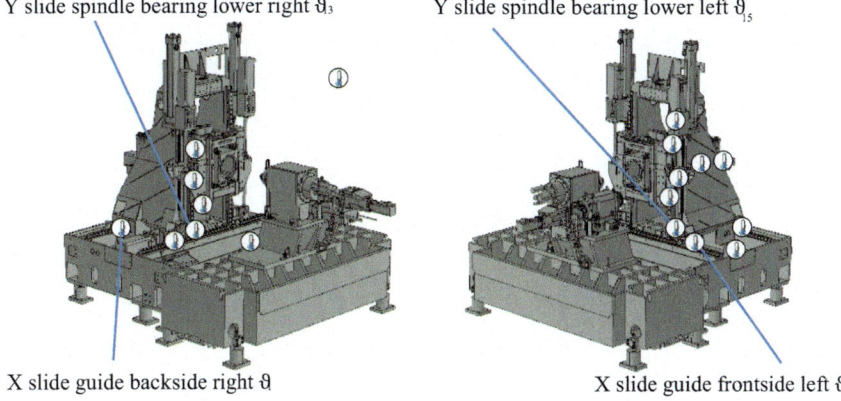

X slide guide backside right ϑ X slide guide frontside left ϑ

Fig. 7 Measurement positions with significant correlations

These specific sensor locations capture critical thermal variations that directly influence geometric deviations in the machining process. The thermal influence on the position deviation of a borehole in the connecting rod is accommodated by the allowance of the pre-machined part, ensuring that only the distance between the two boreholes requires compensation. The following regression approach is used to relate the borehole axis *distance error Δa* to the temperatures.

$$\Delta a = \beta_0 + \beta_1(\vartheta_1 - \vartheta_4) + \beta_2(\vartheta_{15} - \vartheta_{13}) + \varepsilon$$

The dataset of three months of operation from October until December was divided into a test set and a training set. As training data, the first three weeks of the observed period were used. The model was then estimated using the training set to minimize the model error ε and valuated that ε is normally distributed and the test data set corresponds to the model behavior, which leads to the following equation of the prediction model.

$$\Delta a(\vartheta) = 0.0130547 - 0.0111590(\vartheta_1 - \vartheta_4) + 0.0020541(\vartheta_{15} - \vartheta_{13})$$

Figure 8 displays the predicted values of the error (orange) of the borehole distance Δa from the model compared to the measurements from the corresponding production over the whole sample period, including the training data (blue) and the test data set (green).

As shown, the simplified linear model uses only four temperature values and provides reliable results. Particularly during the warm-up phase of the machine tool, continuous compensation could reduce errors down to just a few micrometers.

Figure 9 shows the comparison of predicted values of the borehole distance Δa from the model compared to the measurements from the corresponding production

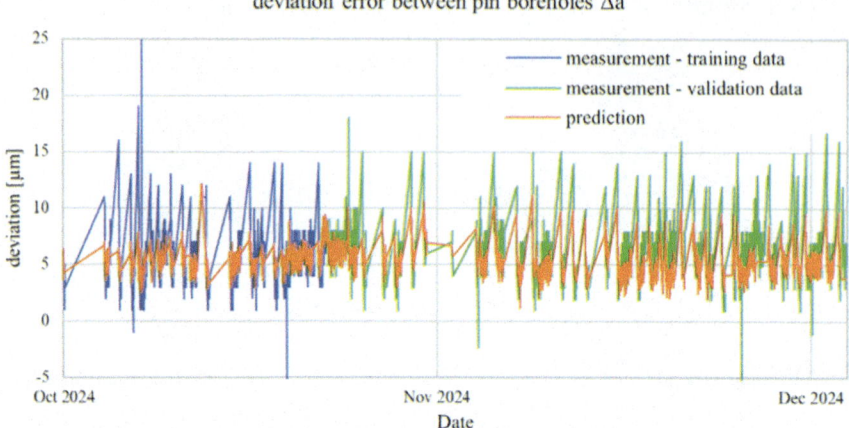

Fig. 8 Predicted values from the model alongside the actual measurements of the deviation error of the borehole distance Δa from the corresponding production

on a test data set of a seven days period. Additionally, the theoretical error of a compensated production batch is depicted. This is defined by the difference between the measurement and the predicted compensation.

Based on process capability analysis, the Six Sigma value of actual production of this period is determined to be 10.614 μm while the theoretically compensated values yield a Six Sigma level of 7.520 μm. This corresponds to an improvement of 29%, demonstrating the effectiveness of such a compensation strategy.

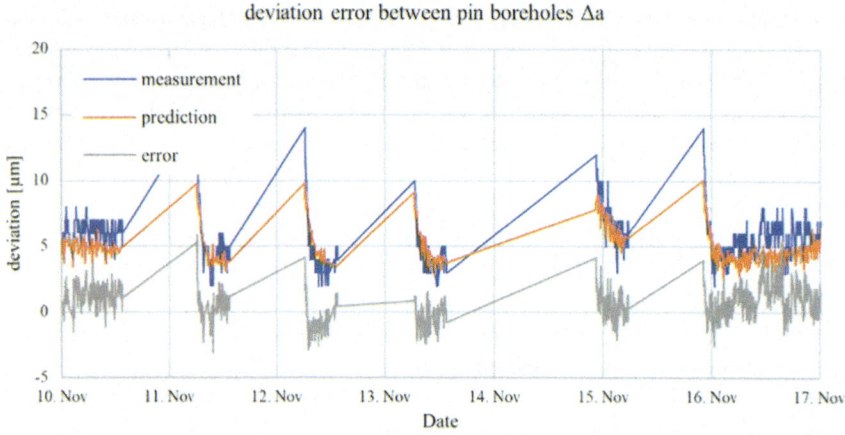

Fig. 9 Measurement values, predicted values and theoretical compensated values of the deviation error of the borehole distance Δa

5 Conclusion and Outlook

The initial analysis of data sequences of about three months of a real-production application demonstrates that reliable results in the prediction of the geometric deformation of a machining center in high precision drilling application—for one crucial measure in a predominant machine axis—can already be obtained using a linear correlation. However, for a more comprehensive understanding of thermal effects and their impact on machining accuracy, further in-depth evaluations of in-process data are necessary. It can be envisioned that the work setup is rotated by 90°, allowing the machining operation to take place e.g. along the X-axis direction. This approach enables single-axis machining to be utilized for model training and can later be evaluated in test setups, such as the machining of a test part featuring a triangular arrangement of three boreholes for spatial accuracy investigations. These evaluations should not only focus on the correlations between temperature and geometric deviations but also incorporate long-term data to capture trends and patterns over extended operating periods. Factors such as seasonal influences (summer–winter cycles), cooling system performance, wear of components in the machine structure, and other operational anomalies must be systematically examined. These external conditions can significantly affect thermal behavior and, consequently, the overall accuracy of the machining process. An additional aspect of future work will be the independent analysis of temperature variations based on the machining program with the aim to reduce the number of involved temperature sensors and substitute them by drive and spindle power or even a NC-code based analysis.

In addition to temperature monitoring, high-frequency data, including spindle and drive power consumption, should be analyzed and used in a data fusion approach. These parameters provide valuable insights into machine dynamics and can help to detect early signs of wear in both tool and machine components. Furthermore, analyzing high-frequency data of the CNC may reveal batch-to-batch variations in material properties or process stability, which could further enhance predictive models for thermal and load induced error compensation.

By integrating these extended data evaluations, a more robust and adaptive approach to thermal error compensation can be developed, ultimately leading to improved machining precision and process reliability in particular for the precision machining applications. Future research will focus on refining the data analysis methodology and integrating advanced machine learning techniques to enhance predictive accuracy and finally enable real-time adaptive control.

Acknowledgements This research was conducted during the CORNET project "GeoComp" 322 EBR funded by the Österreichische Forschungsförderungsgesellschaft mbH (FFG).

References

1. Kuznetsov, A.P., Koriath, H.J., Dorozhko, A.O.: The methods for controlled thermal deformation in machine tools. In: 1st conference on thermal issues in machine tool, Dresden, pp. 117–126 (2018)
2. Zimmermann, N., Müller, E., Lang, S., Mayr, J., Wegener, K.: Thermally compensated 5-axis machine tools evaluated with impeller machining tests. CIRP J. Manuf. Sci. Technol. **46**, 19–35 (2023)
3. Mori, K., Bergmann, B., Kono, D., Denkena, B., Matsubara, A.: Energy efficiency improvement of machine tool spindle cooling system with on–off control. CIRP J. Manuf. Sci. Technol. **25**, 14–21 (2019)
4. Mayr, J., Gebhardt, M., Massow, B.B., Weikert, S., Wegener, K.: Cutting fluid influence on thermal behavior of 5-axis machine tools. Procedia. CIRP **14**, 395–400 (2014)
5. Wiessner, M., Blaser, P., Böhl, S., Mayr, J., Knapp, W., Wegener, K.: Thermal test piece for 5-axis machine tools. Precis. Eng. **52**, 407–417 (2018)
6. Gebhardt, M.: Thermal behavior and compensation of rotary axes in 5-axis machine tools. In: PhD Thesis, institue machine tools and manufacturing, ETH Zürich (2014)
7. Baum, C., Brecher, C., Klatte, M., Hun Lee, T., Tzanetos, F.: Thermally induced volumetric error compensation by means of integral deformation sensors. Procedia. CIRP. **72**, 1148–1153 (2018)
8. Pavlíček, F.: Parametrierbare Metamodelle zur Berechnung des Wärmeübergangs in Hohlräumen. PhD Thesis, institue machine tools and manufacturing, ETH Zürich (2019)
9. Brecher, C., Neus, S., Dehn, M.: Efficient FE-modelling of the transient thermo-elastic machine behaviour of 5-axes machine tools. Thermal Issues, euspen, Northampton, 148–149 (2020)
10. Seber, G.A.F.: Multivariate observations. John Wiley & Sons, Inc., New Jersey (2004)
11. Hotelling, H.: Relations between two sets of variables. Biometrika. **28**(3/4), 321–377 (1936)
12. Faraway, J.J.: Linear models with R, 3rd edn. Chapman & Hall/CRC, London (2025)
13. Gurker. W.: Lecture notes in general regression models. Vienna, Institute of Statistics and Probability Theory (2013)

Experimental Methods and Measurement Techniques in Machine Tools

Ambient Influences on Thermal Errors in Machine Tools: Challenges, Modelling and Compensation

Philipp Klimant⑩, Christian Naumann⑩, Alexander Geist⑩, and Janine Glänzel⑩

Abstract Thermally induced errors are one of the main factors contributing to positioning inaccuracies of cutting machine tools. They are caused by machine-internal, process-related and external heat sources and sinks. Of these three causes, ambient effects pose a particular challenge, because these effects are slow, but large-scale and thus very strong. This paper presents various methodologies to accurately and efficiently model ambient effects in simulation-based thermal machine tool studies. This includes a simplified method for categorizing natural convection based on the surface orientation, methods for obtaining localized heat transfer coefficients via computational fluid dynamics (CFD) simulations and for applying them in real-time to thermal FEM simulations and methods for modelling the complex conditions in the machine tool workspace during cutting operations. The paper also presents some best practices for measuring ambient effects and for the parametrization of CFD simulation models. For a specific milling machine tool, various relevant load scenarios were simulated as well as measured and the results will be shown to demonstrate the efficacy of the simulation models and the scope of the ambient effects. Aside from the mere study and modelling of ambient effects, their reduction or compensation will also be addressed. Different methods for predicting the environment-induced thermal error will be presented, including the use of data-driven black-box methods, transfer functions and simulation-based methods.

P. Klimant · C. Naumann (✉) · A. Geist · J. Glänzel
Fraunhofer Institute for Machine Tools and Forming Technology IWU, Chemnitz, Germany
e-mail: christian.naumann@iwu.fraunhofer.de

P. Klimant
e-mail: philipp.klimant@iwu.fraunhofer.de

A. Geist
e-mail: alexander.geist@iwu.fraunhofer.de

J. Glänzel
e-mail: janine.glaenzel@iwu.fraunhofer.de

P. Klimant
Professorship Virtual Technologies, Hochschule Mittweida–University of Applied Sciences, Mittweida, Germany

© The Author(s) 2026
K. Wegener and M. Bambach (eds.), *4th International Conference on Thermal Issues in Machine Tools (ICTIMT2025)*, Lecture Notes in Production Engineering,
https://doi.org/10.1007/978-3-032-01194-7_16

239

Keywords Thermal error · Machine tool precision · Ambient influences · Error compensation · Error avoidance · Measurement · Simulation

1 Thermal Errors in Machine Tools and the Role of Ambient Influences

Positioning inaccuracy due to thermal influences, referred to as thermal errors, remain a major problem for machine tool manufacturers as well as users, according to an industrial survey by Bräunig et al. [6]. The causes of these errors can be divided into machine-internal, process-related and external heat sources and sinks.

1.1 Thermal Errors in Machine Tools: Sources and Importance

Machine-internal are heat sources such as friction in guides and bearings, electrical losses in drives, waste heat from hydraulic and pneumatic units and the various cooling systems, which may act as both heat sources and heat sinks. Process-related are the friction from the cutting process, cutting fluid, the workpiece and chip nests. Ambient effects comprise the convection of ambient air, radiation, e.g. from sunlight, and conduction with the foundation.

Of these three causes, ambient effects pose a particular challenge, because these effects are slow, but large-scale and thus very strong. From the viewpoint of a machine tool manufacturer, ambient effects are difficult to control, because every machine tool they sell can be placed in a different environment, and difficult to study, because a controlled change of ambient conditions is difficult to achieve without expensive infrastructure. There is also a secondary ambient component of free and forced convection within the workspace due to moving machine tool axes, cutting fluid usage and air/fog evacuation units.

The study of thermal machine tool behavior always starts with the purpose. How to conduct the study can be quite different for someone how wants to optimize the thermal behavior of a new machine tool design, compared to someone who experiences problems with thermal precision only on hot summer days or after production pauses or after a very long, intense milling operation with large cutting volume.

There is also the question of whether the three sources for thermal effects are linked or if they can or should be studied separately. Whether or not they are linked depends on the machine tool design, the answer is often yes. A change in the ambient temperature does not usually change the way internal heat sources affect the machine tool. Electrical losses or friction create waste heat, which causes a similar temperature increase, whether the outside air is at 20 °C or 25 °C. What is different, is the rate at which these heated components transfer heat to the ambient air. In a confined space, this waste heat can also change the ambient temperature and thus create an ambient heat source. This is most obvious in the workspace of a housed machine tool, where air temperature and cutting fluid mist can have a large effect. Likewise, process-related effects influence ambient effects in the workspace, but not the internal sources. When guides, bearings and motors are water-cooled, then the temperature of the coolant system is often either actively controlled to match the ambient temperature or passively influenced by it via heat exchangers. This way, the ambient temperature may also have an influence on how the internal cooling systems affect the machine tool. Therefore, depending on the purpose of the investigation, the three sources of thermal errors should be studied both together and separately to determine how they are linked and to enable easier model parametrization through the isolated study of thermal effects.

1.2 Methods for Thermal Error Investigation

Machine tool thermal behavior can be studied via thermal or thermo-elastic measurement or via simulation. Measurements can be done in the form of local temperature measurement (fixed temperature sensors, e.g. PT100) alongside displacement measurements (position error at the tool center point (TCP)), by infrared camera observation of the machine tool or its components or by quantifying the effects indirectly through test workpieces.

Simulation-based studies are most often done by finite element (FE) simulation, but there have also been some flexible multibody simulation models. Particularly for the study of ambient effects, computational fluid dynamics (CFD) simulations are important tools for the investigation of the heat exchange between machine tool and environment and to compute exact heat transfer coefficients (HTCs). Particle simulations can similarly be used for ambient studies. CFD or particle simulations can also model cooling systems and provide valuable information on how the design and flow rate of cooling systems affect the cooling (or warming) of individual components.

The international standard on thermal error investigation is specified in the ISO 230-3, where three tests are required [27]:

- An environmental temperature variation (ETVE) test;
- A test for thermal distortion caused by rotating spindles;
- A test for thermal distortion caused by moving linear axes.

Together, they can be used to calculate the combined standard thermal uncertainty of a machine tool. This does not include process-related errors, but certainly places emphasis on internal and ambient thermal effects. The ISO norm states that the ETVE test may be run in standby or completely off. There is no clear instruction on how to introduce the ambient changes, so anything from day-night cycles to actively induced ambient temperature changes in a climate chamber are possible. However, for the purpose of determining the ambient thermal precision of a machine tool, the operating conditions, including during ETVE tests, must be in accordance with the manufacturers operating instructions. These instructions may include requirements regarding operating temperature ranges or maximum temperature fluctuations.

1.3 Methods for Thermal Error Reduction

The two general categories for thermal error reduction are error avoidance and error compensation. Error avoidance reduces the thermal deformation or its causes, e.g. by reducing or removing waste heat or by reducing the influence of thermal deformations on the relative position between cutting tool and machining table. Error compensation doesn't change the thermal deformation but rather measures or predicts it in order to apply corresponding positioning offsets. Model-based prediction takes many forms, from purely data-driven regression models mapping sensor data to corresponding TCP displacements, via phenomenological models using simplified thermo-elastic models to realistic FEM simulations creating real-time digital twins of the machine tool.

Which thermal error reduction models are effective and suitable for a given machine tool and application scenario depends on several factors including kinematic structure, position and effect of heat sources and sinks and the environment, in which the machine tool is used.

A specific evaluation of thermal error reduction methods for ambient thermal effects will be given in Chaps. 4 and 5. Before that, Chaps. 2 and 3 provide some fundamentals of measurement and simulation of ambient effects in machine tools, which form the basis for any thermal error reduction scheme. A conclusion on the current state of the art on modelling and reduction of ambient thermal effects alongside some recommended future avenues of research conclude the paper.

2 Measurement of Ambient Influences

The international standard on measuring the impact of ambient effects on the thermal error (ISO 230-3) was already discussed in Chap. 1. This is, however, not always a useful guideline, when it comes to optimizing the machine tool design or creating effective thermal error reduction strategies.

The first step for measuring ambient effects is determining what load cases are relevant and measurable. Typical ambient effects are:

- complete ambient temperature changes, e.g. a longer-lasting temperature increase/decrease
- cyclic ambient temperature changes, e.g. day-night cycles
- exposure to sunlight or other strong radiation sources
- opening of workspace door with significant temperature/moisture gradient
- external forced convection, e.g. from an open factory door/window

- forced convection from moving assemblies or suction units
- cooling fluid in the workspace (cutting fluid, pressured air, cryogenic liquid)

Some of these ambient effects cannot be produced at will, especially strong, controlled temperature changes or forced convection effects. The next step is determining what to measure and how to measure it. The following section will explain this in detail.

2.1 Measurement Procedures

In order to measure the impact of ambient thermal effects on a machine tool, it is generally recommended to measure three things:

- the temperature at various locations on/in the machine tool,
- the resulting displacement at the TCP,
- the ambient influence.

In some cases, an additional measurement of local deformations on various points of the machine tool may also be sensible. Measuring the ambient influence depends on what influence is investigated. This almost always includes the ambient air temperature. It may be useful to measure the air temperature in different heights above ground or in different locations within the workspace. Particularly in the workspace, a measurement of the relative humidity and perhaps even air pressure may also be required. For forced convection experiments, the wind speed should be measured. For solar radiation, the time and angle of exposure should be documented, along with obstructions like cloud cover. For external heat sources like the foundation or other hot machines, ovens, heating units, etc., which may have an effect on the investigated machine tool, any relevant parameters like temperature, relative location or operating schedule should be observed and documented.

For an optimal study of external ambient influences, a climate chamber should be used. This is essentially a climate-controlled room, where the room temperature and sometimes even the relative humidity can be precisely controlled. Figure 1 shows an optimized example of such a climate chamber, which is being used at the Fraunhofer IWU in Chemnitz. The chart on the right shows measured air and ground temperatures for a stepwise ambient air change between 10 and 40 °C. Note, that the air temperature increases nonlinearly with height, particularly for high ambient temperatures. In the low ambient temperature sections, the foundation acts as a heat source, so that the lower air layers are warmer than the layers close to the ceiling.

The climate chamber can be used to measure long load cycles for internal sources, while ensuring no external temperature change occurs. More importantly, day-night-cycles and even seasonal effects can be created in a deliberate and reproducible manner. Different climatic conditions from hot and moist to cold and dry can also be produced with this specific chamber. One important feature in this regard, is the ability to exchange (and heat/cool) large volumes of air without introducing

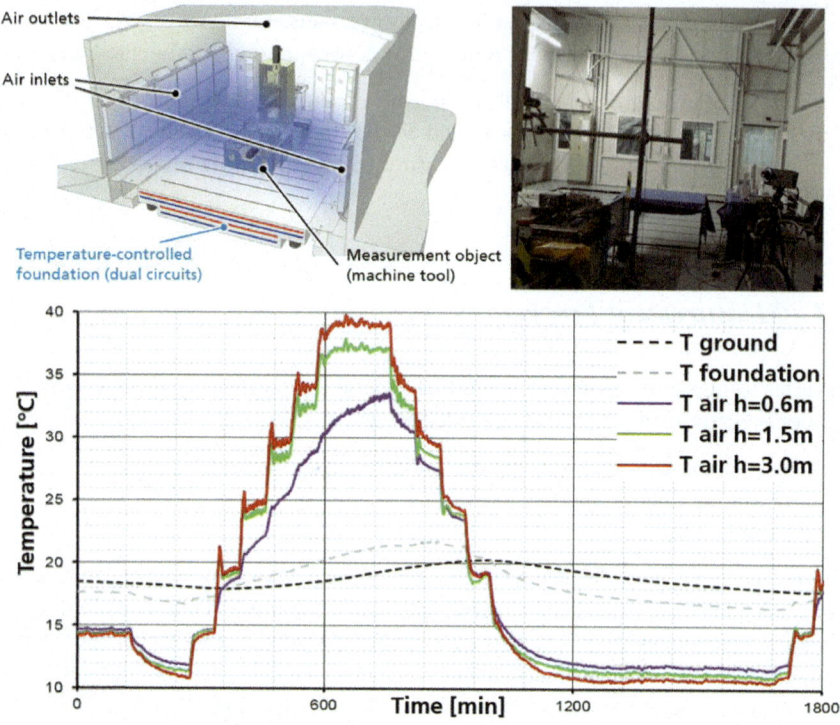

Fig. 1 Example of a climate chamber with sample air temperature measurement

unwanted forced convection. This is achieved though very large wall inlets with many small, downward-facing openings and large, distributed outlets on the ceiling. The Fraunhofer climate chamber can also separately control the floor temperature in order to simulate the effects of the foundation as a heat source or heat sink.

Which internal temperatures of the machine tool need to be measured, depends on the purpose of the investigation. It is usually required to record the temperatures at all integrated temperature sensors. In most cases, this requires for the machine tool to be powered up and in standby. If this is the case, it is usually prudent to place the machine in standby, ideally at constant ambient temperature, for at least one day before starting the measurement, so that the machine tool's internal warm-up does not overlay the actual external load profile. If possible, additional temperature sensors on relevant components, inside the workspace, under the housing, in coolant tanks, etc. may be placed to gain addition information on the temperature distribution and possible other influences on the thermal behavior.

Measurements of ambient effects in the workspace typically require a more thorough coverage of machining table and workspace walls by temperature sensors and may be amplified by infrared (IR) cameras or regular cameras for recording the flow of coolant within the workspace. IR cameras are difficult to employ, however, since they are usually too delicate to place within an enclosed workspace. Reflections and the need for specific emissivity coefficients also require some preparations. Figure 2 shows an example of a temperature measurement setup for the investigation of ambient temperature changes.

Measuring the TCP displacement or thermal deformation is mostly the same as in any thermal error investigation. It can be done by relative measurement or through the use of external displacement sensors, see Fig. 3. The relative TCP error is usually measured with calibrated measurement artifacts mounted on the machining table and a touch-trigger-probe attached as tool. Alternatively, the ISO 230-3 recommends a measurement nest placed on the table with five indicating calipers. Absolute measurements can be done using indicating calipers or equivalent non-touching displacement sensors [27].

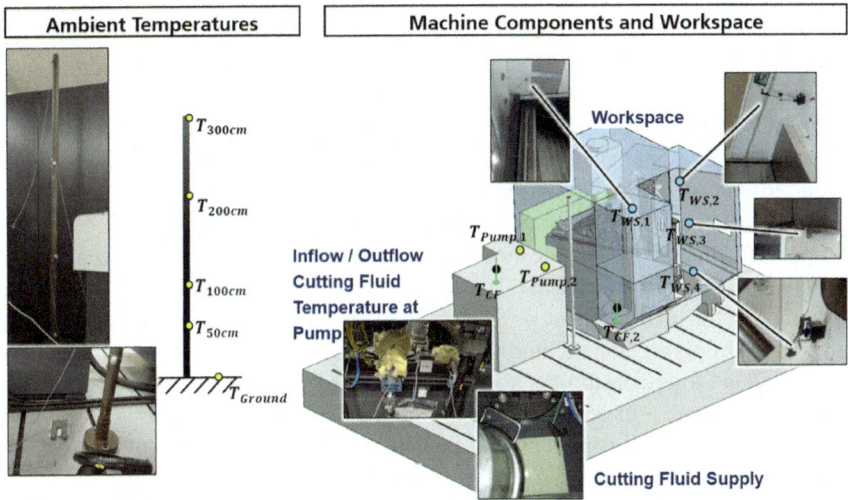

Fig. 2 Example of a temperature measurement setup

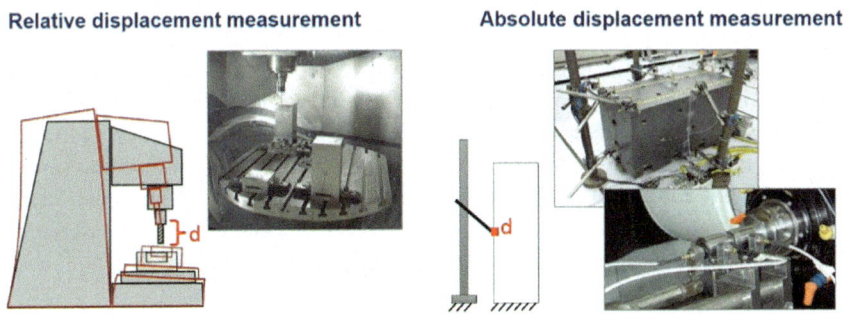

Fig. 3 Relative vs. absolute displacement measurement

Important for absolute displacement measurements, is to properly plan the setup of the measurement frame, which hold these sensors. Under ambient temperature changes, the frame will deform. In order to predict this deformation more easily, a mix of fixed and loose bearings is used. Frames made from Invar steel, ZeroDur or carbon fiber reinforced polymer (CFRP), which have a low thermal coefficient of expansion can also minimize the total length change of the frame. Measuring the frame temperature directly, will also help in exactly accounting for their influence on the measured displacement. Frames are also often sensitive to vibrations, which should be avoided during measurements [19]. Displacement measurements for ambient effects are sufficient every 10–15 min. For absolute measurements with a measurement frame, each measurement should record a series of values in order to filter out vibration-based disturbances.

A very precise method for determining thermal displacements is via the use of laser interferometers. Brecher et al. used four LaserTracers to determine the 3D displacement of the TCP with less than 10 μm uncertainty [8]. The LaserTracers are non-contacting and only need line-of-sight to a reflector placed near the target (e.g. TCP, machine table, structure). This allows the separate study of the thermal error of the machine tool column and the machining table, which is otherwise difficult. LaserTracers can compensate for ambient temperature changes, but deformations of the ground or their tripod/stand may still influence measurements. These position shifts of the LaserTracers must be monitored using additional reflectors and compensated manually after the measurement. A very early work by Schwenke et al. demonstrated that just one LaserTracer can also accurately measure 3D machine tool errors if the measurement is repeated in (near) identical conditions at least four times [50]. Applying this method for ambient temperature variation measurements is difficult and time-consuming, however. One such investigation was performed by Groos et al., where four sequential LaserTracer measurements were used on a 5-axis machine tool under constant ambient temperature, which was repeated for 15, 20, 25 and 30 °C [23].

A new approach using a laser tracker LEICA AT960 mounted to the motor spindle, i.e. near the TCP, and four or more retro-reflectors distributed across the machining table, allowed Egaña et al. to measure ambient thermal effects across the workspace of a very large machine tool (workspace 3.5 × 3.1 × 1.1 m). Instead of using an expensive climate chamber, which is difficult to find for a machine tool of this size, they conducted experiments intermittently over several months in order to capture meaningful ambient temperature variations within their non-climatized production hall. Their method reached a repeatability of 10 μm in X- and Y-direction and 15 μm in Z-direction under constant ambient conditions ($\Delta T_{env} < 1K$). The laser tracker (ca. 14 kg) did not add waste heat relevant to the experiments [12]. The drawbacks of this approach are the limited precision and long time needed to gather measurement data with uncontrolled ambient temperature variation.

2.2 Indirect Thermal Error Measurement

Indirect thermal error measurement refers to the manufacturing and measurement of thermal test workpieces. These are typically produced by long-running NC programs, which are designed to introduce waste heat into the machine tool. Manufacturing the test workpiece with active thermal compensation then shows how effective the compensation is and provides a good method for benchmarking compensation methods on different machine tools.

There are two types of thermal test workpieces. The first type uses the machining operation necessary to produce the workpiece as the source of the thermal error. The second type only uses the workpiece as a measuring instrument and introduces the thermal error, e.g., from long cyclic air cutting operations. The first type therefore represents the thermal precision of the machine tool during production more realistically, while the second type allows a more broad investigation of different thermal influences including slow external ambient effects.

The standard ISO 10791-7 contains different test pieces for the evaluation of machine tool accuracy [28]. Wiessner et al. developed a thermal test piece for measuring the position and orientation errors of five-axis machine tools, which can subsequently be obtained from the workpiece via a coordinate measurement machine [62]. Fickert et al. have similarly developed a test workpiece and were able to show that the obtained thermal error was equivalent to touch probe measurements with a difference of only 2 μm [15]. Zimmermann et al. have developed a new thermal test workpiece, which can identify ten thermal errors over eight time steps including the separate detection of table errors [64].

Aside from the benchmarking aspect and the benefit from having a tangible proof of the achieved thermal precision, indirect thermal error measurements provide no added information compared to the direct measurements described in Sect. 2.1. They are also not well suited for ambient effects, save for those in the workspace.

2.3 Sample Ambient Machine Tool Investigation

Some measurements with practical relevance may be:

1. Production interruptions with open workspace door
2. Rapid temperature increase, e.g. from an open factory door in summer
3. Day-night cycles (sine-like ambient temperature variation)

The following measurement results were recorded on the DMU 80 eVo inside the Fraunhofer IWU climate chamber. Of the 40 installed temperature sensors, three internal and one ambient sensor were chosen to show the machine and air temperatures, see Fig. 4, left. The relative TCP displacement was measured above the center of the machining table, see Fig. 4, right. For this, a small steel sphere and a Heidenhain touch probe were used. The shown displacements were recorded without any active thermal error compensation.

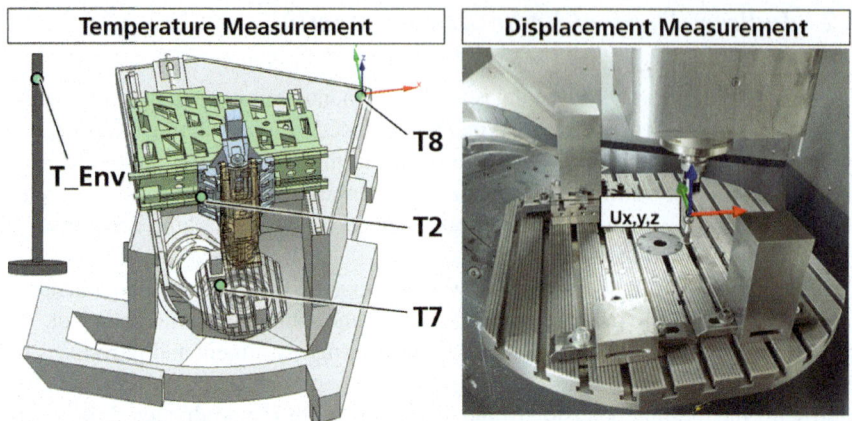

Fig. 4 Selected temperature and displacement measurement positions on the DMU 80 eVo

The first measurement shown here starts with cyclic X- and C-axis movements, which gradually warm up both the machine tool and the air in the workspace. After thermal stability is reached, a 30-min production pause with opened workspace door, followed by continued warm-up cycles and another production pause complete the 24-h load case. The workspace air is approximately 5 K warmer than the constant outside air. Both temperature as well as displacement measurements show an immediate response to the change in load and the interaction with the colder outside air from the climate chamber (Fig. 5).

The second measurement shows the response of the machine tool, placed in stand-by, to a quick rise in ambient temperature from 20 to 30 °C in the space of about one hour (Fig. 6). The machine, having a housing and a high thermal capacity, warms up more slowly and reaches thermal stability only after around two days. The thermal time constant of the machine tool is circa ten hours, which is the time it takes for the machine tool to reach around two thirds of the new final stable temperature under natural convection. There is some debate on whether large time constants are positive or negative. In most cases, large time constants improve thermal stability under short ambient temperature variations and dampen the effects of day-night cycles. The flipside is that thermal energy already in the machine tool also takes a long time to dissipate, even when rigorous cooling is applied.

The observed reversal in the TCP displacement of measurement 2 is owed to the different speed at which the individual components warm up. In the Y-direction, where two reversals occur, there is an interplay of the expanding machine tool column and Y-slide, the expanding table and a tilting motion of the Z-slide around X.

The third measurement simulates an idealized day-night cycle with an amplitude of 16 K within 24 h. In total, three of these day-night cycles were measured, see Fig. 7 [16]. It is interesting to observe, how the machine tool reaches its own maximum and minimum temperatures about three hours after the ambient temperature does and once again not in every part of the machine tool at the exact same time. If a factory

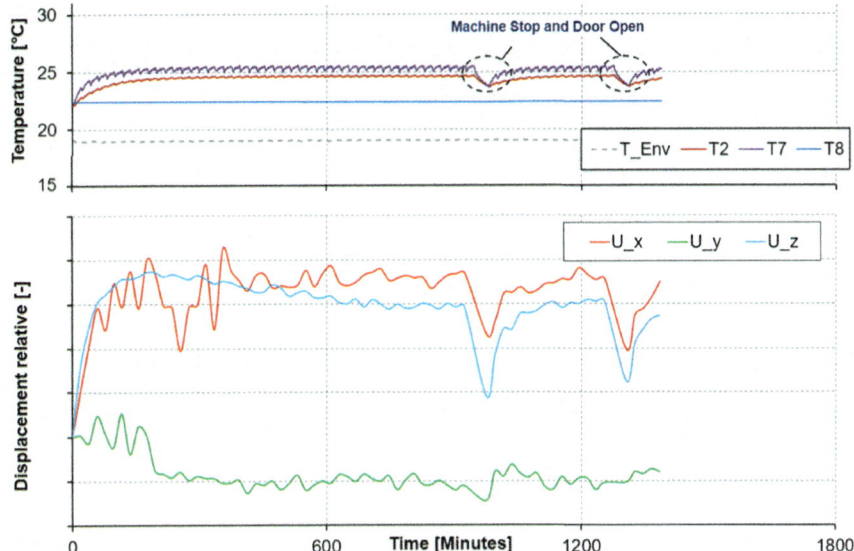

Fig. 5 Measured temperatures and relative TCP displacement for sample measurement 1 "production breaks"

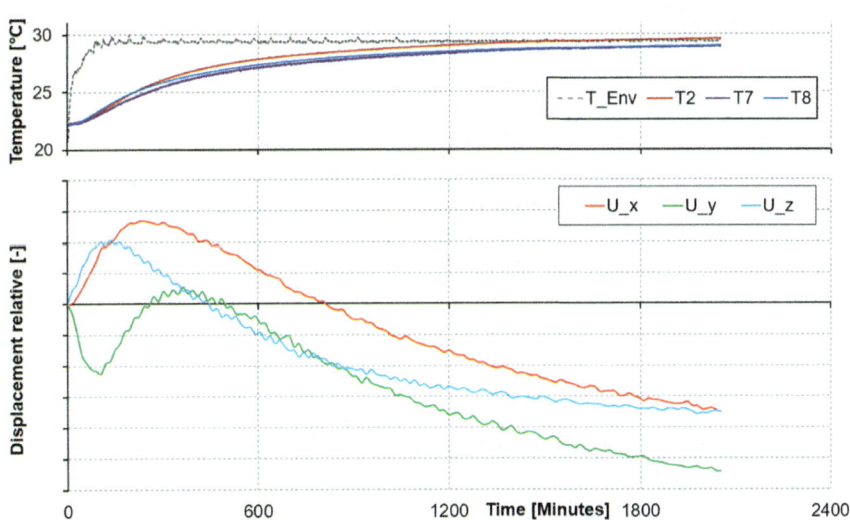

Fig. 6 Measured temperatures and relative TCP displacement for sample measurement 2 "temperature jump"

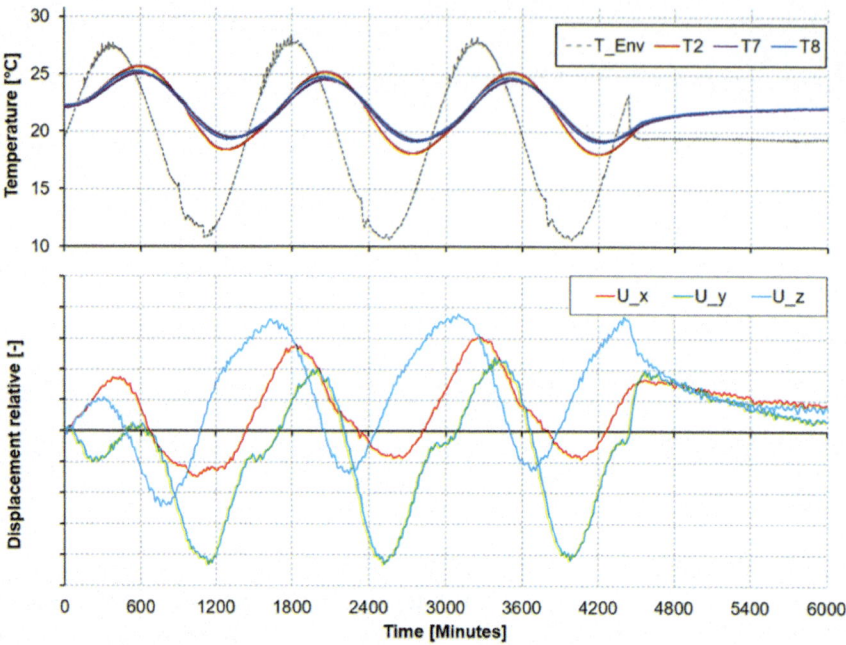

Fig. 7 Measured temperatures and relative TCP displacement for sample measurement 3 "day-night cycles" (modified from [16] licensed under CC-BY 4.0)

door is opened in winter near a machine tool, the effects can still be observed for hours after the door has been closed again, even if the air temperature has already returned to its previous level.

Similarly interesting is the measured thermal error. Under nearly identical day-night-cycles, the displacement also follows a nearly identical curve. Settling into this curve, however, takes almost a full day. Therefore, when measuring day-night cycles, a two-day measurement (at least) is advisable.

3 Simulation of Ambient Influences

Thermal and thermo-elastic simulations are an excellent way for investigating the response of machine tools to ambient effects. Setting up such a simulation requires a number of steps, which are detailed in the next section. Another important prerequisite is model parametrization, which is ideally supported by equivalent measurements. Only in the case of new machine tool design, where no prototype for measurement is available, should such a validation step be omitted. Ambient simulations of external conditions are typically much simpler than similar investigations of internal thermal error sources. Different free or forced convection effects, foundation temperatures or solar radiation are easy to simulate using FEM simulations. Far more complex, however, are realistic simulations of ambient, process-related effects in the workspace. Here, dynamic conditions from moving assemblies, cutting fluid (both liquid and in mist form), tool rotation and suction units present a major challenge for transient thermal and thermo-fluidic simulations.

3.1 Simulation Procedures

Discussing simulation procedures must be divided into thermal simulations and fluidic simulations. The study of the thermo-elastic behavior mostly requires thermal and subsequently thermo-elastic simulations. Achieving accurate results for the thermal simulations, however, often requires additional computational fluid dynamics (CFD) or equivalent particle simulations in order to provide correct boundary condition parameters, such as convective heat transfer coefficients (HTCs).

On a very basic level, thermal simulations can be steady-state or transient. Steady-state simulations calculate the thermal equilibrium after an infinite time has passed. This can be used to gauge how susceptible a machine tool design is to ambient temperature changes or how thermo-symmetric it behaves. While the progression of the temperature change can also be observed, it does not directly correlate to the elapsed time. Transient simulations calculate individual solutions for defined time steps and allow for changing boundary conditions. This can be used to determine how long it takes the machine tool after an ambient temperature change to settle into a new equilibrium or how much a machine tool is affected by day-night cycles. More importantly, short load cases, such as those in the workspace, should be simulated time-dependent, because they usually do not reach any steady state.

The workflow for thermal simulations is:

- CAD model preparation and simplification,
- Meshing with finite elements,
- Setting mechanical and thermal boundary conditions,
- Specifying the load case (e.g. power of heat sources, ambient temperature) and solver parameters,
- Solving and analyzing the simulation results.

Geist et al. demonstrate this workflow for the DMU 80 eVo, which is featured several times in this article [19]. The workflow is the same for ambient influences as it is for internal thermal influences. Perhaps the most important part of this process is the definition of the boundary conditions.

Convective boundary conditions govern the heat or mass transfer at the boundaries of a physical system [2]. There are two main types of convective boundary conditions:

1. Dirichlet boundary condition: Here, the temperature or concentration at the boundary is kept constant, e.g.: $T(x = 0) \equiv T_0$

2. Neumann boundary condition: In this case, the heat flow or mass flow at the boundary is kept constant, e.g.: $\partial n / \partial T (x = 0) = -\alpha \cdot (T - T_\infty)$

Here, α represents the heat transfer coefficient, and T_∞ is the ambient temperature. Convection, whether natural or forced, is specified via the HTC, which is in the range of 2–35 W/(m^{2}K) for natural convection and 10–250 W/m^{2}K for forced convection. Forced convection for cutting fluid can reach HTCs of up to 10^4 W/(m^{2}K) and even higher values when phase changes are involved [19].

Another important mechanism in thermodynamics is heat conduction, which describes the process by which thermal energy is transferred through a material without the material itself moving [2]. Heat flows from areas of higher temperature (heat source) to areas of lower temperature (heat sink). Mathematically, heat conduction is described by Fourier's Law:

$$\dot{q} = -k \cdot \partial T / \partial x$$

where $\dot{q}$ is the heat flow rate intensity or heat flux density, k is the thermal conductivity of the material, and $\partial T / \partial x$ is the temperature gradient.

Thermal radiation is defined by the Stefan-Boltzmann law, which describes the relationship between the temperature of a black body and the emitted radiative power [2]:

$$P = \sigma \cdot A \cdot T^4$$

where P is the total emitted power, σ the Stefan-Boltzmann constant, A the surface area of the body and T the absolute temperature in K. All bodies emit and absorb thermal radiation at the same time so that only radiation from objects with a significantly higher temperature will have a noticeable effect. Industrial ovens and, of course, the sun are obvious sources, though line of sight or direct reflection are necessary for the radiation to reach a machine tool.

3.2 Parametrization of Simulation Models

Parametrization of thermal models, i.e. quantifying thermal boundary conditions, is a complex and time-consuming process. A good machine tool CAD model already has the correct material properties assigned to its various components. This includes both mechanical (density, stiffness, etc.) as well as thermal properties (thermal conductivity, thermal capacity, thermal expansion coefficient, etc.). However, these are not necessarily complete, correct or precise, so that they should be questioned, if implausible simulation results are observed.

Next, there are the HTCs between joined or touching solids. These depend on the contact surface and the contact pressure. Often these contact interfaces are modelled without an additional thermal resistance, which may present a source of simulation error. Lastly, the HTCs with the environment are very important, because they specify at what rate the heat, which builds up inside the machine tool during its operation, gets transferred to the environment. Of course, they also play a major role in machine tool warm-up under high ambient temperatures.

These convective HTCs can be calculated from simple equations, which take the surface properties and its orientation and the surrounding liquid (usually air) into consideration [57].

More precise, especially for the more complex conditions inside the workspace, is the use of CFD or particle simulation. They present two distinct approaches for analyzing heat transfer processes. Both methods have their own strengths and weaknesses that must be considered in practice.

CFD is a numerical method that describes fluid dynamics and heat transfer in liquids and gases. By solving the Navier–Stokes equations, CFD allows for a detailed analysis of flow fields, temperature distributions, and consequently the HTC. The advantage of CFD lies in its ability to account for complex geometries and various operating conditions. However, performing CFD simulations can be computationally intensive and requires extensive knowledge of numerical modeling [58].

In contrast, particle simulation focuses on tracking particles within a fluid field. This method is particularly useful for investigating interactions between particles and fluids, such as in colloid chemistry or sedimentation processes [35]. Particle simulations can be less computationally intensive and are often more efficient when modeling specific transport phenomena. However, the downside is that they may not provide the same level of detail as CFD simulations when examining flow and heat transfer processes.

Overall, the choice between CFD and particle simulation for determining the HTC depends on the specific requirements of the application, available resources, and the desired level of detail. While CFD enables a more comprehensive analysis, particle simulation may be more advantageous in certain scenarios.

An early work by Jedrzejewski et al. described an analytical method for determining heat transfer coefficients for forced convection on rotating machine elements and verified the results experimentally [29].

Zwingenberger developed simple analytical surrogate models for boundary conditions of the thermo-elastic displacements in FEM analyses [65]. CFD simulations were employed to determine the HTCs and an FEM model was created to integrate the developed surrogate models. Analytical and numerical calculation methods were compared to accurately capture the heat transfer from both convection and radiation, which was subsequently confirmed in measurements.

Simulating the complex, often even turbulent conditions in the machine tool workspace during cutting operations requires more detailed modelling techniques. A general comparison of different CFD simulation parameters including computation times and the resulting cooling effect on the cutting tool was performed by Naumann et al. on a test bench. Simulations and corresponding measurements with and without cutting fluid showed the resulting HTCs and the effects of tool rotation on the cooling effect [46].

Liu et al. used the Coupled Eulerian Lagrangian (CEL) method to simulate the cutting fluid's impact on chip formation and the tool temperature in fluid–structure interaction (FSI) simulations. With it, they were able to show that the distribution of the convective HTCs on the tool-rake interface was similar for different cutting parameters and that the CEL tended to overestimate the convective cooling effect due to inaccurate flow velocities in the boundary layer [38].

Coupled numerical approaches like FEM and CFD exhibit substantial resource requirements, which is why mesh-free methods, such as the finite pointset method (FPM) are frequently utilized. The FPM uses a Lagrangian approach, in which numerical points move with the material flow, unrestricted by any mesh. This presents the advantage that the computational domain self-adapts to moving geometries, free surfaces and phase boundaries without the need for costly remeshing. Uhlmann et al. use the FPM to simulate the effects of jet cooling during metal cutting operations [56].

A more precise method for simulating both the heat transfer near the rotating tool and in the rest of the workspace during cutting fluid usage was presented by Brier et al. The developed model uses two coupled CFD models for a control volume in the immediate vicinity of the tool, where a steady-state fluid–structure interface model is used. This model transfers the air and cutting fluid speeds and their volume fractions to a larger workspace model, which can compute the resulting transient thermal effects on the machine tool [9].

3.3 Model Simplification and Surrogate Models

Coupled fluid–structure simulations are highly computationally intensive and complex. This is particularly problematic, when moving components are involved, because these cause transient changes of the HTCs on their surface. To supply the appropriate set of HTCs for every time step of a transient thermal simulation therefore takes long delays to run the corresponding CFD simulations. To eliminate these delays, a decoupling approach using characteristic diagrams was presented in the works of Glänzel [20, 22]. This approach describes the convective HTCs as a function of ambient load parameters such as air temperature, air velocity, and air flow direction. Interpolation between these load cases is performed using characteristic diagrams. This requires the creation of a database with a moderate number of training CFD simulations for computing the characteristic diagrams. It subsequently enables the real-time estimation of HTCs for all relevant ambient scenarios with an accuracy similar to that of a load case-specific CFD simulation [21].

Jungnickel has developed a node model, which is able to simulate enclosed spaces like the workspace housing. The model represents the enclosed space by a node. The heat exchange with the walls results from the spatial orientation as well as air and wall temperatures. The method has proven successful for low temperature gradients between wall and air [30]. Radiation was also modelled and was computed with an error of up to 16%. The model does not explicitly consider the size of the enclosure, nor any additional heat source therein [31].

Pavliček has improved the robustness and precision of machine tool simulations through the development of metamodels for structural cavities in the machine frame. Methods for estimating heat generation are employed, combined with new model reduction techniques. One of the greatest challenges lies in describing the various dynamics involved in the processes, including electrodynamics, fluid dynamics, mechanics, and thermodynamics. To minimize the computation times of conventional approaches and bridge the research gap between thermal and energetic machine models, a custom simulation program was developed in Matlab. The model was subsequently validated with empirical measurements [48].

3.4 Sample Ambient Simulation Results

For housed machine tools, it simplifies the study of environmental interactions, to create separate CFD simulation models for the workspace within the enclosure and the surfaces in contact with the external environment. Considering both environments in a single model is very time-consuming due to the large combined fluid volume. There is also the fact, that the simpler external model can often use steady-state simulations, while the workspace model should usually be time-dependent.

An exemplified CFD simulation result with HTC contour plots is illustrated in Fig. 8. It shows the workspace of the DMU 80 eVo with external cutting fluid supply in subfigures a and b.

A measurement-based validation of the FEM simulation model for the DMU 80 eVo for an ambient temperature increase from 20 °C to 30 °C showed good agreement with a maximum residual error of 1 K, see Fig. 9.

The ambient temperature in the climate chamber (grey dashed curve) rose to 27 °C within minutes and then settled to the final stable temperature of 29 °C within the first hour. The tested model was simulated without a housing and with the machine in standby. There is thus only natural convection with HTCs of circa 3 $W/(m^2K)$ on average. The remaining differences likely result from uncertainties in the cooling system temperatures and the surface HTCs, e.g. as a result of the neglected housing in the FEM simulation.

4 Avoidance of Thermal Errors from Ambient Sources

Thermal error avoidance can be accomplished in many different ways. Some effective methods are climatization, housings, thermo-symmetric design, material selection, cooling/heating, latent heat storage and heat pipes.

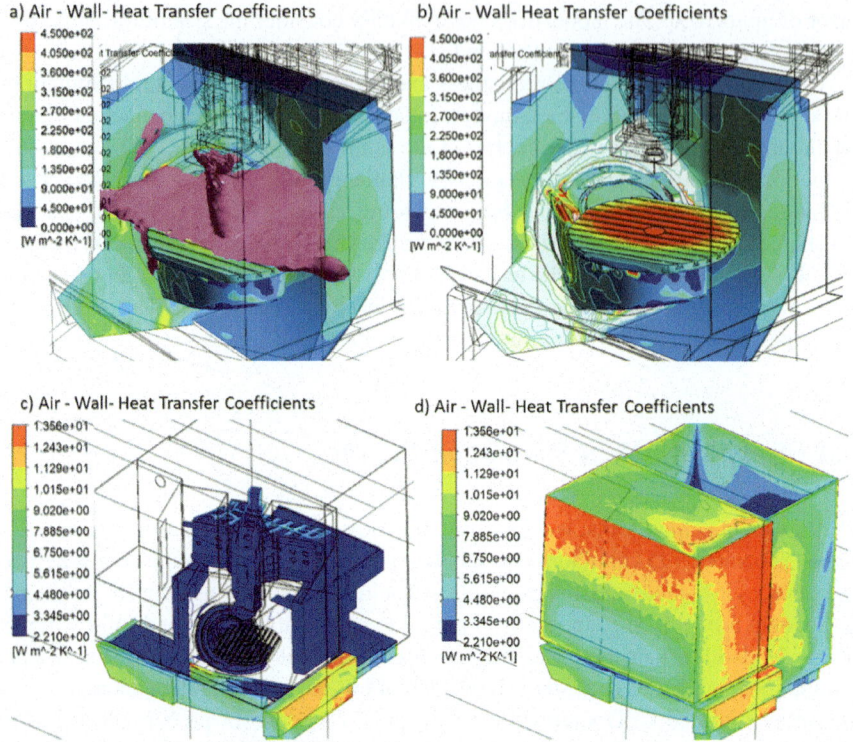

Fig. 8 Sample CFD simulation results of workspace and external environment for DMU 80 eVo (from [36] licensed under CC-BY 4.0)

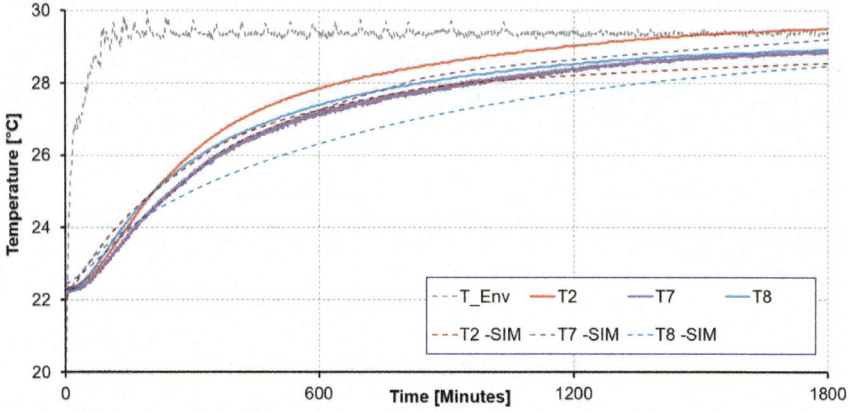

Fig. 9 Validation measurement of DMU 80 eVo for a 10 K temperature jump

4.1   Climatization of Factory and Workspace

Factory climatization is one of the most effective methods for preventing large ambient temperature changes and the thermal error associated with them. This is also why many machine tool manufacturers limit the allowable ambient temperature variation range, when precision machining is desired. Machine tools operating in small rooms can also significantly warm up the ambient air and will thus require climatization even more than ones operating in large open factories.

Another aspect of factory climatization is that it often creates a large difference in temperature and humidity between housed workspaces and the environment. During workpiece exchanges, this difference will balance out, leading to large heat transfers, usually strong cooling effects on the workspace walls. This may be enhanced by evaporative cooling from evaporating cutting fluid, as the workspace air dries rapidly. Employing air evacuation units in the workspace reduces the heat transfer during workpiece exchanges but can similarly produce cooling effects.

Climatization can also be used inside the enclosed workspace. This helps remove process heat from tool, workpiece and chips and also regulates the humidity in the workspace. The housing itself, while primarily for safety and the containment of cutting fluid and chips, is also a means of insulating the machine tool from external ambient temperature changes.

In her Bachelor's thesis, Franke performed a simulation-based study of single and double-walled machine tool housings with different wall thicknesses for a 3-axis demonstration machine. A CFD simulation was run with the machine tool at initially 20 °C under 30 °C ambient air temperature with 1 m/s forced convection from an open factory door on the left and 110-235 W solar radiation from a hypothetical factory window on the top right of the machine tool. The foundation temperature was kept at a constant 20 °C. Variants of the double wall with thicknesses of 10, 30 and 40 mm were investigated, using polyurethane (PUR) as the insulating material between the walls. Figure 10 shows the machine tool and the resulting temperature fields in steady state both outside and inside the housing for the double-walled enclosure. The TCP temperature of the single-walled model was 26.6 °C, while the double-walled model reached 25.7 °C. It was concluded that, although the use of a double-walled housing might be advantageous in cases of extreme external temperature fluctuations, the single wall was sufficient to limit the external heat and its effects on the machine tool structure [17]. Additionally, stronger insulation not only keeps warm air out more effectively, it also slows heat transfer to the outside air, when this air acts a heat sink for the warm machine tool.

Another related aspect is shielding the machine tool from external heat and radiation sources. Very hot nearby machines, radiators or windows letting in direct sunlight can typically be blocked by adding a drywall or jalousie.

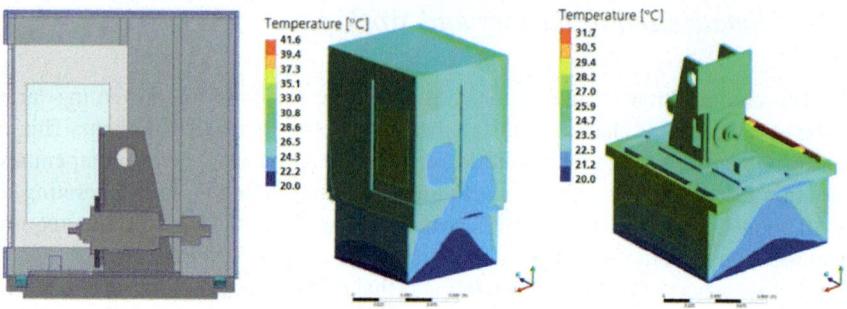

Fig. 10 Simulation of the effects of using an enclosure under ambient loads [17]

4.2 Thermal Symmetry and Machine Tool Design

Thermal symmetry is another method for error avoidance, though it is usually limited. For one thing, it is most effective, when the heat sources and sinks also have locally symmetric effects. For ambient temperature changes, this may be the case. For forced convection or exposure to solar radiation, this will generally not work. Housing a thermo-symmetric machine tool may, however, shield the kinematically relevant assemblies from asymmetric external influences.

Internal heat sources and sinks are often not thermo-symmetric either. This includes process-related effects in the workspace. Thermo-optimized machine design may still help make machine tools more thermally inert, e.g. through high thermal capacity, or more easily predictable by compensation methods, or more easily thermally managed by cooling systems or other heat flow manipulation systems.

Weng et al. have developed an analytical approach for characterizing the equivalent cuboid thermal response of structural components, which can be used to create thermal balancing strategies according to specific thermal conditions [61].

Even if a complete thermal symmetry is often difficult to achieve, some localized thermal symmetry can still improve the thermal precision of the machine tool. One example of this is the use of two fixed bearings for guiding a ball screw, which generally reduces the thermal error compared to one fixed and one floating bearing [40].

CFRP has a much lower thermal coefficient of expansion (≈ 1 μm/mK) than steel or concrete (both ≈ 12 μm/mK). Kato et al. have tested a machine tool employing CFRP for all elements and compared it to a conventional five-axis machine tool. They found that the displacement in the Z-axis was decreased by 37% on cold start and was able to maintain a stable displacement of less than 5 μm after warm-up. However, a small degradation of the thermal precision was detected after long operations due to the mechanical anisotropy of the CFRP [32]. While these results are promising, CFRP machine tools are still far from industrial application.

4.3 Cooling Systems and Heat Flow Manipulation

The most obvious method for reducing thermal errors is cooling. Most thermal errors come from heat sources distributed across the machine tool, so removing this heat as soon as it is introduced, effectively eliminates the source of the thermal error. Cooling to remove ambient effects is more difficult, because they affect all of the large outer surfaces of the machine tool. There may also be a secondary effect via the foundation, which typically follows ambient temperature changes with some delay. Some machine tools have integrated cooling systems to actively cool the machine bed, which are suitable for maintaining a stable temperature under ambient temperature changes. Hellmich et al. have designed and tested a Hydropol machine frame with several cooling circuits and investigated different control strategies such as single- vs. multi-loop control and static vs. harmonized distribution of actuating variables, where the harmonized multi-loop control achieved the best homogeneity of the temperature field under heat exposure [24]. A similar investigation by Steiert et al. also showed the advantages of a decentralized load-dependent machine frame tempering, where a multi-variable control achieved an 18% higher homogeneity under asymmetric thermal loads, when compared to a centralized two-point control [53].

Winiarski et al. have investigated the effectiveness of different liquid cooling strategies under ambient temperature variation for a machining center column. The measured air temperatures changed from around 19 °C to 25 °C/29 °C and back to 20 °C for the lower and upper air of the machining room, respectively. A developed simulation model enabled the investigation and optimization of the cooling system. Spray cooling inside the column channels with gravity flow reduced the maximum displacements from 130 μm to 52 μm. Cooling with a 5 mm thick layer of cooling channels reached 82 μm and cooling through the full cross section still 102 μm maximum displacement. To ensure a fair comparison, all cooling rates were equally set to 6.6 l/min [63].

Another aspect to consider is the coolant temperature. Depending on the machine tool, the coolant may be actively temperature-controlled or it may simply be fed by a large tank, which follows the ambient temperature. In either case, the coolant typically helps to stabilize and homogenize the temperature of the machine tool, but the resulting thermal deformation may still be significant. This is especially relevant, when a machine is freshly turned on after a longer shut-down period. In those cases, a machine frame cooling system, may take many hours to achieve a stable temperature field, during which time the thermal deformation slowly shifts.

Latent heat storage (LHS) provides a method for temporarily storing thermal energy and releasing it at a later time. How this works, depends on the setup and the (phase-change) material (PCM) used. In theory, this can be used to store the ambient heat during mid-day hours and release it at night, when the low ambient temperatures would otherwise cause thermal shrinkage. In practice, however, slow large-scale warming processes are not well suited for LHS. They also have a defined operating point, where the phase-change takes place, which lacks flexibility. In addition, the amount of storable heat is determined by the material and the volume, which limits its applicability. Once an LHS device is full, it transmits heat relatively unimpeded. Nevertheless, LHS show some potential for building envelopes [51]. On the one hand, this may facilitate more energy-efficient climatization of factories and on the other hand, further research may translate these principles to useful applications for machine tool housings.

Heat pipes are highly efficient thermal conductors, which can quickly transfer heat to heat exchangers or to other parts of a machine tool. This passive structure can be used for cooling as well as temperature field homogenization. In contrast to cooling systems, heat pipes require no pumps or electrical energy to operate. For ambient effects, however, they are not well suited, since they operate best with larger temperature gradients such as with local internal heat sources.

Voigt et al. have tested the combined application of latent heat storage and heat pipes on a demonstration machine, which allowed the easy integration of these components near the heat sources. They used PCM-filled steel bricks for ease of installation. The PCM was the paraffin-based RT28HC, which stores 220 kJ/kg with a phase transition between 24 °C and 29 °C. Conventional heat pipes with sintered capillary structures were used for heat transfer to passive metal cooling fins. They managed a temperature reduction of 50% and a thermal error lowered by 20% in the dominant direction [59]. Ambient effects have, however, not been considered and the setup used by Voigt et al. would not have been suitable for ambient loads.

An innovative thermal error control system for motor spindles consisting of CFRP bars, thermo-electric modules and cooling units with liquid coolant was developed and tested by Ge and Ding. The system was integrated into the spindle housing, where a total of eight systems was arranged in a circular configuration. In a test, the axial displacement was reduced from 214 to 5 μm [18]. The system was tested under constant ambient temperature but seems inherently able to handle ambient temperature changes. Nevertheless, extending this principle to an entire machine tool is likely too costly and not nearly as effective.

5 Compensation of Thermal Errors from Ambient Sources

Thermal error compensation involves the prediction and subsequent correction of thermal errors via offsets in the machine tool control. As previous chapters have shown, the relevant thermal displacement at the TCP is an accumulation of all the local deformations along the machine tool's kinematic chain.

Linear scales measure the actually traversed distance on a linear axis and therefore have a large influence on the thermal error. In modern machine tools, most linear axes have such scales. However, even with a linear scale, the thermal error in the respective direction is not automatically compensated. Depending on the location and type of the scale, there are usually still parts of the machine tool, whose thermal deformation is not detected by the scale. As an example, the Y-scale of the DMU 80 evo is placed on the machine frame under the Y-slide, which holds the corresponding reader. This eliminates much of the Y-error from the Y-slide deformation. Deformations of the X- and Z-slide and also from the machining table are, however, not detectable and thus require separate compensation.

The simplest method of doing this, is by producing a part, measuring its geometry, comparing that to the CAD geometry and making the necessary corrections directly in the CAD model. While this may produce one faulty workpiece, it is a legitimate approach for serial production. Changing ambient influences, such as strong day-night cycles may, however, invalidate this approach. This is because, while the production-relevant internal thermal influences remain relatively constant, the ambient influences still change the thermal error between the otherwise identical load cases.

Another similarly simple method is the measurement of the thermal error (relative TCP displacement) in regular intervals with an extrapolation from the measured errors. For this, the thermal error must typically be measured for various kinematic poses and then interpolated between them. The obvious downside to this method is the need for production interruptions and efficient measurement equipment, such as the Etalon multiline. The upside is, however, that the method handles all thermal influences equally well and compared to internal influences, ambient effects are slower and thus require less frequent recalibration measurements. Deutsch et al. demonstrate how to use a multi-channel absolute laser interferometer to efficiently measure the thermal error in the entire workspace [11]. The measurement system is well suited for large machine tools without a housing, while small housed 5-axis machining centers can be problematic for the sensitive optical equipment. Similarly, the earlier-mentioned measurement systems using laser trackers [12], one or more LaserTracers [8, 23] or photogrammetric systems [49] can be used for regular recalibration measurements.

More sophisticated compensation approaches use mathematical models, such as phenomenological models, data-driven models or simulation-based models. Most of these have ways of adjusting their predictions to changing ambient conditions.

5.1 Phenomenological Error Models

There are different types of phenomenological models. Most of these use transfer functions to map the time-dependent thermo-elastic response (TCP error) onto the heat sources and sinks, which cause them. Since there is no limitation on the number and strength of these heat sources or sinks, all ambient effects can likewise be modelled. One necessary requirement is, however, that the heat loads are measurable, which may be done, e.g., via motor current, feed rate, a temperature sensor directly at the heat source or an ambient temperature sensor. Model training is typically done from measurements, which allow the determination of the transfer function coefficients.

Brecher et al. developed and tested such a phenomenological model for compensating both internal and ambient thermal influences. Their model uses control data from the drives and spindle (torque and speed) plus one ambient temperature sensor, while process heat and cutting fluid are neglected. The transfer functions comprise first and second order time delay functions and are computed for all error directions and all thermal effects separately and then added up. The training requires the measurement of constant loads (25%, 50%, 75% and 100%) for all drives and the spindle and an ambient temperature jump of $+ 1$ K and -1 K, where all experiments are recorded until thermal equilibrium is reached [7].

Horejš et al. have likewise used transfer functions to compensate the thermal error in a horizontal milling center, where the focus was initially on spindle heat and ambient temperature variation. The model proved to reduce the thermal error from 47 μm (y) and 93 μm (z) to around 5 μm, which significantly outperformed an also tested linear regression model [26].

A method integrating grey relational and thermal sensitivity analyses and fuzzy C-means clustering (GTF) was used by Cao et al. to reduce the average thermal error on two tested machine tools by 28 and 26%, respectively. The measurements were performed in a climate chamber and contained varying spindle speeds (2000–20,000 rpm) and varying ambient temperatures (2–30 °C). The GTF method proved superior to a similar method without thermal sensitivity analysis [10].

5.2 Data-Driven Models and Artificial Neural Networks

Data-driven models are most often regression models or neural networks. In either case, their ability to handle ambient influences depends on their model structure and on their training data. Generally, at least one ambient temperature sensor is required in order to allow the model to detect ambient changes.

Naumann and Priber included ambient effects in a high-dimensional characteristic diagram. They use temperature sensors inside the machine tool to detect the impact of the ambient temperature change without the need for an ambient temperature sensor. They were able to reduce the thermal error by over 80% (RMSE) [44]. While the test included some ambient temperature change, their machine tool had no housing so that the internal sensors more directly experienced the ambient effects. Especially for complex machine tools, the number of sensors necessary for this approach can become impractical and creating the required training data too costly.

In a later work, Naumann et al. have developed a composite model, which splits the thermal influences into internal, short-term ambient and long-term ambient in order to improve the thermal error predictions for many simultaneously active thermal influences, while at the same time limiting the number of training simulations required for data acquisition [45]. Due to assumptions necessary for splitting the thermal error, this method may not work for every machine tool or it may have larger modelling errors for specific combinations of thermal influences.

Groos et al. have investigated the thermal error of a five-axis machine tool for discrete ambient temperatures of 15, 20, 25 and 30 °C in a climate chamber. They were able to reduce the maximum thermal error by 80% from 147 μm to 28 μm using a simple linear regression model. The authors admit, however, that the model is likely not transferable to continuously changing ambient temperatures in this large temperature range. They also found that squareness and linear positioning errors are dominantly affected by ambient temperatures, while straightness and rotational errors are less sensitive [23].

Blaser et al. and Mayr et al. have developed the thermal adaptive learning control (TALC), which uses an auto-regressive model with exogeneous inputs (ARX) [5, 42]. The ARX model uses both present and past data and can thus handle time-delayed correlations. This ability makes it suitable for handling both ambient and internal effects, which has been demonstrated by Lang et al., who achieved a volumetric error reduction of around 72%. The investigation was performed in a non-climatized facility with a measured ambient temperature span of around 4 K. The TALC model uses intermittent displacement measurements to recalibrate itself, as soon as larger accuracy losses are detected. This happened, e.g., when significant ambient temperature changes occurred, and the TALC automatically recalibrated to account for the new thermal conditions [37].

Wei et al. use principal component regression and ambient temperature intervals (ATIs) to compensate spindle thermal errors in a Victor Taichung Vcenter-55. In total, 46 measurements were performed across an entire year, where the initial ambient temperature varied between 4 and 33 °C due to seasonal temperature changes. By splitting the total ambient temperature range into four intervals (4.1–8.1, 8.1–17.2, 17.2–24.0, 24.0–32.2)[°C], they were able to create more reliable error predictions, improving accuracy by over 20% and robustness by over 48%. The ATIs were determined by C-means clustering. A separate model was then calculated for each ATI and during use, the currently applicable model is selected based on the measured ambient temperature [60].

Tan et al. compared an analytical compensation model with multiple regression analysis (MRA) for a large vertical 3-axis machine tool. Measurements were performed with variable spindle rotations in different seasons with average ambient temperatures of 20, 32, 23 and 7 °C for spring, summer, autumn and winter, respectively, and overlayed by day-night cycles of about 3 K amplitude. They achieved error reductions of more than 85% with their analytical model, which was superior to the MRA model. In terms of amplitude, the ambient effects contributed 40–60% to the total measured thermal error [54].

Feng et al. have used a radial basis function (RBF) neural network optimized by improved particle swarm optimization (IPSO) to compensate thermal errors on a DMG DMU 65 machine tool. They varied spindle speeds (800–12,000 rpm) and ambient temperatures (14–22 °C) in 20 measurements and were able to reduce the thermal error in Z-direction to 2 μm, which was superior to four alternatively tested ANN methods [14].

Recurrent neural networks (RNN) such as long short-term memory (LSTM) neural networks also allow for computing correlations on different time scales. Ngoc et al. used stacked LSTM models and, for comparison, also gated recurrent unit (GRU) models with power consumptions as input variables. Tests of both model types on the rotary axes of a five-axis machine tool yielded high accuracies of over 70%, though at controlled ambient temperatures between 22 °C and 23.5 °C [47]. Liu et al. used a different LSTM variant, which includes variational mode decomposition (VMD) to remove the coupling effect of high- and low-frequency data and grey wolf algorithms for hyper-parameter optimization. In their tests on spindle systems, over 75% of the thermal error was compensated, which exceeded the results for VMD-LSTM and basic RNN models [39].

Detecting solar radiation influence is only possible via temperature sensors across the machine surface. In theory, photo-electric sensors could be employed to measure the strength and direction of the incoming sunlight and use these inputs in the compensation model, but no such elaborate applications are currently known.

Another important question is what training data or more specifically training load cases to use. This depends on the conditions under which the machine tool is operated. Unfortunately, for a machine tool manufacturer, this is not always known. It also depends on what types of ambient effects can be produced for a thermal measurement or if perhaps a validated thermal simulation model is available. As was already mentioned, Brecher et al. have used a single ambient temperature jump of + 1K and −1K to train their transfer function model [7].

The composite regression model by Naumann et al. uses several ambient temperature variation simulations to train their model, specifically −10K, −5K, −2K, − 1K, + 0.2K, + 1K, + 2K, + 5K and + 15K. In two validation measurements with varying internal thermal loads at 20 and at 30 °C ambient temperature, a thermal error reduction of 56–72% and 42–52% was achieved, respectively [45]. Using so many ambient load cases should not always be necessary, depending on the operating conditions of the machine tool, but for the DMU 80 eVo, where a strong interaction between cooling system and ambient temperature was seen, the response to ambient changes varied strongly.

Compensation based on local deformation measurements, such as using CFRP rods in/on machine tool assemblies, is another effective strategy for thermal error reduction. By measuring the deformation of individual assemblies and adding up those errors along the kinematic chain, the total relative error can be computed. Bertaggia et al. used this method to reduce the thermal error of a 3-axis machine tool by between 50 and 75% [4]. This method is inherently able to handle all thermal effects, including any ambient influences. Nevertheless, explicit investigations on the effectiveness of this approach under ambient temperature variation are not known. The downside of the method is its reliance on the placement of a sufficient number of sensors in all the necessary locations. While one CFRP rod placed along the length of a component can measure its thermal elongation in one direction, component bending or partial elongation due to different kinematic poses are not directly evident without additional sensors.

5.3 Simulation-Based Compensation

Simulation-based compensation, as the name suggests, uses thermo-elastic FEM simulations or more rarely flexible multibody simulations to simulate the thermo-elastic deformation of the machine tool. By simulating the current state of the machine tool, i.e. computing both temperature and deformation field, the displacement of tool and workpiece give a prediction of the current relative TCP error. Such a simulation model necessarily has convection boundary conditions, which can (theoretically) simulate most ambient effects. Nevertheless, the method has many weak points, specifically the need for accurate parameters and long computation times. For simple load cases, the computation times can be reduced drastically through model order reduction [3, 52].

Simulating the complex conditions in the machine tool workspace with a high degree of realism is, however, still very time-consuming. Obtaining accurate parameters requires component and material data, control data as well as numerous parametrization measurements. Parameter determination is an iterative optimization process, which is performed by running many simulations and comparing the results to actual sensor data from equivalent measurements. There are, however, a number of methods and software tools, which can reduce the effort for parameter optimization [1, 25, 34].

Simulating ambient effects thus requires measurements with ambient temperature changes and a subsequent optimization of the relevant convective HTCs in the simulation model.

Ess presents a thorough description of thermal machine tool simulation and error compensation in his dissertation, where ambient temperature changes including the effects of the housing are considered. A 20 h validation using a measurement of the machine tool warm-up showed an error reduction of 50–75% [13]. The effectiveness of the method under changing ambient effects was, however, not tested.

Mian et al. developed an offline thermal error modelling methodology in an FE framework for the characterization of the thermo-mechanical behavior of a machine tool subjected to ambient temperature perturbations. The method was tested on a 3-axis vertical milling machine using two ETVE tests performed during summer (23–28 °C) and winter (20–25 °C) and showed good agreement of 60–70% between measurement and simulation [43].

Mayr et al. modelled the thermal behavior using FEA in the frequency domain and validated their methodology using the experimental data. They used their model to investigated different insulation thicknesses for their ability to reduce the thermal sensitivity of the structure. They found that the thickest tested insulation of 40 mm almost completely dampened the thermal resonance frequencies [41].

Kaulagi and Sonawane used a network-based compensation model with six nodes to predict the thermal errors of a vertical machining center in Y- and Z-direction when subjected to ambient temperature fluctuations. They investigated a near linear temperature increase from 20 to 40 °C first during machine tool shutdown, second with machine tool idle and spindle at 5600 rpm and third with 5 m/min X-, Y- and

Z-axis movements and spindle off. The model achieved mean error reductions of 69–92% on the training data and 36–88% on the test data [33].

In his dissertation, Thiem has created a thermo-elastic simulation model for structure-model based error compensation. In order to better predict thermal errors from an unknown initial thermal machine tool state, he has developed methods for estimating the initial temperature field based on temperature sensors. Using different load jumps, the time constants of all components are calculated. The smallest time constant determines the maximum duration of an interruption in the simulation, after which the simulation can be resumed without significant error (threshold set to 20%). The largest time constant determines how much data is needed to synchronize a newly started simulation to a running machine tool [55].

6 Summary and Critical Assessment of the State of the Art

The paper provides a broad overview of thermal error modelling, avoidance and compensation with a focus on ambient influences. The general methodology for thermo-elastic measurement and simulation are presented with some exemplified results for ambient thermal load cases. In the second half of the paper, methods for avoiding thermal errors, such as climatization, enclosures, heat manipulation devices, thermo-symmetric design and cooling are described with references to many relevant works in the field. The final chapter describes thermal error compensation strategies that can cope with ambient influences, e.g. via the use of transfer functions, regression models or finite element simulations.

Thermal error research has been a hot topic for many decades. The involved thermodynamic and thermo-mechanical processes are therefore quite well understood. Both measurement procedures as well as simulation models are available for a large number of different internal as well as ambient thermal loads. Measurement-based investigations of ambient effects typically require specialized climate chambers, but are otherwise state of the art. Alternative approaches, which gather data over long periods of time in order to capture meaningful ambient temperature variations, have likewise been investigated by several research groups.

Simulation-based studies for external ambient effects are well known and commercial simulation tools are, e.g. reliably able to predict the thermal response of a machine tool to ambient thermal loads. One aspect, however, which requires further study, is the determination of HTCs for machine tools with an enclosure. While studies on hull design options and their effects exist, the discrepancies between simulated and observed thermal behavior is often still significant and clear parameter identification methods for this specific issue are not known.

The other and perhaps larger issue is the simulation of the dynamic effects within the workspace during cutting operations with cutting fluid and moving assemblies. Here too, there have been studies, but accurate transient simulations are still too time-consuming, so that effective surrogate models are needed. There also need to be experimental validations of these models, which are difficult in practice due to the highly dynamic conditions in the workspace. This often only allows for local temperature measurements, which only reveal some aspects of the thermodynamic events in the workspace.

Thermal error avoidance is generally very effective at managing ambient effects. Climatization, the use of enclosures and thermo-symmetry can eliminate most of the ambient thermal errors. Devices like latent heat storage and heat pipes are not well suited for ambient influences and there are no known industrial applications of these devices in general. Since they are, however, very energy-efficient, a new concept for ambient effects followed by a thorough prototype test, could find successful commercial applications in the future.

Compensation methods are often designed to account for both internal as well as external thermal influences. For phenomenological models, an ambient temperature sensor can be used to account for air temperature fluctuations. Here, new model variants, which include ambient effects in the workspace, including process-related effects, would be a good avenue for future research.

Data-driven models have similar shortcomings with regard to process-related influences. They do, however, typically only need appropriate training data and suit-ably placed sensors to account for workspace effects. As this data becomes available and more precise, these models can also advance. One additional problem in this regard, is that most data-driven models ideally require training data with all realistic combinations of all loads, including ambient loads. Therefore, the more details of the workspace condition are represented in the model, the more data is required for a complete compensation model. Methods for reducing this growing data hunger or methods for efficiently supplying large batches of training data would help advance these correlative approaches.

Simulation-based compensation can handle external ambient effects well in general, but HTC identification and the accurate measurement of the ambient load in real-time are difficult and lead to discrepancies between simulation results and the actual thermal error. Simulations of the dynamic effects in the workspace are possible with some simplifications but are still too time-consuming for integration into online compensation routines. Here suitable surrogate models or moder order reduction techniques are needed to enhance the simulation performance. Thermo-elastic simulations are also still very time-consuming in their model preparation and parametrization. Methods for simplifying this procedure could lead to a more wide-spread usage of such simulations among machine tool manufacturers and users.

Acknowledgements This research was created in part during the Cornet project "GeoComp" 322 EBR funded by Germany's Federal Ministry for Economic Affairs and Climate Action (BMWK). The research was also cofinanced by German Research Foundation (DFG) for funding the project "Modelling of dual-phase fluid-solid interactions in the workspace of cutting machine tools to enable process-specific thermal error compensation" (ID 174223256).

References

1. Abuaniza, A., Fletcher, S., Simon, M., Naeem, S., Longstaff, A.P.: Thermal error modelling of a CNC machine tool feed drive system using FEA method. Int. J. Eng. Res. Techn. **5**(3), 118–126 (2016)
2. Baehr, H.D., Stephan, K.: Wärme- und Stoffübertragung–Grundlagen und Praxis. Springer Verlag Berlin Heidelberg, 9th edition (2016)
3. Benner, P., Herzog, R., Lang, N., Riedel, I., Saak, J.: Comparison of model order reduction methods for optimal sensor placement for thermo-elastic models. Eng. Optimization **51**(3), 465–483 (2019). https://doi.org/10.1080/0305215X.2018.1469133
4. Bertaggia, N., Tzanetos, F., Zontar, D., Brecher, C.: Investigation of thermally induced TCP-displacement under load of the machine axes in different areas. Procedia CIRP **107**, 600–604 (2022). https://doi.org/10.1016/j.procir.2022.05.032
5. Blaser, P., Pavliček, F., Mori, K., Mayr, J., Weikert, S., Wegener, K.: Adaptive learning control for thermal error compensation of 5-axis machine tools. J. Manuf. Sys. **44**(2), 302–309 (2017). https://doi.org/10.1016/j.jmsy.2017.04.011
6. Bräunig, M., Regel, J., Richter, C., Putz, M.: Industrial relevance and causes of thermal issues in machine tools. In: Conference proceedings, 1st conference on thermal issues in machine tools. Wissenschaftliche Scripten, Auerbach/Vogtland, pp. 127–139 (2018)
7. Brecher, C., Hirsch, P., Weck, M.: Compensation of thermo-elastic machine tool deformation based on control internal data. CIRP Ann. Manuf. Technol. **53**(1), 299–304 (2004). https://doi.org/10.1016/S0007-8506(07)60702-1
8. Brecher, C., Behrens, J., Flore, J., Wenzel, C.: Comprehensive calibration of robots and large machine tools using high precision laser-multilateration. Euspen Laser Metrology and Machine Performance X (2013)
9. Brier, S., Geist, A., Glänzel, J., Naumann, C., Regel, J., Dix, M., Ihlenfeldt, S.: Coupled CFD model of tool environment and workspace to determine the convective heat transfer in jet cooling of milling processes in machine tools. In Procedia CIRP, 20th CMMO (2025)
10. Cao, L., Khim, G., Baek, S.G., Chung, S.C., Park, C.H.: A selection method of key temperature points for thermal error modeling of machine tools featuring multiple heat sources. Precis. Eng. **93**, 528–539 (2025). https://doi.org/10.1016/j.precisioneng.2025.01.021
11. Deutsch, J., Albrecht, T., Riedel, M., Penter, L., Wiemer, H., Müller, J., Ihlenfeldt, S.: Thermo-elastic structural analysis of a machine tool using a multi-channel absolute laser interferometer. J. Machine Eng. **20**(3), 63–75 (2020). https://doi.org/10.36897/jme/12712
12. Egaña, F., Mutliba, U., Yagüe-Fabra, J.A., Gomez-Acedo, E.: A novel methodology for measuring ambient thermal effects on machine tools. Sensors **24**, 2380 (2024). https://doi.org/10.3390/s24072380
13. Ess, M.: Simulation and Compensation of Thermal Errors of Machine Tools. Dissertation ETH Zurich (2012). https://doi.org/10.3929/ethz-a-7357121
14. Feng, Z., Min, X., Jiang, W., Fan, S., Li, X.: Study on thermal error modeling for CNC machine tools based on the improved radial basis function neural network. Appl. Sci. **13**(9), 5299 (2023). https://doi.org/10.3390/app13095299
15. Fickert, A., Wiemer, H., Gißke, C., Penter, L.: Measuring thermally induced tool center point displacements on milling machines using a test workpiece. In: 3rd ICTIMT, Springer LNPE, ed. Ihlenfeldt S, pp. 345–357 (2023). https://doi.org/10.1007/978-3-031-34486-2_25
16. Friedrich, C., Geist, A., Yaqoob, M.F., Hellmich, A., Ihlenfeldt, S.: Correction of thermal errors in machine tools by a hybrid model approach. Appl. Sci. **14**, 671 (2024). https://doi.org/10.3390/app14020671
17. Franke, K.: Optimierte Maschinenumhausung durch Isolationseffekt. Bachelor's thesis, TU Chemnitz, Faculty for mechanical engineering (2018)
18. Ge, Z., Ding, X.: Design of thermal error control system for high-speed motorized spindle based on thermal contraction of CFRP. Int. Journ. Machine Tools Manufacture **125**, 99–111 (2018). https://doi.org/10.1016/j.ijmachtools.2017.11.002

19. Geist, A., Naumann, C., Glänzel, J., Putz, M.: Methods for determining thermal errors in machine tools by thermo-elastic simulation in connection with thermal measurement in a climate chamber. MM Sci. J. pp. 6575–5681 (2023). https://doi.org/10.17973/MMSJ.2023_06_2023049

20. Glänzel, J., Ihlenfeldt, S., Naumann, C., Putz, M.: Decoupling of fluid and thermo-elastic simulations of machine tools using characteristic diagrams. Procedia CIRP **62**, 340–345 (2016). https://doi.org/10.1016/j.procir.2016.06.068

21. Glänzel, J., Naumann, C., Ihlenfeldt, S., Putz, M.: Efficient quantification of free and forced convection via the decoupling of thermo-mechanical and thermo-fluidic simulations of machine tools. J. Machine Eng. **18**(2), 41–53, ISSN 1895–7595 (2018)

22. Glänzel, J., Kumar, T.S., Naumann, C., Putz, M.: Parameterization of environmental influences by automated characteristic diagrams for the decoupled fluid and structural-mechanical simulations. J. Machine Eng. **19**(1), 98–113, ISSN 1895–7595 (2019)

23. Groos, L., Held, C., Keller, F., Wendt, K., Franke, M., Gerwien, N.: Mapping and compensation of geometric errors of a machine tool at different constant ambient temperatures. Precis. Eng. **63**, 10–17 (2020). https://doi.org/10.1016/j.precisioneng.2020.01.001

24. Hellmich, A., Mater, S., Popken, J., Ihlenfeldt, S.: Control approaches for tempering machine tool frames with multiple fluid channels and limited, jointly used actuating variable. MM Sci. J. Special Issue ICTIMT2021 (2021). https://doi.org/10.17973/MMSJ.2021_07_2021074

25. Hensel, B., Schroeder, S., Kabitzsch, K.: Parameter identification software for various thermal model types. In: 1st ICTIMT 2018, Dresden, URN: nbn:de:bsz:14-qucosa2-326441 (2018)

26. Horejš, O., Mareš, M., Kohut, P., Barta, P., Hornych, J.: Compensation of machine tool thermal errors based on transfer functions. MM Sci. J. **162/163** (2010). https://doi.org/10.17973/MMSJ.2010_03_201001

27. ISO.: ISO 230-3—Test code for machine tools–Part 3: determination of thermal effects, ISO, 2020 (2020)

28. ISO.: ISO 10791-7—Test conditions for machining centres (2014)

29. Jedrzejewski, J., Kaczmarek, J., Reifur, B.: Description of the forced convection along the walls of machine tool structures. Ann. CIRP **37**(1), 397–400 (1988). https://doi.org/10.1016/S0007-8506(07)61663-1

30. Jungnickel, G.: Forschung Lehre Praxis–Simulation des thermischen Verhaltens von Werkzeugmaschinen. TU Dresden (2000)

31. Jungnickel, G.: Lehre Forschung Praxis–Simulation des thermischen Verhaltens von Werkzeugmaschinen–Modellierung und Parametrierung. Institut f. Werkzeugmaschinen und Steuerungstechnik (2010)

32. Kato, M., Mizoguchi, Y., Kono, K., Kakinuma, Y.: Machining-based thermal error analysis of CFRP-structured machine tool. In: Proceeding of the Machining Innovations Conference 2020, MIC Procedia, SSRN, pp. 27–33 (2020). https://doi.org/10.2139/ssrn.3724093

33. Kaulagi, M.N., Sonawane, H.A.: Thermal network-based compensation model for a vertical machining center subjected to ambient temperature fluctuations. Int. J. Adv. Manuf. Technol. **124**, 3973–3994 (2022). https://doi.org/10.1007/s00170-021-08241-6

34. Kim, B.-S., Park, J.-K.: Thermal error compensation for a high-precision lathe. In: Proceeding 9th international workshop on micro factories, Honolulu (2014)

35. Klein, S., Kummer, B., Wiegand, K.: Particle simulation in Flu-id mechanics: basics and applications. Wiley-VCH (2016)

36. Kumar, T.S., Naumann, C., Geist, A., Glänzel, J.: Modellierung von Umgebungseinflüssen. In: Thermo-energetische Gestaltung von Werkzeugmaschinen, Springer Vieweg (2025). https://doi.org/10.1007/978-3-658-45180-6_10

37. Lang, S., Zimmermann, N., Mayr, J., Wegener, K., Bambach, M.: Thermal error compensation models utilizing the power consumption of machine tools. In: ICTIMT 2023, Springer LNPE, ed. Ihlenfeldt S, pp. 41–53 (2023). https://doi.org/10.1007/978-3-031-34486-2_4

38. Liu, H., Meurer, M., Schraknepper, D., Bergs, T.: Investigation of the cutting fluid's flow and its thermomechanical effect on the cutting zone based on fluid–structure interaction (FSI) simulation. Int. J. Adv. Manuf. Technol. **121**, 267–281 (2022). https://doi.org/10.1007/s00170-022-09266-1

39. Liu, J., Ma, C., Gui, H., Wang, S.: Thermally-induced error compensation of spindle system based on long short term memory neural networks. Appl. Soft Comput. **102**, 107094 (2021). https://doi.org/10.1016/j.asoc.2021.107094
40. Mayr, J., Jedrzejewski, J., Uhlmann, E., Donmez, M.A., Knapp, W., Härtig, F., Wendt, K., Moriwaki, T., Shore, P., Schmitt, R., Brecher, C., Würz, T., Wegener, K.: Thermal issues in machine tools. CIRP Ann. Manuf. Technol. **61**(2), 771–791 (2012). https://doi.org/10.1016/j.cirp.2012.05.008
41. Mayr, J., Ess, M., Pavlíček, F., Weikert, S., Spescha, D., Knapp, W.: Simulation and measurement of environmental influences on machines in frequency domain. CIRP Ann. **64**(1), 479–482 (2015). https://doi.org/10.1016/j.cirp.2015.04.001
42. Mayr, J., Blaser, P., Ryser, A., Hernandez-Becerro, P.: An Adaptive self-learning compensation approach for thermal errors on 5-axis machine tools handling an arbitrary set of sample rates. CIRP Ann. **67**, 551–554 (2018). https://doi.org/10.1016/j.cirp.2018.04.001
43. Mian, N.S., Fletcher, S., Longstaff, A.P., Myers, A.: Efficient estimation by FEA of machine tool distortion due to environmental temperature perturbations. Precision Eng. **37**(2), 372–379 (2013). https://doi.org/10.1016/j.precisioneng.2012.10.006
44. Naumann, C., Priber, U.: Modellierung des Thermo-Elastischen Verhaltens von Werkzeugmaschinen mittels Hochdimensionaler Kennfelder. In: Proceeding workshop computing intelligence 22, Dortmund, pp. 365–383 (2012)
45. Naumann, C., Geist, A., Putz, M.: Handling ambient temperature changes in correlative thermal error compensation. J. Machine Eng. **23**(4), 43–63 (2023). https://doi.org/10.36897/jme/175397
46. Naumann, C., Schmidt, C.-D., Geist, A., Brier, S., Glänzel, J., Klimant, P., Ihlenfeldt, S., Dix, M.: Investigation of tool cooling and heat transfer using computational fluid dynamics simulations. J. Machine Eng. **24**, 5–26 (2024). https://doi.org/10.36897/jme/196074
47. Ngoc, H.V., Mayer, J.R.R., Bitar-Nehme, E.: Deep learning to directly predict compensation values of thermally induced volumetric errors. Machines **11**, 496 (2023). https://doi.org/10.3390/machines11040496
48. Pavlíček, F.: Parametrierbare Metamodelle zur Berechnung des Wärmeübergangs in Hohlräumen. Dissertation, TU Chemnitz (2017)
49. Riedel, M., Deutsch, J., Müller, J., Ihlenfeldt, S.: Design of a photogrammetric measurement system for displacement and deformation on machine tools. In: Proceedings of the conference on thermal issues in machine tools, Verlag Wiss. Scripten, Zwickau, ISBN: 978-3-95735-085-5 (2018)
50. Schwenke, H., Franke, M., Hannaford, J., Kunzmann, H.: Error mapping of CMMs and machine tools by a single tracking interferometer. CIRP Ann. **54**(1), 475–478 (2005). https://doi.org/10.1016/S0007-8506(07)60148-6
51. Salihi, M., Chhiti, Y., El Fiti, M., Harmen, Y., Chebak, A., Jama, C.: Enhancement of buildings energy efficiency using passive PCM coupled with natural ventilation in the Moroccan climate zones. Energy Build. **315**, 114322 (2024). https://doi.org/10.1016/j.enbuild.2024.114322
52. Spescha, D., Hernandez-Becerro, P., Wegener, K.: Model order reduction of thermo-mechanical models with parametric convective boundary conditions: focus on machine tools. Comput. Mech. **67**, 167–184 (2020). https://doi.org/10.1007/s00466-020-01926-x
53. Steiert, C., Weber, J., Hellmich, A., Geist, A., Mater, S., Glänzel, J., Weber, J., Ihlenfeldt, S.: Optimierte Temperierung von Maschinengestellen für unsymmetrische Lasteinträge. In: Brecher, C. (eds) Thermo-energetische Gestaltung von Werkzeugmaschinen. Springer Vieweg (2025). https://doi.org/10.1007/978-3-658-45180-6_18
54. Tan, B., Mao, X., Liu, H., Li, B., He, S., Peng, F., Yin, L.: A thermal error model for large machine tools that considers environmental thermal hysteresis effects. Int. J. Mach. Tools Manuf **82–83**, 11–20 (2014). https://doi.org/10.1016/j.ijmachtools.2014.03.002
55. Thiem, X.: Ein Beitrag zur strukturmodellbasierten Korrektur thermisch bedingter Fehler an Werkzeugmaschinen. Dissertation, TU Dresden (2024)
56. Uhlmann, E., Barth, E., Seifarth, T., Höchel, M., Kuhnert, J., Eisenträger, A.: Simulation of metal cutting with cutting fluid using the Finite-Pointset-Method. Procedia CIRP **101**, 98–101 (2021). https://doi.org/10.1016/j.procir.2021.02.013

57. Verein Deutscher Ingenieure (VDI).: VDI-Wärmeatlas–Berechnungsblätter für den Wärmeübertrag. VDI Verlag Berlin (2006)
58. Versteeg, H.K., Malalasekera, W.: An introduction to computational fluid dynamics: the finite volume method. Pearson Education (2007)
59. Voigt, I., Fickert, A., Wiemer, H., Drossel, W.-G.: Experimental investigation of passive thermal error compensation approach for machine tools. In: 3rd ICTIMT 2023, Springer LNPE, ed. Ihlenfeldt S, pp. 265–277 (2023). https://doi.org/10.1007/978-3-031-34486-2_19
60. Wei, X., Ye, H., Feng, X.: Year-round thermal error modeling and compensation for the spindle of machine tools based on ambient temperature intervals. Sensors 22, 5085 (2022). https://doi.org/10.3390/s22145085
61. Weng, L., Gao, W., Zhang, D., Huang, T., Liu, T., Li, W., Zheng, Y., Shi, K., Chang, W.: Analytical modelling method for thermal balancing design of machine tool structural components. Int. J. Machine Tools Manuf. 164 (2021). https://doi.org/10.1016/j.ijmachtools.2021.103715
62. Wiessner, M., Blaser, P., Böhl, S., Mayr, J., Knapp, W., Wegener, K.: Thermal test piece for 5-axis machine tools. Precis. Eng. 52, 407–417 (2018). https://doi.org/10.1016/j.precisioneng.2018.01.017
63. Winiarski, Z., Jedrzejewski, J., Kwasny, W., Ha, H.: Reduction of precise machining centre column thermal deformations caused by changes in ambient temperature by means of liquid cooling. J. Manuf. Process. 110, 192–201 (2024). https://doi.org/10.1016/j.jmapro.2023.12.030
64. Zimmermann, N., Lang, S., Mayr, J., Wegener, K.: Validating real time compensation: a thermal test piece for 5-axis machine tools to separate thermal errors in Z-direction. Precis. Eng. 91, 263–277 (2024). https://doi.org/10.1016/j.precisioneng.2024.08.014
65. Zwingenberger, C.: Beitrag zur Verbesserung der Simulationsgenauigkeit bei der Bestimmung des thermischen Verhaltens von Werkzeugmaschine. Dissertation, TU Chemnitz (2014)

Thermally Induced Volumetric Error Modelling on Large-Scale Machining Centre

Álvaro Sáinz de la Maza García ⓘ, **Leonardo Sastoque Pinilla** ⓘ, and **Luis Norberto López de Lacalle Marcaide** ⓘ

Abstract The aerospace sector is in the constant need of large-scale, highly precise components manufactured in challenging materials with tight tolerances by chip removal processes [1]. When machining, ambient temperature changes, the heat generated by machine moving components and motors, and the chip removal process itself, lead to dimensional changes in structural components of machine tools, and a subsequent positioning error. Thermally induced positioning errors of large-scale machines can represent up to a 75% of the total error [2]. To this day, avoiding temperature-induced errors in the machined workpiece remains a challenge. This work presents a methodology for measuring the impact of ambient and machine usage temperature changes, and a simple yet useful relation between temperature or deformation measurements and Tool Centre Point (TCP) positioning errors. The research was experimentally conducted in a large-scale, five-axis machining centre, in which, apart from numerous internal sensor values and CNC variables, a total of 49 temperature and 14 Integral Deformation Sensor measurements were recorded and correlated to tool positioning errors. Errors for different temperature conditions were measured using a calibrated metrological artifact. The obtained model, capable of estimating thermal errors in TCP can be easily applied to compensate them in real time with low computing requirements, as proposed at the end of the article. Therefore, implementing it to compensate thermal deformations in industrial environment is straightforward. In this article, two similar models are presented, one based on the

Á. Sáinz de la Maza García (✉)
Aeronautics Advanced Manufacturing Centre (CFAA), University of the Basque Country (UPV-EHU), Zamudio, Spain
e-mail: alvaro.sainzdelamaza@ehu.eus

L. Sastoque Pinilla
Aeronautics Advanced Manufacturing Centre (CFAA), University of the Basque Country (UPV-EHU), Zamudio, Spain
e-mail: edwarleonardo.sastoque@ehu.eus

L. N. López de Lacalle Marcaide
Aeronautics Advanced Manufacturing Centre (CFAA), University of the Basque Country (UPV-EHU), Zamudio, Spain
e-mail: norberto.lzlacalle@ehu.eus

© The Author(s) 2026   273
K. Wegener and M. Bambach (eds.), *4th International Conference on Thermal Issues in Machine Tools (ICTIMT2025)*, Lecture Notes in Production Engineering,
https://doi.org/10.1007/978-3-032-01194-7_17

temperature measurements, while the other considers deformation measurements. The main objective of this research was not only to predict thermal errors, but to do it with a volumetric approach. Thus, the proposed models, which both show accurate error predictions, consist of two staged non-linear models. Firstly, volumetric error is modelled by a non-linear multivariable fit depending on a low number of parameters. In a second stage, each parameter is fitted by other non-linear multivariable functions depending on the most representative sensors, thus considering thermal effects. The results showed a 77% error reduction when using the model based on temperature measurements.

Keywords Thermal effects · Volumetric error · Prediction model · Machine tool

1 Introduction

Some large-scale components, such as those for the aerospace sector need to be manufactured by chip removal processes in difficult to cut materials with tight tolerances using big machining centres [1]. During the process, machine movements, the chip removal process and the ambient temperature variation tend to modify machine components temperature, affecting their dimensions and inducing Tool Centre Point (TCP) positioning errors that may represent up to a 75% of the total positioning error [2]. Avoiding thermally induced errors remains a challenge but controlling them in small machines may be accomplished using complex cooling systems [3]. In bigger machines however, this solution is unacceptable in terms of cost and energy consumption. Thus, machine tool manufacturers tend to design their machines in such a manner that the machine deforms uniformly, trying to avoid local effects [4].

Modelling thermal errors in machine tools has been a recurrent research topic for the last four decades [5]. Therefore, multiple methods have been studied, and diverse models have proven effective. In the latest years, there is a growing tendency on using complex neural-network-based models, for example those presented by Zang, et al. [6] and Yang, et al. [7], letting aside simple classical models using single variable (like the one by Liu, et al. [8]) or multivariable (such as that by Liu, et al. [9]) regressions. This work uses two staged non-linear multivariable fits, showing better results with less parameters than the usual single or multivariable linear functions.

It is common to model thermal and kinematic errors separately (this is the way both corrections are usually introduced using simple models natively in CNCs). For this research, a different approach was used, considering the positioning error of the TCP as a single value for each XYZ direction depending on the coordinates of the TCP (kinematic approach) and on the temperature of the components (thermal approach). A single model that predicts the total error was obtained, not considering the source of it, similarly as studied by Iñigo et al. [10]. In this sense, it is common to study machine errors using the decomposition showed on ISO 230-7 standard [11], determining the origin of each error, as done by Iñigo, et al. [12]. In this research, although the source of the error was analysed to understand if the results make sense, the exposed model does not consider it, it estimates the error from a set of calibration data as a "black box", keeping the model simple enough for real-time execution.

Working in this way, without needing to understand the effect of each heat source on all components allows using sensor information in each moment (or average a few number of measurements to avoid noise), not needing to collect the historical values to do a prediction. For a certain combination of sensor measurements, no matter how that situation was reached, the predicted error would be the same. This behaviour, consequence of the machine high thermal inertia, was detected during previous tests.

This research resulted in a model that estimates positioning errors decomposed in XYZ components depending on theoretical TCP coordinates and temperature or deformation measurements. A mix of temperatures and deformations could also be used. At the same time, the model must be simple and light enough to be implemented in real time. Both the experimental testing and the proposed model are described.

This work, giving a similar approach to that by Lang, et al. [13] but using temperature and deformation measurements instead of power consumption, covers the gap of data-driven thermal error prediction models requiring low training time. The proposed model, gives virtually the same improvement as S. Lang, et al.'s, requiring approximately 10 times less experimental parameter adjustment time, making it easier to use in industrial environments, and covering the proposed future research.

2 Experimental Measurements

For measuring the deformation of the main structural components of the machine, 14 Integral Deformation Sensors (IDS) were installed. To obtain a temperature map of the machine, 49 thermocouples were also mounted all over the machine, 3 of which were installed in the workshop to measure ambient temperature at different heights. The location of the different IDSs and thermocouples was chosen based on the results of exhaustive simulations using the finite element method and multiple experimental warming tests, which are out of the scope of this research. To determine the best possible locations the effect of ambient temperature and heat generated by the moving components on each structural element was studied, similarly as done by Naumann et al. [14]. Both structure deformations and temperature maps were analysed, and the most representative positions were selected. Final sensor placement locations were chosen to ensure that all significant heat sources effects in machine components were measured. Sensors are located where components showed greater temperature and length variations as consequence of both ambient temperature and internal heat generation. During this stage, in each support used to fix the IDSs a thermocouple was installed, to study if the use of IDSs could be avoided for future industrial environments and substituted by much cheaper thermocouples.

To measure positioning errors, after analysing different alternatives [15] and considering the machine configuration and limitations, the use of a touch probe installed in the machine against a calibrated metrological artifact like the one used by Gomez-Acedo et al. [16] was chosen. The artifact geometry is shown in Fig. 2a.

Both before starting the research and during the analysis of the data, relations between sensors (thermocouples and IDSs) and positioning errors were studied aiming at simplifying the mathematical model. These readings showed great correlation, allowing to reduce the number of sensors to three selected IDSs or thermocouples. These most relevant sensors were IDS bars 2, 5 and 9 and thermocouples 25, 43 and 50, which correspond to the bars coloured in red in Fig. 1 and thermocouples coloured in blue in this same figure. It should be noted that the value named "thermocouple 50" is not a real thermocouple, but the average value of thermocouples 47, 48 and 49, measuring the workshop ambient temperature at different heights. Note that the numbering used for both types of sensors is not consecutive as these are the numbers used internally to identify all sensors, avoiding possible mistakes. Figure 1 also shows the main structural elements, and the configuration of the machine used for this research. Using IDSs, it was seen that the biggest components suffered thermal deformations of just over 200 μm during the pretesting, which indicates the expected magnitude of errors.

It is important to point out that although these sensors showed more relevant data and are inputs for the model, during the design and adjustment stage all sensors and variables were considered, and their correlation with TCP errors was analysed. For clearness, in this article sensors not included in the final model are omitted.

The calibrated artifact (Fig. 2a), with a carbon fibre structure to assure low thermal deformations (negligible compared to machine deformations), has 11 highly precise spheres located at 3 different angular positions (0°, 45° and −45° respect to the vertical plane) around the axis of the main bar and uniformly distributed all along its 1500 mm main dimension to make it three-dimensional, being the one used by Iñigo, et al. [10]. During calibration, sphere centre distances were measured with an

Fig. 1 Machine structure and location of used thermocouples and IDSs

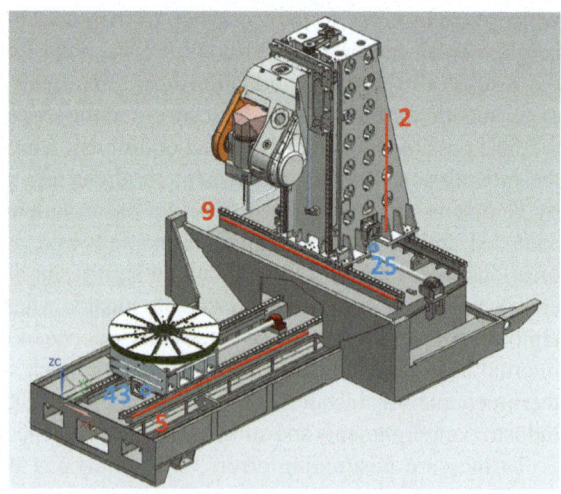

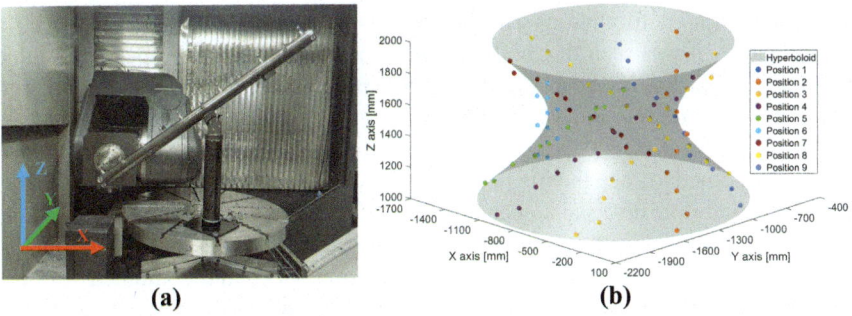

Fig. 2 **a** Carbon fibre artifact with 11 spheres inside the machine for measurement. **b** Hyperboloid generated by the line connecting spheres 1 and 11 when rotating the table and location of all measuring points

uncertainty of 9 μm in a Mitutoyo Crysta Apex S162012 CMM. To achieve a full coverage of the machining centre working volume, the artifact was located in the turntable, near the perimeter of it and perpendicular to its radius, so that rotating the C axis led to a single sheet hyperboloid as shown in Fig. 2b, where the theoretical location of all spheres in each position is also shown. It was also considered that for the machine construction, both when milling and when turning, the periphery of the turntable is used most often.

To generate all measurement points, 8 different positions of the turntable were used, adding a ninth one coincident with the first to quantify the effect of the temperature change during the measuring process (which took around 45 min). In some of these positions, due to the size of the artifact, it was not possible to measure all the spheres. To work in a representative situation and, considering that moving also the A axis would increase the collision chance and excessively lengthen the measurement time, positioning the spindle only in vertical or horizontal directions was accepted (emulating vertical or horizontal machining centres configuration). To maximize the number of measurements per test without colliding, of those two options, the horizontal spindle position was selected. 76 sphere positions could be reached without interferences, and thus, measured (X, Y and Z position measurements for each sphere (228 data values)). The measurement method is explained in detail by Sáinz de la Maza García, et al. [17]. Note that in Fig. 2b, position 1 cannot be seen as position 9 is drawn on top. To increase the significance of the results, instead of absolute positions, distances between spheres were measured and compared to the calibrated values, obtaining the error values used in this work. Although absolute positions were not analysed, in position 1 for the reference ambient temperature measurement, sphere 1 was set as local coordinate system origin, locating the artifact so that sphere 9 was also in XZ plane. However, for all calculations, machine coordinates (and errors calculated from distances between sphere centres) were used, obtaining a more useful result.

Before the positioning error measurements, during all sensor and acquisition system testing, the most important heat sources were identified and the relation of temperatures and deformations with a combination of heat sources was analysed. In these pretests, it was seen that both the main spindle and the high-speed turn table had very fast dynamics, reaching the maximum temperature and cooling back to ambient temperature in approximately 5 minutes. A similar behaviour was seen in the torque motor of the A axis. This can be explained because all three mentioned motors are cooled to keep their temperatures as low and constant as possible. Thanks to this refrigeration, it was also seen that the movement of these motors did not have a noticeable effect on other sensors than those installed in the motors themselves, even when their movement was made at maximum speed and combined with the axes movements. The measurement of all spheres took longer than the cooling of these components, so it was decided to avoid their movement during heating cycles. If a faster acquisition method could have been used, it would have been interesting to analyse also their influence on errors.

To study the effects of both ambient and internal temperature changes, limited to the time availability of the machine, a total of 10 tests were designed, in which all selected axes were moved individually and in the combinations that have most cross influence. Some additional measurements without internal warming and after the heating cycles of all axes together, during the cooling stage, were performed. Except from these tests, all measurements were started 5 min after the heating cycle had finished to keep consistency and avoid technical limitations immediately after heating cycle finished. Table 1 shows briefly the tests performed and the ID used in the rest of the article to refer to them. Ambient tests allowed modelling uniform temperature changes, while heating tests were specially designed to study time-varying, non-uniform temperature changes due to moving parts friction and axes heat generation.

Table 1 Performed tests' warming speeds of axes

Id	Test	Linear speeds (m/min)			Rotating speeds (rpm)			Movement range	Comments
		X	Y	Z	A	C	Spindle		
1	X warming	30	0	0	0	0	0	Full	–
2	Y warming	0	25	0	0	0	0	Full	–
3	Z warming	0	0	30	0	0	0	Full	–
4	XZ warming	30	0	30	0	0	0	Full	–
5	XYZ warming	30	25	30	0	0	0	Full	–
6	XYZ cooling 1	0	0	0	0	0	0	–	After 5
7	XYZ cooling 2	0	0	0	0	0	0	–	After 6
8	Half X warming	30	0	0	0	0	0	Last half	–
9	Ambient 1	0	0	0	0	0	0	–	–
10	Ambient 2	0	0	0	0	0	0	–	–

3 Proposed Model

As the main objective of this method is to calculate expected errors to allow a real-time error compensation implementation, it was chosen to keep the model as light as possible without severely compromising its precision. To effectively estimate positioning errors, as these depend not only on measured temperatures but also on the position of the TCP (due to non-uniform deformations), a thermo-volumetric error model is proposed, estimating the effect of temperature changes in a volumetric manner.

Function 1 represents the model of the volumetric error, where x, y, z and c are the machine coordinates of all three linear axes and the angular position of the rotating table measured in mm and degrees respectively. K_i for $i = 1 \ldots 8$ are the coefficients that define the error in the studied direction. Function 1 was obtained from each individual axis contribution shape and the most representative cross-correlations of the different axes. It was seen that axes are not very coupled for calculating positioning error. The effect of C rotation axis was modelled using sinusoidal terms, while linear axes effect showed better fitting to polynomials. As this equation is valid for the modelling of errors in the whole volume for the three XYZ components, K_i coefficients are K_{ij}, where $j = x, y, z$ represents the error calculation component, and f_{Vol} is in fact f_{Vol_j}.

$$f_{Vol}(x, y, z, c) = K_1 \cdot x + K_2 \cdot y^2 + K_3 \cdot z + K_4 + K_5 \cdot \cos\left(\frac{\pi \cdot c}{180}\right) \quad (1)$$
$$+ K_6 \cdot \sin\left(\frac{\pi \cdot c}{180}\right) + K_7 \cdot x \cdot y^2 \cdot z + K_8 \cdot \cos\left(\frac{\pi \cdot c}{180}\right) \cdot \sin\left(\frac{\pi \cdot c}{180}\right)$$

To completely model the error for each test k during the calibration stage, the coefficients K_{ij}^k were calculated based on Function 1, and the measured errors by using the least squares method. As already exposed, during the calibration stage 10 tests were performed, leading to a total of $8 * 3 * 10 = 240$ K_i coefficients.

Once the volumetric error was fitted, the second step in the calibration stage is to fit each K_{ij} for all k values as a function of a reduced number of temperatures or IDS readings. To achieve this task, the least squares method was used again using Functions 2 and 3 with the temperature or deformations measurements respectively, applying the Levenberg-Marquardt [18] algorithm in this case, as the data available is scarce.

$$f_{Temp}(T_{25}, T_{43}, T_{50}) = K_{T1} \cdot T_{25}^2 + K_{T2} \cdot T_{50}^{\frac{2}{3}} + K_{T3} + K_{T4} \cdot T_{43}^2 \quad (2)$$

Function 2 represents the values of the K_i coefficients of Function 1 depending on the temperature of thermocouples 25 and 43 and the mean ambient temperature (T_t for $t = 25, 43, 50$) in °C. This function is defined by four K_{Ti} coefficients that take different values for each positioning error component, thus being $4 * 8 * 3 = 96$ K_{Tij}

coefficients that define the errors in a volumetric way as function of the selected temperatures.

Similarly Function 3 represents the values of the K_i coefficients of Function 1 depending on the deformation measurements of IDS bars 2, 5 and 9 (B_b for $b = 2, 5, 9$) in μm. Like in Function 2, K_i coefficients are fitted by the coefficients K_{Bi}, which also take different values for each direction of the error (K_{Bij}). To successfully fit the model, ($\overline{B}_b^*$), mean values of the deformations during the calibration stage were used as centring values, mathematically needed because of the small deformations compared to initial bar length measurements.

$$f_{Bar}(B_2, B_5, B_9) = K_{B1} \cdot \left(\overline{B}_2^* - B_2\right)^{\frac{2}{5}} + K_{B2} \cdot \left(\overline{B}_9^* - B_9\right)^3 + K_{B3} \cdot \left(\overline{B}_5^* - B_5\right)^{\frac{2}{5}} + K_{B4} \quad (3)$$

Using Functions 2 and 3 and the least squares method, the result of the calibration stage is the value of all K_{Bij} or K_{Tij} coefficients. These values are considered to keep constant during the lifetime of the machine, unless significant structural modifications are made, although periodic calibrations are recommended. During the design of the model, numerous functions were analysed and compared to choose the most appropriate ones in terms of fitting quality and stability.

During the stage of use of the model, using the temperatures or the IDSs measurements and Functions 2 or 3, the coefficients for Function 1 can be calculated virtually in real time. These coefficients completely define the value of the error components using Function 1 for the theoretical value of X, Y, Z and C axes.

4 Experimental Error Prediction

As stated above, during the experimental measurements, 10 different conditions were studied. In each test, the temperature and deformation measurements were seen to keep virtually constant thanks to the great thermal inertia of the machine, which allowed to consider the mean measurement of each sensor as representative of the whole cycle. Using this data and the model exposed in the above section, XYZ error components for each measurement point were estimated and compared with the performed measurements. The results for the model using the temperature measurements are summarized in Fig. 3, in which for all 10 test situations, the real measured errors and the estimations are shown. Results using deformation measurements are similar to those obtained with temperatures, but as the cost of IDSs is much higher than that of thermocouples, this model is industrially less useful. Note that the horizontal axis does not have any physical meaning, it represents all measurement points. Some real errors are missing as there was no chance to reach them with the touch-probe without colliding with the artifact.

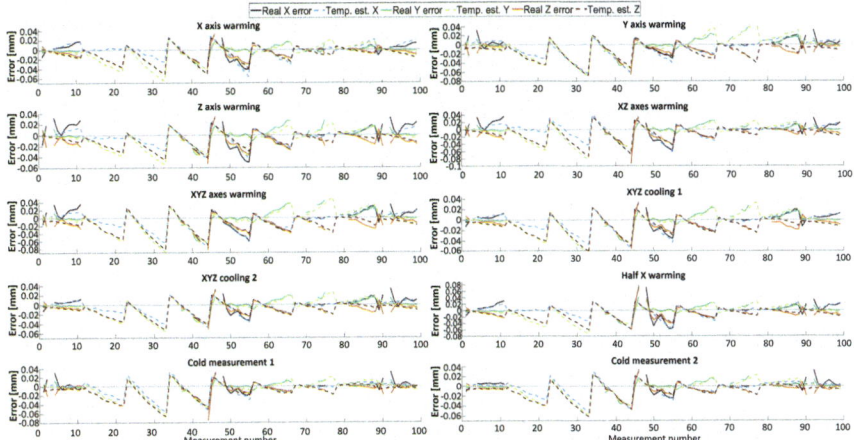

Fig. 3 Measured error and temperature-based estimation of positioning errors for all measurement points in all performed tests by X, Y and Z components

As it can be seen in Fig. 3, the error predicted in each direction fits adequately the measurement of the real error in most measurement points, except in those where in a short variation of axis positions the positioning error changed fast. It can also be seen that the global shape of the error is similar in all cases, showing the importance of the volumetric error of the machine, but not all predictions are equal, each is adjusted to the measured temperatures, which slightly modifies the amplitude of errors.

Using the temperature measurements as input for the model, the mean absolute error of the predictions is approximately of 5.17 μm, while using the deformation measurements, this error is slightly bigger, of 5.48 μm. In both cases, the error is mainly due to a low number of measurement points. These estimation errors, although still significant, are considerably smaller than the amplitude of the measured errors, which reached up to 171 μm, with a mean value of about 22.4 μm. Furthermore, these modelling errors have a similar magnitude to the measurement uncertainty of the touch probe and encoders of the machine used during the tests. It should also be noted that in the majority of measurement points, the prediction underestimates the magnitude of the error, which means that if implemented to correct machine movements, the error would be reduced approximately by a 77%, avoiding overcorrection issues.

Experimental results showed that, as predicted, large-scale machine tool thermally-induced-positioning-errors may be modelled using rather simple equations that depend on a low number of temperature or deformation measurements. This result implies that thermally induced positioning error can be predicted using adequate discrete measurements, not being necessary to study temperature evolution and temperature gradients in machine components: no matter how a certain thermal situation is reached, if it is consequence of environment and machine axes movements, positioning error can be predicted with a few local temperature real-time measurements.

Once the model was designed and after seeing the promising results showed by it, the statistical meaningfulness was analysed. As the number of tests performed was very limited, possible overfitting was studied using the "Leave one out cross validation (LOOCV)" method [19]. This way, overfitting of any of the two stages of the model could be identified, for which the calculation of the expected errors was made for each test using the data of all other 9 tests, leaving its own data out of the calibration stage (10 calibrated models were used, one for each test predictions). The final prediction errors for all measurement points in each of the 10 tests were calculated and compared with those obtained using the data of all 10 tests for all predictions. The result of this validation showed little difference between using the data of each test for the prediction of the test itself or removing this data, with a total difference of a 1.5% on the coefficients of the first stage, and just a 0.23% in the final error prediction. Using the data of each test for its own prediction slightly improves the estimation, as expected for having more data, but not enough to be caused by overfitting of the model. Thus, the model, with the data available, can be considered statistically meaningful, but should be further validated before using it in a production environment.

Therefore, using larger datasets to adjust the model is expected to improve the representation of the real behaviour of the machine and improve the model as a whole, reducing the prediction error and as a consequence, allowing better compensation of tool paths to minimize positioning errors.

5 Possible Implementation of the Model

The proposed method is not only capable of modelling thermo-volumetric errors of large-scale machines with low control on ambient and internal temperatures, but it may also be applied on thermally highly stable machines reducing the model to the first step and avoiding the second, in which the variation of the coefficients with the temperature is estimated. In this way, thermal effects would not be considered by the model, reaching a simple yet useful volumetric error model, that compensates kinematic errors. Although the implementation and validation of a full model to compensate thermal errors is out of the scope of the research presented here, a methodology to compensate thermal errors in real time using any of the proposed models (with temperature or deformation measurements) is briefly exposed hereafter.

Aiming at implementing the positioning error model, the first step is to calculate K_{Bij} or K_{Tij} coefficients performing multiple measurements during a calibration stage, which should be carried out at least once after the installation of the machine. The computation of the coefficients during calibration could be programmed both in the CNC or in a computer. For this research, it was carried out externally in MatLab.

After that, during the normal use of the machine, using Function 2 or Function 3, the coefficients K_{ij} must be determined, which could be calculated either inside the CNC of the machine, or in the data acquisition system used to measure the temperatures or deformations [20]. For the system used in this research, the preferred option would be to determine these coefficients in the edge computing system used to measure all variables. Considering the large thermal inertia of machine tools, the update of K_{ij} values can be done at low frequency, for example, a value between 0.02 Hz and 0.1 Hz is recommended.

Having all K_{ij} coefficients calculated, the following step is to estimate the expected error using Function 1, which should be performed inside the CNC at a higher frequency. For simplicity, a parallel execution of a synchronous function (function that runs in the CNC parallel to the main program and synchronized with it) in each interpolation step is proposed, so that the end point of each interpolation of the machining path is substituted by the theoretical coordinate minus the result of Function 1 with the coefficients associated with each XYZ component of the error (three evaluations of the function). Using this methodology, the thermo-volumetric error is modelled with high accuracy, at a low computational cost using a simple calibration that adjusts the model to different machines by obtaining different K_{Bij} or K_{Tij} coefficients.

6 Limitations and Further Research

It should be noted that the models here exposed predict machine positioning error at different temperatures, but as the temperature changes, the workpiece is also affected by this variation, changing its dimensions. During the experimental testing, the ambient temperature was seen to vary approximately by 5.3°C, which, for a workpiece made of steel (linear expansion coefficient of around 10^{-5} °C^{-1} [21]) with a length of 1.25 m (similar to the measured range of each axis), would imply a change of 66 μm. This result is comparable to the amplitude of the measured errors during the tests, and explains why even with the identified errors, workpieces were being manufactured without significant errors in this machine. Consequently, using the proposed model, the TCP positioning errors could be drastically reduced, but the improvement in the manufactured components is expected to be of the same magnitude only when the workpiece material has a low thermal expansion. When the workpiece is expected to significantly change its dimensions as consequence of temperature changes, the path followed by the tool should be corrected not only by the model proposed in this article (considering machine deformations), but also to compensate workpiece deformations.

One of the main objectives of this research was to design a general methodology to easily, and without the need of simulations or complex algorithms, estimate the positioning errors of a machine tool. This work is limited to the modelling of the errors of a certain machine, but it could be expected to have a similar result on other machines of the same model, which would probably use the same functions, but the coefficients calculated during the calibration stage would be different. To apply the model exposed in this article to machines with different but similar structure, the proposed functions would probably be adequate to fit the measurements. However, for machines with a completely different configuration, these functions should be adjusted to achieve a correct fitting. Despite the need of slight adjustments for different machines, the proposed method showed precise error estimations with a low computational cost, even with a small dataset, which makes this approach interesting for further development.

In this research, using the three-dimensional artifact and rotating the turntable, a reduced number of tests could be made as a consequence of the slow measurement method. To increase the number of measurements and to make possible the analysis and modelling of the effect of the spindle and both C and A axes torque motors (with high thermal dynamics), a faster method for measuring TCP positioning errors should be used. The employed measurement method is limited in the maximum accuracy that can be achieved as well, which also suggests this change. The use of optical or laser-based technologies is recommended (such as that proposed by Gomez-Acedo, et al. [22]), which would also allow to introduce the movement of the A axis in the thermo-volumetric error prediction.

7 Conclusions

Thermal deformations of machine tools because of ambient temperature changes and internally generated heat, added to other sources of volumetric errors lead to insufficient tool positioning accuracy. In this article, a simple yet useful and precise model to estimate thermo-volumetric positioning errors of a five-axis large-scale machine tool based on internal component's deformation or temperature measurements is exposed. Moreover, it was seen that the complete thermal behaviour of the machine could be accurately studied using only three adequate temperature or deformation measurements. Although results are promising, further research is needed to completely validate the model and to determine the changes to be made to adjust it to machines of different configurations. In addition, to manufacture in-tolerance components, the effect of temperature changes in the workpiece should be considered and added to this model that is limited to the deformations of the machine structure. The proposed model and implementation method were specially designed to be simple enough to allow compensating positioning errors running in real-time in the CNC of the machine, opening new research paths towards thermally immune large-scale machine tools without needing ambient and internal temperature controls. Thermo-volumetric positioning errors of machine tools are still a challenge but controlling and fully compensating them is becoming progressively closer.

Acknowledgements We would like to acknowledge Francisco Javier Amigo Fuertes from CFAA for his valuable help during the experimental testing, Nagore Villarrazo Rubia from CFAA for her calibration of the artifact and Beñat Iñigo from Ideko for lending us the metrological arti-fact. The authors gratefully acknowledge the funding support received from the Centre for the Development of Industrial Technology (CDTI), entity of the Ministry of Science and Innovation, though MHAYA project, REF. MIG-20221059. Part of analysis was funded by The Basque Government, Spain, supported University research groups, IT1573-22. Thanks are addressed to PID2022-137380OB-I00, NEOPHYM project, funded by Spanish Ministry of Science and innovation with the help of MCIN/AEI/https://doi.org/10.13039/501100011033/ and European Union Next Generation EU/PRTR.

References

1. Klocke, F., Soo, S.L., Karpuschewski, B., Webster, J.A., Novovic, D., Elfizy, A., Axinte, D.A., Tönissen, S.: Abrasive machining of advanced aerospace alloys and composites. CIRP Annals. **64**(2), 581–604 (2015). ISSN: 0007-8506. https://doi.org/10.1016/j.cirp.2015.05.004

2. Mayr, J., Jedrzejewski, J., Uhlmann, E., Alkan Donmez, M., Knapp, W., Härtig, F., Wendt, K., Moriwaki, T., Shore, P., Schmitt, R., Brecher, C., Würz, T., Wegener, K.: Thermal issues in machine tools. CIRP Annals. **61**(2), 771–791 (2012), ISSN: 0007-8506. https://doi.org/10.1016/j.cirp.2012.05.008

3. Mares, M., Horejs, O., Fiala, S., Havlik, L., Stritesky, P.: Effects of cooling systems on the thermal behaviour of machine tools and thermal error models. J. Mach. Eng. **20**(4), 5–27 (2020), ISSN: 1895–7595. https://doi.org/10.36897/jme/128144

4. Gomez-Acedo, E., Olarra, A., Lopez de Lacalle, L.N.: A method for thermal characterization and modeling of large gantry-type machine tools. Int. J. Adv. Manuf. Technol. **62**(9), 875–886 (2012), ISSN: 1433-3015. https://doi.org/10.1007/s00170-011-3879-0

5. Li, Y., Yu, M., Bai, Y., Hou, Z., Wu, W.: A review of thermal error modeling methods for machine tools. App. Sci. **11**(11) (2021), ISSN: 2076-3417. https://doi.org/10.3390/app11115216

6. Zhang, J., Li, Y., Wang, S.T., Gou, W.D.: High-speed motorized spindle thermal error modeling based on genetic RBF neural network. J. Huazhong Univ. Sci. Technol. (Nat. Sci. Ed.) **46**(07), 73–77 (2018)

7. Yang, T., Sun, X., Yang, H., Liu, Y., Zhao, H., Dong, Z., Mu, S.: Integrated thermal error modeling and compensation of machine tool feed system using subtraction-average-based optimizer-based CNN-GRU neural network. Int. J. Adv. Manuf. Technol. **131**(12), 6075–6089 (2024), ISSN: 1433-3015. https://doi.org/10.1007/s00170-024-13369-2

8. Liu, H.W., Yang, Y., Xiang, H., Wang, J.P., Chen, G.H.: Research on thermal error compensation technology of machine tool spindle on least square method. Mach. Des. Res **36**, 130–133 (2020)

9. Liu, J., Ma, C., Wang, S.: Data-driven thermally-induced error compensation method of high-speed and precision five-axis machine tools. Mech. Syst. Signal Process. **138**, 106–538 (2020), ISSN: 0888-3270. https://doi.org/10.1016/j.ymssp.2019.106538

10. Iñigo, B., Colinas-Armijo, N., López de Lacalle, L.N., Aguirre, G.: Digital twin for volumetric thermal error compensation of large machine tools. Sensors. **24**(19) (2024), ISSN: 1424-8220. https://doi.org/10.3390/s24196196

11. ISO 230-7, test code for machine tools, part 7: Geometric accuracy of axes of rotation (2006)

12. Iñigo, B., Colinas-Armijo, N., López de Lacalle, L.N., Aguirre, G.: Digital twin-based analysis of volumetric error mapping procedures. Precis. Eng. **72**, 823–836 (2021), ISSN: 0141-6359. https://doi.org/10.1016/j.precisioneng.2021.07.017

13. Lang, S., Zimmermann, N., Mayr, J., Wegener, K., Bambach, M.: Thermal error compensation models utilizing the power consumption of machine tools. In: 3rd International conference on thermal issues in machine tools (ICTIMT2023), Ihlenfeldt, S. Ed. Cham: Springer International Publishing, pp 41–53 (2023), ISBN: 978-3-031-34486-2

14. Naumann, C., Naumann, A., Bertaggia, N., Geist, A., Glänzel, J., Herzog, R., Zontar, D., Brecher, C., Dix, M.: Hybrid thermal error compensation combining integrated deformation sensor and regression analysis based models for complex machine tool designs. In: 3rd International conference on thermal issues in machine tools (ICTIMT2023) Ihlenfeldt, S. Ed., Cham: Springer International Publishing, pp. 28–40 (2023), ISBN: 978-3-031-34486-2

15. Kwasny, W., Turek, P., Jedrzejewski, J.: Survey of machine tool error measuring methods. J. Mach. Eng. (2011)

16. Gomez-Acedo, E., Olarra, A., Orive, J., Lopez de Lacalle, L.N.: Methodology for the design of a thermal distortion compensation for large machine tools based in state-space representation with kalman filter. Int. J. Mach. Tools Manuf. **75**, 100–108 (2013), ISSN: 0890-6955. https://doi.org/10.1016/j.ijmachtools.2013.09.005

17. Sáinz de la Maza García, A., Sastoque Pinilla, L., López de Lacalle, L.N.: Measuring thermally induced volumetric error of large-scale machining centre (manuscript submitted for publication) (2025)

18. Troiano, M., Nobile, E., Mangini, F., Mastrogiuseppe, M., Conati Barbaro, C., Frezza, F.: A comparative analysis of the Bayesian regularization and Levenberg–Marquardt training algorithms in neural networks for small datasets: a metrics prediction of neolithic laminar artefacts. Information. **15**(5) (2024), ISSN: 2078-2489. https://doi.org/10.3390/info15050270

19. Yates, L.A., Aandahl, Z., Richards, S.A., Brook, B.W.: Cross validation for model selection: a review with examples from ecology. Ecol. Monogr. **93**(1), e1557 (2023). https://doi.org/10.1002/ecm.1557

20. Tapia, E., Lopez-Novoa, U., Sastoque-Pinilla, L., L. de Lacalle, L.N.: Implementation of a scalable platform for real-time monitoring of machine tools. Comput. Ind. **155**, 104065 (2024), ISSN: 0166-3615. https://doi.org/10.1016/j.compind.2023.104065

21. Nedoseka, A.: 2-welding stresses and strains. In: Fundamentals of evaluation and diagnostics of welded structures, ser. Woodhead Publishing Series in Welding and Other Joining Technologies, Nedoseka, A. Ed., Woodhead Publishing (2012), pp. 72–182, ISBN: 978-0-85709-531-2. https://doi.org/10.1533/9780857097576.72

22. Gomez-Acedo, E., Olarra, A., Zubieta, M., Kortaberria, G., Ariznabarreta, E., López de Lacalle, L.N.: Method for measuring thermal distortion in large machine tools by means of laser multilateration. Int. J. Adv. Manuf. Technol. **80**(1), 523–534 (2015), ISSN: 1433-3015. https://doi.org/10.1007/s00170-015-7000-y

Extension of the Torque Limit Skip Method to Multi-spindle Thermal Error Compensation

Daniel Divisek, Petr Kaftan, Josef Mayr, Markus Bambach, and Konrad Wegener

Abstract This paper introduces an extension of the thermal error compensation based on the Torque Limit Skip (TLS) to two independent spindle systems of a multi-spindle Swiss-type lathe. The TLS method is a recently introduced approach for the measurement and compensation of thermal errors of precision machine tools. Until now, this method had been used for probing at a single location and to create a single compensation model, specifically on the main spindle of a Swiss-type lathe. The challenge of a second counter-spindle system is that it introduces a second tool-center-point into the working space, the thermal behaviour of which is different compared to that on the main spindle. Furthermore, on Swiss-type lathes, there is no possibility to measure with a movable touch-trigger probe or displacement sensors. This work therefore extends the TLS compensation to the counter-spindle, by devising a second probing location and implementing a second compensation model. It is demonstrated that the two measurement cycles and compensation models can run independently and that each model can compensate the thermal errors of its corresponding spindle system. Results show that with out-of-sync operation of the two spindles, the TLS measured thermal errors are reduced by 80% when using a compensation model based on transfer functions.

Keywords Machine tool · Thermal error · Torque limit skip · Thermal compensation

D. Divisek (✉)
Department of Production Machines and Equipment (RCMT), Czech Technical University, Prague, Czech Republic
e-mail: D.Divisek@rcmt.cvut.cz

D. Divisek · P. Kaftan · M. Bambach
Advanced Manufacturing Laboratory (AMLZ), ETH Zürich, Zurich, Switzerland

P. Kaftan · J. Mayr · K. Wegener
Inspire AG, Zurich, Switzerland

© The Author(s) 2026
K. Wegener and M. Bambach (eds.), *4th International Conference on Thermal Issues in Machine Tools (ICTIMT2025)*, Lecture Notes in Production Engineering,
https://doi.org/10.1007/978-3-032-01194-7_18

287

1 Introduction

Machine tools (MT) can be considered as a rigid system. However, during the working process, various phenomena, such as vibrations, chatter, gravitational forces and thermal effects etc., load the machine, resulting in different types of errors. Errors caused by non-stationary thermal effects can be described as displacements from the desired tool and workpiece path. According to research by Bryan [1] and Mayr [2], these errors account for 40–70% of the total MT inaccuracy. Due to the significance of this error source, minimizing thermal errors remains a long-term goal of MT designers and programmers. One possible solution is to compensate MTs through correction of the NC-axes or the tool position. Data-based thermal compensation is a gray/black box approach based on correlating machine tool variables, such as temperature, speed, position etc., with thermal errors and then compensating them.

There are many types of data-based compensation models, ranging from simple regression models such as multiple-linear regression analysis (MLRA) [3], to transfer functions (TF) [4], to more complex models such as convolutional neural networks (CNNs) [5] or long short-term memory neural networks (LSTM) [6]. As stated by Wegener et al. [7], a major aspect of any compensation model is the type of sensor technology used to measure the thermal error itself.

The most widely used standardized direct method for measuring thermal errors is based on ISO 230-3: Test code for machine tools - Part 3: Determination of thermal effects [8]. The basis of this method is the measurement of a precision mandrel using 5 probes that measure three linear errors and two angular errors. This setup is used in many works, e.g. [5, 9, 10].

A requirement for data-based compensation models is that the model parameters or even the inputs can be adaptively updated if conditions change, and a certain tolerance band is exceeded. In order to meet this condition and to achieve high accuracy over long time scales, it is necessary to measure the thermal errors periodically with optimal time spacing [11].

Currently, many machine tool manufacturers are addressing the implementation of measuring devices in real production. Static solutions defined in ISO standards are not suitable for these cases. One suitable solution is touch-trigger probes (TTP), which however require clamping the artifact on the table and also the machine must be equipped with this feature [12].

This publication focuses on the Swiss-type lathe. This specialized machine tool has several features that distinguish it from a standard lathe:

I.   a guide bush near the cutting tool, which supports the workpiece;
II. an automatic bar feeder that feeds the workpiece through the guide bushing into the work area;
III. a sliding spindle that allows the workpiece to move along the Z-axis;
IV. a larger number of tools and axes;
V.   a counter-spindle that allows operations to be performed independently.

In general, Swiss-type lathes are not equipped with a TTP. Some of the larger and more complex Swiss-type lathes are equipped with a rotating tool tower which may contain a TTP in one of the tool positions. However, basic Swiss-type lathes designed for mass production of less complex parts usually do not have a TTP. Kaftan et al. [13] developed a new thermal error measurement method based on torque limit skip (TLS) integrated into the machine control system without additional specialized measuring equipment. In this paper, this method is applied to a counter spindle that is also subjected to thermal effects. Based on the measured data, two independent compensation models for individual spindles are developed.

2 Methodology

2.1 Torque Limit Skip Method

The torque limit skip (TLS) is a special type of a linear motion command. If, during the movement with this command, the torque of the servo-motor reaches the torque limit value due to pressing or other causes, any remaining move commands will be canceled, and the next block will be executed. The torque limit value is equal to the maximum torque on the servo motor multiplied by the override; the override value can be specified by the operator as a percentage value from 0 to 100% directly in the CNC code. The operation during the TLS command is shown in Fig. 1 and the probing consists of a TLS probe and the artefact.

During the probing, the TLS probe moves along a linear axis from the starting position O towards the artefact to a user-specified coordinate C. The probe comes into contact with the artefact at point A but its movement continues in the direction of the coordinate C since the torque limit on the axis servo motor had not yet been reached at point A. The movement is subsequently terminated at point B where the torque limit value is reached and the skip signal is triggered. The set torque limit at point B corresponds to a current limit on the axis motor, as also shown in Fig. 1. The probe then returns to its starting position or another position in the working space. As

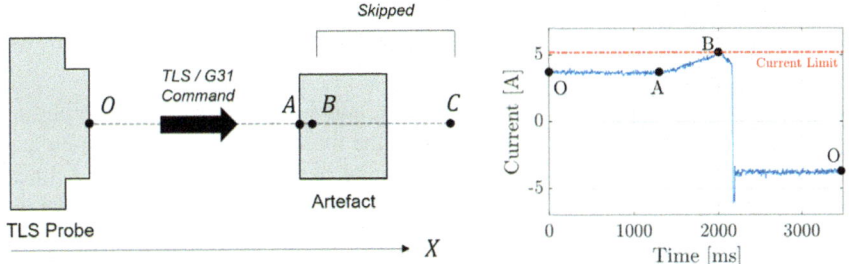

Fig. 1 Left: operation during the execution of the TLS command with a TLS probe. O: axis starting position in the working space, A: first point of contact, B: position when torque limit is reached, C: target axis position. Right: plot of the motor current during the TLS probing sequence—the torque limit is equivalent to a current limit in the axis motor

described in the work of Kaftan et al. [13], the repeated application of the movement sequence as described, i.e. the repeated triggering of the TLS over a time horizon of several hours, can be used to track the relative change in position of the axis and therefore the relative change in position between the probe and the artefact, which is mostly due to the thermal error.

2.2 Compensation Model

The compensation strategy based on TF theory is widely used in electrical and mechanical systems [14]. It is a dynamic method that can describe the heat transfer in the MT structure with sufficient analytical accuracy [15]. A discrete TF is used to describe the coupling between the so-called excitation, here the temperature difference, and its response, in this case the deformation:

$$y(t) = \varepsilon \cdot u(t) + e(t). \tag{1}$$

The vector $u(t)$ in Eq. (1) is the input of the TF and $y(t)$ is the output in the time domain, ε represents the general TF in the time domain and $e(t)$ is the interference value which is neglected further. The thermal compensation model used in this paper is based on an AutoRegressive model with eXogenous inputs (ARX). The ARX model considers the past and present inputs of the system but also the past outputs of the system. Based on this, it calculates a prediction for the current system output. The ARX model is suitable for time series data processing and therefore can capture the thermal dynamics of the machine. The mathematical structure of the model is expressed in the form of a discrete TF in the time domain using differential Eq. (2). This form is generally suitable for modern MT control systems using their programming languages.

$$y(k) = \frac{u(k-n) \cdot a_n + \ldots + u(k-1) \cdot a_1 + u(k) \cdot a_0}{b_0}$$
$$- \frac{y(k-m) \cdot b_m + \ldots + y(k-1) \cdot b_1}{b_0}, \tag{2}$$

where k represents the examined time instant, and k-n (k-m) is the n-multiple (m-multiple) delay in the sampling frequency of the measured input vector (simulated output vector). The sampling frequency is equal to 1 s^{-1} in the article. A linear parametric model ARX is used with the help of *Matlab Identification Toolbox* [16]. The ARX as an optimal model structure with the best fit in quality and robustness is also discussed in [17].

The approximation quality of the simulated behavior is expressed by a global approach based on a normalized root mean squared error (NRMSE) method *fit* [16] as expressed in Eq. (3). The *fit* is a percentage value of model efficiency where 100%

would equal a perfect match of the measured and simulated behaviors:

$$fit = \left(1 - \frac{\|y_{mea} - y_{sim}\|}{\|y_{mea} - \bar{y}_{mea}\|}\right) \cdot 100. \tag{3}$$

The y_{mea} value in Eq. (3) is the measured output - displacements in the machine direction X in this case, y_{sim} is the simulated model output, and $\bar{y}_{mea}$ expresses the arithmetic mean of the measured output over time.

3 Experimental Setup

3.1 Investigated Machine Tool

The target machine for this research is a Swiss-type lathe. This machine has two spindles that have a maximum speed of 8000 rpm. Their kinematic chains are shown schematically in Fig. 2. According to ISO 10791-1:2015 [18] the kinematic chains of main and counter spindle can be described as H[w-(C')-Z'-b-X-Y-t] and H[w-(C')-Z'-X'-b-Y-t]. The temperature is measured using Pt100 temperature sensors, which are mounted on both spindles at thermally relevant locations (motors, linear guide etc.) and on the machine structure. The TLS method, described in the previous section, was used to measure the thermal error on both the main and counter spindles.

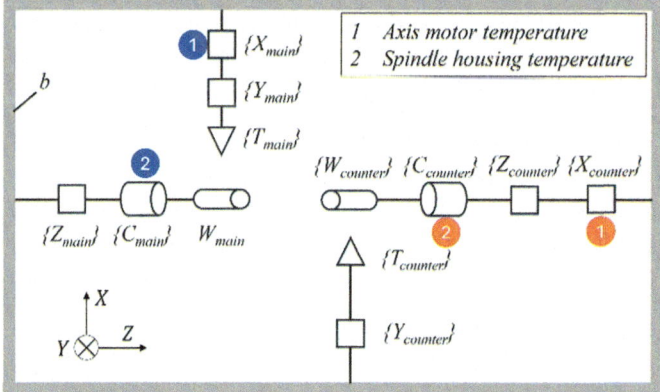

Fig. 2 The axis configuration of the Swiss-type Lathe. Curly brackets indicate NC controlled axes in the coordinate system as shown. W denotes the workpiece and T denotes the tool, b is the machine bed. The coloured points show the position of the temperature sensors used for the model in the following section (blue dots are for the main spindle and red for the counter spindle)

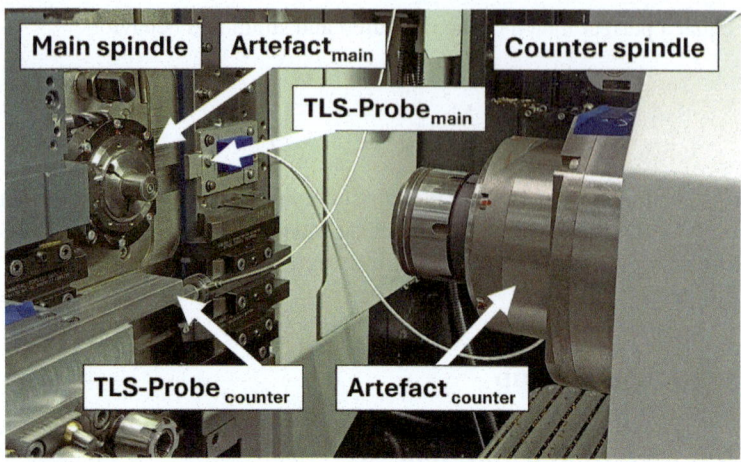

Fig. 3 Experimental setup of measurements on a Swiss-type lathe

3.2 Experimental Setup and Load Case

The experimental setup consists of two TLS probes, also called 'reference blocks' in earlier terminology, and two corresponding artefacts. The first TLS probe, which is used to capture the thermal error of the main spindle (shown as TLS_{main} in the plots below), is located on the vertical table on the right hand side of the main spindle. The second TLS probe, which is used for the measurement of the thermal error on the counter spindle side (shown as $TLS_{counter}$), is located on the horizontal tool table next to the counter spindle. Both artefact surfaces are also shown in Fig. 3. On the main spindle side, the guide bush serves as the artefact surface, while on the counter spindle side, it is the outside metal cover of the counter spindle. As described above, the contact between the TLS probe and the artefact on each of the spindle sides triggers the torque limit skip signal.

The experiment is based on independent loading of both spindles. The cycle for the main spindle consists of a loading phase—rotating the spindle at 8000 rpm and moving in the X-axis for a period of 0.75 h and a cooling phase - spindle stopped without movement for 0.75 h. The cycle for the counter spindle is the same but each phase lasts 1 h. 'One cycle' refers to one repetition heating and cooling. The cycles are repeated for both spindles for 21 h. Every minute the relative error in the X-axis is measured by TLS. Recording of the measured relative errors and temperature are shown in Fig. 4. Both spindles are stopped during the probing and the TLS measurements take place on the main and counter spindles simultaneously. The TLS measurements take ca. 5 s and the cooling during this time is therefore negligible.

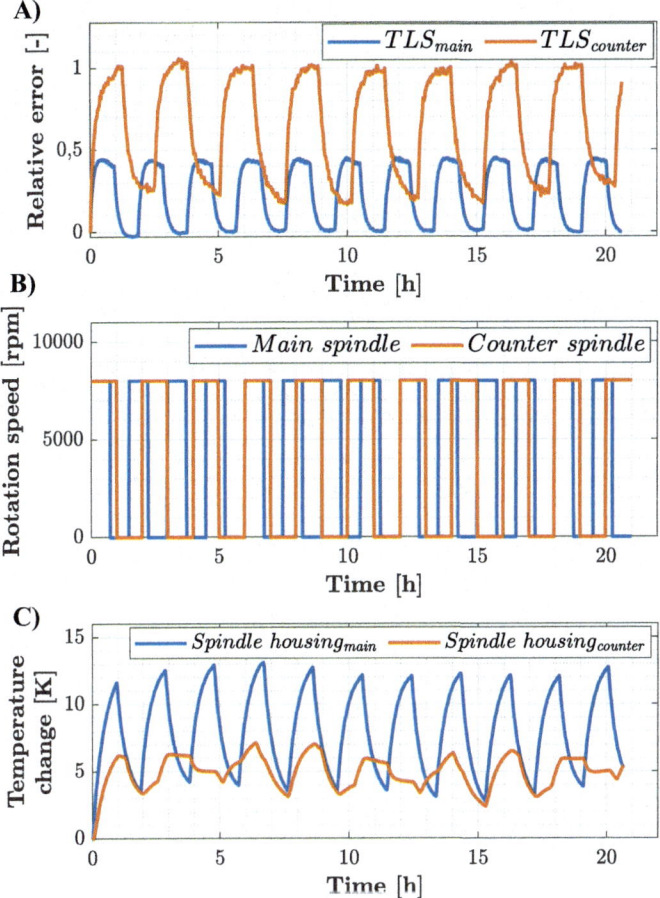

Fig. 4 Load case—**a** measured X-axis relative errors for both spindles; **b** spindle speed setting; **c** measured temperature change of spindle housing—model input

4 Results

The thermal error measured by the TLS is used to calibrate (train) two separate ARX thermal compensation models. Both models are of the Multiple Input—Single Output (MISO) type, where the inputs are two temperatures measured close to the respective spindles and the output is a prediction of the relative displacement. The scheme of the compensation models is shown in Fig. 5 and the position of inputs is marked in Fig. 2.

The calibration of the compensation model was carried out on the data from the first 7.5 h: 4 cycles for the main spindle and 3 cycles for the counter spindle; and the remaining part of the time series data was used for model validation, as indicated by the vertical solid line in Figs. 6 and 7.

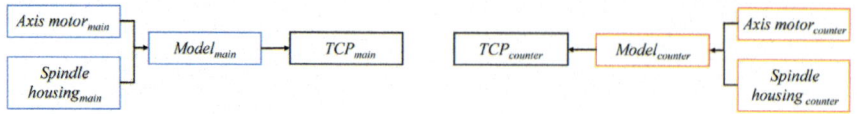

Fig. 5 Scheme of compensation models

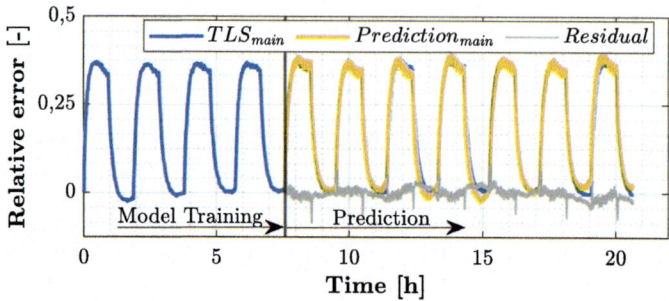

Fig. 6 Compensation of the thermal error at the main spindle

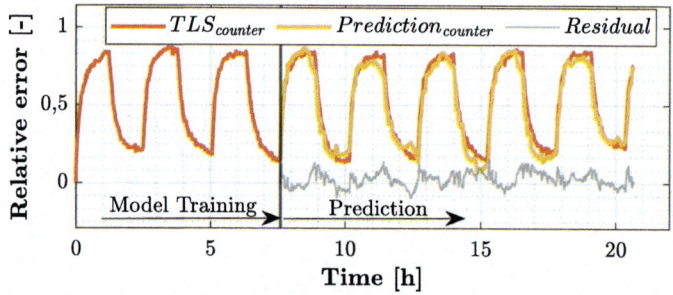

Fig. 7 Compensation of the thermal error at the counter spindle

The model is applied offline and the compensation results are shown in Fig. 6 for main spindle and Fig. 7 for counter spindle. The predicted reductions in thermal errors, expressed as *fit* values, are computed between the predicted model and TLS values and are shown in Table 1. For both models, the magnitude of the thermal error is significantly reduced, by more than 80%. The state after the application of the compensation model is indicated by the residual curve in both graphs.

Table 1 Thermal error reductions for main spindle and counter spindle	Error	TLS_{main}	$TLS_{counter}$
	Fit [%]	84	82

5 Conclusion and Outlook

This paper presents an extension and simultaneous independent use of the TLS method on a multi-spindle precision machine tool. The recently introduced TLS method has so far been used only for single location measurements. The extension of TLS measurements to multiple locations allows the creation of more independent compensation models, for example for different spindles or tools positions. It is demonstrated and experimentally validated on the Swiss-type lathe. Results have shown that a model can be created for each spindle, regardless of the operation of the other spindle, and successfully compensate its thermal errors. Additional models can therefore be created step-by-step to compensate the machine tool without having to revise existing models.

Future research will focus on the implementation of the TLS measurement and compensation for the counter spindle in real workpiece production. With the current configuration, the use of the TLS means occupying the table for the clamping of the tools with a TLS probe. To keep the same number of tools, the tool holder would be redesigned so that the TLS probe becomes part of the tool holder itself. Future research will also extend the introduced compensation approach to a multi-Swiss lathe with more than two spindle systems on both main and counter operations.

Acknowledgements The authors would like to thank the Swiss Innovation Agency Innosuisse for the provided funding of the Innosuisse Project 53644.1 IP-ENG 'Intelligent Precision Swiss-type Automatic Lathe'.

This work was supported by the Grant Agency of the Czech Technical University in Prague, grant no. SGS25/135/OHK2/3T/12.

Competing Interests The authors have no conflicts of interest to declare that are relevant to the content of this paper.

References

1. Bryan, J.: International status of thermal error research. CIRP Ann. Manuf. Technol. **39**(2), 645–656 (1990)
2. Mayr, J., Jedrzejewski, J., Uhlmann, E., Donmez, M., Knapp, W., Härtig, F., Wendt, K., Moriwaki, T., Shore, P., Schmitt, R., Brecher, C., Würz, T., Wegener, K.: Thermal issues in machine tools. CIRP Ann. **61**, 771–791 (2012)
3. Naumann, C., Glänzel, J., Putz, M.: Comparison of basis functions for thermal error compensation based on regression analysis—a simulation based case study. J. Mach. Eng. **20**(4), 28–40 (2020)
4. Brecher, C., Hirsch, P., Weck, M.: Compensation of thermo-elastic machine tool deformation based on control internal data. CIRP Ann. Manuf. Technol. **53**(1), 299–304 (2004)
5. Fujishima, M., Narimatsu, K., Irino, N., Mori, M., Ibaraki, S.: Adaptive thermal displacement compensation method based on deep learning. CIRP J. Manuf. Sci. Technol. **25**, 22–25 (2019)
6. Vu, N.H., Mayer, J.R.R., Bitar-Nehme, E.: Deep learning LSTM for predicting thermally induced geometric errors using rotary axes' powers as input parameters. CIRP J. Manuf. Sci. Technol. **37**, 70–80 (2022)

7. Wegener, K., Weikert, S., Mayr, J.: Age of compensation—challenge and chance for machine tool industry. Int. J. Autom. Technol. **10**, 609–623 (2016)
8. ISO 230-3: Test Code for Machine Tools—Part 3: Determination of Thermal Effects, 3rd ed., p. 50. International Organisation for Standardization, Genf, Switzerland (2020)
9. Wei, X., Ye, H., Miao, E., Pan, Q.: Thermal error modeling and compensation based on Gaussian process regression for CNC machine tools. Precis. Eng. **77**, 65–76 (2022)
10. Divisek, D., Mares, M., Horejs, O.: Post-process updating of model parameters to approximate thermal errors of machine tools operating in different configurations. Precis. Eng. **88**, 241–250 (2024)
11. Blaser, P., Pavlíček, F., Mori, K., Mayr, J., Weikert, S., Wegener, K.: Adaptive learning control for thermal error compensation of 5-axis machine tools. J. Manuf. Syst. **44**, 302–309 (2017). Special Issue on Latest advancements in manufacturing systems at NAMRC 45
12. Lin, C.J., Su, X.Y., Hu, C.H., Jian, B.L., Wu, L.W., Yau, H.T.: A linear regression thermal displacement lathe spindle model. Energies **13**(4), 949 (2020)
13. Kaftan, P., Porquez, F., Mayr, J., Pomodoro, K., Keel, M., Trombert, D., Wegener, K.: Thermal error measurement and compensation with torque limit skip in Swiss-type lathe manufacturing. Precis. Eng. **88**, 315–323 (2024)
14. Bart, H.: Transfer functions and operator theory. Lin. Algebra Appl. **84**, 33–61 (1986)
15. Fraser, S., Attia, M.H., Osman, M.O.M.: Modelling, identification and control of thermal deformation of machine tool structures, part 1: concept of generalized modelling. J. Manuf. Sci. Eng. Trans. ASME **120**(3), 623–631 (1998)
16. Ljung, L.: System Identification Toolbox™ User's Guide. The MathWorks (2015)
17. Mayr, J., Blaser, P., Ryser, A., Hernandez-Becerro, P.: An adaptive self-learning compensation approach for thermal errors on 5-axis machine tools handling an arbitrary set of sample rates. CIRP Ann. Manuf. Technol., 551–554
18. ISO 10791-1:2015: Test Conditions for Machining Centres—Part 10: Evaluation of Thermal Distortions. Standard, International Organization for Standardization (2015)

Necessary Camera Calibration Accuracy for Vision-Based Thermal Error Measurement in Machine Tools

Eri Tsuchiya, Shunsuke Ohmae, and Daisuke Kono

Abstract Thermal error is one of the significant motion errors in machine tools. It is often evaluated in terms of the relative displacement between the tool and the workpiece. However, relying solely on relative displacement measurements fails to capture the individual thermal deformations of the tool and workpiece, which can lead to incomplete or inaccurate understanding of the thermal error mechanisms. In contrast, vision-based measurements, using multiple target markers, allow for the evaluation of the absolute displacements of both the tool and workpiece. This approach provides a more comprehensive understanding of thermal error mechanisms. However, achieving sufficient accurate measurements in machine tools requires highly accurate camera calibration. In vision-based measurements, to ensure accurate transformations between world and image coordinates, the camera's intrinsic and extrinsic parameters must first be determined before measuring 3D positions. This study aims to determine the necessary calibration accuracy for vision-based thermal error measurements. The impact of error factors in camera calibration on measurement accuracy is investigated by Monte Carlo simulations and experiments. The simulations assessed the influence of the calibration board accuracy and the accuracy in the board pattern detection. Furthermore, experiments were conducted using different calibration patterns to estimate camera parameters, and the resulting measurement accuracy with calibrated images was compared. The Monte Carlo simulation results revealed that a board error of less than 1 μm and a detection error of less than 0.01 pixel are necessary to achieve the required measurement accuracy. Furthermore, the simulation results indicated that using more than 200 feature points is preferable to achieve a sufficient averaging effect in calibration. Experimental results showed that the calibration boards used in the experiment could not achieve the target measurement accuracy due to insufficient accuracy in both the board error and detection error. Additionally, enlarging the board size relative to the field of view improved the estimation accuracy of the higher-order coefficients of radial distortion, contributing to higher calibration accuracy.

E. Tsuchiya (✉) · S. Ohmae · D. Kono
Department of Micro Engineering, Kyoto University, Kyoto, Japan
e-mail: tsuchiya.eri.55r@st.kyoto-u.ac.jp

© The Author(s) 2026

K. Wegener and M. Bambach (eds.), *4th International Conference on Thermal Issues in Machine Tools (ICTIMT2025)*, Lecture Notes in Production Engineering,
https://doi.org/10.1007/978-3-032-01194-7_19

Keywords Machine tools · Thermal errors · Vision-based measurement · Camera calibration

1 Introduction

Thermal error is a major source of machine tool position errors [1, 2]. Positioning uncertainty of machine tools immediately affects the dimensional accuracy of manufactured parts, and, up to 75% of the overall geometrical errors of machined workpieces can be induced by the thermal error [3].

The measurement of the thermal error is important to analyze and develop the thermal error model. Vision-based measurement has attracted attention in the motion error measurement of machine tools. This is because it has advantages in three-dimensional measurement in large spaces. Ibaraki et al. proposed a two-dimensional machine tool error measurement method based on monocular vision [4]. Irino et al. reported that an uncertainty of the sub-micrometer level could be achieved in the vision-based measurement of static positioning accuracy [5]. However, the usability of these reported methodologies is limited in the measurement in large spaces because they require a reference artifact with high accuracy.

Simple motion tracking has higher usability because its measurement apace is more flexible by scaling with optical systems. In motion tracking, images captured by cameras are subject to distortion due to the characteristics of lenses and image sensors. Therefore, achieving sufficient measurement accuracy in vision-based measurement requires high-precision camera calibration [6]. Camera calibration is a fundamental technique in computer vision, and numerous calibration methods based on camera models have been established [7–10]. There have been various studies on camera calibration in vision-based measurements for applications such as robotic positioning [11] and structural health monitoring [12]. These studies generally require calibration accuracy in the order of millimeters or sub-millimeters. However, in the field of precision measurement, particularly for machine tools, significantly higher accuracy is required, often from sub-micron to micron level [13]. Despite this need, the requirement of calibration accuracy in machine tools remains unexplored.

In this study, the calibration accuracy required for two-dimensional vision-based thermal error measurements in machine tools using a single camera was investigated. Guidelines for realizing thermal displacement evaluation through vision-based measurement are provided. The target accuracy in this study is set to the order of 1 μm within a measurement range of 1 m. Monte Carlo simulations and experiments were conducted to investigate the impact of error factors in camera calibration on measurement accuracy for machine tools.

2 Uncertainty Evaluation

The uncertainty in the conventional camera calibration using the calibration board is discussed. Neglecting the modeling error of the camera model, errors in camera calibration can be attributed to two factors: manufacturing errors of the calibration board (board error) and detection errors when detecting feature points in the calibration board (detection error). Therefore, to investigate the allowable error to achieve the desired accuracy in vision-based measurement, we investigated the impact of both errors on calibration accuracy through simulations.

2.1 Camera Model

The camera model used in the simulation for uncertainty evaluation is described in this section. The conventional model for perspective projection is used in this study. Figure 1 shows a schematic of the projection in the camera. The relationship between the target marker position (X_W, Y_W, Z_W) in the world coordinates and the corresponding position (x, y) in the image coordinates is formulated as follows [7]:

$$\lambda \begin{bmatrix} x \\ y \\ 1 \end{bmatrix} = \begin{bmatrix} f_x & s & c_x \\ 0 & f_y & c_y \\ 0 & 0 & 1 \end{bmatrix} \begin{bmatrix} R_{11} & R_{12} & R_{13} & T_x \\ R_{21} & R_{22} & R_{23} & T_y \\ R_{31} & R_{32} & R_{33} & T_z \end{bmatrix} \begin{bmatrix} X_W \\ Y_W \\ Z_W \\ 1 \end{bmatrix} = A[\mathbf{Rt}] \begin{bmatrix} X_W \\ Y_W \\ Z_W \\ 1 \end{bmatrix} \quad (1)$$

where R_{ij} ($i = 1 - 3, j = 1 - 3$) expresses the three-dimensional rotation of the camera; and T_x, T_y, and T_z denote the translation of the camera. R_{ij} is obtained by multiplying the rotation matrices. These parameters are called extrinsic parameters;

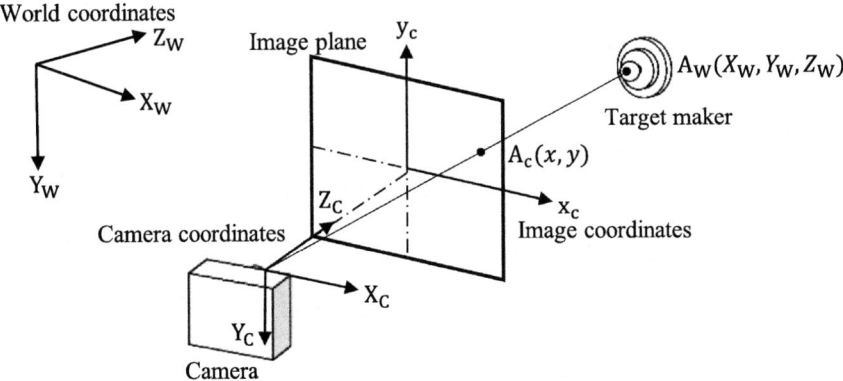

Fig. 1 Coordinate systems and perspective projection model

f_x and f_y are focal length in the x and y directions; s is the skew factor; c_x and c_y are principal points. The parameters f_x, f_y, s, c_x and c_y are called intrinsic parameters.

In an actual camera lens, distortion occurs both in the radial and tangential direction, making it necessary to account for lens distortion. The image coordinates (x_d, y_d), considering lens distortion, are expressed by the following equation using the radial distortion coefficients k_1, k_2, k_3 and the tangential distortion coefficients p_1 and p_2[14]:

$$
\begin{bmatrix} x_d \\ y_d \\ 1 \end{bmatrix} = A \begin{bmatrix} x_n\left(1 + k_1 r^2 + k_2 r^4 + k_3 r^6\right) + 2p_1 x_n y_n + p_2\left(r^2 + 2x_n^2\right) \\ y_n\left(1 + k_1 r^2 + k_2 r^4 + k_3 r^6\right) + 2p_2 x_n y_n + p_1\left(r^2 + 2y_n^2\right) \\ 1 \end{bmatrix} \quad (2)
$$

where $r = \sqrt{x_n^2 + y_n^2}$, and (x_n, y_n) are the normalized image coordinates without distortion.

2.2 Influence of Calibration Board and Detection Errors on Calibration Uncertainty

Based on the camera model described in Sect. 2.1, extrinsic parameters, intrinsic parameters, and lens distortion parameters are determined using the calibration board before performing measurements. These camera parameters are estimated from the relationship between the known position of feature points on the calibration board and their corresponding projected position in the image. In a common approach, the camera parameters are determined through nonlinear optimization based on the camera model, by minimizing the error between the observed position in the image and the reprojected position calculated using the estimated camera parameters [7].

In camera calibration with board error, when the world coordinates of each feature point on the calibration board (X_W, Y_W) have errors (r_X, r_Y), the image coordinates of the corresponding point are calculated as follows based on Eq. (1):

$$
\lambda \begin{bmatrix} x \\ y \\ 1 \end{bmatrix} = A[\boldsymbol{Rt}] \left(\begin{bmatrix} X_W \\ Y_W \\ 0 \\ 1 \end{bmatrix} + \begin{bmatrix} r_X \\ r_Y \\ 0 \\ 0 \end{bmatrix} \right). \quad (3)
$$

In camera calibration with detection error, assuming the detection error in identifying the image coordinates is (r_x, r_y), the detected image coordinates are expressed by the following equation based on Eq. (2):

$$
\begin{bmatrix} x_d + r_x \\ y_d + r_y \\ 1 \end{bmatrix} = A \begin{bmatrix} x_n\left(1 + k_1 r^2 + k_2 r^4 + k_3 r^6\right) + 2p_1 x_n y_n + p_2\left(r^2 + 2x_n^2\right) \\ y_n\left(1 + k_1 r^2 + k_2 r^4 + k_3 r^6\right) + 2p_2 x_n y_n + p_1\left(r^2 + 2y_n^2\right) \\ 1 \end{bmatrix}. \quad (4)
$$

In camera calibration incorporating both board error and detection error, the camera parameters are estimated through nonlinear optimization using Eqs. (3) and (4).

In the simulation, the board error (r_X, r_Y) is assumed to follow a uniform distribution $U(-\sigma_W, \sigma_W)$, and the detection error (r_x, r_y) is assumed to follow a uniform distribution $U(-\sigma_i, \sigma_i)$. Here, σ_W and σ_i represent the noise scales. The parameters (r_X, r_Y) and (r_x, r_y) were randomly selected, and parameter estimation was repeated 1000 times to examine the uncertainty of the estimated parameters. The same evaluation was performed with varying noise scales to investigate the requirement of the noise scale to achieve the desired parameter uncertainty. The intrinsic parameters and lens distortion parameters, assumed as the true values, were kept the same, while the extrinsic parameters were randomly selected from 70 to 15 and projected onto the image coordinates.

The simulation results showed similar results qualitatively across all parameters. Therefore, the evaluated uncertainty of the focal length f_x is presented in Fig. 2 as a representative example. In Fig. 2, the color bar indicates the coefficient of variation of the parameter, which is defined as the ratio of the standard deviation to the mean value. It was observed that the coefficient of variation increases with both larger board error and detection error. Given that the target accuracy in this study is set to the order of 1 μm within a measurement range of 1 m, the coefficient of variation in the focal length must be less than 0.001%. To achieve this, from Fig. 2, the board error of the calibration board needs to be below 1 μm, and the detection error must be below 0.01 pixel.

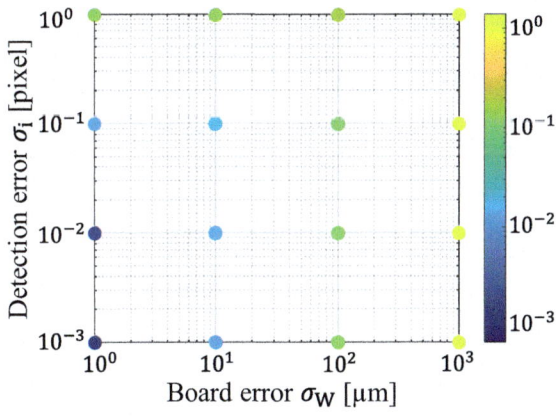

Fig. 2 Coefficient of variation of f_x in various conditions of board error and detection error

2.3 Influence of Number of Calibration Feature Points on Calibration Uncertainty

The number of feature points on the calibration board has an influence on the calibration accuracy because the larger number of feature points enables more robust calibration by averaging. To determine the required number of feature points on the calibration board, the effect of the number of feature points on calibration accuracy was evaluated by simulation. The camera parameters used in the simulation are the same as those in Sect. 2.2. Simulations were repeated for 1000 times with different number of feature points from 5 to 930. The feature points used for the calibration were selected randomly. The noise scale of the errors was set to 10 μm for board error and 0.1 pixel for detection error.

Figure 3 shows the relationship between the number of feature points and the coefficient of variation of focal length, which is used as a representative value of camera parameters. It can be observed that the uncertainty of the camera parameters decreases as the number of feature points increases. Notably, the uncertainty decreases rapidly with an increase in feature points until the number exceeds 200. It was found that the number of feature points should be preferably more than 200 to achieve sufficient averaging effect.

3 Experiment

The simulation results described in Sect. 2 were verified through experiments. Multiple types of calibration boards were prepared for camera calibration, and the effect of calibration accuracy on the measurement accuracy for machine tools was evaluated experimentally.

Fig. 3 Relationship between uncertainty of f_x and number of feature points

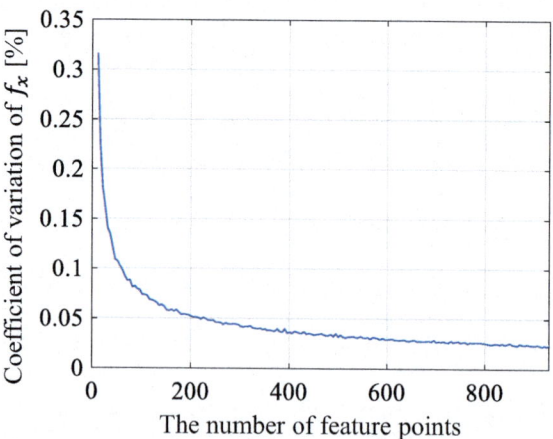

3.1 Experimental Method

Figure 4a shows the schematic of the experiment method. The camera and luminous target marker were fixed on the machine tool table and spindle, respectively. Camera calibration was performed in advance using multiple calibration boards, and the position determined by machine tool positioning on the same plane was measured using the camera parameters estimated from each calibration pattern. The target marker was moved in a grid pattern within the YZ plane, as shown in Fig. 4b. The marker was held stationary for two seconds at each commanded position, while the camera captured. The ambient temperature was maintained at approximately $20 \pm 2\ °C$ using an air conditioning system.

To determine the actual length in world coordinates from images captured by a single camera, it is necessary to know the working distance between the camera and the measured object. However, measuring the accurate working distance is not feasible. Therefore, the vision-based measured position has scaling errors. To address this issue, the maker's motion plane was shifted two times in parallel to the initial motion plane. The distance between each plane was controlled to 100 mm by the feed drive. This distance was estimated from the vision-based measured points and compared with commanded distance to evaluate the accuracy of the vision-based measurement.

The relationship between the image coordinates and world coordinates in the experiment is illustrated in Fig. 5. From this figure, working distance differences $W_1 - W_2$ and $W_2 - W_3$, focal length f, length in world coordinates P, and lengths in image coordinates p_1, p_2, p_3 have the following relationship:

$$W_1 - W_2 = \left(\frac{1}{p_1} - \frac{1}{p_2}\right)Pf \tag{5}$$

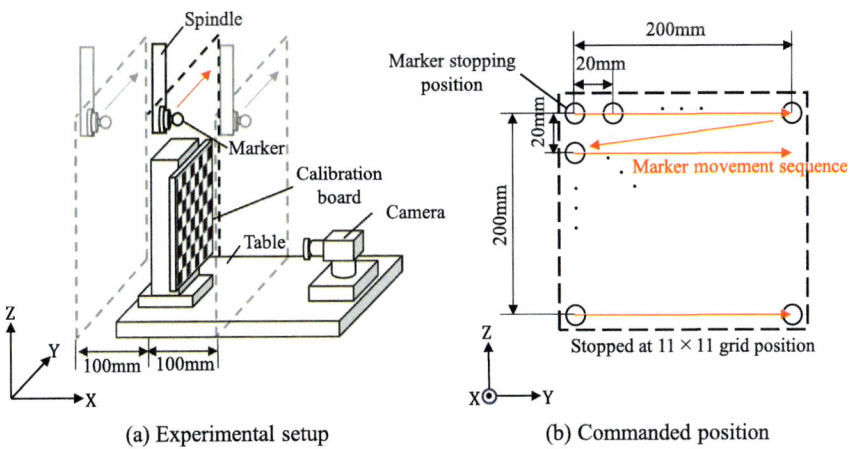

(a) Experimental setup   (b) Commanded position

Fig. 4 Schematic of experiment

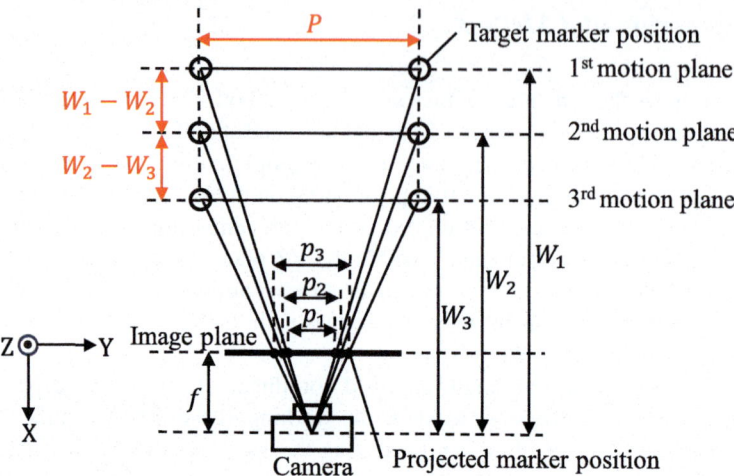

Fig. 5 Relationship between image coordinates and world coordinates

$$W_2 - W_3 = \left(\frac{1}{p_2} - \frac{1}{p_3}\right)Pf \qquad (6)$$

By comparing the estimated working distance differences obtained using these equations with the commanded values, the camera calibration accuracy is evaluated.

The shooting conditions are shown in Table 1. A high-intensity LED was used as the marker to achieve high resolution by enhancing the signal-to-noise ratio. The marker position was obtained by fitting the sigmoid function to the intensity distribution of the marker [15]. The subpixel resolution of 1/100 pixel can be obtained by this method. The marker position was obtained by averaging the detected positions across the frames where the marker remained stationary.

Table 2 provides the specifications of the calibration boards used for camera calibration. In this experiment, three types of calibration patterns were used to estimate the intrinsic parameters and lens distortion parameters. The checkerboard and circle grid are commercially available calibration boards with 10 μm accuracy. For the selected circle grid, 20 feature points on the edge of the circle grid were selected and used in the calibration. The positional accuracy of these selected feature points

Table 1 Image capture condition

Frame rate [fps]	10
Shutter speed [s]	1/1000
Image size [pixels]	4096 × 2160
Field of view [mm × mm]	655 × 345
Focal length [mm]	8
Bit depth [bit]	8

Table 2 Characteristics of calibration patterns

	Checkerboard	Circle grid	Selected circle grid
	Square size: 20 mm	Circle diameter: 2.5 mm	Grid spacing: 5 mm
Number of detected points	132 (11 × 12)	930 (30 × 31)	20
Board error (μm)	10	10	1

was known with 1 μm accuracy because their position was measured by a high-precision 2D image measuring instrument. For each calibration pattern, 15 images were captured at different positions and angles relative to the camera. The parameters were then estimated based on these images. The estimated intrinsic parameters and lens distortion parameters were subsequently used to perform image coordinate calibration.

The extrinsic parameters, which represent the positional relationship between the camera and the maker's motion plane, were determined using the detected grid points of the marker as a substitute for the calibration board. The inverse matrix of the obtained extrinsic parameters was applied to correct the maker's grid point coordinates, aligning the motion plane to be perpendicular to the optical axis of the camera.

3.2 Detection Error Evaluation

The detection error of the feature point position can be influenced by the environmental lighting conditions, an investigation of the detection errors was conducted in the experimental setup. Specifically, the variation in the detected positions was evaluated by capturing image of stationary boards. 100 images of the stationary boards were taken while varying the position of the lighting. The feature points in each image were detected using the detection function of commercial data analysis software (MATLAB ver. R2024b, detectPatternPoints). The shooting conditions are the same as described in Table 1.

Figure 6 shows the standard deviation of detected positions for each feature point on the checkerboard and circle grid. In this figure, the color bar indicates the magnitude of the standard deviation. The circle grid shows a biased distribution of the standard deviation, while no such trend is observed for the checkerboard. This difference is likely due to the use of different algorithms for checkerboard detection and circle detection in MATLAB, as well as differences in how each calibration board's surface

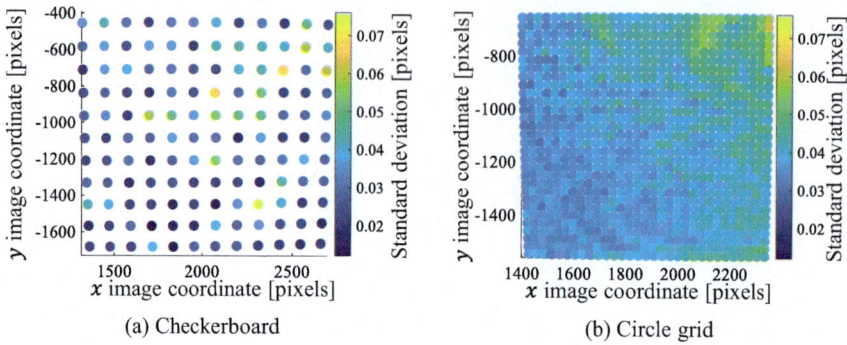

Fig. 6 Standard deviation of detected position in static state

treatment affects the sensitivity to lighting condition variations. The maximum standard deviations for the checkerboard and circle grid were 0.076 and 0.062 pixels, respectively. Thus, their expanded uncertainties are 0.152 and 0.124 pixels, respectively, when the coverage factor is two. Thus, both results exceed the required accuracy of 0.01 pixel for achieving the target measurement accuracy. The uncertainty of the circle grid is smaller by 0.01 pixel at the best than that of checkerboard in a stationary state.

3.3 Uncertainty Evaluation Method by Simulation

The obtained detection error and the board error for each calibration board were used as the noise scale in the uncertainty simulation described in Sect. 2, allowing the estimation of the working distance uncertainty under the experimental conditions. Specifically, the uncertainty of the working distance was evaluated through Monte Carlo simulations by assuming errors in the estimation of camera parameters and the detection positions of the marker under the experimental conditions. The simulation iteration was set to 10,000 times.

The uncertainty of the intrinsic parameters and lens distortion parameters was assumed to follow a normal distribution with the scale factor obtained in the uncertainty simulation in Sect. 2. To account for the influence of the number of feature points on the calibration patterns, the number and spacing of the feature points used in the simulation were adjusted to match those of each calibration pattern. The resolution uncertainty of the marker position detection was assumed to have a uniform distribution with $\pm$ 1/100 pixel limit, which was estimated from the experimental variation of the detected position.

3.4   Evaluation of Calibration Uncertainty

The deviations of the distances between the marker's motion planes from the commanded values are shown in Fig. 7. Additionally, Fig. 7 also compares the combined standard uncertainties of the working distances obtained from the uncertainty simulation. The combined standard uncertainty was obtained from the standard uncertainties of two distant positions. For all calibration patterns, the experimental results show that the deviations are larger than the target accuracy by a factor of one hundred. Although the selected circle grid should have sufficiently small board error according to the uncertainty simulation in Sect. 2, it showed the largest deviations. Since the detection errors of the circle grid and the selected circle grid are comparable, it is considered that other error factors, such as the number of feature points, contribute to this result.

The uncertainty obtained by simulation indicates that the selected circle grid has the largest standard deviation, similar to the experimental results. While the distance deviation is similar for the checkerboard and the circle grid in the experiment, the uncertainty of the checkerboard was approximately half that of the circle grid in the simulation. The deviations in the experimental results are smaller than the uncertainties obtained from the simulation, indicating that the experimental results are consistent with the simulation.

Further investigation revealed that the radial distortion coefficients k_2 and k_3 had the most significant impact on the uncertainty of the measured working distance difference. The checkerboard showed smaller uncertainty, likely due to its larger size, which reduced variations in k_2 and k_3. Higher order coefficients k_2 and k_3 are more sensitive to the size of calibration area.

Fig. 7 Deviation of working distance difference from commanded value

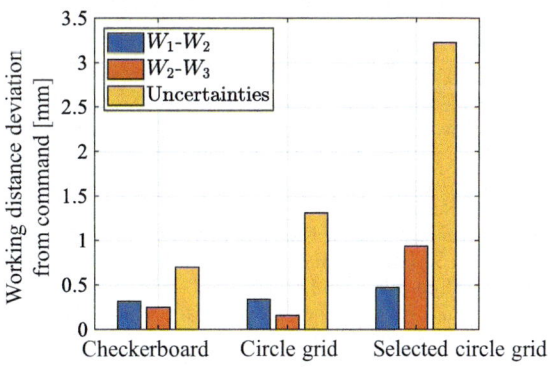

4   Conclusion

The calibration accuracy required for vision-based thermal error measurements was evaluated through simulations and experiments. The results of the Monte Carlo simulation revealed that achieving the target accuracy of 1 μm within a measurement range of 1 m requires a board error of less than 1 μm and a detection error of less than 0.01 pixels.

Experimental results showed that the calibration boards used in the experiment could not achieve the target measurement accuracy due to insufficient accuracy in both the board error and detection error. Both errors were approximately an order of magnitude larger than the required accuracy. Furthermore, for accurate calibration, the number of feature points on the calibration board should be more than 200. Enlarging the calibration board size relative to the field of view was also effective for higher accurate calibration by enhancing the estimation accuracy of the higher-order coefficients of radial distortion.

Acknowledgements This study was supported by The Die and Mould Technology Promotion Foundation and JSPS KAKENHI Grant Number JP25K01132. The authors express their gratitude for the support.

References

1. Bryan, J.: International status of thermal error research. CIRP Ann. **39**, 645–656 (1990). https://doi.org/10.1016/S0007-8506(07)63001-7
2. Ramesh, R., Mannan, M.A., Poo, A.N.: Thermal error measurement and modelling in machine tools: Part I. Influence of varying operating conditions. Int. J. Mach. Tools Manuf **43**, 391–404 (2003)
3. Mayr, J., Jedrzejewski, J., Uhlmann, E., Donmez, M.A., Knapp, W., Härtig, F., Wendt, K., Moriwaki, T., Shore, P., Schmitt, R., Brecher, C., Würz, T., Wegener, K.: Thermal issues in machine tools. CIRP Ann. **61**, 771–791 (2012). https://doi.org/10.1016/J.CIRP.2012.05.008
4. Ibaraki, S., Tanizawa, Y.: Vision-based measurement of two-dimensional posional positioning errors of machine tools. J. Adv. Mech. Design Syst. Manuf. **5**, 315–328 (2011)
5. Irino, N., Shimoike, M., Mori, K., Yamaji, I., Mori, M.: A vision-based machine accuracy measurement method. CIRP Ann. **69**, 445–448 (2020)
6. Zhang, S.: High-speed 3D shape measurement with structured light methods: a review. Opt. Lasers Eng. **106**, 119–131 (2018)
7. Zhang, Z.: A flexible new technique for camera calibration. IEEE Trans. Pattern Anal. Mach. Intell. **22**, 1330–1334 (2000). https://doi.org/10.1109/34.888718
8. Duane, C.B.: Close-range camera calibration. Photogramm. Eng. **37**(8), 855–866 (1971)
9. Tsai, R.Y.: A versatile camera calibration technique for high-accuracy 3D Machine vision metrology using off-the-shelf TV cameras and lenses. IEEE J. Robot. Autom. **3**(4), 323–344 (1987)
10. Sobel, I.: On calibrating computer controlled cameras for perceiving 3-D scenes. Artif. Intell. **5**, 185–198 (1974). https://doi.org/10.1016/0004-3702(74)90029-0
11. Motta, J.M.S.T., De Carvalho, G.C., McMaster, R.S.: Robot calibration using a 3D vision-based measurement system with a single camera. Robot Comput. Integr. Manuf. **17**, 487–497 (2001). https://doi.org/10.1016/S0736-5845(01)00024-2

12. Azimbeik, K., Hossein Mahdavi, S., Rahimzadeh Rofooei, F.: Improved image-based, full-field structural displacement measurement using template matching and camera calibration methods. Measurement **211**, 112650 (2023)
13. Ding, L., Xu, H., Xu, B., Du, P.: A review of Research on machine vision-based dimensional measurement methods for parts. In: 2024 5th International Conference on Machine Learning and Computer Application (ICMLCA) **5**, 376–381 (2024)
14. Weng, J., Coher, P., Herniou, M.: Camera calibration with distortion models and accuracy evaluation. IEEE Trans. Pattern Anal. Mach. Intell. **14**, 965–980 (1992). https://doi.org/10.1109/34.159901
15. Mori, K., Kono, D., Matsubara, A.: Vision-based volumetric displacement measurement with a self-illuminating target. CIRP Ann. Manuf. Technol. **72**, 305–308 (2023). https://doi.org/10.1016/j.cirp.2023.04.059

Utility of Radio Transmission Probes for Identifying Non-stationary Thermal Errors of Spindle Units

Michal Straka, Martin Mareš, Hyeok Kim, and Horejš Otakar

Abstract Thermal errors caused by internal and external heat sources constitute a significant part of the geometric errors in machine tools (MTs). With machining accuracy and MT utilisation rates continuously rising, the thermal behaviour of MT structures research is crucial for advancing manufacturing efficiency. To mitigate costly structural redesigns, thermal errors can be reduced through two primary strategies: direct methods (real-time measurement of the tool-workpiece contact point) or indirect approaches (predictive models). These methods avoid the need for active thermal control mechanisms, which are often costly to implement and operate, especially when using climate chambers. As MT sensory equipment continues to improve, predictive models adaptable to areas of higher nonlinearity and inhomogeneity in thermo-mechanical systems are becoming more prevalent, leveraging real-time feedback from intermittent probing of manufactured parts. This study investigates the utility of radio transmission part probe (RMP) measurements for non-stationary thermal error identification on MTs. Experimental validation involves exposing the target MT to controlled conditions in a climate chamber, including different spindle speeds and ambient temperature settings.

Keywords Machine tool · Thermal error · On-machine measurement · Non-stationary thermal errors · Radio transmission part probe · RMP · Climate chamber experiments

M. Straka (✉) · M. Mareš · H. Otakar
Faculty of Mechanical Engineering, Department of Production Machines and Equipment, Czech Technical University in Prague, RCMT, Prague, Czech Republic
e-mail: M.Straka@rcmt.cvut.cz

H. Kim
DN Solutions Co., Ltd, Seongsan-Gu, Changwon-Si, Gyeongsangnam-Do, Korea

© The Author(s) 2026   311
K. Wegener and M. Bambach (eds.), *4th International Conference on Thermal Issues in Machine Tools (ICTIMT2025)*, Lecture Notes in Production Engineering,
https://doi.org/10.1007/978-3-032-01194-7_20

1   Introduction

Precision manufacturing continues to demand ever-tighter tolerances. However, machine tool (MT) accuracy remains fundamentally constrained by thermal errors. This is a persistent challenge that is responsible for 40–70% of total machining inaccuracies [1–3]. The state-of-the-art approach for minimising thermal errors in CNC MTs relies on real-time error compensation using thermal error estimation models, also known as indirect compensation [2]. The practical utility of these models depends on their predictive accuracy and robustness, with robustness referring to a model's ability to maintain accuracy under varying operational and external conditions.

Nevertheless, existing thermal error compensation approaches demonstrate limited generalisability across diverse manufacturing environments, with degraded prediction accuracy and poor robustness observed under dynamic thermal loads [4, 5]. A fundamental constraint pertains to the model's reliance on training data that inadequately reflects the complete spectrum of MT operational states. This often leads to incomplete physical representations and reduced adaptability. Consequently, these models demonstrate a decline in predictive performance when confronted with scenarios that extend beyond the confines of their initial training set.

In order to enhance robustness, recent studies suggest for the implementation of adaptive parameter updating through process-intermittent probing (on-machine measurement). This technique enables real-time identification of thermal error variations at the tool contact point (TCP), which helps to maintain the model accuracy over time [6–11].

Various on-machine measurement techniques can provide thermal displacement feedback for adaptive thermal error compensation models. A comprehensive review of state-of-the-art on-machine measurement systems was conducted by Gao et al. [12].

Radio transmission part probe (RMP) systems, which have become standard equipment in precision MTs, offer a particularly valuable solution for real-time thermal error compensation model adaptation. The utilisation of on-machine measurement as a method for in-situ verification of workpiece dimensions has become a preferred approach, as it eliminates the necessity for external metrology equipment and enhances manufacturing efficiency [12, 13]. Beyond their primary applications, on-machine measurements with RMPs enable direct compensation of thermal errors at the TCP, as demonstrated in [14].

Several research teams have successfully demonstrated approaches that use process-intermittent measurements to update compensation models, as evidenced in several studies [7–11]. These methods typically use RMPs in combination with precision calibration artefacts positioned on the MT table. Intermittent probing of these artefacts during machining cycles allows dynamic updates of compensation model parameters or their entire structure.

The main research question in this study is how reliable is RMP as feedback for (adaptive) thermal error compensation models? The objective is to evaluate the reliability of non-stationary thermal error (relative displacement between the TCP and the MTs table) identification on MT equipped with spindle unit using RMP during thermo-mechanical transient behaviour due to spindle rotation and under different ambient temperatures.

The paper is organized into five main sections. Following the introduction, Sect. 2 details the experimental methodology and testing conditions. Section 3 presents the experimental results, followed by a comprehensive analysis and discussion in Sect. 4. The final section provides conclusions and outlines future research directions related to this approach.

2 Experimental Set-Up and Conditions

This study focuses on a vertical 5-axis milling centre with a workspace of $625 \times 450 \times 400$ mm ($X \times Y \times Z$) placed in a climate chamber and equipped with an RMP measurement system. The MT is designed with three linear axes dedicated to tool holder movement and two rotary axes for workpiece manipulation. According to ISO 10791-2 [15], the kinematic chain can be represented as V[w-C'-B'-b-Y-X-Z-(C)-t].

For reliability assessment purposes, a reference displacement measuring method is implemented. This reference system comprises non-contact eddy-current sensors (PR6423) arranged in a five-point configuration and mounted in specialized fixtures to the MT table and a test mandrel clamped in the spindle, following the ISO 230-3 standard [16], illustrated in Fig. 1, left. This measurement approach is characterised by its precision and consistency. Consequently, it is regarded as the reference standard and is designated as ECS throughout this study.

The reliability assessment of RMP displacement measurements is conducted using the Renishaw RMP 400 system, which has a manufacturer-certified repeatability of $2\sigma \leq 0.25$ μm. This system is deployed in conjunction with a datum ball (DB) with a sphere diameter of 22 mm, depicted in Fig. 1, right.

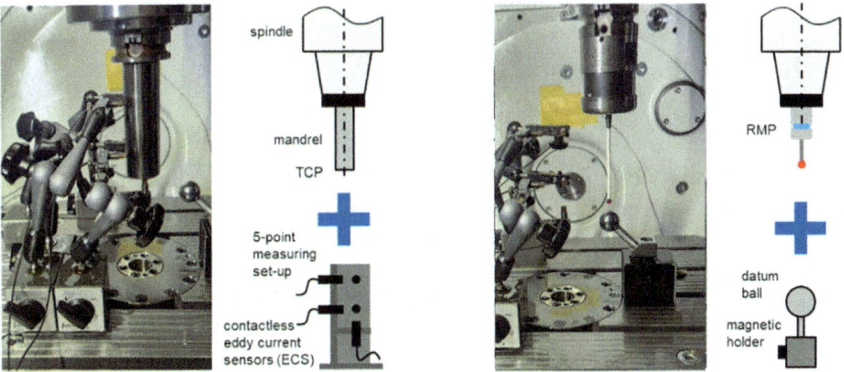

Fig. 1 Experimental configurations for the ECS (left) and the TMP (right) approach

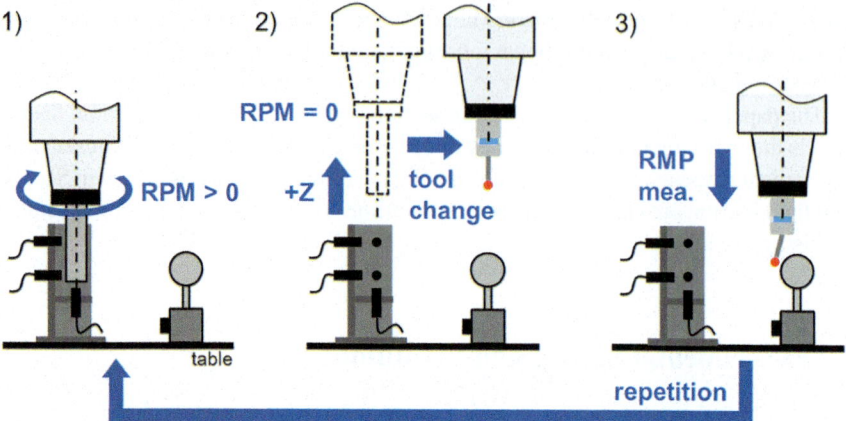

Fig. 2 Measurement cycle scheme

In all conducted experiments in which the spindle speed is varied and the room temperature is controlled to different ambient temperature levels, an identical measurement cycle is periodically performed. Initially, see part 1) in Fig. 2, the test mandrel clamped in the spindle is positioned at the measuring position, and then the spindle rotation is initiated for a period of 15 min. Subsequently, see part 2) in Fig. 2, the spindle rotation is stopped, the test mandrel is removed from the fixture, and an exchange with the RMP is performed, which allowed measurement of the DB in the X, Y, and Z-directions, see part 3) in Fig. 2. Following DB measurements, the RMP is replaced back with the test mandrel, and the entire process is repeated from part 1) until the end of the experiment. The interval between spindle stop and subsequent reactivation is five minutes. A reduction in feed rate is applied to mitigate potential thermal influences on the MT caused by linear axis movements, thereby isolating the thermal effects attributable solely to spindle rotation.

3   Test Results

A total of four experiments were conducted, see Table 1. One experiment was performed at a constant spindle rotation speed of 6700 rpm (1/3 of maximum spindle speed) and three at a constant spindle speed of 15,000 rpm (3/4 of maximum spindle speed), each under different controlled ambient temperatures that were maintained at 18, 22, and 26 °C within the climate chamber throughout the entire experimental duration. The temperature sensor is installed at the rear of the machine bed, and the measured bed temperature is considered the ambient temperature $T_{amb.}$. The difference between the set-up temperature in the climate chamber and measured ambient temperature $T_{amb.}$ values is attributable to the absence of machine's temperature sensor calibration for absolute values, as this study prioritises relative changes in measurements rather than precise absolute values.

By ensuring a constant ambient temperature during each experiment, thermal errors attributable to environmental temperature changes were minimised to a negligible level. The implementation of different ambient temperature conditions across experiments enables an assessment of the reliability of the RMP under varying environmental conditions and thereby evaluating the robustness of this approach for deployment in standard industrial applications. Two distinct spindle speed levels were selected to verify the base robustness of the RMP approach when measuring thermal errors under different magnitudes of thermal load induced by different spindle activity. After each ambient temperature adjustment in the climate chamber, the MT was thermally stabilised at the new temperature for 14 h before the subsequent experiment was conducted.

The temperature variation measured at the spindle bearing ($T_{bear.}$) and the ambient environment ($T_{amb.}$) during the first hour of each experiment is illustrated in Fig. 3.

This time frame was selected as the transient thermal phase induced by spindle rotation had stabilised by this point, and it provided a clearer demonstration of the influence of the individual parts of the measurement cycle (see Fig. 2) on the measured

Table 1 Overview of test conditions

Test no.	Spindle speed [rpm]	Ambient temp. set-up [°C]	Ambient temp. measured $T_{amb.}$ [°C]
1	6700	22	22.1
2	15,000	18	19.3
3	15,000	22	22.8
4	15,000	26	25.6

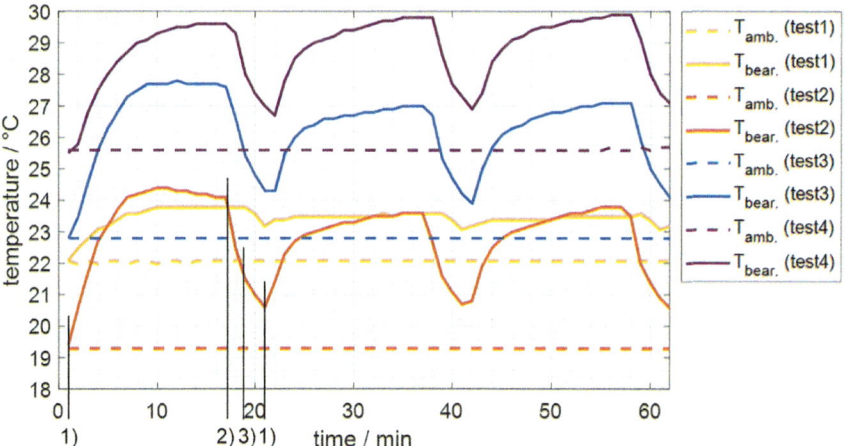

Fig. 3 Measured temperatures for tests no. 1–4

temperatures—the time points 1), 2) and 3) corresponding to the initiation of each part of the measurement cycle are indicated below the timeline in Fig. 3. A significant observation is that the MT activity did not appreciably affect the ambient temperature, which remained constant throughout each experiment.

The variation of measured TCP displacement relative to the MT's table during Test 3 (see Table 1) is illustrated in Fig. 4. The graph shows the TCP displacement measured via the ECS approach, denoted as (Z_{ECS}); missing values resulting from measurement cycle execution were supplemented through linear interpolation. The displacement values that were measured at time points 1) and 2) of the measurement cycle illustrated in Fig. 2 are represented by the variables $Z_{ECS\ 1)}$ for time point 1) and $Z_{ECS\ 2)}$ for time point 2). The difference between values $Z_{ECS\ 1)}$ and $Z_{ECS\ 2)}$ is attributable to spindle cooldown during the measurement cycle. The $Z_{ECS\ 2)}$ measurement is almost equivalent to the displacement that would be measured if the MT heating process through spindle rotation were not interrupted by the measurement cycle. The progression of values measured using RMP is represented in the graph as $Z_{RMP\ 3)}$; linear interpolation was applied to fill in missing values in discrete measurement data.

In consideration of the objectives of this study, the key parameter illustrated in Fig. 4 is the difference between the measured displacement using the ECS and RMP approaches in the Z-direction, denoted as *Residue Z: $Z_{ECS\ 2)}$-$Z_{RMP\ 3)}$*, as it represents the magnitude of the difference between the displacement measured via the RMP approach and the value obtained through the reference ECS method. Furthermore,

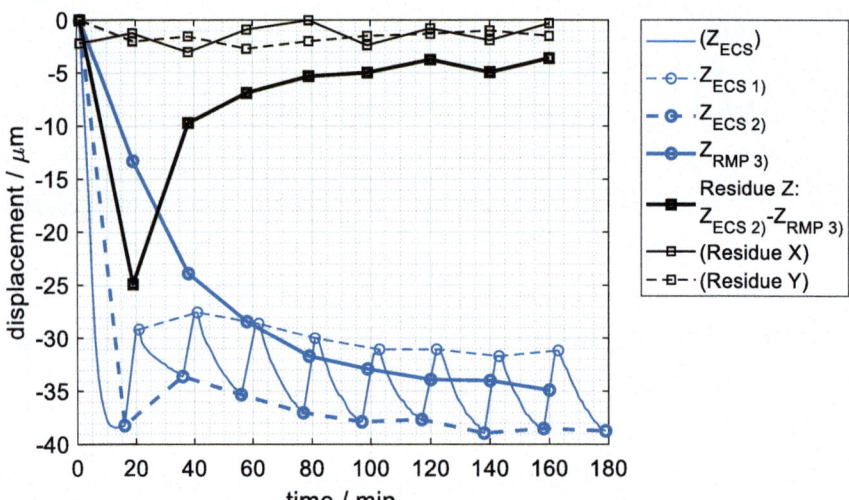

Fig. 4 Measured TCP displacement in the Z-direction and residual values for test no. 3

residual values demonstrating the measurement differences between both measurement approaches in the X and Y-directions are presented. It is noted that the magnitude of the difference for both these evaluated directions is insignificant compared to differences in the Z-direction and will therefore not be further presented.

4 Discussion

The residuals, representing the difference between the reference ECS and the assessed RMP approach in the Z-direction, are illustrated in Fig. 5 for individual tests no. 1–4 (see Table 1), which differ in the constant values of the set spindle speed and ambient temperature. The observed trends indicate that, after approximately two hours of spindle rotation at constant speed, the transient phenomenon induced by spindle activity had almost stabilised, regardless of the specific spindle speed or ambient temperature applied. This observation confirms the repeatable behaviour of the system and the reliability of the RMP approach in its typical application.

However, it is evident that there is a substantial impact of a change in spindle speed from 1/3 of max. rpm, as in test no. 1, to 3/4 of max. rpm, as in test no. 3, on the residual value, in both transient and steady-state phases, thus on the difference in the measured value of TCP displacement between the reference ECS and assessed RMP approach. Similarly, the impact of ambient temperature changes on residual values is detected during the transient phase following the first spindle speed cycle in tests no. 2, 3, and 4. This may indicate a limited reliability of the RMP approach in obtaining feedback from the thermo-mechanical system of the MT in terms of

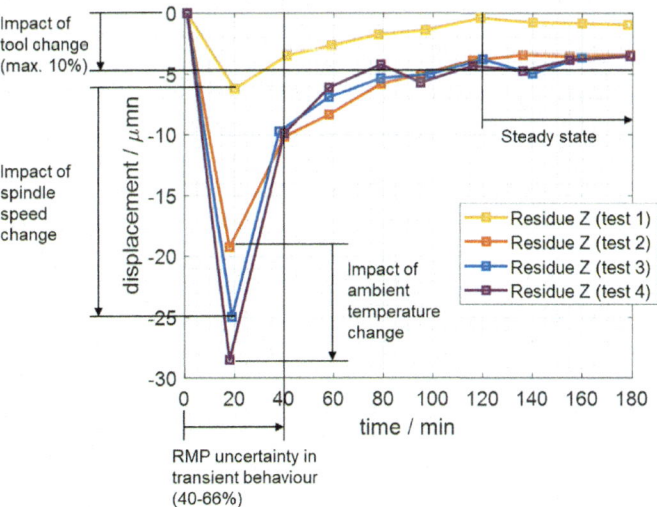

Fig. 5 Residual values for tests no. 1–4

TCP displacement during the transient phase. A comparison of $Z_{ECS\ 2)}$ and $Z_{RMP\ 3)}$, see Fig. 4, revealed a 40–66% loss of displacement value information with the RMP approach during the transient phase, relative to the ECS approach. The deviation is likely attributable to process interruptions caused by test mandrel—RMP exchange and to effective spindle cooling during the measurement cycle. Furthermore, the experimental conditions applied in this study do not fully reflect standard machine tool operation, which could also influence the observed results. However, a detailed investigation of these influences lies beyond the scope of the present study.

The residual value observed during the steady-state phase is most likely caused by the inability to perform both measurements simultaneously and is attributed to the operational necessity of replacing the test mandrel with the RMP. During the steady-state phase, this results in a maximum deviation of 10% relative to the ECS approach. It is hypothesized that this deviation could be reduced by shortening the operational time interval between ECS and RMP measurements.

5   Conclusions and Outlook

This research evaluated the reliability of RMP measurements for providing accurate feedback on thermally induced TCP displacement for potential application in adaptive thermal error compensation in MTs. Through a series of experiments comparing RMP with the reference ECS method, in accordance with ISO 230-3 [16], under different spindle speeds and ambient temperatures, there were identified critical limitations of the RMP approach in non-stationary thermal error identification on MT equipped with spindle unit.

Analysis of the measurements showed that a potentially significant reliability issue occurs during thermal transient phases, as RMP measurements missed 40–66% of the thermal displacement detected by the reference ECS method in the Z-direction. The residual value, the difference between ECS and RMP measurements, varies substantially with both spindle speed and ambient temperature, being most pronounced at higher rpm levels. In contrast, once thermal steady-state is achieved, for this MT approximately two hours after spindle activity, the RMP approach demonstrates acceptable reliability with deviations limited to 10%. Furthermore, displacement measurement deviations in the X- and Y-directions were found to be insignificant in comparison to those in the Z-direction.

The findings of this study indicate that the RMP approach may not be suitable for capturing rapid transient thermal phenomena. Nevertheless, its performance during thermally stable operational phases confirms its potential for use in adaptive thermal error compensation models. To ensure robust implementation in industrial environments, it is advisable to verify the performance of the RMP system against a reference measurement method on the specific MT and RMP configuration and to incorporate identified system limitations into the design of compensation models. This research provides a basis for developing more reliable adaptive thermal error compensation strategies that account for the measurement characteristics of commonly available RMP systems in practical MT applications.

Future work should extend this study to include cutting processes, allowing the evaluation of RMP reliability under real machining conditions. This extension will provide critical insights into the practical implementation of RMP feedback for adaptive compensation approaches in production environments. Additionally, to establish the generalisability of these findings, tests should be conducted on different MT configurations, including MTs of the same type but equipped with different spindle units, as well as MTs of different types and sizes. Moreover, a comparative analysis of RMP performance against alternative measurement methods, such as tool probe systems or acoustic emission techniques, will provide a more comprehensive understanding of the relative advantages and limitations of different measurement approaches for real-time on-machine thermal error identification. These expanded investigations will contribute to the development of more flexible and robust adaptive compensation strategies suitable for diverse manufacturing applications.

Acknowledgements This work was supported by the Grant Agency of the Czech Technical University in Prague, grant no. SGS25/135/OHK2/3T/12. This work was supported by the Industrial Technology Innovation Program (20023418, Development of Machine Tool Digital Twin and Intelligence Technology) funded by the Ministry of Trade, industry & Energy of the Republic of Korea.

References

1. Bryan, J.: International status of thermal error research. CIRP Ann. Manuf. Technol. **39**(2), 645–656 (1990)
2. Mayr, J., Jedrzejewski, J., Uhlmann, E., Alkan Donmez, M., Knapp, W., Härtig, F., Wendt, K., Moriwaki, T., Shore, P., Schmitt, R., Brecher, C., Würz, T., Wegener, K.:: Thermal issues in machine tools. CIRP Ann. Manuf. Technol. **61**(2), 771–791 (2012)
3. Putz, M., Richter, C., Regel, J., Bräunig, M.: Industrial relevance and causes of thermal issues in machine tools. In: Proceedings of the Conference on thermal issues in Machine Tools, Dresden (2018)
4. Mareš, M., Horejš, O., Hornych, J., Smolík, J.: Robustness and portability of machine tool thermal error compensation model based on control of participating thermal sources. J. Mach. Eng. **13**(1), 24–36 (2013)
5. Miao, E.M., Gong, Y., Niu, P.C., Ji, C.Y., Chen, H.D.: Robustness of thermal error compensation modeling models of CNC machine tools. Int. J. Adv. Manuf. Technol. **69**(2013), 2593–2603
6. Horejš, O., Mareš, M., Straka, M., Švéda, J., Kozlok, T.: Adaptive thermal error compensation model of a horizontal machining centre. In: Ihlenfeldt, S. (ed.) 3rd International Conference on Thermal Issues in Machine Tools (ICTIMT2023), ICTIMT 2023.Lecture Notes in Production Engineering. Springer, Berlin, pp 83–98 (2023)
7. Mou, J.: A method of using neural networks and inverse kinematics for machine tools error estimation and correction. J. Manuf. Sci. Eng. **119**(2), 247–254 (1997)
8. Yang, H., Ni, J.: Adaptive model estimation of machine-tool thermal errors based on recursive dynamic modeling strategy. Int. J. Mach. Tools Manuf. **45**(1), 1–11 (2005)
9. Blaser, P., Pavlíček, F., Mori, K., Mayr, J., Weikert, S. and K. Wegener. Adaptive learning control for thermal error compensation of 5-axis machine tools. J. Manuf. Syst. **44**(2), 302–309 (2017)

10. Zimmermann, N., Breu, M., Mayr, J., Wegener, K.: Autonomously triggered model updates for self-learning thermal error compensation. CIRP Ann. **70**(1), 431–434 (2021)
11. Zimmermann, N., Müller, E., Lang, S., Mayr, J., Wegener, K.: Thermally compensated 5-axis machine tools evaluated with impeller machining tests. CIRP J. Manuf. Sci. Technol. **46**(2023), 19–35 (2023)
12. Gao, W., Haitjema, H., Fang, F., Leach, R., Cheung, C., Savio, E., Linares, J.: On-machine and in-process surface metrology for precision manufacturing. CIRP Ann. **68**(2), 843–866 (2019)
13. Zhuang, Q., Wan, N., Guo, Y., Zhu, G., Qian, D.: A state-of-the-art review on the research and application of on-machine measurement with a touch-trigger probe. Measurement **224** (2024)
14. Brecher, C., Hirsch, P., Weck, M.: Compensation of thermo-elastic machine tool deformation based on control internal data. CIRP Ann. Manuf. Technol. **53**(1), 299–304 (2004)
15. ISO 10791-2: 2015: Test conditions for machining centres—Part 2: Geometric tests for machines with vertical spindle (vertical Z-axis) (2015)
16. ISO 230-3: Test code for machine tools—Part 3: determination of thermal effects (2020)

Effect of Thermal Changes on the Squareness Error Between Linear Axes of Machine Tools

Morteza Dashtizadeh, Andrew Longstaff, and Simon Fletcher

Abstract The international standard ISO 230-3 generally specifies four different tests to characterise the thermal behaviour of machine tools. All these tests quantify performance of a single linear or a single rotary axis/component of the machine tool under certain thermal conditions. However, this standard does not address the influence of thermal changes in the assembly of machine tools' linear axes, namely all their mutual squareness, E_{C0X}, E_{A0Y}, and E_{B0Z}. This study examines the variation of squareness error between each pair of linear axes of machine tools. The mutual squareness errors between all three linear axes of a 3-axis vertical machining centre (VMC) with kinematic chain of [w X' Y' b Z (C) t] were measured using the Renishaw QC-20W ballbar. Reasons for this choice of instrument, supported by an analysis of the associated uncertainty contributors are provided to show that this is a robust and convenient method to measure these thermal effects. The measurements were repeated over a period of two weeks, where the ambient temperature naturally varied by 4.5 °C. Findings of this study indicate significant changes of 12 µm/m and 14 µm/m in squareness error in two vertical planes, YZ and ZX, respectively, whereas the squareness in the horizontal XY plane shows negligible variation. The reasons for these differences are explained by the mechanical construction of the machine. Monitoring the temperature of the machine reveals a significant correlation between the temperature of elements of the Z-axis and the squareness of this axis with respect to both horizontal axes, X and Y of this machine. An uncertainty analysis based on GUM for squareness measurements using a ballbar in the three principal planes reveals that the expanded uncertainties are ± 2.2 µm/m, ± 7.1 µm/m, and ± 7.5 µm/m in the XY, YZ, and ZX planes, respectively. The paper concludes by proposing a suitable outline method of test for the thermal influences on the squareness error of precision machines, in line with other parts of the ISO 230 series of standards.

M. Dashtizadeh (✉) · A. Longstaff · S. Fletcher
Centre for Precision Technologies, Future Advanced Metrology Hub for Sustainable Manufacturing, University of Huddersfield, Huddersfield, UK
e-mail: m.dashtizadeh2@hud.ac.uk

© The Author(s) 2026

K. Wegener and M. Bambach (eds.), *4th International Conference on Thermal Issues in Machine Tools (ICTIMT2025)*, Lecture Notes in Production Engineering,
https://doi.org/10.1007/978-3-032-01194-7_21

1 Introduction

ISO 230-3:2020 [1] standardises four sets of tests on machine tools to evaluate their thermal distortions under different conditions. Additionally, ISO 10791-10:2022 [2] and ISO 13041-8:2004 [3] specify the same tests on machining centres and turning centres, respectively, without providing acceptable tolerances for these machine tools.

Squareness error as an index of the assembly quality of two linear axes of a machine tool is missing in ISO standards specifying thermal tests. This research addresses this gap experimentally.

Conducting squareness test to obtain reliable results in industrial environment is generally challenging. The availability of appropriate measuring instruments with low uncertainty capable of covering the full axis travel of the axes under test is a major concern. Several measuring instruments and techniques are commonly used in industrial environments to measure the squareness error between two linear axes. ISO/TR 230-11:2018 [4] outlines suitable measuring instruments for testing the squareness between linear axes of a machine tool. Table 1 lists these measuring instruments along with their typical measurement ranges for checking the squareness between two linear axes. Reference squares are one of the most commonly used on the list but the ballbar enables convenient automated capture with simple manual setup for new test positions.

Szipka et al. [5] showed that the squareness errors can be measured efficiently using circular motion and measurement of inertia. The study validated performance against traditional measurements and included an uncertainty analysis. The method was not applied to longer term testing of thermal variation. However, the ability to perform such measurements with sensors integrated into machine tool will enable higher levels of measurement automation and increase the efficiency of obtaining the level of data presented in our study. Tang et al. [6] also showed that on-machine probing can be used to measure squareness with low uncertainty. While the instrument is often readily available for automated measurement, an artefact is required similar to the ISO 10791-10:2022 standard and the approach by Ibaraki and Okumura [7], which

Table 1 Suitable measuring instruments for squareness tests with their range according to the guidelines of ISO/TR 230-11:2018 [4]

Squareness measurement	Typical applicable working range (length)			
	10 mm	100 mm	1 m	10 m
Reference squares with linear displacement sensor				
Reference cube				
Laser with squareness optics				
Optical square with angle reading device				
Index table with straightedge				
Three planes laser scanning device				
Ballbar application				
2D ball array				
Two dimensional digital scale				

may require loading/unloading and re-orienting for different axis combinations. The influence of thermal changes on the machine squareness was not included in either of the artefact studies but the method could be fully automated on some machines such as those with pallet loading for example.

Sarvas et al. [8] employed ISO 230-4 procedures utilising QC-20W ballbar to examine how different operating conditions affect circular test results, from which squareness error can be estimated. Tests were repeated at various machine states (cold start vs. warmed up) and showed variability in scaling errors and the perpendicular (squareness) error between the X–Y squareness of more than 10 μm/m as the machine warmed. Unfortunately, only XY results were presented, but the Z position was changed. Temperature sensors were only placed on the scales with no assessment of the causes of the variation.

In a production setting, frequent manual squareness checks are often impractical therefore few thermal studies exist due to the difficulty of performing them repeatedly at different thermal states. This work emphasises the significant impact of thermal effects on a popular machine configuration and automating the reviewed methods could reduce these barriers significantly.

2 Experiments

For this research, experiments were conducted on a vertical 3-axis Cincinnati Arrow500 machining centre (VMC) with the kinematic chain of [w X' Y' b Z (C) t] as shown in Fig. 1(left). In this machine configuration, the Z-axis assembled on the column carrying the tool while both X and Y-axes are located under the table moving the workpiece. Squareness between the linear axes of this VMC was measured using Renishaw ballbar QC-20W as depicted in Fig. 1(right).

Technically, Renishaw software calculates the squareness error by analysing the variations in radius sensed by its transducer at different angle with respect to the initial position of the transducer. Therefore, this provides a direct measurement of squareness derived from simultaneous interpolation of the two axes under test as they deviate from a nominal circular path.

Figure 2 illustrates the experimental setups for measuring the squareness errors between linear axes of the machine under test in the three principal planes of XY, YZ and ZX. Fig. 2 illustrates the experimental setups for measuring the squareness error between linear axes of the machine under test in the three principal planes of XY, YZ and ZX. To examine how squareness values are influenced by heat generated from axes motions, the squareness measurements cycle was repeated over time. No separate heating up cycle was applied between two consecutive circular tests. In other words, the heat was generated purely from the circular interpolation of the two axes under test. The position of the third axis of the machine remained unchanged throughout the entire experiment and between subsequent tests. By monitoring and

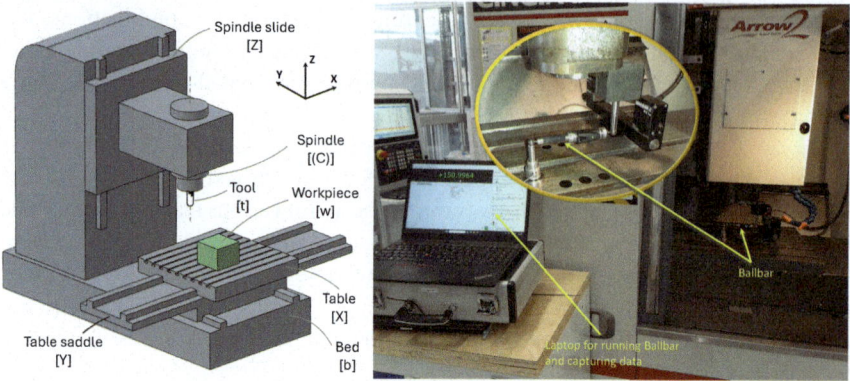

Fig. 1 VMC with kinematic chain of [w X' Y' b Z (C) t] according to ISO 10791-2 [9] (left), Setup of a circular test using Renishaw QC-20W Ballbar in XY plane on Cincinnati Arrow 500 VMC (right)

recording the temperature at various locations on the machine and its main components, the relationship between squareness changes and temperature variations can be identified.

Squareness measurements in the XY plane were conducted over two consecutive days while the experiments in the ZX and YZ planes were carried out over a period of 11 days. Greater variations were observed in the vertical ZX and YZ planes compared to the horizontal XY plane, which is why these experiments were conducted over a longer period.

There are approximately 40 locations on the VMC under test where the temperature sensors were installed. Each of these sensors recorded temperature data approximately every 5 s. It allows detailed investigation of temperature changes in both the machine and its ambient environment over very short time intervals.

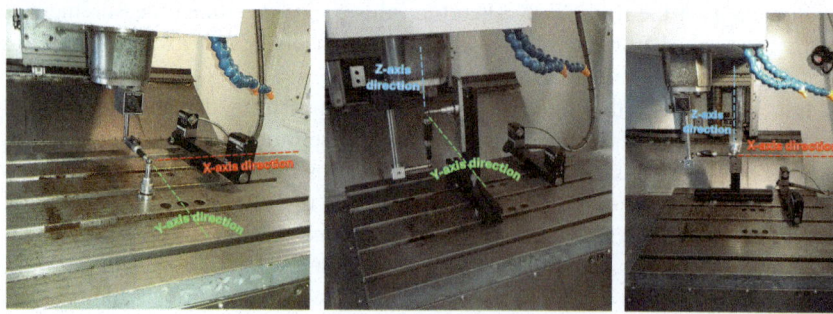

Fig. 2 Squareness measurement by Renishaw ballbar in different planes, XY (left), YZ (middle), and ZX (right) planes

3 Results of the Experiments

In this section, the results of the tests carried out in different planes are presented. Figure 3 presents the measured squareness error of X-axis to Y-axis, E_{COX}, over time. It shows that the squareness error fluctuated within $\pm$ 1 µm/m over the two-day experiment, during which 370 circular tests with feed speed of 2500 mm/min were performed in the XY plane. This figure also illustrates the internal ambient temperature alongside the temperature of the main components of the X and Y-axis, both located under the table of the VMC during the 2-day experiments in XY plane. This plot shows that the internal ambient temperature varied by 1.8 °C throughout the experiment while the temperature of X-axis and Y-axis ballnuts varied by 1.8 °C and 5.4 °C, respectively. Temperature of motors of X and Y-axes exhibited a variation of 5.2 °C and 5.1 °C, respectively.

An assessment of the correlation between recordings of each temperature sensor and the measured squareness in the XY plane reveals that most sensors have a correlation coefficient (R) between 0.7 and 0.8. Since nearly all locations of the machine display similar temperature trends with almost identical correlation coefficients, identifying a specific cause for the squareness error variations is challenging. It can be concluded that the heat generated in the horizontal axes of this VMC, when operating at a feed speed of 2500 mm/min, has no significant impact on the squareness variations between the axes under test.

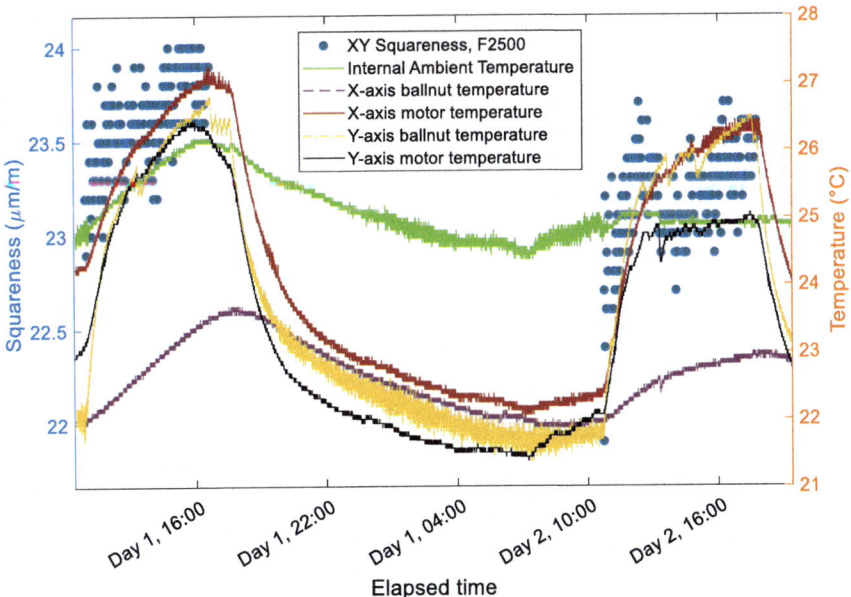

Fig. 3 Result of squareness measurement in XY plane using ballbar and monitored temperature

Figure 4 presents the monitored temperatures by the sensors mounted on the three axes of the VMC as well as the internal ambient temperature during the experiments in the YZ and ZX vertical planes. During the first five days, almost the same pattern is observed. After finishing the tests at the end of each shift, the machine was switched off and upon restarting the next day, all component temperatures increased. Z-axis motor shows the largest temperature increase, rising by 35 °C from cold condition to its warm condition, followed by a sharp drop when the machine was switched off. Day 6 was a weekend and no experiments were conducted. In the second phase of testing, the machine remained switched on continuously from days 7 until 15:30 on Day 10, when the machine was turned off again. After a cooldown period of almost 2.5 h, the machine was switched on again at around 18:00 for another set of tests. It remained switched on overnight between Day 10 and Day 11, with the final portion of the experiment carried out on Day 11. The experiments were finished at approximately 17:00.

Throughout the 11-day experiment, the internal ambient temperature of the machine varied by about 4 °C, ranging from 24 to 28 °C.

From Day 3 to the end of Day 10, approximately 400 circular tests were conducted in the ZX plane with two feed speeds of 270 and 2500 mm/min. All these tests were performed without interruption using a single setup. Figure 5 shows the test results of squareness of Z-axis to X-axis, E_{B0Z}, from Day 3 to Day 10 inclusive. It also demonstrates monitored temperature of Z-axis and X-axis components alongside the internal ambient temperature (represented on the secondary vertical axis). This plot reveals a strong correlation between the measured squareness and the Z-axis motor temperature.

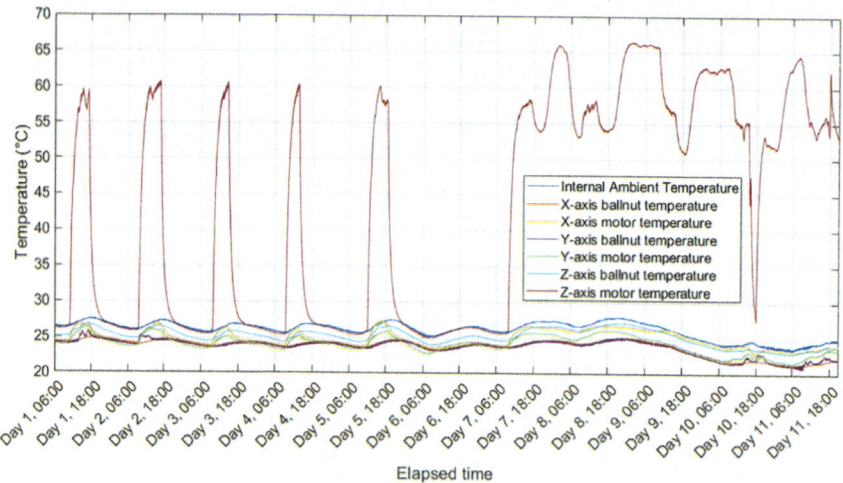

Fig. 4 Machine's internal ambient temperature and main components temperature during squareness tests in vertical YZ and ZX planes over 11 days

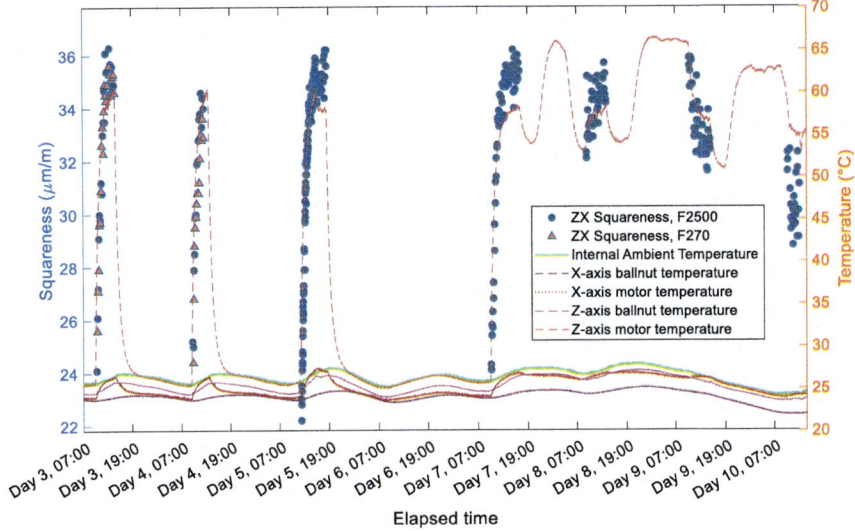

Fig. 5 Result of squareness measurement in ZX plane using ballbar and monitored temperature

Tests performed on Days 3 and 4 with both feed speeds of 270 and 2500 mm/min showed no significant difference in squareness results. This suggests that conducting circular tests at low and high feed speeds on this machine does not produce a distinguishable effect. Therefore, to speed up the process of data collection, all subsequent tests from Day 4 onwards were carried out with only the higher feed speed of 2500 mm/min.

Correlation analysis between measured squareness and temperature at different machine locations supports these findings. As shown in Fig. 6, both the Z-axis motor and the Z-axis motor plate have the highest correlation coefficient of more than 0.9 indicating that the heat generated by the Z-axis motor is the primary factor influencing the column on which the Z-axis guideway is installed.

Figure 7 illustrates squareness in the ZX plane against temperature at the locations with the highest correlation among all the 40 sensors. Besides the Z-axis motor and the column (due to heat transfer from the motor to column body), the X-axis motor temperature also shows a high correlation of 0.82 with squareness errors. Similar to the results observed in the XY plane, higher temperatures are associated with larger squareness errors.

Approximately 140 tests were performed in the YZ plane to evaluate the squareness of Y-axis to Z-axis, E_{A0Y}, and its changes due to different feed speeds as well as the heat generated by the two axes under test. Figure 8 presents the results of squareness measurement in the YZ plane. These tests were carried out on Days 1, 2, 10 and 11.

In the first two days of the experiments, the effect of two feed speeds of 270 and 2500 mm/min on the measured squareness was examined. Like the results in ZX

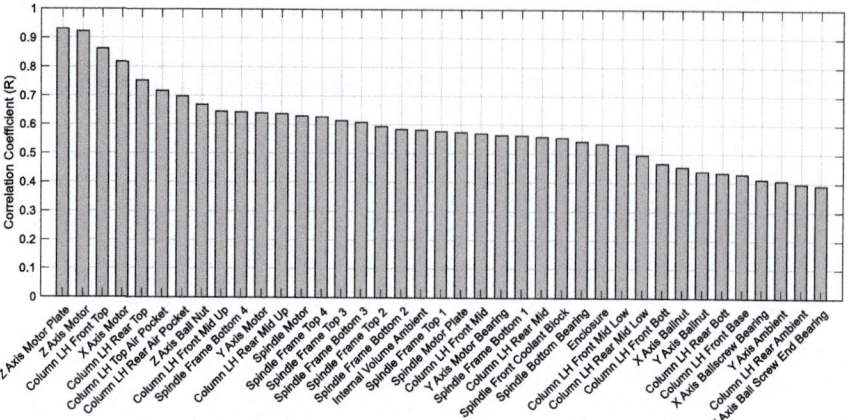

Fig. 6 Correlation coefficient (R) between monitored temperature and measured squareness in ZX plane

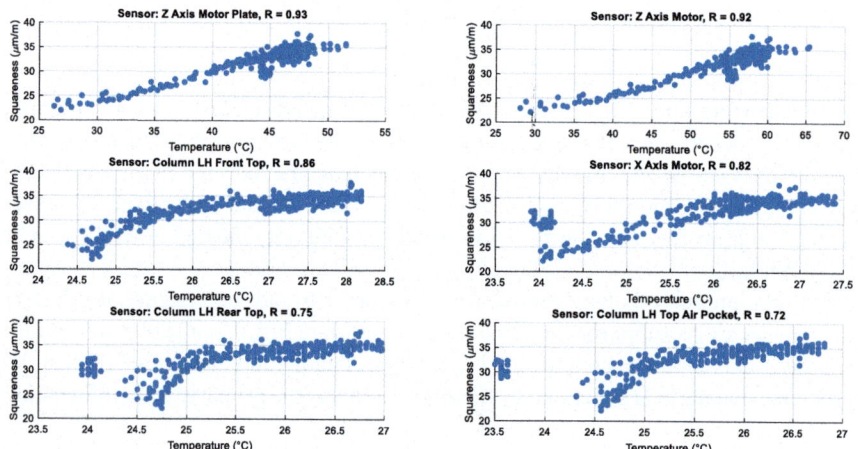

Fig. 7 Squareness error in ZX plane against recorded temperature with the highest correlations

plane, these measurements do not show a significant difference between these two feed speeds on the squareness.

In the gap between Day 3 and Day 9 inclusive, the experiments in the ZX plane were conducted meaning that the tests in the YZ plane were carried out with two different setups in two separate phases. However, the tests were performed exactly at the same centre position of Y and Z-axes as well as at the same position of the X-axis to maintain the same test condition during these two phases. It was also monitored to have very small centre offset along both Y and Z-axes less than 25 μm to minimise the effect of setup on the test results in the first test of both phases of

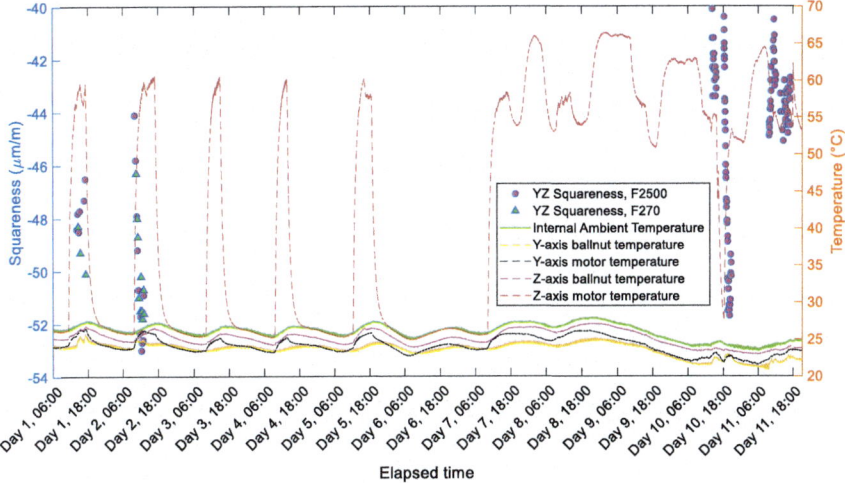

Fig. 8 Result of squareness measurement in YZ plane using ballbar and monitored temperature

the experiments, although Renishaw recommends that the tests with offsets below 100 μm are acceptable.

The results in the first 2 days in Fig. 8 indicate that as the Z-axis motor temperature increases over time, the squareness error shifts from almost −44 to −53 μm/m. In other words, the squareness error increases towards negative direction which means that there is an inverse correlation between the Z-axis motor temperature and squareness error in the YZ plane.

As mentioned before, from Day 7 onwards, the machine remained switched on, and the Z-axis motor reached a peak temperature of over 65 °C. To assess the machine's behaviour during cooldown from the warm condition, it was switched off at 15:30 on Day 10 and restarted at about 18:00 the same day. Figure 9 highlights the switching off and on moments in which Z-axis motor temperature dropping from almost 55–27 °C. This figure also shows that upon restarting, the squareness error was −40 μm/m, increasing negatively to −52 μm/m as the motor warmed back up to almost 53 °C. This corresponds to a squareness variation of 12 μm/m over a 28 °C temperature gradient of the Z-axis motor, recorded at the maximum feed speed of 2500 mm/min.

Given the limited squareness data in the YZ plane compared to the ZX plane, correlations analysis indicates that only the Z-axis motor and its plate exhibit strong correlation coefficients of −0.86 and −0.84, respectively. Figure 10 presents these correlation coefficients with the largest values, along with squareness versus temperature plots for the two locations with the highest correlations.

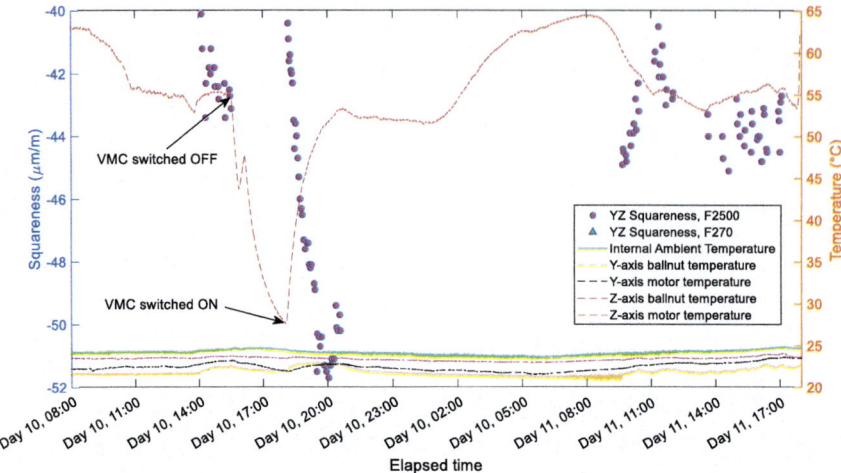

Fig. 9 Result of squareness measurement in YZ plane and monitored temperature over the last two days of experiments

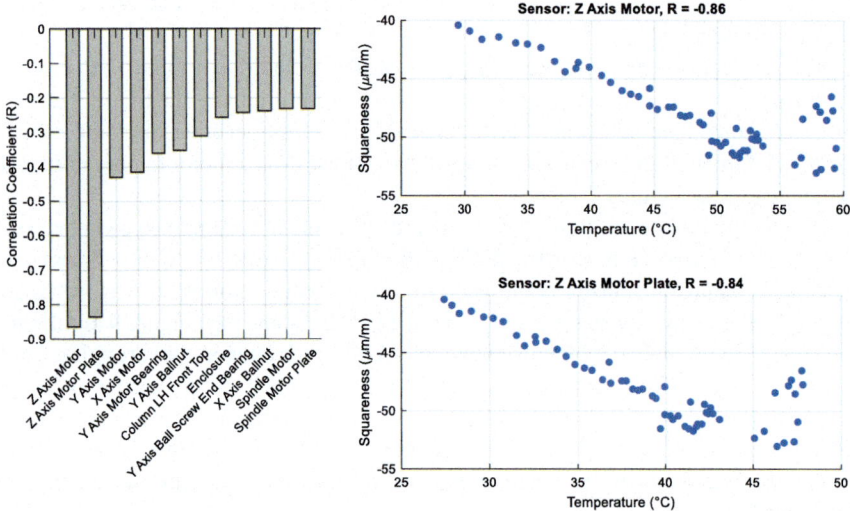

Fig. 10 Correlation coefficient (R) between monitored temperature and measured squareness in YZ plane (left) Squareness error against temperature of Z-axis motor and its plate (right)

4 Uncertainty of Squareness Measurement with a Ballbar

To evaluate the uncertainty of squareness measurement using a ballbar, firstly, the measurement principle should be addressed. According to ISO 230-1:2012 [10], subclause 10.3.2.6, the squareness error derived from a circular test is equal to the

Fig. 11 Concept of squareness measurement with ballbar

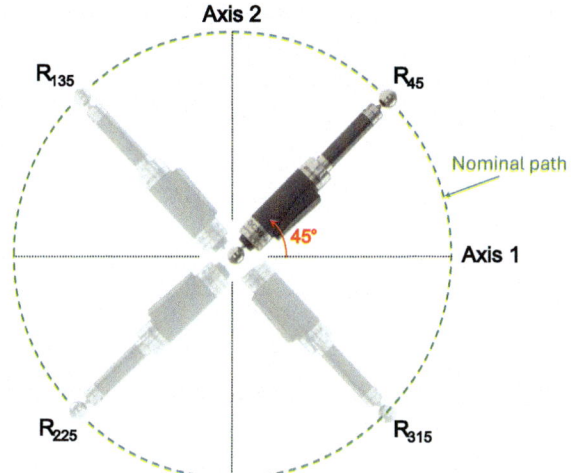

difference of the two diameters at $\pm 45°$ divided by the diameter of the nominal path. Figure 11 illustrates the concept of squareness measurement using a ballbar.

For the initial uncertainty estimation, it is assumed that Renishaw software follows the same concept. Therefore, according to the explanations of ISO 230–1 and the schematic shown in Fig. 11, the squareness error between Axis 1 and Axis 2, Sq in µm/m, is given by Eq. (1):

$$Sq = \frac{(R_{45} + R_{225}) - (R_{135} + R_{315})}{D_{Nom}} \times 1000 \qquad (1)$$

where D_{Nom} is the nominal diameter of the programmed path on the machine tool under test in mm, and R_{45}, R_{135}, R_{225}, and R_{315} are the actual length of the ballbar in mm at angles 45°, 135°, 225°, and 315°, respectively. For the tests conducted on the Cincinnati VMC, D_{Nom} was set to 300 mm.

Based on Eq. (1), the standard instrumental uncertainty of squareness measurement for a single setup at any centre position is calculated using Eq. (2):

$$u_{Sq} = \sqrt{\begin{array}{l} \left(\dfrac{\partial Sq}{\partial R_{45}}\right)^2 u_{R_{45}}^2 + \left(\dfrac{\partial Sq}{\partial R_{225}}\right)^2 u_{R_{225}}^2 \\[2mm] + \left(\dfrac{\partial Sq}{\partial R_{135}}\right)^2 u_{R_{135}}^2 + \left(\dfrac{\partial Sq}{\partial R_{315}}\right)^2 u_{R_{315}}^2 \end{array}} \qquad (2)$$

where u_{R_θ} (for any θ angle) is the standard instrumental uncertainty of the ballbar's linear measurement, which can be extracted from the calibration certificate of this device. Figure 12 shows the main technical information provided in the calibration certificate of the Renishaw QC20-W ballbar used for these experiments. As shown in this certificate, the instrumental expanded uncertainty under laboratory condition

Specification *Positional error* ±0.5 µm

Measured values and uncertainties of calibration

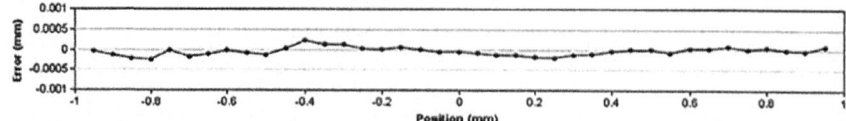

Results	Value
Maximum error:	±0.24 µm
Uncertainty of measurement (95% confidence level, k=2):	±0.24 µm

Fig. 12 Main technical data from the calibration certificate of the Renishaw QC20-W ballbar used for experiments

is reported as ± 0.24 µm over the full range of ± 1 mm of the ballbar. Therefore, u_{R_θ} is taken as 0.24 µm for all four readings at angles 45°, 135°, 225°, and 315°.

In Eq. (2), all derivatives of Sq with respect to the ballbar readings at different angles, R_θ, are equal and can be obtained from Eq. (3):

$$\frac{\partial Sq}{\partial R_\theta} = \frac{\partial Sq}{\partial R_{45}} = \frac{\partial Sq}{\partial R_{225}} = \frac{\partial Sq}{\partial R_{135}} = \frac{\partial Sq}{\partial R_{315}} = \frac{1000}{D_{Nom}} = \frac{1000}{300} \approx 3.3 \qquad (3)$$

Substituting $\frac{\partial Sq}{\partial R_\theta} = 3.3$ and $u_{R_\theta} = \pm 0.12$ into Eq. (2) gives the standard instrumental uncertainty of squareness measurement using a ballbar for a single measurement under laboratory conditions as shown in Eq. (4).

$$u_{Sq} = \sqrt[2]{\begin{array}{c}(3.3)^2(0.12)^2 + (3.3)^2(0.12)^2 \\ +(3.3)^2(0.12)^2 + (3.3)^2(0.12)^2\end{array}} \approx \pm 0.8 \, \mu m/m \qquad (4)$$

During the measurement of squareness, using either a calibrated ballbar with a calibrator or an uncalibrated ballbar does not have a significant impact, as any small deviation from the nominal length cancels out across all readings. This is because the ballbar length remains constant throughout the measurement process, particularly when conducted at higher feed speeds, which reduces the measurement duration.

Another uncertainty contributor is the ballbar's resolution. According to Renishaw software, the ballbar's resolution in squareness measurement is 0.1 µm/m. Assuming a rectangular distribution for this uncertainty contributor results in a divisor of $\sqrt{3}$. Therefore, the standard uncertainty of the resolution is given by Eq. (5):

$$u_{res} = \frac{0.5 \times \text{Resolusion}}{\text{Divisor}} = \frac{0.5 \times 0.1}{\sqrt{3}} \approx \pm 0.03 \, \mu m/m \qquad (5)$$

Equation (6) gives the combined standard uncertainty of the ballbar in squareness measurement:

$$u_{Device} = \sqrt{u_{Sq}^2 + u_{res}^2} = \sqrt{0.8^2 + 0.03^2} \approx \pm 0.8 \mu m/m \qquad (6)$$

Summarises of the sources of uncertainty and their contributions to the squareness measurement using a ballbar according to the standard method of GUM [11] for all three principal planes of XY, YZ and ZX are presented in Table 2.

5 Discussion on the Results

Based on the data collected throughout the experiment for all three planes, Table 3 summarises the squareness measurement results, considering only geometric uncertainty (assuming no thermal changes) and incorporating thermal distortions.

According to the summarised results, thermal distortions of the machine does not significantly affect the squareness of X-axis to Y-axis, E_{C0X}. However, the squareness in the vertical planes is meaningfully influenced due to the thermal effects. A key factor contributing to the observed squareness variations in the vertical planes of YZ and ZX is the heat generated by the Z-axis motor. Unlike other axes, the Z-axis motor must continuously overcome the weight of the Z-axis slider, spindle and its attachments, leading to a rise in temperature even when the axis remains stationary. This phenomenon was observed during several nights when the machine remained switched on. Since there is no cooling mechanism on this motor, the heat transfers into the column and surrounding structures, causing additional distortions. These thermal effects influence the squareness measurements, particularly in the vertical planes, where the accumulated heat can induce structural deformations over time. This highlights the significance of thermal considerations when assessing the geometric accuracy of machine tools, especially in precision applications. Understanding the changing geometry of a machine tool is the first step to compensation, but also vital in avoiding overcompensation if inadequate data is taken.

Considering the 40 μm/m tolerances specified in ISO 10791 series for the squareness of linear axes in general-purpose machining centres with normal accuracy, the obtained results in this research can be evaluated. However, it is common for customers, even for general-purpose machines to request tighter tolerances when negotiating and exchanging technical annexes of contracts, which serve as criteria for acceptance tests. The primary reason expressed by the customers or end users is that the machine accuracies worsen over time due to wear and tear; therefore, they prefer to set stricter tolerances during commissioning.

For the specific VMC used for the experiments, the results indicate that under certain conditions, squareness in the vertical planes can exceed the 40 μm/m tolerance. This means that if squareness were checked at such a moment, the machine might fail its acceptance test. It is worth noting that these measurements were

Table 2 Uncertainty budget table for squareness measurement between linear axes of the Cincinnati VMC by a ballbar

Nr	Source of uncertainty	Standard uncertainty, u
1	Measuring device (Ballbar)	0.8 µm/m
2	Machine X-axis repeatability at 20°C (obtained from a separate laser test on this axis)	0.3 µm/508 mm = 0.59 µm/m
3	Machine Y-axis repeatability at 20°C (obtained from a separate laser test on this axis)	0.2 µm/508 mm = 0.39 µm/m
4	Machine Z-axis repeatability at 20°C (obtained from a separate laser test on this axis)	0.2 µm/464 mm = 0.43 µm/m
5	Machine's structural changes due to thermal distortions in the XY plane	0.3 µm/m
6	Machine's structural changes due to thermal distortions in the YZ plane	3.4 µm/m
7	Machine's structural changes due to thermal distortions in the ZX plane	3.6 µm/m
Nr	Squareness uncertainty in different planes	Expanded uncertainty, U
I	Squareness in the XY plane, E_{C0X} (Geometric, excluding thermal)	$2 \times \sqrt[2]{(0.8)^2 + (0.59)^2 + (0.39)^2} \approx \pm 2.1 \mu m/m$
II	Squareness in the YZ plane, E_{A0Y} (Geometric, excluding thermal)	$2 \times \sqrt[2]{(0.8)^2 + (0.39)^2 + (0.43)^2} \approx \pm 2.0 \mu m/m$
III	Squareness in the ZX plane, E_{B0Z} (Geometric, excluding thermal)	$2 \times \sqrt[2]{(0.8)^2 + (0.59)^2 + (0.43)^2} \approx \pm 2.2 \mu m/m$
IV	Squareness in the XY plane, E_{C0X} (Geometric and thermal)	$2 \times \sqrt[2]{(0.8)^2 + (0.59)^2 + (0.39)^2 + (0.3)^2} \approx \pm 2.2 \mu m/m$
V	Squareness in the YZ plane, E_{A0Y} (Geometric and thermal)	$2 \times \sqrt[2]{(0.8)^2 + (0.39)^2 + (0.43)^2 + (3.4)^2} \approx \pm 7.1 \mu m/m$
VI	Squareness in the ZX plane, E_{B0Z} (Geometric and thermal)	$2 \times \sqrt[2]{(0.8)^2 + (0.59)^2 + (0.43)^2 + (3.6)^2} \approx \pm 7.5 \mu m/m$

Table 3 Summary of squareness measurement results for all three planes on Cincinnati VMC using a ballbar

Parameter	Result		
	Average of repeated measured squareness	Geometric squareness	Geometric and thermal squareness
Squareness in the XY plane, E_{COX}	23.4 μm/m	23.4 ± 2.1 μm/m	23.4 ± 2.2 μm/m
Squareness in the YZ plane, E_{A0Y}	−45.3 μm/m	−45.3 ± 2.0 μm/m	−45.3 ± 7.1 μm/m
Squareness in the ZX plane, E_{B0Z}	32.4 μm/m	32.4 ± 2.2 μm/m	32.4 ± 7.5 μm/m

conducted over a 300 mm diameter, while the machine's axis travel is approximately 500 mm. Therefore, when examining these results more closely, the acceptance test of such a machine during official inspections could be subject to scrutiny.

6 Summary and Conclusions

This paper explored the importance of squareness measurements and their variations due to thermal distortions in machine tools. Experimental results on a vertical machining centre showed that the squareness uncertainty in the horizontal XY plane was ± 2.2 μm/m, while the uncertainties in the vertical YZ and ZX planes were ± 7.1 μm/m and ± 7.5 μm/m, respectively.

Compared to the squareness tolerances specified in ISO standards and considering the uncertainty analysis, the ballbar can be effectively used for these measurements. However, its ability to cover the full axis travel is limited by the commercially available device lengths. In other words, this device could serve as a convenient method for assessing local squareness measurements over a diameter of 300 or 500 mm as specified in different ISO standards.

This research also presented an uncertainty analysis based on GUM for the squareness measurement of linear axes in machine tools using a ballbar. The main uncertainty contributors were identified and listed in the uncertainty budget table along with their respective contribution.

Based on the discussions presented in this paper, it is recommended to consider thermal squareness measurements, particularly for the machine tools used in precision applications. This test might reveal hidden aspects related to assembly, design and installed main components of the machine which may not be detectable with conventional squareness tests. It is also concluded that thermal squareness variations in some principal planes can be considerably high where under certain circumstances the squareness error may exceed the specified tolerance for a machine tool.

7 Future Work

In future work, similar tests will be conducted on another machining centre with a different kinematic chain to investigate the effect of thermal changes on squareness variations. Furthermore, angular errors of both linear axes will be assessed over a longer period than the standard tests. The results of these thermal tests will be assessed alongside the measured squareness variations. Although the heating cycle of the two axes under test differs between circular tests and testing each axis individually, this comparison could provide insight into the machine's behaviour. Specifically, it could help determine whether the squareness changes are caused by changes of the angular errors or by other underlying factors.

Furthermore, different warm-up cycles on the machine tool such as rotating the spindle and moving the linear axes at higher feed speeds will be applied between some of circular tests using the ballbar. This approach is achievable because of the ease with which the measurement instrument can be reintroduced. In such cases, an additional uncertainty contributor that must be taken into account is the setup uncertainty introduced by the relocation of the ballbar.

Acknowledgements The authors gratefully acknowledge the UK's Engineering and Physical Sciences Research Council (EPSRC) funding of the Future Metrology Hub (Grant Ref: EP/P006930/1) and UKRI-funded Advanced Machinery and Productivity Initiative (84646). The authors also extend their gratitude to Andrew Bell and Ashley Cusack, colleagues at CPT (University of Huddersfield), for their valuable contribution to the experiments and data processing.

References

1. ISO 230-3:2020, Test code for machine tools—Part 3: determination of thermal effects
2. ISO 10791-10:2022, Test conditions for machining centres—Part 10: evaluation of thermal distortions
3. ISO 13041-8:2004, Test conditions for numerically controlled turning machines and turning centres—Part 8: evaluation of thermal distortions
4. ISO/TR 230-11:2018, Test code for machine tools—Part 11: measuring instruments suitable for machine tool geometry tests
5. Szipka, K., Archenti, A., Vogl, G.W., Donmez, M.A.: Identification of machine tool squareness errors via inertial measurements. CIRP Annals. **68**(1), 547–550 (2019). https://doi.org/10.1016/j.cirp.2019.04.070
6. Tang, Y., Feng, X., Ge, G., Du, Z., Lv, J.: Uncertainty assessment of machine tool squareness error identification using on-machine measurement. Meas. Sci. Technol. **35**(3), 035022 (2024). https://doi.org/10.1088/1361-6501/ad1368
7. Ibaraki, S., Okumura, R.: A machining test to evaluate thermal influence on the kinematics of a five-axis machine tool. Int. J. Mach. Tools Manuf. **163**, 103702 (2021). https://doi.org/10.1016/j.ijmachtools.2021.103702
8. Sarvas, M., Holub, M., Marek, T., Prochazka, J., Bradac, F., Blecha, P.: Influence of machine tool operating conditions on the resulting circularity and positioning accuracy. Machines. **12**(5). https://doi.org/10.3390/machines12050352
9. ISO 10791-2:2023, Test conditions for machining centres—Part 2: geometric tests for machines with vertical spindle (vertical Z-axis)

10. ISO 230-1:2012, test code for machine tools—Part 1: geometric accuracy of machines operating under no-load or quasi-static conditions
11. JCGM 100:2008, Evaluation of measurement data–Guide to the expression of uncertainty in measurement (GUM)

Enhancing Thermal Precision of Coordinate Measuring Machine (CMM) Through Implementation of Temperature Correction for Shopfloor Metrology

Gaurav Abhay Kulkarni, Rajeev Rajampeta, Steffen Strauss, and Malte Langmack

Abstract For monitoring the quality of manufactured components, reliable and precise measurements are critical. These measurements inform decisions regarding compliance with predefined tolerances and adjustments to manufacturing process parameters. In a production environment, a shopfloor metrology-compatible Coordinate Measuring Machine (CMM) is essential for providing quick feedback and is economically viable. Therefore, CMMs must deliver precise measurements across a wider temperature range and withstand higher temperature gradients. This paper presents a novel feedforward thermal correction technique designed to enhance the measurement accuracy of CMMs in dynamic production environments. The approach emphasizes seamless implementation and robustness throughout the entire lifecycle of the CMM, ensuring consistent performance despite environmental variations and wear over time. A key focus is on robust real-time data collection from temperature sensors and the implementation of a temperature correction algorithm. This advancement not only improves quality control but also extends the operational longevity and reliability of CMMs in manufacturing settings.

1 Introduction

In the domain of modern manufacturing processes, the desired precision and the required reliability of measurements are critical for ensuring the quality of produced components. CMMs play a pivotal role in this realm, enabling for the accurate assessment of dimensional tolerances and the adjustment of manufacturing parameters. However, the performance of the CMMs is significantly influenced by thermal conditions, which can lead to measurement errors due to thermal drift. Mayr and Wegener [1] has investigated the impact of thermal error on the precision of machine tools and measurement systems. They report that thermal errors can contribute to

G. A. Kulkarni (✉) · R. Rajampeta · S. Strauss · M. Langmack
Carl Zeiss Industrielle Messtechnik GmbH, Oberkochen, Germany
e-mail: gaurav-abhay.kulkarni@zeiss.com

© The Author(s) 2026
K. Wegener and M. Bambach (eds.), *4th International Conference on Thermal Issues in Machine Tools (ICTIMT2025)*, Lecture Notes in Production Engineering,
https://doi.org/10.1007/978-3-032-01194-7_22

up to 75% of the total measurement deviations observed. As manufacturing environments often experience fluctuations in temperature, it is imperative to develop strategies that enhance the thermal stability of CMMs to maintain their accuracy.

Shopfloor metrology is an essential component of modern manufacturing processes, playing a critical role in ensuring the quality and precision of manufactured components. Unlike traditional metrology, which is often conducted in controlled laboratory environments, shopfloor metrology is integrated directly into the production environment, allowing for real-time measurement and feedback. This integration is crucial for maintaining tight tolerances and achieving high-quality standards, as it enables immediate detection and correction of deviations during the manufacturing process. By providing quick and accurate measurements, shopfloor metrology helps reduce waste, minimize rework, and improve overall production efficiency. Furthermore, it supports the implementation of advanced manufacturing strategies, such as just-in-time production and lean manufacturing, by ensuring that components meet specifications without the need for extensive post-production inspection. As manufacturing environments become increasingly dynamic, with varying temperature and humidity conditions, the importance of robust shopfloor metrology systems that can adapt to these changes becomes even more pronounced. This adaptability ensures that measurement precision is maintained, regardless of environmental fluctuations, thereby enhancing the reliability and consistency of the manufacturing process.

Recent advancements in thermal management techniques have highlighted the potential for improving measurement precision through innovative approaches. One such approach is the implementation of feedforward thermal correction techniques, which aim to predict and compensate for thermal errors before they affect measurement outcomes. This paper presents a novel methodology for enhancing the thermal precision of CMMs by integrating a comprehensive thermo-mechanical finite element (FE) model with experimental validation. The proposed method not only addresses the challenges posed by varying thermal conditions but also seeks to optimize the placement of temperature sensors to effectively monitor and mitigate thermal drift.

The significance of this research lies in its potential to provide quick feedback for manufacturing processes, enabling CMMs to deliver precise measurements over a wider temperature range and higher temperature gradients. By establishing a robust thermal correction workflow, this study aims to contribute to the development of a 'Design for Correction' approach, fostering the design of more correctable machines that can adapt to the dynamic conditions of shopfloor metrology.

Through a series of controlled experiments, the effectiveness of the thermal correction technique will be validated, adhering to established standards such as DIN EN ISO 10360-2. The results of this research are expected to demonstrate a significant reduction in measurement errors, thereby enhancing the overall accuracy and reliability of CMMs in production environments. This paper ultimately aims to advance the field of shopfloor metrology by providing a comprehensive framework for thermal correction in CMMs, paving the way for future innovations in precision engineering.

2 Literature Review and State of the Art

Thermal error in CMMs and in machine tools is a long known phenomenon and therefore has been examined extensively. Budak and Ozturk [2] has highlighted the impact of change in temperature on the measurement accuracy, emphasizing the need for effective thermal management strategies in manufacturing environments. Additionally, the authors have proposed a framework for integrating thermal correction strategy into CMM for enhancing thermal stability. An important component for implementation of correction strategy is an accurate a thermo-mechanical Finite Element (FE). This model serves as a foundational tool for simulating the thermal behavior and mechanical response of the CMM under varying environmental conditions. By accurately capturing the relationship between temperature fluctuations and measurement deviations, the FE model enables the development of precise thermal correction algorithms. The development of an FE model is crucial for understanding and in future predicting the thermal behavior of CMMs. Bambach and Zäh [3] present a comprehensive FE modeling approach to predict thermal effects in machine tools. Their findings support the methodology employed in the current research, where a validated FE model is used to simulate thermal responses under various load conditions. Experimental validation is very critical for ensuring the accuracy of the FE model. CMMs being complicated systems with many components, verifyng the FE model is an important step to enhance the precision of the system. Ihlenfeldt and Weigold [4] conducted experiments to validate their thermal models of CMMs, demonstrating the importance of empirical data in refining simulation results. Their work underscores the necessity of rigorous testing protocols.

3 Requirements for Implementation of Thermal Correction

Various thermal correction approaches are implemented since many years to improve performance of the CMMs. These correction methods were implemented based on analytical approaches or based on trial and error. For successful implementation of feedforward correction in CMMs, based on a physics informed approach following critical technologies are required, as described in Fig. 1.

- FE modeling
- Developing an observer
- Correction algorithm

FE modeling is critical for simulating the thermal behavior of CMMs under various operating conditions. By creating a detailed model of the CMM's structure and materials, engineers can predict how temperature changes will affect the machine's components. This understanding is crucial for identifying potential sources of thermal drift and quantifying their impact on measurement accuracy. With the help of FE modeling the CMM design can be optimized for minimizing thermal errors in the

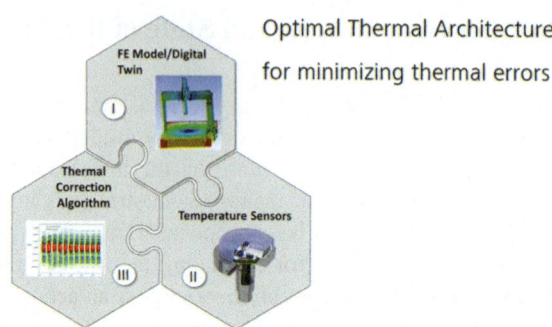

Fig. 1 Technologies required for implementation of thermal correction in CMMs

system. One can test different materials, geometries, and cooling strategies in a virtual environment before implementing changes in the physical machine. This leads to more efficient designs that are inherently more stable under thermal fluctuations.

For an observer, temperature sensors need to be integrated in the system at predefined positions and are essential for accurately monitoring the temperatures affecting the CMMs. Precise temperature data is crucial for understanding the thermal conditions that lead to measurement errors. Without reliable sensors, it is challenging to assess the extent of thermal drift and implement effective correction strategies.

The correction algorithm uses physics-based models to simulate the thermal behavior of the CMM. These models incorporate principles of thermodynamics and heat transfer to predict how temperature variations affect the machine. Based on a validated FE model, training data for different load cases is generated for training the thermal correction algorithm.

4 Development of Temperature Sensors

As pointed out in Sect. 1, for observing the thermal state space of the CMM during the operation, reliable temperature sensors based on the design requirements are critical. Developing advanced temperature sensors allows for real-time data collection, which is vital for dynamic thermal correction techniques like feedforward correction. Continuous temperature monitoring enables immediate adjustments to the CMM's operation, reducing the impact of temperature fluctuations on measurement accuracy. Temperature sensors provide the necessary data input for the digital twin of the CMM. By feeding real-time temperature data into the correction algorithm, the thermal error at the measurement position can be predicted.

4.1 Testing of Temperature Sensors

Two PT100 sensors, classified under accuracy class 1/10 DIN B, are employed as reference sensors, as described in Fig. 2. Temperature adjustments are made at specific points: 10, 20, and 30 °C, to ensure high precision measurements. The reference sensor, accompanied by a calibration certificate, exhibits a measurement uncertainty of 30 mK, relative to the true value. This entire setup, including the calibration sensors enclosed within a plastic bag, is immersed in a calibration bath to avoid contact of water with the sensor. The calibration bath operates within a range of 5–80 °C and maintains a bath stability of ±10 mK. Figure 2 describes the water bath test setup for testing of the temperature sensors.

4.2 Data Acquisition for Sensor Testing

Unimoc (Universal Measurement Data Observer and Controller) An internal software developed by Carl Zeiss AG facilitates both the logging of measurement data and the automated control of the calibration bath. Within the software, all conditions pertinent to the calibration process are specified, including the maximum reference range, reference lower limit, reference upper limit, calibration duration, tolerance limit, and other relevant parameters.

4.3 Parameters to be Tested

Following are the parameters that are identified as critical and evaluated for testing of the temperature sensors:

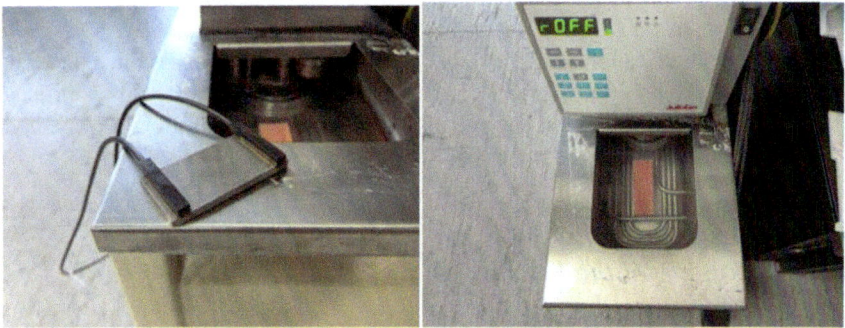

Fig. 2 Reference PT100 temperature sensors and water bath for testing of newly developed temperature sensors

- Precision
- Stability
- Linearity
- Hysteresis
- Reproducibility of measurement for same sensor
- Reproducibility of measurement for same sensor type
- Thermal conductivity of the carrier material

The precision of measurement describes the accuracy of the sensor measurement through filtering and averaging. The linearity of measurement describes the proportional deviation of the sensor from the true value over the entire temperature range. Hysteresis in a measurement describes a system behaviour in which the direction and therefore the previous state of the input variable have an influence on the output variable. Hysteresis cannot be corrected. Reproducibility of measurement describes the ability to always achieve the same results under the same conditions. A lack of or low reproducibility cannot be corrected. The thermal conductivity of FR4, which is commonly used as a base material for PCBs, is significantly lower compared to ceramics. It is therefore necessary to check whether this influences the accuracy and response time of the newly developed sensor. Along with sensor testing, an additional test is also conducted to evaluate the performance of the sensors under transport conditions. This test was performed across an extreme temperature range, specifically from -40 to $+70\,°C$. The purpose of this test was to quantify the storage temperature range for both reference sensors and calibration sensors.

4.4 Internal ID and Digital Calibration Certificate

The Unique Internal ID (UID) for sensors serves as a distinctive identifier assigned to each individual sensor, playing a crucial role in sensor management, quality control, and traceability. Each sensor is equipped with a QR code that functions as its UID, which is also stored within the sensor's PCB for database management purposes. Carl Zeiss operates a German Accreditation Body *Deutsche Akkreditierungsstelle* (DAkks)-accredited calibration laboratory where all sensor testing is conducted. The data from the calibration certificate is stored within each sensor. This certificate includes essential information such as the calibration object, calibration method, measuring conditions and ambient conditions.

4.5 Packaging of the Sensors

The casing of the sensors is constructed from aluminum, which offers high thermal conductivity, thereby enhancing the sensors' ability to accurately detect temperature. In terms of design, the sensors feature a bolt mechanism that allows them to be directly screwed into the designated slots. At the top of each sensor, an LED is integrated into the sensor PCB, providing visible confirmation of the sensor's positioning and cabeling. Regarding connectivity, the sensors are equipped with a USB-C port, to which a cable is connected. At the other end of this cable is the data acquisition box, which collects data from the individual sensors and transmits it to the controller for further control and processing (Fig. 3).

4.6 Sensor Helth Check

During the assembly of the CMMs, the temperature sensors can be installed at pre-defined locations. By integrating the sensors in a bolt, an accurate sensor position can be ensured and the temperature of the body and not the surface is measured. This design also minizes the enviormental impact on sensor measurement. Additionally, by having a QR code with calibration certificate saved internally along with a unique ID, it can be ensured that corrected sensors are mounted. LED lights ensured that the sensor cables are in correct ports at the data acquisition unit. However, it is possible that the sensor is not mounted correctly. It could happen that the sensor is not tightened correctly or presence of contaminants between sensor and the system. This will cause additional thermal resistance between the system and the temperature sensor, causing incorrect temperature measurement as displayed in Fig. 4. In order to validate the assembly of the temperature sensor, a heating element is also integrated in the sensor. This heating element can be switched on, generating a heat load in the sensor head. Based on the response of the measured temperature the contact status of the sensor can be checked. In case the sensor is mounted correctly, in other words the contact between the sensor and structure is good, the heat generated from the temperature sensor will be transfered to the structure and the temperature will increase slowly andwill reamin low at steadystate, as described in Fig. 4. On the other hand, if the sensor is not mounted correctly and the contact is open, the heat capacity of the structure is missing and therefore, the measured temperature will jump quickly and

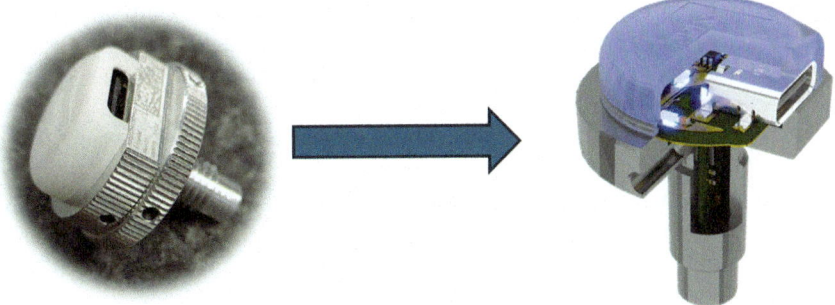

Fig. 3 Temperature sensors developed for CMMs

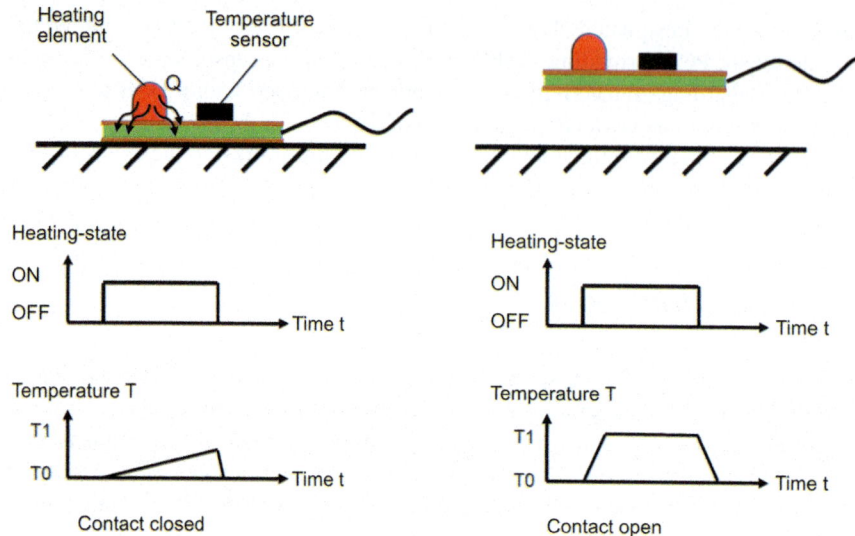

Fig. 4 Temperature responce of the heating element in temperature sensor casing to ensure accurate mounting of the temperature sensor

at steady state will be higher, as described in Fig. 4. With this approach the sensor health can be ensured in the field as well.

5 Experimental Setup

The experimental setup was designed to evaluate the performance of the feedforward thermal correction technique for length measurements using a Zeiss Micura CMM in a real-world environment as described in Fig. 5. This setup aimed to simulate typical industrial conditions, which often involve variable temperatures and other environmental factors that can affect measurement accuracy. This machine is equipped with a probe that can accurately measure the dimensions of the Zeiss calibration artifact as described in Fig. 6. The artifact served as a reference standard for length measurements. Its known dimensions allowed for the validation of the CMM's accuracy and the effectiveness of the thermal correction algorithm.

Multiple internally developed temperature sensors as described in Sect. 4 were selected for their high sensitivity and accuracy. These sensors were capable of providing real-time temperature readings in the shopfloor environment. The sensors were mounted on the CMM at critical locations to capture temperature variations that could influence measurement accuracy. The placement was determined using an internally developed algorithm that optimized sensor locations based on expected

Fig. 5 Zeiss MICURA
CMM with temperature
sensors and thermal
correction algorithm
implemented

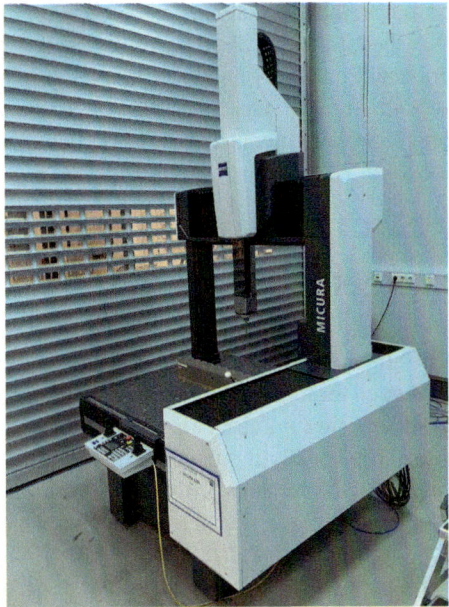

Fig. 6 Zeiss caliberation
artificat for a stable length
measurement

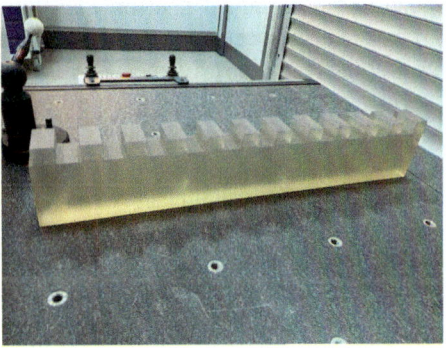

thermal gradients and airflow patterns in the enviorment. Unlike traditional laboratory settings or climate chambers, the measurements were conducted outside a climate chamber environment. Throughout the measurement period of nearly 70 days,
the enviorment experienced significant temperature fluctuations, which were continuously monitored by the installed sensors. This variability posed a challenge for
maintaining measurement accuracy, making it an ideal scenario to test the robustness
of the thermal correction technique. Length measurements of the Zeiss calibration
artifact were taken at regular intervals over the 70-day period. Each measurement ses-
sion included recording the corresponding temperature data from the sensors. The
internally developed algorithm processed the real-time temperature data to apply
corrections to the length measurements. This algorithm adjusted the measurements

based on the thermal expansion characteristics of the materials involved, ensuring that the CMM delivered accurate results despite the environmental variations. The collected data was analyzed to evaluate the effectiveness of the temperature correction strategy. The focus was on assessing the consistency and reliability of the length measurements over time, particularly in the context of the dynamic shopfloor environment.

6 Results

The effectiveness of the implemented feedforward thermal correction technique was evaluated through a series of length measurements conducted over a period of nearly 70 days using a Zeiss calibration artifact. During this period, temperature sensors were strategically mounted on the CMM to monitor environmental conditions continuously. The sensors provided real-time data that informed the internally developed temperature correction algorithm, which was designed to optimize sensor placement for maximum accuracy.

Figure 7 illustrates the length measurements obtained from the CMM over the 70-day period. The plot displays the error in measured lengths alongside the corresponding temperature data recorded by the sensors. Notably, the results demonstrate a robust temperature correction capability, as evidenced by the reduced deviations in error of the length measurements.

The analysis reveals that the temperature correction algorithm effectively compensated for thermal expansion and contraction, resulting in reliable measurements that remained within predefined tolerances. The standard deviation of the length measurements was significantly reduced after the application of the temperature correction, indicating enhanced measurement stability and accuracy.

The combination of strategically placed temperature sensors and the internally developed temperature correction algorithm has proven to deliver consistent and reliable length measurements over an extended period. This robustness underscores the potential of the proposed approach in maintaining measurement integrity in dynamic production environments.

7 Conclusion

The integration of the thermal correction approach into the Zeiss MICURA CMM represents a significant advancement in the field of precision metrology. This innovative solution addresses one of the critical challenges in measurement accuracy–thermal variations. By correcting for temperature fluctuations, consistent and reliable mesurement, regardless of the environmental conditions can be delivered. Following are some of the key benefits observed by integrating thermal correction in CMM:

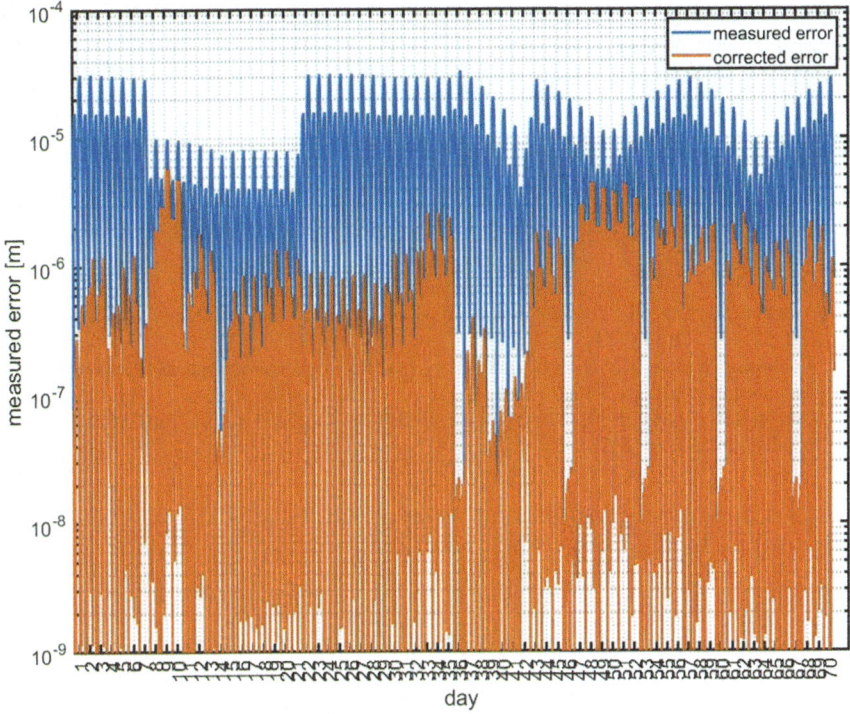

Fig. 7 Measurement error and the corrected error with implementation of temperature correction algorithm

- Enhanced Thermal Stability: The thermal correction approach significantly improves the thermal stability of the system. This enhancement means that the machine can maintain its high level of performance and accuracy even when subjected to varying temperatures. This is particularly important in industries where environmental control may not be perfect, yet precision is non-negotiable.
- Robust Measurements: The ability to produce robust measurements in dynamic thermal environments makes the system an invaluable tool for field applications. This robustness ensures that the data collected is reliable and can be trusted for critical decision-making processes, whether in quality control, product development, or research.
- High Precision and Reliability: The machine's ability to perform reliably in fluctuating thermal conditions enhances its usability in real-world applications. This high precision is crucial for industries such as aerospace, automotive, and manufacturing, where even minor deviations can lead to significant issues.

The implementation of thermal correction technology in the CMM represents a significant advancement in precision measurement solutions. This innovation addresses the critical challenge of maintaining measurement accuracy in varying thermal environments, ensuring enhanced thermal stability, improved measurement robustness, and increased precision.

References

1. Mayr, J., Jedrzejewski, J., Uhlmann, E., Donmez, M.A., Knapp, W., Härtig, F., Wendt, K., Moriwaki, T., Shore, P., Schmitt, R., Brecher, C., Würz, T., Wegener, K.: Thermal issues in machine tools. CIRP Ann. Manufact. Technol. **62**, 771–791 (2012)
2. Budak, E., Ozturk, M.: Thermal effects on coordinate measuring machines: a review. Precis. Eng. **52**, 1–12 (2018)
3. Bambach, M., Zäh, M.: Thermal analysis of machine tools using finite element methods. J. Manufact. Sci. Eng. **52**, 141 (2019)
4. Ihlenfeldt, S., Weigold, M.: Experimental validation of thermal models for CMMs. Measur. Sci. Technol. **53**, 31 (2020)

Simulation Strategies and Digital Twins
of Machine Tools

Advancing Thermal Control in Machine Tools: The Role of Digital Twins

Lars Penter and Steffen Ihlenfeldt

Abstract Thermally induced deviations are a leading cause of geometric error in machine tools and present persistent challenges for high-precision manufacturing. This review explores how digital twins (DTs) have advanced thermal management by integrating real-time data, advanced simulation models, and adaptive control strategies. Recent developments in thermal digital twins are categorized as physics-based, data-driven, hybrid, or adaptive modeling approaches. The review highlights their contributions to the prediction and compensation of thermally induced deviations. The emphasis is on model order reduction, machine learning integration, and sensor technologies. Finally, we discuss open challenges, such as model generalization, calibration robustness, early-stage design integration, and lifecycle-wide applications, to guide future research.

Keywords Thermal deviations · Hybrid simulation · Machine tools · Lumped-parameter thermal network

1 Introduction

Thermal issues in machine tools have long been recognized as a major source of manufacturing errors, with thermal deformations contributing up to 70% of total machining inaccuracies [1]. Despite decades of research into compensation of thermally induced deviation and temperature control, significant challenges remain in achieving consistent, high-precision manufacturing. Early methods relied heavily on manual compensation techniques and basic control of internal temperature sources. However, as noted in [1], the complex origins of thermo-mechanical deviations continue to impact the accuracy and stability of machine tools. These tool center

L. Penter (✉) · S. Ihlenfeldt
Chair of Machine Tool Development and Adaptive Controls, Dresden University of Technology TUD, Dresden, Germany
e-mail: lars.penter@tu-dresden.de

Fraunhofer Institute of Machine Tools and Forming Technology IWU, Chemnitz, Germany

© The Author(s) 2026
K. Wegener and M. Bambach (eds.), *4th International Conference on Thermal Issues in Machine Tools (ICTIMT2025)*, Lecture Notes in Production Engineering,
https://doi.org/10.1007/978-3-032-01194-7_23

353

point (TCP) deviations are influenced by internal heat sources like spindles, motors, and guideways, as well as external factors like environmental temperature fluctuations along the thermo-elastic functional chain, see Fig. 1. In recent years, digital twins (DTs) have emerged as a promising solution to this challenge.

A DT is a representation of a physical entity, system, or process. It is enabled by real-time data, simulation models, and integration technologies. The goal is to mirror, predict, and optimize the performance and behavior of its physical counterpart throughout its lifecycle [3]. According to Tao et al., the core elements of a DT are the physical and virtual entities which are connected by a real-time data link (see Fig. 2).

This link enables bidirectional interaction and real-time synchronization. DTs are updated and adapted to changes in the real world.

Physical and virtual entities are connected to the service layer that facilitates key functionalities such as:

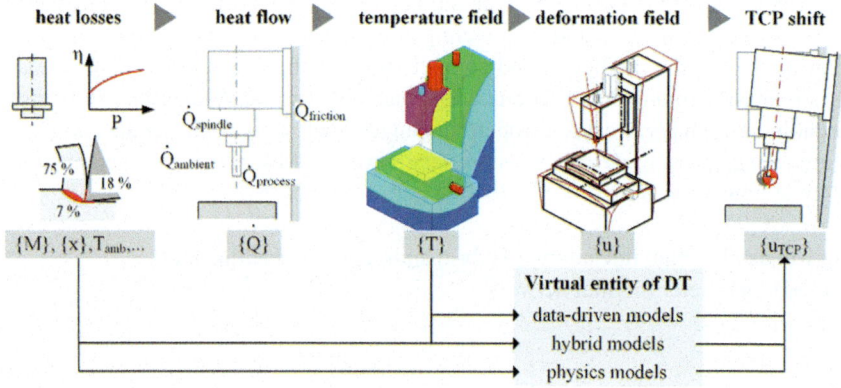

Fig. 1 Thermo-elastic functional chain and model types in accordance with [2]

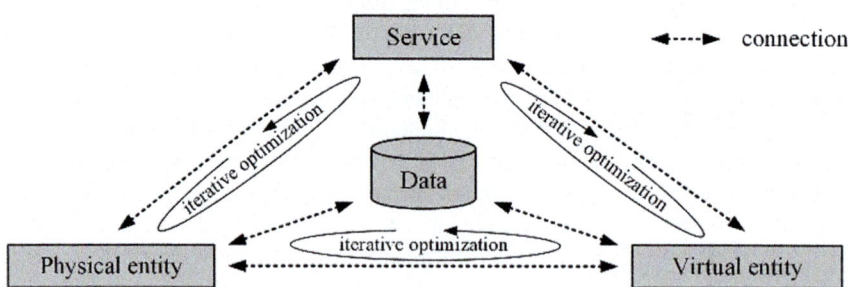

Fig. 2 Definition of a DT according to [3]

- **Real-time monitoring and prediction services**, such as predicting of thermal deformations or temperature changes (e.g. spindle thermal growth, overall thermal drift and TCP deviations)
- **Optimization and correction services**, such as optimizing thermal management strategies such as cooling strategies [4, 5], and dynamically adjusting the tool path by incorporating real-time sensor data.
- **Maintenance services**: Predicting potential temperature-related part failures and suggesting preemptive repairs [6].
- **Digital design service**, such as engineering support for thermo-energetic design

DTs, with their ability to simulate real-time conditions and integrate predictive analytics, offer promising solutions to these ongoing challenges. By leveraging advanced simulation models, real-time data, and continuous feedback, DTs can predict, prevent, and compensate for thermal issues as they happen. The next sections will talk about how DTs revolutionized thermal management in machine tools over the last decade, improving precision, reducing energy consumption, and enhancing operational efficiency.

2 Physical Entities and Their Contribution to Thermally Induced TCP Deviations

To effectively mitigate the thermal challenges faced by machine tools, it is essential to understand the various physical sources of thermally induced deviations. The following section delves deeper into the complexities and challenges of thermal management in machine tools.

The sources of heat in machine tools include spindles, servo drives, friction in linear guides, ball screws, the cutting process, and ambient temperature fluctuations. These factors contribute to significant thermally induced TCP deviations, impacting machining precision. Brecher et al. [7] highlight the substantial impact of thermal deformations on the performance of spindles, noting that motorized spindles in particular are highly susceptible to thermo-elastic errors. The thermal expansion of spindle components—such as the shaft and bearings—can lead to significant deflections in the tool center point (TCP), directly affecting machining accuracy. The study underlines how spindle thermal growth is one of the primary contributors to errors in high-speed milling and precision machining, where even minor changes in spindle dimensions can result in substantial positioning errors. Friction and motion in linear guideways, ball-screws, and feed drive motors produce heat that causes these components to expand. This thermal expansion leads to axis positioning errors and structural deformations. For example, a heated ball screw will elongate and can shift the tool axis, directly impacting positioning precision. The cutting process generates intense localized heat from tool–workpiece friction and plastic deformation in the chip formation zone. A large fraction of the mechanical energy (often ~ 90%) converts into heat [8], raising temperatures of the cutting tool, workpiece, chips, and nearby machine elements. This causes thermal expansion of the tool and workpiece, leading to dimensional inaccuracies or thermal drift during machining. Ambient temperature changes and external heat sources (sunlight on the machine, HVAC drafts, etc.) induce thermal expansion or contraction of the machine structure. If the workshop temperature fluctuates, the machine's frames and tables may bend or drift out of alignment as gradients form, leading to positioning errors. Even in a controlled factory, day/night temperature swings or an unbalanced thermal environment will cause a calibrated machine to lose accuracy over time unless corrected. The research has expanded from specific machine components, such as main spindles [7], single coordinate axes [9] and cooling systems [10] to entire machine tools such as 3- and 5- axis milling machines [11, 12], lathes [13, 14] and grinding machines [15] as well as entire machine tool fleets [16, 17].

3   Advancements in the Virtual Entity of the Thermal Digital Twin

The DT's virtual entity provides the necessary computational models and simulations that allow for real-time thermal management. The simulation models used can be categorized based on the source of system knowledge into physics-based models [18, 19], data-driven models [15, 17, 20, 21] and hybrid models [22]. In addition, models can be either static or adaptive [13], depending on whether they adjust their parameters based on new data.

3.1   Physics Based Models

Physics-based models provide a deep understanding of the underlying physical processes and relationships. They are highly interpretable and can model complex thermal and mechanical interactions accurately, but they may be computationally expensive and struggle with real-time predictions. Physics based models comprise finite element analysis (FEA), lumped parameter thermal networks (LPTN) or transfer functions. Specifically model order reduction (MOR) approaches applied to FE models where focus of research in the last 15 years as they promise faster estimates for real-time control and compensation. Galant et al. [23] apply MOR to a machine tool column using a Krylov subspace projection method (KMS), specifically implemented via the Arnoldi process. Their approach reduced the FE model from 16,626 to 100° of freedom (DOF), cutting simulation time from 18,300 to 18 s, with a maximum temperature deviation of 0.6% and displacement error below 4.4%. Brecher et al. [7] applied this MOR-FE technique to a motorize milling spindle and achieved real-time capability, with total computation time per step below 0.5 s. Hernández Becerro et al. [24] develop a physics-based thermo-mechanical model

using FEA with KMS to simulate thermal deformations in machine tools. Vettermann et al. [25] focuses on balanced truncation combined with structured decomposition to treat nonlinearities, which seem more suitable for systems like machine
tools with localized nonlinear behavior. Their MOR approach reduces a thermo-
mechanical model of a feed axis from 340,000 DOFs to 81 DOFs. Cao et al. [26]
achieve 97.60% accuracy in predicting thermal deformations at spindle speeds of
1500–2500 rpm using a combination of Hybrid BC updates and MOR-FE. The
real-time model allows for thermal deformation estimation within 30 s, enabling
quasi-static compensation for thermally induced TCP deviations once per minute.
Irino et al. [27] developed a DT framework that integrates two physics-based model
types: a reduced-order thermal model using KMS methods, and a dynamic model
based on modal analysis of machine substructures. The method was validated on a
5-axis machining center, with thermal deformation reduced from up to 15 μm to just
1–3 μm after compensation.

3.2 Data-Driven Modelling

Data-driven models, such as regression equations and neural networks, are widely
employed in thermal management. These models rely on measured data to capture
complex relationships between system variables, allowing for real-time predictions
and adjustments. Although some models are initially trained using simulation data,
they often require further calibration with real-world measurements to improve
their accuracy. However, data-driven models can be difficult to interpret, especially
when complex machine interactions are involved. They may also perform poorly
when faced with unknown situations or conditions that differ significantly from the
training data, which limits their effectiveness in dynamic or unanticipated operational
scenarios. Creighton et al. [28] developed a thermal displacement model using FEA to
simulate spindle thermal growth and associated deformations. They then correlated
this model with real-time temperature data to predict thermal-induced deviations.
Using this correlation, the DT can provide real-time thermal compensation during
machining, eliminating the need for real-time FEA simulations. Brecher et al. [9]
investigated backpropagation neural networks (BP-ANN), long short-term memory
(LSTM) networks and bidirectional LSTM (BiLSTM) networks to predict and correct
thermo-elastic errors. Their approach uses encoder difference data as input, enabling
the virtual model to self-adapt to thermal deviations in real-time. A key performance
comparison of the LSTM model was conducted during realistic machining tasks. It
achieved an average reduction in thermo-elastic TCP deviation of up to 88.1% when
the environmental temperature was included as an input. Boos et al. [20] demonstrate how integrating domain randomization (DR) with deep learning models can
significantly improve the accuracy of the virtual entities in DTs. They demonstrate
that a temporal fusion transformer (TFT) model trained on DR simulation data can
outperform a model trained on limited real data by a significant margin.

3.3 Hybrid Models

Hybrid models are a synthesis of physics-based and data-driven approaches, leveraging physics-based knowledge to inform or constrain data-driven models, and data-driven models to enhance and adapt the physics-based model over time. Dahlem et al. [29] superpose the output of three model component. A rigid body model calculates geometric-kinematic deviations. A white-box model expresses the relative deviation to the rigid body model. Finally, machine learning algorithms are used to create a model of the remaining difference between the other two models and the actual reference sensor outputs. Iñigo et al. [30] combine kinematic modeling, phenomenological thermal models, and an ARX regression model in their DT framework. The kinematic model simulates machine movement, the phenomenological model predicts thermal behavior, and the ARX model links past data to predict future errors. This combination allows the DT to predict and correct thermal deformations dynamically, improving the machine's positioning accuracy and overall compensation for thermally induced TCP deviations. Kumar et al. [31] present a split computational fluid dynamics (CFD) simulation approach that decouples the internal and external convective domains of a 5-axis machine tool, enabling efficient estimation of heat transfer coefficients (HTCs). A multilayer feed-forward neural network with backpropagation, trained on the decoupled CFD data, can predict HTCs based on machine configurations and environmental inputs. They authors show that the predicted temperature fields from the coupled CFD-FE model deviated by only ± 0.8 °C from actual machine sensor readings. Geist et al. [32] introduce a hybrid thermal compensation model that combines data-driven Characteristic Diagrams (CD) compensation with structure model-based compensation. Their system uses real-time data to continuously update compensation models and dynamically adjust for thermal changes during machine operation, ensuring consistent machining precision.

3.4 Adaptive Models for Thermal Management

Mayr et al. [1] discussed traditional thermal compensation methods in 2012, including manual adjustments and pre-programmed corrections, which are static and unable to respond to real-time changes in the machine's environment. In contrast, adaptive models provide greater flexibility by continuously adjusting to dynamic conditions such as ambient temperature fluctuations, machine wear, and varying machining parameters, making them essential for advancing thermal management. Fujishima et al. [33] propose an adaptive thermal displacement compensation method using a convolutional neural network (CNN). The model adapts to real-time temperature data and adjusts the compensation values based on the prediction uncertainty calculated by Bayesian dropout. This adaptive mechanism provides dynamic adjustment to changing operating conditions, improving the accuracy of thermal error compensation compared to static models. Blaser et al. [34] integrate a Thermal ARX model (Autoregressive with Exogenous inputs) in adaptive learning control (ALC), a machine learning-based approach that dynamically updates thermal models based on real-time data. The ALC system was implemented on a five-axis machine tool. It adjusts measurement intervals, model parameters, and NC code during production. With ALC, deviations stayed within the target range (typically $< 5\ \mu$m), confirming the systems long-term stability and effectiveness. Naumann et al. [35] propose a compensation framework for thermal error correction using regression-based characteristic diagrams. Their update method uses weighted training data to fine-tune the model for specific load cases without having to fully retrain it. Mareš et al. [36] introduce an adaptive thermal error compensation system that uses on-machine measurements to continuously update a transfer function-based model. Their system significantly improves thermal management by adapting the model to real-time machine conditions, enhancing prediction accuracy and robustness through periodic updates. Thiem et al. [12] suggest an adaptive thermal MOR-FE model that dynamically updates thermal error compensation parameters based on real-time sensor data. Their approach, which accounts for long-term wear and operational changes, continuously improves model accuracy. Zimmermann et al. [11] propose a self-learning system using Thermal Adaptive Learning Control (TALC), which combines machine learning algorithms and real-time thermal data to dynamically adjust compensation for thermal errors. This adaptive model outperforms static models by continuously refining compensation strategies, ensuring high precision in 5-axis machining over long-term operations. Lang et al. [22] enhance the DT framework by applying Kalman filter-based state observers with MOR-FE models. This allows for real-time, accurate prediction of thermal errors in machine tools with minimal sensor requirements. The advancements in adaptive models for thermal management, particularly those using machine learning algorithms, sensor fusion, and model-based updates, are paving the way for the DTs. By continuously adapting to real-time data, these models provide significant improvements in thermal error prediction and compensation, enabling high-precision machining and reducing downtime in manufacturing environments.

4   Sensor Integration and Advanced Measurement Techniques for Thermal Control

Accurate measurement of thermo-elastic errors in machine tools remains a significant challenge, primarily due to the complexity and time constraints of current measurement techniques as well as necessary interruptions of production. However, recent innovations in sensor technology and measurement systems are working to address these issues by enabling more efficient, precise, and real-time thermal displacement measurements. They rely either on measuring the temperature field or internal structural deformations to indirectly calculate the deviations at the TCP or directly measure the TCP deviations.

Horejš et al. [37] develop a smart sensor for measuring spindle drum temperature, improving the thermal error compensation model in multi-spindle lathes. Their sensor provides real-time data that increases thermal displacement prediction accuracy by 16% and reduces displacements by 65%, enabling real-time thermal compensation and enhancing machining precision. Baum et al. [38] propose the use of integral deformation sensors to measure thermally induced deformations in machine components. These sensors capture real-time data about structural deformations caused by thermal loads, providing more detailed information than traditional temperature sensors. Bertaggia et al. [39] improve these so-called Integrated Deformation Sensors (IDS) to enabling accurate predictions of Tool Center Point (TCP) displacement. Deutsch et al. [40] introduce a multichannel laser interferometer for simultaneous thermal displacement measurements at multiple points on the machine tool. This sub-micron accuracy system allows for the detection of thermal deformation across the machine's entire structure, providing data for thermal compensation and also for improving the machine design. Brecher et al. [41] introduce a machine-integrated thermal error measurement system that uses pentaprisms to create a laser reference frame for multi-axis thermal displacement measurements. While offering comparable accuracy to interferometric methods, their approach provides significant cost savings. Erben et al. [42] developed a smart pressure film sensor that uses shape memory alloy (SMA) thin films. This sensor can simultaneously measure ultrafine pressure variations and surface temperature with high spatial resolution on machine tool surfaces. This design is significantly more sensitive to low-pressure changes in the single-digit Pascal range than conventional sensors. It enables the real-time correlation of pressure, temperature, and flow velocity data to determine convective heat transfer. Ota et al. [43] developed advanced thermal deformation measurement methods using laser trackers and vision cameras to visualize thermal displacements. They also introduced a deep learning-based thermal displacement prediction model using CNNs that outperforms conventional methods, achieving up to 81% improvement in thermal error compensation accuracy.

5 Services Provided by Thermal Digital Twins

In the context of machine tools, DTs offer a revolutionary approach to addressing thermal management issues that are critical to achieving and maintaining precision, efficiency, and reliability in manufacturing processes. DTs provide could not only by during the operation, but also throughout its entire lifecycle of a machine tool, from design and manufacturing to real-time production monitoring.

During the design phase, a DT is essentially a virtual model of the product, built from CAD data and simulation models. It represents the intended geometry, materials, and thermal behavior. Based on this virtual prototype, thermal simulations are performed to predict how heat will flow through the system, where potential thermal problems may occur, and how they may affect machine performance. FEA or LPTN simulate the machine's behavior under various thermal loads, essentially creating a "twin" that reflects the expected behavior of the future real product. These early simulations help engineers to optimize the machine design to mitigate these issues.

In the virtual commissioning phase, which occurs before or while the physical machine is being built, a DT is used to simulate and validate the machine's performance, including its thermal behavior, in a virtual environment. The goal is to ensure that all systems, including thermal management, function as intended and that the machine design meets production requirements. This minimizes the risk of delays once the machine is physically commissioned. Testing the machine's behavior virtually allows potential thermal issues to be identified and resolved early on, ensuring a smoother transition into the actual commissioning phase.

Once the machine moves into the production phase, it enters active use in a manufacturing environment. Here, the DT integrates real-time data, such as temperature readings and operational parameters, to monitor and adjust the machine tool's thermal performance during actual production.

During the operational phase, the machine tool is continuously monitored and maintained. The DT accumulates data, refining its predictive capabilities and adapting to changes such as wear and tear, environmental factors, and operational adjustments. Maintenance becomes an integral, ongoing process, as the DT helps predict when components will require attention, anticipate degradation indicated by abnormal temperature increases, and provide insights into necessary repairs or upgrades to ensure optimal performance. Effective compensation models depend on the deterministic behavior of the machine–where predictable inputs lead to consistent, repeatable outputs. However, as degradation progresses, it can introduce variability or scatter into the system's response, which undermines the accuracy of compensation. However, this increased scatter itself becomes a valuable indicator of degradation, enabling the DT to detect subtle changes in machine behavior before major faults occur. With a dual focus on both trend prediction and statistical variability, the DT supports proactive maintenance strategies that enhance reliability, minimize unplanned downtime, and extend the machine tool's lifespan.

6   Challenges and Open Research Questions

Despite significant advancements in adaptive learning control for DTs, challenges remain in addressing model uncertainties due to ever-changing boundary conditions. As DT systems evolve with the integration of machine learning models, future research should focus on optimizing calibration methods and enhancing the robustness of these models in dynamic production environments. Although DTs offer the advantage of real-time adaptability, current methods still struggle to handle extreme thermal fluctuations and ensure consistency across different machine configurations. Further improvements are needed to account for unpredictable thermal influences with high temperature gradients, such as sudden cooling or heating events.

Alongside efforts to leverage DTs for the correction of thermally induced deviation, there is a clear need for more tools that enable designers to assess and adapt machine concepts for improved thermal management in the early design stages. Transferring or generalizing thermal models between different machines remains challenging, as existing thermal simulation models often lack the parametric flexibility needed for direct integration into design workflows. Moreover, there is no guarantee that a model developed for one machine will perform reliably for another. While LPTNs, as proposed by [44], show promise in early-stage design, they have yet to be fully integrated into design processes. Furthermore, integrating topology optimization with thermal analysis provides significant opportunities to improve thermal efficiency in the early stages of machine design. Key challenges include the need for accurate thermal boundary conditions, the high computational cost of coupled simulations, and maintaining mechanical performance while optimizing for heat flow.

The relationship between component temperature and part failure is well established. However, the effective use of DTs for temperature-based anomaly detection is still in development. Key challenges include modeling accuracy, variability across machine types, and limited fault data. One promising approach is to use hybrid anomaly detection methods that compare simulated and measured temperature data. Calibrated physics-based simulation models predict expected behavior based on internal control data. Then data-driven models such as variational autoencoders (VAEs) and generative adversarial networks (GANs), evaluate deviations using sensor readings and simulation outputs [45]. Other emerging strategies include deep learning methods based on multi-sensor data fusion, synthetic fault data generation via generative AI [46], application of transformer models [47], transfer learning [48], and federated learning [49]. These strategies will help to overcome data scarcity and improve generalizability.

References

1. Mayr, J., Jedrzejewski, J., Uhlmann, E., Donmez, M.A., Knapp, W., Härtig, F., Wendt, K., Moriwaki, T., Shore, P., Schmitt, R., Brecher, C., Würz, T., Wegener, K.: Thermal issues in machine tools. CIRP Ann. **61**, 771–791 (2012). https://doi.org/10.1016/j.cirp.2012.05.008
2. Schroeder, S., Kauschinger, B., Hellmich, A., Ihlenfeldt, S, Phetsinorath, D.: Identification of relevant parameters for the metrological adjustment of thermal machine models. Int. J. Interact. Des. Manuf. **13**, 873–883 (2019). https://doi.org/10.1007/s12008-019-00529-y
3. Tao, F., Cheng, J., Qi, Q., Zhang, M., Zhang, H., Sui Zhang, F., Sui, F.: Digital twin-driven product design, manufacturing and service with big data. Int. J. Adv. Manuf. Technol. **94**, 3563–3576 (2018). https://doi.org/10.1007/s00170-017-0233-1
4. Wenkler, E., Steiert, C., Ihlenfeldt, S., Weber, J.: Application and evaluation of classic and demand oriented cooling strategies in context of machine tools. Int. J. Adv. Manuf. Technol. **130**, 1451–1463 (2024). https://doi.org/10.1007/s00170-023-12467-x
5. Bani-Hani, M., Hanenkamp, N.: Enhancing energy efficiency in machining through digital twin technology: predictive modeling of thermal loads in machine tool spindles. Procedia CIRP **133**, 662–667 (2025). https://doi.org/10.1016/j.procir.2025.02.113

6. Touret, T., Changenet, C., Ville, F., Lalmi, M., Becquerelle, S.: On the use of temperature for online condition monitoring of geared systems—A review. Mech. Syst. Signal Process. **101**, 197–210 (2018). https://doi.org/10.1016/j.ymssp.2017.07.044

7. Brecher, C., Ihlenfeldt, S., Neus, S., Steinert, A., Galant, A.: Thermal condition monitoring of a motorized milling spindle. Prod. Eng. Res. Devel. **13**, 539–546 (2019). https://doi.org/10.1007/s11740-019-00905-3

8. Liu, H., Meurer, M., Bergs, T.: Modeling and monitoring of the tool temperature during continuous and interrupted turning with cutting fluid. Metals **14**, 1292 (2024). https://doi.org/10.3390/met14111292

9. Brecher, C., Dehn, M., Neus, S.: A data-based model of the thermo-elastic TCP error using the encoder difference and neural networks. In: Ihlenfeldt, S. (ed) 3rd international conference on thermal issues in machine tools (ICTIMT2023). Springer Nature, Cham, pp. 119–131 (2023)

10. Horejš, O., Mareš, M., Fiala, Š. Havlík, L., Stříteský, P.: Effects of cooling systems on the thermal behaviour of machine tools and thermal error models. J. Mach. Eng. **20**, 5–27 (2020). https://doi.org/10.36897/jme/128144

11. Zimmermann, N., Müller, E., Lang, S., Mayr, J., Wegener, K.: Thermally compensated 5-axis machine tools evaluated with impeller machining tests. CIRP J. Manuf. Sci. Technol. **46**, 19–35 (2023). https://doi.org/10.1016/j.cirpj.2023.07.005

12. Thiem, X., Rudolph, H., Krahn, R., Ihlenfeldt, S., Fetzer, C., Müller, J.: Adaptive thermal model for structure model based correction. In: Ihlenfeldt, S. (ed) 3rd International conference on thermal issues in machine tools (ICTIMT2023). Springer Nature, Cham, pp 67–82 (2023)

13. Horejš, O., Mareš, M., Straka, M., Švéda, J., Kozlok, T.: Adaptive thermal error compensation model of a horizontal machining centre. In: Ihlenfeldt, S. (ed) 3rd International conference on thermal issues in machine tools (ICTIMT2023). Springer Nature, Cham, pp 83–98 (2023)

14. Kaftan, P., Mayr, J., Wegener, K.: Thermal compensation of sudden working space condition changes in Swiss-type lathe machining. In: Ihlenfeldt, S. (ed) 3rd International conference on thermal issues in machine tools (ICTIMT2023). Springer Nature, Cham, pp 15–27 (2023)

15. Mareš, M., Horejš, O., Nykodym, P.: An indicative model considering part of the thermo-mechanical behaviour of a large grinding machine. In: Ihlenfeldt, S. (ed) 3rd International conference on thermal issues in machine tools (ICTIMT2023). Springer Nature, Cham, pp 54–66) (2023)

16. Stoop, F., Mayr, J., Sulz, C., Bleicher, F., Wegener, K.: Fleet learning of thermal error compensation in machine tools. In: 2021 26th IEEE international conference on emerging technologies and factory automation (ETFA). IEEE, pp 1–4 (2021)

17. Stoop, F., Mayr, J., Sulz, C., Kaftan, P., Bleicher, F., Yamazaki, K., Wegener, K.: Cloud-based thermal error compensation with a federated learning approach. Precis. Eng. **79**, 135–145 (2023). https://doi.org/10.1016/j.precisioneng.2022.09.013

18. Thiem, X., Kauschinger, B., Ihlenfeldt, S.: Online correction of thermal errors based on a structure model. IJMMS **12**, 49 (2019). https://doi.org/10.1504/IJMMS.2019.097852

19. Thiem, X., Kauschinger, B., Müller, J., Ihlenfeldt, S.: Estimation of the influence of volumetric correction approaches on the thermo-elastic correction accuracy. In: Behrens, B.-A., Brosius, A., Hintze, W. et al. (eds) Production at the leading edge of technology: proceedings of the 10th. Springer-Verlag Berlin AN, [S.l.], pp 324–333 (2020)

20. Boos, E., Thiem, X., Wiemer, H., Ihlenfeldt, S.: Improving a deep learning temperature-forecasting model of a 3-axis precision machine with domain randomized thermal simulation data. In: Liewald, M., Verl, A., Bauernhansl, T. et al. (eds) Production at the leading edge of technology: Proceedings of the 12th. Springer International PU, [S.l.], pp 574–584 (2023)

21. Lang, S., Zimmermann, N., Mayr, J., Wegener, K., Bambach, M.: Thermal error compensation models utilizing the power consumption of machine tools. In: Ihlenfeldt, S. (ed) 3rd International conference on thermal issues in machine tools (ICTIMT2023). Springer Nature, Cham, pp 41–53 (2023)

22. Lang, S., Talleri, S., Mayr, J., Wegener, K., Bambach, M.: Kalman filter-driven state observer for thermal error compensation in machine tool digital twins. Manuf. Lett. **41**, 208–218 (2024). https://doi.org/10.1016/j.mfglet.2024.09.025

23. Galant, A., Beitelschmidt, M., Großmann, K.: Fast high-resolution FE-based simulation of thermo-elastic behaviour of machine tool structures. Procedia CIRP **46**, 627–630 (2016). https://doi.org/10.1016/j.procir.2016.04.020

24. Hernández Becerro, P., Mayr, J., Wegener, C.: Efficient thermo-mechanical model of a precision 5-axis machine tool. In: Euspen special interest group meeting: thermal issues (2020)

25. Vettermann, J., Steinert, A., Brecher, C., Benner, P., Saak, J.: Compact thermo-mechanical models for the fast simulation of machine tools with nonlinear component behavior. at–Automatisierungstechnik. **70**, 692–704 (2022). https://doi.org/10.1515/auto-2022-0029

26. Cao, L., Park, C.-H., Chung, S.-C.: Real-time thermal error prediction and compensation of ball screw feed systems via model order reduction and hybrid boundary condition update. Precis. Eng. **77**, 227–240 (2022). https://doi.org/10.1016/j.precisioneng.2022.05.017

27. Irino, N., Kobayashi, A., Shinba, Y., Kawai, K., Spescha, D., Wegener, K.: Digital twin based accuracy compensation. CIRP Ann. **72**, 345–348 (2023). https://doi.org/10.1016/j.cirp.2023.04.088

28. Creighton, E., Honegger, A., Tulsian, A., Mukhopadhyay, D.: Analysis of thermal errors in a high-speed micro-milling spindle. Int. J. Mach. Tools Manuf **50**, 386–393 (2010). https://doi.org/10.1016/j.ijmachtools.2009.11.002

29. Dahlem, P., Sanders, M.P., Birck Fröhlich, H., Schmitt, R.H.: Hybrid model approaches for compensating environmental influences in machine tools using integrated sensors. At–Automatisierungstechnik. **68**, 465–476 (2020). https://doi.org/10.1515/auto-2020-0007

30. Iñigo, B., Colinas-Armijo, N., López de Lacalle, L.N., Gorka, A.: Digital twin for volumetric thermal error compensation of large machine tools. Sensors (Basel). **24** (2024). https://doi.org/10.3390/s24196196

31. Kumar, T.S., Geist, A., Naumann, C., Glänzel, J., Ihlenfeldt, S.: Split CFD-simulation approach for effective quantification of mixed convective heat transfer coefficients on complex machine tool models. Procedia CIRP **118**, 199–204 (2023). https://doi.org/10.1016/j.procir.2023.06.035

32. Geist, A., Yaqoob, M.F., Friedrich, C., Naumann, C., Ihlenfeldt, S.: Concept of integrating a hybrid thermal error compensation into an existing machine tool control architecture. J. Mach. Eng. **24**, 32–46 (2024). https://doi.org/10.36897/jme/192866

33. Fujishima, M., Narimatsu, K., Irino, N., Mori, M., Ibaraki, S.: Adaptive thermal displacement compensation method based on deep learning. CIRP J. Manuf. Sci. Technol. **25**, 22–25 (2019). https://doi.org/10.1016/j.cirpj.2019.04.002

34. Blaser, P., Mayr, J., Wegener, K.: Long-term thermal compensation of 5-axis machine tools due to thermal adaptive learning control. MM SJ 2019:3164–3171 (2019). https://doi.org/10.17973/MMSJ.2019_11_2019066

35. Naumann, C., Glänzel, J., Dix, M., Ihlenfeldt, S., Klimant, P.: Optimization of characteristic diagram based thermal error compensation via load case dependent model updates. J. Mach. Eng. (2022). https://doi.org/10.36897/jme/148181

36. Mareš, M., Horejš, O., Straka, M., Kozlok, T.: An update of thermal error compsation model via on-machine measurement. MM SJ 2022 (2022). https://doi.org/10.17973/MMSJ.2022_12_2022150

37. Horejš, O., Mareš, M., Mlcoch, A.: Smart sensor for enhancement of a multi-spindle automatic lathe thermal error compensation model. MM SJ 2021:4706–4712 (2021). https://doi.org/10.17973/MMSJ.2021_7_2021079

38. Baum, C., Brecher, C., Klatte, M., Hun Lee, T., Tsanetos, F.: Thermally induced volumetric error compensation by means of integral deformation sensors. Procedia CIRP **72**, 1148–1153 (2018). https://doi.org/10.1016/j.procir.2018.03.045

39. Bertaggia, N., Tzanetos, F., Zontar, D., Brecher, C.: Investigation of thermally induced TCP-displacement under load of the machine axes in different areas. Procedia CIRP **107**, 600–604 (2022). https://doi.org/10.1016/j.procir.2022.05.032

40. Deutsch, J., Albrecht, T., Riedel, M., Lenter, L., Wiemer, H. Müller, J., Ihlenfeldt, S.: Thermo-elastic structural analysis of a machine tool using a multi-channel absolute laser interferometer. J. Mach. Eng. **20**, 63–75 (2020). https://doi.org/10.36897/jme/127128

41. Brecher, C., Spierling, R., Fey, M., Neus, S.: Direct measurement of thermo-elastic errors of a machine tool. CIRP Ann. **70**, 333–336 (2021). https://doi.org/10.1016/j.cirp.2021.04.084
42. Erben, A., Geist, A., Voigt, I., Senf, B., Mäder, T., Glänzel, J., Ihlenfeldt, S., Drossel, W-G.: Smart pressure film sensor for machine tool optimization and characterization of the dynamic pressure field on machine surfaces. In: Ihlenfeldt, S. (ed) 3rd International conference on thermal issues in machine tools (ICTIMT2023). Springer Nature, Cham, pp 179–191 (2023)
43. Ota, K., Mori, M., Irino, N.: Development of thermal displacement prediction model and thermal deformation measurement methods. In: Ihlenfeldt, S. (ed) 3rd International conference on thermal issues in machine tools (ICTIMT2023). Springer Nature, Cham, pp 3–14 (2023)
44. Schroeder, S., Ihlenfeldt, S., Penter, L.: Efficient and robust creation of structural component models for thermo-elastic analysis of machine tools. MM SJ. **2021**, 4644–4651 (2021). https://doi.org/10.17973/MMSJ.2021_7_2021071
45. Mikołajewska, E., Mikołajewski, D., Mikołajczyk, T., Paczkowski, T.: Generative AI in AI-based digital twins for fault diagnosis for predictive maintenance in industry 4.0/5.0. App. Sci. **15**, 3166 (2025). https://doi.org/10.3390/app15063166
46. Shao, S., Wang, P., Yan, R.: Generative adversarial networks for data augmentation in machine fault diagnosis. Comput. Ind. **106**, 85–93 (2019). https://doi.org/10.1016/j.compind.2019.01.001
47. Boos, E., Zimmermann, J., Wiemer, H., Ihlenfeldt, S.: Investigation of alternative attention modules in transformer models for remaining useful life predictions: addressing challenges in high-frequency time-series data. Procedia CIRP **122**, 85–90 (2024). https://doi.org/10.1016/j.procir.2024.01.012
48. Schwendemann, S., Amjad, Z., Sikora, A.: Bearing fault diagnosis with intermediate domain based layered maximum mean discrepancy: a new transfer learning approach. Eng. Appl. Artif. Intell. **105**, 104415 (2021). https://doi.org/10.1016/j.engappai.2021.104415
49. Singh, A., Sampath, R., Raj, S.A.D., Mary, G.I., Aarthi, G., Kumar, R.M.: Federated learning for predictive maintenance: model comparisons and privacy advantages. In: Proceedings of international conference on IoT based control networks and intelligent systems (ICICNIS 2024); conference location: Bengaluru, India. IEEE, Piscataway, NJ, pp 1408–1415 (2024)

Model-Reduction-Based Temperature Field Reconstruction for Volumetric Error Compensation

Daniel Spescha⑩, Nino Ceresa, and Mayra Hoppstädter

Abstract Thermally induced deviations are a major contributor to positioning errors in machine tools, significantly impacting machining precision. This paper presents a model-reduction-based temperature field reconstruction method for volumetric error compensation. The temperature field is interpolated between temperature probe locations using projection basis vectors obtained through model order reduction of finite element models. Thermal effects such as convection are modeled as external heat loads, allowing their influence to be incorporated without requiring exact heat transfer coefficients or fluid temperatures. The approach allows thermal errors to be evaluated at arbitrary positions in the machine's kinematic range, facilitating volumetric error compensation and integration with numerical compensation strategies. Experimental validation shows a good agreement with measured displacements. Maximum deviations of $19\,\mu m$ were observed, while the maximum reconstruction error is $1.1\,\mu m$. The method's computational efficiency and real-time capability make it a scalable and robust solution for model-based thermal error compensation in modern manufacturing machinery.

Keywords Thermal error compensation · Model order reduction (MOR) · Finite Element Method (FEM) · Machine tools · Digital twin · Volumetric error compensation

D. Spescha (✉) · N. Ceresa · M. Hoppstädter
inspire AG, Zurich, Switzerland
e-mail: daniel.spescha@inspire.ch

N. Ceresa
e-mail: nino.ceresa@inspire.ch

M. Hoppstädter
e-mail: mayra.hoppstaedter@inspire.ch

K. Wegener and M. Bambach (eds.), *4th International Conference on Thermal Issues in Machine Tools (ICTIMT2025)*, Lecture Notes in Production Engineering,
https://doi.org/10.1007/978-3-032-01194-7_24

1   Introduction

Maintaining high precision in machine tools is becoming increasingly challenging as accuracy requirements continue to increase and machines are expected to operate in environments with less stringent temperature stability. As machining tolerances reach the micrometer scale, even minor thermal variations can lead to significant deviations in the machine structure, leading to relative displacements between the tool and the workpiece at the Tool Center Point (TCP).

Conventional mitigation strategies, such as air conditioning, cooling systems, and steady-state production methods, often require substantial infrastructure investment and energy consumption, as well as warm-up cycles before production. To overcome these limitations, model-based and data-driven compensation strategies have gained increasing attention.

In recent years, advancements in machine learning and deep learning approaches, Finite Element Method (FEM)-based models, Model Order Reduction (MOR), and hybrid compensation methods have demonstrated the potential for real-time volumetric thermal error compensation.

The following sections categorize and discuss state-of-the-art compensation methods.

1.1   Traditional Regression Models for Thermal Compensation

Regression models are widely used for thermal compensation due to their simplicity and efficiency. Linear regression assumes a direct correlation between temperature variations and thermal displacement and includes, e.g., the least squares method or multivariate regression analysis, as presented by Li et al. [9]. Naumann et al. [10] proposed a hybrid thermal error compensation approach integrating deformation sensors and multiple regression analysis for five-axis machine tools. Blaser et al. [1] presented a method based on Auto-Regressive models with eXogenous inputs (ARX) with temperature probe data inputs, leveraging historical data to improve predictions by considering the dynamic behaviour of thermal systems.

1.2   Machine Learning and Adaptive Compensation Models

Machine learning techniques have emerged as powerful tools for thermal error compensation. Blaser et al. [1] introduced a Thermal Adaptive Learning Control (TALC) strategy for 5-axis machine tools, dynamically adjusting ARX-models and compensation parameters based on periodic process feedback. Similarly, Zimmermann et al.

[12] developed a data-driven thermal error prediction model utilizing machine learning techniques to enhance compensation accuracy in complex and varying thermal conditions. Lang et al. [7] demonstrated that power consumption data, an alternative to pure temperature sensor networks, can be used to predict and compensate thermal errors. Kaftan et al. [6] applied an approach combining ARX and Random Forest Regression (RFR) to compensate for sudden thermal boundary condition changes in Swiss-type lathe machining.

Ota et al. [11] developed a Convolutional Neural Network (CNN)-based prediction model that outperformed traditional regression models in estimating thermal displacements.

1.3 Finite Element Method and Model Order Reduction

In contrast to data-driven approaches, FEM-based models provide physics-based thermal predictions but are computationally expensive. Hernández-Becerro et al. [2] introduced a parametric MOR technique, enabling real-time thermal deformation predictions, while maintaining model accuracy also with varying convective boundary conditions. With this, Hernández-Becerro et al. [3] presented a thermo-mechanical model of a five-axis precision machine tool, demonstrating that MOR enables the efficient time-integration of thermal responses under dynamic conditions.

Irino et al. [5] demonstrated the use of MOR techniques in a digital twin framework for predicting thermal displacements and compensating machining accuracy under varying ambient temperature conditions. The approach integrates reduced-order finite element models with dynamic system simulations, enabling efficient estimation and correction of volumetric errors in machine tools.

A recent study by Lang et al. [8] introduced a Kalman filter-driven state observer for thermal error compensation in digital twins of machine tools. This approach combines reduced-order finite element models with Kalman filtering to estimate the entire thermal field in real time.

1.4 Discussion of the State of the Art

Traditional regression models for thermal compensation rely on measured data to establish relationships between temperature variations and tool center point displacement. Although these models are effective under constant operating conditions, they are difficult to generalize to scenarios that differ from the conditions during training. Unforeseen temperature fluctuations, environmental changes, or varying machining processes can result in significant deviations, rendering regression-based models ineffective for varying machining conditions.

Adaptive learning methods, including machine learning-based compensation strategies, provide a solution by dynamically updating model parameters to account for evolving conditions. However, these approaches require continuous or periodic measurements, which may not always be feasible due to displacement sensor limitations, machine accessibility, or productivity constraints. The necessity for frequent recalibration introduces downtime, thereby reducing overall machine efficiency.

Despite their advantages, all data-driven approaches, whether regression-based or machine learning-based, are fundamentally limited to what can be measured. Depending on machine kinematics and the specific machining process, certain displacements affecting workpiece accuracy within the working envelope cannot be directly measured during machining. This constraint limits the applicability of purely data-driven methods to scenarios where comprehensive in-process measurements are available.

FEM-based approaches, in contrast, offer high flexibility by enabling thermal analysis at any point within the complete kinematic range of the machine structure. These models allow for the assessment of thermal effects across various configurations, making them highly versatile for compensation applications. However, the accuracy of thermal FEM models strongly depends on boundary conditions, material properties, and contact conditions. While material properties and contact conditions can be determined with reasonable effort and typically remain constant, boundary conditions vary over time, introducing significant uncertainties. Heat transfer coefficients for free and forced convection at different surfaces, the influence of enclosures, and the proximity of heat sources or ventilation systems remain difficult to predict with high accuracy, reducing the reliability of purely physics-based models in real-world applications.

To address the issue of unknown and varying thermal loads, the Kalman-filter approach offers a promising solution. By continuously updating thermal state predictions based on limited sensor inputs, this approach compensates for unmodeled disturbances and enhances robustness against uncertain boundary conditions. An FEM model with moving axes, however, has a nonlinear dependence on axis positions and thus poses additional challenges to the concept of Kalman filtering.

This paper presents a method for direct observation of the temperature state of a reduced-order FEM model using temperature probes, ensuring effective FEM-based compensation with minimal complexity while maintaining maximum flexibility. The proposed method is introduced in Sect. 2, followed by an experimental validation in Sect. 3. Section 4 concludes the paper with a discussion and outlook.

2 Temperature Field Reconstruction Methodology

The aim is to reconstruct the temperature field of a structure at any time without the need for any temporal information. Subsequently, such a temperature field can then be applied as a load for static thermoechanical calculations in order to assess volumetric errors of a machine tool. Temperature probes acquire temperatures at multiple locations on the structure. Following this, a method is required that allows for a physically meaningful interpolation between these known temperatures.

2.1　Fundamental Principle

In this section, a method for direct estimation of the temperature state of a thermal model by means of a limited number of temperature probes is presented. The starting point is the generalized state space equation of a linear system according to

$$E\dot{x} = Ax + Bu \tag{1}$$
$$y = Cx \tag{2}$$

with the thermal state vector x, the thermal capacity matrix E, the conductivity matrix A, the input vector u, the input matrix B, the temperature output vector y, and the output matrix C. More information on the derivation of this equation from a finite element model is given by Hernández-Becerro et al. [2].

Let y be the vector of temperatures known from temperature measurements and C be an invertible matrix. Then, the state vector x can directly be calculated by multiplying Eq. 2 with the inverse of C according to

$$x = C^{-1} y. \tag{3}$$

In order for the matrix C to be invertible, it has to be square and have full rank. This implies some requirements on the setup:

1. For a square matrix, there must be the same number of outputs (temperature probes) as system states.
2. For the matrix to be of full rank, the columns and rows have to be linearly independent, which poses constraints on where the temperature probes can be placed.

To obtain a system of sufficiently low order that allows placing one probe for each state, an efficient model reduction method is required.

2.2　Model Order Reduction

Hernández-Becerro et al. [2] presented a Krylov and Modal Subspace (KMS) based reduction method, which leads to an efficient reduction. This method ensures that the input-output behavior for every thermal load and evaluation interface is matched exactly at steady state and that an error limit estimation is provided which monotonically increases with increasing frequency within a frequency range of interest. This model reduction technique provides a projection basis

$$V \in \mathbb{R}^{N \times n} \tag{4}$$

for projection from a subspace of dimension N, typically the nodal degrees of freedom of a FEM model, onto a KMS subspace of lower dimension n. Consider a large-scale finite element model

$$\hat{\mathbf{E}}\dot{\hat{\mathbf{x}}} = \hat{\mathbf{A}}\hat{\mathbf{x}} + \hat{\mathbf{B}}\mathbf{u} \tag{5}$$

$$\mathbf{y} = \hat{\mathbf{C}}\hat{\mathbf{x}}. \tag{6}$$

Projecting this system to the reduced coordinates $\mathbf{x} = \mathbf{V}^T\hat{\mathbf{x}}$ with the projection basis $\mathbf{V}$ leads to

$$\mathbf{V}^T\hat{\mathbf{E}}\mathbf{V}\dot{\mathbf{x}} = \mathbf{V}^T\hat{\mathbf{A}}\mathbf{V}\mathbf{x} + \mathbf{V}^T\hat{\mathbf{B}}\mathbf{u} \tag{7}$$

$$\mathbf{y} = \hat{\mathbf{C}}\mathbf{V}\mathbf{x} \tag{8}$$

Writing this equation by means of the reduced system matrices

$$\mathbf{E} = \mathbf{V}^T\hat{\mathbf{E}}\mathbf{V} \quad \in \mathbb{R}^{n\times n} \tag{9}$$

$$\mathbf{A} = \mathbf{V}^T\hat{\mathbf{A}}\mathbf{V} \quad \in \mathbb{R}^{n\times n} \tag{10}$$

$$\mathbf{B} = \mathbf{V}^T\hat{\mathbf{B}} \quad \in \mathbb{R}^{n\times p} \tag{11}$$

$$\mathbf{C} = \hat{\mathbf{C}}\mathbf{V} \quad \in \mathbb{R}^{q\times n} \tag{12}$$

leads to Eqs. 1 and 2. Here, p is the number of inputs and q is the number of outputs.

With the proposed KMS reduction, the dimension n is determined by the number of inputs to the system and the number of thermal modes considered. In the case of Krylov-only reduction, i.e. without consideration of thermal modes, the order of the reduced system is equal to the number of inputs, i.e. $n = p$.

From this and the condition that the matrix $\mathbf{C}$ must be square from Sect. 2.1, it follows that the number of probes must be equal to the order of the reduced system which is strongly related to the number of considered heat loads.

2.3 Sensor Placement

This section examines the conditions under which the matrix $\mathbf{C} = \hat{\mathbf{C}}\mathbf{V}$ attains full rank, with the goal of understanding the requirements for sensor placement in MOR-based temperature-field reconstruction.

Given that $\mathbf{V} \in \mathbb{R}^{N\times n}$ is an orthogonal matrix satisfying $\mathbf{V}^T\mathbf{V} = \mathbf{I}_n$, its columns constitute an orthonormal basis. The rank satisfies

$$\text{rank}(\hat{\mathbf{C}}\mathbf{V}) \leq \min(\text{rank}(\hat{\mathbf{C}}), n). \tag{13}$$

Thus, a necessary condition for full rank is that $\hat{\mathbf{C}}$ has at least n linearly independent rows

$$\text{rank}(\hat{\mathbf{C}}) = n. \tag{14}$$

While full row rank is necessary, it is not sufficient. Two scenarios can cause rank loss: First, the j-th column of $\mathbf{V}$, denoted by $\mathbf{v}_j$, lies in the null space of $\hat{\mathbf{C}}$, which means that

$$\hat{\mathbf{C}}\mathbf{v}_j = \mathbf{0}. \tag{15}$$

This implies that the corresponding column in $\hat{\mathbf{C}}\mathbf{V}$ is zero, reducing the rank.

Second, a row of $\hat{\mathbf{C}}$ is in the null space of $\mathbf{V}^T$, meaning there exists a row $\mathbf{c}_i$ of $\hat{\mathbf{C}}$ such that

$$\mathbf{c}_i\mathbf{V} = \mathbf{0}. \tag{16}$$

This leads to a loss of independent rows in $\hat{\mathbf{C}}\mathbf{V}$.

For $\hat{\mathbf{C}}\mathbf{V}$ to have full rank n, the following conditions are both necessary and sufficient:

1. $\hat{\mathbf{C}}$ has full row rank:

$$\text{rank}(\hat{\mathbf{C}}) = n. \tag{17}$$

2. No column of $\mathbf{V}$ is in the null space of $\hat{\mathbf{C}}$:

$$\forall j, \quad \hat{\mathbf{C}}\mathbf{v}_j \neq \mathbf{0}. \tag{18}$$

3. No row of $\hat{\mathbf{C}}$ is in the null space of $\mathbf{V}^T$:

$$\forall i, \quad \mathbf{c}_i\mathbf{V} \neq \mathbf{0}. \tag{19}$$

These conditions ensure that $\text{rank}(\hat{\mathbf{C}}\mathbf{V}) = n$.

In practical terms, the matrix $\hat{\mathbf{C}}$ represents the sensor readout locations, where each row corresponds to a sensor measuring temperature at a specific position. The matrix $\mathbf{V}$ defines the reduction basis, with each column representing a basis temperature field.

The necessary and sufficient conditions for $\hat{\mathbf{C}}\mathbf{V}$ to have full rank can be interpreted as follows:

1. **Distinguishable sensor locations**: The first condition, $\text{rank}(\hat{\mathbf{C}}) = n$, ensures that the sensor locations are sufficiently independent to provide unique measurements. If the sensors were too close together or redundantly positioned, the measurements would be linearly dependent, reducing the rank.
2. **Detectability of basis vectors**: The second condition, $\forall j, \quad \hat{\mathbf{C}}\mathbf{v}_j \neq \mathbf{0}$, ensures that each basis temperature field can be detected by at least one sensor. If a basis vector is entirely undetectable, the system cannot capture its contribution, leading to rank loss.

3. **Sensor sensitivity to basis vectors**: The third condition, $\forall i, \quad \mathbf{c}_i \mathbf{V} \neq 0$, guarantees that each sensor captures information from at least one of the basis temperature fields. If a sensor measures only zero values across all basis fields, it does not contribute to the reconstruction and effectively becomes redundant.

These conditions collectively ensure that the reduced-order model remains fully observable, meaning that the sensor placement and basis selection allow for accurate reconstruction of the temperature field.

2.4 Overdetermined System with Additional Sensors

In practice, it can be beneficial to add more sensors than the number of thermal states. By increasing the number of temperature probes, the system of equations

$$\mathbf{y} = \mathbf{Cx} \tag{20}$$

becomes overdetermined, meaning that there is no unique solution for $\mathbf{x}$ in terms of $\mathbf{C}^{-1}$. Instead, an optimal solution can be obtained in the least-squares sense using the Moore-Penrose pseudo-inverse $\mathbf{C}^{+}$:

$$\mathbf{x} = \mathbf{C}^{+}\mathbf{y}. \tag{21}$$

The pseudo-inverse ensures that the estimated state $\mathbf{x}$ minimizes the squared error in the measurement equation.

After computing $\mathbf{x}$, the temperature values at the sensor locations can be re-evaluated using

$$\tilde{\mathbf{y}} = \mathbf{Cx}. \tag{22}$$

If the reduction basis perfectly captures the temperature field, the estimated sensor temperatures $\tilde{\mathbf{y}}$ will match the measured values $\mathbf{y}$. However, in the presence of modeling errors or unforeseen thermal loads, a discrepancy

$$\Delta\mathbf{y} = \mathbf{y} - \tilde{\mathbf{y}} \tag{23}$$

will appear. This residual provides valuable information about deviations from the assumed thermal model and can be used to detect unexpected thermal influences in the system.

2.5   Practical Implications

A fundamental property of the MOR-based temperature field reconstruction is that the absolute magnitude of thermal loads is irrelevant, as the input u is not part of Eq. 3. However, their spatial distribution is crucial because it defines the input matrix $\hat{B}$, which in turn determines the reduction basis V, and thus also the reduced output matrix C. This means that this approach accounts for the effects of thermal loads without requiring knowledge of their absolute values.

A key feature of this approach is the use of a measurement that eliminates the need to specify convective heat load parameters. In thermal simulations, convection is typically defined by two parameters: the heat transfer coefficient and the fluid temperature. These parameters are often uncertain. Instead of specifying the heat transfer coefficient and fluid temperature, the resulting heat load can be used directly as a single non-constant parameter that produces the same physical effect on the structure. The MOR-based temperature-field reconstruction incorporates the spatial distribution of the heat load as a thermal input. This directly influences the reduction basis and is inherently integrated into the reconstruction approach. This treatment of convection introduces an additional degree of freedom for each convection load defined in the model, each of which necessitates an extra sensor on the structure. However, it eliminates the need to specify the two constant convection parameters. The user only needs to define the spatial distribution of the load.

By introducing thermal load cases considering the heat flow in insufficiently known thermal couplings, this approach allows the system to account for previously unmeasurable thermal interactions and reconstruct the resulting temperature field.

3   Experimental Validation

3.1   Implementation

The implementation of the proposed method has been carried out using the software MORe [4]. A dedicated plugin has been developed to facilitate sensor placement by visualizing reduction basis vectors and enabling temperature field estimation from temperature probe data.

The estimated temperature field is then used as input for subsequent static thermomechanical calculations. This integration enables the full feature set of MORe to be leveraged, using the estimated temperature field as a load, e.g., to compute thermally induced displacements at the TCP and workpiece.

Notably, this includes axis controllers that use motor actuation to correct position errors at the encoder, automatically adjusting for any shifts in the encoder head or linear scale.

Furthermore, the ability to reposition axes according to the machine's kinematics allows the evaluation of thermal errors at arbitrary positions within the working envelope, providing a comprehensive assessment of thermal effects on machine accuracy.

3.2 Experimental Setup

The experimental validation was conducted on a two-axis test bench equipped with a linear direct drive on the X axis and an externally applied heat load using a heat pad on the Y axis. The measurement system employed was a cross grid encoder from Heidenhain, enabling high-precision measurement of displacements at the TCP between encoder head, mounted on the X axis, and cross grid, mounted on the Y axis. Figure 1 provides an overview of the test bench used in the experiment.

The thermal loads considered for MOR-based temperature field reconstruction are as follows:

- X motor primary part heat
- X motor secondary part heat
- Heat pad on the Y axis
- Two convection heat loads on the X slider (upper/lower part)
- Two convection heat loads on the Y slider (upper/lower part)
- Two convection heat loads on the frame (upper/lower part)
- Convection heat load on the ball-screw spindle.

Fig. 1 Photograph of the two-axis test bench

The convection heat loads account for both fluctuations in ambient temperature and auxiliary heat sources within the lower frame, such as drive electronics. In total, 10 distinct heat loads were considered. The order of the reduced system was 12. Therefore, to ensure a well-conditioned system of equations, 12 temperature probes were placed according to the guidelines in Sect. 2.3.

Figure 2 shows a visualization of an example of a basis vector. To fulfill the detectability condition for basis vectors derived in Sect. 2.3, a good location for a temperature probe is the orange region near the motor of the X axis. Figure 3 provides an overview of all chosen probe locations.

Fig. 2 Visualization of the temperature field for an example basis vector

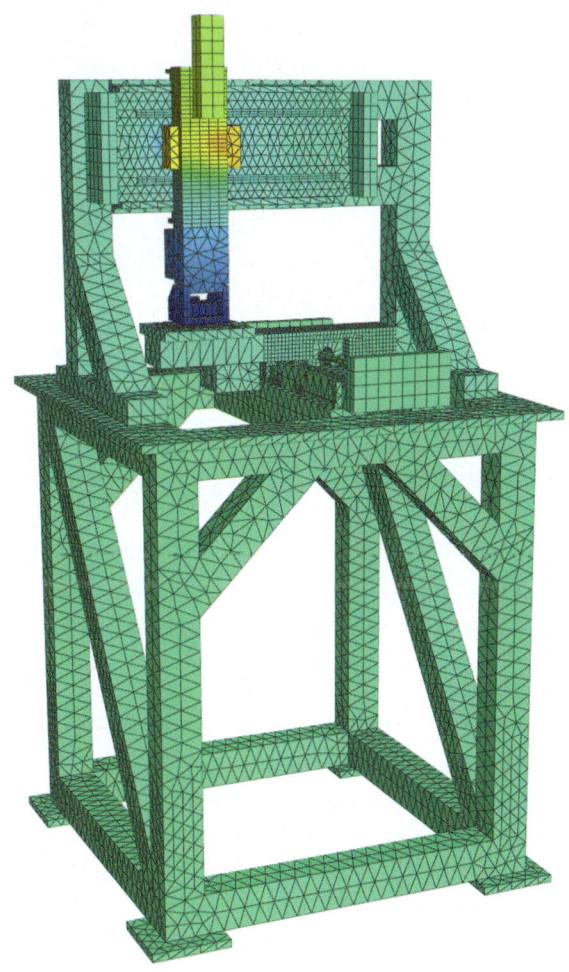

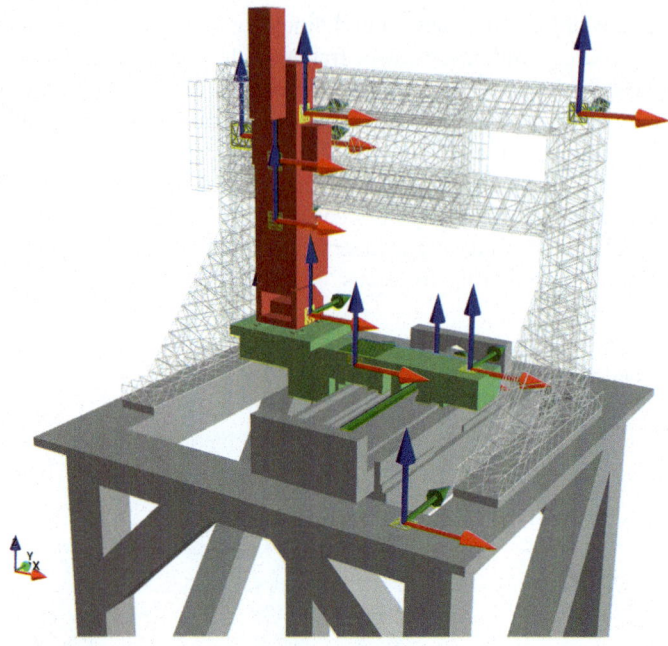

Fig. 3 Overview of the chosen probe locations. Selected probe locations are highlighted with yellow mesh areas and coordinate systems. The location of the heat pad is shown as a green mesh on the Y slide (green component)

3.3 Load Cycle

The experiment was conducted in four sequential phases:

1. **Phase 1**: High-acceleration periodic motion of the X axis for one hour.
2. **Phase 2**: Cooldown phase with no external loads for one hour.
3. **Phase 3**: Heat pad activation with approximately 40 W applied to the Y axis for one hour.
4. **Phase 4**: Second cooldown phase lasting 1.5 h.

Throughout the experiment, displacement measurements were taken every two minutes. Each measurement cycle included the evaluation of a circular trajectory to determine a shift of the center position in X and Y direction. This measurement required approximately 30 s, during which the periodic motion in Phase 1 had to be paused.

3.4 Results and Discussion

Figure 4 shows the comparison between the estimated displacement at the TCP in the X and Y directions over the entire measurement period. The plots exhibit strong agreement between the measured and predicted values. While the maximum displacement in the X direction reaches $19\,\mu m$, the maximum estimation residual is $1.1\,\mu m$. Similarly, the maximum displacement in the Y direction is $11\,\mu m$, with a maximum estimation residual of $0.85\,\mu m$, demonstrating the high accuracy of the proposed reconstruction method.

3.5 Applications

A key advantage of the proposed approach is the ability to evaluate thermal errors at arbitrary axis positions. This flexibility allows for the computation of various error measures beyond local displacements, including comprehensive evaluations such as volumetric errors of a machine. By assessing thermally induced deviations across different positions within the working envelope, a more complete characterization of machine accuracy can be obtained. The visualization of these deformations further enhances the understanding of machine behavior under thermal influences. An example of this evaluation is illustrated in Fig. 5.

The knowledge of thermal errors can be leveraged for real-time monitoring and error compensation in machine tools through a digital twin application. In this approach, the digital twin continuously evaluates thermally induced deviations and provides corrective measures to enhance machining accuracy. The interface between the machine's numerical control system and the digital twin remains minimal, as it only requires the transmission of temperature data to the digital twin and the reception of computed thermal error measures. This streamlined integration ensures low implementation complexity while enabling adaptive compensation, ultimately improving machining precision without significant modifications to the existing control architecture.

Figure 6 presents a visualization of the deformed machine structure, with scaled deformation and color mapping representing the temperature field. This visualization not only illustrates the extent of thermal deformation but also provides valuable insights into the thermal behavior of the machine. By animating and analyzing these visualizations, potential thermal issues can be identified, and the effects of temperature changes on the stability and performance of the structure can be better understood. This capability aids in diagnosing sources of thermal errors and improving overall machine performance with appropriate design measures.

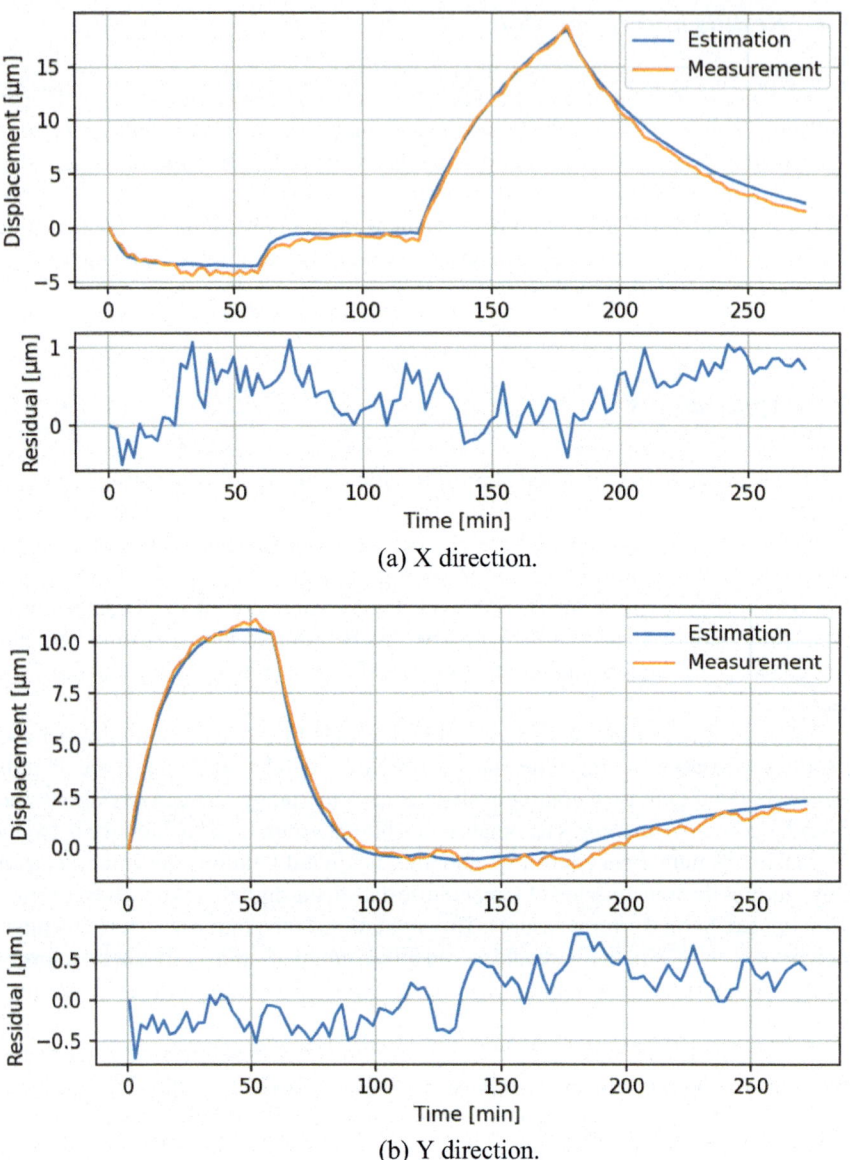

(a) X direction.

(b) Y direction.

Fig. 4 Comparison of estimated and measured displacements at the tool center point for X and Y directions throughout the complete load cycle (phases 1–4) defined in Sect. 3.3

Fig. 5 Visualization of an
undeformed and a deformed
grid on the X-Y plane at
56 min after the start of the
experiment. The color scale
indicates the total
displacement of each grid
point in the X-Y-plane

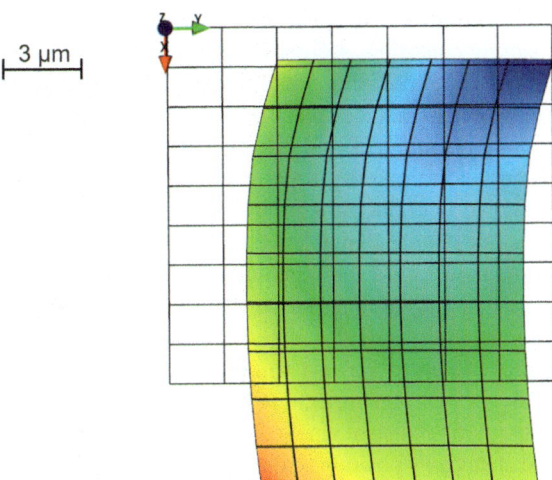

Fig. 6 Visualization of the
deformed machine structure
with scaled deformation and
temperature field
representation at 56 min after
the start of the experiment

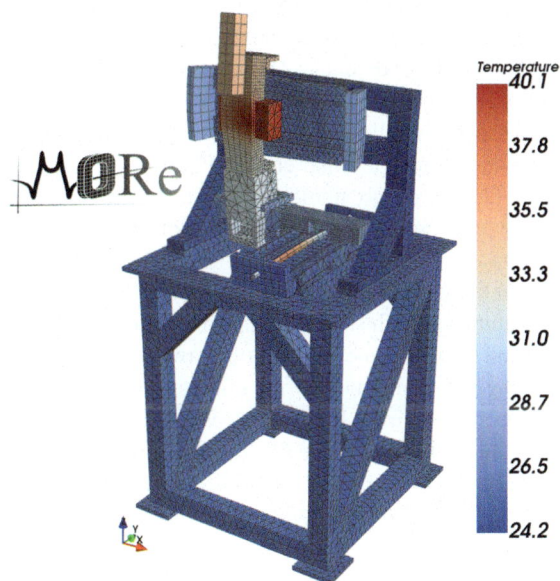

4 Conclusion

The presented method for MOR-based temperature field reconstruction provides a
robust and flexible approach to the identification of temperature fields, which, when
combined with thermomechanical computations, allow for thermal error compensation. Since this method uses a reduced-order FEM model driven by temperature probe

data, it eliminates the need for extensive measurement campaigns and recalibration, ensuring consistent accuracy across varying operating conditions.

The experimental validation confirms that the proposed method is able to accurately reconstruct thermally induced displacements with sub-micron residual errors, making it suitable for precise error compensation.

Furthermore, this direct reconstruction approach minimizes dependency on unknown boundary conditions. Rather than requiring precise heat transfer coefficients or detailed knowledge of convection dynamics, convection is treated as an external heat load and inherently incorporated into the reconstruction process. This makes the estimation more robust to environmental variations and ensures reliability even in cases where thermal interactions with enclosures, ventilation, or nearby heat sources would otherwise introduce significant uncertainties.

By separating temperature field estimation from static thermomechanical calculations, thermal displacement estimation becomes possible at any point on the structure and at any axis position within the machine's kinematic range, even in regions where direct measurements are not feasible. This enables comprehensive thermal compensation beyond localized sensor data, ensuring that thermal deviations affecting workpiece accuracy are accounted for across the entire working envelope.

Another key advantage of the model-based approach is its integration of axis control, allowing for the consideration of measurement system locations and displacements. Additionally, repositioning the axes according to machine kinematics enables the evaluation of thermal errors at arbitrary positions, facilitating the generation of specific data for any numerical compensation strategy.

By leveraging reduced-order models and computationally efficient calculations, this approach enables real-time capability, even when evaluating multiple points within the working volume to determine time-dependent volumetric errors.

Future research should focus on the applicability and integration of the proposed method for real-time interaction between digital twins and machine tools, demonstrating that volumetric error compensation and error monitoring during machining operations are achievable. This includes validation under machining conditions with the application of cutting fluid. Furthermore, adopting a modular approach that accounts for errors from the machine, tool holder, tool, and workpiece would enable a more comprehensive compensation strategy.

References

1. Blaser, P., Pavlíček, F., Mori, K., Mayr, J., Weikert, S., Wegener, K.: Adaptive learning control for thermal error compensation of 5-axis machine tools. J. Manuf. Syst. **44**, 302–309 (2017). ISSN 0278-6125. https://doi.org/10.1016/j.jmsy.2017.04.011. Special Issue on Latest advancements in manufacturing systems at NAMRC 45
2. Hernández-Becerro, P., Spescha, D., Wegener, K.: Model order reduction of thermo-mechanical models with parametric convective boundary conditions: focus on machine tools. Comput. Mech. **67**(1), 167–184 (2021)

3. Hernández-Becerro, P., Mayr, J., Wegener, K.: Efficient thermo-mechanical model of a precision 5-axis machine tool. In: euspen Special Interest Group Meeting on Thermal Issues, Aachen, Germany (2020)
4. inspire AG: MORe Simulations (2024). https://www.more-simulations.ch. Accessed 26 Feb 2025
5. Irino, N., Kobayashi, A., Shinba, Y., Kawai, K., Spescha, D., Wegener, K.: Digital twin based accuracy compensation. CIRP Ann.—Manuf. Technol. **72**, 345–348 (2023). https://doi.org/10. 1016/j.cirp.2023.04.088
6. Kaftan, P., Mayr, J., Wegener, K.: Thermal compensation of sudden working space condition changes in swiss-type lathe machining. In: Ihlenfeldt, S. (ed.) 3rd International Conference on Thermal Issues in Machine Tools (ICTIMT2023), pp. 15–27. Springer International Publishing, Cham (2023). ISBN 978-3-031-34486-2
7. Lang, S., Zimmermann, N., Mayr, J., Wegener, K., Bambach, M.: Thermal error compensation models utilizing the power consumption of machine tools. In: Ihlenfeldt, S. (ed.) 3rd International Conference on Thermal Issues in Machine Tools (ICTIMT2023), pp. 41–53. Springer International Publishing, Cham (2023). ISBN 978-3-031-34486-2
8. Lang, S., Talleri, S., Mayr, J., Wegener, K., Bambach, M.: Kalman filter-driven state observer for thermal error compensation in machine tool digital twins. Manuf. Lett. **41**, 208–218 (2024). https://doi.org/10.1016/j.mfglet.2024.09.025. ISSN 2213-846. 52nd SME North American Manufacturing Research Conference (NAMRC 52)
9. Li, Y., Yu, M., Bai, Y., Hou, Z., Wu, W.: A review of thermal error modeling methods for machine tools. Appl. Sci. **11**(11) (2021). https://doi.org/10.3390/app11115216. ISSN 2076-341
10. Naumann, C., Naumann, A., Bertaggia, N., Geist, A., Glänzel, J., Herzog, R., Zontar, D., Brecher, C., Dix, M.: Hybrid thermal error compensation combining integrated deformation sensor and regression analysis based models for complex machine tool designs. In: Ihlenfeldt, S. (ed.) 3rd International Conference on Thermal Issues in Machine Tools (ICTIMT2023), pp. 28–40. Springer International Publishing, Cham (2023). ISBN 978-3-031-34486-2
11. Ota, K., Mori, M., Irino, N.: Development of thermal displacement prediction model and thermal deformation measurement methods. In: Ihlenfeldt, S. (ed.) 3rd International Conference on Thermal Issues in Machine Tools (ICTIMT2023), pp. 3–14. Springer International Publishing, Cham (2023). ISBN 978-3-031-34486-2
12. Zimmermann, N., Büchi, T., Mayr, J., Wegener, K.: Self-optimizing thermal error compensation models with adaptive inputs using Group-LASSO for ARX-models. J. Manuf. Syst. **64**, 615–625 (2022). https://doi.org/10.1016/j.jmsy.2022.04.015. ISSN 0278-612

Overcoming Uncertainty With an Ensemble of Physical Models and Real-Time Measurements: Thermal Error Compensation Using Kalman Filters In a Digital Twin

Sebastian Lang⦿, Marco Schneider, Lenny Rhiner, and Markus Bambach⦿

Abstract Thermally induced errors critically limit precision in machine tools (MT), affecting dimensional accuracy in high-precision manufacturing. Physics-based models offer a sustainable solution and can do without training measurements, unlike data-driven models, but are often unsuitable for real-time thermal error compensation due to computational demands and boundary condition uncertainties. This paper presents an enhanced ensemble Kalman filter approach to accurately and efficiently reconstruct the thermal state from limited sensor data. Utilizing a reduced finite element model generated via Krylov modal subspace techniques, the proposed method integrates real-time sensor measurements with an ensemble of perturbed models to robustly estimate thermal states and mechanical deformations. Validation experiments on a thermal test bench with 10 out of 40 temperature sensors observed demonstrate significant improvements in accuracy achieving an RMSE of 1.5 °C. The computational speed is over $45\times$ faster than real-time allowing the implementation of the digital twin for any MT. The reduction of the volumetric thermal error by consistently more than 70% confirm the potential of this digital twin-based approach for real-time thermal error compensation in precision manufacturing.

Keywords Thermal error · Machine tool · Kalman filter · Digital twin · Model reduction

---

Sebastian Lang and Marco Schneider: These authors contributed equally to this work.

S. Lang (✉) · M. Schneider · L. Rhiner · M. Bambach
Advanced Manufacturing Laboratory, ETH Zurich, Zurich, Switzerland
e-mail: selang@ethz.ch

S. Lang · L. Rhiner
inspire AG, Zurich, Switzerland

K. Wegener and M. Bambach (eds.), *4th International Conference on Thermal Issues in Machine Tools (ICTIMT2025)*, Lecture Notes in Production Engineering,
https://doi.org/10.1007/978-3-032-01194-7_25

1 Introduction

Production accuracy in machine tools (MT) is heavily impacted by thermal effects. According to Mayr et al. [21] up to 75% of all errors on workpieces can be induced by effects of temperature. To alleviate these thermal effects in a MT, considerable effort is necessary. For example, design efforts such as thermal symmetry can minimize the impact of temperature changes [29]. Another approach is the use of extensive cooling of the machine components [5], structure [9] and environment [22]. However, these approaches are typically resource intensive and leave a considerable residual error.

Approaches which determine the error and compensate for it in the MT control, promise to be a more effective and resource efficient solution. However, this requires a timely prediction or measurement of the thermal error. For these predictions, data-driven models can be used, which often use temperature sensors as model inputs [20]. In addition, the power of the rotary drives [28] or the power of the key components of the machine [17] can be used. The crucial factor is the sufficient availability of representative training data [15]. Once a data-driven model is no longer performing well, it can be retrained, for example, by using the thermal adaptive learning control approach introduced by Blaser et al. [3]. Zimmermann et al. [30] extended this to include a re-selection of the most suitable temperature sensors for every model. The drawback to these approaches, however, is the availability of measurements at the required machine states, which adversely impacts productivity. Furthermore, it has to be assumed, that during a measurement the error is constant, such that no thermal error changes the machine during the measurement which is often not the case [7]. On machine measurement cycles agree well with manufactured workpieces, Zimmermann et al. [31] introduced a novel thermal test piece for 5-axis machine tools, enabling the isolation and analysis of axis-specific thermal errors. Zimmermann et al. [33] also demonstrated a 73% reduction in machining errors on workpieces by implementing thermal compensation during the manufacturing of impellers.

The drawback of these data-driven approaches, however, is their reliance on measurement data and the black box nature of the underlying models. An alternative approach is the use of physics-based models, such as finite element methods (FEM) or computational fluid dynamic (CFD) models. If a robust data-driven model can be found to map between the temperatures or power inputs and thermal errors a physics-based model can be found for the same relationship without requiring training data. However, allowing compensation on machine tools necessitates fast computational speeds, which are capable of simulations in thermal real-time, which is prohibitive to many physics-based approaches. Digital twins can be used to fuse real-time simulations with measurements and subsequent compensation on the MT [19]. Irino et al. [13] utilized a digital twin based on a model order reduced (MOR) finite element (FE) model to compensate the thermal errors and another model to compensate the tool displacement induced by the cutting force. Kizaki et al. [14] placed 284 temperature sensors on a machine tool and applied linear interpolation to obtain the temperature field for an FE model, which enabled a direct solution of the thermo-mechanical model and therefore the calculation of the thermally-induced errors. However, this approach requires a high number of temperature sensors and uses little physical knowledge about the MT behavior. Tanaka et al. [27] extended this approach and developed a temperature measurement system for the temperature measurement and thermal compensation of a ballscrew. Lang et al. [18] introduced a digital twin based approach for selecting the most suitable temperature sensor positions. Another challenge associated to physics based models is the uncertainty faced in the model generation as many parameters are difficult to determine exactly and some are dynamic or dependent on other variables [12].

Kalman filters are a widely used tool for state estimation, using a model and measurement feedback [2]. Gomez-Acedo et al. [8] used a Kalman filter with a lumped mass model of a MT to develop a physics-based compensation model for thermal errors based on two temperature sensors and the spindle speed. Lang et al. [16] introduced the first Kalman filter for state estimation of the temperature field of an entire MT using reduced FE models. This approach enabled the recreation of the entire temperature field from only 13 to 16 temperature sensors, which in turn allowed the computation of thermal errors by using a thermo-mechanical FE model.

This paper aims to extend the previous work in physics based modeling of thermal errors in MTs by enhancing the machine model accuracy through the use of ensemble approaches, which contain many different models. This has the advantage that the uncertainty in model parametrization becomes a strength as the true system is expected to be within the ensemble. First, the measurement setup is introduced before the thermo-mechanical model and the sampling of 500 such models is introduced that form the basis of the digital twin for compensation. Section 4 introduces the Kalman filter and the ensemble that is used to determine the estimated thermal state so the heart of the digital twin as it combines the measurement data with the simulation models. Subsequently the results are presented before a conclusion and outlook are given.

2 Experimental Setup

This section introduces the experimental setup on which the ensemble approach is tested and evaluated.

Figure 1 illustrates the finite element (FE) model of the thermal test bench from two perspectives (front and rear). For simplicity, this test setup will be referred to as the machine in the following. The machine is designed as a scaled-down representation of a typical C-frame MT, consisting of three hollow structural aluminum sections attached to a base plate, as well as a mounting for the displacement probes. This configuration enables continuous thermal error measurement, while retaining the general characteristics of a representative MT [32].

A total of 40 temperature sensors (type-K thermocouples and Pt100 resistance temperature detectors) are installed on the machine. Their attachment points in the FE model are marked in green in Fig. 1. Each sensor is affixed directly to the structure using thermally conductive paste, while being insulated from the ambient air. These precautions reduce measurement noise arising from ambient influences. The data acquisition system records sensor readings at a sampling rate of 0.4 Hz. Although additional sensors measure ambient temperature, only the 40 sensors physically attached to the structure can be directly compared to the FE mesh and therefore used for the thermomechanical model.

Four heating pads, highlighted in yellow in Fig. 1 and each rated at 50 W, provide controlled heat inputs to excite a thermal load. In this study, only three of the four pads are used. The experimental procedure consists of sequentially activating each pad for six hours, followed by a ten-hour cool-down period, before proceeding to the next. After all three pads (back, front, and top) have been individually cycled, they are simultaneously switched on for six hours, followed by a final 15-hour cool-down phase. In total, the experiment spans 70 hours. This heating protocol ensures both spatial and temporal temperature gradients that closely mimic real operational thermal loads in machine tools.

Thermally induced displacements are typically measured between the tool center point (TCP) and the workpiece center point (WCP). For this measurement setup the TCP is represented by a stylus mounted on the spindle equivalent body while the WCP is approximated by a static probe. Five incremental displacement sensors (Heidenhain Specto ST1288, rated accuracy 1 μm) measure these relative displacements at 0.8

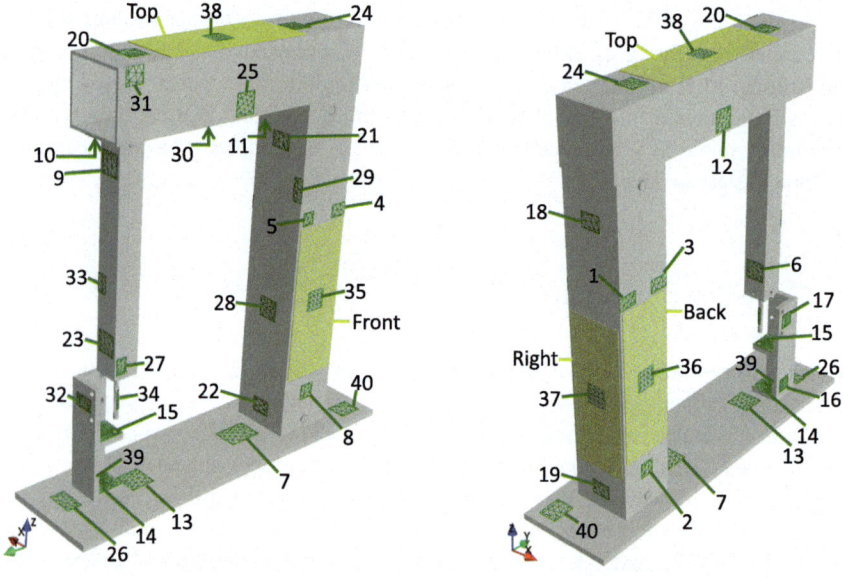

(a) Front view of the machine's FE model.   (b) Rear view of the machine's FE model.

Fig. 1 The machine's FE mesh shown from front and rear views. It consists of three hollow structural aluminum parts, a base plate, and the displacement probe holder. Four heat pads are labeled and highlighted in yellow. Temperature sensor locations are highlighted in green and labeled with a number. The average temperature of the green area is compared to the physical temperature sensor mounted in the center of this location

Hz. One sensor tracks the displacement in the Z-direction, while four additional sensors measure X- and Y-direction deformations at two different heights near the spindle tip. This arrangement isolates out-of-plane distortions and permits an accurate characterization of the three-dimensional thermal behavior of the bench.

Finally, the entire structure is enclosed to reduce external air drafts and temperature fluctuations, facilitating more consistent results. With this setup, the machine successfully emulates essential features of a MT's thermal behavior and offers a valuable test environment for validating real-time thermal error compensation strategies based on the proposed ensemble Kalman filter approach.

3 Thermomechanical Model for a Digital Twin

This section presents the thermomechanical modeling framework that powers the digital twin used for real-time temperature and thermal error estimation in a machine tool (MT). The digital twin consists of two core components. A physical MT equipped with 40 temperature sensors as well as 5 displacement probes, as introduced in Sect. 2 and an ensemble of 500 reduced-order thermomechanical simulation models (ROM) that share the same geometry but differ in (the uncertain) boundary conditions that affect the resulting ROM. The underlying assumption is that the true system behavior can be captured by one of these simulation models or approximated as a weighted combination of several models in the ensemble. To fuse live sensor data with model predictions, a separate Kalman filter is run for each ROM. At every timestep, each filter computes a model-specific likelihood based on the agreement between predicted and measured temperatures, resulting in a dynamic weight distribution across the ensemble. Because each ROM is based on a different Krylov subspace, the thermal states cannot be directly averaged. Instead, each predicted state is mapped to measurement space and the ensemble prediction is formed as a weighted sum of model outputs. This applies to both the temperature at validation sensor locations and the thermally induced deformation, computed via each model's mechanical coupling. Figure 2 illustrates the full digital twin workflow, from model generation and filtering to probabilistic fusion and final output estimation. The remainder of this section is organized as follows: Sect. 3.1 details the thermal and mechanical simulation setup. Section 3.2 outlines the sampling strategy used to generate the model ensemble. Section 3.3 presents the discretization of the continuous-time state-space models. Finally, Sect. 3.4 describes the input sensor selection procedure used to define the observation and validation temperature sensor sets.

3.1 Thermal and Mechanical Simulation Setup

The initial step of the simulation setup is to convert the 3D computer-aided design (CAD) geometry of each machine component into a finite element (FE) mesh. In

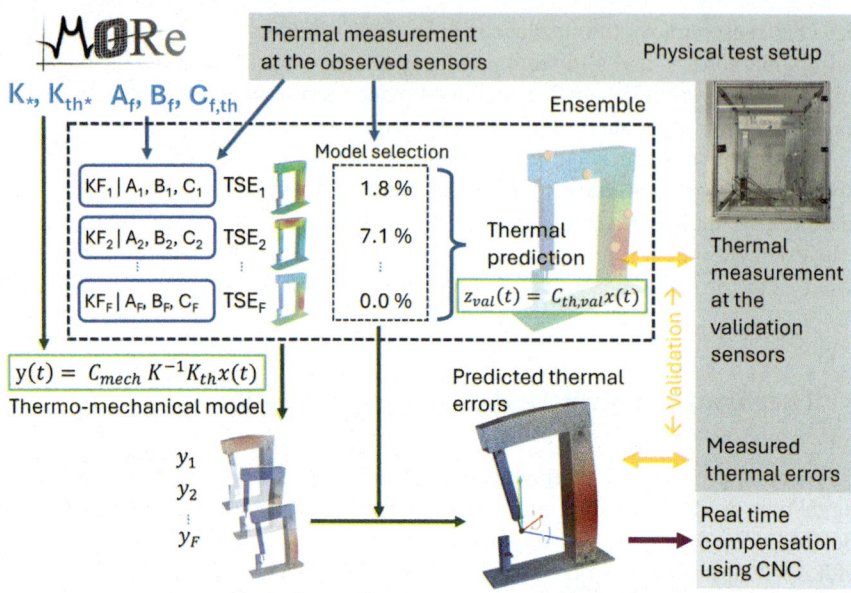

Fig. 2 Overview of the ensemble methodology fueling the digital twin. Multiple state-space representations with varying boundary conditions are generated and used in separate Kalman filters. Each filter computes its own thermal state estimation and at each timestep, model probabilities are inferred based on the difference between predicted and measured sensor data. Since not all sensors are used for state observation, the remaining ones are used for validation of the thermal behavior. Thermally induced deformations and thermal errors at the TCP are computed for each estimated thermal state (TSE) and aggregated using the ensemble weighting strategy. Ultimately, the deviations at the TCP are compared to the physical test setup for performance evaluation of the ensemble in the thermomechanical domain

this work, ANSYS is used to generate detailed thermal and structural meshes. Each mesh is then exported to the model-order reduction environment MORe [10, 26] to create a state-space representation:

$$\mathbf{E}\dot{\mathbf{x}}(t) = \mathbf{A}\mathbf{x}(t) + \mathbf{B}\mathbf{u}(t), \tag{1}$$

$$\mathbf{z}(t) = \mathbf{C}_{\text{th}}\mathbf{x}(t) \tag{2}$$

where $\mathbf{x}(t)$ is the thermal state vector, $\mathbf{A}$ is the system matrix encoding conduction and convection, $\mathbf{E}$ is the mass matrix reflecting thermal capacitance's and $\mathbf{B}$ couples exogenous thermal inputs $\mathbf{u}(t)$ (e.g. heat fluxes). The thermal observability matrix $\mathbf{C}_{\text{th}}$ maps the thermal state to the temperature sensors $\mathbf{z}(t)$, which can be observed in the experimental setup (see Sect. 2, Fig. 1).

Measuring or modeling exact heat fluxes in a production environment is often challenging for a MT. Hence, one way to deal with $\mathbf{u}(t)$ is setting it to zero, leaving the filter to capture these unmeasured heat fluxes via process noise [16]. In this work,

Table 1 Model size of the FE-model and modes after reduction for all components of the machine

	FE nodes	Reduced thermal modes	Reduced thermomechanical modes
Frame	186,339	55	91
Meas. holder	22,602	58	88
Total	208,941	113	179

nominal heat flows from the heat pads can be estimated and $\mathbf{u}(t)$ is set accordingly. For all other boundaries, a constant convection temperature of 25 °C is applied. To reduce computation time while preserving key thermomechanical dynamics, a Krylov Modal Subspace (KMS) reduction [11, 25] is applied to each individual component using the MORe software framework. This enables real-time estimation of the thermal state at high sampling rates such as every ten seconds and also improves the performance of the Kalman filter, as the model is much smaller and better observable. Table 1 lists the mesh sizes and the number of reduced modes used for the machine, which is divided into its components.

Table 1 details the model size of the machine before and after the KMS-based reduction. The reduction is significant and the system is compressed around ~ 1849 fold, which leads to even larger computational gains, as the scaling is not linear.

The thermomechanical system consists of the coupling of the thermal and mechanical system, which allows the calculation of the thermally induced deformation through a virtual force.

$$\mathbf{y}(t) = \mathbf{C}_{\text{mech}} \mathbf{K}^{-1} \mathbf{K}_{\text{th}} \mathbf{x}(t) \tag{3}$$

Equation 3 represents the mechanical part of the system. The vector $\mathbf{y}(t)$ denotes the thermally induced deformation at the TCP at time t. The mechanical observability matrix $\mathbf{C}_{\text{mech}}$ is assembled from the individual components and maps the deformation between the TCP and workpiece holder for the five displacement sensors. Meanwhile, $\mathbf{K}^{-1}$ captures the structural stiffness, and $\mathbf{K}_{\text{th}}$ translates temperature changes into volumetric forces via the thermal expansion coefficient α that is defined as a material property before system reduction.

3.2 Model Sampling

Due to the significant uncertainty and unknown exact values of the model parameters, an ensemble approach comprising multiple models is introduced in this paper. One key aspect is that model parameters, such as convection coefficients, change during an experiment or during a machine's operation and therefore cannot be assumed to remain constant. This means that finding a single linear model with the functions

Table 2 Bounds within which the 500 models are sampled using LHS

Model parameter	Lower bound	Upper bound
Column inside (CI) HTC	1 W/(m^2 K)	7.5 W/(m^2 K)
Beam inside (BI) HTC	1 W/(m^2 K)	7.5 W/(m^2 K)
Spindle inside (SI) HTC	1 W/(m^2 K)	7.5 W/(m^2 K)
Working space (WS) HTC	1 W/(m^2 K)	10 W/(m^2 K)
Baseplate conduction (BC)	0.5 W/(m^2 K)	20 W/(m^2 K)
TCC	100 W/K	1000 W/K

describing the transient behavior of all parameters is generally impossible. However, by sampling a large enough space of models, the actual system can be represented by one of the sampled models or a combination of them.

For this purpose, 500 models are sampled using Latin Hypercube Sampling (LHS), as this is a very suitable method for creating a generalized representation of the sampling space [4, 24]. The parameter bounds listed in Table 2, within which the models are constructed, are informed by experimental data and extensive CFD simulations. The parameter bounds are chosen to encompass the expected physical values, based on the assumption that only natural convection occurs within the sealed enclosure. The numerical limits are informed by general engineering approximations and are consistent with the operators experience. The heat transfer coefficients (HTCs) of every surface are incorporated in the reduced order model, which is why they are a crucial model parameter. Different HTCs are defined for the hollow spaces within the aluminum tubes compared to the HTC of the working space for all outside facing parts. The same holds true for the thermal contact conductivity (TCC) between all metal-metal contacts. The baseplate is modeled as transferring heat to the underlying table, which is treated as an infinite thermal reservoir.

3.3 Discretization

The system equation introduced in Eq. 1 describes the system in a continuous-time framework. In practical applications, temperature measurements are only available at discrete time intervals. To enable state estimation using a Kalman filter, the system must be discretized. Since the thermo-mechanical dynamics evolve significantly more slowly than the available sampling rate, the discretization method is chosen to ensure an accurate representation of the system while minimizing numerical artifacts. Various discretization techniques, including zero-order hold (ZOH), first-order hold (FOH), and impulse-invariant discretization were evaluated, revealing negligible differences in performance. For this study, an exact discretization based on the matrix exponential method [6] is employed. The thermal state-space equations are transformed into their discrete counterparts using $\Delta t = 10$s:

$$\mathbf{A}_d = e^{\mathbf{E}^{-1}\mathbf{A}\Delta t}$$
$$\mathbf{B}_d = \mathbf{E}^{-1}\mathbf{A}^{-1}(\mathbf{A}_d - \mathbf{I})\mathbf{E}^{-1}\mathbf{B} \qquad (4)$$
$$\mathbf{C}_d = \mathbf{C}_{th}$$

This results in the discrete-time state-space equations with $\mathbf{u}$ set to the experimental heat input as described in Sect. 3.1:

$$\mathbf{x}[k] = \mathbf{A}_d\mathbf{x}[k-1] + \mathbf{B}_d\mathbf{u}[k-1]$$
$$\mathbf{z}[k] = \mathbf{C}_d\mathbf{x}[k] \qquad (5)$$

Equation 5 is used in the ensemble of Kalman filters described in Sect. 4.

3.4 Input Sensor Selection

Sensor selection plays a critical role in Kalman filtering, as it impacts the accuracy of state estimation. The objective is to identify a subset of sensors, $M_{obs} \subset M$, from the full set of available sensors $M = \{1, \ldots, 40\}$ (as shown in Fig. 1), in order to maximize system observability. The remaining sensors, denoted by M_{val}, are used exclusively for validation, satisfying:

$$M_{val} = M \setminus M_{obs}. \qquad (6)$$

Since multiple state-space models are used in the ensemble Kalman filter, the selected sensor subset must remain consistent across all models. Observability is quantified, according to Preumont [23], using the observability Gramian $\mathbf{W}_{o,f}$ for each model f, obtained by solving the discrete Lyapunov equation:

$$\mathbf{A}_{d,f}^T \mathbf{W}_{o,f} \mathbf{A}_{d,f} - \mathbf{W}_{o,f} + \mathbf{C}_{d,f}^T \mathbf{C}_{d,f} = 0. \qquad (7)$$

A widely used criterion to maximize system observability is to use the smallest eigenvalue of the observability Gramian $\lambda_{min}(\mathbf{W}_{o,f})$, which reflects the least observable state direction. Accordingly, sensor selection is formulated as an optimization problem. Let $S \subset M$ be the selected sensor subset for filtering. The objective is to maximize the mean of the smallest eigenvalues across all models, subject to the constraint that the number of selected sensors does not exceed the predefined limit m_{max}:

$$M_{obs} = \arg\max_{S \subset M, \, |S| \leq m_{max}} \frac{1}{F} \sum_{f=1}^{F} \lambda_{min}(\mathbf{W}_{o,f}(S)). \qquad (8)$$

Here, F denotes the number of models in the ensemble. The optimization problem in (8) is inherently non-convex and combinatorial in nature. Exhaustively evaluating all possible sensor subsets becomes computationally prohibitive for $m_{max} \gtrsim 5$. Since m_{max} is planned to be in between 10 and 20, greedy algorithms (GR), genetic algorithms (GA) and particle swarm optimization (PSO) are employed to efficiently approximate the optimal sensor selection while maintaining computational feasibility. Among the tested approaches, the GA yielded the highest objective function value for most selections of m_{max}. Consequently, the sensor subset obtained via GA is used for ensemble Kalman filtering. After evaluation of the ensemble results for $m_{max} \in \{10, 15, 20\}$, m_{max} was set to ten. The ten sensors selected by GA, used for state observation are highlighted in red in Fig. 3 while the remaining 30 sensors are used for validation of the predictions on the thermal state are shown in green.

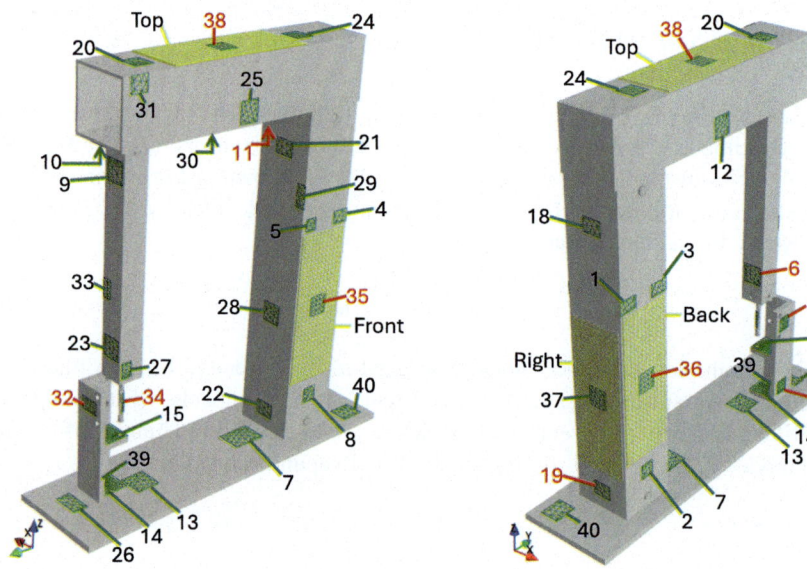

(a) Front view of the machine's FE model with the sensors used for state estimation (observation sensors) highlighted with red numbering.

(b) Rear view of the machine's FE model with the sensors used for state estimation (observation sensors) highlighted with red numbering.

Fig. 3 Visualization of the sensor set M_{obs} obtained by solving the optimization problem in Eq. 8 using genetic algorithms. This selected sensor set (marked in red) is used for state estimation, while the unselected ones (green) are used only for validation of the accuracy of the thermal state

4 Ensemble Kalman Filtering for Thermo-Mechanical Systems

State estimation in thermo-mechanical systems presents significant challenges due to uncertainties in boundary conditions and model parameters. Traditional Kalman filtering relies on a single fixed model, which may not adequately capture real-world variations. To address this, an ensemble-based Kalman filtering approach is employed, leveraging multiple reduced-order models that represent different boundary conditions, as outlined in Sect. 3. By dynamically adjusting model weights at each timestep k based on the current discrepancy between prediction and measurement, the ensemble filter continuously identifies the most probable models, enhancing estimation accuracy. The section is organized as follows. Section 4.1 provides the formulation of the filtering process for each model in the ensemble and gives an overview of how the ensemble process works. Sections 4.2 and 4.3 detail the prediction and measurement update steps, including the choice of noise and measurement covariance matrices $\mathbf{Q}_f$ and $\mathbf{R}_f$. Section 4.4 describes the likelihood-based model selection, which assigns weights $w_f[k]$ to each model at every timestep. Section 4.5 introduces the weighted model combination process, which integrates information from multiple models based on the computed model weights $w_f[k]$.

4.1 Mathematical Foundation of Ensemble Kalman Filtering

A Kalman filter is a recursive estimation algorithm that reconstructs state variables of dynamical systems from noisy measurements. In this study, the classical Kalman filter framework is extended to an ensemble-based approach to improve the estimation of the thermal state in reduced-order models. Figure 4 gives an overview of the ensemble Kalman-filtering approach: Instead of relying on a single state-space representation, an ensemble of reduced-order models is employed, each corresponding to a unique set of boundary conditions sampled according to the method described in Sect. 3.2. At each time step k, multiple Kalman filters run in parallel, each using a different reduced-order model f. The system dynamics for each model in the ensemble is represented in discrete-time state-space form:

$$\begin{aligned}
\mathbf{x}_f[k] &= \mathbf{A}_f \mathbf{x}_f[k-1] + \mathbf{B}_f \mathbf{u}[k-1] + \mathbf{v}_f[k-1] \\
\mathbf{z}_{\text{obs}}[k] &= \mathbf{C}_{\text{obs},f} \mathbf{x}_f[k] + \mathbf{w}_f[k]
\end{aligned} \tag{9}$$

The subscript d indicating discretized quantities is omitted in the notation of Eq. 9 but it represents the discretized thermal system outlined in Sect. 3.3, Eq. 5. $\mathbf{x}_f[k]$ represents the thermal state vector for model f at time k and $\mathbf{A}_f$ is the state transition matrix of model f. $\mathbf{C}_{\text{obs},f}$ is the observation matrix for model f and $\mathbf{z}_{\text{obs}}[k]$ denotes the temperature measurements of the observed sensors at time step k. The process noise $\mathbf{v}_f[k] \sim \mathcal{N}(0, \mathbf{Q}_f)$ and measurement noise $\mathbf{w}_f[k] \sim \mathcal{N}(0, \mathbf{R}_f)$ are assumed

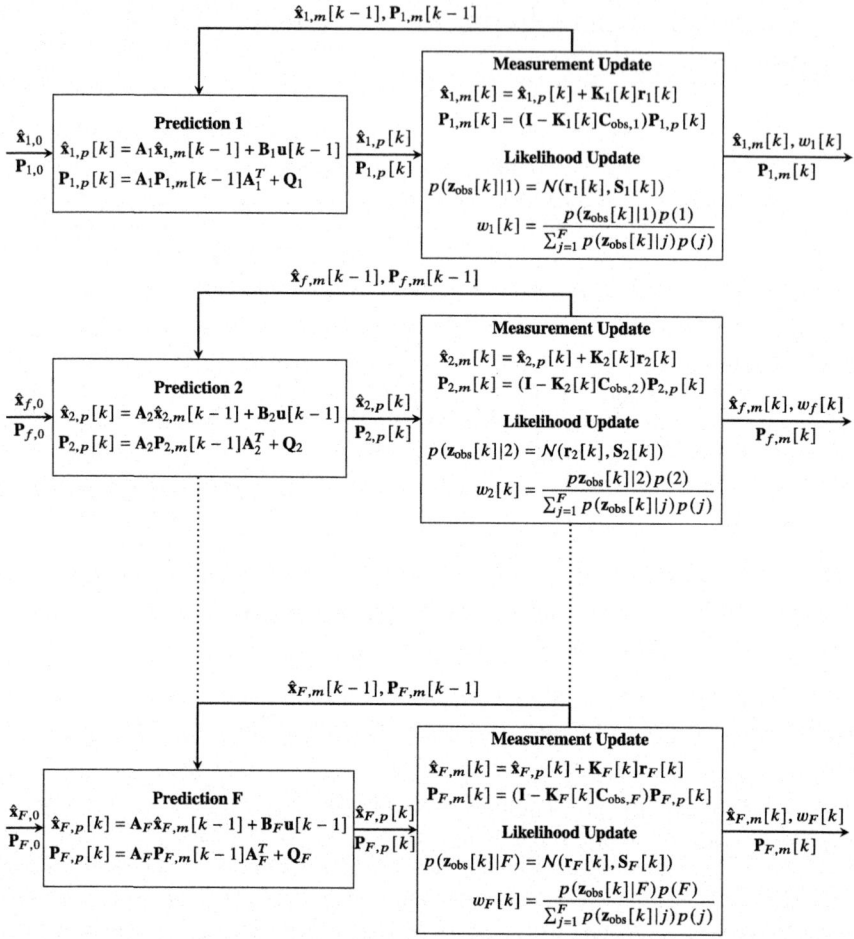

Fig. 4 Overview of the ensemble Kalman filtering approach. Each model in the ensemble is used in a separate Kalman filter, which consists of a prediction and a measurement update step. To decide which model best explains the current measurement data, a weight $w_f[k]$ is assigned to each model f at each time step k

to be Gaussian. Note the difference between Eqs. 9 and 5, where $\mathbf{C}_d$ is replaced by $\mathbf{C}_{\text{obs},f}$ and $\mathbf{z}[k]$ by $\mathbf{z}_{\text{obs}}[k]$. This is due to the fact that in the Kalman filter only the sensor subset M_{obs} is used. $\mathbf{C}_{\text{obs},f}$ therefore maps the thermal state to the observed sensors. Further note that $\mathbf{z}_{\text{obs}}[k]$ and $\mathbf{u}[k-1]$ do not change over the models in the ensemble as the measurements and the applied input stays the same.

The ensemble-based Kalman filter process operates as follows. At each timestep k, all F reduced-order models evolve their state estimates using their respective state-space matrices $\mathbf{A}_f$. Each model predicts a temperature based on its state estimate and observation matrix $\mathbf{C}_{\text{obs},f}$. The temperature measurements $\mathbf{z}_{\text{obs}}[k]$ are compared to each model's predicted temperature and a residual $\mathbf{r}_f[k]$ is computed. Based on this residual, a likelihood is computed for each model, assigning a weight $w_f[k]$ that represents how well each model explains the data. To apply the Kalman filter methodology to the system model f described in Eq. 9, each model follows the standard two-step Kalman update procedure: the a priori prediction step, which estimates the next state based on the system model, and the a posteriori measurement update step, which corrects the state estimate based on observations [1]. The notation $\hat{\mathbf{x}}$ is used to indicate the expected or predicted value of a variable, whereas P denotes its corresponding covariance matrix.

4.2 Prediction Step

The prediction step estimates the system's evolution by incorporating prior knowledge of the dynamics while accounting for process noise. The estimated mean state and covariance matrix before incorporating new measurements are given by:

$$
\begin{aligned}
\hat{\mathbf{x}}_{f,p}[k] &= \mathbf{A}_f \hat{\mathbf{x}}_{f,m}[k-1] + \mathbf{B}_f \mathbf{u}[k-1] \\
\mathbf{P}_{f,p}[k] &= \mathbf{A}_f \mathbf{P}_{f,m}[k-1] \mathbf{A}_f^T + \mathbf{Q}_f
\end{aligned}
\tag{10}
$$

In a probabilistic sense, the prediction step updates the prior probability distribution of the state vector. The predicted mean $\hat{\mathbf{x}}_{f,p}[k]$ represents the expected state given all past measurements. The predicted covariance $\mathbf{P}_{f,p}[k]$ quantifies the uncertainty in this estimate. The process noise covariance matrix $\mathbf{Q}_f$ is assumed to be diagonal and the same for each model f.

4.3 Measurement Update Step

The measurement update step refines the predicted state using new sensor data. The Kalman gain $\mathbf{K}_f[k]$ is computed to optimally blend the predicted estimate with the measurement:

$$
\begin{aligned}
\mathbf{r}_f[k] &= \mathbf{z}_{\text{obs}}[k] - \mathbf{C}_{\text{obs},f} \hat{\mathbf{x}}_{f,p}[k] \\
\mathbf{S}_f[k] &= \mathbf{C}_{\text{obs},f} \mathbf{P}_{f,p}[k] \mathbf{C}_{\text{obs},f}^T + \mathbf{R}_f \\
\mathbf{K}_f[k] &= \mathbf{P}_{f,p}[k] \mathbf{C}_{\text{obs},f}^T \mathbf{S}_f^{-1}[k] \\
\hat{\mathbf{x}}_{f,m}[k] &= \hat{\mathbf{x}}_{f,p}[k] + \mathbf{K}_f[k] \mathbf{r}_f[k] \\
\mathbf{P}_{f,m}[k] &= (\mathbf{I} - \mathbf{K}_f[k] \mathbf{C}_{\text{obs},f}) \mathbf{P}_{f,p}[k]
\end{aligned}
\tag{11}
$$

In this step, $\mathbf{r}_f[k]$ represents the innovation or measurement residual, which quantifies the discrepancy between the predicted measurement and the actual observation. The residual covariance $\mathbf{S}_f[k]$ determines the uncertainty in this residual. The

Kalman gain $\mathbf{K}_f[k]$ then adjusts the state estimate based on the reliability of the measurements. The measurement covariance matrix $\mathbf{R}_f$ is assumed to be diagonal and the same for each model f. From a probabilistic perspective, the measurement update step applies Bayes' theorem, updating the prior probability distribution with the new measurement to obtain the posterior probability distribution. The updated state $\hat{\mathbf{x}}_{f,m}[k]$ represents the maximum a posteriori estimate, and the updated covariance $\mathbf{P}_{f,m}[k]$ reflects the reduced uncertainty after incorporating measurement information. By iterating through the prediction and measurement update steps, the Kalman filter continuously refines the state estimate, improving accuracy as more measurements become available.

4.4 Model Weighting Via Likelihood Estimation

At each timestep k, the goal is to determine which model in the ensemble best explains the measurement data. To do so, the likelihood function $L(f|\mathbf{z}_{\text{obs}}[k])$ is defined, which represents the relative plausibility of model f given the temperature measurement $\mathbf{z}_{\text{obs}}[k]$. The likelihood is proportional to the probability density function (PDF) of the measurement given the model. Using Bayes' rule, it follows:

$$L(f|\mathbf{z}_{\text{obs}}[k]) \propto p(\mathbf{z}_{\text{obs}}[k]|f)p(f) \tag{12}$$

Here, $p(\mathbf{z}_{\text{obs}}[k]|f)$ is the PDF of the measurement $\mathbf{z}_{\text{obs}}[k]$ given model f, and $p(f)$ is the prior probability of selecting model f. This is assumed to be uniform over all models, i.e. $p(f) = \frac{1}{F}$ as all models are equally likely to be selected. The PDF $p(\mathbf{z}_{\text{obs}}[k]|f)$ arises naturally in the measurement update step of the Kalman filter, as it is computed based on the residual $\mathbf{r}_f[k]$ and the associated residual covariance $\mathbf{S}_f[k]$. Assuming Gaussian measurement noise, $p(\mathbf{z}_{\text{obs}}[k]|f)$ follows a multivariate normal distribution:

$$p(\mathbf{z}_{\text{obs}}[k]|f) = \frac{1}{(2\pi)^{m/2}|\mathbf{S}_f[k]|^{1/2}} \exp\left(-\frac{1}{2}\mathbf{r}_f[k]^T\mathbf{S}_f[k]^{-1}\mathbf{r}_f[k]\right) \tag{13}$$

To obtain model weights that can be interpreted as proper probabilities, likelihoods across all models are normalized at each time step:

$$w_f[k] = \frac{p(\mathbf{z}_{\text{obs}}[k]|f)p(f)}{\sum_{f=1}^{F} p(\mathbf{z}_{\text{obs}}[k]|f)p(f)}. \tag{14}$$

The interpretation of $w_f[k]$ is as follows: At each timestep k, a weight factor is computed for each model f, which represents its posterior probability, given the measurement data. The weights are relative probabilities, meaning that a model with $w_f[k] = 0.5$ is twice as likely as a model with $w_f[k] = 0.25$. However, it does not mean that model f explains 50% of the measurement data $\mathbf{z}_{\text{obs}}[k]$. Rather, it indicates that this model is statistically more plausible compared to other models in the ensemble.

4.5 Weighted Model Combination

Once the model weights $w_f[k]$ are computed, they are used to obtain an ensemble-based estimate for both temperature and thermal error. Since each state-space model f corresponds to a reduced order model with a different basis, direct combination in state space is not meaningful. The state vectors $\mathbf{x}_f[k]$ belong to different reduced spaces, making a direct weighted sum in the reduced-order domain mathematically inconsistent. Instead, the combination is performed in measurement space, where all models provide predictions in terms of the same physical sensor outputs. Each model maps its estimated state to measurement space using its corresponding validation matrix $\mathbf{C}_{\text{val}, f}$, yielding a predicted measurement for the temperature of the validation sensors $\hat{\mathbf{z}}_{\text{val}}[k]$ and displacement $\hat{\mathbf{y}}_f[k]$:

$$
\begin{aligned}
\hat{\mathbf{z}}_{\text{val}, f}[k] &= \mathbf{C}_{\text{val}, f}\hat{\mathbf{x}}_f[k] \\
\hat{\mathbf{y}}_f[k] &= \mathbf{C}_{\text{mech}, f}\mathbf{K}_f^{-1}\mathbf{K}_{\text{th}, f}\hat{\mathbf{x}}_f[k]
\end{aligned}
\tag{15}
$$

Different methods can be used to compute the final ensemble predictions $\hat{\mathbf{z}}_{\text{val, ens}}[k]$ (temperature) and $\hat{\mathbf{y}}_{\text{ens}}[k]$ (displacement):

1. Model combination using overall top weights:
    In this approach, ensembling is performed over the N models that have the highest time-averaged weights across the entire measurement duration. This stabilizes the ensembling process by favoring models that consistently explain the data well over time, rather than reacting to momentary fluctuations. Mathematically, this is represented as a weighted sum:

    $$
    \begin{aligned}
    \hat{\mathbf{z}}_{\text{val, ens}}[k] &= \sum_{f \in \mathcal{F}_N} \tilde{w}_f[k]\hat{\mathbf{z}}_{\text{val}, f}[k] \\
    \hat{\mathbf{y}}_{\text{ens}}[k] &= \sum_{f \in \mathcal{F}_N} \tilde{w}_f[k]\hat{\mathbf{y}}_f[k]
    \end{aligned}
    \tag{16}
    $$

    $\mathcal{F}_N$ is the set of the top N models with the highest time-averaged weights. The weights $\tilde{w}_f[k]$ must be re-normalized at each timestep to ensure that they sum to one:

    $$
    \tilde{w}_f[k] = \frac{w_f[k]}{\sum_{j \in \mathcal{F}_N} w_j[k]}
    \tag{17}
    $$

    The extreme cases $\mathcal{F}_1$ and $\mathcal{F}_F$ correspond to selecting only the best model and computing a weighted average over all models, respectively.

2. Model combination using timestep-based top weights:
   In this approach, ensembling is performed over the N models with the highest weights at each individual timestep k. This method enables rapid adaptation to changing boundary conditions by dynamically selecting the most probable models at each timestep. However, it may introduce more frequent switching between the models which can cause a jumping behavior in the temperature and deformation prediction. Mathematically, this is again computed as a weighted sum. However, this time, the set of top models $\mathcal{F}_F[k]$ changes with each timestep k:

   $$\hat{\mathbf{z}}_{\text{val, ens}}[k] = \sum_{f \in \mathcal{F}_N[k]} \tilde{w}_f[k]\hat{\mathbf{z}}_{\text{val}, f}[k],$$
   $$\mathbf{y}_{\text{ens}}[k] = \sum_{f \in \mathcal{F}_N[k]} \tilde{w}_f[k]\hat{\mathbf{y}}_f[k]. \tag{18}$$

   Also here, the weights need to be re-normalized at each timestep such that they sum up to one:

   $$\tilde{w}_f[k] = \frac{w_f[k]}{\sum_{j \in \mathcal{F}_N[k]} w_j[k]}. \tag{19}$$

   The extreme cases $\mathcal{F}_1[k]$ and $\mathcal{F}_F[k]$ correspond to selecting only the best model at each timestep k and performing a weighted average over all models.
3. Model combination using overall top models with equal weights (unweighted):
   In this approach, the ensemble prediction is computed using the N models with the highest time-averaged weights, similar to approach 1. However, instead of applying the time-varying weights $w_f[k]$, each of the selected models is given equal importance in the ensemble. That is, the contribution of each model is uniformly weighted at each timestep. Let $\mathcal{F}_N$ denote the set of the N models with the highest time-averaged weights over the full time horizon. Then, the temperature and deformation predictions are computed as:

   $$\hat{\mathbf{z}}_{\text{val, ens}}[k] = \frac{1}{N} \sum_{f \in \mathcal{F}_N} \hat{\mathbf{z}}_{\text{val}, f}[k]$$
   $$\hat{\mathbf{y}}_{\text{ens}}[k] = \frac{1}{N} \sum_{f \in \mathcal{F}_N} \hat{\mathbf{y}}_f[k] \tag{20}$$

   This method ignores the computed model weights $w_f[k]$. Instead, it assumes that the top N models, based on their overall performance, are equally informative. For $N = 1$, this approach yields the same prediction as approach 1. When $N \to F$ (i.e., using all models in the ensemble), this unweighted averaging gives equal influence to poorly performing models, which should significantly degrade prediction accuracy. Hence, this approach serves as a naive baseline, helping to evaluate the effectiveness of the dynamic weighting strategies in Approaches 1 and 2.

5   Results

To evaluate the performance of the proposed ensemble Kalman filter, the method is applied to the 70-hour measurement dataset introduced in Sect. 2. The evaluation is structured as follows. First, the computed model weights $w_f[k]$ are analyzed over time. Then, the different model combination strategies introduced in Sect. 4.5 are compared in terms of prediction accuracy and number of models used. Finally, the estimated and measured temperatures at the validation sensor locations, as well as the corresponding thermal errors, are assessed.

The ensemble evaluation involves 500 reduced-order Kalman filter models over a duration of 70 h (25'200 timesteps with $\Delta t = 10$ s), which requires approximately 1.6 hours on a Intel Core i7-13700K CPU. This corresponds to a speedup factor of $\sim 47\times$ compared to real-time, demonstrating that even the complete ensemble filtering process remains highly computationally efficient and therefore suitable for real-time application with online sensor measurements even for a large number of ensemble models.

5.1   Analysis of Weights

Figure 5 shows the evolution of the weights $w_f[k]$ for the top ten most probable models (based on time-averaged weight) alongside the mean temperature of all sensors. The mean temperature solely serves as a reference for the start of a loadcase and is not directly correlated to the weight selection.

The weights vary strongly across the four thermal load cases, demonstrating the ensemble Kalman filter's ability to dynamically select models that best explain the observed thermal behavior. During stationary (non-heating) phases, all model weights converge toward a uniform distribution, consistent with the physical intuition that no thermal loads are present and it is in a steady state equilibrium with the environment. In these phases, all models are equally likely with a baseline prior of $p(f) = \frac{1}{500} = 0.2\%$, and the likelihood does not distinguish between them. When heating is activated, the model probabilities shift significantly, indicating that the measurement data contain enough information to favor certain models. Notably, the first two heating cycles result in very similar weight evolution, where the same set of models dominate the ensemble. This can easily be explained by the fact that it is a mirrored heat load along the symmetry plane of the system. Therefore all system models (which inherently are defined symmetrically) should perform equally for both loads as only minor sensor locations are not perfectly symmetrical. In contrast, heating cycles three and four show distinct weight profiles. Cycle three activates the top heating pad, targeting a different structural region, while cycle four activates all pads simultaneously. The resulting thermal states differ significantly from cycles one and two, and the ensemble correspondingly shifts its model preference. It can be noted that especially for these, the preference does not only follow the activation

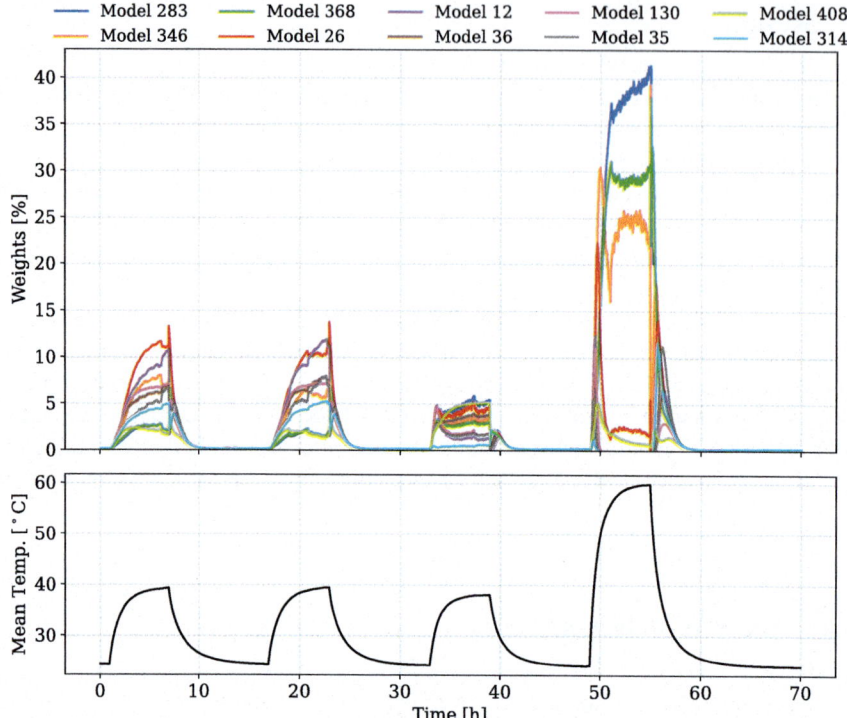

Fig. 5 Time evolution of the top 10 model weights and the mean sensor temperature. The first two thermal loads of the front and back heat pad produce similar weight patterns as the heat input is mirrored along the symmetry plane. The third loadcase activates the top heat pad and in the fourth loadcase all previous heating pads are active which leads to different model preferences. During stationary phases without any heat input, all weights converge to the same value of 0.2%

profile, but also the temperature profile as the HTC's are functions of temperature. For example, Model 130 shows initially a higher weight which then decreases as the overall temperature rises and temperature gradients decrease, similar to the bending behavior of the thermomechanical system.

To gain insight into which physical parameters influence the model selection, Table 3 lists the sampled parameters of the top 10 models. Among all parameters, the heat transfer coefficient between the structure and the working space (WS HTC) appears to have the strongest influence and all top models have WS HTC values near the lower bound of the sampled range. This suggests that models with lower heat losses to the environment are favored and might be due to the fact that the convection temperature is assumed to be constant at 25 °C, but actually increases with sustained heat input.

Table 3 Sampling parameters of the top 10 models based on time-averaged weight

Model	CI HTC [W/(m² K)]	BI HTC [W/(m² K)]	SI HTC [W/(m² K)]	WS HTC [W/(m² K)]	Baseplate [W/(m² K)]	TCC (W/K)
283	6.850	1.342	3.940	1.348	1.120	533.395
346	5.306	1.773	3.459	1.135	6.343	948.419
368	5.450	1.415	1.714	1.768	5.266	960.560
26	4.034	1.308	4.426	1.448	6.413	626.308
12	2.366	1.372	4.822	1.238	8.790	336.348
36	2.775	2.251	6.237	1.276	5.023	199.145
130	2.192	1.053	4.073	1.559	12.944	873.100
35	1.206	1.799	2.506	1.674	6.928	566.616
408	4.945	2.640	6.508	1.152	8.132	641.223
314	2.288	4.620	1.952	1.434	7.056	324.625

Models with low WS HTC values are consistently preferred, indicating that models with reduced heat loss to the working space better match the measurements

5.2 Analysis of Ensembling Strategies

To analyze the results of the ensembling strategies, the methods outlined in Sect. 4.5 are used:

1. Combining N models with the overall top weights (time-averaged)
2. Combining N models with the highest weights at each timestep
3. Combining N models with the overall top weights but using equal weights (unweighted)

Each combination strategy yields a time-dependent ensemble prediction for the temperature $\hat{z}_{ens}[k]$ at the validation sensor locations and for the thermal errors $\hat{y}_{ens}[k]$. In the configuration considered here, 10 temperature sensors are used for state estimation, and the remaining 30 sensors are used for validation. Hence, $\hat{z}_{ens}[k]$ is a vector of length 30, containing the predicted temperatures at the validation sensor location. The thermal deformation is characterized by five components (X-up, X-down, Y-up, Y-down, and Z), resulting in a corresponding 5-dimensional vector $\hat{y}_{ens}[k]$.

Two standard metrics are used to quantify the accuracy of the temperature and deformation predictions: the Root Mean Square Error (RMSE) and the Peak-to-Peak (P2P) error. The RMSE for a signal $a[k]$ relative to a reference $b[k]$ over T timesteps is defined as:

$$\text{RMSE} = \sqrt{\frac{1}{T}\sum_{k=1}^{T}(a[k] - b[k])^2}. \tag{21}$$

The Peak-to-Peak (P2P) error is the range of the signal:

$$\text{P2P} = \max_k(a[k] - b[k]) - \min_k(a[k] - b[k]). \tag{22}$$

The predicted temperature $\hat{z}_{val, ens,i}[k]$ is compared to the measured value $z_{val,i}[k]$ for each validation sensor i. The RMSE is computed per sensor, and the total RMSE over all validation sensors is then defined as:

$$\text{RMSE}_{\text{tot}} = \sqrt{\frac{1}{30} \sum_{i=1}^{30} \text{RMSE}_{\text{temp},i}^2}, \tag{23}$$

providing a single aggregated metric of temperature prediction accuracy.
For each of the five deformation components, the predicted thermal error $\hat{y}_{ens,i}[k]$ is compared to the measured error $y_i[k]$, and the RMSE and P2P values are computed. To evaluate the benefit of the prediction, the relative error reduction is calculated:

$$\text{RED}_i = 100 \cdot \left(1 - \frac{\text{Error}_{\text{res},i}}{\text{Error}_{\text{meas},i}}\right) \tag{24}$$

where $\text{Error}_{\text{res},i}$ is computed from the residuals between predicted and measured thermal error. This can be both applied to individual displacement sensors or the volumetric error and both the RMSE and P2P metric. A reduction of 100% would correspond to a residual of zero, so the prediction matching the measured displacement at every timestep.

To obtain a single characteristic value for the overall deformation error, a volumetric error metric is introduced. It combines the five thermal deformation components into a scalar magnitude per timestep using the following definition:

$$e[k] = \sqrt{\left(\frac{y_1[k] + y_2[k]}{2}\right)^2 + \left(\frac{y_3[k] + y_4[k]}{2}\right)^2 + y_5[k]^2}. \tag{25}$$

Therefore the average euclidean displacement of the midpoint of the tool is considered as the volumetric error. This is applied both to the measured deformation $y_i[k]$ and to the residual $y_i[k] - \hat{y}_{ens,i}[k]$ to compute RMSE and P2P metrics for the volumetric deformation. The corresponding reductions follow the same form as Equation (24) but applied to the volumetric error.

To assess the impact of the different model combination strategies introduced in Sect. 4.5 on temperature and thermal error prediction, the aggregated quantities total temperature RMSE and volumetric RMSE / P2P reduction are used. Figure 6 shows the total RMSE between predicted and measured temperatures at the validation sensors as a function of the number of models N used in the ensemble. The timestep-based combination method (2) consistently achieves the lowest RMSE across all values of N. particularly for small ensembles. A minimum in total RMSE is observed for all three strategies at approximately $N = 15$, indicating that combining the top 15 models yields the best temperature prediction performance. For small ensemble sizes $N < 20$, both the weighted (1) and unweighted (3) approaches yield nearly identical performance. However, as N increases, the RMSE of the unweighted strategy

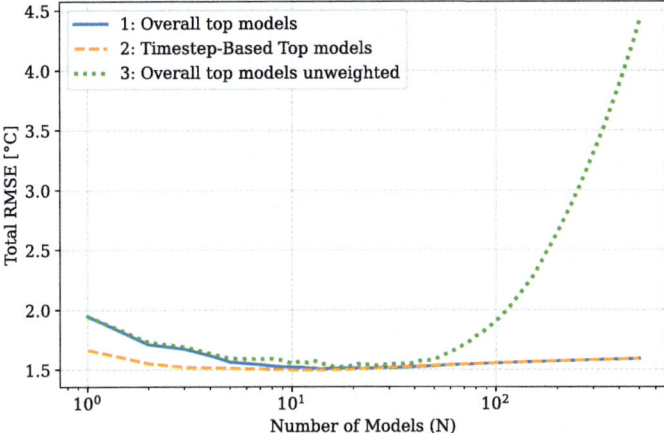

Fig. 6 Total RMSE between predicted and measured validation temperatures as a function of the number of models N in the ensemble, shown for the three model combination strategies. The RMSE is minimized around $N = 15$. The unweighted strategy degrades significantly as more models are included, while the weighted approaches remain stable

increases sharply, while the weighted strategies maintain low error. This demonstrates the effectiveness of the weight-based filtering in suppressing poorly performing models: even when a large number of models is included, their influence remains minimal if they are assigned appropriately low weights.

Figure 7 shows the volumetric RMSE and P2P reduction as a function of N. Similar to the temperature results, all methods achieve maximum error reduction around $N = 15$. This suggests that the model weights obtained from thermal estimation are also informative for mechanical predictions. Interestingly, unlike the temperature case, the unweighted model combination (3) outperforms the weighted methods for small to moderate values of N, achieving up to 10 percentage point higher reduction in both volumetric RMSE and P2P metrics. This indicates that while the relative ranking of the models (i.e. their weight ordering) is meaningful for mechanical accuracy, the absolute values of the weights are less aligned with mechanical performance. This is consistent with the fact that the model weights are computed based solely on thermal quantities and not on mechanical error. However, for values of $N \gtrsim 200$, the unweighted approach begins to underperform compared to the weighted methods. This again highlights the advantage of weighting strategies in reducing the influence of poorly performing models, even for a quantity (mechanical deformation) that was not directly used to compute the weights.

Overall, the results demonstrate that ensembling significantly improves both temperature and thermal error prediction as opposed to using one single model. In particular, using around 15 models in the ensemble provides a good trade-off between accuracy and robustness. It can be expected, that the optimal number of model increases with the complexity of the load cycles and the complexity of the machine, as more

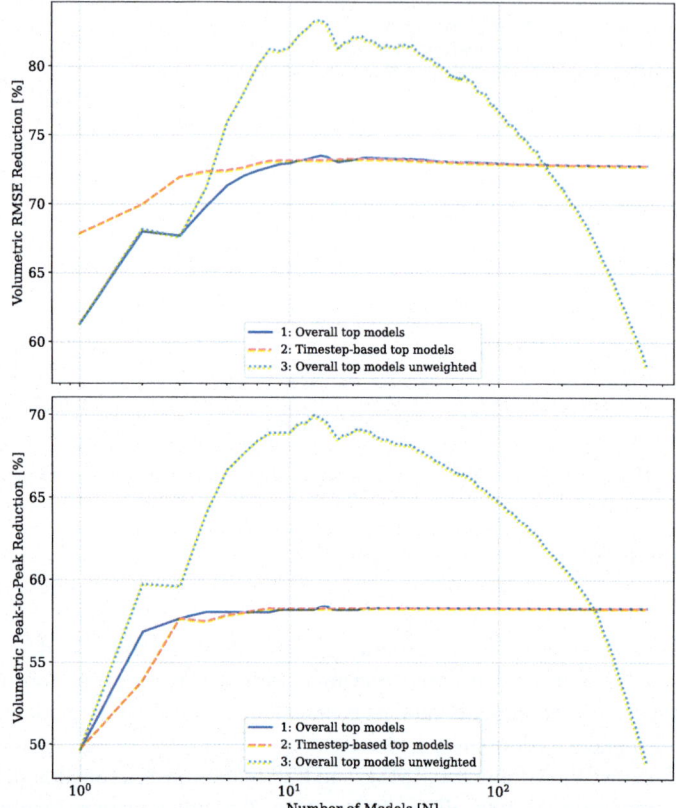

Fig. 7 Volumetric RMSE and P2P reduction as a function of the number of models N used in the ensemble. While all approaches achieve maximum reduction around $N = 15$, the unweighted strategy outperforms the weighted ones for small to moderate N, but declines at large N

models capture different aspects of the thermal behavior. Weight-based model combination effectively suppresses the influence of poor models, maintaining performance even as the ensemble size increases.

5.3   Compensation Results of the Machine

According to Fig. 6, the ensemble matches the measurement validation data the best for 15 models, when the overall top modeling strategy (1) is used. Figure 8 visualizes the performance of the ensemble with that configuration in more detail, by displaying the 3 sensors with the highest and lowest RMSE, respectively.

The ensemble matches the measurement at locations 14, 28 and 7 and is quite far off at locations 8, 37 and 15 (in those orders). The mismatch at location 15 can

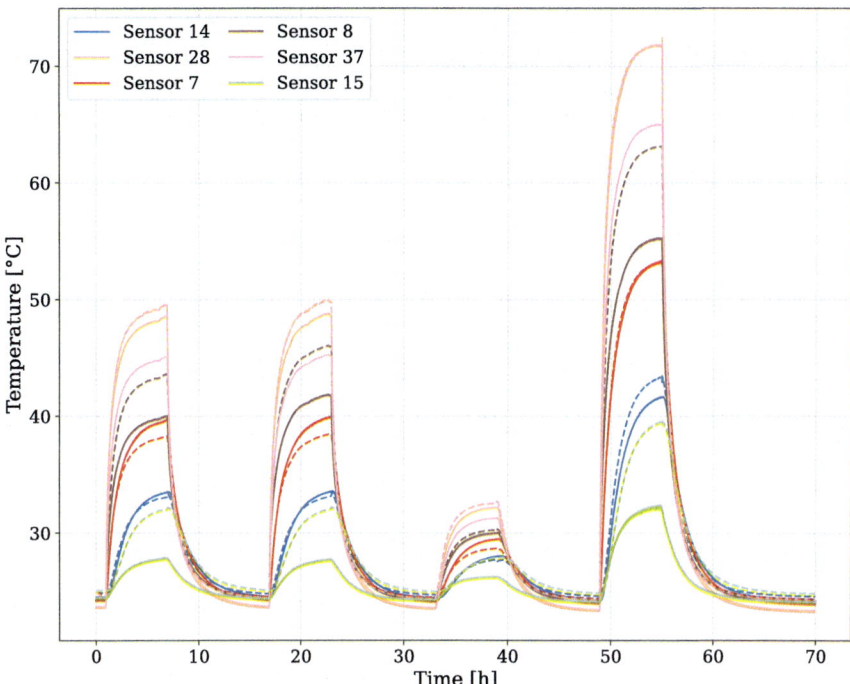

Fig. 8 Visualized are three validation sensors respectively, for which the prediction of the ensemble with the overall top 15 models matches the measurement setup the best and the worst. The thick line corresponds to the measured value and the dashed line to the predicted value of the corresponding sensor

be attributed to a strong modeling simplification. The plate on which that sensor is mounted in the experiment has only loose contact with the displacement sensor holder, while in the simulation, it is assumed, that the entire displacement sensor holder is a single piece of steel and thus the thermal connection is far greater. The overestimation in sensors 8 and 37, which are both located close to the front and back heating pads are likely due to an overestimation of the heat flux.

Ultimately, however, the goal is not to predict the temperatures, but to predict the displacement, which is caused by the thermally invariant behavior. To that end, the ensemble ranking and weighing of the thermal behavior is applied to the mechanical deformation at the TCP in the same fashion. Figure 7 also indicates an optimum at 15 models, although here using the overall ranked top models with equal weighting promises better performance than when the weighting is considered. Figure 9 visualizes the predictive performance of the ensemble to the thermally induced displacement. The ensemble captures the dynamic behavior of the thermally induced displacement very well, going so far as to capture minute shifts of low magnitude. This is very difficult to replicate when not using live sensor data, as would be the case in conventional modeling approaches. Furthermore, the ensemble also correctly

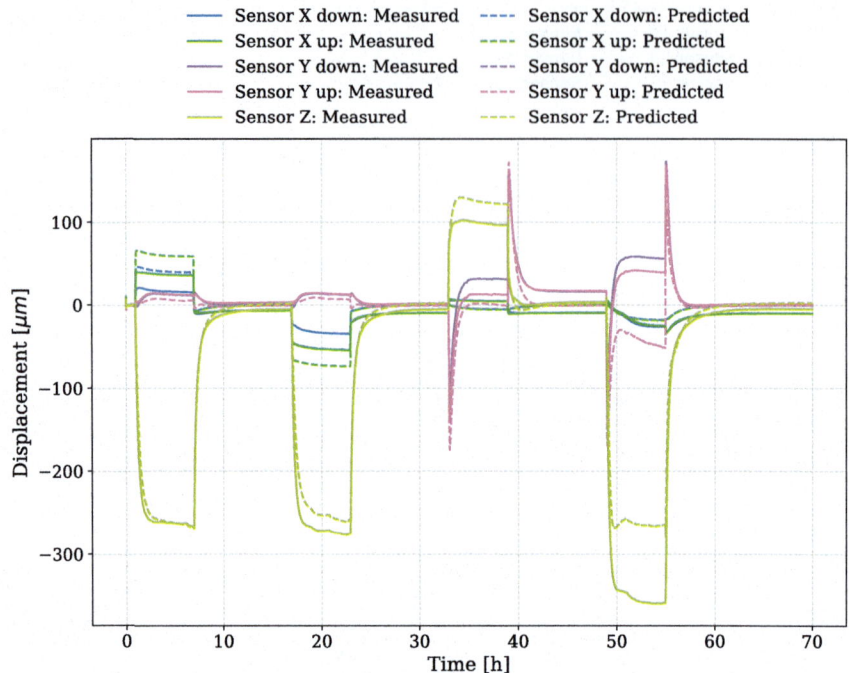

Fig. 9 Displacement of the machine at the TCP positions described in Sect. 2, using the ensemble with the overall top 15 models

estimates the magnitudes of the displacements in many cases, reducing RMSE's by more than 70%.

The ensemble never quite captures the magnitude of the X-axes displacements of the measurement setup. However, this might not actually be an issue of the models or the ensemble, but rather due to a lack of repeatability in the measurement setup, as especially the upper X displacements do not return to zero in the steady-state phases indicating a shift in one of the connecting joints.

6   Conclusion and Outlook

This paper proposes an approach which unites physics-based ROMs with on-machine measurements to achieve real-time thermal error compensation in machine tools. The efficacy of the Kalman filter in estimating thermal states is enhanced by utilizing an ensemble of models. The ensemble approach assigns weights based on the likelihood estimated for each model given the measurement data. This leads to a change of model preference depending on the active load case and temperature distribution which in turn affects the physical system as for example HTCs are generally a function of temperature which is prohibitively hard to model a priori as boundary conditions. The method ensures robust and precise reconstruction of thermal states from a limited number of temperature sensors and handles inherent uncertainties in boundary conditions and system parameters well. Through the use of ROMs the computational time was ~ 47 times faster than the measurement allowing for real-time physics based thermal error compensation.

Validation on a real machine demonstrated significant accuracy in predicting thermal states and associated thermomechanical deformations significantly outperforming the state of the art for physics based models. A RMSE of 1.5 °C was achieved for the best models compared to 2.7 °C reported by state of the art models. This translates into a reduction of the volumetric thermal error between ~ 61 to $\sim 82\%$. The ensemble sampling also improves this significantly as even the model with the best fitting boundary conditions yields a $\sim 8\%$ lower volumetric RMSE reduction. Therefore, an ensemble sampling approach combined with measured temperatures in the Kalman filter promises an avenue for a digital twin with real-time thermal compensation capabilities that are not adversely affected by uncertainty in the boundary conditions of the physics-based simulation models.

Looking forward, several avenues merit exploration to further improve and generalize the proposed methodology. Firstly, a feedback process between model selection and model generation (boundary condition sampling) could be established. Moving the sampling space in real time depending on the Kalman filter. Furthermore, sensor selection strategies could be expanded to consider the thermomechanical system as well. Adaptively estimating heat input and process noise in real-time could also improve prediction accuracy and model weights. Finally, exploring the scalability of this ensemble approach across a wider range of machine tool configurations and industrial applications will help realize the full potential of digital twin technologies in achieving sustainable and precision-driven manufacturing processes through thermal error compensation.

Acknowledgements This work was co-financed by Innosuisse—Swiss Innovation Agency, Project no. 27835.1 IP-ENG. The authors wish to thank Prof. Stephen Duncan, Sofia Talleri and Dr. Nico Zimmermann for the helpful discussions and their insights as well as the MORe team for support and Dr. Josef Mayr for proofreading.

CRediT Authorship Contribution Statement Sebastian Lang: Conceptualization, Data curation, Investigation, Methodology, Software, Writing—original draft. Marco Schneider: Conceptualization, Investigation, Methodology, Software, Writing—original draft. Lenny Rhiner: Conceptualization, Software, Writing—review and editing. Markus Bambach: Supervision, Writing—review and editing.

Declaration of Competing Interest The authors declare that they have no known competing financial interests or personal relationships that could have appeared to influence the work reported in this paper.

References

1. Alsadik, B.: Kalman filter. Adjustment Models in 3D Geomatics and Computational Geophysics, pp. 299–326 (2019). https://doi.org/10.1016/b978-0-12-817588-0.00010-6
2. Auger, F., Hilairet, M., Guerrero, J.M., Monmasson, E., Orlowska-Kowalska, T., Katsura, S.: Industrial applications of the Kalman filter: a review. IEEE Trans. Ind. Electron. **60**(12), 5458–5471 (2013). https://doi.org/10.1109/TIE.2012.2236994
3. Blaser, P., Pavliček, F., Mori, K., Mayr, J., Weikert, S., Wegener, K.: Adaptive learning control for thermal error compensation of 5-axis machine tools. J. Manuf. Syst. **44**, 302–309 (2017). https://doi.org/10.1016/j.jmsy.2017.04.011
4. Damblin, G., Couplet, M., Iooss, B.: Numerical studies of space-filling designs: optimization of Latin hypercube samples and subprojection properties. J. Simul. **7**(4), 276–289 (2013). https://doi.org/10.1057/jos.2013.16
5. Denkena, B., Bergmann, B., Klemme, H.: Cooling of motor spindles—A review. Int. J. Adv. Manuf. Technol. **110**(11–12), 3273–3294 (2020). https://doi.org/10.1007/s00170-020-06069-0
6. Franklin, G.F., Powell, J.D., Workman, M.L.: Digital Control of Dynamic Systems, vol. 3. Addison-Wesley, Menlo Park (1998). ISBN: 0201820544
7. Gao, W., Ibaraki, S., Donmez, M.A., Kono, D., Mayer, J.R., Chen, Y.L., Szipka, K., Archenti, A., Linares, J.M., Suzuki, N.: Machine tool calibration: measurement, modeling, and compensation of machine tool errors. Int. J. Mach. Tools Manuf. **187**(December 2022) (2023). https://doi.org/10.1016/j.ijmachtools.2023.104017
8. Gomez-Acedo, E., Olarra, A., Orive, J., Lopez De La Calle, L.N.: Methodology for the design of a thermal distortion compensation for large machine tools based in state-space representation with Kalman filter. Int. J. Mach. Tools Manuf. **75**, 100–108 (2013). https://doi.org/10.1016/j.ijmachtools.2013.09.005
9. Hellmich, A., Glänzel, J., Pierer, A.: Analyzing and Optimizing the Fluidic Tempering of Machine Tool Frames. In: Proceedings of Conference on Thermal Issues in Machine Tools, Dresden, June 2018
10. Hernández Becerro, P.: Efficient thermal error models of machine tools. Ph.D. thesis, ETH Zurich (2020). https://doi.org/10.3929/ethz-b-000449279
11. Hernández-Becerro, P., Purtschert, J., Konvicka, J., Buesser, C., Schranz, D., Mayr, J., Wegener, K.: Reduced-order model of the environmental variation error of a precision five-axis machine tool. J. Manuf. Sci. Eng. Trans. ASME **143**(2) (2021). https://doi.org/10.1115/1.4047739
12. Ihlenfeldt, S., Schroeder, S., Penter, L., Hellmich, A., Kauschinger, B.: Adjustment of uncertain model parameters to improve the prediction of the thermal behavior of machine tools. CIRP Ann. **69**(1), 329–332 (2020). https://doi.org/10.1016/j.cirp.2020.04.056
13. Irino, N., Kobayashi, A., Shinba, Y., Kawai, K., Spescha, D., Wegener, K.: Digital twin based accuracy compensation. CIRP Ann. **72**(1), 345–348 (2023). https://doi.org/10.1016/j.cirp.2023.04.088
14. Kizaki, T., Tsujimura, S., Marukawa, Y., Morimoto, S., Kobayashi, H.: Robust and accurate prediction of thermal error of machining centers under operations with cutting fluid supply. CIRP Ann. **70**(1), 1–4 (2021). https://doi.org/10.1016/j.cirp.2021.04.074
15. Lang, S., Lampert, N., Mayr, J., Wegener, K., Bambach, M.: Training efficient and compensating fast : Data augmentation for thermal error compensation models of machine tools . In: EUSPEN Special Interest Group Meeting on Thermal Issues, March. EUSPEN, Eindhoven (2024). https://doi.org/10.3929/ethz-b-000669528
16. Lang, S., Talleri, S., Mayr, J., Wegener, K., Bambach, M.: Kalman filter-driven state observer for thermal error compensation in machine tool digital twins. Manuf. Lett. **41**, 208–218 (2024). https://doi.org/10.1016/j.mfglet.2024.09.025
17. Lang, S., Zimmermann, N., Mayr, J., Wegener, K., Bambach, M.: Thermal error compensation models utilizing the power consumption of machine tools. In: International Conference on Thermal Issues in Machine Tools, vol. 3, pp. 41–53 (2023). https://doi.org/10.1007/978-3-031-34486-2_4

18. Lang, S., Zorzini, M., Scholze, S., Mayr, J., Bambach, M.: Sensor placement utilizing a digital twin for thermal error compensation of machine tools. J. Manuf. Syst. **80**(June), 243–257 (2025). https://doi.org/10.1016/j.jmsy.2025.03.003

19. Liu, M., Fang, S., Dong, H., Xu, C.: Review of digital twin about concepts, technologies, and industrial applications. J. Manuf. Syst. **58**(PB), 346–361 (2021). https://doi.org/10.1016/j.jmsy.2020.06.017

20. Mareš, M., Horejš, O., Havlík, L.: Thermal error compensation of a 5-axis machine tool using indigenous temperature sensors and CNC integrated python code validated with a machined test piece. Precis. Eng. **66**(July), 21–30 (2020). https://doi.org/10.1016/j.precisioneng.2020.06.010

21. Mayr, J., Jedrzejewski, J., Uhlmann, E., Alkan Donmez, M., Knapp, W., Härtig, F., Wendt, K., Moriwaki, T., Shore, P., Schmitt, R., Brecher, C., Würz, T., Wegener, K.: Thermal issues in machine tools. CIRP Ann. **61**(2), 771–791 (2012). https://doi.org/10.1016/j.cirp.2012.05.008

22. Mori, K., Ogura, D., Matsubara, A.: Energy-efficient manufacturing with indoor conditions offset considering weather conditions (2022). https://doi.org/10.1016/j.cirpj.2022.02.014

23. Preumont, A.: Controllability and observability. In: Vibration Control of Active Structures: An Introduction, pp. 289–312. Springer, Cham (2018). https://doi.org/10.1007/978-3-319-72296-2_12

24. Shields, M.D., Zhang, J.: The generalization of Latin hypercube sampling. Reliab. Eng. Syst. Saf. **148**, 96–108 (2016). https://doi.org/10.1016/j.ress.2015.12.002

25. Spescha, D., Weikert, S., Retka, S., Wegener, K.: Krylov and modal subspace based model order reduction with A-priori error estimation (2018). https://doi.org/10.3929/ethz-b-000284435

26. Spescha, D., Weikert, S., Wegener, K.: Simulation in the design of machine tools. In: Reinventing Mechatronics, pp. 163–177. Springer, Cham (2020). https://doi.org/10.1007/978-3-030-29131-0_11

27. Tanaka, S., Kizaki, T., Tomita, K., Tsujimura, S., Kobayashi, H., Sugita, N.: Robust thermal error estimation for machine tools based on in-process multi-point temperature measurement of a single axis actuated by a ball screw feed drive system. J. Manuf. Processes **85**(July 2022), 262–271 (2023). https://doi.org/10.1016/j.jmapro.2022.11.037

28. Vu Ngoc, H., Mayer, J.R., Bitar-Nehme, E.: Deep learning LSTM for predicting thermally induced geometric errors using rotary axes' powers as input parameters. CIRP J. Manuf. Sci. Technol. **37**, 70–80 (2022). https://doi.org/10.1016/j.cirpj.2021.12.009

29. Weng, L., Gao, W., Zhang, D., Huang, T., Liu, T., Li, W., Zheng, Y., Shi, K., Chang, W.: Analytical modelling method for thermal balancing design of machine tool structural components. Int. J. Mach. Tools Manuf **164**(February), 103715 (2021). https://doi.org/10.1016/j.ijmachtools.2021.103715

30. Zimmermann, N., Lang, S., Blaser, P., Mayr, J.: Adaptive input selection for thermal error compensation models. CIRP Ann. **69**(1), 485–488 (2020). https://doi.org/10.1016/j.cirp.2020.03.017

31. Zimmermann, N., Lang, S., Mayr, J., Wegener, K.: Validating real time compensation: a thermal test piece for 5-axis machine tools to separate thermal errors in z-direction. Precis. Eng. **91**(March), 263–277 (2024). https://doi.org/10.1016/j.precisioneng.2024.08.014

32. Zimmermann, N., Mayr, J., Wegener, K.: An action-oriented teaching approach for intelligent and energy efficient precision manufacturing. Manuf. Lett. **33**, 961–969 (2022). https://doi.org/10.1016/j.mfglet.2022.07.117

33. Zimmermann, N., Müller, E., Lang, S., Mayr, J., Wegener, K.: Thermally compensated 5-axis machine tools evaluated with impeller machining tests. CIRP J. Manuf. Sci. Technol. **46**, 19–35 (2023). https://doi.org/10.1016/j.cirpj.2023.07.005

Simulation-Based Design of a Thermoelectrically Temperature-Controlled Motorized Spindle

Roland Binninger, Hans-Fridtjof Pernau, Florian Triebel, Katrin Schmitt, and Jürgen Wöllenstein

Abstract The precision and efficiency of modern manufacturing processes are critically dependent on the thermal stability of the machine tool. As one of the most crucial components, the motorized spindle is of particular importance. By thermally stabilizing the bearings, the thermal expansion of the shaft can be reduced. A second advantage of the thermal stabilization of motorized spindles is the significant reduction of warm-up times after tool change down-times and the warm-up time at the start of the machining process. This study focuses on the simulation-based design of a thermoelectrically tempered motorized spindle that is using a novel design of tubular Peltier modules. The tubular design was chosen for its efficiency in providing a uniform and precise temperature control around the cylindrical surface of the bearing. In a first step, the thermal behavior of a functional prototype of a motorized spindle was measured under various operating conditions and compared to a simulation model using COMSOL MULTIPHYICS™. The FEM simulation validated by measurements was subsequently applied to the geometry of a real spindle to determine the optimized shape of the thermoelectric modules as well as their optimal number and placement around the bearings.

1 Introduction

In the fields of high and ultra-precision machining, the requirements for the geometric accuracy of components are extremely high, reaching down to the sub-micrometer level [1]. In the last decades, due to technological advancement and improvement of

R. Binninger (✉) · H.-F. Pernau · K. Schmitt · J. Wöllenstein
Fraunhofer Institute for Physical Measurement Techniques, Freiburg, Germany
e-mail: roland.binninger@ipm.fraunhofer.de

F. Triebel
Institute for Machine Tools and Factory Management (IWF), Technische Universität Berlin, Berlin, Germany

K. Schmitt · J. Wöllenstein
Institute of Microsystems Engineering—IMTEK, Freiburg, Germany

© The Author(s) 2026
K. Wegener and M. Bambach (eds.), *4th International Conference on Thermal Issues in Machine Tools (ICTIMT2025)*, Lecture Notes in Production Engineering,
https://doi.org/10.1007/978-3-032-01194-7_26

materials, the achievable accuracy of machine tools has continuously increased [2]. This trend is driven by the growing demand for accuracy of workpieces. Consequently, the level of achievable accuracy has become a key selling point for machine manufacturers. Thereby, the thermal behavior of the machine tools plays a crucial role in achieving these necessary operational accuracy. It is estimated that up to 75% of geometric component errors are caused by thermally induced displacements of the Tool Center Point (TCP) [3]. Due to electrical and mechanical losses, heat flows are induced in the components of the machine tool, such as the frame, guides, or spindle. These heat flows lead to increased temperatures and subsequently to thermally induced deformations of the respective components. If this occurs in the accuracy-determining components of the machine tool, it results in a displacement of the TCP and thus a reduction in achievable working accuracy. To mitigate the effects of the described thermal phenomena, complex cooling concepts are implemented in machine tools, and minimal changes in thermal loads during the process are pursued. This necessitates long warm-up times for the machine tool to reach a thermal equilibrium state, which in turn negatively affects the productivity and energy efficiency of the machine tool. Motorized spindles play a special role regarding the thermal behavior of machine tools, as high thermal flows are directly induced into the spindle due to their functionality. Thus, thermally induced deformations in motorized spindles have an immediate effect on the position of the TCP. At this point, it should be noted that this paper focuses solely on the effects of bearing temperature as part of the chain of effects. The influence of the tools and tool holders, which are used untempered during tool changes, is not addressed in this paper. Furthermore, the temperature of the bearings and their lubrication—especially for permanently grease-lubricated bearings—is important. At low Temperatures the viscosity of the oil content in the grease increases, leading to a change in its consistence and becoming sticky. At high temperatures, the viscosity of the grease becomes too low to generate a continuous oil film under the load of the bearings, leading to a reduction of its lubrication ability. Moreover, the lifespan of lubricants depends on the temperature. Lubricants have specific temperature limits. If these are exceeded or not reached, it can affect their lifespan or the lifetime of the entire bearing. A typical temperature range for the use of bearings in high-speed spindles is between 30 and 70 °C.

Uhlmann et al. [4] have shown that the concept of a thermoelectrically temperature controlled motorized spindle is capable regulating temperatures at various points within the motorized spindle. This can significantly reduce warm-up times and thermally induced deformations, regardless of varying induced thermal flows. As already stated in Uhlmann et al. [5] the use of prismatic Peltier modules for applications in motorized spindles is not optimal. On the one hand, the outer diameter of the spindle significantly increases due to the geometric arrangement. This implies certain power classes of spindles increasing in size, which in turn reduces their usability and thus negatively impacts application use. Furthermore, the components of the spindle become more expensive due to the complex geometry, as not only turned parts but also milled components, necessary for good thermal coupling, must be used. In general, Peltier modules can be effectively used to transfer heat from or to flat surfaces and offer several advantages over alternative heat pumps.

They are compact, operate silently, and have no moving parts. With small temperature differences, they can achieve a COP (coefficient of performance) significantly larger than 2, which is the ratio of supplied electrical to pumped thermal power. Furthermore, their pumping power and direction depend solely on the current intensity and direction. Therefore, they can be used for both cooling and heating of structures. To address the deficit that thermoelectric modules are not suitable for use on cylindrical structures such as pipes, several different concepts have already been pursued. Min et al. [6] present for the first time a concept in which, instead of the typically used pellets, discs made of thermoelectric material are utilized. The resulting structure is thus ideally suited for use on pipes, provided the thermal contact is well designed. However, this approach is limited both in the manufacturable diameter and, from an electrotechnical perspective, is not suitable for larger systems [7]. Tian et al. [8] demonstrate another approach in their publication. Instead of entire discs, individual pellets are assembled into a round structure. Both the contact pads made of copper and the thermoelectric material itself are rounded. They also simulated different geometries of the pellets for comparison. There are also several other publications that deal with optimizing the pellet geometry for cylindrical constructions [9–11] as well as publications that specifically address the simulation of leg geometries [12–14]. There is a significant advantage in Peltier-modules designed in this way. They can adapt very well to cylindrical surfaces, and thus the thermal contact resistance can be reduced. However, the manufacturing costs of such pellets and modules are significantly higher compared to commercial ones.

With the intention of reducing production complexities and the elevated manufacturing costs associated with mechanical imperfections of segmented pellets, Zoui et al. [15] proposed a ring-shaped module with quadratic profiled thermoelectric pellets as they are used in commercial flat modules. The pellets are placed, alternating n- and p- material, on copper stripes to create an electrical series connection between the pellets. These thermoelectric stripes are glued to an outer copper tube. The inner tube is made of individual strips of aluminum glued on the lower side of the thermoelectric stripes and connected by a sealing resin. Schäfer-Welsen et al. [16] pursued a similar approach to reduce the costs of a tubular thermoelectric module, thereby enabling commercialization. Unlike Zoui et al. the module is not glued, but a press fit is established with the thermoelectric modules positioned in between the inner and outer tubes to achieve good thermal and mechanical bonding. For this press fit, the strip-shaped thermoelectric modules presented by Binninger et al. [7] are suitable. For the fabrication of these strips, commercially available ceramic plates with copper pads are utilized. After positioning and soldering the thermoelectric pellets, the resulting square module is separated into individual strips. Since the width of each strip is limited to that of the thermoelectric pellets, which is only a few millimeters, they can be adapted to curved thermally connectable surfaces at large radii. To ensure optimal thermal contact, the strips are connected on both sides with graphite foil during assembly. This foil also compensates the differences between curved and flat connection surfaces. With this design, the number of strips and their electrical configuration can be tailored to the specific operating conditions. Reducing the number of strips decreases the thermal pumping power and increases the thermal resistance between the two sides. If high pumping power is not required in the system, the design can be simplified, resulting in cost savings. A second approach to optimize the tubular thermoelectric module to the necessary thermoelectric power is to change the footprint and height of the pellets.

This work presents the results of an FEM simulation set up using COMSOL MULTIPHYICS^{TM}, aimed at optimizing tubular thermoelectric modules for use in a motorized spindle. The initial simulation of a functional prototype of a motorized spindle allows for a comparison between the simulation results and actual measured values from the functional prototype. This allows for the adjustment of the thermal resistances in the simulation to correspond with those of the measurements. These thermal resistances will be used in the subsequent simulations. In a second step, the resulting FEM model is adapted to a real motorized spindle setup, to optimize the pellet geometry and thermoelectric strip number for this application.

2 Functional Prototype Configuration and Simulation

The functional prototype of a motorized spindle was developed and designed, based on the studies of Uhlmann et al. [5], to demonstrate the benefits and performance of tubular thermoelectric modules for temperature control of a motorized spindle bearing. This concept assumes a thermal separation between the front bearing and the motor, which results in the high thermal output of the motor having only a minimal impact on the temperature of the bearing.

The setup of the functional prototype is shown in Fig. 1a. It consists of a simplified model of a bearing seat, surrounded by 20 thermoelectric strips. The thermoelectric strips are designed with four pellet pairs. Each pellet (n- and p- doped material) has the dimensions of $0.8 \times 0.8 \times 2.5$ mm. The strips are connected to the inner and outer structure by thin graphite foil to create a good thermal and mechanical connection. The surrounding water cooling is attached by shrinking. It has two counter-flow cooling circuits to achieve high temperature homogeneity. The thermal behavior in different stages can be monitored by implemented temperature sensors in the cooling structure and the bearing seat. To replicate the heat input of the rotating bearings, a heating coil is pressed into the bearing seat of the test stand. This allows to induce thermal energy corresponding to different rotational speeds of the shaft. With this functional prototype, measurements were carried out at varying heater and Peltier powers. The measurements and their evaluation are presented in Sect. 3.

To validate the simulation model based on the measurements, the first model was created according to the functional prototype. To save computational time and performance, only a single slice with one thermoelectric strip consisting of 8 pellets was simulated, as shown in Fig. 1b. The boundary conditions were chosen to ensure that the energy levels in the slice correspond to those of the corresponding part of the complete ring. Thus, the thermal behavior in the system can be replicated by

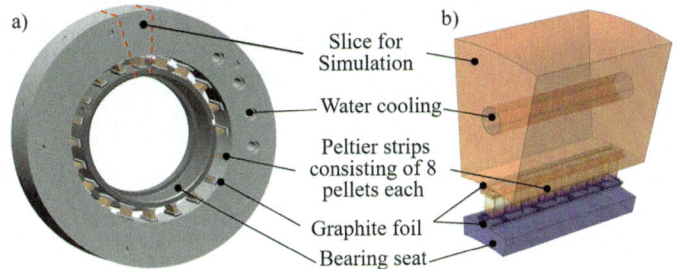

Fig. 1 a CAD model of the functional prototype. The bearing seat was realized with a sleeve on which 20 thermoelectric strips consisting of 8 pellets are mounted. Around it, a water cooling system made of Alumina has been shrunk. **b** The simulation model corresponds to one slice of the functional prototype

the simulation. Based on this, the pellet geometry and number of strips used in the system can then be optimized.

3 Prototype Measurement and Simulation Validation

Various measurements were conducted using the functional prototype presented in Sect. 2 to investigate the thermal behavior of the system. In the initial investigations, the heating power of the heating coil was set to 0, 5, 10, and 20 W at constant cooling liquid temperature of 20 °C. The heat input into the bearing of a rotating spindle can be much higher than 20 W. Depending on the type and size of the spindle, this study anticipates a maximum heat input of 50 W when utilizing a Z62 spindle from NAKANISHI JAEGER GMBH, Ober-Mörlcn, Germany, featuring a front bearing composed of two high-speed angular ball bearings with an inner diameter of 20 mm and an outer diameter of 42 mm [7]. However, with higher heat inputs than 20 W, the system with 20 strips reaches its cooling limit. With this prototype system and the set cooling liquid temperature, temperatures lower than 30 °C cannot be reached with max. 7 V applied Peltier voltage, but the measurements can still be used for the validation of the simulation. For these measurements, the electrical power of the Peltier module was varied by increasing the applied voltage in 1 V steps from 0.5 V respectively 1 V up to 7 V and the steady-state temperature at the bearing seat was measured.

Figure 2 shows the measurement at a heating power of 20 W. At the beginning of the measurement, the temperature of the bearing seat rises up to nearly 110 °C. By applying a Peltier power of only 0.4 W (0.5 V), the temperature at the bearing seat drops significantly to approximately 78 °C. As the Peltier power increases, the temperature continues to decrease. At an electrical power of 20 W (7 V), a bearing seat temperature of 38 °C can be achieved. As can be seen in this graph, the temperature drop becomes smaller with increasing electrical Peltier power. This is

Fig. 2 Results of the bearing seat temperature measurement with a heating power of 20 W

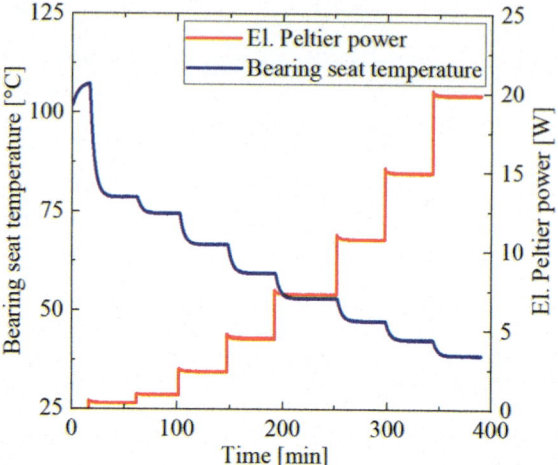

attributed to the increase in Joule heating with higher electrical power and to the decreasing temperature difference between the bearing seat and the water cooling, which subsequently reduces the pumping performance of the modules. The steady-state temperatures measured in this experiment were used to validate the simulation model.

For this purpose, the same thermal and electrical powers and conditions were applied as in the measurement. Periodic boundary conditions (PBC) were applied to the sides of the slice, defined as a unit cell, in Fig. 1b. The unit cell is replicated around itself, resulting in the formation of a closed ring. A heat input of 50 W was simulated on the inner side of the bearing shield. The COMSOL MULTIPHYICS™ standard module "Heat Transfer in Solids (ht)" interface was employed for heat conduction, while the thermoelectric module was utilized for the thermoelectric characteristics. Both the cooling fluid and the surrounding air are set to a temperature of 20 °C in the simulation. Free convection with a heat transfer coefficient of 20 W/m² K was assumed for the surrounding environment. The results of this comparison are presented in Fig. 3.

It can be observed that an increase in electrical Peltier power leads to a further reduction in the temperature of the bearing seat beyond the measurements recorded. The slight differences between simulation and measurement can be attributed to the thermal connection of the measurement probe as well as its position. In the simulation, the entire surface of the bearing seat was used for temperature measurement, while in the measurement of the functional prototype only a point measurement with a shielded PT1000 temperature sensor was performed. At high temperatures, this has a greater impact than at low temperatures. As a conclusion, it can be stated that the simulation sufficiently represents the measurement, and the identified boundary conditions will therefore be used for further simulations.

Fig. 3 Comparison of the measured and simulated minimum bearing seat temperature

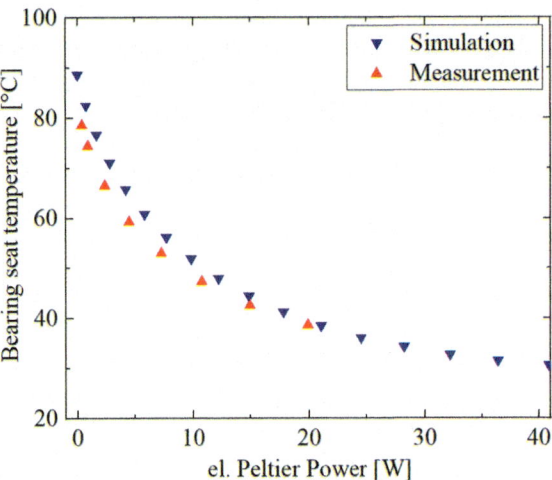

4   Spindle Simulation

The simulation model, validated based on the measurements, will further be used to optimize the number of built-in Peltier strips as well as the pellet height for use in a commercial spindle. For this purpose, as shown in Fig. 4, the geometry of the bearing seat was replaced with a bearing shield of a commercial spindle, and the water cooling of the prototype was replaced by the one inbuilt into the spindle. Thereby the Peltier strip is placed on the bearing shield above the position of the two angular contact ball bearings of the front bearing. To optimize computational efficiency and resource allocation, a single slice was simulated, with boundary conditions selected to replicate the temperature distribution of an entire bearing shield.

While the measurement in Fig. 2 was performed with a heat input of 20 W, the heat input in the outer bearing ring of the analyzed motorized spindle can be up to 50 W. To estimate the number of strips required to achieve bearing temperatures of less than 50°C for ensuring a long lifespan in a real motorized spindle, the heat input in this simulation was set to 50 W.

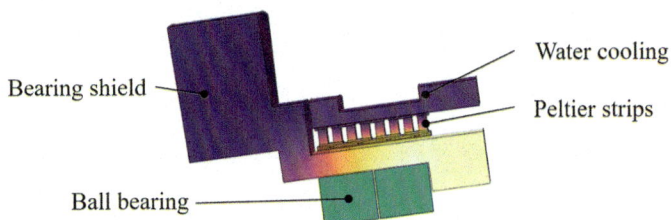

Fig. 4 Simulation model of a front bearing shield for a commercial spindle, incorporating Peltier strips and water cooling

Figure 5 depicts the simulation results with varying numbers of strips and the respective achievable temperature of the bearing shield at different Peltier voltages.

An increase in the number of strips in the system leads to a reduction in the temperature of the bearing shield. The temperature at the bearing seat strongly depends on the number of strips. The higher their number, the lower the thermal resistance of the system. However, the increase in cooling performance becomes smaller with a rising number of strips, meaning that an additional increase in strips, if space permits, does not necessarily lead to a significant improvement in the system performance.

Figure 6 shows the achievable minimum temperatures at the bearing shield per number of thermoelectric strips.

The curve flattens as the number of strips increases. With 10 strips in the system, a minimum temperature of 75 °C can be achieved, while with 20 strips, it drops to 59 °C, resulting in a temperature reduction of 16 K. When the number of strips is further increased to 30, the temperature decreases to 49 °C, and with an additional 10 strips, it only reduces to 43 °C. The system with 50 strips has a minimum temperature of 39 °C. Therefore, in optimizing the system not only the maximum possible number of strips is crucial, but a trade-off between benefits, costs, and installation effort must also be considered. Furthermore, it is evident that there is also an optimum with respect to the Peltier voltage. As previously described, this is due to Joule heating within the Peltier modules at increasing electrical power. To achieve a low minimum temperature while keeping the installation effort of the strips manageable, a total of 40 strips will be utilized in the subsequent analysis.

The optimization of the pellet geometry offers further potential to increase the cooling performance of the module. By lengthening the pellets at constant cross-section, the maximal achievable temperature difference increases but in the same time, the maximal heat transfer decreases [17]. To evaluate the impact of the pellet hight in the described setup, the hight was varied in the simulation from 1.0 to 2.5 mm

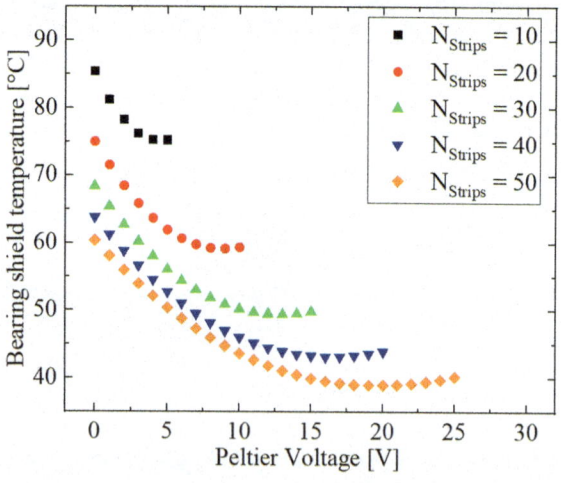

Fig. 5 Simulation results of the bearing shield temperature for heat input of 50 W to the outer bearing ring dependent on Peltier voltage for various numbers of strips

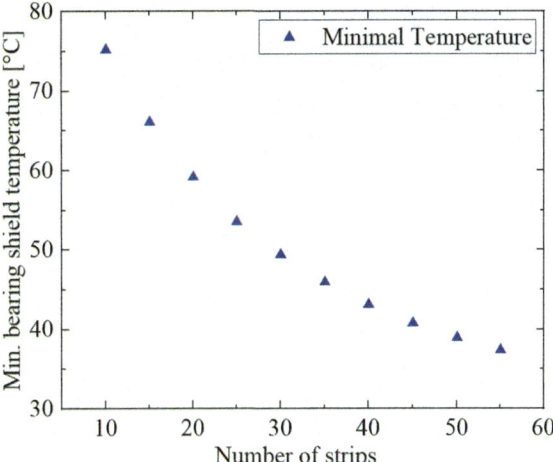

Fig. 6 Representation of the minimum achievable bearing shield temperature under 50 W of induced heat as a function of the number of strips

in 0.25 mm steps. The result of this simulation is shown in Fig. 7. A temperature difference depending on the pellet heights can be observed when no voltage is applied. With rising height of the pellets, the thermal resistance between the bearing shield and the water cooling increases, resulting in a higher temperature at the bearings. By increasing the Peltier voltage, the temperature decreases until the thermal pumping power and the Joule heating are equal. Beyond this point, the temperature rises again.

For shorter pellets, the reachable minimum temperature is at smaller voltages than for longer ones, this is due to the impact of the electrical resistance. The shorter the pellets, the smaller the electrical resistance and thus the higher the driving current. This leads to higher electrical powers even for smaller voltages as can be seen in

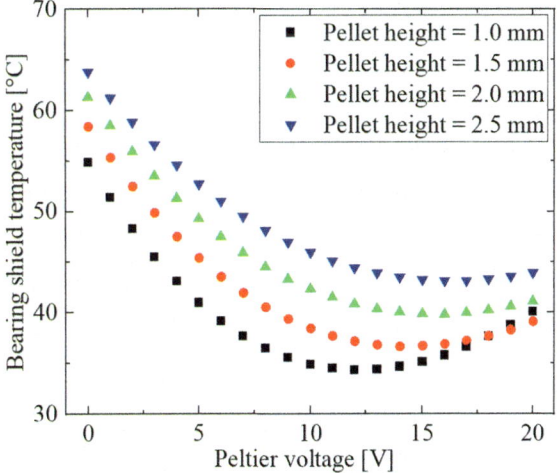

Fig. 7 Simulation results illustrating the relationship between bearing shield temperature, Peltier voltage, and pellet height

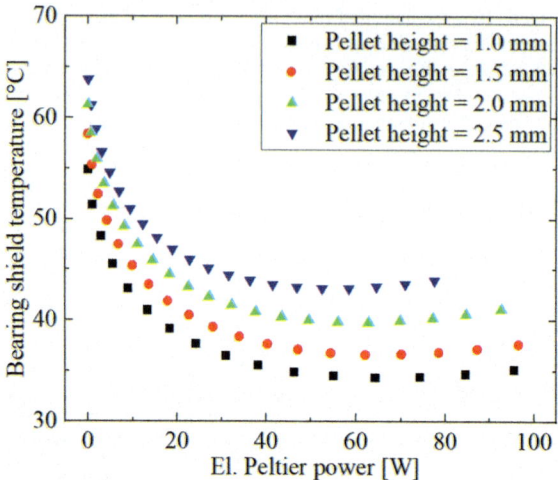

Fig. 8 Graphical representation of bearing shield temperature as a function of electrical Peltier power for pellets of varying heights

Fig. 8. However, the energy required to reach a target temperature is lower for smaller pellets than for long ones. To reach a target temperature of 50°C at 50 W heat input, the system with 40 strips and a pellet hight of 1 mm requires 2.3 W while a system with a pellet height of 2.5 mm requires 11.6 W.

Figure 9 shows the minimal reachable temperatures of the cooling systems with varying pellet heights from 1.0 to 2.5 mm and different number of stripes.

This graph, as previously shown in Figs. 5 and 6, illustrates that the minimum achievable temperature decreases with an increasing number of strips. It is also evident that the temperature reduction from 30 to 40 strips is greater than that from 40 to 50 strips. Additionally, the influence of pellet height is clearly observable in

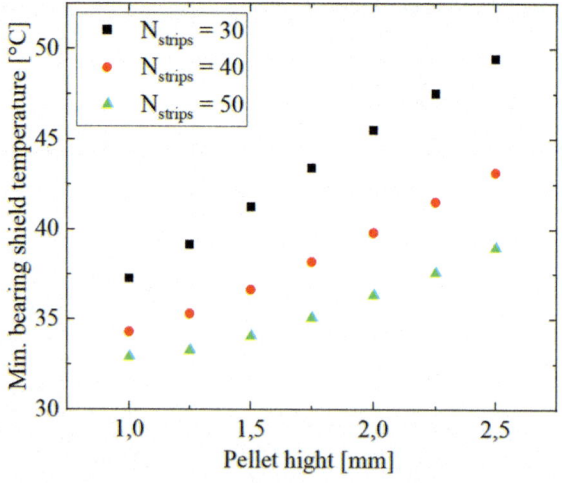

Fig. 9 Analysis of the correlation between simulation results for minimum bearing shield temperature, pellet height, and the number of strips in the system

this graph. However, a flattening of the curve is also apparent with decreased pellet height, most notably in the graph with 50 strips. This phenomenon is caused by the presence of materials with a low or even neglegtable thermoelectric strength as the contacts and the solder (see also Fig. 7). The shorter the pellets, the higher the current density, resulting in a greater voltage drop across the contacts due to their reduced effective cross-sectional area, which consequently leads to increased Joule heating. Each strip also contributes to this amount of "dead" material for its in- and outlet, thus the effect is more pronounced for the 50 strips than for the 30 strips configuration.

5 Conclusion

The behavior of the bearing temperature of motorized spindles has a large impact on the machining accuracy due to the thermal displacement of the TCP. The investigated tempering system, based on Peltier modules placed in between the bearing shield and the water cooling of the spindle, offers the advantage of versatile temperature control of the bearings by electrical induced heat transfer. This method allows for heating and cooling that is faster and easier to adjust compared to conventional water-based systems. In this work, a simulation based optimization of the used Peltier module, consisting of single Peltier strips, is introduced. Therefor the number of strips used and the height of the thermoelectric pellets was analyzed.

By increasing the number of strips, the thermal resistance and thus the bearing shield temperature, without applying an electrical power to the module, decreases. Beside this, for larger number of strips, the heat transfer increases. But the increase of thermal output is not linear to the number of strips. Thus the optimal number is a trade-off between the required thermal power and the increasing costs of the thermoelectric system. By analyzing the optimized hight of the pellets it can be observed, the lower the pellets are, the more thermal energy can be transferred. Thereby the thermal resistance of the system decreases for lower pellets which offers an additional benefit for the system.

Acknowledgements This work was funded within the Fraunhofer-DFG transfer program by Fraunhofer Society and Deutsche Forschungsgemeinschaft (DFG, German Research Foundation)—529738427.
Artificial intelligence was employed to improve the accuracy and efficiency of the translation process, ensuring a precise representation of the research findings.

References

1. Creighton, E., Honegger, A., Tulsian, A., Mukhopadhyay, D.: Analysis of thermal errors in a high-speed micro-milling spindle. Int. J. Mach. Tools Manuf **50**(4), 386–393 (2010)
2. Taniguchi, N.: Current status in, and future trends of, ultraprecision machining and ultrafine materials processing. CIRP Ann. **32**(2), 573–582 (1983)

3. Mayr, J., Jedrzejewski, J., Uhlmann, E., Donmez, M.A., Knapp, W., Härtig, F., Wendt, K., Moriwaki, T., Shore, P., Schmitt, R., Brecher, C., Würz, T., Wegener, K.: Thermal issues in machine tools. CIRP Ann. **61**(2), 771–791 (2012)
4. Uhlmann, E., Polte, M., Triebel, F., Salein, S., Temme, P., Hartung, D., Perschewski, S.: Reduction of warm-up period after machine downtime by means of a thermoelectric tempered motorized milling spindle. Special Interest Group Meeting on Thermal Issues, ETH Zurich, Switzerland (2022)
5. Uhlmann, E., Polte, J., Salein, S., Iden, N., Temme, P., Hartung, D., Perschewski, S.: Entwicklung einer thermoelektrisch temperierten Motorspindel. wt Werkstattstechnik online **110**, 299–305 (2020)
6. Min, G., Rowe, D.M.: Ring-structured thermoelectric module. Semicond. Sci. Technol. **22**, 880
7. Binninger, R., Triebel, F., Pernau, H.-F., Polte, M., Schäfer-Welsen, O., Uhlmann, E.: Simulativer Vergleich tubularer Peltierelemente/FEM simulation of tubular Peltier modules for tempering of machine tool components—Simulative comparison of tubular peltier modules. wt Werkstattstechnik online **112**(7/8), 451–457 (2022)
8. Tian, X.-X., Asaadi, S., Moria, H., Kaood, A., Pourhedayat, S., Jermsittiparsert, K.: Proposing tube-bundle arrangement of tubular thermoelectric module as a novel air cooler. Energy **208**(1), 118428 (2020)
9. Zhanga, A.B., Wangb, B.L., Pangc, D.D., Chena, J.B., Wanga, J., Dua, J.K.: Influence of leg geometry configuration and contact resistance on the performance of annular thermoelectric generators. Energy Convers. Manage. **166**, 337–342 (2018)
10. Bauknecht, A., Steinert, T., Spengler, C., Suck, G.: Analysis of annular thermoelectric couples with nonuniform temperature distribution by means of 3-D multiphysics simulation. J. Electron. Mater. **42**, 1641–6 (2013)
11. Shen, Z.G., Wu, S.Y., Xiao, L.: Theoretical analysis on the performance of annular thermoelectric couple. Energy Convers. Manage. **89**, 244–250
12. Ebling, D., Bartholomé, K., Bartel, M., Jägle, M.: Module geometry and contact resistance of thermoelectric generators analyzed by multiphysics simulation. J. Electron. Mater. **39**, 1376–80 (2010)
13. Sahin, A.Z., Yilbas, B.S.: The thermoelement as thermoelectric power generator: effect of leg geometry on the efficiency and power generation. Energy Convers. Manage. **65**, 26–32 (2013)
14. Lamba, R., Kaushik, S.C.: Thermodynamic analysis of thermoelectric generator including influence of Thomson effect and leg geometry configuration. Energy Convers. Manage. **144**, 388–98 (2017)
15. Zoui, M.A., Bentouba, S., Velauthapillai, D., Zioui, N., Bourouis, M.: Design and characterization of a novel finned tubular thermoelectric generator for waste heat recovery. Energy **253**(15), 124083 (2022)
16. Schäfer-Welsen, O., Vetter, U., Vergez, M., König, J.: Method for manufacturing a thermoelectric module and thermoelectric mudle as a press band (EP 3 933 946 B1) (2022). https://register.epo.org/application?number=EP20183520&lng=de&tab=doclist
17. Rowe, D.M., Bhandari, C.M.: Modern Thermoelectrics. Rinehart and Winston Ltd., Eastbourne (1983)

Mixed-Dimensional Finite Element Modelling of Passive Thermal Error Compensation Systems in Machine Tools

Immanuel Voigt◉ and Welf-Guntram Drossel◉

Abstract Due to the high relevance of thermally induced errors in machine tools, numerous thermal error compensation methods are discussed in literature. While most of the approaches demand significant cooling, measurement or simulation effort, design-related strategies contribute to the reduction of thermal errors by convenient heat distribution within the machine tool. This is achieved by reducing the sensitivity of the machine structure to thermal loads as, for example, realized through thermal-symmetric design. By providing more design-related strategies, the compensation efforts during machine operation may be reduced. A currently investigated approach is the integration of passive thermal error compensation systems that inherently enhance the heat transfer and storage capabilities of the machine structure. As demonstrated in recent studies, latent heat storage systems can reduce temperature fluctuations close to machine-internal heat sources of high temporal variability. This approach is based on the high effective heat capacity of phase change materials during the phase transition. Another approach is to add heat pipe heat sink assemblies to passively increase the convective heat dissipation. Both phase change materials and heat pipes are characterized by nonlinear thermal properties that require a thorough understanding of the underlying heat transfer and storage mechanisms. At the same time, efficient modelling techniques are needed to provide convenient design tools that are compatible with common machine tool models. This work illustrates the application of mixed-dimensional finite element models to describe the effect of latent heat storage systems and heat pipe assemblies integrated into machine tool structures. It is demonstrated how thermal resistance networks composed of point and line elements can be implemented within three-dimensional finite element models to represent heat pipe assemblies. Furthermore, the use of volume-shell elements is proposed as an efficient way to model latent heat storage systems. Both approaches are

I. Voigt (✉) · W.-G. Drossel
Technische Universität Chemnitz, Chemnitz, Germany
e-mail: immanuel.voigt@iwu-extern.fraunhofer.de

W.-G. Drossel
e-mail: welf-guntram.drossel@iwu.fraunhofer.de

W.-G. Drossel
Fraunhofer Institute for Machine Tools and Forming Technology, Chemnitz, Germany

© The Author(s) 2026   427
K. Wegener and M. Bambach (eds.), *4th International Conference on Thermal Issues in Machine Tools (ICTIMT2025)*, Lecture Notes in Production Engineering,
https://doi.org/10.1007/978-3-032-01194-7_27

evaluated using different numerical scenarios and validated by comparing the simulation results to experimental data. The results imply that the mixed-dimensional finite element modelling provides high accuracy at considerably lowered modelling and simulation effort compared to three-dimensional modelling. The proposed method hence allows for an increased efficiency during the design phase of the corresponding compensation components.

1   Introduction

As the industrial energy demand is globally increasing, the energy efficiency of machine tools is gaining relevance as a central aspect in the research area of production engineering. Cooling systems are among the major consumers of electrical energy in machine tools [1]. Considering an exemplary production process, Denkena et al. demonstrate that the cooling system contributes to around 27% of the overall power consumption of the machine tool [2]. Accordingly, solutions that lead to a reduction of active cooling activities exhibit a considerable energy saving potential.

The significance of thermal aspects on the accuracy of machine tools has led to various thermal error compensation approaches. According to Ramesh, methods for thermal error reduction can be divided into three categories [3]:

- control of heat flow into the machine tool environment (*regulatory approaches*),
- compensation through controlled movement (*control-based approaches*) and
- redesign of the machine tool system to reduce sensitivity to heat flow (*design-related approaches*).

Regulatory approaches aim for an active influence on the occuring heat flow, especially through active cooling systemes. Control-based approaches are based on predicting the thermal error on the tool center point (TCP) by using sensors and computational models. The determined error value is then used for a position correction between the tool and the workpiece [4, 5]. Design-related approaches comprise methods that lead to increased thermal stability of machine tools such as convenient positioning and isolation of heat sources as well as the thermosymmetric design of the machine [6]. In addition, the selection of suited material properties is of high importance. Materials of low thermal expansion coefficients can be used to obtain a low sensitivity of thermoelastic displacements to occuring temperature fields.

It can be stated that regulatory and control-based approaches usually lead to an increased electrical energy demand or considerable effort due to additional measurement and simulation tools. As opposed to that, design-related approaches are realized during the design phase of the machine without causing additional efforts during operation. To enable the development of further design-related compensation strategies, the present work deals with passive systems that are able to reduce the thermal error without needing additionaly power supply. As shown in previous works by the authors, this can be realized by adding latent heat storage (LHS) systems and heat pipe heat sink assemblies to the machine tool structure [7].

To enable the design of corresponding passive compensation systems, an efficient modelling approach is proposed by the authors. It is based on the use of lower-dimensional elements within three-dimensional (3D) finite element (FE) models. Such mixed-dimensional FE models are described by different authors addressing, among others, mechanical boundary problems containing slender beams or thin sheets that are modelled by one-dimensional (1D) or two-dimensional (2D) elements coupled with adjacent 3D elements [8–10]. The present work demonstrates, how mixed-dimensional FE models can be implemented to efficiently describe LHS systems and heat pipe heat sink assemblies within 3D thermal FE models. It is illustrated, how the coupling between elements of different order can be achieved. Furthermore, the needed characteristics of lower-dimensional elements are elaborated to provide sufficient simulation accuracy. The implementation of mixed-dimensional thermal FE modelling is illustrated considering an experimental set-up previously described. By comparing the experimentally and numerically obtained results, the precision of the proposed modelling concept is evaluated.

2 Passive Systems for Thermal Error Reduction

The underlying compensation concept is based on the structural integration of passive compensation systems that alter the heat flow within the machine tool in a way that causes a reduction of the thermal error (see Fig. 1). This can be achieved by two different approaches. The first approach is the integration of LHS systems close to lossy machine components. LHS systems make use of the high effective heat capacity of phase change materials (PCM) during melting and solidification aiming for an increased thermal stability within the corresponding phase transition temperature range.

The second approach considers the integration of heat pipe systems to enhance the heat transport within the machine structure or between the machine and the environment. An exemplary compensation scenario is shown in Fig. 2. An LHS component

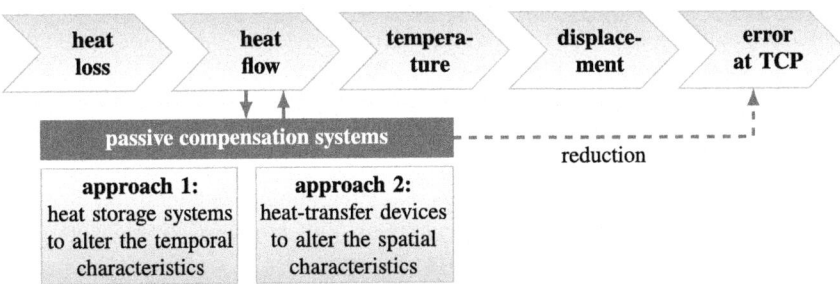

Fig. 1 Intended alteration of heat flows within machine tools to reduce the thermal error by the integration of passive compensation systems

Fig. 2 Exemplary
integration of heat pipes, heat
sinks and LHS components
between the primary part of a
linear direct drive and the
adjacent slide structure

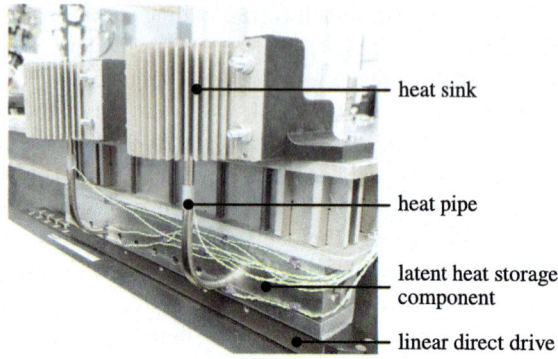

heat sink

heat pipe

latent heat storage
component

linear direct drive

based on an organic PCM with a phase change temperature range around 28 °C was
placed between the primary part of a linear direct drive and the adjacent slide structure. Additionally, heat pipes were integrated to transfer the heat losses to heat sinks
enhancing the convective heat dissipation on the moving slide.

3   Mixed-Dimensional Finite Element Modelling

To design passive compensation systems for a given machine and a defined load scenario, thermal FE models of the machine tool and the intended components are used.
Hereby, the non-linear thermal properties of the introduced compensation systems
must be taken into account. While PCM can be described as a continuum assigned
with a temperature-dependent enthalpy curve, the description of the complex fluid
flow inside a heat pipe requires detailed fluid mechanical models. However, to avoid
excessive modelling effort, the behaviour of heat pipes as well as heat sinks can be
described by simplified models considering a thermal resistance curve depending on
heat flow, orientation and motion state [11].

The present work aims to provide an efficient way of modelling passive compensation systems by means of lower-dimensional elements. The desired modelling
approach features a fixed 3D mesh of the reference machine that is not altered during the design of the passive compensation systems to avoid iterative geometry and
mesh updates while simulating the thermal impact of the compensation system for
different component locations and characteristics. In the following, the possibility
of the integration of point (zero-dimensional, 0D) elements and line (1D) elements
into 3D models is discussed. Afterwards, the integration of volume-shell elements
(2.5D) into 3D models is evaluated. Any simulations presented in this work were
conducted using the commercial FE software Ansys.

3.1 Integration of Point and Line Elements

Solving thermal boundary problems using FEM requires a geometry discretization. To define an adequate mesh the inhomogenity of the expected temperature field must be considered. In some cases it is acceptable to model a structure of low temperature inhomogenity using the lumped capacitance method. In such cases the corresponding component can be represented by 0D elements that neglect any temperature gradients. To evaluate, if this abstraction is a valid option, the Biot number Bi as a measure of the resistance to heat conduction within the solid R_{cond} relative to the resistance to convection across the surface R_{conv} can be used. As found in literature, for solids with a Biot number below a critical value Bi_{crit} the lumped capacitance method is generally considered a valid approach:

$$Bi = \frac{R_{\mathrm{cond}}}{R_{\mathrm{conv}}} < 0.1 = Bi_{\mathrm{crit}}. \tag{1}$$

High-conductive components such as heat pipes and heat sinks usually satisfy the criterion stated by Eq. 1 and can hence be modelled by means of 0D elements representing lumped capacitances C_i. The heat conduction between distributed 0D elements can be modelled by means of resistances R_i as 1D elements. Thereby, a thermal resistance network is created that describes the thermal behaviour of the corresponding components at the interface of interest, e.g. between two machine tool components Ω_A and Ω_B as shown in Fig. 3.

To integrate the thermal resistance network into a 3D FE machine tool model, the nodes $0'$ and n' must be coupled to the nodes of the corresponding faces Γ_A and Γ_B, respectively. The simplest method to provide a 3D/1D interface is to couple the

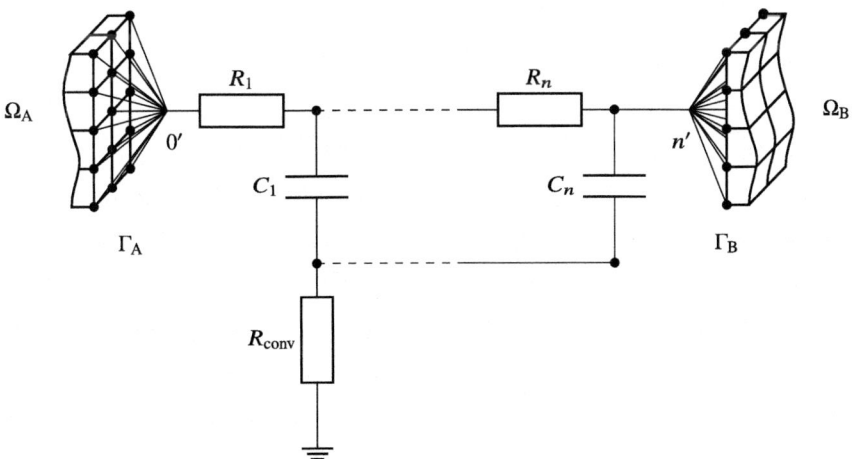

Fig. 3 Thermal resistance network representation of a structural component integrated between two solids for transient FE analyses

degrees of freedom on Γ_A and Γ_B (*temperature coupling*). Another approach is the use of *constraint equations* that define a linear relationship between the temperature of the nodes of lower-dimensional elements T' and the temperatures of the nodes of higher-dimensional elements T_i using constraint coefficients X_i:

$$T' = \sum_{i=1}^{n} X_i T_i. \tag{2}$$

The validity of both coupling methods is evaluated considering an exemplary 2D thermal boundary problem as shown in Fig. 4. The boundary problem is solved in three different ways: analytically (as illustrated in detail in the appendix), numerically using FEM with constraint equations on Γ_A and numerically using FEM with temperature coupling on Γ_A. The comparison of the calculated stationary temperature fields reveals that using temperature coupling leads to a significant error close to the edge Γ_A. Conclusively, 1D/3D coupling should be carried out using constraint equations.

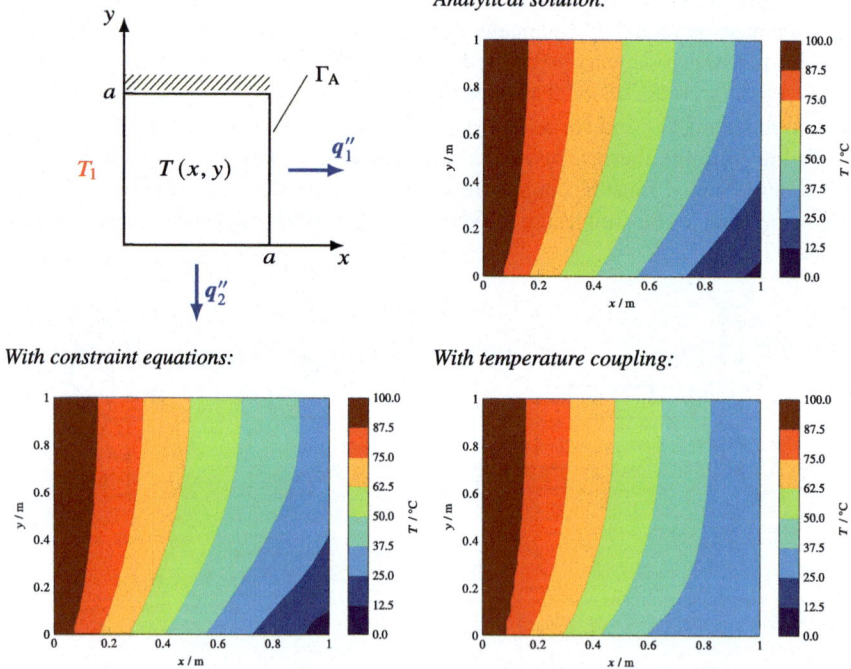

Fig. 4 Two-dimensional heat transfer problem with analytical solution and numerical solutions using constraint equations and temperature coupling with $a = 1$ m, $k = 10$ W/(m K), $q_1'' = q_2'' = 500$ W/m^2

3.2 Integration of Volume-Shell Elements

The criterion stated by Eq. 1 is usually not satisfied by LHS systems due to the low thermal conductivity of the underlying PCM (below $1\,\mathrm{W/(m\,K)}$ for organic PCM). Therefore, thermal resistance networks are not a suitable modelling type for LHS systems. As a novel approach, the use of volume-shell (2.5D) elements is investigated. 2.5D elements are composed of nodes aligned within a plane. As opposed to shell elements, the nodes of 2.5D elements can have multiple temperature degrees of freedom to account for through-thickness heat conduction. The number of degrees of freedom depends on the number of element layers.

Two FE models were used to compare the performance of 2.5D elements and 3D elements. The first model represents an LHS component between two solids where all bodies are discretized by 3D elements. The second model represents the LHS component as 2.5D elements (using the Ansys-specific element type SHELL131 [12]) with one element in through-thickness direction (see Fig. 5). As proposed by other authors [9, 10], multipoint constraint (MPC) equations were applied to connect the 2.5D elements to the 3D elements. Both models were used to calculate transient temperature profiles under variation of physical parameters within an LHS-typical parameter range (length of LHS component size l, thermal conductivity k, enthalpy H, heat flows q_{in} and q_{conv}) and numerical parameters (element size, number of layers n_{lay}). Excluding contact elements, the 3D model is composed of a minimum of 25576 elements, while the mixed-dimensional model contains 3615 elements.

By comparing the results of the 3D model and the mixed-dimensional model, the necessary number of layers needed to guarantee a defined maximal deviation of 1 K was obtained. A regression was conducted to mathematically describe the required number of layers as a function of the dimensionless entity $\beta = l\, q_{\mathrm{in}}\, k^{-1} \mathrm{K}^{-1} \mathrm{m}^{-2}$

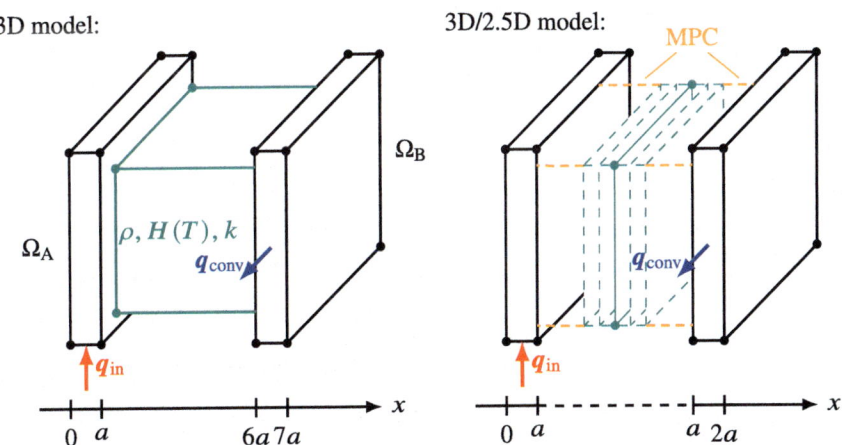

Fig. 5 3D and 3D/2.5D thermal FE models of an LHS component between two solids

$$n_{\text{lay,reg}}(\beta) = a - b\ln(\beta + c) \tag{3}$$

with the regression coefficients $a = -5.7872$, $b = -3.4837$ und $c = 5.0705$ leading to a coefficient of determination of $Z^2 = 0.99422$. The data obtained by the simulations is illustrated alongside the regression function in Fig. 6.

The results indicate that the use of 2.5D elements is valid for a broad range of physical parameters. Even when using one element across the thickness, a convenient estimation of the thermal LHS impact on the adjacent structure is achieved.

4 Validation

To evaluate the accuracy of the proposed mixed-dimensional modelling approach, a compensation scenario, that was previously experimentally investigated [7], was modelled. The model represents the y-slide of the experimental machine MAX developed by the Chair of Machine Tools Development and Adaptive Controls of Dresden University of Technology (see Fig. 7) [13].

In the previous work, LHS components and heat pipe heat sink assemblies were integrated between the primary parts of the two linear direct drives Y1 and Y2 (face Γ_A) and the slide (face Γ_B). Temperatures were measured using 16 thermocouples

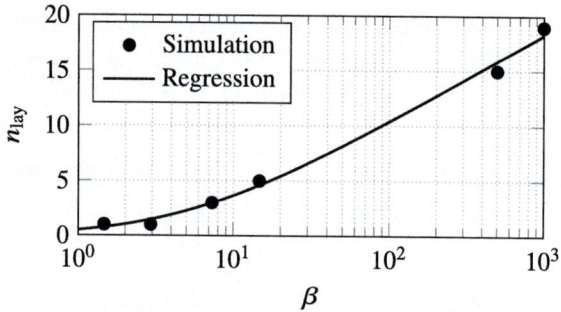

Fig. 6 Number of layers required to achieve the target accuracy as obtained by the parametric study and estimated using a regression function

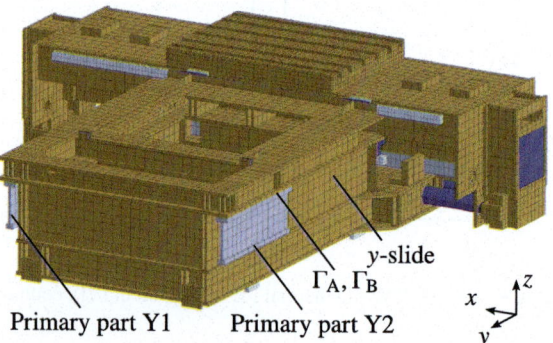

Fig. 7 FE model of the y-slide of the experimental machine MAX

on the compensation components and ten Pt100 sensors on the slide. At the corresponding faces of the 3D model, the LHS components were added as 2.5D elements. The heat pipe heat sink assemblies were modelled be means of 0D and 1D elements. The resulting mixed-dimensional model is shown in Fig. 8. The 2.5D elements are assigned the density ρ, heat capacity c_p, enthalpy H and thermal conductivity k of the PCM added by nanographite particles (C − PCM), the storage container (St) and a copper coupling element (Cu). In accordance to Eq. 3, three layers were defined. The enthalpy curve for the PCM RT28HC used in the experimental investigations was determined via differential scanning calorimetry (see Fig. 9). The heat pipes were modelled by 1D elements characterized by a non-linear relationship between the thermal resistance and the heat flow according to previous tests [11].

Figure 10 shows the experimentally obtained and simulated temperatures at the slide location that corresponds to the measurement point close to Γ_B. Both for the reference configuration (without compensation components) Ω and the compensation configuration $\tilde{\Omega}$ the simulated temperatures are found to be in good agreement

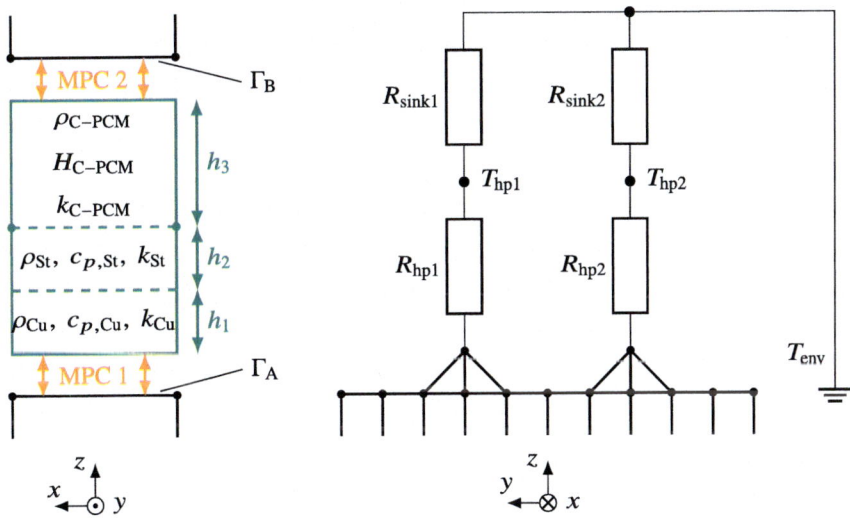

Fig. 8 2.5D representation of the LHS system and 1D representation of the heat pipes and heat sinks

Fig. 9 Temperature-dependent volumetric enthalpy curve of PCM RT28HC obtained by differential scanning calorimetry

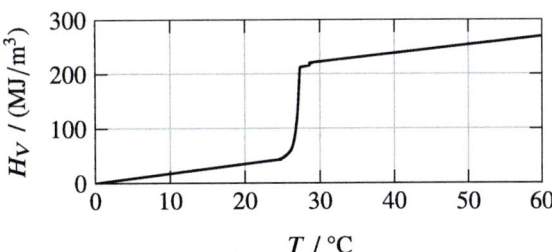

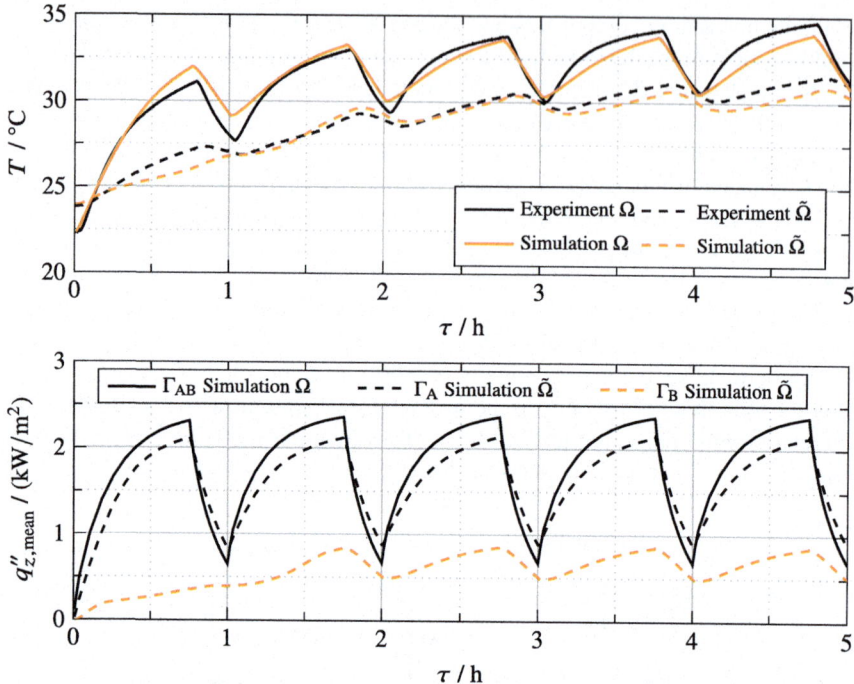

Fig. 10 Simulated and measured temperatures as well as heat flux at Γ_A and Γ_B in reference configuration Ω and compensation configuration $\tilde{\Omega}$ of the experimental machine MAX

with the measurement data. The maximum deviation amounts to 1.2 K. The deviations may be related to the model simplification by only considering the y-slide without accounting for the adjacent machine tool structure. Furthermore, the flow conditions of the environmental air during the experiments and the corresponding convective heat transfer were modelled using simplified convection boundary conditions assuming constant convective heat transfer coefficients that potentially contribute to increased simulation inaccuracy. While the enclosure of the machine by surrounding walls limited undesirable forced convection effects, the environment temperature inside the enclosure increased by approximately 2 K during each measurement. However, constant environment temperature was assumed in the boundary conditions.

The calculated mean heat flux $q''_{z,\mathrm{mean}}$ at Γ_A between the primary part and the LHS component as well as at Γ_B between the LHS component and the adjacent slide structure demonstrates how the LHS component alters the heat input into the slide. The maximum heat flux in the compensation configuration is reduced by 9% compared to the reference configuration. This corresponds to the heat dissipation caused by the heat pipe heat sink assembly. Significant differences become apparent between the heat flow into the LHS component and the heat flow out of the LHS component. By means of the LHS system, the heat flow into the slide is reduced by approximately 60%. This corresponds to the phase change enthalpy of the underlying PCM as well as the convective heat dissipation on the surfaces of the LHS component.

5 Conclusions

Passive compensation systems can be integrated into machine tools to reduce temperature fluctuations aiming for a reduction of thermal errors. This work proposes an efficient modelling approach to support the design of corresponding components. The modelling approach is based on the use of lower-dimensional elements within 3D FE elements. By analyzing different thermal boundary problems, coupling methods between elements of different dimensions as well as the accuracy of lower-dimensional models compared to 3D models were investigated. The findings of the investigations can be summarized as follows:

- heat pipes and heat sinks can be modelled by means of thermal resistance networks composed of 0D and 1D elements that can be integrated into 3D FE models using constraint equations,
- LHS systems can be modelled by 2.5D elements whereas the adecuate number of layers can be calculated through a regression function formulated based on a parametric simulation study and
- the mixed-dimensional modelling of compensation components within a machine tool was found to be in good agreement with experimental data.

The proposed modelling method allows for an efficient simulation-assisted design of LHS and heat pipe systems by providing an estimation of the expected compensation effect for different compensation scenarios without the need to iteratively alterate the 3D geometry and mesh of the reference machine tool. However, the mixed-dimensional modelling as demonstrated in this work is restricted to the evaluation of the compensation effect on a thermal level. To provide a comprehensive prediction on the impact of passive compensation systems on the thermoelastic deformation state, it is necessary to expand the current approach by adding a thermoelastic coupling. It is suggested to extend the mixed-dimensional FE modelling approach onto a thermoelastic level to provide a design tool that delivers direct conclusions regarding the expected reduction of the thermal error due to passive compensation systems.

Acknowledgements The authors would like to thank the German Research Foundation (DFG) for financial support within the Collaborative Research Centre Transregio 96. Furthermore, the authors would like to thank the Chair of Machine Tools Development and Adaptive Controls of Dresden University of Technology for providing the reference model of the experimental machine MAX.

Appendix

The boundary problem discussed in Sect. 3.1 is divided into three problems with one non-homogeneous boundary condition each using the superposition principle (see Fig. 11).

Problem A comprises three homogeneous heat flux boundary conditions and hence gives the homogeneous temperature field

$$T_A(x, y) = T_1. \tag{4}$$

Problem B does not exhibit heat flow in y-direction. The temperature field can hence be described through

$$T_B(x, y) = C_1 x + C_2. \tag{5}$$

The boundary condition $T_B(x = 0) = 0$ gives $C_2 = 0$ leading to

$$T_B(x) = \frac{q_1''}{k} x. \tag{6}$$

Problem C

$$\frac{\partial^2 T_C}{\partial x} + \frac{\partial^2 T_C}{\partial y} = 0 \quad \text{in} \quad 0 < x < a, \, 0 < y < a \tag{7a}$$

$$T_C(x = 0) = 0 \tag{7b}$$

$$\left. \frac{\partial T_C}{\partial x} \right|_{x=a} = 0 \tag{7c}$$

$$-k \left. \frac{\partial T_C}{\partial y} \right|_{y=0} = -q_2'' \tag{7d}$$

$$\left. \frac{\partial T_C}{\partial y} \right|_{y=a} = 0 \tag{7e}$$

is solved by separation of variables.

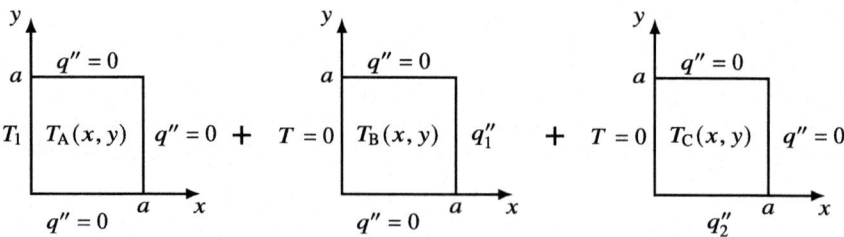

Fig. 11 Superposition of the thermal boundary problem

$$T_C(x, y) = X(x)Y(y) \tag{8}$$

gives two ordinary differential equations

$$\frac{d^2 X}{dx^2} + \lambda^2 X = 0 \tag{9a}$$

$$X(x = 0) = 0 \tag{9b}$$

$$\left.\frac{dX}{dx}\right|_{x=a} = 0 \tag{9c}$$

and

$$\frac{d^2 Y}{dy^2} - \lambda^2 Y = 0 \tag{10a}$$

$$\left.\frac{dY}{dy}\right|_{y=a} = 0. \tag{10b}$$

Equation 9 gives

$$X(x) = C_1 \cos(\lambda x) + C_2 \sin(\lambda x), \tag{11}$$

whereby the boundary conditions lead to $C_1 = 0$ as well as the eigen values

$$\lambda_n = 1/a \, (\pi/2 + n\pi) \tag{12}$$

and

$$X_n(x) = C_2 \sin(\lambda_n x). \tag{13}$$

The result of Eq. 10 is

$$Y(y) = C_3 \cosh(\lambda y) + C_4 \sinh(\lambda y) \tag{14}$$

with $C_4 = 0$ resulting from the homogenous boundary condition. The composition of $X(x)$ and $Y(y)$ gives

$$T_C(x, y) = X_n(x)Y_n(y) = C_2 \sin(\lambda_n x)C_3 \cosh(\lambda y) = C_n \sin(\lambda_n x) \cosh(\lambda y) \tag{15}$$

leading to

$$T(x, y) = T_1 + \frac{q_1''}{k}x - \sum_{n=0}^{\infty} \frac{\frac{q_2''}{k} \int_{x=0}^{a} \sin(\lambda_n x) dx}{\sinh \lambda_n a \int_{x=0}^{a} \lambda_n \sin^2(\lambda_n x) dx} \sin(\lambda_n x) \cosh(\lambda_n y). \tag{16}$$

References

1. Brecher, C., Bäumler, S., Jasper, D., Triebs, J.: Energy efficient cooling systems for machine tools. In: Leveraging Technology for a Sustainable World, pp. 239–244. Springer, Berlin, Heidelberg (2012). https://doi.org/10.1007/978-3-642-29069-5_41
2. Denkena, B., Möhring, H.-C., Hackelöer, F., Hülsemeyer, L., Dahlmann, D.: Effiziente Fluidtechnik für Werkzeugmaschinen. wt Werkstatttechnik online 101(5), 347–352 (2011)
3. Ramesh, R., Mannan, M. Poo, A.: Error compensation in machine tools—a review. Int. J. Mach. Tools Manuf. 40(9), 1257–1284 (2000). https://doi.org/10.1016/s0890-6955(00)00010-9
4. Brecher, C., Fey, M., Wennemer, M.: Volumetric measurement of the transient thermo-elastic machine tool behavior. Prod. Eng. 10(3), 345–350 (2016). https://doi.org/10.1007/s11740-016-0666-7
5. Thiem, X., Kauschinger, B., Ihlenfeldt, S.: Online correction of thermal errors based on a structure model. Int. J. Mechatron. Manuf. Syst. 12(1), 49 (2019). https://doi.org/10.1504/ijmms.2019.097852
6. Weck, M.: Werkzeugmaschinen 2. Konstruktion und Berechnung. Springer GmbH, (2006). 701 pp. ISBN: 9783540304388
7. Voigt, I., Fickert, A., Wiemer, H., Drossel, W.-G.: Experimental investigation of passive thermal error compensation approach for machine tools. In: Lecture Notes in Production Engineering, pp. 265–277. Springer International Publishing (2023). https://doi.org/10.1007/978-3-031-34486-2_19
8. Reissner. E.: On bending of elastic plates. Q. Appl. Math. 5(1), 55–68 (1947). https://doi.org/10.1090/qam/20440
9. McCune, R.W., Armstrong, C.G., Robinson, D.J.: Mixed-dimensional coupling in finite element models. Int. J. Numer. Methods Eng. 49(6), 725–750 (2000). https://doi.org/10.1002/1097-0207(20001030)49:6<725::aid-nme967>3.0.co;2-w
10. Yue, J.: Micro-macro simulation technique combined with multilevel damage assessment methodology for RC building structures. Adv. Mater. Sci. Eng., pp 1–13 (2015). https://doi.org/10.1155/2015/764517
11. Voigt, I., Drossel, W.-G.: Experimental investigation of heat pipe performance under translational acceleration. Heat Mass Transf. (2021). https://doi.org/10.1007/s00231-021-03106-w
12. Ansys Academic Research Mechanical: Release 2023 R2. Element Reference, ANSYS, Inc., Help System (2023)
13. Peukert, C., et al.: Efficient FE-modelling of the thermo-elastic behaviour of a machine tool slide in lightweight design. In: Proceedings of the Conference on Thermal Issues in Machine Tools. Auerbach/Vogtl, pp. 61–71 (2018)

Modeling for Thermal Error of Slant Bed Lathe Based on Error Decomposition and Differential Equations

Meng Bai, Tao Tao, Hu Shi, and Yingqiang Zheng

Abstract To address the growing demands for precision machining, this paper proposes a novel thermal error compensation method for slant-bed lathes that combines mechanism-based thermal error decoupling methods with differential equation thermal error models. In the first step, the approach primarily decouples thermal errors of the slant bed lathe into component-level thermal errors through a movement measurement method improved on the basis of the experimental setup in ISO230-2 standards. Based on the thermal errors at the level of components, the method further decouples component-level thermal errors with the help of differential equations. The differential equations explicitly model distinct heat sources, including spindle rotation and feed motor operation, enabling precise decomposition of their individual contributions to overall thermal errors. To validate the robustness of the thermal error compensation model, an experiment simulating real processing was conducted, which emphasized that these models achieved a significant improvement in the experimental results by up to 50%. The mechanism-driven approach not only designs thermal error compensation models of the slant bed lathe's main components but also provides a unique insight into thermal error origins through the thermal errors decoupling method.

Keywords Slant bed lathe · Thermal error compensation model · Error decomposition · Differential equations

1 Introduction

Thermal errors account for an increasing proportion of errors in machine tools up to approximately 40–70% [1, 2], which substantially degrade the machining accuracy of machine tools. To eliminate the compact of thermal errors to machining accuracy, comprehension about origins machine tools' thermal errors is necessary. The total

M. Bai · T. Tao (✉) · H. Shi · Y. Zheng
School of Mechanical Engineering, Xi'an Jiaotong University, Xi'an, China
e-mail: taotao@xjtu.edu.cn

© The Author(s) 2026 441
K. Wegener and M. Bambach (eds.), *4th International Conference on Thermal Issues in Machine Tools (ICTIMT2025)*, Lecture Notes in Production Engineering,
https://doi.org/10.1007/978-3-032-01194-7_28

thermal errors of machine tools, affecting the tool-workpiece position deviations, arise from the combined effects of thermal deformation of various components. And thermal errors of CNC machine tools' components are influenced by various internal and external heat sources jointly. Therefore, to improve robustness of the thermal error compensation model decoupling machine tools' thermal errors by components and heat sources is an important prerequisite before establishing a high-precision thermal error compensation model, which remains a challenging and urgent problem to be addressed. Due to the complex coupling factors of thermal errors in machine tools, thermal error compensation methods with decoupling thermal errors by components and heat sources exhibit broad applicability and generalization. Jiri Vyroubal [3] presents a method focused on compensation of the machine's thermal deformation in the spindle axis direction based on decomposition analysis, which is carried out with the help of a specially developed measuring frame that measures deformation of the machine column, headstock, spindle and tool simultaneously. The decomposition process allows describing each machine part's thermal dynamics more precisely than the usual deformation curve describing the complete machine thermal error. Hongyang Du et al. [4] present a new measuring method used in a slant bed lathe to decouple the thermal linear and angular distortions of the spindle and the turret. During the machine tools' working period, positioning errors in feed systems are also combined with geometric errors and thermal errors. It is a fact that machine tools' geometric errors are inevitable results of the inaccuracies of manufacturing and assembling processes of machine tools and their components [5].

The approaches to improve machining accuracy are error avoidance and error compensation generally applied in the field of reducing thermal errors [6]. And thermal error compensation has become the primary approach for enhancing the machining accuracy of machine tools due to its high applicability and low cost [7]. Nowadays, data-mechanism fusion framework modeling has become a trend in thermal error compensation modeling methods. Ziquan Zhan et al. [8] introduce a data-model hybrid-driven framework based on sensor optimization placement for accurate thermal error prediction of Spindle-Bearing systems. To solve challenges poor generalization of data-driven models lacking of structural thermal deformation mechanism, Yingqiang Zheng et al. [9] establish a data-mechanism fusion digital twin (DT) system for spindle thermal errors modeling and compensation, which encompasses the physical entity layer (PEL), DT prediction layer (DT-PL), and DT interaction service layer (DT-ISL). Yingqiang Zheng et al. [10] establish a hybrid model integrating physical and data-driven approaches to forecast the spindle's thermal error by analyzing the thermal deformation law within the spindle assembly.

A thermal error compensation approach targeting a slant bed lathe is presented in this research, which utilizes a move measuring method evolved from ISO230-3 standards and differential equations to decouple thermal errors of the machine tool on the level of components and heat sources.

2 Lathe Structure and Thermal Characteristics Analysis

In the paper, the object of the research is a slant bed lathe machine tool as shown in Fig. 1. The main components of this machine tool include the spindle, headstock, turret, triangular saddle, slideway, tailstock, lathe bed and other auxiliary accessories. The left-view simplified structure and thermal deformation profile of the T65-750 machine tool are shown in Fig. 2. Each component exhibits distinct thermal error under the influence of different heat sources. Heat generated by spindle motor and friction in the front and rear bearings contributes to thermal expansion of the head-stock in the positive H direction and negative V direction due to the asymmetrical construction of the headstock and thermal expansion of the lathe bed in both H direction and V direction. And the thermal expansion of the lathe bed has lag effect because of its high thermal capacity. The slideway installed at the lathe bed also deforms under the thermal deformation of the lathe bed, which causes the motion errors of the turret along the Z-axis. High-speed rotation of the spindle also leads to a sharp increase in the temperature of the hydraulic cylinder and heat sources coming from the hydraulic cylinder would be transformed to the turret component of the lathe through the oil of the hydraulic station, which is the reason of the turret's thermal deformation in the positive X direction and negative Y direction.

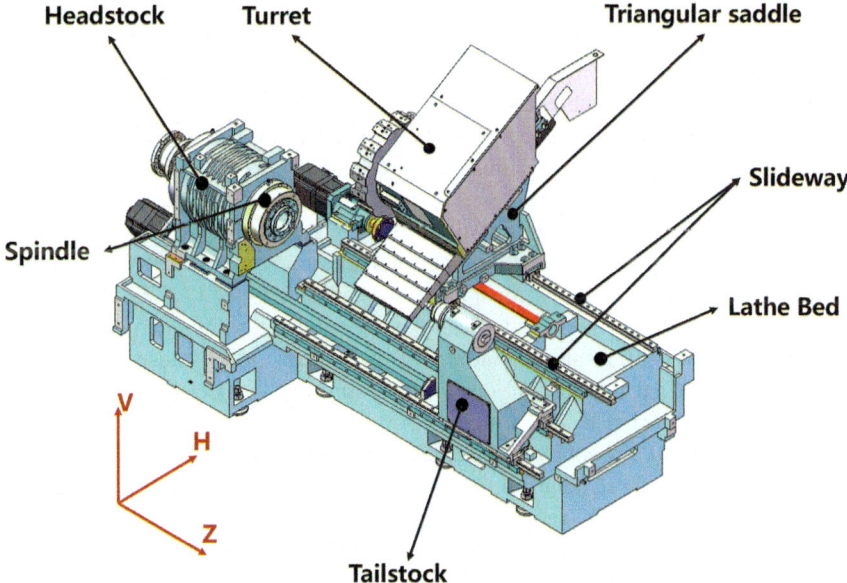

Fig. 1 The schematic diagram of the structure of T65-750

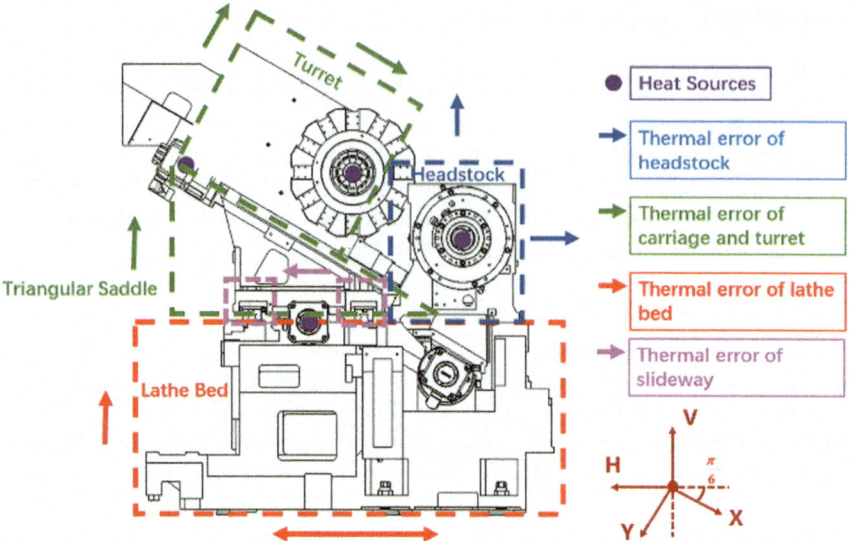

Fig. 2 The left view of T65-750: structural simplification and thermal deformation

3   Thermal Error Decoupling Method of the Lathe

In the paper, the thermal errors of the slant bed lathe result from the combined thermal errors of components caused by different heat sources in Sect. 2. As suggested in ISO230-3 [11], the typical setup for thermal distortion test of structure caused by rotating spindle and moving linear axis on a slant bed lathe is mainly for the measurement of the thermal derivations of the relative distance between cutting tools and workpieces. To achieve robust thermal error compensation, the absolute thermal errors of each component are requirements that are impossible to be measured only by typical experimental setups for a slant bed lathe in ISO230-3. To address the struggle this paper improves the typical experimental setup for thermal distortion of structure caused by rotating spindle and thermal distortion caused by moving linear axis on a slant bed turning center according to the special structure of the slant lathe, as shown in Fig. 3. Through the optimization of the experimental setup, the absolute thermal errors of components are relatively easy to be obtained, which will be explained in more details below. As to the experimental setup evolved from typical setup in ISO230-3 standards, the first mandrel is clamped in the spindle unit and the second mandrel is clamped in the tailstock unit. Four contact displacement Sensors (two aligned with the X-axis and two with the Y-axis) are mounted on the turret unit via a sensor holding bracket, which can measure the change of relative distance in both X and Y directions between test mandrels and the turret unit through the contact between the sensors and mandrels. The sensor probes must contact the test mandrels while maintaining precise alignment to ensure their measurement axes intersect the mandrels' central axes.

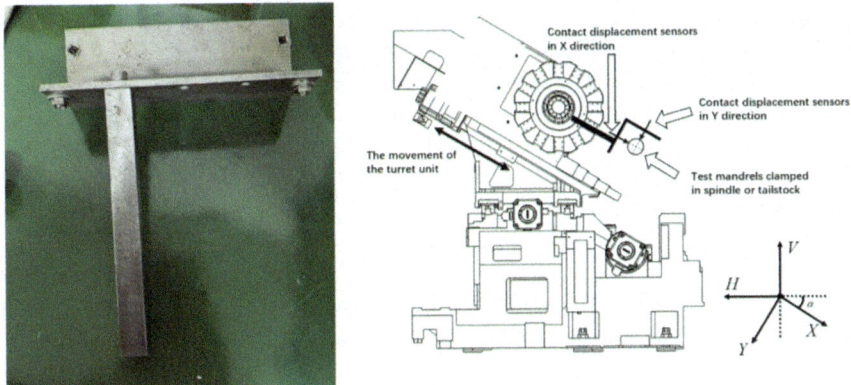

Fig. 3 The experimental nest and schematic diagram of the measurement method

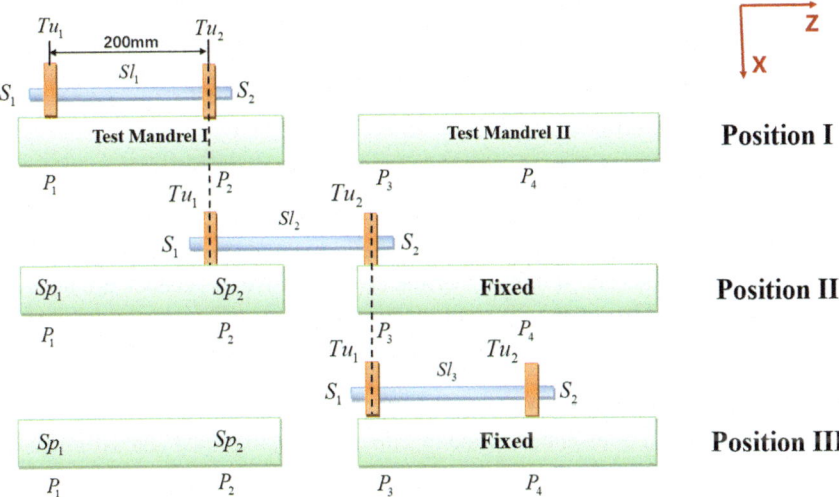

Fig. 4 The schematic diagram of the experimental procedure

The target machine tool was placed in a laboratory where the environmental temperature is controlled within a certain range. The environmental temperature of the experiments on the target machine tool is considered stable during the tests. The whole experiment repeats the measurement process at different positions multiple times, as shown in Fig. 4. The test mandrel I is clamped in the spindle unit, while the test mandrel II is clamped in the tailstock unit. The distance in Z direction between sensors is 200mm, which is the same as the distance the turret unit moves in every measurement cycle. The contact displacement sensors denoted as S_1, S_2 are utilized to measure thermal error data affected by thermal deformation of spindle, turret and slideway in both X and Y direction, which can be represented by Eq. (1). The whole experiment procedure and equations are the same for both X and Y directions so the following details about experiments would use the measuring steps in the X direction as an example. The first step of the experiment is to measure the relative distance between the turret unit and the spindle unit on the Position I, which are denoted as $\Delta X_{P_1,S_1}$, $\Delta X_{P_1,S_2}$, and then the measurement equipment move 200mm along Z direction by the feed drive of the slant lathe so the Sensor I reaches to the last measuring point measured by S_2 to measure $\Delta X_{P_2,S_1}$ and the S_2 obtains $\Delta X_{P_2,S_2}$ symbolizing the deviations between the tailstock and the turret. The final procedure is same as last operation so sensors can measure $\Delta X_{P_3,S_1}$, $\Delta X_{P_3,S_2}$ individually between the turret and the tailstock. And the most important detail about the measurement of thermal errors of spindles is spindle orientation to eliminate data jump due to the roughness of the test mandrels in every measurement for the spindle unit.

$$
\begin{cases}
\Delta X_{P_1,S_1} = -\Delta X_{Sp_1} + \Delta X_{Sl_1} + \Delta X_{Tu_1} \\
\Delta X_{P_1,S_2} = -\Delta X_{Sp_2} + \Delta X_{Sl_1} + \Delta X_{Tu_2} \\
\Delta X_{P_2,S_1} = -\Delta X_{Sp_2} + \Delta X_{Sl_2} + \Delta X_{Tu_1} \\
\Delta X_{P_2,S_2} = \Delta X_{Sl_2} + \Delta X_{Tu_2} \\
\Delta X_{P_3,S_1} = \Delta X_{Sl_3} + \Delta X_{Tu_1} \\
\Delta X_{P_3,S_2} = \Delta X_{Sl_3} + \Delta X_{Tu_2}
\end{cases}
\tag{1}
$$

The measured values are denoted as $\Delta X_{P_1,S_1}$, $\Delta X_{P_1,S_2}$, $\Delta X_{P_2,S_1}$, $\Delta X_{P_2,S_2}$, $\Delta X_{P_3,S_1}$, $\Delta X_{P_3,S_2}$ where $P_i(i = 1,2,3)$ represents measured positions and $S_i(i = 1,2)$ represents the contact displacement sensors. The spindle's thermal errors relative to tailstock are denoted as ΔX_{Sp_1}, ΔX_{Sp_2}. The turret's thermal errors relative to tailstock are denoted as ΔX_{Tu_1}, ΔX_{Tu_2}. The slideway's thermal errors are denoted as ΔX_{Sl_1}, ΔX_{Sl_2}, ΔX_{Sl_3}.

Based on the above chapter about the analysis of the lathe's structure and thermal characteristics and the graphical relationship of these thermal errors, the actual thermal deformation of each unit of the machine tool is simplified and assumed:

(1) All thermal errors measured are coupled with thermal errors of the spindle, turret and slideway.
(2) All thermal errors are relative to the position of test mandrel II clamped in the tailstock. Therefore $\Delta X_{P_2,S_2}$ $\Delta X_{P_3,S_1}$ $\Delta X_{P_3,S_2}$, measured from tailstock, are simplified as only thermal errors caused by the turret.
(3) The lathe has no thermal deformation in a cycle of measurement on the basis of the fact that the measuring time is too short and the spindle is spineless when measuring thermal errors.

According to the experiments procedure and the equations of the measurement results as shown in Eq. (1), thermal errors of the lathe are decoupled into the thermal errors of spindle, slideway and turret based on the analysis of the structure and the thermal characteristics of the lathe and the assumptions mentioned above, as shown in Eq. (2).

$$\begin{cases} \Delta X_{Sl_1} - \Delta X_{Sl_3} = \Delta X_{P_1,S_2} - \Delta X_{P_2,S_1} + \Delta X_{P_2,S_2} + \Delta X_{P_3,S_1} - 2 \times \Delta X_{P_3,S_2} \\ \Delta X_{Tu_1} + \Delta X_{Sl_3} = \Delta X_{P_3,S_1} \\ \Delta X_{Tu_2} + \Delta X_{Sl_3} = \Delta X_{P_3,S_2} \\ \Delta X_{Sp_1} = \Delta X_{Sl_1} + \Delta X_{Tu_1} - \Delta X_{P_1,S_1} \\ \Delta X_{Sp_2} = \Delta X_{Sl_1} + \Delta X_{Tu_2} - \Delta X_{P_1,S_2} \end{cases} \quad (2)$$

4 Experimental Verification

Four contact displacement sensors were employed to measure the thermal errors in the X and Y directions. The distance between sensors I (III) and II(IV) was set to 200 mm. The test mandrels are clamped individually in the spindle and the tailstock as illustrated in Fig. 5. The following experiments were carried out within some configurations in power status and speed of the spindle, as shown in Table 1. Every Test consists of the warming and cooling phase of the lathe. Test 5 and 6 were separated by a power-off interval, during which no data was collected.

There are 12 PT100 temperature sensors deployed to measure the temperatures of the main units of the slant lathe and the ambient temperature. The locations of the temperature sensors are detailed in Fig. 6.

The lathe's status and raw temperature data were collected, as shown in Fig. 7, where the ambient temperature was controlled in $\pm 2.5°C$ within a day. From the results of PT-100 sensors, the analysis about the thermal characteristics of the slant lathe in the previous section has been proved to be correct and reasonable. The higher temperature of T3 than T1 confirms the analysis that the temperature of rear

Fig. 5 Experimental setting

Table 1 Experimental setup of all tests

Test	Duration (h)	Power status	Spindle speed (rpm)
1	0–24	Off	3000
2	24–48	On	3000
3	48–72	On	3000
4	72–96	On	2500
5	96–168	On	2000
6	192–264	Off	2000
7	264–336	On	2000

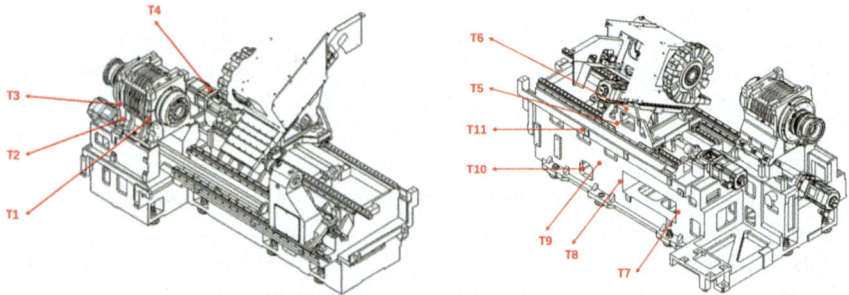

Fig. 6 The schematic diagram of the location of temperature sensors

bearings is much higher than the front bearings because the hydraulic cylinder which generates lots of heat energy locates at the rear side of the spindle unit. T9-T11 which locate at the right side of lathe bed far away from spindle unit follow fluctuations of ambient temperature while the temperature tendencies of T4, T7 and T8 which locate at relative closer to spindle correlate with T1, which demonstrates that the slideway far away from the spindle is unaffected whereas the slideway near the spindle is heated by heat sources from the spindle through contact thermal transfer. So the assumption that $\Delta X_{Sl_3} = 0$ is accepted because the position III of the slideway is far away from the spindle unit and the temperature of the slideway at position III is similar to environmental temperature, so that the variation of slideway at position III is considered to be a relatively small quantity [12]. The Eq. (2) can be simplified as

$$
\begin{cases}
\Delta X_{Sl_1} = \Delta X_{P_1,S_2} - \Delta X_{P_2,S_1} + \Delta X_{P_2,S_2} + \Delta X_{P_3,S_1} - 2 \times \Delta X_{P_3,S_2} \\
\Delta X_{Tu_1} = \Delta X_{P_3,S_1} \\
\Delta X_{Sp_1} = \Delta X_{Sl_1} + \Delta X_{Tu_1} - \Delta X_{P_1,S_1} \\
\Delta X_{Tu_2} = \Delta X_{P_3,S_2} \\
\Delta X_{Sp_2} = \Delta X_{Sl_1} + \Delta X_{Tu_2} - \Delta X_{P_1,S_2}
\end{cases}
\tag{3}
$$

Compared to the temperature of T4, the high amplitude value of T6 is caused by the special structure of the triangular saddle, which leads to high current in the

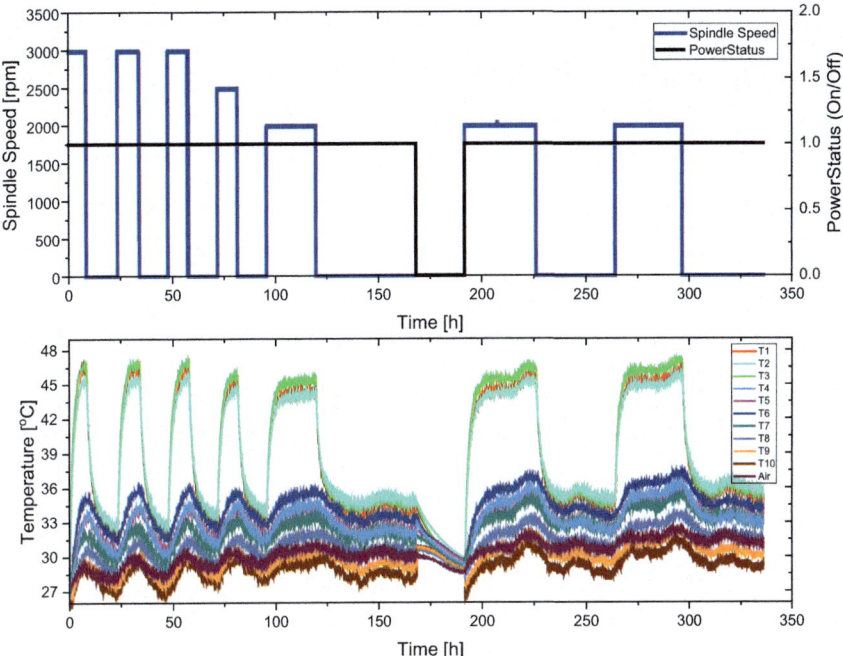

Fig. 7 The status and the temperature of the lathe T65-750

X-axis feed motor aiming to overcome the gravity of the turret unit and maintain the position of the turret unit in the X direction [13]. The spindle unit's time to equilibrium is much less than the turret and the lathe bed. The turret's lag effect stems from heat transfer delays through long hydraulic oil pipes. The factor of the lathe bed's lag effect is mainly its large size and its large heat capacity mentioned in Sect. 2. Because of the lag effect of the turret and the lathe bed thermal errors in the X direction, which are the main factors influencing the accuracy of the machine tool, have peak-like curves that are complicated to be compensated only by temperature data.

Based on the decoupling method mentioned in Sect. 3, thermal errors of each unit of the lathe were individually decoupled from the thermal errors of Sensor I and Sensor III measured on the Position I, and in both X and Y directions the error data for each decoupled unit is shown in Fig. 8. In the X–Y coordinate system, the curve of the turret unit's displacement initially increased during the early machining phase due to the turret's thermal expansion in the X direction. This phenomenon that the displacement of the turret gradually decreased over time after reaching its peak value, suggesting an increasing distance between the turret and the tailstock, can be attributed to the lag effect of the lathe bed. Since the headstock holding the spindle is installed in the H-V coordinate system, modeling the spindle unit's thermal errors in the X–Y coordinate system may not be fully accurate. Because the contact surface between the headstock and the lathe bed is composed of numerous asperities with different scales, the thermal conductivity of the filling material in the cavities (such as air or oil) is much lower than that of the parts and the thermal contact resistance (TCR) and temperature jumps are formed at the joint surface. Namely, TCR restricts the heat flowing from the heat source to another object contacting the heat source, which results in a high local temperature and high local thermal displacements of the heat source [14]. This limitation arises because the spindle unit's thermal deformation is influenced by its unique mounting configuration. Because of the installation methods of the headstock unit and slideway unit and the thermal contact resistance formed by the mounting contact surface, it is better to decompose the thermal errors of the spindle and the slideway in H-V coordinates. Further analysis is required to determine the actual deformation of the spindle and the slideway by applying coordinate transformation. The mapping relationship between the H-V coordinates and the X–Y coordinates is available at each measuring point on the mandrel:

$$\begin{cases} \Delta H = -\Delta X \cos\alpha + \Delta Y \sin\alpha \\ \Delta V = -\Delta X \sin\alpha - \Delta Y \cos\alpha \end{cases} \tag{4}$$

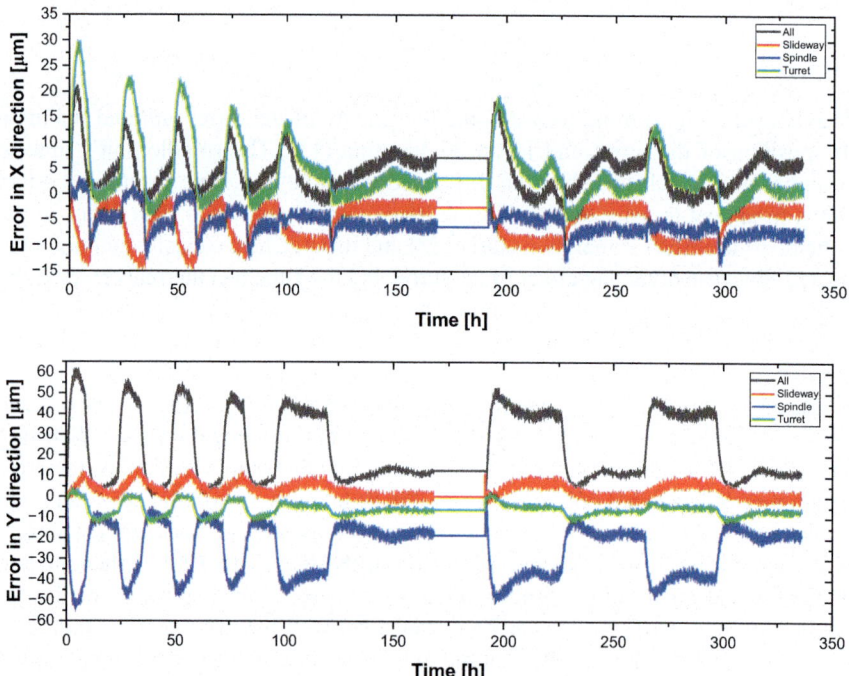

Fig. 8 Results of decoupling thermal errors on P_1 in X–Y coordinate system

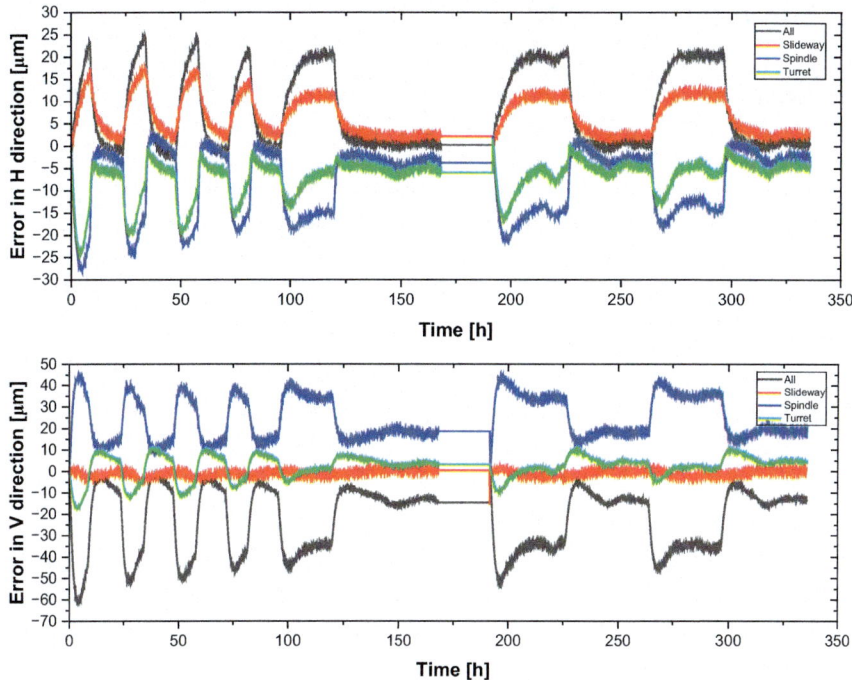

Fig. 9 Results of decoupling thermal errors on P_1 in H-V coordinate

where ΔX, ΔY, ΔH, ΔV symbolize the thermal deformation of the same component in X direction, Y direction, H direction and V direction. Through Eq. (4) above decoupled thermal errors in the H-V coordinate are shown in Fig. 9. Thermal errors of the slideway and spindle are the main focus in the H-V coordinate. The thermal errors of slideway in the H direction are the main thermal errors because of the truth that the thermal displacement of the slideway in the V direction is nearly 0, which validates the structural and thermal analysis of the lathe bed. The trends of spindle's thermal errors in both H and V directions are also consistent with the analysis of the spindle in Sect. 2. According to structural and thermal characteristics analysis of the machine tool and installation of the main components thermal error compensation models for the spindle, slideway in H-V coordinate and the turret in the X direction would be investigated in the following chapters.

5 Thermal Error Modeling

The thermal error of machine tools is caused by a system of heat transfer and thermal expansion. The description of the thermal displacements depending on an underlying thermal load is done by a system of differential equations. For each thermal error identified as significant, a differential equation is used to predict the error depending on a system input and specific parameters found by a parameter identification procedure [15]. The thermal deformation of each unit of machine tools can be regarded as the mutually coupled influence of each heat source from the machine tool. Through a series of simplifications, a system of differential equations used for modeling thermal errors in [15, 16] is optimized as:

$$
\begin{cases}
\dfrac{dy}{dt} = p_1 \cdot (u - y), u \neq 0 \\
\dfrac{dy}{dt} = -p_2 \cdot y, u = 0
\end{cases}
\tag{5}
$$

where t denotes as duration of the heat source impact, the unit of t is seconds, y represents the thermal error caused by the heat source at t time, and u symbolizes the amplitude value of thermal error caused by the heat source. The term dy/dt describes the change over time of a thermally induced error, which is a function of either the axis speed or the power status of the machine tool, read-out from a programmable logic controller of the machine tool. Parameters p_1, p_2 are constants only associated with heat sources and thermal convection, where p_1 is used for the warming period and p_2 is used for the cooling period. The target machine tool is experimented in the same environment so the equation above neglects the influence of the environmental temperature during the measurement. The constant parameters p_1, p_2 are derived from measurement. In the working period of the machine tool each unit is affected by various heat sources in the machine tool, and thermal displacements of different units are also described as the combination of a series of differential equations mentioned above. Focusing on the slant bed lathe, the heat sources mainly divided into two types, the heat sources from the rotation of the spindle and the heat sources caused by the power of the machine tool. Through the different equations mentioned above, the thermal errors of main components are decoupled into thermal errors caused by different heat sources. The performance of the compensation model is evaluated by three different criteria. In order to further quantify the fitting performance of the compensation model, this paper used metrics such as Mean Absolute Error (MAE), Root Mean Square Error (RMSE) and R-squared as performance evaluation indicators. The compensation model of each unit on P_1 in different directions based on the differential equations is shown as Figs. 11 and 12. Because of the installation of the spindle unit and the slideway unit, the compensation model of thermal errors of the spindle unit is carried out in H-V coordinate, while the thermal errors of the turret unit which is moved in the X direction by the X-axis feed motor are compensated by the model based on differential equations in the X direction. At the same time thermal errors of the slideway in the V direction are too small to be compensated limited by the resolution of axis positioning of the machine tool.

According to Table 2, the MAE and RMSE are both relatively small for each unit of the machine tool. For the spindle unit, the turret unit and the slideway unit, the R-squared are 0.9526, 0.7060, 0.9720, 0.8831 in order, which indicates that the compensation models based on the differential equation above have a good fit and can meet the requirements for practical compensation (Figs. 10 and 11).

The thermal error compensation model for the coupled thermal errors of the slant bed lathe at measured point P_1 in the X-direction is illustrated in Fig. 12. Based

Table 2 Criteria of main components' thermal error compensation models

Error	MAE [μm]	RMSE [μm]	R^2
Spindle in H direction	1.1505	1.5614	0.9526
Spindle in V direction	2.6631	5.2872	0.7060
Slideway in H direction	0.5930	0.7305	0.9720
Turret in X direction	1.6151	2.0158	0.8831

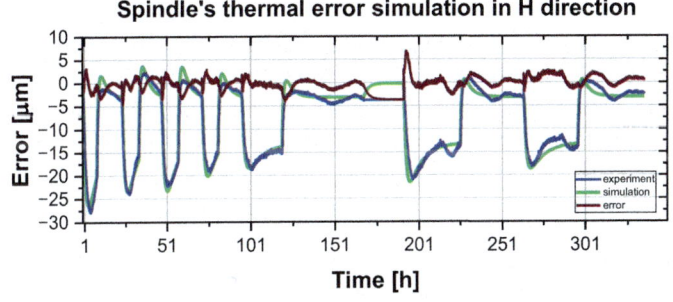

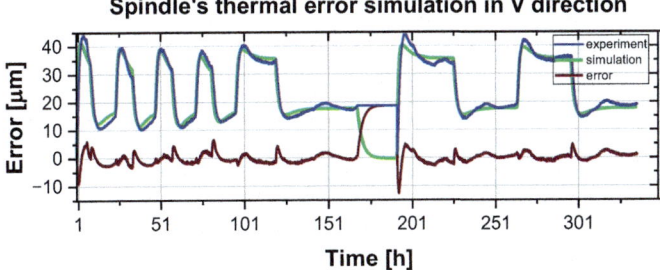

Fig. 10 The thermal error compensation model on P_1 of the spindle in the H-V coordinate system

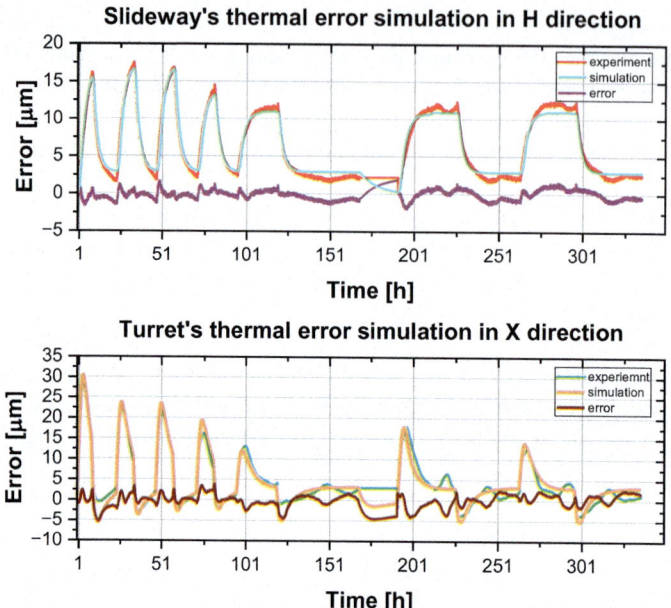

Fig. 11 The thermal error compensation model on P_1 of slideway in H direction and turret in X direction

on the experimental setting of contact displacement sensors, the performance of the compensation model for $\Delta X_{P_1,S_1}$ was validated by the experiment to simulate the real manufacturing procedure of the machine tool, as shown in Fig. 13. The thermal error compensation function was activated 40 min after the experiment began. To simulate actual operating conditions, the machine tool initially ran at a spindle speed of 2000 rpm for 3 h, followed by a 2-h power-off cooling period. Subsequently, the machine was restarted and operated continuously for 9 h at 2000 rpm. As evident from the figure, the measured value from the experiment exhibit fluctuations within a small range, while the compensation model's performance metrics presented in Table 3 confirm the robustness and effectiveness of the proposed thermal error compensation approach for the slant-bed lathe.

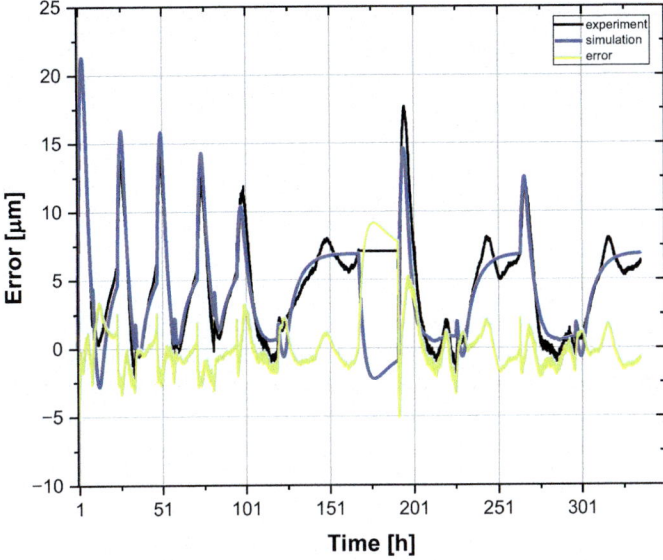

Fig. 12　The compensation model of coupled thermal error on P_1 of the slant bed lathe

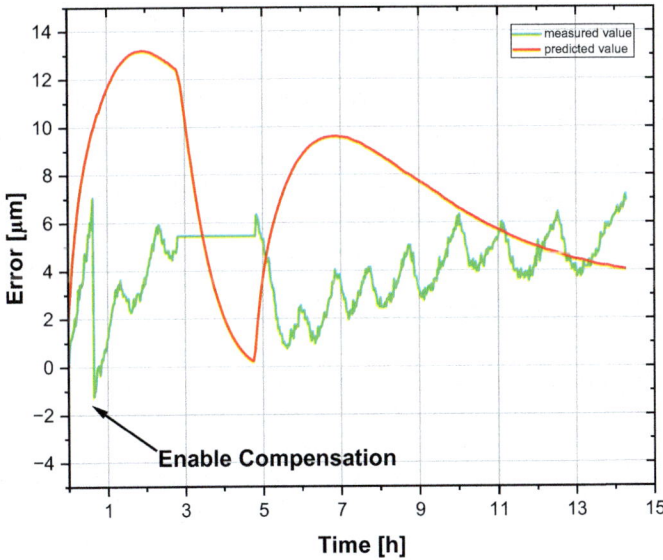

Fig. 13　Thermal error compensation model validation on P_1 in X direction

Table 3 Criteria of coupled thermal error compensation model in X direction

Criteria	Peak value [μm]	MAE [μm]	RMSE [μm]
Uncompensated	17.2	11.3	11.7
Compensated	8.4	4.2	4.4
Improvement [%]	48.94	63.10	62.04

6 Conclusion

This paper proposes a thermal error compensation approach for slant-bed lathes based on the decoupling thermal errors by movement measurement improved from the test setup in ISO230-3 standards and differential equations. Through thermal and structural analysis of the slant-bed lathe, absolute thermal errors of its main components are analyzed, and the lag effects of both the turret and lathe bed are identified as one of the primary contributors to the machine tool's thermal displacements. Using experimental setups optimized from ISO 230–3 standards and the thermal error decoupling method based on movement measurement presented in Sect. 3, thermal errors are decoupled into independent component-level thermal errors, ensuring unbiased separation of individual components' effects. And these models also account for the distinct mounting configurations of the components in different coordinate systems. Based on the decoupled thermal errors of each component from experimental results, this study further decouples thermal errors of main components through differential equations on the level of heat sources, which reveals the underlying thermal deformation mechanisms of the machine tool. The input parameters of models are only the speed of the spindle and the power status of the machine tool, which are the main factors affecting the temperature field of the machine tool and can be easily obtained from the NC system of machine tools. The thermal error compensation models for each component of the slant-bed lathe demonstrate strong individual performance based on evaluation criteria. When applied to experiments simulating actual machining conditions, these models achieved a significant reduction in thermal errors along the X-direction, improving measuring accuracy by up to 50%, validating the robustness and effectiveness of the proposed compensation methodology.

This decoupling method demonstrates potential applicability to other structurally similar systems under appropriate operating conditions. The decoupling method only requires two test mandrels and a special sensor holder and involves three measurements within one experimental cycle to effectively utilize experimental results for thermal errors decomposition, which is really helpful to find out absolute displacements of main components of machine tools to further predict and compensate thermal errors of each unit. However, the thermal error compensation model presented in this paper still has some refinement for real processing compensation, for example, the thermal angular errors of the spindle and the turret. The thermal error decoupling method mentioned in Sect. 3, a newly proposed experimental method, still has some aspects to validate by some extra experiments, such as the effect of turret movement on measurements. Although the thermal error compensation model has some refinement to do, it is still worthwhile to explore the mechanism of thermal errors in machine tools which are similar to the slant bed lathe.

Acknowledgements The presented findings are funded by the National Key Research and Development Program of China and Zhejiang Headman Machinery Co., Ltd. The authors would like to express their sincere gratitude for the financial support.

References

1. Mayr, J., Jedrzejewski, J., Uhlmann, E., Alkan Donmez, M., Knapp, W., Härtig, F., Wendt, K., Moriwaki, T., Shore, P., Schmitt, R., Brecher, C., Würz, T., Wegener, K.: Thermal issues in machine tools. CIRP Ann. **61**(2), 771–791 (2012). https://doi.org/10.1016/j.cirp.2012.05.008
2. Gao, W., Ibaraki, S., Donmez, M.A., Kono, D., Mayer, J.R.R., Chen, Y.-L., Szipka, K., Archenti, A., Linares, J.-M., Suzuki, N.: Machine tool calibration: measurement, modeling, and compensation of machine tool errors. Int. J. Mach. Tools Manuf **187**, 104017 (2023). https://doi.org/10.1016/j.ijmachtools.2023.104017
3. Vyroubal, J.: Compensation of machine tool thermal deformation in spindle axis direction based on decomposition method. Precis. Eng. **36**(1), 121–127 (2012). https://doi.org/10.1016/j.precisioneng.2011.07.013
4. Du, H., Jiang, G., Tao, T., Hou, R., Yan, Z., Mei, X.: A thermal error modeling method for CNC lathes based on thermal distortion decoupling and nonlinear programming. Int. J. Adv. Manuf. Technol. **128**(5–6), 2599–2612 (2023). https://doi.org/10.1007/s00170-023-12038-0
5. Huang, Y.B., Fan, K.C., Lou, Z.F., Sun, W.: A novel modeling of volumetric errors of three-axis machine tools based on Abbe and Bryan principles. Int. J. Mach. Tools Manuf **151**, 103527 (2020). https://doi.org/10.1016/j.ijmachtools.2020.103527
6. Liu, K., Han, W., Wang, Y., Liu, H., Song, L.: Review on thermal error compensation for feed axes of CNC machine tools. J. Mech. Eng. **57**(3), 156–173 (2021). https://doi.org/10.3901/JME.2021.03.156
7. Fu, G., Zheng, Y., Zhou, L., Lu, C., Zhang, L., Wang, X., Wang, T.: Look-ahead prediction of spindle thermal errors with on-machine measurement and the cubic exponential smoothing-unscented Kalman filtering-based temperature prediction model of the machine tools. Measurement **210**, 112536 (2023). https://doi.org/10.1016/j.measurement.2023.112536
8. Zhan, Z., Fang, B., Wan, S., Bai, Y., Hong, J., Li, X.: Application of a hybrid-driven framework based on sensor optimization placement for the thermal error prediction of the spindle-bearing system. Precis. Eng. **89**, 174–189 (2024). https://doi.org/10.1016/j.precisioneng.2024.06.011
9. Zheng, Y., Yang, H., Jiang, G., Hu, S., Tao, T., Mei, X.: Data-mechanism fusion modeling and compensation for the spindle thermal error of machining center based on digital twin. Measurement **250**, 117152 (2025). https://doi.org/10.1016/j.measurement.2025.117152
10. Zheng, Y., Tao, T., Mei, X., Tian, W., Hu, S., Du, H.: Modeling of thermal errors in electric spindle based on a hybrid approach of thermal deformation theory and data drive. J. Intell. Manuf. (2025). https://doi.org/10.1007/s10845-025-02565-w
11. ISO 230–3.: Test code for machine tools–Part 3: determination of thermal effects, international organization for standardization, Geneva, Switzerland (2007)
12. Shi, H., Qu, Q., Mei, X., Tao, T., Wang, H.: Robust modeling for thermal error of spindle of slant bed lathe based on error decomposition. Case Stud. Therm. Eng. **51**, 103564 (2023). https://doi.org/10.1016/j.csite.2023.103564
13. Yan, Z., Tao, T., Du, H., Shi, H., Mei, X.: An experiment-based multi-objective modeling method for thermal errors of slant bed CNC lathes. Int. J. Adv. Manufact. Technol. **120**(9–10), 6565–6583 (2022). https://doi.org/10.1007/s00170-022-09158-4

14. Ma, C., Mei, X., Yang, J., Zhao, L., Shi, H.: Thermal characteristics analysis and experimental study on the high-speed spindle system. Int. J. Adv. Manufact. Technol. **79**(1–4), 469–489 (2015). https://doi.org/10.1007/s00170-015-6821-z
15. Mayr, J., Egeter, M., Weikert, S., Wegener, K.: Thermal error compensation of rotary axes and main spindles using cooling power as input parameter. J. Manuf. Syst. **37**, 542–549 (2015). https://doi.org/10.1016/j.jmsy.2015.04.003
16. Wegener, K., Gebhardt, M., Mayr, J., Knapp, W., Blaser, P.: Thermal compensation for 5-axis machine tools with physical model (2014). https://doi.org/10.3929/ETHZ-A-010268930

Simulation Approach for Considering the Thermal Effects of Cutting Fluid Use and Axis Movements in the Machine Tool Workspace via Coupled CFD Models

Alexander Geist⬤, Steffen Brier, Janine Glänzel⬤, Christian Naumann⬤, Joachim Regel, Martin Dix, Steffen Ihlenfeldt, Marc-André Dittrich, Matthias Brand, and Denis Ulrich

Abstract Advances in hardware performance and modern simulation software tools have enabled significant improvements in the capabilities and precision of machine tool thermal error simulations. This in turn allows for better predictions of thermal errors and more effective machine tool design optimization. One area, which still

---

A. Geist (✉) · J. Glänzel · C. Naumann · M. Dix · S. Ihlenfeldt
Fraunhofer Institute for Machine Tools and Forming Technology IWU, Reichenhainer Straße 88, 09126 Chemnitz, Germany
e-mail: alexander.geist@iwu.fraunhofer.de

J. Glänzel
e-mail: janine.glaenzel@iwu.fraunhofer.de

C. Naumann
e-mail: christian.naumann@iwu.fraunhofer.de

M. Dix
e-mail: martin.dix@mb.tu-chemnitz.de

S. Ihlenfeldt
e-mail: steffen.ihlenfeldt@tu-dresden.de

S. Brier · J. Regel · M. Dix
Professorship Production Systems and Processes, Chemnitz University of Technology, Reichenhainer Straße 70, 09126 Chemnitz, Germany
e-mail: steffen.brier@mb.tu-chemnitz.de

J. Regel
e-mail: joachim.regel@mb.tu-chemnitz.de

S. Ihlenfeldt
Institute for Machine Tools Development and Adaptive Controls, Technical University Dresden, Helmholtzstraße 10, Dresden 01069, Germany

M.-A. Dittrich · M. Brand · D. Ulrich
DMG MORI Seebach GmbH, Neue Straße 61, Seebach 99846, Germany
e-mail: m.dittrich@dmgmori.com

© The Author(s) 2026   459
K. Wegener and M. Bambach (eds.), *4th International Conference on Thermal Issues in Machine Tools (ICTIMT2025)*, Lecture Notes in Production Engineering,
https://doi.org/10.1007/978-3-032-01194-7_29

holds large uncertainties, is the workspace of machine tools, where moving assemblies, the cutting process, cooling systems and other support systems create highly dynamic situational conditions, which are hard to predict, model or even measure. To account for these situational thermal interactions in error compensation models, this paper presents a coupled CFD modeling approach, which computes realistic transient heat transfer coefficients (HTCs) for improved thermal FEM analyses. This new composite model includes a tool-adjacent model equipped with internal coolant supply and integrated linear axis movements. The second submodel considers forced convection from axis movements in the workspace, the air evacuation unit and also from the near-tool model. Subsequently, the steady-state tool-adjacent CFD model is integrated into the transient workspace CFD model by transferring flow velocities and volume fractions as boundary conditions. Steps three and four use the computed HTCs on the machine tool surface to simulate the resulting temperature and deformation fields. By combining all four models, the influence of different process scenarios on the resulting thermal error of the machine tool can be obtained. Initial simulations for selected load cases reveal load-dependent differences in the resulting HTCs and temperature fields.

Keywords Thermal error · Modelling · Computational fluid dynamics · Simulation · Heat transfer coefficients · Cutting fluid

1 Introduction and State of the Art

Ambient factors and the cutting process are significant contributors to thermal errors in machine tools [1]. The convective heat transfer coefficients (HTCs) on machine surfaces, particularly during the use of cutting fluid, play a crucial role. Since HTCs are challenging to measure directly and analytical calculations are only feasible for simple cases, numerical calculations typically represent the most accurate approach. Ideally, a combination of computational fluid dynamics (CFD) simulation for determining the HTCs and finite element method (FEM) simulations for calculating temperature fields and thermo-elastic deformations would be employed. However, such coupled simulations are resource- and time-intensive [2]. To address this challenge, Kumar et al. [3] have developed a method for decoupling the CFD from the subsequent thermo-elastic simulation via the use of characteristic diagram-based HTC interpolation.

A noteworthy aspect is the heat transfer at the tool-chip interface [4]. High HTCs exceeding 400 kW/(m^{2}K) have been observed at this interface. Coupled numerical

M. Brand
e-mail: matthias.brand@dmgmori.com

D. Ulrich
e-mail: denis.ulrich@dmgmori.com

approaches like FEM and CFD exhibit substantial resource requirements, which is why mesh-free methods, such as the finite pointset method [5], are frequently utilized.

Brier et al. (2024) simulated a macroscopic model of the tool under coolant influence, using only a heat source term to represent process energy, thereby omitting details from the chip formation zone.

Fluid–structure interaction (FSI) simulations demonstrate that local HTCs are heavily dependent on the wetting status of the tool surface and the turbulence level of the coolant flow [6]. However, these simulations are time-consuming, especially for complex geometries. Therefore, a new, simplified simulation approach is required.

A multi-domain simulation model for deep hole drilling was developed by Baumann et al. [7] utilizing coupled particle simulations to calculate chip positions, which serve as a basis for CFD simulations of fluid flows.

While the aforementioned factors complicate the precise determination of local HTCs in the workspace, they are crucial for the accurate prediction and subsequent compensation of thermal errors. Data-driven methods require a comprehensive and accurate representation of both the machine tool and the relevant thermal influences [8].

This paper presents the necessary CFD submodels to couple a near-tool CFD model with an encompassing workspace model, allowing for the description of airflow and coolant flow throughout the entire workspace of the machine tool. This enables the determination of the effects of multi-phase flow on the local HTCs at the machine surfaces across the entire workspace. The geometrically simplified near-tool model captures the detailed coolant distribution resulting from contact with the rotating tool surface and is based on previous numerical investigations [3], [9]. The relevant flow field data from the near-tool CFD model are transferred to the workspace model to subsequently calculate the HTCs in the workspace, which is the focus of this paper. Based on the detailed description of the submodels used here, which were explained in detail in [10], the coupling of the submodels was supplemented and complemented with different load cases for milling and drilling operations.

The following section provides a detailed description of a new coupled CFD model consisting of a near-tool model A and a larger workspace model B. Section 3 then defines some relevant thermal load cases, for which the composite simulation model is suitable. The results of these simulations are shown in Sect. 4. A summary and outlook on subsequent research conclude the paper.

2 Description of the Coupled CFD Workspace Model

This chapter describes the implementation of the two coupled submodels for the tool environment and the workspace. The near-tool CFD model is a static, isothermal FSI model, which determines the air and cutting fluid movement into the wider workspace, which result from the tool rotation, cutting fluid sprayed onto the tool from external supply systems and the tool-internal supply channels. This submodel has a cube-shaped fluid domain containing the tool (and workpiece) and provides the boundary conditions for the workspace simulation. The workspace model is the second submodel, which treats the near-tool domain as an opening and calculates the transient effect of cutting fluid and moving air on the heat transfer with the workspace walls.

2.1 Near-Tool Environment Model (Model A)

Both presented CFD models A and B are based on the governing equations of continuity (Eq. 1) and momentum (Eq. 2), and for the workspace model also energy (Eq. 3) [11]:

$$\nabla v = \frac{\partial v_x}{\partial x} + \frac{\partial v_y}{\partial y} + \frac{\partial v_z}{\partial z} \tag{1}$$

$$\rho \left(\frac{\partial v_x}{\partial x} + (v\nabla)v \right) = f - \nabla p + \eta \Delta v \tag{2}$$

$$\rho c_p \left(\frac{\partial T}{\partial t} + (u\nabla)T \right) = \lambda - \nabla^2 T \tag{3}$$

with ρ (density), f (volume force), t (time), p (pressure), v (velocity), η (dynamic viscosity) and x, y, z as spatial directions. To calculate the HTCs on the surface in the workspace model, the energy equation is solved with c_p (specific heat capacity), λ (thermal conductivity), u (internal energy) and T (temperature). The turbulence of the flow field was included via the k-ω SST model [12] and additional transport equations for the turbulent kinetic energy k and the dissipation rate ε were solved.

As previously mentioned, the developed coupled CFD model is particularly useful for simulating the effects of complex thermo-fluidic conditions during cutting operations. One such scenario is hole drilling with tool-internal cutting fluid supply. In order to simulate the flow field for drilling with internal cooling, it is necessary to utilise a split model approach. The flow field before the tool engagement differs significantly compared to the situation with an engaged drill. In open space, the coolant flow is unobstructed in the absence of a workpiece. Model A_{free} contains only the rotating tool with the coolant flow due to the internal cooling channels and A_{wp} holds the configuration of an engaged rotating tool for different drilling depths.

Submodel A_{free} is based on a static isothermal multi-phase (volume-of-fluid) CFD model implemented in ANSYS CFX (ANSYS 2021 R2), see Fig. 1. The simulation domain consists of a static cube-shaped volume ($0.2 \times 0.2 \times 0.2$ m) with a rotating cylinder inside. The rotation of the drill (here: Kentip FS GTP by Kennametal) was implemented via a frozen-rotor approach and all surrounding exterior faces of the cube were defined as openings with the corresponding ambient pressure level. The multi-phase simulation uses a Euler-Euler formulation (volume-of-fluid) with a k-ω SST turbulence model to describe the flow behaviour near the walls and in a free space. Additionally, the surface tension ($\sigma_{coolant} = 0.073$ N/m) and continuum surface force model were also included.

The near-tool model in free space has $\approx 2.8 \times 10^6$ cells. This approach delivers the coolant distribution prior to the drill engagement and the obstructing workpiece. The drill engagement requires the inclusion of the tool holder and workpiece geometries. The simulated drilling depth was 16 mm (Fig. 2). In this model, the tool and tool holder walls were rotating, realized by the frozen-rotor method in ANSYS CFX. The relevant model and tool parameters are shown in Fig. 1. The utilized numerical methods are identical to model A_{free}. Model A_{wp} comprises $\approx 2.3 \times 10^6$ cells.

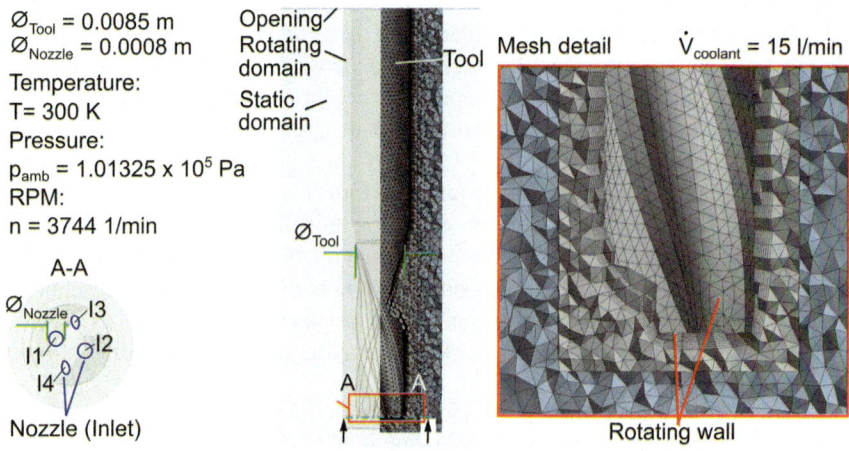

Fig. 1 Geometry and setup of the free stream model A_{free}

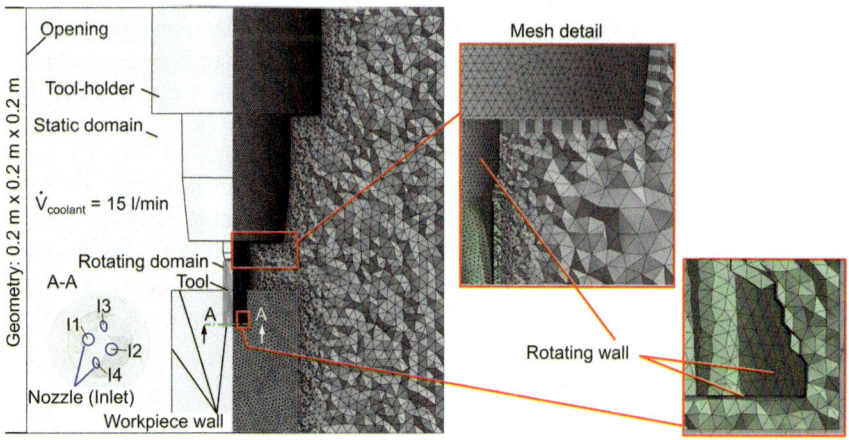

Fig. 2 Geometry and setup of the engaged drill model

Table 1 Transferred flow field data

Variable	Unit	Description
α	–	Volume fraction
v_x	m/s	Velocity component in x-direction
v_y	m/s	Velocity component in y-direction
v_z	m/s	Velocity component in z-direction
k	m^2/s^2	Turbulent kinetic energy
ω	m^2/s^3	Specific turbulent dissipation rate

In order simulate the workspace conditions during a drilling process in a transient CFD simulation, some additional simulations of the near-tool model may be required. Models A_{free} and A_{wp} provide different input to the workspace model depending on the considered situation. A drilling operation comprises the free space situation, then an approach phase, where the internal coolant is reflected from the approaching workpiece surface, then several stages of the drill engaged at different depths inside the workpiece and finally the withdrawal of the tool. The withdrawal should be very similar to the penetration phase in terms of cutting fluid dispersal. In total, a set of 4–5 different steady-state near-tool CFD simulations with the workpiece, between which interpolation may be used, should be sufficient to obtain an accurate representation of the near-tool effects of drilling on the workspace and the machine tool as a whole. Similarly, the near-tool model with workpiece may be adjusted to fit different types of milling operations.

All exterior surfaces of the cubic simulation domain are interfaces to the workspace model B. On these surfaces, the flow field variables are determined and transferred to model B (Table 1).

The near-tool model assumes a time-independent, local flow field in the control volume, which is why a stationary calculation is possible. This assumption is valid for fast rotating tools and a sufficiently small control volume. The temperature hardly plays a role for the flow field, which is why the energy equation is neglected here.

The model of the entire workspace, on the other hand, must consider the flow resolved in time and describes the space outside the control volume.

The energy equation is taken into account for the heat transfer. This model can be used to calculate the HTCs for assemblies in direct contact with the cooling lubricant and supply them to subsequent thermal simulations of the entire machine tool. As the linear slides are covered by a protective cover, there is no direct heat transfer to them. To reduce the model size, these are thus neglected in this model. A second model coupling, which would be required to include these assemblies, is planned for the future.

2.2 Machine Tool Workspace Model (Model B)

The following section describes the workspace model (Model B) on the example of the machine tool DMU 80 eVo by DMG MORI, which is a five-axis CNC machine with a swivel rotary table (B and C axis) for simultaneous five-axis milling. The machine has a compact shape, which has led to a thermally asymmetrical design. As a result, the workspace of the DMU 80 eVo has a complex and irregular geometry. Figure 3 shows a model of the machine tool with the machine frame, swivel rotary table and its workspace.

The three linear slides have been omitted because they are separated from the workspace by the workspace housing. The fluid domain of the workspace is created from the CAD data of the machine tool and the enclosure as a filling volume and meshed with tetrahedral volume cells. In total, 8.1×10^6 volume cells were generated, which contain the elements of the fluid region, the solids and the boundary layer cells (inflation). The model allows HTC calculation, because the machine surfaces contacting the workspace are taken into account.

Fluid–solid interfaces with heat transfer are defined between the contact surfaces of the fluid region and the solid surfaces. The linear slides and the headstock were omitted in this model as there is still a partition wall between them in the form of a protective covering. There is thus no direct heat transfer with these assemblies. This model therefore focuses on the workspace walls and the swivel rotary table. The mesh density is refined locally in areas, where a high flow gradient and turbulence are expected. This is done by partitioning the fluid workspace into multiple parts and assigning the appropriate mesh density based on aspects like surface topology, expected turbulence or proximity to relevant flow sources like the cutting tool, coolant nozzles or evacuation unit. Observations of actual machine tool operations can help to determine these partitions. In particular, the interface to the near-tool volume is finely and evenly meshed to enable a clean transfer of the solution data from model

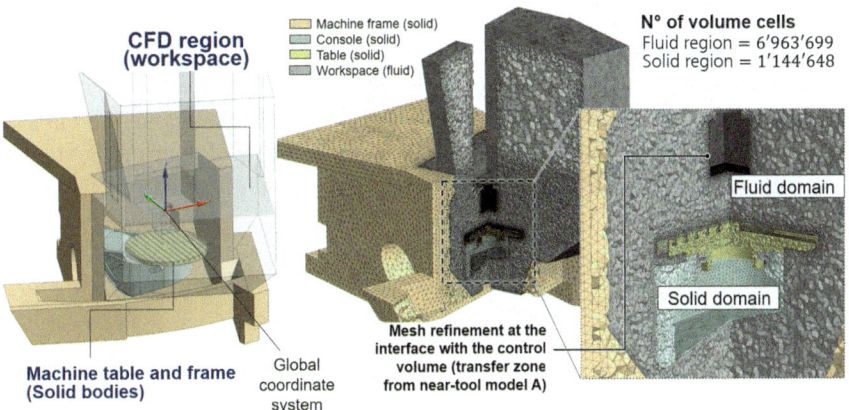

Fig. 3 CAD model and volume mesh of the entire workspace model

A. The same applies to the table surface, in order to account for the complex table geometry with its T-slots. In the effective range of the multi-phase flow, the volume mesh is generally highly refined in order to resolve the phase boundaries in more detail and thus increase the accuracy of the solution.

Unlike the near-tool model, the workspace model calculates a transient solution with defined time steps. With the implicit solver, each time step is solved with a maximum number of intermediate iterations. The settings in this model allow a maximum of 12 substeps per time step. The convergence between the individual intermediate iterations is monitored via defined temperature monitor points. These monitor points correspond to the actual temperature sensors in the machine. The time step size can be selected adaptively depending on the load case. Table 2 provides values recommended by the authors for the time step sizes suitable for different scenarios. The model is therefore very flexible, allowing both very short time periods (e.g. for fast axis movements) and very long, quasi-stationary processes to be simulated.

In addition to the solution of the energy conservation equation, the multi-phase model differs slightly from that in model A. The phase boundary between gaseous (air) and liquid (coolant) is mapped via the free surface instead of the mixture model. This allows phase boundaries to be clearly separated, so that wetting of the workspace surface with liquid cooling lubricant can be simulated. Adhesion effects and contact angles (surface normals) are not yet taken into account in this model. The entire fluid in the workspace is considered a continuum, i.e. the fluid particles inside are continuously distributed and fill the entire region of the space.

In order to simulate axis movements, a distinction is made between rotational and translational axes, where the latter are assigned a time-defined axis position for each time step (specified boundary wall displacement). For the C-axis, a pitch angle of $360°$ is assumed (i.e. one full rotation per step). With the applied transient rotor–stator mixing model, however, partial rotations can also be investigated.

The axis movements affect the flow field via moving wall assemblies or the rotation of the table, so that air is displaced and circulated. There is also the secondary influence of the (possibly changing) position of the control volume and its movement in the workspace. The position and movement of the control volume is coupled to the translational axis movement. This coupling is, however, only one-way—from model A to model B, because the linear axis movements are very slow compared to the speed of the rotating tool. Figure 4 shows an example of the consideration of translational (Y-axis) and rotational (C-axis) axis movements.

Table 2 Time step sizes for various thermal issues

Time step [s]	CFD load case description
0.01 … 1	Axis movement investigation
10...60	Usage of cutting fluid without axis motion
180 … 600	Constant heat load; natural convection only, without axis motion or coolant flows

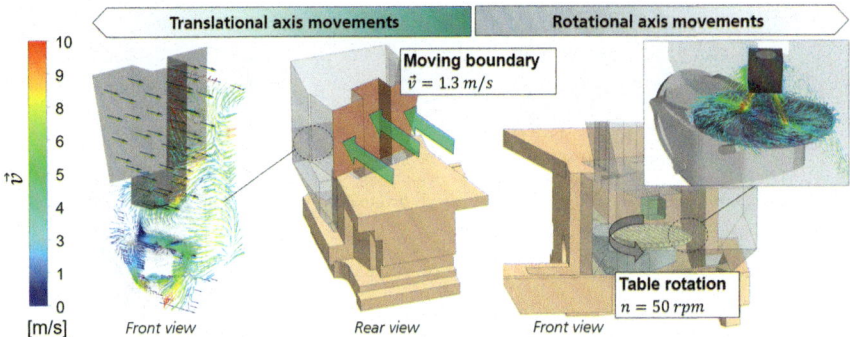

Fig. 4 Representation and modelling of moving machine components

The movement of the Y-axis is particularly dominant in this type of machine, as a lot of air is circulated due to the large wall surface. This has a strong influence on the global flow field in the entire workspace. The rotation of the C-axis causes the impacting cooling lubricant to be ejected tangentially, which has a major influence on the wetting of the workspace walls. Naturally, this influence depends heavily on the C-axis rotation speed and on the B-axis position, which determines the table surface inclination.

3 Investigated Load Cases

The method of cutting fluid supply (internal or external) and the consideration of a possible cutting process have a decisive influence on the flow field and the fluid distribution of the coolant in the workspace. The coolant is most often supplied directly to the cutting zone of the tool via external nozzles on the headstock (external coolant supply). In addition to cooling and lubricating both tool and workpiece, this primarily serves to remove chips. Inside specialized tools, coolant is supplied at high pressure via internal supply channels (internal coolant supply). The coolant reaches the outside as a jet or spray via small holes near the tool tip (see Fig. 2).

In order to demonstrate the influences of these operating/cooling modes on the heat transfer in the workspace, three different process situations, referred to as load cases, are investigated here (Fig. 5).

In the first case, cooling lubricant is supplied both externally and internally. No cutting process takes place yet. This leads to wide-spread wetting in large sections of the workspace. In the second case, only the tool-internal coolant supply is active. A distinction is made between no load (2a) and the drilling process (2b). In all of these load cases, the tool is directly over the table center position. Translational axis movements (including Z-axis) as well as table rotations are neglected here.

The third load case (3) is used to validate the simulation results using real measurement data. For this purpose, the table C-axis is rotated cyclically at a fixed speed.

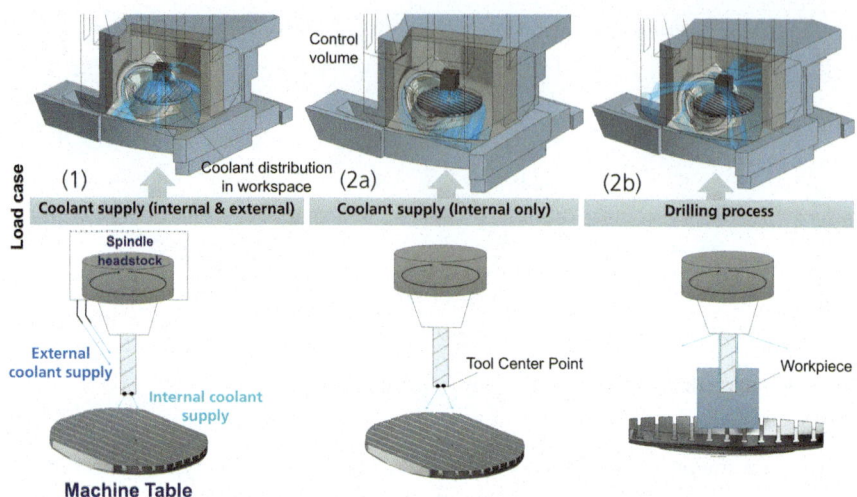

Fig. 5 Graphical representation of the analysed load cases for the calculations

Since the investigated machine tool model has no turning option, the C-axis was rotated by 360°, halted and then reversed 360°, which was repeated cyclically. The heat loss due to friction in the roller bearing of the table and the drive losses of the C-motor cause the swivel rotary table to heat up. In the first part of load case 3, rotation takes place without the use of cooling lubricant until the steady state is reached. After approx. eight hours, the internal and external cutting fluid is switched on while the rotation continues, resulting in rapid cooling of the assembly. The temperature is recorded using the three temperature sensors shown in Fig. 6, which are all located inside the table, so that no direct exposure to the workspace fluid occurs. The recorded temperatures are compared with the results of the simulation described in the following section.

Fig. 6 Selected temperature sensor positions for model validation

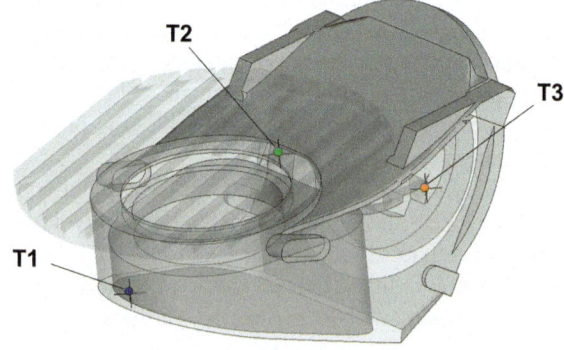

4 Simulation Results and Discussion

The subsequent sections show the calculation results for the previously specified three load cases, as they were obtained for the individual submodels in ANSYS CFX. The results for the near-tool model were computed as a static solution and passed to the transient workspace model in order to obtain the HTCs on the workspace surfaces.

4.1 Results of the Near-Tool Model A

The fluid distribution of the coolant jet from the tool-internal supply system shows the characteristic conical expansion due to a jet with high fluid velocity entering a space filled with a stationary gas. A jet is a flow from a nozzle into a free environment without any wall boundary and the conical expansion results from velocity differences for the fluids involved. As a result, a shear layer is created between air and coolant, from which a free jet develops, and the surrounding fluid is sucked in and entrained. Immediately after the free jet exits the coolant holes at a defined speed, it begins to disperse. The deflection of the fluid flow from the two inner coolant nozzles (I1 and I2 in Fig. 1) to the outside is noticeable, although the direction of the coolant outlet is parallel to the axis of rotation. This is caused by the rotation of the drill. An investigation of the local Reynolds number (R_e) for the analysed process configuration shows a value of $R_e \approx 17925$, a highly turbulent flow, for a high velocity $v_{res} \approx 30$ m/s (Fig. 7a). The deflection of the two central coolant nozzles can be seen in the plot of the volume fraction (Fig. 7b).

If a workpiece is present (Fig. 8a), the fluid flow shows a different behaviour compared to a free stream. The coolant flow is redirected through the drill hole and

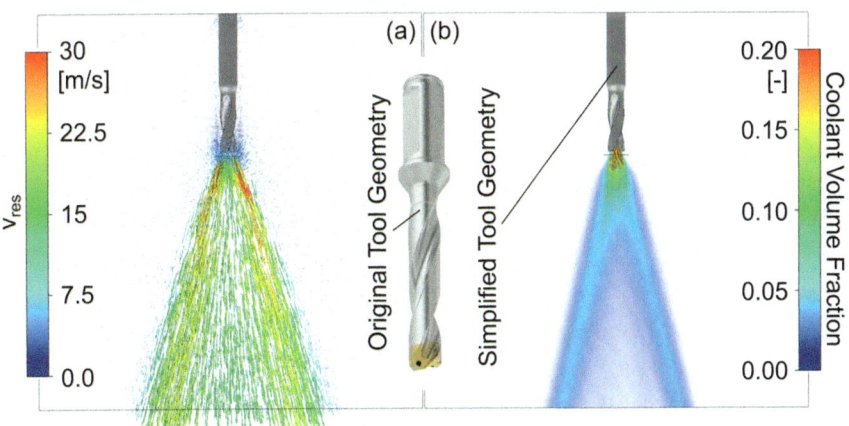

Fig. 7 Velocity and coolant volume fraction for unobstructed coolant jet flow

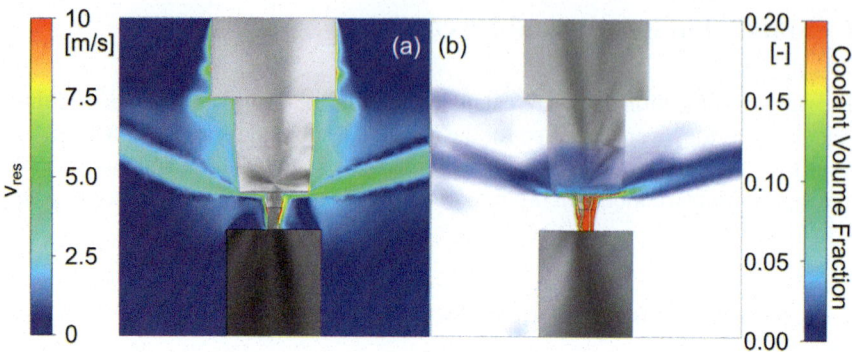

Fig. 8 Side view of the obstructed coolant flow field velocity and volume fraction

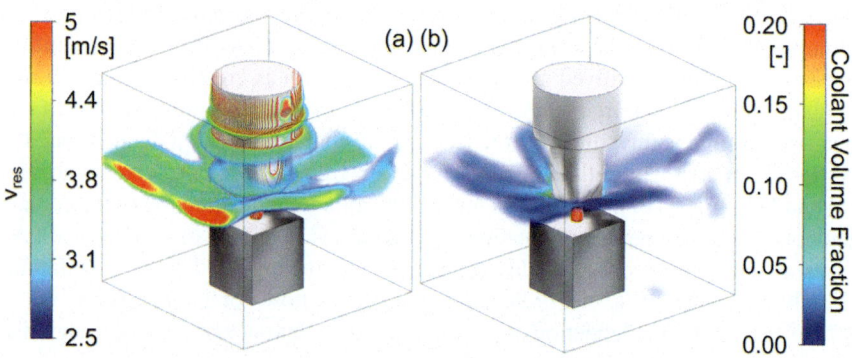

Fig. 9 Spatial velocity and volume fraction of the coolant flow field

significantly slowed down. Compared to the unobstructed state, the speed is only a fraction of the value of the free-stream jet (Fig. 7a). A significant deflection of the coolant flow is present in the images of cutting fluid velocity (Fig. 7a) and coolant volume fraction (Fig. 7b). Figure 9 shows, how the internal cutting fluid supply is dispersed mostly horizontally and slightly upward after being pushed out of the bore hole. The low volume fractions indicate that the cutting fluid disperses as a fine mist into the workspace. This coincides with observations from real drilling tests on the DMU 80 eVo.

4.2 Results of the Workspace Model B

The workspace model B uses the results of the near-tool model A to calculate the effect of the different cutting fluid supply systems for various application scenarios. The goal is to determine the time-dependent temperature field and the heat transfer coefficients on the workspace walls. The latter are the most important results of the model, as they are required as boundary conditions for thermo-elastic FE simulations, which provide the response of the machine tool to different conditions within its workspace. Understanding and quantifying this response can then be used to optimize the cooling systems and train thermal error compensation methods.

4.2.1 Calculation of the HTCs for Load Cases 1 and 2

The heat transfer coefficient was calculated for the three load cases presented above. For heat transfer to occur, the solid surface must have a temperature difference compared to the (local) fluid temperature. In order to demonstrate the effect, the solids in the model were preheated, as might be the case after intensive machine tool usage. A starting temperature of 30 °C was assigned to all solids in order to better visualize the quantitative cooling effect of the investigated scenarios. The fluid temperature and the supplied cooling lubricant temperature are both 20 °C. It is assumed that the ambient temperature and the coolant temperature do not change in the period under consideration.

The result (Fig. 10) shows that the HTC strongly depends on the wetting with cooling lubricant, as the liquid absorbs the heat significantly faster than air. High HTCs lead to faster cooling rates of the assemblies. Comparing the results of load cases (2a) and (2b) shows that using internal cutting fluid supply can have a strong cooling effect on the machining table before and after the drill engages the workpiece. During the actual drilling, however, the coolant disperses in the workspace, which leads to stronger cooling of the workspace walls and almost no cooling of the table. It is important to mention that load case 2 has no external cutting fluid active and that the workpiece is not included in the workspace simulation, only in the near-tool model. If external cutting fluid was active, the effects of the internal fluid on the table would be much less noticeable. Likewise, if the workpiece was present in the workspace model, then depending on the size and shape of the workpiece, the internal coolant jet in load case (2a) may not have much effect on the table either.

4.2.2 Calculation of Temperatures and Estimation of Cooling Periods

To determine the cooling speed, three virtual sensor positions in the swivel rotary table are compared in Fig. 11. These can be found at different positions within the table assembly, as shown in Fig. 6, and also correspond to the real sensor positions in the machine. In addition, the top part of Fig. 11 shows a temperature contour plot of the simulated solids for a fixed point in time (1200s). Depending on the load case and the results from the near-tool model A, the curves in the lower part of Fig. 11 show different thermal time constants, i.e. different cooling speeds.

The transient workspace simulation covered two hours. The highest cooling rate is achieved in load case 1, where the external cutting fluid supply was active. This is mainly due to the extensive wetting of the entire table assembly and the high thermal conductivity of the components. As could already be seen in Fig. 10 (1), the highest

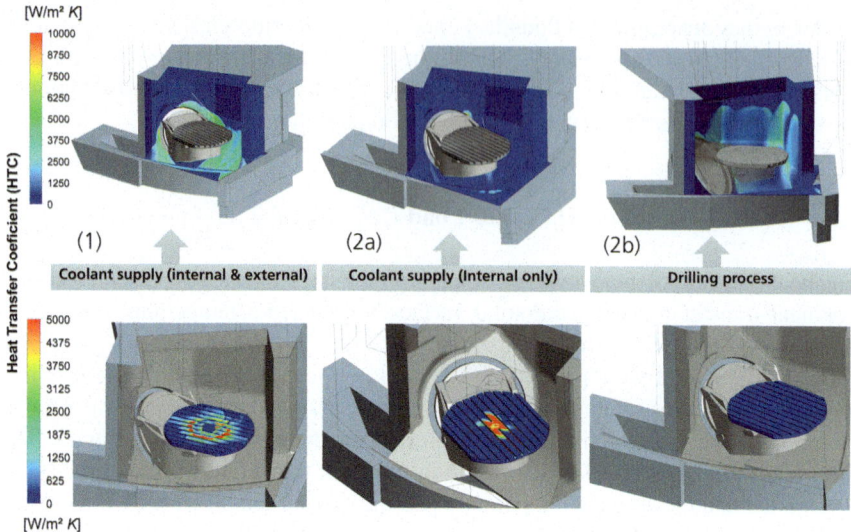

Fig. 10 Contour plot of the HTCs for the table surface and workspace wall

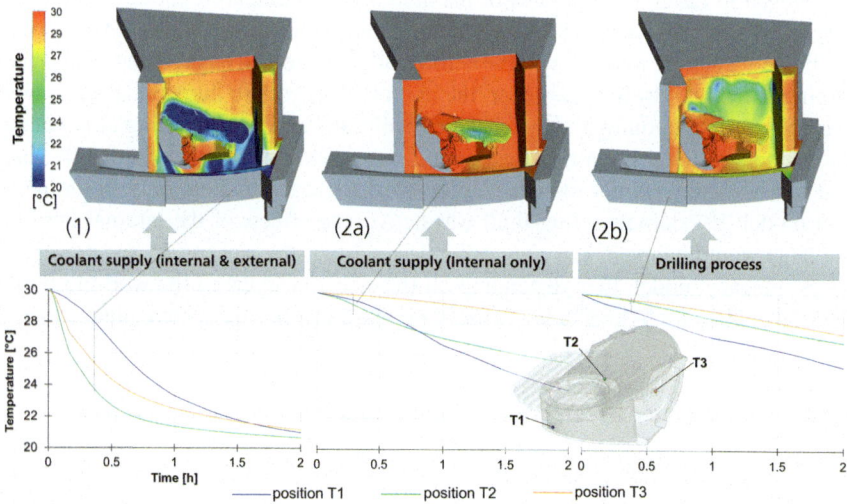

Fig. 11 Simulated temperatures for load cases 1, 2a and 2b, top: temperature contour plot of solids at t = 1200 s, bottom: temperatures at three sensor locations within the table

HTCs are achieved there. For this simulation, the machine frame is thermally inert and thus shown in grey. In load case (2a), the cooling process takes longer, as the high volume flow of the external coolant supply is missing. In an exemplary drilling process (2b), the tool-internal coolant sprays almost entirely onto the workspace wall, while the table is only wetted slightly. This results in a correspondingly slower cool-down of the table.

4.3   Validation of Simulation Results with Measurement Data

To validate the workspace model, the thermal load profile (3) was experimentally investigated by rotating the C-axis cyclically with different cooling systems in an air-conditioned climate chamber. The table rotation heats up the machine tool as a result of dissipation losses in the swivel rotary table. In the simulation model, the heating was simulated by heat flows in the table bearing and the interface boundary of the drive motor. In order to represent the influence of machining with cutting fluid, the axis was first rotated cyclically dry (without cutting fluid) until a steady state was reached after approximately eight hours. At this point, cooling lubricant was added via both external and internal supply systems using a test tool. The cooling lubricant causes rapid cooling, as shown in Fig. 12. The oscillations of the temperature curve in the thermal measurement are caused by the measuring cycles during which the table warm-up is interrupted.

There is a good agreement between the measured and simulated temperature curves. The measuring points are located in different positions within the machining table, see Fig. 6. The measurement cycles of the displacement measurement and thus the load interruptions were not taken into account in the simulation, which is why the curves are smooth. Sensor T3 is located near the C-motor, which introduces waste heat during axis movements and to a smaller degree during stand-by. This sensor shows the highest temperature rise. However, T3 is far away from the influence of the cooling lubricant, which is why the cooling effect is lower in the middle section of the load case.

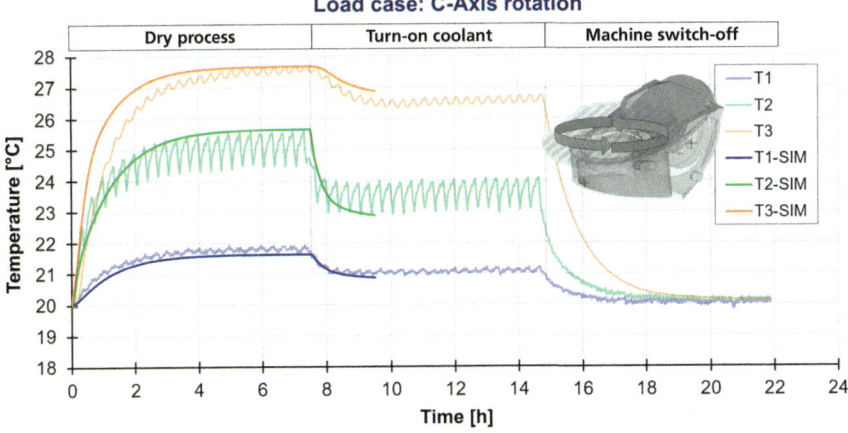

Fig. 12 Comparison of measured and simulated temperature profile for C-axis rotation

The situation is similar with sensor T1, which is located on the underside of the console. It is far away from the heat sources and therefore exhibits a lower heating effect. Sensor T2 is located very close to the C-axis bearing of the table in the upper area of the console, where heating occurs there due to bearing friction. It is, however, closer to the influence of the cooling lubricant, which is why the largest temperature drop is registered there.

Overall, both heat-up and cooling in the performed measurements match the simulated curves quite well. For sensor T3, the heating occurs slightly faster in the simulation, though the steady-state temperature matches almost perfectly. This may be due to inaccuracies of the sensor placement or incorrect HTCs within the table. For sensor T2, the simulated steady-state temperature is slightly too high, possibly from a too high coefficient of friction in the bearing. After the activation of cutting fluid, the cooling is also a bit stronger than the measured value, which hints at inaccuracies in the simulated HTC distribution of the table.

Generally, the experiments show that the presented modeling approach gives an accurate representation of the qualitative effects of different production-relevant thermal load scenarios. Once further parameter optimization has been realized, additional load cases will similarly be validated using measurements.

5   Summary and Outlook

The article presents a method using coupled CFD simulations to investigate the thermal effects of cooling lubricant usage in the machine tool workspace. The heat transfer during wet machining is determined via a multi-phase flow in the workspace. The immediate environment around the tool and near the cutting process plays a decisive role in the distribution of the coolant. The models are coupled indirectly by transferring the results from the area close to the tool to the surfaces of a control volume, where they serve as input boundary conditions to the workspace model.

The two models (near-tool model A and workspace model B) can be calculated independently of each other and with the mesh fineness and model type most suitable for the local conditions. The near-tool model, e.g., is very finely meshed, but isothermal and steady-state, while the workspace model is mostly much coarser and transient thermo-fluidic. This enables transient workspace simulations of larger time frames without losing accuracy at the tool interface and without the otherwise unacceptably high computation times.

For the near-tool model A, two specific load cases were simulated, which are relevant for drilling and various other cutting operations. The first is the flow of coolant from the tool-internal supply channels in free space, such as may occur before and after tool engagement with the workpiece. The second is the spreading of that same coolant throughout the workspace during drilling operations. In this case, the tool and drill hole redirect the coolant flow upward and out into the workspace. The resulting coolant flow speed and distribution are shown for the two scenarios.

For the workspace model B, the thermal effects of the aforementioned coolant for the different scenarios are simulated. There, a strong cooling effect on the machining table can be seen for the free tool and on the workspace walls for the engaged tool during the simulated drilling operation. Another operation with and without coolant use during constant C-axis rotation shows the different heating and cooling effects on different parts of the machining table, which was validated by an equivalent machine tool measurement on the DMU 80 eVo. The good agreement at the three temperature monitor points validates the modelling approach for the investigated scenario.

Subsequent research will expand the number of different cooling scenarios, also in combination with axis movements, to get a more complete overview of the processes in the workspace. Once this is done, these computed HTCs will be used to compare the resulting temperature fields of the machine tool and use these to train thermal error compensation methods.

Acknowledgements The authors thank the German Research Foundation (DFG) for funding the project "Modelling of dual-phase fluid-solid interactions in the workspace of cutting machine tools to enable process-specific thermal error compensation" (ID 174223256), which enabled this investigation.

References

1. Putz, M., Richter, C., Regel, J., Bräunig, M.: Industrial relevance and causes of thermal issues in machine tools. In: 1st International Conference on Thermal Issues in Machine Tools, March 2018, pp. 127–139, Dresden, Verlag Wiss. Scripten (2018)
2. Kumar, T.S., Glänzel, J., Bergmann, M., Putz, M.: Prediction of thermal errors in machine tools through decoupled simulations using genetic algorithm and artificial neural networks. MM Sci. J. 4683–4691 (2021). https://doi.org/10.17973/MMSJ.2021_7_2021076
3. Kumar, T.S., Geist, A., Naumann, C., Glänzel, J., Ihlenfeldt, S.: Split CFD-simulation approach for effective quantification of mixed convective heat transfer coefficients on complex machine tool models. Procedia CIRP **118**, 199–204 (2023). https://doi.org/10.1016/j.procir.2023.06.035
4. Liu, H., Meurer, M., Schraknepper, D., Bergs, T.: Investigation of the cutting fluid's flow and its thermomechanical effect on the cutting zone based on fluid-structure interaction (FSI) simulation. Int. J. Adv. Manuf. Techn. **121**, 267–281 (2022). https://doi.org/10.1007/s00170-022-09266-1
5. Nabbout, K., Sommerfeld, M., Barth, E., Uhlmann, E., Bock-Marbach, B., Kuhnert, J.: Cooling capacity of oil-in-water emulsion under wet machining conditions. Procedia CIRP **117**, 74–79 (2023). https://doi.org/10.1016/j.procir.2023.03.014
6. Brier, S., Regel, J., Putz, M., Dix, M.: Unidirectional coupled finite element simulation of thermoelastic TCP-displacement through milling process caused heat load. MM. Sci. J, 4534–4539 (2021). https://doi.org/10.17973/MMSJ.2021_7_2021056
7. Baumann, A., Oezkaya, E., Schnabel, D., Biermann, D., Eberhard, P.: Cutting-fluid flow with chip evacuation during deep-hole drilling with twist drills. Euro. J. Mech. B/Fluids **89**, 473–484 (2021). https://doi.org/10.1016/j.euromechflu.2021.07.003
8. Naumann, C.: Kennfeldbasierte Korrektur thermo-elastischer Verformungen an spanenden Werkzeugmaschinen. Dissertation. TU Chemnitz, URN: nbn:de:bsz:ch1-qucosa2–906384 (2024)

9. Brier, S., Topinka, L., Regel, J., Dix, M.: Coupled fluid-structure-interaction simulation approach for correction of the thermal tool elongation to improve the milling precision. Forschung im Ingenieurwesen **88**(31) (2024). https://doi.org/10.1007/s10010-024-00751-5

10. Brier, S., Geist, A., Glänzel, J., Naumann, C., Regel, J., Dix, M., Ihlenfeldt, S.: Coupled CFD model of tool environment and workspace to determine the convective heat transfer in jet cooling of milling processes in machine tools. In: 20th CIRP Conference on Modeling of Machining Operations, Mons (2025)

11. Chorin, A.J., Marsden, J.E.: A Mathematical Introduction to Fluid Mechanics, 3rd edn, Springer, New York (2000). https://doi.org/10.1007/978-1-4612-0883-9

12. Menter, F.R., Ferreira, J.C., Esch, T.: The SST turbulence model with improved wall treatment for heat transfer predictions in gas turbines. Int. Gas Turbine Congr. **2003**, 1–7 (2003)

13. Anderson, P., Culley, P., Parker, T.J.: Marketing Research. Hansen Publisher, London. ISBN: 09525242 (2003)

14. Smith, D., et al.: Specification formulation. J. Eng. **2**(2), 223228. ISSN: 09544828

Hybrid Thermal Modeling Approach for the Early Stages of the Machine Tool Design Process

Lars Penter, Steffen Schroeder, Holger Rudolph, Matthias Brand, and Steffen Ihlenfeldt

Abstract The thermally optimized design of machine tools is a prerequisite for both high accuracy and high productivity. Numerical prediction of the temperature field is therefore an essential piece of the puzzle in the development and design process of the machine. Especially in the early stages of development, little data is available and only rough assembly dimensions are known. As the design process progresses, the level of detail increases and thus the model requirements change, but so does the information for more detailed modeling. Neither high-resolution FE models nor data-driven models can meet these evolving requirements. This article presents a hybrid approach based on a lumped parameter thermal network and the incremental integration of high-resolution FEM assemblies. The model approach allows both the qualified analysis and optimization of the temperature field of the machine tool in the early project phase by using simple pre-parameterized basic elements, and the consideration of fully designed components in the later course of the project. The model was verified experimentally on a rotary-tilt table and used for design considerations regarding the layout of cooling channels. The study showed that the steady-state temperature of the assembly could be reached much faster and reduced by up to 38%.

Keywords Thermal deviations · Hybrid simulation · Machine tools · Lumped-parameter thermal network

L. Penter (✉) · S. Schroeder · H. Rudolph · S. Ihlenfeldt
Chair of Machine Tool Development and Adaptive Controls, Dresden University of Technology TUD, Dresden, Germany
e-mail: lars.penter@tu-dresden.de

L. Penter · S. Ihlenfeldt
Fraunhofer Institute of Machine Tools and Forming Technology, Chemnitz, Germany

M. Brand
DMG MORI Seebach GmbH, Seebach, Germany

© The Author(s) 2026   477
K. Wegener and M. Bambach (eds.), *4th International Conference on Thermal Issues in Machine Tools (ICTIMT2025)*, Lecture Notes in Production Engineering,
https://doi.org/10.1007/978-3-032-01194-7_30

1 Introduction

Rotary-tilt tables and rotary heads of machine tools (MT) for 5-axis machining are compact assemblies characterized by a high degree of integration of mechanical and electrical functions. These units contribute significantly to motion errors of the MT and limit the achievable manufacturing accuracy due to their complex thermal behavior. Internal heat sources cause heating of external structural components, which in turn undergo thermoelastic deformation. Through the kinematic chain, these deformations result in tool center point (TCP) deviations that are amplified by lever effects.

To mitigate thermal effects, countermeasures such as insulation or cooling of motors and bearings are commonly applied. The effective use of these measures requires a thorough understanding of the underlying thermal interactions. These are typically obtained by measurement and numerical analysis. However, the ability to analyze compact assemblies remains challenging due to high costs and limited accessibility for metrological investigations.

In the thermal analysis of compact assemblies, the relationship between the internal processes and their effect on the external structural components is of interest. Simulation-based studies have great potential for determining these relationships. The simulation models used can be categorized based on the source of system knowledge into physics-based models [1, 2], data-driven models [3–6] and hybrid models [7]. In addition, models can be either static or adaptive [8], depending on whether they adjust their parameters based on new data.

However, building these models is currently very time consuming. This is due to the stringent requirements for accurately capturing the thermal and thermoelastic behavior under various operational scenarios, as well as the individual characteristics of system components within assemblies. In particular, data-driven and adaptive models require extensive experimental data for training, which is often not available during the early design phase, making their application even more challenging at that stage.

Given these diverse modeling challenges, either Lumped Parameter Thermal Networks (LPTN) or FE models are typically employed in the initial development stages. Existing tools are tailored to one of these model types. As a result, some system components cannot be integrated into current analyses in an appropriate model form. They must first be transformed or even simplified, which leads to significant implementation effort or a considerable reduction in accuracy. To address these challenges, Schroeder et al. [9] developed a methodology that enables the seamless combination of LPTN and FE submodels, thereby improving modeling flexibility and preserving system accuracy. This allows a more agile response to different design stages and component specifics.

In this paper, this method has been further developed and applied to the calculation of the thermo-elastic behavior of the rotary-tilt table of a 5-axis MT. A key innovation over previous approaches is the integration of MOR-based simulation technology [10] to obtain sufficiently accurate results on the transient thermal behavior of typical

MT structures with high structural resolution in a short time. This advancement allows for efficient modelling of the thermal behavior of compact assemblies with dense internal layouts, delicate structural elements, and relative axis motion.

The paper describes the methods for creating the multiphysics network models in Sect. 2. The transfer and application of the simulation technology was carried out in Sect. 3 for qualitative analysis and in Sect. 4 for quantitative analysis on the example of a MT rotary-tilt table.

2 Modelling Approach and Experimental Setup

This section describes the approach for the modular model approach and the element types used. Furthermore, the procedure for the semi-automated creation of subcomponents and their connection is discussed. Finally, the experimental setup is presented.

2.1 Approach and Element Types

A network model approach is used to model and calculate the thermal behavior of the MT. It allows the integration of submodels from different physical domains and a flexible handling of the available information in the different development stages of a project. The modeling approach allows both the qualified analysis and optimization of the MT temperature field in the early stages of the project by using simple pre-parameterized basic elements, as well as fully designed components in the later stages. MT models for the thermal analysis require parameterized instances of these model elements, which are then assembled to a complete model.

MT assemblies consist of a large number of components with varying heat transfer conditions and sources for energy losses. This results in a high degree of manual effort in model creation and validation. To reduce the manual modeling effort, a model kit of reusable domain-specific model elements is proposed. Established, open and documented model approaches as well as standardized interfaces are used to enable an easy transferability of the technology into industrial practice. Two different approaches can be used to model structural components.

- Coarse models for early design stages: The structural models and their corresponding element types have a coarse geometric resolution and can be created by combining simple basic shapes such as cuboids and cylinders. LPTNs are derived from them. These models are suitable for qualitative analysis or for components with little overall impact on the model accuracy. The parameters are determined using well-known analytical modeling approaches. The models have very few degrees of freedom and are therefore computationally efficient.

- Detailed models for advanced design: When detailed geometric designs are available, the FE method is used to describe the complex components. The resulting system degrees of freedom are usually very large. To reduce the computational effort, methods of geometric simplification and structure- and parameter-preserving Krylov-type Model Order Reduction (MOR) [11, 12] are used. These methods provide an efficient way to generate computationally efficient models of the structural components with good approximation properties directly from an FE model. A modeling strategy has been designed and element types have been developed to allow continuous relative movements between individual assemblies. As a result, compact reduced models can be generated semi-automatically from high-resolution FE models. With the implemented MOR-FEM combination, the high geometric complexity typical of machine tools can be modeled efficiently.

Physical and empirical models are used to describe heat transfer and power dissipation, which can be temperature, velocity, position and/or load dependent over time. Due to the large number and complexity of the relevant thermal effects, a library of reusable model elements covering all typical heat transfer mechanisms has been developed. This library is continuously extended and provides standardized interfaces for integration into the network model.

2.1.1 Implementation of Simple Structural Parts

The structural components of the machines are modeled by combining basic bodies such as cuboids and cylinders. These primitive shapes allow for an accurate approximation of most machine geometries. The thermal network model is then derived from the combined geometries, see Fig. 1.

An automated method is used to calculate thermal coupling coefficients between adjacent bodies. This is done by combining elementary coupling coefficients, each representing a constrained heat flow in a specific direction or through a particular interface (as shown in Fig. 2). These partial conductances capture directional dependencies and are derived from analytical and numerical models developed for rectangular and cylindrical elements. Coupling coefficients are precomputed at discrete reference points across the relevant range of geometric parameters. During model generation, the appropriate conductance values for arbitrary geometries are obtained by interpolation between these precomputed reference points.

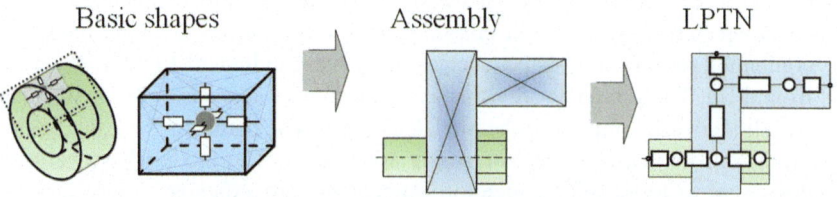

Fig. 1 Combination of basic shapes to machine assemblies and derivation of LPTN

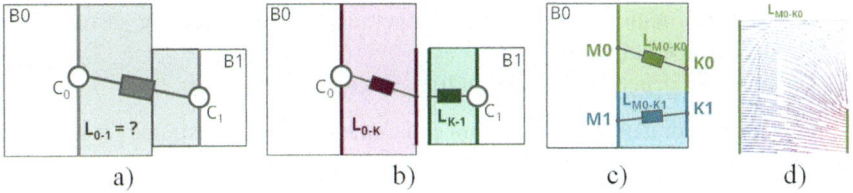

Fig. 2 Automated thermal coupling between two contacting bodies: **a** Thermal conductance between the centers of mass of two adjacent bodies; **b** Thermal conductance between the center of mass and the center of the contact surface; **c** Use of two partial conductances to model directionally constrained heat flow; **d** Example of a temperature field used to compute a single thermal conductance based on defined boundary conditions

2.1.2 Implementation of Detailed Structural Parts

The modeling process begins with CAD representations of the subassemblies, each consisting of rigidly connected components. These are combined into closed geometries within the CAD system. A simplified "de-featuring" step is recommended to remove elements such as chamfers or engravings that do not influence thermal behavior but would unnecessarily increase mesh density.

The geometric discretization is then performed using either tetrahedral or hexahedral elements, depending on the geometry and meshing strategy. This mesh forms the basis for generating the FE system of equations. Optionally, the resulting system can be reduced using a model order reduction (MOR) method to improve computational efficiency. While unreduced models benefit from sparse matrix solvers, MOR-based models typically yield dense matrices and require transformations between virtual and physical domains for temperature and heat flow data. The final FE subassembly model is connected to other components of the thermal network via coupling surfaces. These interfaces may represent different heat transfer mechanisms (e.g., conduction, convection) and typically consist of multiple selected contact faces, which must be defined manually.

2.1.3 Thermal Network Model

The overall thermal network model is constructed by assembling all relevant component models—reduced FE assemblies, library solids, and heat transfer functionals—and connecting them at their defined coupling surfaces. This integration allows the simulation of transient temperature fields by considering both localized physical behavior and global thermal interactions. Time-domain simulations are performed across discrete time steps and may require iterative convergence. Each time step consists of the following four stages:

1. Computation of temperature gradients within each subsystem,
2. Determining the temperature fields by numerical integration for each system of equations,

3. Back-transformation of the virtual temperatures of the coupling surfaces into physical temperatures, and

4. Updating the dependencies of heat transfers and power losses.

A detailed formulation of the system equations and coupling strategy is provided in [9].

2.2 Experimental Setup

The presented component is a rotary-tilt table of the 5-axis milling machine DMU 80 eVo. In order to identify the uncertain parameters and to quantify their effect on the thermal behavior of the whole system, the procedure for parameterizing thermal models according to [13] was used. A series of measurements were made while the MT was operating with constant motion of the rotary and tilting axis at 75% of the maximum speed until the approximate thermal steady state was reached after about 8 h. Cooling after heating to equilibrium was performed both with and without active cooling.

Temperature measurements were made using 13 wireless, contact-type PT100 sensors integrated into the Saveris PTD system and 2 ambient air sensors. The sensors were equipped with high-precision amplifiers to achieve an overall measurement uncertainty of $\pm$ 0.1 K. The contact sensors were installed on both the inner and outer surfaces of the structural components of the rotary-tilt table, as well as on the drive units (Fig. 3), in order to capture relevant thermal gradients and identify major heat sources. The two ambient sensors were placed in the machine workspace and the surrounding environment, in order to monitor the air temperature and characterize the thermal boundary conditions during the experiments. Wireless data transmission was used to collect readings from the moving assemblies.

3 Qualitative Analysis of the Rotary-Tilt Table

For the qualitative analyses, a simulation model was created that roughly represents the internal assembly structures. It contains three moving structural components (rotary table, console, drive housing) as fine-meshed FE models and twelve components of the drive trains, simplified by LPTNs. The external boundary conditions (ambient air, frame, cooling water) are represented by eight temperature nodes, while the behavior of the enclosed air is modeled by three concentrated heat capacities, see Fig. 4.

The components and capacities are linked in the overall model via 129 conductance values for heat transfer in joints, roller guides, and gear contacts, as well as for heat transfer in interior spaces and to the environment through convection (free and forced) and heat radiation. A total of 35 heat sources represent the internal loads of

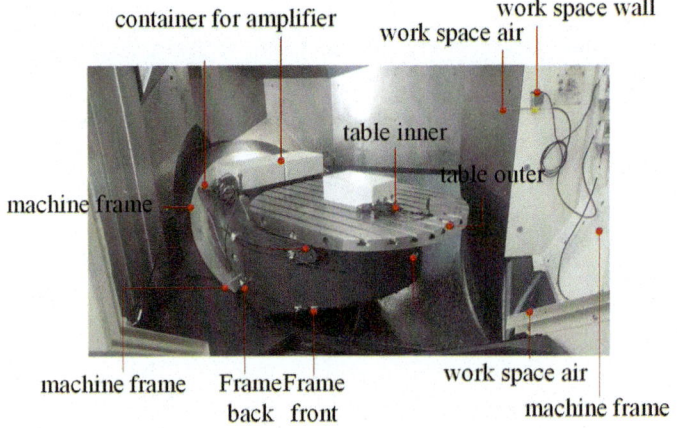

Fig. 3 Sensors for temperature measurement on rotary-tilt table of DMU 80 eVo

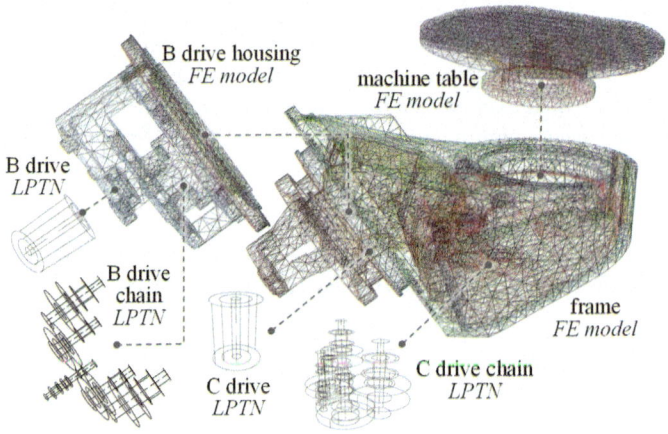

Fig. 4 Thermal network model with LPTN and FE components for rotary-tilt table

the rotary-tilt table. They represent the friction losses in the bearings of the rotary axes and the shafts of the drive trains, the electrical losses of the motors, and the friction losses at the tooth contacts of the drive shafts. The parameter in the early phase are based on literature values [1, 2, 9, 10].

A thermal network diagram was developed (see Fig. 5) to visualize assembly behavior through lumped parameters and variables. In this model, heat capacities, conductances, power losses, and heat flows are represented by scaled symbols. For LPTN, conductances and capacities can be taken directly from the diagram, whereas for FE models they have to be condensed over larger areas. The condensed coupling coefficients are determined numerically and calculated by a least-squares method. This simplifies the representation of the model properties and enables qualitative

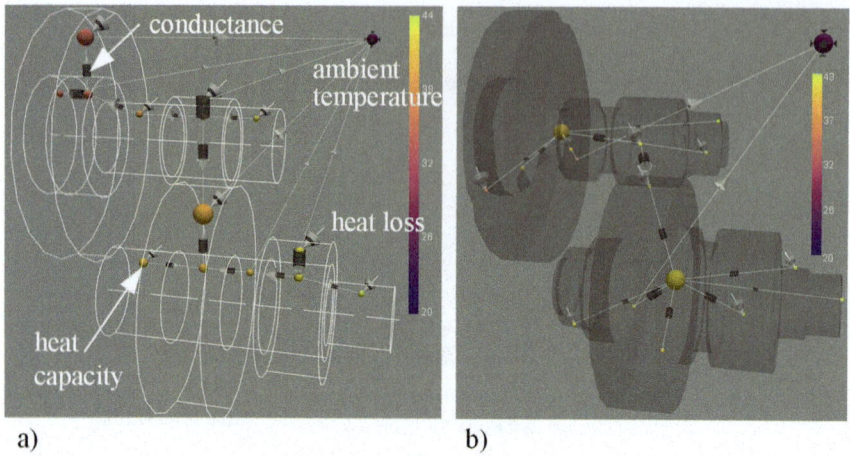

Fig. 5 Thermal network model with **a** LPTN and **b** FE-submodels

behavior analyses, such as in transmission stages. An analysis showed that heat accumulation in the transmission shafts occurs due to low coupling coefficients for heat dissipation, which explains the high simulated temperatures. The transmission shafts therefore have a significant influence on the thermal behavior (as shown in Fig. 5).

4 Quantitative Analysis of the Rotary-Tilt Table

4.1 Model Structure

For the quantitative analyses, the model of the rotary-tilt table was further detailed. The internal components and heat transfer were described in greater physical and geometric resolution. For power train components, the original LPTNs were replaced by FE models. A methodology for MOR was implemented in the workflow of the FEM-based models and its suitability for the analysis of compact assemblies was examined. Finally, to improve the model quality, the uncertain parameters of the model were ultimately adjusted with the support of measurement technology.

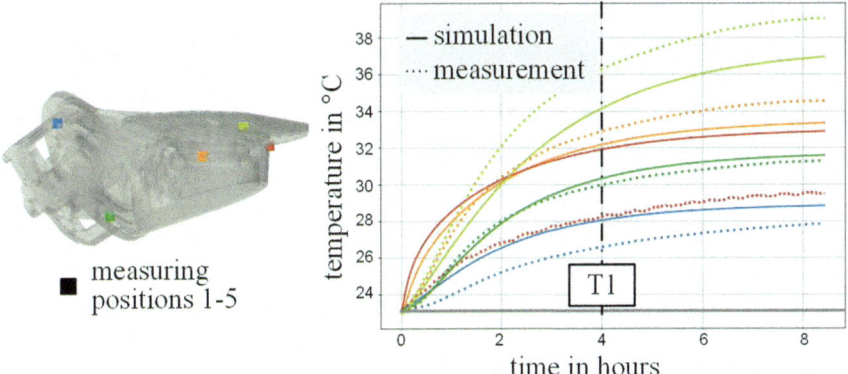

Fig. 6 Comparison of measured values for 75% maximum speed with simulation

4.2 Parameter Identification

The model quality was improved by parameter adjustment for the quantitative analyses, since some parameters of the thermal models have high uncertainties. Uncertainty analysis according to [13] identified parameters with high influence and uncertainties, in particular the frictional losses of the pivot bearings, preloaded bearings and gears. The parameter identification procedure is based on minimizing the difference between measured and simulated local temperatures through iterative simulation-based optimization. To support this, temperature measurements were obtained at selected locations on the rotary-tilt table under controlled operating conditions. These measurements were first used to calibrate the internal heat source parameters. Once these were matched, the heat transfer coefficients were derived. Further uncertainties result from rough model approaches, scatter of the prestressing, and heat transfer in the cooling channels, which were determined only approximately.

The comparison is based on measured temperature curves, with uncertain parameters adjusted within their uncertainties. As an example, Fig. 6 shows deviations of 8 to 42% between simulation and measurement at T1 = 4 h for the table rotation load case (75% maximum speed). The causes are simplified model structures, missing smaller components, and inaccurate approaches to heat transfer and power dissipation. Despite these residual errors, the model quality is sufficient for quantitative influence analysis.

4.3 Model Application in Design Phase

The detailed analysis was based on additional cooling measures on the rotary-tilt table. Measurements and simulations showed uncooled areas with high temperature increases in the rotary table and console, which could cause significant displacements.

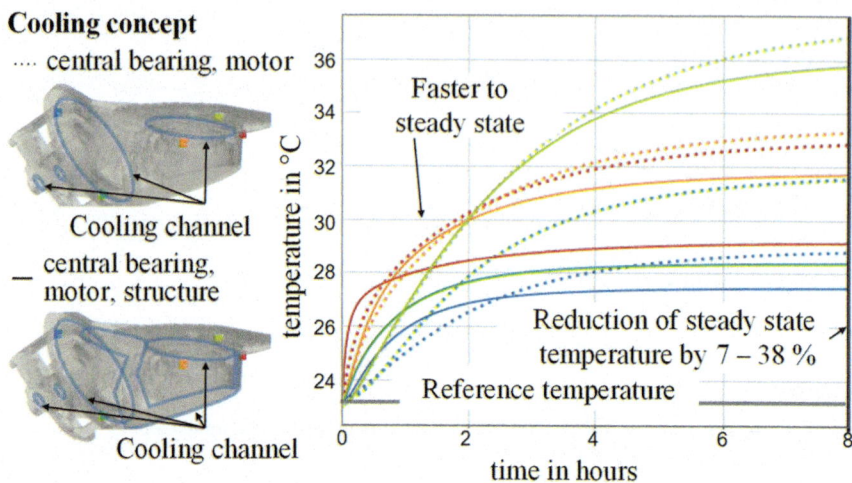

Fig. 7 Comparison of cooling channal concepts for rotary-tilt table at 75% n_{max}

As a result, new cooling structures were designed with a compromise between flow through critical structures and manufacturing effort.

The model was extended to include cooling channels, cooling water temperature boundary conditions, and convective heat transfer. The analysis focused on the console with the internal C-axis drive train. With additional cooling, the time constant is significantly reduced, allowing the assembly to reach steady state about twice as fast. In addition, the steady-state temperatures (in °C) are reduced by 7% to 38% (as shown in Fig. 7).

5   Conclusion

Rough modeling of structural components using geometric primitives is suitable for conceptual and qualitative analysis of internal components of the axes. The resulting LPTN models have a low degree of freedom, which significantly reduces computation time. The Advanced Network Diagram provides a rapid assessment of the thermal behavior of larger assembly models through the three-dimensional representation of model elements and their condensed parameters. It also provides quick insight into the thermal relationships within the assemblies. Compared to traditional simulation tools, the developed domain-specific modeling toolbox allows for faster creation and clearer model representation. In addition, a high degree of automation of the model creation process has been achieved. This significantly reduces manual effort. The transfer of the MOR methodology to compact assemblies has been successful in principle. However, the modeling of an active water cooling system leads to larger deviations from the real behavior, so that only rough qualitative analyses are possible. There is potential to improve the quality of the model by applying alternative approaches and using other known reduction methods such as balanced truncation. The modular structure of the workflow tool chain for the generation of FEM-based models of the structural components allows an easy integration of the workflow into the user's existing tool chains.

Acknowledgements The authors would like to thank the German Research Foundation (DFG) for funding the SFB/TR96 (project number 174223256) transfer project T05, and DMG Mori for providing the machine tool and supporting the execution of the tests.

References

1. Thiem, X., Kauschinger, B., Ihlenfeldt, S.: Online correction of thermal errors based on a structure model. Int. J. Mechatron. Manufact. Syst. **12**, Nr. 1, S. 49 (2019)
2. Thiem, X., Kauschinger, B., Müller, J., Ihlenfeldt, S.: Estimation of the influence of volumetric correction approaches on the thermo-elastic correction accuracy. In: Behrens, B.A., Brosius, A., Hintze, W., Ihlenfeldt, S., Wulfsberg, J.P.: (Hrsg.): Production at the leading edge of technology. In: Proceedings of the 10th. [S.l.]. Springer-Verlag Berlin AN (Lecture Notes in Production Engineering), S. 324–333 (2020)
3. Stoop, F., Mayr, J., Sulz, C., Kaftan, P., Bleicher, F., Yamazaki, K., Wegener, K.: Cloud-based thermal error compensation with a federated learning approach. Precis. Eng. **79**, S. 135–145 (2023)
4. Boos, E., Thiem, X., Wiemer, H., Ihlenfeldt, S.: Improving a deep learning temperature-forecasting model of a 3-axis precision machine with domain randomized thermal simulation data. In: Liewald, M., Verl, A., Bauernhansl, T., Möhring, H.C. (Hrsg.): Production at the leading edge of technology: Proceedings of the 12th. [S.l.] : SPRINGER INTERNATIONAL PU, 2023 (Lecture Notes in Production Engineering), S. 574–584 (2023)
5. Lang, S., Zimmermann, N., Mayr, J., Wegener, K., Bambach, Markus: Thermal Error Compensation Models Utilizing the Power Consumption of Machine Tools. In: Ihlenfeldt, Steffen (Hrsg.): 3rd International Conference on Thermal Issues in Machine Tools (ICTIMT2023). Cham : Springer Nature (Lecture Notes in Production Engineering), S. 41–53 (2023)
6. Mareš, M., Horejš, O., Nykodym, P.: An indicative model considering part of the thermo-mechanical behaviour of a large grinding machine. In: Ihlenfeldt, S.teffen (Hrsg.): 3rd International Conference on Thermal Issues in Machine Tools (ICTIMT2023). Springer Nature, Cham. (Lecture Notes in Production Engineering), S. 54–66 (2023)
7. Lang, S., Talleri, S., Mayr, J., Wegener, K., Bambach, M.: Kalman filter-driven state observer for thermal error compensation in machine tool digital twins. Manufact. Lett. **41**, S. 208–218 (2024)
8. Horejš, O., Mareš, M., Straka, M., Švéda, J., Kozlok, T.: Adaptive thermal error compensation model of a horizontal machining centre. In: Ihlenfeldt, S. (Hrsg.): 3rd International Conference on Thermal Issues in Machine Tools (ICTIMT2023), S. 83–98. Springer Nature, Cham (Lecture Notes in Production Engineering) (2023)
9. Schroeder, S., Ihlenfeldt, S., Penter, L.: Efficient and robust creation of structural component models for thermo-elastic analysis of machine tools. MM Sci. J. **2021**, Nr. 3, S. 4644–4651 (2021)
10. Galant, A., Beitelschmidt, M., Großmann, K.: Fast high-resolution fe-based simulation of thermo-elastic behaviour of machine tool structures. Procedia CIRP **46**, S. 627–630 (2016)
11. Bock, H-G., Hoog, F., Friedman, A., Gupta, A., Neunzert, H., Pulleyblank, W.R., Rusten, T., Santosa, F., Tornberg, A-K., Bonilla, LL., Mattheij, R., Scherzer, O., Schilders, W.H.A., van der

Vorst, H.A., Rommes, J.: Model Order Reduction : Theory, Research Aspects and Applications ; with 9 Tables (Mathematics in industry The European Consortium for Mathematics in Industry 13). Springer, Berlin, Heidelberg

12. Benner, P(Hrsg.); Sorensen, D.C.(Hrsg.); Mehrmann, Volker (Hrsg.): Dimension Reduction of Large-Scale Systems: Proceedings of a Workshop held in Oberwolfach, Germany, October 19–25. 2005 (Springer eBook Collection Mathematics and Statistics 45). Springer Berlin Heidelberg, Berlin, Heidelberg (2005)

13. Ihlenfeldt, S., Schroeder, S., Penter, L., Hellmich, A., Kauschinger, B.: Adjustment of uncertain model parameters to improve the prediction of the thermal behavior of machine tools. CIRP Annals **69**(1), S. 329–332 (2020)

Strategy for Sensor Placement to Estimate Thermal Errors Using Temperature-Sensitivity Distribution Based on a Reduced-Order Model of Machine Tools

Satsuma Ando, Shun Tanaka, Yuta Teshima, Jun Morishita, and Toru Kizaki

Abstract Thermal errors in machine tools account for up to 70% of machining errors, making it critical. Thermal displacement estimation using a reduced-order model (ROM) correlates temperature inputs with tool center point (TCP) displacement, where regression coefficients represent temperature sensitivity at each measurement point. Increasing the number of temperature input points in the ROM improves accuracy, but no method exists for determining the optimal number and placement of measurement points. This study provides guidelines for optimal sensor placement. An objective function was designed to explore temperature sensor placement, balancing two key factors: (1) ensuring the accuracy of thermal error estimation and (2) mitigating the impact of variability in temperature measurements. The objective function was defined as a linear combination of the absolute sum of temperature-sensitivity, representing estimation accuracy, and the squared sum of temperature-sensitivity, representing measurement uncertainty. The search for optimal sensor placement was formulated as a minimization problem of this objective function. Temperature-sensitivity distributions were utilized to visualize sensitivity across different areas of the machine tool to explore sensor placement. A sensor placement was achieved by quantitatively evaluating and appropriately weighing the two factors, which balance estimation accuracy with robustness against measurement errors.

Keywords Machine tool · Thermal deformation · Temperature sensor · Sensor placement

S. Ando · S. Tanaka · Y. Teshima · J. Morishita · T. Kizaki (✉)
Department of Mechanical Engineering, The University of Tokyo, Tokyo, Japan
e-mail: kizaki@g.ecc.u-tokyo.ac.jp

K. Wegener and M. Bambach (eds.), *4th International Conference on Thermal Issues in Machine Tools (ICTIMT2025)*, Lecture Notes in Production Engineering,
https://doi.org/10.1007/978-3-032-01194-7_31

1   Introduction

Many studies have explored more rigorous "Temperature-Sensitive Points" (TSPs) selection. Statistical methods and clustering techniques have been employed to identify sensor placements that minimize redundancy while capturing key thermal characteristics [1]. Dimension-reduction algorithms such as principal component analysis and partial least squares regression address multicollinearity among temperature signals, enabling more compact representations of the thermal field [2]. Yet, most of these methods still assume a relatively small total sensor count, frequently under 30, due to the complexity of wiring, hardware constraints, and cost considerations [3]. As a result, large-scale machinery with extensive structures may remain under-instrumented, leaving regions of high thermal gradient unmonitored [4].

Insufficient spatial coverage of temperature measurement can lead to inaccuracies in thermal compensation when operating conditions shift or previously unobserved heating patterns arise [5]. Conversely, installing many sensors without a proper strategy can inflate system costs, increase measurement variability, and reduce reliability if any single sensor fails [6]. This tension between coverage and cost points to a key gap in the literature: no established guideline for determining the necessary and sufficient number of temperature sensors and their optimal placement to achieve robust and high-accuracy thermal error compensation [7].

To address this gap, we propose a new approach that leverages a large-scale array of temperature sensors interconnected in series (LATSIS) with over 100 temperature measurement points. By gathering dense thermal data from across the entire machine structure, we construct a reduced-order model (ROM) that captures the relationship between local temperature fluctuations and overall thermal displacement of the TCP. The ROM, derived from a finite element model, preserves essential dynamics while remaining computationally tractable. We obtain a temperature-sensitivity distribution from this ROM that quantifies how each potential sensor location contributes to the net thermal error.

Using these sensitivity values, we design an objective function that balances two requirements: (1) maximizing the overall magnitude of temperature sensitivity so that critical hotspots are not overlooked, and (2) minimizing the variance of sensitivity among sensors to avoid placing too much reliance on any single measuring point. Formulated as an optimization problem, our method systematically searches for sensor placements that satisfy these criteria. The resulting configuration provides practical guidelines on how many sensors are needed and where they should be placed to capture thermal behaviour effectively.

This paper is organized as follows. Section 2 describes the construction of the reduced-order model and explains how the temperature-sensitivity distribution is derived. Section 3 presents our objective function and discusses the sensor placement optimization procedure. Section 4 provides validation results using a simplified machine tool model and discusses the trade-off between accuracy and robustness. Finally, Sect. 5 concludes with a summary of our findings and suggestions for future research.

2 Methods

2.1 Sensor Sensitivity Coefficient from Reduced Order Model

Finite element analysis (FEA) is widely used in model-based machine tool analysis due to its high accuracy with 3D models [8]. However, computing TCP displacement is time-consuming, and large matrix sizes make interpretation difficult. Model reduction techniques have been developed to simplify finite element models while retaining essential information and accuracy [9–11].

The sensitivity of machine tools to temperature variations has also been investigated using reduced-order models. Teshima et al. [11] defined a transfer function matrix and sensor sensitivity functions based on a reduced-order model of a machine tool. The transfer function represents how TCP displacement responds to temperature variations across the entire structure and is defined by [12, 13]:

$$d = S\Delta T \tag{1}$$

Here, $d = \begin{bmatrix} dx\ dy\ dz \end{bmatrix}^T$ is the relative displacement between the TCP and workpiece, ΔT the temperature change at N sensors, and $S \in \mathbb{R}^{3 \times N}$ the transfer matrix relating them. Each row of S corresponds to an axis, and each column to a sensor, as defined in Eq. (2) via sensitivity coefficients s_i.

$$S = [s_1\ s_2\ \cdots\ s_N] \tag{2}$$

Here, s_i denotes the TCP displacement caused by a 1 K increase at the ith sensor, making 1 K. In this sense, the transfer function matrix S a representation of the machine's thermal sensitivity. Evaluating s_i, allows quantitative assessment of each sensor's contribution to thermal error.

2.2 Sensitivity Density Function

The sensor sensitivity coefficient s_i, reflects each sensor's impact on TCP displacement, but is affected by sensor layout. To isolate structural effects, a new sensitivity density function $S_d(r)$ is introduced.

The sensitivity density function is a vector-valued function that satisfies Eq. (3).

$$s_i = \int_{V_i} S_d(r)dv \tag{3}$$

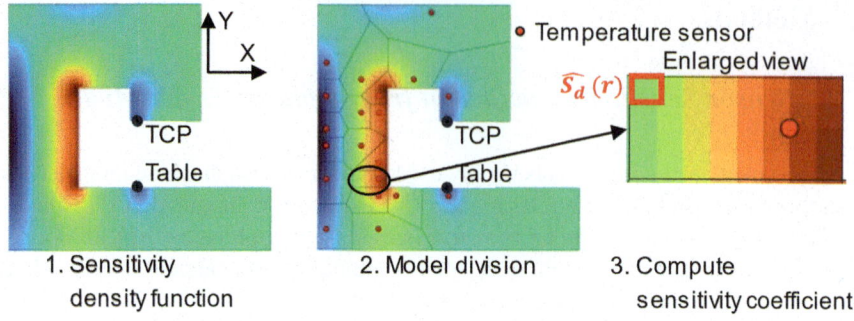

Fig. 1 Computation of sensor sensitivity coefficients

Let $r = [xyz]^T$ be the position vector, and V_i denotes the Voronoi region of the ith sensor. Each area is assigned to its nearest sensor, and s_i is computed by integrating over V_i.

2.3 Computation of the Sensitivity Density Function and Sensor Sensitivity Coefficient

This study aims to determine the optimal number and placement of sensors. Given a sensor arrangement, sensitivity coefficients s_i are computed as shown in Fig. 1.

1. $\hat{S}_d(r)$ is computed from a fine mesh using FEM.

The sensitivity density function $S_d(r)$, defined from structural characteristics, is continuous; however, due to discretization in finite element analysis, it is approximated as $\hat{S}_d(r)$ by sensor sensitivity coefficients s_i at mesh nodes.

2. For a given arrangement of sensors, the integration region V_i for each sensor is determined using the Voronoi partitioning method.
3. The sensor sensitivity coefficient is approximated using Eq. (4):

$$s_i = \int_{V_i} \hat{S}_d(r) dv \ (i = 1, 2, \ldots, M) \tag{4}$$

2.4 Definition of the Objective Function for Sensor Placement

This study defines two requirements for sensor placement. First, sensors should be placed in high-sensitivity regions where small temperature changes cause large

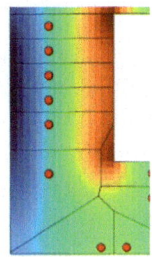
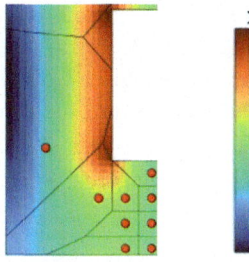

Sensors are placed near a boundary where the sign of $\hat{S}_d(r)$ changes.

No sensor is placed in the region where $\hat{S}_d(r)$ reverses.

Fig. 2 Sensor positions where sensitivity cancellation occurs

displacements. From Eq. (3), s_i is computed by integrating $\hat{S}_d(r)$ over V_i. However, if $\hat{S}_d(r)$ contains opposing sings within V_i, cancellation may occur, reducing s_i. Therefore, in Eq. (5), the following inequality holds:

$$\sum_{i=1}^{N} |s_{i,k}| \leq \int_V \left| \hat{S}_{d,k}(r) \right| dv, \quad k \in \{x, y, z\} \tag{5}$$

Sensitivity cancellation occurs when opposite-sign regions of $\hat{S}_d(r)$ exist within a sensor's Voronoi region V_i, often near sign-change boundaries or unmonitored reversals. This reduces the sensor coefficient s_i due to partial or complete integration cancellation. Effective placement requires locating sensors where $\left| \hat{S}_d(r) \right|$ is high and avoiding mixed-sign regions within V_i. As shown in Fig. 2, poor placement leads to low sensitivity coverage. Thus, evaluating cancellation helps assess placement quality.

Thus, the amount of information loss due to sensitivity cancellation is defined as L and expressed by the following equation:

$$L_k = \int_V \left| \hat{S}_{d,k}(r) \right| dv - \sum_{i=1}^{N} |s_{i,k}|, \quad k \in \{x, y, z\} \tag{6}$$

The first term on the right represents the total sensitivity density, while the second indicates the sensitivity captured by the sensors. Their difference, L, quantifies information loss due to sensor placement. A smaller L suggests better sensor positioning in high-sensitivity regions. The second requirement prevents excessive sensitivity coefficients by avoiding sensor clustering. A large coefficient increases vulnerability to disturbances, leading to significant errors. High-density regions naturally produce larger coefficients, so placing multiple sensors in these areas reduces each sensor's Voronoi region size, balancing the coefficients.

Sensor readings include errors from accuracy and installation, causing uncertainty in estimated displacement. This is expressed as $T_i = T_i' + \Delta T_i$, where T_i is the measured temperature, T_i' the actual temperature, and ΔT_i the measurement error. Using this temperature, the thermal displacement is shown in Eqs. (7) and (8).

$$d = s_1\left(T_1' + \Delta T_1\right) + s_2\left(T_2' + \Delta T_2\right) + \cdots + s_N\left(T_N' + \Delta T_N\right) \tag{7}$$

$$= \sum_{i=1}^{N} s_i T_i' + \sum_{i=1}^{N} s_i \Delta T_i \tag{8}$$

The second term in Eq. (8) represents uncertainty in displacement due to temperature variations. Assuming $\Delta T_i \sim N\left(0, \sigma^2\right)$, the resulting uncertainty is shown in Eq. (9):

$$\sum_{i=1}^{N} s_{i,k} \Delta T_i \sim N\left(0, \left(\sum_{i=1}^{N} s_{i,k}^2\right)\sigma^2\right), \quad k \in \{x, y, z\} \tag{9}$$

This expression shows that displacement uncertainty follows a normal distribution with variance dependent on the squared sensor sensitivities and temperature measurement error variance. To minimize uncertainty, the following quantity must be minimized, as shown in Eq. (10):

$$S_k = \sum_{i=1}^{N} \left|s_{i,k}\right|^2, \quad k \in \{x, y, z\} \tag{10}$$

Minimizing S reduces measurement error propagation in displacement estimation, ensuring robustness and accuracy. S evaluates the impact of temperature errors, and minimizing it distributes sensitivity evenly, reducing localized errors. Based on these requirements, the objective function J for sensor placement optimization is defined in Eq. (11):

$$J = \lambda S + (1 - \lambda)L, \quad 0 \le \lambda \le 1 \tag{11}$$

The objective function J optimizes sensor placement by allocating sensors in high-sensitivity regions. The weighting coefficient λ, between 0 and 1, balances error suppression ($\lambda = 1$) and sensitivity representation ($\lambda = 0$). Sensor placements minimizing J are explored for different numbers of sensors and various values of λ.

3 Sensor Placement Optimization in a Simplified Model

3.1 Details of the Simplified Model

In this study, a simplified and scaled-down model of a three-axis vertical machining center, projected from the X-axis direction as shown in Fig. 3, is employed to evaluate the influence of sensor placement clearly The material properties used in the model are listed in Table 1. Some material properties were assigned dummy values for the creation of the ROM. The analysis focuses solely on thermal displacement in the Z-axis direction between the TCP and the table. The candidate sensor placement positions are the 147 points indicated in the figure, from which multiple points are selected for sensor placement. Instead of using the full sensitivity density function $S_d(r)$, the sensitivity distribution is approximated using 2292 uniformly distributed virtual sensors. The sensor sensitivity coefficients are efficiently computed by applying the Krylov subspace method to a ROM.

Fig. 3 Mesh diagram of the simplified model

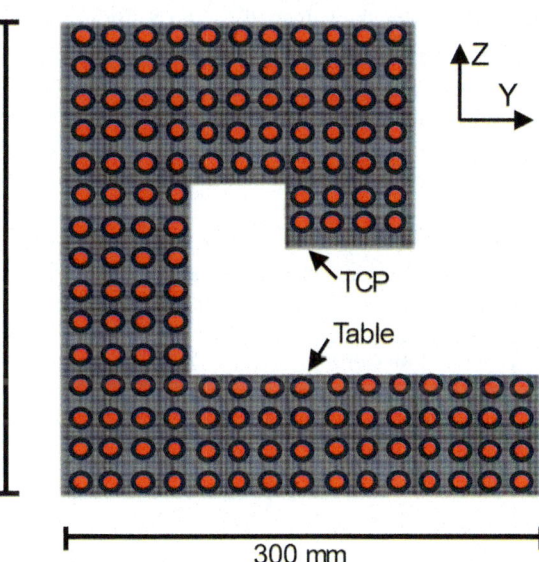

300 mm

Table 1 Physical properties

Material	A5056
Thermal conductivity	137 W/(m · K)
Specific heat	0.001 J/(kg ·° C)
Density	0.001 kg/m^3
Thermal expansion coefficient	2.3e−5 °C

A linear hexahedral mesh with an edge length of 2.5 mm was applied to the model, resulting in 36,672 elements.

3.2 Parameters

The objective function J in Eq. (11) includes λ, which balances the minimization of S and L. Increasing λ reduces sensor sensitivity imbalance, while decreasing λ minimizes information loss. Optimization is performed for different λ values to select the sensor placement that best balances S and L. The optimization process, shown in Fig. 4, considers thermal displacement in the Z-axis direction.

A genetic algorithm was used to minimize the objective function J. The details of the genetic algorithm are shown in Table 2.

Figure 5 presents the results of the optimization process, where the number of sensors was varied from 10 to 80, and the parameter λ was incrementally increased from 0 to 1 in steps of 0.05, minimizing the objective function at each step. The figure illustrates the values of L_z and S_y for each optimized sensor arrangement.

When λ is large, the optimization prioritizes minimizing S_z, causing the plotted points to shift toward the left side of the graph. However, this also increases L_z. Conversely, when λ is small, the optimization prioritizes minimizing L_z, leading to a distribution of points toward the lower part of the graph. As the number of sensors increases, the individual sensor sensitivity coefficients tend to decrease, resulting in a reduction in both L_z and S_z.

Figure 6 shows the optimized 20-sensor layouts for $\lambda = 0$, 0.15, 0.5, and 0.9. At $\lambda = 0$, the sensors are spread across the entire model to prevent sensitivity

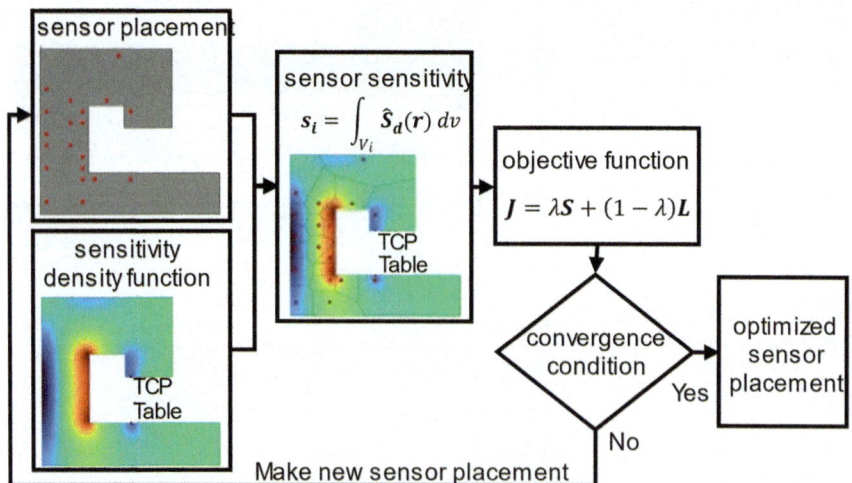

Fig. 4 Flowchart for sensor placement explore

Table 2 Genetic algorithm parameters

Setting	Value
Population size	200
Generations	300
Elite individuals	10
Crossover rate	0.7
Mutation rate	Applied to crossover offspring
Stopping criteria	– No improvement for 50 generations (StallGenLimit = 50) – Objective function change < 10^{-6} (TolFun = 1e−6)
Parent selection	Stochastic Universal Sampling (SUS)
Crossover method	Order Crossover (OX)
Post-crossover correction	Ensures valid sensor arrangement

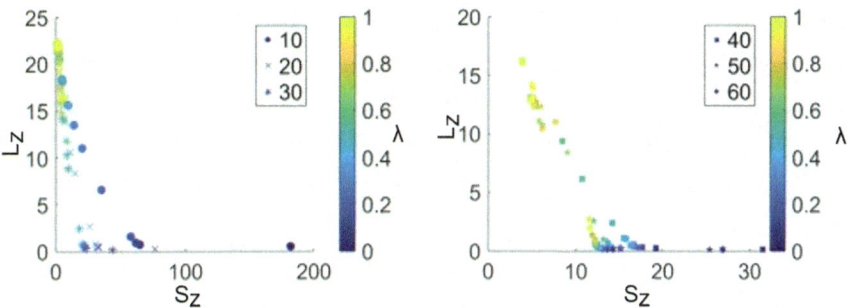

Fig. 5 Parameters of each optimized sensor placement

cancellation. When λ is set to 0.15, in addition to reducing sensitivity cancellation, more sensors are placed in regions with high sensitivity density. For $\lambda = 0.5$, the sensors are positioned so that their Voronoi regions contain both positive and negative sensitivity areas, leading to some degree of sensitivity cancellation. At $\lambda = 0.9$, fewer sensors are near the column, causing strong sensitivity cancellation within single regions.

3.3 Experimental Setup

This experiment evaluates the sensor placements obtained through the optimization process by comparing them with the results of an actual heating experiment.

Figure 7 shows the experimental setup. Temperature sensors were installed at all 147 candidate sensor positions. Heating was performed by applying a flame from a long lighter to the heating point for 30 s, repeated three times. During heating,

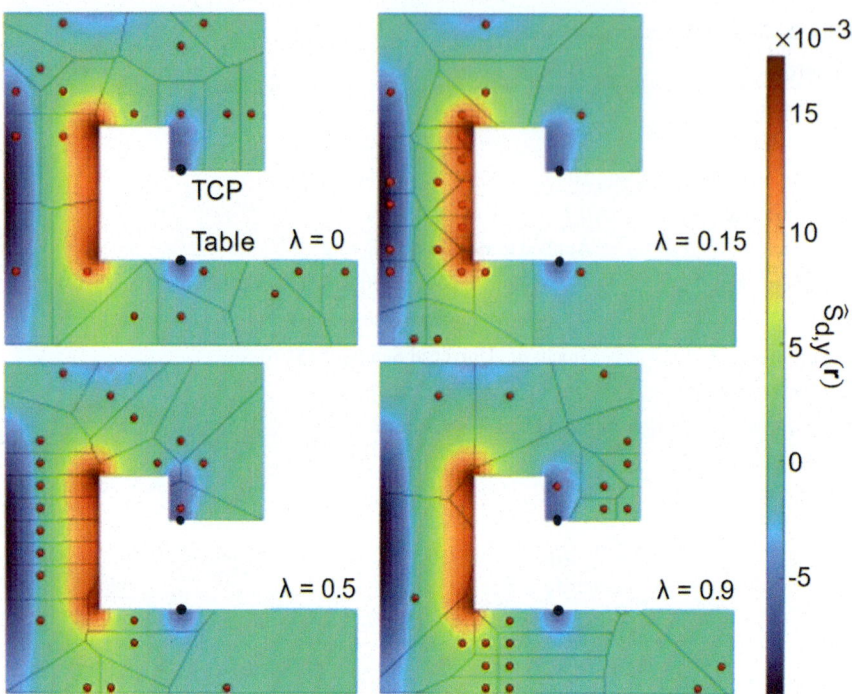

Fig. 6 Sensor placement obtained through optimization process

the relative displacement in the Z-direction between the TCP and the table was measured using an eddy current sensor. In this study, the estimated thermal displacement obtained using all 147 sensors is compared with the estimated displacement obtained using only the optimized sensor placements from the exploration process. Additionally, 50 temperature patterns were generated by adding randomly generated errors to the measured temperatures at each time point to evaluate the sensitivity of the estimated values to temperature measurement errors. These errors followed a normal distribution with a standard deviation of 0.3 °C.

4   Results and Discussion

Figure 8 shows the results of the displacement measurements obtained using a displacement sensor and the estimated thermal displacement calculated based on the sensor sensitivity coefficients for all 147 sensor positions. The time lag is due to infrequent temperature measurements and sensor thermal capacity.

Figure 9 plots the Root Mean Square Error (RMSE) between the estimated thermal displacement using different sensor configurations and the estimation using all sensors, with λ on the horizontal axis and RMSE on the vertical axis. The maximum

Fig. 7 Experimental setup
to measure thermal deviation

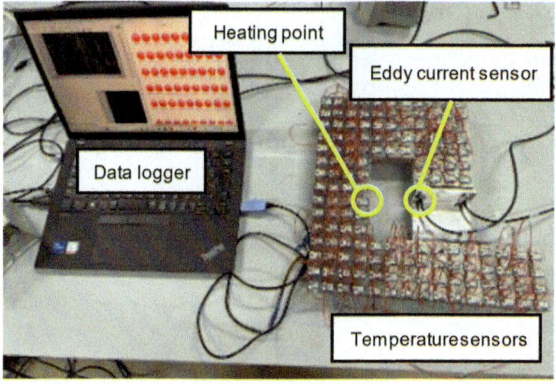

Fig. 8 Thermal
displacement of TCP of a
simplified model

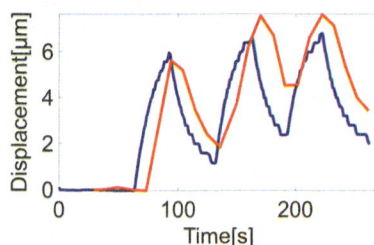

value of λ at which the RMSE falls below 0.5 μm increases with the number of sensors as follows: 0.15 for 10 and 20 sensors, 1 for 30, 0.5 for 40, 0.6 for 50, and 0.95 for 60 sensors.

Figure 10 shows a strong correlation (r = 0.910) between L_z and RMSE. Minimizing I_z improves estimation accuracy, comparable to denser sensor setups. More sensors facilitate lower L_z by reducing Voronoi region size and sensitivity cancellation. Configurations with 30 and 60 sensors yielded low RMSE even at higher λ. The 30-sensor case had a relatively high L_z, likely due to heating-position dependence and optimization for specific conditions. For 60 sensors at $\lambda = 0.85$ and 0.9, low L_z suggests possible influence of local optima in the genetic algorithm.

Figure 6 shows the optimal 20-sensor layout at $\lambda = 0.15$, yielding the lowest RMSE by avoiding sensitivity cancellation and concentrating sensors in key regions. Figure 11 presents the standard deviation under random temperature errors, revealing greater variability at low sensor counts and small λ. The λ threshold for stable estimates increased with sensor count.

From Eq. (10), estimation variability follows a Gaussian distribution with standard deviation $\sqrt{S_z}\sigma$. Since L_z and S_z are trade-off, increasing λ raises L_y and lowers S_z. To keep RMSE low while reducing S_z, more sensors are needed. Table 3 configurations with RMSE < 0.5μm have lower variability as sensor count increases, improving both accuracy and robustness.

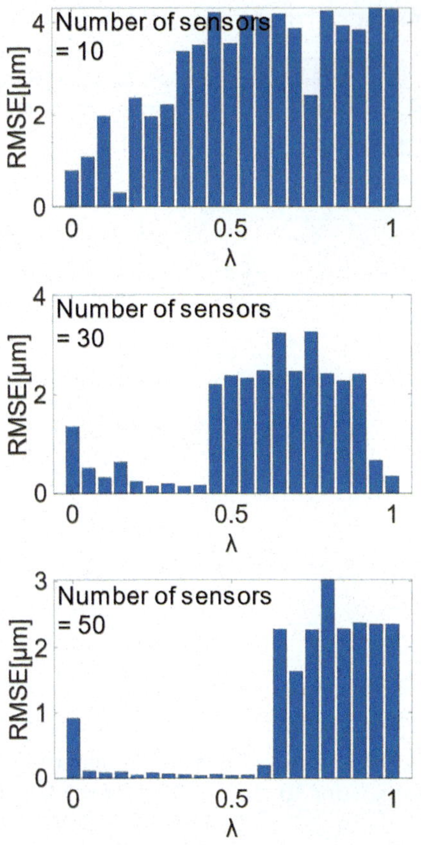

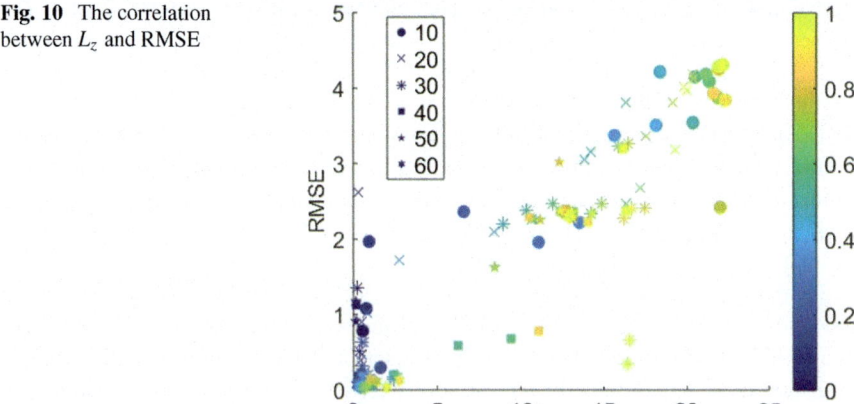

Fig. 9 RMSE between the estimated deviation using the 147 sensors and the optimized sensor placements

Fig. 10 The correlation between L_z and RMSE

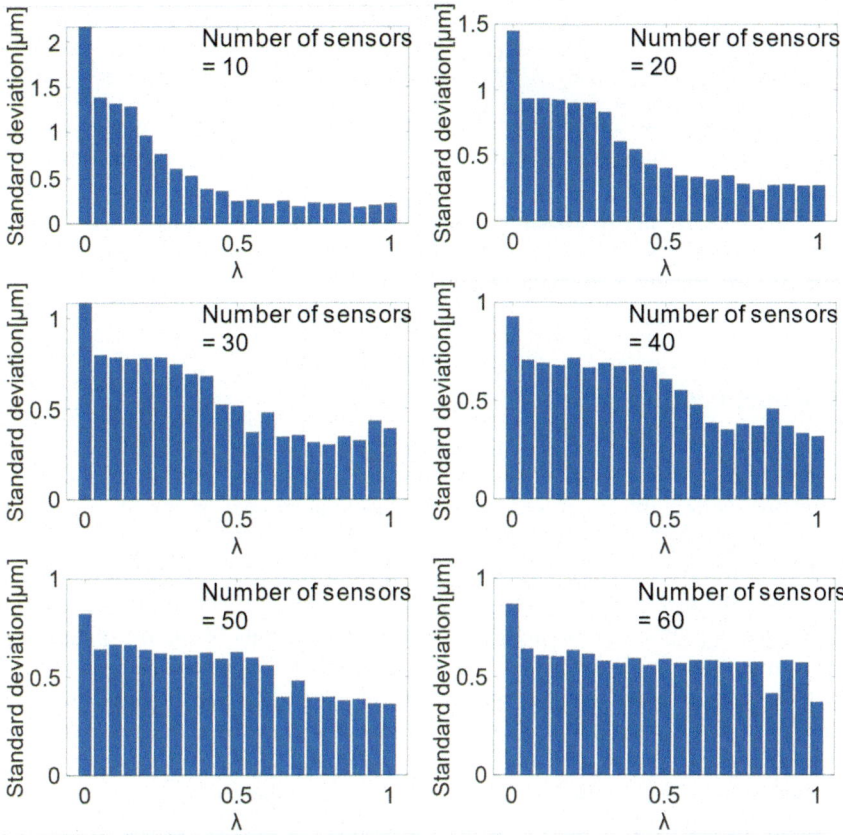

Fig. 11 Standard deviation of the estimated values calculated based on randomly generated errors

Table 3 Standard deviations for each number of sensors

Number of sensors	10	20	30	40	50	60	147
Standard deviation for RMSE < 0.5 μm	1.28	0.929	0.737	0.680	0.621	0.589	0.533

The minimum S_z is constrained by L_z. From Eq. (10), minimizing S_z under constant L_z leads to the condition shown in Eq. (12):

$$s_{i,z} = \frac{\sum_{i=1}^{N} |s_{i,z}|}{N} = \frac{L_z}{N} \tag{12}$$

In this case, the theoretically minimum standard deviation is given by:

$$\sqrt{S_z}\sigma = \sqrt{N \cdot \left(\frac{L_z}{N}\right)^2}\,\sigma = \frac{L_z}{\sqrt{N}}\sigma \tag{13}$$

Due to the inverse square-root relation, accuracy gains diminish as sensor count increases. In this study, a simplified model was used, so practical applications may require more sensors.

Minimizing L_z improves accuracy, minimizing S_z improves robustness. Balancing both requires more sensors, though benefits diminish with count. Ultimately, sensor accuracy and sensitivity must also be tuned.

5 Conclusion

In this study, a sensor placement optimization strategy for thermal displacement estimation using multi-point temperature measurements was proposed, and sensor placement was explored using a simplified model. The sensor placement was determined by introducing a new parameter based on sensor sensitivity coefficients and optimizing the arrangement using a genetic algorithm. As a result, it was demonstrated that appropriately setting the weighting coefficients of the two parameters makes it possible to find a sensor configuration that balances estimation accuracy and robustness against temperature measurement errors. This study focused on sensor placement optimization in a single-axis direction. Future work should extend this to all three axes. In practical machine tools, constraints such as inaccessible sensor locations and thermal contact resistance between components must also be considered. To maintain high estimation accuracy under these conditions, machine learning approaches may be necessary. Additionally, the effectiveness of the proposed sensor placement should be evaluated with estimation methods beyond the ROM.

References

1. Cheng, Q., Qi, Z., Zhang, G., Zhao, Y., Sun, B., Gu, P.: Robust modelling and prediction of thermally induced positional error based on grey rough set theory and neural networks. Int. J. Adv. Manuf. Technol. **83**(5–8), 753–764 (2016)
2. Miao, E.M., Gong, Y.Y., Niu, P.C., Ji, C.Z., Chen, H.D.: Robustness of thermal error compensation modeling models of CNC machine tools. Int. J. Adv. Manuf. Technol. **69**(9–12), 2593–2603 (2013)
3. Kumar, S., Srinivasu, D.S.: Optimal number of thermal hotspots selection on motorized milling spindle to predict its thermal deformation. Mater. Today **62**, 3376–3385 (2022)
4. Zimmermann, N., Lang, S., Blaser, P., Mayr, J.: Adaptive input selection for thermal error compensation models. CIRP Ann. Manuf. Technol. **69**(1), 485–488 (2020)
5. Tan, F., Yin, M., Wang, L., Yin, G.: Spindle thermal error robust modeling using LASSO and LS-SVM. Int. J. Adv. Manuf. Technol. **94**(5–8), 2861–2874 (2018)
6. Li, L., Chen, B., Yu, J.: Thermal error modeling method of truss robot based on GA-LSTM. Ind. Robot. **51**(5), 809–819 (2024)

7. Chen, C., Dai, H., Lee, C., Hsieh, T., Hung, W., Jywe, W.: The development of thermal error compensation on CNC machine tools by combining ridge parameter selection and backward elimination procedure. Int. J. Adv. Manuf. Technol. **130**(5–6), 2423–2442 (2024)
8. Mian, N., Fletcher, S., Longstaff, A., Myers, A.: Efficient estimation by FEA of machine tool distortion due to environmental temperature perturbations. Precis. Eng. **37**(2), 372–379 (2013). https://doi.org/10.1016/j.precisioneng.2012.10.006
9. Srinivasan Puri, R., Morrey, D., Bell, A.J., Durodola, J.F., Rudnyi, E.B., Korvink, J.G.: Reduced order fully coupled structural–acoustic analysis via implicit moment matching. Appl. Math. Model. **33**(11), 4097–4119 (2009). https://doi.org/10.1016/j.apm.2009.02.016
10. Hernández-Becerro, P., Purtschert, J., Konvicka, J., Buesser, C., Schranz, D., Mayr, J., Wegener, K.: Reduced-order model of the environmental variation error of a precision five-axis machine tool. J. Manuf. Sci. Eng. **143**(2), 021005 (2021). https://doi.org/10.1115/1.4047739
11. Teshima, Y., Tanaka, S., Kizaki, T., Sugita, N.: Sensor placement strategy based on reduced-order models for thermal error estimation in machine tools. CIRP J. Manuf. Sci. Technol. **55**, 403–410 (2024). https://doi.org/10.1016/j.cirpj.2024.10.015
12. Mayr, J., Ess, M., Pavlíček, F., Weikert, S., Spescha, D., Knapp, W.: Simulation and measurement of environmental influences on machines in frequency domain. CIRP Ann. **64**(1), 479–482 (2015). https://doi.org/10.1016/j.cirp.2015.04.001
13. Fujishima, M., Narimatsu, K., Irino, N., Ido, Y.: Thermal displacement reduction and compensation of a turning center. CIRP J. Manuf. Sci. Technol. **22**, 111–115 (2018). https://doi.org/10.1016/j.cirpj.2018.04.003

Comparisons of Model Order Reduction Techniques for Efficient Thermal Simulations

Pritam Bari⊙, Holger Rudolph, Lars Penter⊙, and Steffen Ihlenfeldt⊙

Abstract Machine tools are susceptible to a wide range of thermal field variations that result in thermal deformations or positional deviations of the tool center point (TCP), which affect the manufacturing quality, life cycle of the tool, and production time. Therefore, accurate thermal and thermo-elastic predictions play a significant role in satisfying the manufacturing demand for precise machine tools. Finite Element Method (FEM) based correction of the thermal deviations of the mechanical systems is a very energy-efficient strategy. However, those are computationally expensive, especially for large-scale simulations. Model order reduction (MOR) techniques are applied to reduce computational costs. However, little attention is paid to instruct clear guidelines for the optimal MOR technique for specific boundary conditions and types of problems. To address such a problem, this paper compares two Krylov subspace-based MOR techniques regarding their computational efficiency and accuracy. Therefore, the study examines two levels of model complexity: a bar-slider and a test bench. For both cases, the thermal computational time is compared with and without MOR. Finally, proper guidelines are established to select the optimal MOR. It is observed that MOR using the moment matching method is more computationally efficient and accurate than the Arnoldi iteration.

Keywords Thermal deviations · Model order reduction · Krylov subspace · Moment-matching

1 Introduction

In the machine tool industry, the surging demand for products with compact tolerances in the micrometer range requires persistent advancement in manufacturing highly precise machine tools. To enable the production of such machines with a high degree of accuracy, we need to reduce the wide range of deviations that occur in the

P. Bari (✉) · H. Rudolph · L. Penter · S. Ihlenfeldt
Chair of Machine Tools Development and Adaptive Controls, Institute of Mechatronic Engineering, Dresden University of Technology, Dresden, Germany
e-mail: pritam.bari@tu-dresden.de

© The Author(s) 2026
K. Wegener and M. Bambach (eds.), *4th International Conference on Thermal Issues in Machine Tools (ICTIMT2025)*, Lecture Notes in Production Engineering,
https://doi.org/10.1007/978-3-032-01194-7_32

machine tools. The dominant source (up to 70%) of the overall deviations is attributed to the thermal impact [1]. Thermally induced deviations are caused by internal and external thermal influences [2]. The internal effects arise most commonly due to heat generated in electrical components such as motors, or due to mechanical friction in gearboxes, ball screws, linear guides, and bearings, or due to the tool-workpiece interaction in the cutting process. The external factors are typically heat generation due to environmental/personal radiation. The deviations in the temperature field produce thermal deformations in the machine tools, again causing positional uncertainties of the TCP and degrading the precision of the workpiece, manufacturing quality, tool life cycle, and production time [1]. Therefore, accurate thermal and thermo-elastic predictions play a significant role in satisfying the manufacturing demand for precise machine tools.

To diminish the detrimental consequences of thermal deviations, thermal stabilization of machine tools could be done via conventional methods (such as machine cooling or shopfloor climatization) or by correcting the NC axes or tool position. To meet the intensifying demand for the reduction of energy consumption, the model-based correction of the thermal deviations of the mechanical systems plays a very crucial role. The model-based thermal compensation approaches are divided into two categories: data-based and physics-based. The data-based models vary from simple regression models or transfer function-based models to more sophisticated pioneering approaches using memory neural networks, ARX (AutoRegressive with eXogenous inputs), and deep learning [3–10]. The physics-based models [11] compute the thermal deformation of the machine tool by partial differential equations, which are further discretized typically via finite elements [e.g., Finite Element Method (FEM), or Finite Difference Method (FDM), or Finite Volume Method (FVM)] [12]. Although finite element method-based thermal compensation models can predict deformations, but are not fast enough to calculate them in real time, especially for large, complex machine tools.

The computational difficulty of thermal simulations originates largely from the continual decompositions of high-dimensional system matrices of order n (the system's number of degrees of freedom). For large, complex geometric machine tools, n becomes very high, making the thermal FEA model computationally expensive. This, in turn, hinders the efficient design of structure-based correction of thermal deviations. As an optimum solution, the Model Order Reduction (MOR) methods are applied to generate a compact, efficient model from the original FE model [13]. MOR enables the conversion of a high-order thermal model into a reduced-order model by reducing the degrees of freedom of the system while preserving the input-output behaviour of the original model within a given accuracy. Hence, to reduce the computational time, the physics-based models were coupled with the model order reduction (MOR) techniques [14–16]. Typical projection-based MOR techniques that are most promisingly used in thermal problems are the Krylov subspace methods (using the Arnoldi iteration [14, 17, 18] or moment matching method [19]), balanced truncation method [20], and iterative rational Krylov algorithm (IRKA) [21]. For large-scale problems, Krylov-based model reduction methods are efficient. However, the stability of the reduced system is not necessarily preserved, and the global error bound does not exist in contrast to the balanced truncation method [22]. Although MOR techniques have been used in thermal models in the past, negligible contemplation was drawn to provide distinct guidelines for selecting the optimal MOR technique based on specific boundary conditions and the complexity of the problems. To attenuate such a problem, this paper focuses comparison of two widely used Krylov subspace methods (using moment-matching and Arnoldi iteration) in terms of computational efficiency and accuracy.

The technical contributions of this paper are structured as follows: Sect. 2 discusses the thermal modelling of machine tools. The solution methodology is further presented in Sect. 2.1. An equivalent mechanical CAD model is first imported into Ansys® to prepare geometry, material, contact, components definition, volume division, and meshing. The thermal FEM matrices are then exported (using Ansys APDL Macro Programming) to Matlab® for geometry reconstruction and thermal analysis with and without MOR. Following this, the mathematical formulation of the MOR techniques is presented in Sect. 2.2. Thereafter, Sect. 3 presents validation of the thermal results for two difficulty levels of the model. In the first model, we consider a bar-slider where heat generation occurs due to mechanical friction between two moving bodies as discussed in Sect. 3.1. In the second case, we perform the thermal simulation of a real test bench having a rotating shaft and a motor as presented in Sect. 3.2. In this case, the main heat source is the ball bearings and the motor. Furthermore, the thermal computation times for the original system are compared with the reduced models for both cases. Finally, proper guidelines are established to select the optimal MOR. This is followed by the summary of the work as in Sect. 4.

2 Thermal Modeling of Machine Tools

The thermal characteristics of a homogeneous body can be expressed by a partial differential equation (PDE) of the temperature field $T(t, \delta)$ at time t and location δ as follows [20]:

$$\rho c_{\mathrm{p}} \frac{\partial}{\partial t} T(t, \delta) - \nabla \cdot (\lambda \cdot \nabla T(t, \delta)) = \dot{q}(t, \delta), \tag{1}$$

where, the Laplace operator is written by ∇, the external thermal load per unit time is denoted by $\dot{q}$ and the material properties are as follows: density ρ, heat capacity c_{p}, and conductivity λ. To calculate the temperature distribution field, Eq. 1 must be solved. Since the geometries of the systems and the governing equations are very complex, the closed-form solution of Eq. 1 could not be used for most real-world problems. Therefore, the mathematical discretization method (here, FEM) is applied to calculate approximate solutions [12].

2.1   Framework for Thermal Analysis

The framework for thermal analysis is illustrated in Fig. 1. Initially, the mechanical CAD geometry of the interested machine components is prepared in SolidWorks®. Afterward, this is modified into a thermal equivalent CAD model which is imported into Ansys for the discretization of the geometry, material selection, connection of contact and target bodies, component definition, mechanical and thermal boundary conditions (Neumann/Dirichlet/Robin), volume division, projection of interacting bodies, and meshing using FEM. The FEM converts the PDE (Eq. 1) into a system of coupled ordinary differential equations (ODE) that enables the temperature to be solved using numerical integration methods in the time domain. Applying FEM, the following linear system is obtained [11, 14]:

$$\mathbf{C}_{TT} \cdot \dot{\mathbf{T}} + \mathbf{L}_{TT} \cdot \mathbf{T} = \dot{\mathbf{Q}}, \tag{2}$$

where, $\mathbf{C}_{TT}$ —the matrix of heat capacities, $\mathbf{L}_{TT}$—the matrix of heat conduction and heat transfer within the system, $\mathbf{T}$—the vector of temperatures, $\dot{\mathbf{Q}}$—the vector of heat flow rate at the system boundaries. These thermal FEM matrices are computed using Ansys APDL Macro Programming, which are further exported into Matlab for geometry reconstruction, and thermal analysis. All computer programs were executed on Ansys Workbench 2023 R2 and Matlab 2023a, which were implemented on a computer with specifications of: Intel(R) Core(TM) i7 Processor, 3.42 GHz, 32 GB RAM. Since Ansys takes a long time for thermal simulation of large complex machine tools, and Ansys has less flexibility to choose an optimized solver, the programming environment is developed in Matlab. The backward/implicit Euler method is applied in the thermal solver. Furthermore, to enhance the computational efficiency of the thermal calculations, the system of equations from the FEM model having a very large degree of freedom is transferred into a reduced system with substantially fewer degrees of freedom using MOR techniques as discussed next.

2.2   Model Order Reduction for Machine Tools

The MOR methods start with reformulating the thermal problem (Eq. 2) into an entirely equivalent linear dynamic system through state-space representation as follows [20]:

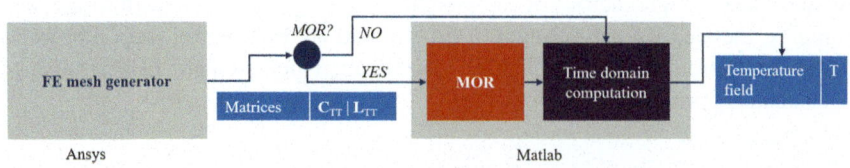

Fig. 1 Strategy for solving the thermal behaviour with/without MOR

$$\Sigma : \begin{cases} \mathbf{E}\dot{\mathbf{x}}(t) = \mathbf{A}\mathbf{x} + \mathbf{B}\mathbf{u}(t) \\ \mathbf{y}(t) = \mathbf{C}\mathbf{x} \end{cases}, \qquad (3)$$

with capacity and conductivity matrices $\mathbf{E}, \mathbf{A} \in \mathbb{R}^{n \times n}$, control and observer matrices $\mathbf{B} \in \mathbb{R}^{n \times m}$, $\mathbf{C} \in \mathbb{R}^{p \times n}$, state $\mathbf{x} = \mathbf{T} \in \mathbb{R}^n$, input $\mathbf{u} \in \mathbb{R}^m$, and output $\mathbf{y} \in \mathbb{R}^p$. The objective of the projection-based MOR method is to construct a reduced-order surrogate model as follows:

$$\widehat{\Sigma} : \begin{cases} \widehat{\mathbf{E}}\dot{\hat{\mathbf{x}}}(t) = \widehat{\mathbf{A}}\hat{\mathbf{x}} + \widehat{\mathbf{B}}\mathbf{u}(t) \\ \hat{\mathbf{y}}(t) = \widehat{\mathbf{C}}\hat{\mathbf{x}} \end{cases}, \qquad (4)$$

where, $\widehat{\mathbf{E}}, \widehat{\mathbf{A}} \in \mathbb{R}^{r \times r}$, $\widehat{\mathbf{B}} \in \mathbb{R}^{r \times m}$, $\widehat{\mathbf{C}} \in \mathbb{R}^{p \times r}$ are the reduced matrices and $\hat{\mathbf{x}} \in \mathbb{R}^r$ is the state in the reduced space of the dimension $r \ll n$. The reduced-order matrices are computed by projecting the original system matrices into a lower-dimensional subspace as follows:

$$\widehat{\mathbf{E}} = \mathbf{W}^T \mathbf{E} \mathbf{V}, \quad \widehat{\mathbf{A}} = \mathbf{W}^T \mathbf{A} \mathbf{V}, \quad \widehat{\mathbf{B}} = \mathbf{W}^T \mathbf{B}, \quad \widehat{\mathbf{C}} = \mathbf{C} \mathbf{V}, \qquad (5)$$

where, $\mathbf{V}, \mathbf{W}$ are called truncation/projection matrices. Finally, the state vector $\hat{\mathbf{x}}$ of the reduced degrees of freedom is re-transformed into the full state/temperature vector of the original FE model as follows:

$$\mathbf{x} = \mathbf{V}\hat{\mathbf{x}}. \qquad (6)$$

The truncation matrices in Eq. 5 could be calculated in different ways using various projection-based algorithms: Krylov subspace methods (using the Arnoldi iteration or moment matching method), balanced truncation method, and iterative rational Krylov algorithm (IRKA) [14, 19–21]. In this work, the thermal behaviour of the original system is compared with two Krylov subspace methods (using the moment matching method and Arnoldi iteration) for two cases as discussed next.

3 Model Validation

3.1 Case Study 1: Bar-Slider

In the first case study, we consider a bar-slider problem. The bar has a dimension of $0.2 \times 0.02 \times 0.005$ m (L $\times$ W $\times$ H) and the slider has a dimension of $0.02 \times 0.02 \times 0.005$ m (L $\times$ W $\times$ H). The material is considered to be Structural steel for both. The Ansys model of the bar-slider is shown in Fig. 2. We consider pressure $P = 1$ MPa over the positive zx surface of the slider. The slider moves over the bar from one end to the other along the positive x direction at a speed of $v = 0.18$ m/s. Hence, the

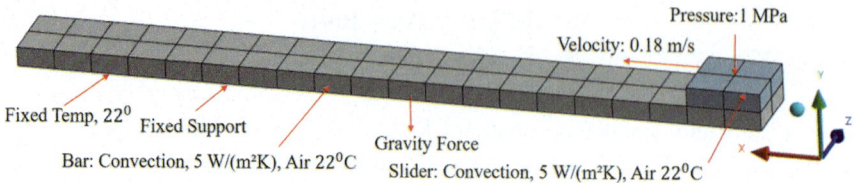

Fig. 2 The bar-slider problem: ansys model

time for each forward/backward stroke is 1 sec. We consider a total of two cycles of movement, which is covered in 4 sec. The main source of heat comes from the friction (coefficient $\mu = 0.2$) between the bar and slider during movement. The total heat ($\dot{Q} = \mu P A v$, where A is the contact area) is equally distributed towards the contact surfaces of the bar and slider. Since the slider moves over the bar, the contact surface over the bar changes over time. Therefore, we discretize the positive zx surface of the bar into 10 segments and develop a weighting function in Matlab that calculates the position-dependent weighting factors of power loss and thermal conductivity over the bar. This enables the heat control matrix **B** to be constant and the input matrix **u**(t) to vary with time as in Eq. 3. The negative zx surface of the bar is at the fixed support and fixed temperature of 22 °C. For the bar, free convection (coefficient of 5 W/m²/K) to the air (at 22 °C) occurs over positive yz surface, negative and positive xy surfaces. For the slider, it occurs over negative and positive xy, yz surfaces. We consider 444 nodes and 44 elements in the Ansys mesh.

The transient thermal simulations of the bar-slider (combined with the linear movement of the slider) are illustrated in Fig. 3. The results for $t = 1.5$ s (1st cycle, middle position of return stroke) are shown in Fig. 3a–d. The temperature behaviour of the original full model (Fig. 3b) shows fair agreement with the Ansys result (as in Fig. 3a) in terms of temperature distribution and maximum/minimum temperature. The results for the full model are further compared with reduced models applying the moment-matching and Arnoldi iteration as depicted in Fig. 3c–d, respectively. The reduced order $r = 10$ is considered for both MOR methods. It is chosen based on the convergence test because there is a minimum value of r, below which the reduced-order solution does not converge with the original solution. We see that the temperature distribution using the moment matching method is quite promising with the original result. However, for the case of the reduced model with Arnoldi iteration, the slider is at a higher temperature. A similar comparison for $t = 4$ s (2nd cycle, end position of return stroke) is shown in Fig. 3e–h. In this case too, the maximum temperature of the slider using the Arnoldi iteration is higher than 23.43 °C, while it is 23.43 °C for both full-order model and moment-matching MOR. To confirm such characteristics, we further perform a robustness analysis by varying the friction coefficient and velocity of the slider. We consider four cases as follows: case 1 ($\mu = 0.2$, $v = 0.18$ m/s), case 2 ($\mu = 0.4$, $v = 0.18$ m/s), case 3 ($\mu = 0.2$, $v = 0.36$ m/s), case 4 ($\mu = 0.4$, $v = 0.36$ m/s). The maximum temperature using full-order and reduced models for all four cases are provided in Table 1. In all cases, we see

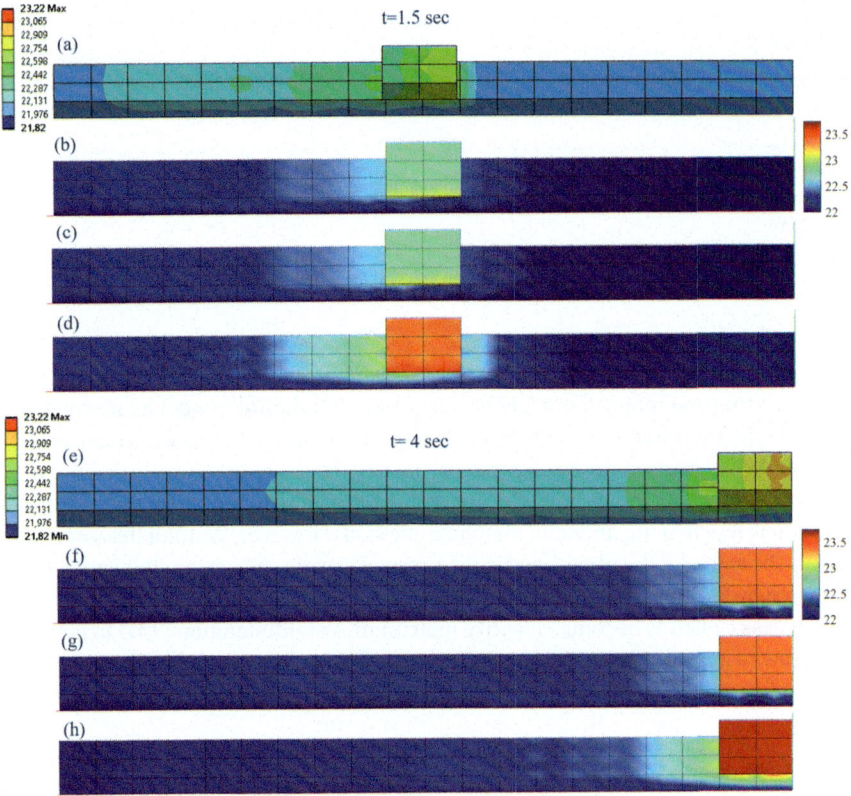

Fig. 3 Transient thermal simulation of the bar-slider (combined with the linear movement of the slider). At $t = 1.5$ s **a** Ansys result; Matlab results for **b** original full model and reduced models applying **c** moment-matching and **d** Arnoldi iteration. At $t = 4$ s **e** ansys result; Matlab results for **f** original full model and reduced models applying **g** moment-matching and **h** Arnoldi iteration. The temperatures in the colorbar are in °C

Table 1 Comparisons of the maximum temperatures using full-order and reduced models for four cases of the bar-slider problem

Different models	Case 1 (°C)	Case 2 (°C)	Case 3 (°C)	Case 4 (°C)
Full order	23.43	24.87	23.98	25.97
Moment matching	23.43	24.87	23.98	25.97
Arnoldi iteration	23.75	25.5	24.49	26.99

that the Arnoldi iteration overestimates the maximum temperature while the full-order model and moment-matching MOR give the same maximum temperature. It confirms that the Arnoldi iteration is not as accurate as moment-matching MOR.

Table 2 Comparison of the transient thermal computational time (CPU time) for the case of the bar-slider

Different models	Geometry reconstruction (s)	Model reduction (s)	ODE solver time (s)	Total computation time (s)
Full order	8.4		3.25	11.65
Moment matching	8.4	0.085	0.785	9.27
Arnoldi iteration	8.4	0.023	3.25	11.67

Furthermore, the computational effort (CPU time) for the transient simulation (up to $t = 4$ s) for the reduced models is compared with the full model as presented in Table 2. The time step in the transient thermal simulations is considered to be 0.082 sec. Total computation time comprises geometry reconstruction time, model reduction time, and ODE solver time. We see that the model reduction time in the Arnoldi iteration is less than the moment matching method. However, Arnoldi iteration takes almost the same ODE solver time as the original system, which could be potentially triggered by the near-singular form of matrix $\mathbf{A} = -\mathbf{C}_{TT}^{-1/2} \mathbf{L}_{TT} \, \mathbf{C}_{TT}^{-1/2}$ due to the lumping process (when $\mathbf{E}$ becomes identity matrix) in Arnoldi iteration [14]. In contrast, the moment-matching method is much more efficient in terms of ODE solver time and total computational time. A test bench is further considered as discussed next.

3.2 Case Study 2: Test Bench

In the second case, a real test bench (see the set-up in the laboratory in Fig. 4a) is prepared for the experimental validation of our thermal model. The primary components of the test bench are as follows: DC motor (max 24 V, 1800 U/min), shaft, base plate, angular contact ball bearings mounted over the shaft, thermal insulation over the bearings, and bearing blocks (see Table 3 for details)[1] The bearing mounted over the motor side is a fixed type, whereas it is a floating type on the opposite end. When we turn on the motor, the shaft rotates, and heat is produced due to frictional power loss in the bearings. The bearing power loss is calculated using the following empirical formula [23]:

$$
\begin{aligned}
P &= (M_0 + M_1 + M_s) \cdot \omega, \\
M_0 &= 4501 \cdot f_0 \cdot d_m^3 \cdot (v \cdot \omega)^{2/3}, \quad \text{for } v \cdot \omega \geq 2 \cdot 10^{-4} \text{ m}^2/\text{s}^2 \\
M_0 &= 16 \cdot f_0 \cdot d_m^3, \qquad\qquad\quad \text{for } v \cdot \omega < 2 \cdot 10^{-4} \text{ m}^2/\text{s}^2 \\
M_1 &= f_1 \cdot F_r \cdot d_m, \\
F_r &= 0.5 m_s g + F_a / \sin \alpha,
\end{aligned}
\tag{7}
$$

---

[1] The comprehensive dimensions of all components can be found online at https://gitlab.hrz.tu-chemnitz.de/oshi/bearing-test-rig/design-data.

Table 3 The material properties and dimensions of major components of the test bench

Components	Materials	Dimensions
Stator (motor)	Structural steel	Major and minor diameter-0.095, 0.0855 m
Rotor (motor)	Structural steel (inner part), Copper (outer part)	Major diameter-0.0635 m (inner), major and minor diameter-0.0835, 0.0635 m (outer)
Radial ball bearings (motor)	Bearing steel 100Cr6	Major and minor diameter-0.035, 0.017 m
Shaft	Aluminium alloy AlMgSi0.5	Diameter-0.05 m, length-0.4 m
Base plate	Structural steel	Length-0.6 m, width-0.2, height-0.03 m
Angular contact ball bearings	Bearing steel 100Cr6	Major & minor diameter-0.072, 0.035 m
Thermal insulation	Phenolic foam	Major and minor diameter-0.096, 0.06 m
Bearing blocks	Structural steel	Width-0.162, height-0.115 m

where, M_0 [N.m] is the speed-dependent frictional torque, M_1 [N.m] load-dependent frictional torque, M_s [N.m] seal frictional torque, ν [m²/s] is kinematic viscosity, ω [rad/s] is rotational speed, d_m [m] is pitch circle diameter, F_r [N] is radial bearing load, m_s [Kg] is mass of the shaft, g [m/s²] is the gravitational acceleration, F_a [N] is the axial bearing load, and α [°] is the contact angle of the bearing. F_r has two components because each bearing (out of two) carries half of the weight of the shaft and the component of F_a. The values of coefficients f_0, f_1 depend on the type of lubrication and bearings as provided in [23].

A pre-tension device provides an axial force $F_a = 1000$ N and compensates for the thermal deformation of the shaft. The constructional mechanical CAD model is shown in Fig. 4b. Since some parts (e.g., small holes, screws, pre-tension device, etc.) are geometrically complex and do not significantly influence the thermal calculations, we ignore them in the equivalent thermal CAD model (see Fig. 4c). As evident in Fig. 4c, we also make an equivalent angular contact bearing ($\alpha = 40°$) composed of a single solid ring (removing the balls) cut around the diagonal plane to represent the inner and outer rings. We further prepare an equivalent motor divided into a rotor and stator separated by a thin air gap. We also consider two radial contact ball bearings mounted over the rotor shaft to connect with the stator of the motor. Heat generation inside the motor originates from two parts: mechanical power loss occurs in radial bearings, and the electrical power loss occurs in the copper windings of the rotor. Ansys mesh model of the test bench and reconstructed geometry in Matlab are shown in Fig. 4d–e respectively. We consider 64,142 nodes and 63,464 linear elements in the Ansys mesh. The lower part of the base plate of the test bench is fixedly mounted over a machine bed through the clamping claws. We neglect the radiation effect in

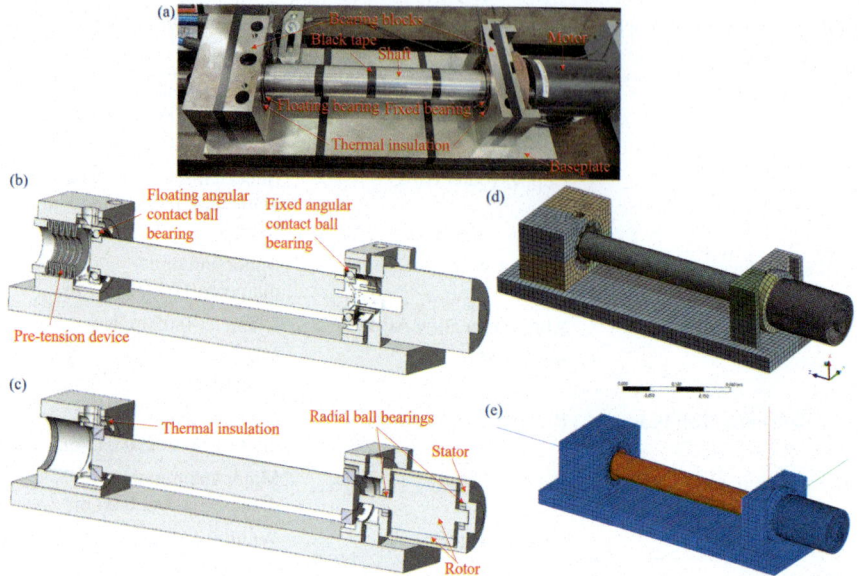

Fig. 4 The test bench: **a** set-up in the laboratory, **b** mechanical CAD model, **c** thermal equivalent CAD model, **d** Ansys model, and **e** reconstructed geometry in Matlab

this study. However, we consider free convection (coefficient of 5 W/m²/K) for all surfaces that are exposed to the air (measured at 16° C) except the open surfaces of the rotating parts where forced convection is considered. The forced convection coefficient α_k [W/m²/K], which depends on the speed ω [rad/s] and radius of rotating shaft r [m], is evaluated as follows [23]:

$$\alpha_k = 3.75 \cdot \omega^{0.7} \cdot r^{0.4}. \tag{8}$$

The experimental setup in the laboratory is shown in Fig. 5a. We ran the test rig at 1505 rpm for 2 h to obtain the steady-state results. The rotational speed of the shaft was measured with a tachometer (VOLTCRAFT DT-10L). The power in the motor was controlled by a variable DC Power Supply (VOLTCRAFT VSP 1410). The measured steady-state voltage and current were 18.9 V and 4.94 A, respectively. To precisely measure the temperature field over the test bench with time, we used an IR-TCM HD Infrared thermal camera (IR 1.0/15 LW Jenoptik VarioCAM® HD lens). The camera was focused over the test bench, where a total of 18 measurement points (see the point locations in Fig. 5b) were selected. We added coatings that behave like a black body to get the correct temperatures over those points. We used black tape as the coating material. Before starting the measurement, we performed some preliminary tests to calibrate the emissivity value of the coating materials, which was eventually tuned to 1, that is not unusual in other studies too [24]. The frame rate of the camera was 22 Hz, and the resolution of the photos was 768x1024 (height x width) pixels. We used thermography software IRBIS® 3.1 for live data recording, acquisition, and post-processing. We covered the whole setup area with a thermal shield so that no reflection of any external moving bodies occurred. We also used four Testo Saveris PtD radio probe sensors to cross-check the results of the thermal camera. Three radio probe sensors were attached to the motor, floating and fixed bearing covers, and one more was used for the measurement of the air temperature. The probe displays were mounted on the wall as visible in Fig. 5a. A manual Temperature logger (ALMEMO® 2290-2) was also used to double-check the results before starting and after stopping the motor. The snapshot of the real-time steady-state temperature field of the test bench captured by the camera (at $t = 120$ min) is shown in Fig. 5b. Time histories of temperatures measured by the radio probe sensors are depicted in Fig. 5c. Similarly, the temperature-time histories of all 18 measurement points captured by the camera are shown in Fig. 5d. We notice that the motor, bearings, and shaft are heated the most (the maximum temp is around $50\,°C$ captured by the camera and $51\,°C$ measured by the probe sensor). However, the base plate and bearing blocks are the coolest parts (temperatures are close to air temperature) even after 2 hours of running the setup due to thermal insulation.

Next, we compare these experimental results with our predictions based on thermal models as illustrated in Fig. 6. Comparisons of the time histories of the temperature field are analyzed for the following measured points only: P1: Motor Front, P2: Fixed Bearing Block Front, P3: Fixed Bearing Insulation Front, P4: Fixed Bearing Cover Front, P7: Shaft Fixed Bearing Side, P9: Shaft Floating Bearing Side, as shown in Fig. 6a–f respectively. For all cases, we see that the transient and steady-state responses from our thermal models show a fair agreement with the experimental results. However, for the case of P7 and P9 points over the shaft (see Fig. 6e–f), the steady-state temperature from our thermal model slightly deviates downwards ($< 6\,°C$) from the experimental results. The potential reason could be the uncertainties in the forced convection formula, due to which strong heat dissipation occurs in the air from the large surface area of the shaft. The time histories of the full thermal model are also compared with the reduced models applying moment-matching and Arnoldi iteration. We see that the full model and reduced model, applying the moment-matching method, almost overlap with each other. Only for the case of the reduced model with Arnoldi iteration, the initial temperature gradient is quite high, which makes a notable difference ($< 15\,°C$) in the transient response. However, that difference minimizes over time and converges with the original model at the steady state condition. Furthermore, the comparisons of temperature deviations (in %) of the full and reduced models with respect to experimental results using the thermal camera are listed in Table 4. The temperature deviations ($e = (T - T_e)/(T_e + 273.15) \times 100$, where T, T_e are the simulated and experimental (with camera) temperatures in $°C$ respectively) are shown for the following points over the test bench: P1, P2, P3, P4, P7, and P9 at $t = 20$ min (where the maximum differences occur) taken from Fig. 6). A negative deviation indicates that the model temperature is lower than the measured temperature with the camera. As evident in Table 4, a high temperature discrepancy (around 4.33%) occurs for the P1 point

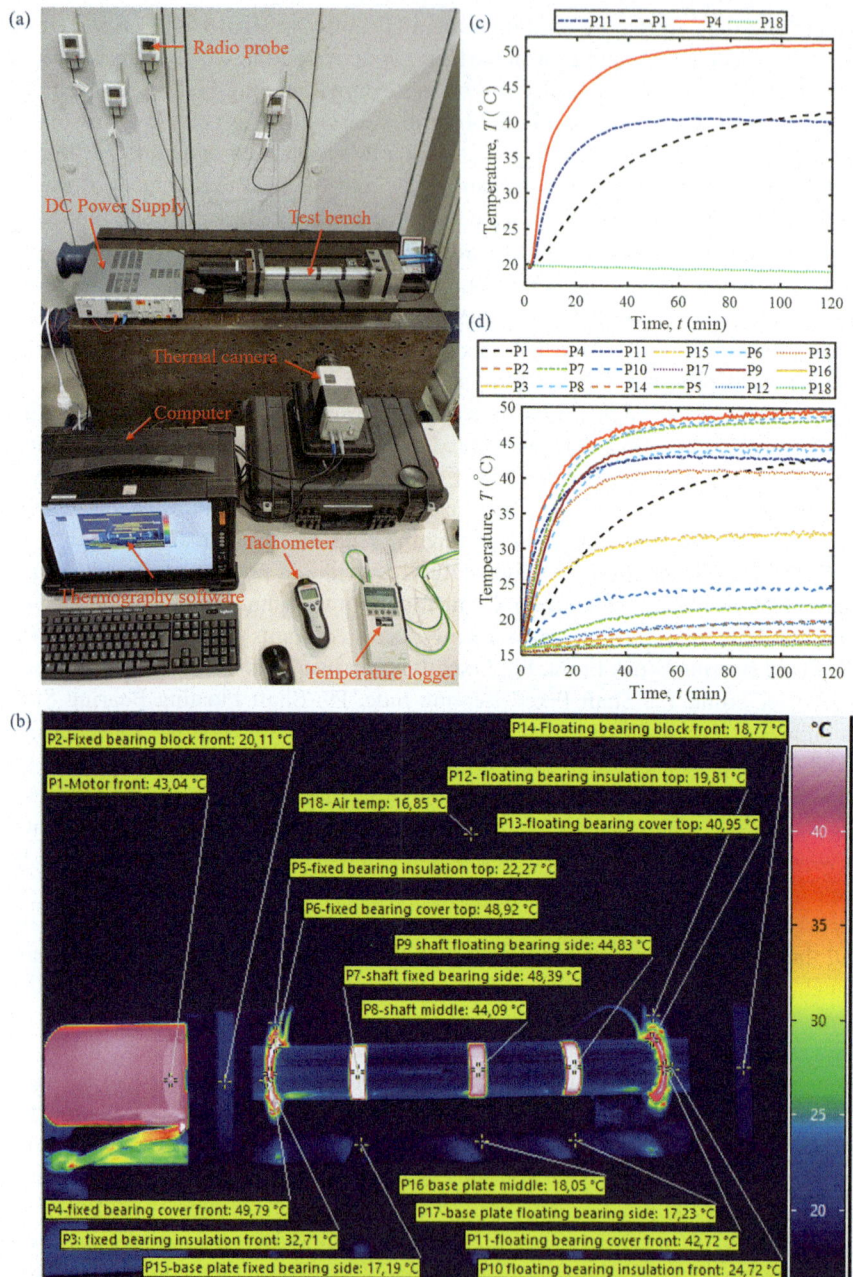

Fig. 5 **a** The experimental setup of thermal measurements in the laboratory, **b** snapshot of real-time steady-state temperature field of the test bench captured by the camera at $t = 120$ min (comma denotes dot here), **c** time histories of temperatures measured by radio probe sensors, **d** temperature-time histories captured by the thermal camera

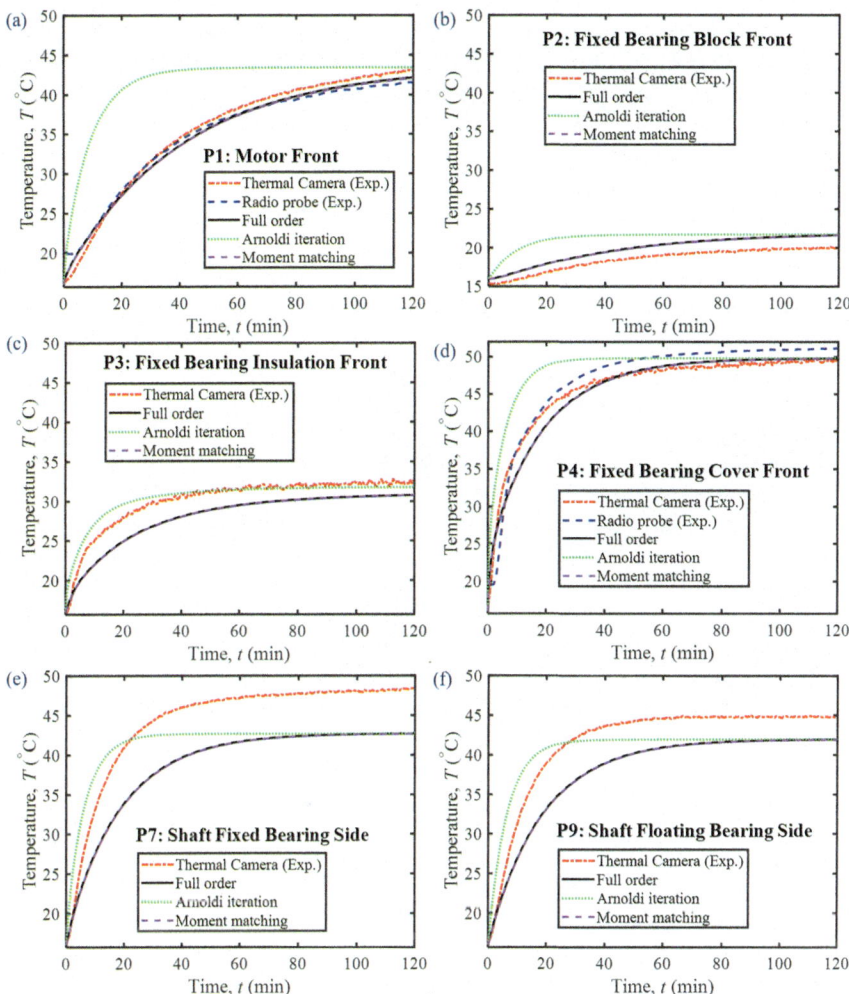

Fig. 6 Experimental validation of the full order thermal model and comparison with the reduced models applying moment matching and Arnoldi iteration. The comparisons of the temperature-time histories are shown for the following measured points: **a** P1: motor front, **b** P2: fixed bearing block front, **c** P3: fixed bearing insulation front, **d** P4: fixed bearing cover front, **e** P7: shaft fixed bearing side, **f** P9: shaft floating bearing side

(also seen in Fig. 6a) for the case of the reduced model with Arnoldi iteration. The absolute temperature deviations of the full-order model and moment matching MOR are the same, but lower than the Arnoldi iteration for most of the points. Therefore, we can conclude that the Arnoldi iteration is not as accurate as moment-matching MOR at the transient zone, which further validates the robustness analysis as in not as accurate as moment-matching MOR at Sect. 3.1. Furthermore, the temperature field of the whole test bench (for both cases of the full model and reduced models)

Table 4 Comparison of temperature deviations with respect to experimental result using the thermal camera at $t = 20$ min from Fig. 6 for the following points over the test bench: P1, P2, P3, P4, P7, and P9

Different models	P1 (%)	P2 (%)	P3 (%)	P4 (%)	P7 (%)	P9 (%)
Full order	−0.12	0.36	−1	−0.70	−2.26	−1.87
Moment matching	−0.12	0.36	−1	−0.70	−2.26	−1.87
Arnoldi iteration	4.33	1.45	0.59	1.84	0.23	0.61

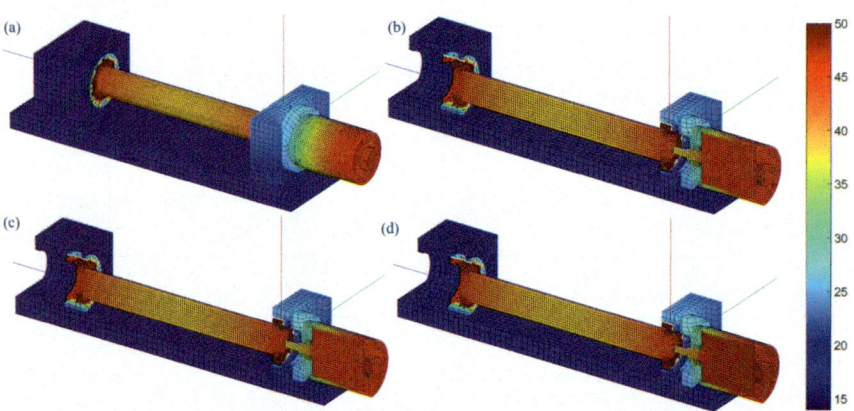

Fig. 7 The full thermal model and comparison with the reduced models for the whole test bench at $t = 120$ min. Original model: **a** full view, **b** cross-sectional view. The cross-sectional view of the reduced models applying **c** moment matching and **d** Arnoldi iteration. The temperatures in the colorbar are in °C

at $t = 120$ min is shown in Fig. 7. For the original model, the full view and cross-sectional view are shown in Fig. 7a, b. The cross-sectional views of the reduced models applying moment matching and Arnoldi iteration are illustrated in Fig. 7c, d respectively. We observe that the heat distribution pattern and maximum-minimum temperatures for the reduced models are almost similar to the original system under steady-state conditions.

Finally, the computation time (CPU time) of the transient simulation (up to $t = 120$ min) for the reduced models is compared with the full model as listed Table 5. We take the reduced order $r = 13$ in both MOR methods based on the convergence test of the reduced-order solution. As in Sect. 3.1, we observe here also that the Arnoldi iteration reduces the system more quickly than the moment-matching method. However, the ODE solver time for the moment-matching method is far less than the Arnoldi iteration. We observe here that the Arnoldi iteration takes more ODE solver time than the original system. It is potentially attributed to the lower time step value in the Arnoldi iteration. The simulation time reduces with a higher time step.

Table 5 Comparison of the transient thermal computational time (CPU time) for the case of the test bench

Different models	Geometry reconstruction (min)	Model reduction (min)	ODE solver time (min)	Total computation time (min)
Full order	3.16		3.84	7
Moment matching	3.16	2.6	1.2	6.96
Arnoldi iteration	3.16	0.15	9.2	12.51

However, above a certain time step, the solution becomes infeasible. The maximum value of the time step is found from the convergence test. This value is found to be 10 sec for the original system and the moment-matching reduced model. However, it is 3 sec for the Arnoldi iteration, which increases the ODE solver and total computation time. Hence, the moment-matching method is more efficient and accurate for the overall thermal simulations. Therefore, the moment-matching method will be more suitable for the time-efficient, precise thermal calculation for large machine tools.

4 Conclusions

This paper explores two model order reduction techniques for two different levels of thermal problems (bar-slider and test bench). MOR tools offer a potential opportunity to generate a computationally effective compact model from a high-dimensional complex FE model. Therefore, MOR methods are eminently used in model-based compensation models to expeditiously reduce the thermal deviations of precise machine tools, which helps to improve the manufacturing quality, life cycle of the tool, and production time.

For both cases, the thermal FEA matrices are exported from Ansys to Matlab to perform the geometry reconstruction and thermal analysis (with/without MOR) using our optimized codes. For the case of bar-slider problems, the proposed thermal model shows good numerical verification with the Ansys simulation result. The proposed thermal model is further experimentally validated for the case of the test bench, which indicates good agreement between experiments and model, which could be enhanced further, including the radiation effect, in future work. For both cases, the result of the original system is compared with the Krylov subspace-based reduced models. Due to the limitation of computational load, the balanced truncation method and the iterative rational Krylov algorithm are not compared in the current work, which could be further considered in future studies. It is noticed that the moment-matching method executes more efficiently and accurately than the Arnoldi iteration in terms of the overall thermal simulations for all cases. Therefore, our proposed numerical thermal correction model with the moment-matching method will be applicable to accurate time-saving thermal calculations for the next generation of large and complex machine tools.

Acknowledgements The authors gratefully acknowledge BMWK's funding as part of the Central Innovation Program for SMEs (ZIM).

References

1. Mayr, J., Jedrzejewski, J., Uhlmann, E., Alkan Donmez, M., Knapp, W., Härtig, F., Wendt, K., Moriwaki, T., Shore, P., Schmitt, R., Brecher, C., Würz, T., Wegener, K.: Thermal issues in machine tools. CIRP Ann. **61**(2), 771–791 (2012)
2. Wegener, K., Weikert, S., Mayr, J.: Age of compensation—challenge and chance for machine tool industry. Int. J. Autom. Technol. **10**(4), 609–623 (2016)
3. Mareš, M., Horejš, O., Nykodym, P.: An indicative model considering part of the thermo-mechanical behaviour of a large grinding machine. In: Ihlenfeldt, S. (ed.) 3rd International Conference on Thermal Issues in Machine Tools (ICTIMT2023), pp. 54–66. Springer, Cham (2023)
4. Horejš, O., Mareš, M., Straka, M., Švéda, J., Kozlok, T.: Adaptive thermal error compensation model of a horizontal machining centre. In: Ihlenfeldt, S. (ed.) 3rd International Conference on Thermal Issues in Machine Tools (ICTIMT2023), pp. 83–98. Springer, Cham (2023)
5. Stoop, F., Mayr, J., Sulz, C., Bleicher, F., Wegener, K.: Fleet learning of thermal error compensation in machine tools. In: 2021 26th IEEE International Conference on Emerging Technologies and Factory Automation (ETFA). pp. 1–4 (2021)
6. Boos, E., Thiem, X., Wiemer, H., Ihlenfeldt, S.: Improving a deep learning temperature-forecasting model of a 3-axis precision machine with domain randomized thermal simulation data. In: Liewald, M., Verl, A., Bauernhansl, T., Möhring, H.C. (eds.) Production at the Leading Edge of Technology, pp. 574–584. Springer, Cham (2023)
7. Lang, S., Zimmermann, N., Mayr, J., Wegener, K., Bambach, M.: Thermal error compensation models utilizing the power consumption of machine tools. In: Ihlenfeldt, S. (ed.) 3rd International Conference on Thermal Issues in Machine Tools (ICTIMT2023), pp. 41–53. Springer, Cham (2023)
8. Zimmermann, N., Müller, E., Lang, S., Mayr, J., Wegener, K.: Thermally compensated 5-axis machine tools evaluated with impeller machining tests. CIRP J. Manuf. Sci. Technol. **46**, 19–35 (2023)
9. Blaser, P., Pavlíček, F., Mori, K., Mayr, J., Weikert, S., Wegener, K.: Adaptive learning control for thermal error compensation of 5-axis machine tools. J. Manuf. Syst. **44**, 302–309 (2017) (Special Issue on Latest advancements in manufacturing systems at NAMRC 45)
10. Fujishima, M., Narimatsu, K., Irino, N., Mori, M., Ibaraki, S.: Adaptive thermal displacement compensation method based on deep learning. CIRP J. Manuf. Sci. Technol. **25**, 22–25 (2019)
11. Thiem, X., Rudolph, H., Krahn, R., Ihlenfeldt, S., Fetzer, C., Müller, J.: Adaptive thermal model for structure model based correction. In: International Conference on Thermal Issues in Machine Tools. pp. 67–82. Springer (2023)
12. Lui, S.: Numerical Analysis of Partial Differential Equations. Pure and applied mathematics, Wiley (2011), https://books.google.de/books?id=6c7OvQEACAAJ
13. Schilders, W., Vorst, van der, H., Rommes, J. (eds.): Model order reduction: theory, research aspects and applications. In: Mathematics in Industry. Springer, Germany (2008)
14. Galant, A., Großmann, K., Mühl, A.: Thermo-elastic simulation of entire machine tool. In: Großmann, K. (ed.) Thermo-energetic Design of Machine Tools. Lecture Notes in Production Engineering. pp. 69–84. Springer, Cham (2014)

15. Galant, A., Beitelschmidt, M., Großmann, K.: Fast high-resolution FE-based simulation of thermo-elastic behaviour of machine tool structures. Procedia CIRP **46**, 627–630 (2016)
16. Hernández-Becerro, P., Spescha, D., Wegener, K.: Model order reduction of thermo-mechanical models with parametric convective boundary conditions: focus on machine tools. Comput. Mech. **67**, 167–184 (2020)
17. Arnoldi, W.E.: The principle of minimized iterations in the solution of the matrix eigenvalue problem. Quart Appl Math **9**(1), 17–29 (1951). http://www.jstor.org/stable/43633863
18. Großmann, K., Mühl, A.: Reduktion strukturdynamischer und thermoelastischer fe-modelle. Zeitschrift für wirtschaftlichen Fabrikbetrieb **105**(6), 594–599 (2010). https://doi.org/10.3139/104.110340
19. Model Order Reduction: Volume 1: System- and Data-Driven Methods and Algorithms. De Gruyter, Berlin, Boston (2021)
20. Aumann, Q., Benner, P., Saak, J., Vettermann, J.: Model order reduction strategies for the computation of compact machine tool models. In: International Conference on Thermal Issues in Machine Tools. pp. 132–145. Springer (2023)
21. Lang, N., Saak, J., Benner, P.: Model order reduction for thermo-elastic assembly group models. In: Thermo-energetic Design of Machine Tools: A Systemic Approach to Solve the Conflict Between Power Efficiency, Accuracy and Productivity Demonstrated at the Example of Machining Production, pp. 85–93 (2015)
22. Reis, T., Stykel, T.: Balanced truncation model reduction of second-order systems. Math. Comput. Model. Dyn. Syst. **14**(5), 391–406 (2008)
23. Jungnickel, G.: Simulation des thermischen Verhaltens von Werkzeugmaschinen. Inst. für Werkzeugmaschinen und Steuerungstechnik, Lehrstuhl für Werkzeugmaschinen, TU Dresden (2000)
24. Ibos, L., Marchetti, M., Boudenne, A., Datcu, S., Candau, Y., Livet, J.: Infrared emissivity measurement device: principle and applications. Meas. Sci. Technol. **17**(11), 2950 (2006)

Finite Element Analysis of Residual Stress Variations in Prestressed Fiber-Reinforced Polymer Concrete Induced by Temperature Changes

Michelle Engert, Kim Torben Werkle, and Hans-Christian Möhring

Abstract The novel composite material, prestressed fiber-reinforced polymer concrete, is distinguished by its low weight and exceptional vibration-damping properties, making it highly suitable for use in heavily loaded structural components in machine tools. Additionally, the material exhibits high thermal stability. Preliminary bending tests on simple prismatic test samples have demonstrated that preheating prior to testing increases the bending stiffness. However, the influence of heating on the static stiffness of a complex machine component with varying fiber orientations remains unexplored. This study aims to investigate this influence through experimental analysis. Focusing the further application of polymer concrete and the associated need for feasible design and calculation methods additionally a method for numerical investigation is developed. The simulation models the interactions between the bonding behavior of carbon fiber rovings with polymer concrete, the residual stresses induced by prestressing during curing, and the effects of ambient temperature. The accuracy of the simulation results is validated through comparison with experimental investigation.

Keywords Composite · Residual stress · Thermal effect

1 Introduction

The novel material prestressed fiber-reinforced polymer concrete (PFRPC) combines the vibration-damping properties of its polymer concrete matrix material [1, 2] with the high rigidity of the integrated, prestressed carbon fiber rovings [3, 4]. Consequently, the lightweight material can be used in machine tool components [1] for the first time (see Fig. 1).

In previous studies, the thermal sensitivity of the material was examined using prismatic test samples with unidirectional roving orientation. An increased bending

M. Engert (✉) · K. T. Werkle · H.-C. Möhring
Institute for Machine Tools (IfW), University of Stuttgart, Stuttgart, Germany
e-mail: michelle.engert@ifw.uni-stuttgart.de

© The Author(s) 2026   523
K. Wegener and M. Bambach (eds.), *4th International Conference on Thermal Issues in Machine Tools (ICTIMT2025)*, Lecture Notes in Production Engineering,
https://doi.org/10.1007/978-3-032-01194-7_33

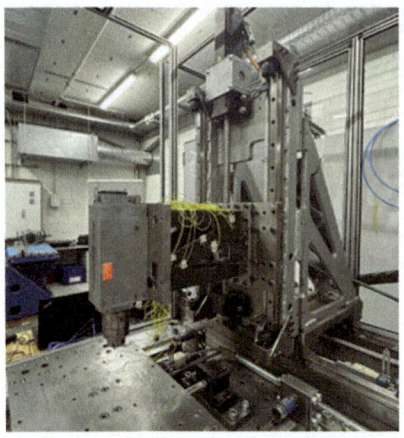

Fig. 1 Machine arm made out of PFRPC and roving orientation in the component

strength was found compared to samples tested at room temperature, which can be attributed to the heating of the samples [3]. It is hypothesized that the differing signs of the coefficients of thermal expansion between the polymer concrete and the carbon fiber rovings [5, 6], contribute to an increase in the residual stress field, as demonstrated in [4] when the material is heated [3]. Although the component considered in literature has already been investigated with regard to its thermal inertia [1], the influence of temperature on the static stiffness of the component was not examined in detail.

Within the scope of this paper, the influence of temperature on the stiffness of the PFRPC component presented in [1] is initially determined experimentally. This is followed by a numerical representation of the experimental setup and a numerical assessment of component stiffness as a function temperature. This analysis builds upon the findings reported in [7] and establishes a foundation for the design and optimization of components made from PFRPC.

2 Experimental Determination of Static Stiffness as a Function of Component Temperature

For the subsequent validation of the numerical model, the static stiffness of the component is first determined as a function of temperature. For this purpose, the machine arm is clamped to its base plate on a machine bed, analogous to the test procedure outlined in [1]. A force of 500 N is applied centrically on the spindle plate in two separate tests in the Y and Z directions. The applied force is recorded using a load cell of type KM26-10kN (by ME Meßsysteme), while the deformation of the cantilever is recorded using two laser triangulators (LK-H052 by Keyence). One of the sensors is positioned directly opposite the force application point to capture displacement, while the second sensor is arranged orthogonally to the force direction to ensure accurate centering of the applied force (see Fig. 2).

The component is exposed to a temperature distribution box maintained at 40 °C for a period of two hours. This is the optimal thermal treatment to replicate the heating up of a machine on a day with elevated temperatures in a machine hall. Immediately following the heating process, the static stiffness of the component is determined. The test results are compared in Fig. 3.

In analogy to the prismatic samples investigated in [3], the heating results in an increase in static stiffness (+ 13% in Y-direction, + 5% in Z-direction). However, this increase exhibits a significant dependence on the load direction, which may be attributed to the multidirectional arrangement of the roving bundles. Twelve of the integrated roving bundles are situated in the XZ plane, thereby contributing to the strengthening effect in the Z direction. In comparison, 16 roving bundles are located in the XY plane (see Fig. 4).

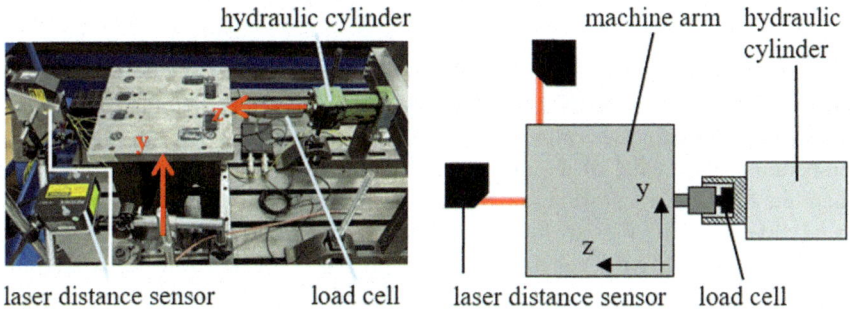

Fig. 2 Test setup for the determination of the static stiffness of the machine arm

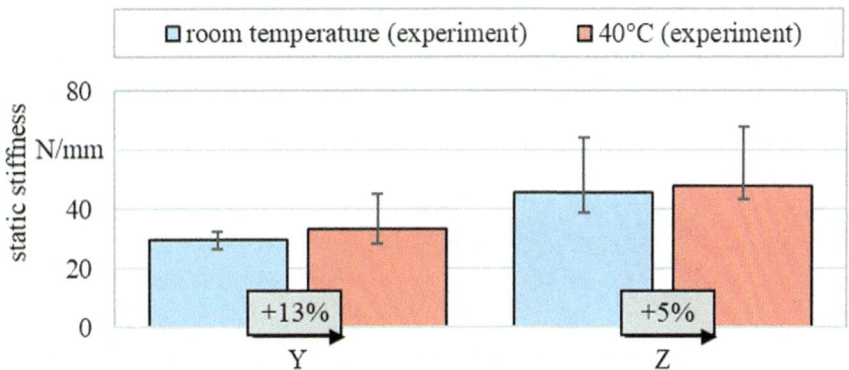

Fig. 3 Static stiffness of the component at different temperatures (test results)

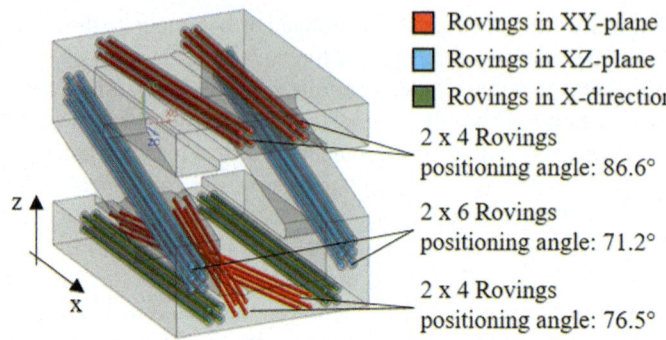

Fig. 4 Roving bundles and their position angle in relation to the base plate

3   Numerical Determination of Static Stiffness as a Function of Component Temperature

In order to be able to design future components without carrying out time-consuming and cost-intensive experiments, the experiment is replicated in a simulation environment. To numerically determine the static stiffness as a function of the component temperature, the model of the machine arm is first simplified. All drill holes and chamfers on the two connecting plates are removed for this purpose. The impregnated carbon fiber rovings are represented as rods with a diameter of 4 mm according to [4, 7]. To simplify the generation of a fine mesh in the area directly around the carbon fiber rovings, which is particularly sensitive for the representation of the residual stress field, areas with a diameter of 9 mm are separated from the rest of the polymer concrete body wherever possible (see Fig. 5).

The assumed material parameters are based on the results in [7] and are summarized in Table 1. The Young's modulus assumed for the polymer concrete EPUMENT 130/3 (by RAMPF) was calculated from the results of uniaxial bending tests on prismatic test samples presented in [3]. Following the assumption in [8], the polymer

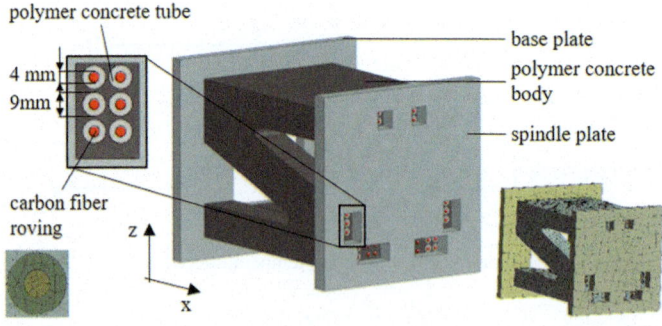

Fig. 5 Adapted model of the polymer concrete machine arm

concrete is assumed to be homogeneous. The contact are between the rovings and the polymer concrete is assumed to be ideally stiff which might not correspond to reality.

The standard procedure for mapping residual stresses in PFRPC resulting from the prestressing of the carbon fiber rovings, which supposes the application of the residual stresses uniformly in the whole PFRPC body, is not applicable due to the multidirectional roving orientation. The study in [7] posits that the residual stresses are uniform throughout the material, irrespective of the distance from the carbon fiber rovings. The present study proposes an alternative method, whereby the rovings' contraction, as measured in [4] by integrating strain gauges placed on the carbon fiber rovings, is simulated. The resulting residual stress field is then transferred as initial state to the component in the following simulations (see Fig. 6).

As can be seen in Fig. 6, a decreasing residual stress field forms around the carbon fiber rovings. This is independent of the roving orientation and is characterized by an abrupt change in sign compared to the residual stresses remaining in the carbon fiber rovings. In the immediate vicinity of the carbon fiber roving, the residual stresses have a value of about − 12 MPa. With increasing distance from the carbon fiber rovings, the value stabilizes at about − 4 MPa, which is only a slight deviation from the value of − 5.72 MPa known from the literature [7, 10]. The stresses in the carbon fiber rovings are largely uniform, but at about 10 MPa they are well below the residual stress value calculated in [7, 10]. It should be noted, however, that this value still requires experimental confirmation, but it could nevertheless be a potential reason for deviations in the calculation.

In the ensuing phase, the entire base plate of the machine arm is provided with a fixation. The load is realistically applied to the center of the sided surface area of the spindle plate as a concentrated load. Subsequent to the calculation being complete, the displacement along the respective axis direction is read out and used to calculate the static stiffness. The calculation shows that the stiffness is initially overestimated. Specifically, the numerically determined static stiffness in the Y direction is approximately 323% higher than the experimentally determined stiffness, while in the Z direction, the overestimation is around 155%. This discrepancy might be attributed either to the described comparatively low stresses in the carbon fiber rovings or to a reduced Young's modulus compared to the prismatic samples considered so far. This could be caused by internal defects resulting from the complex manual manufacturing

Table 1 Material characteristics known from literature

	Density (g/cm^3)	Young's modulus (GPa)	Poisson's ratio (-)	Coefficient of thermal expansion (1/K)
Polymer concrete	2.3	18.5	0.222	$20 \cdot 10^{-6}$
	[6]	[3]	[9]	[6]
Carbon fiber rovings	1.8	96.3	0.317	$-0.1 \cdot 10^{-6}$
	[7]	[7]	[7]	[5]

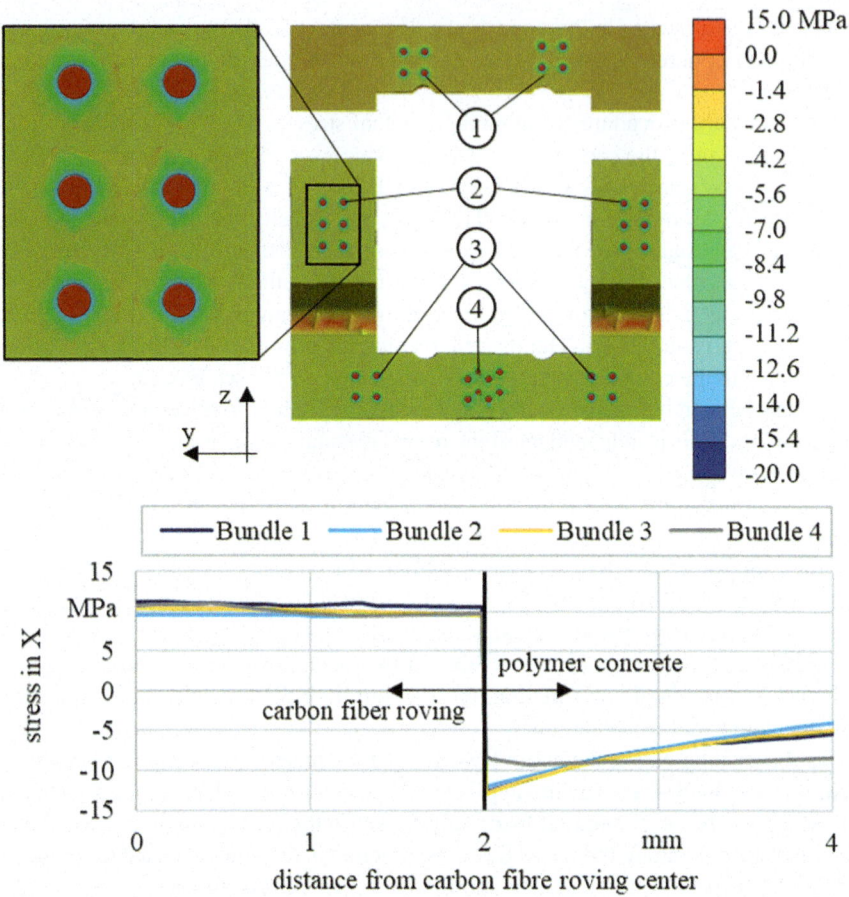

Fig. 6 Simulated residual stresses in X-direction

of the component. Damage to the carbon fiber rovings, on the other hand, is ruled out. To facilitate a numerical analysis that does not necessitate precise knowledge of the component's condition and a resulting resource-intensive simulation of the internal defects, an ideal Young's modulus of the polymer concrete is first determined in an iterative process (see Fig. 7).

During the iterative adjustment of the Young's modulus, it was observed that the dependence of static stiffness on the Young's modulus exhibits directional sensitivity. This phenomenon is likely attributable to variations in the polymer concrete cross-sections along different loading directions, in addition to the previously discussed influence of roving orientation. Consequently, the simulated static stiffness in Y direction is more affected by changes in the Young's modulus compared to the stiffness in Z direction. The ideal Young's modulus of approximately 5 GPa in Y direction is contrasted with its value of around 11 GPa in Z direction. To establish a suitable

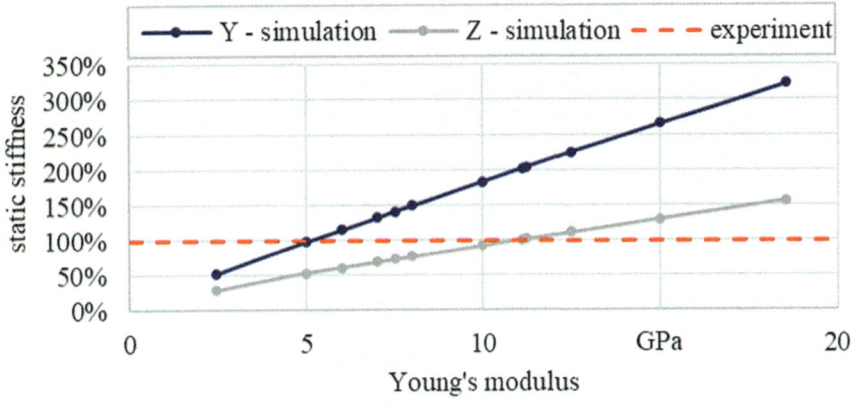

Fig. 7 Iterative determination of an ideal Young's modulus for the numerical determination of static stiffness

compromise for further analysis, a Young's modulus of 7 GPa is proposed, resulting in an error margin of approximately 32%, irrespective of the load direction.

To simulate the heating process, a thermal load is applied to all components. To avoid rigid body movements, the base plate is fixed anew, even if this does not conform to the actual conditions. The residual stress field resulting from the heating process is shown in Fig. 8. For clarity, only the thermally induced residual stress field is shown. However, in subsequent simulations, both residual stress fields are superimposed.

The thermally induced residual stress field shows a similar pattern to the initially imposed residual stress field. The stress in the rovings is nearly constant at approximately 3 MPa. In the polymer concrete, a dependence of the stress field on the distance to the roving can be observed. In close proximity of the rovings, the stress is approximately − 2.7 MPa, gradually stabilizing at around − 0.9 MPa. Contact with the plates causes a slight increase in stress, but it remains at a maximum absolute value of − 0.009 MPa. A slightly positive influence on the result of the static stiffness is to be expected. The static stiffness as a function of the component temperature is shown in Fig. 9. The incorporation of the thermally induced residual stress field into the simulation has a beneficial effect on the predicted static stiffness. Due to the previously observed deviation from the test results, it is difficult to make a clear comparison with the test result of the heated-up component.

However, when evaluating the relative increase in static stiffness with rising component temperature, the simulation results show a comparable trend to the experimental findings.

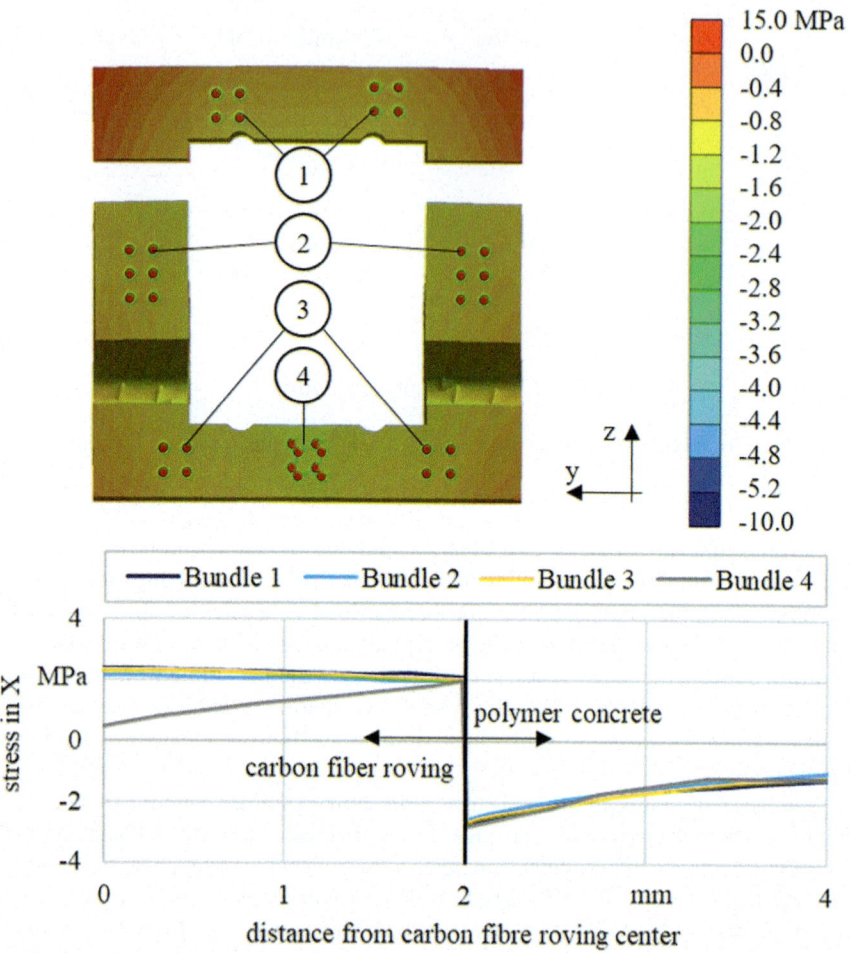

Fig. 8 Numerical estimated thermal induced residual stress field as a function of the distance from the roving center point

4   Conclusion and Outlook

This study investigated the influence of component temperature on the static stiffness of a prestressed fiber-reinforced polymer concrete (PFRPC) component through both experimental and numerical analysis. For the numerical determination of the static stiffness, a new concept was first developed to map the residual stress field induced by the integration of the prestressed carbon fiber rovings. A key challenge in this process was the multidirectional fiber arrangement, as existing methodologies in the literature are only applicable to unidirectional rovings. However, a significant discrepancy was identified between the residual stress remaining in the rovings,

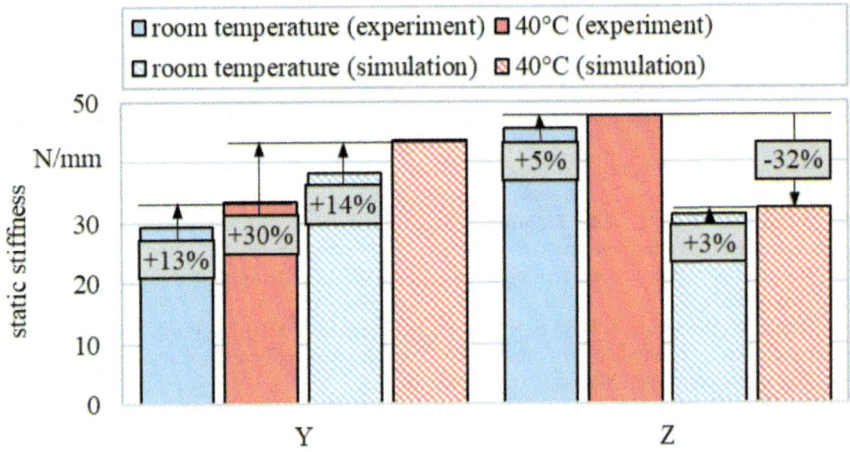

Fig. 9 Experimentally and numerically determined static stiffness as a function of temperature

which could be a contributing factor to the observed deviation of about 30% in numerically determined static stiffness. To accurately ascertain the underlying cause of this discrepancy, further experimental investigations into the residual stress field are required.

In order to ascertain the effect of temperature on the component, it was heated to a temperature of 40 °C. Similar to previous tests on prismatic test samples, an increase in component temperature led to an increase in static stiffness. However, the effect demonstrated a high directional dependence, which could be due to the multidirectional roving orientation. Both effects were similarly reflected in the numerical analysis.

Future research will focus on the fundamental investigation of the residual stress field in PFRPC, with particular emphasis on the intensity and distribution of residual stresses. Furthermore, additional methods for mapping the resulting residual stress field with multidirectional roving guidance will also be developed. These investigations are of decisive importance for the subsequent design of components made of PFRPC.

Acknowledgements The authors would like to thank the DFG (Deutsche Forschungsgemeinschaft) for funding the research within the project MO 2091/15-1.

Competing Interests The authors have no conflicts of interest to declare that are relevant to the content of this chapter.

References

1. Engert, M., Werkle, K.T., Lazic, A., Möhring, H.-C.: Development, manufacturing and analysis of a machine arm made out of prestressed fiber-reinforced polymer concrete. Procedia CIRP 1790–1794 (2024)
2. Möhring, H.-C., Brecher, C., Abele, E., Fleischer, J., Bleicher, F.: Materials in machine tool structures. CIRP Ann. **2**, 725–748 (2015)
3. Engert, M., Werkle, K.T., Möhring, H.-C.: Determination of the thermal properties of prestressed fiber-reinforced polymer concrete. In: Ihlenfeldt, S. (Ed.) 3rd International Conference on Thermal Issues in Machine Tools (ICTIMT2023). Springer International Publishing, Cham (2023)
4. Engert, M., Werkle, K.T., Wegner, R., Born, L., Gresser, G.T., Möhring, H.-C.: Validation of the manufacturing methodology of prestressed fiber-reinforced polymer concrete by the variation of process parameters. Materials **16** (2023)
5. Mitsubishi Chemical Carbon Fiber and Composites, Inc.: GRAFIL 34-700 12K & 24K— Product Data Sheet (2019)
6. Rampf Machine Systems GmbH & Co. KG: Technisches Merkblatt - EPUSELF. Göppingen (2017). Zuletzt geprüft am 18 Apr 2023
7. Engert, M., Frankenbach, K., Werkle, K.T., Möhring, H.-C.: Simulation of the flexural behavior of prestressed fiber-reinforced polymer concrete. J. Adv. Manuf. Technol. (2025)
8. Relea, E., Pfyffer, B., Weiss, L., Wegener, K.: Experimental comparative investigation on creep behavior of mineral cast, ultra-high-performance concrete, and natural stone for precision machinery structures. Int. J. Adv. Manuf. Technol. **7–8**, 2073–2081 (2021)
9. Kepczak, N., Pawlowski, W., Blazejewski, W.: The study of the mechanical properties of the mineral cast material. Arch. Mech. Technol. Autom. **2**, 25–32 (2014)
10. Mostafa, N.H., Ismarrubie, Z.N., Sapuan, S.M., Sultan, M.: Fibre prestressed composites: theoretical and numerical modelling of unidirectional and plain-weave fibre reinforcement forms. Compos. Struct. 410–423 (2017)

Modeling and Simulation of the Thermal Behavior of an Electromechanical Compact Axis

Darío Fernández, Holger Rudolph, Lars Penter, and Steffen Ihlenfeldt

Abstract Electromechanical compact axes, which integrate electric motors with screw drives, are becoming increasingly popular in decentralized drive systems due to their high power density and precise control. These systems are emerging as effective alternatives to hydraulic systems, driven by the growing demand for more compact and efficient drive solutions. However, the compact architecture of electro-cylinders poses critical thermal management challenges. High power density and the proximity of heat-generating components within the design often result in heat accumulation, leading to various performance degradations. Limited pathways for heat dissipation create the risk of thermal buildup, which can degrade motion accuracy through thermo-elastic effects and compromise long-term reliability as heat-related stress affects components such as bearings and nuts. Predicting these effects with current simulation techniques is possible but comes at a high computational cost or accuracy loss, making design optimizations challenging and highlighting the essential role of Model order reduction (MOR) in improving simulation efficiency and enabling faster, more cost-effective design analysis without sacrificing accuracy. While MOR can reduce computation time, its integration into thermal Finite Element Method (FEM) models is not yet standardized. This requires addressing non-linearities arising from heat generation patterns, specifically mechanical friction, electrical resistance, and heat radiation, since conductivity and specific heat capacity remain constant across the temperature range considered. The challenge intensifies in mobile systems with continuously changing convective and conductive heat transfer conditions. To address these aspects, this paper presents a framework that integrates MOR into thermal FEM

D. Fernández (✉) · H. Rudolph · L. Penter · S. Ihlenfeldt
Department Machine Tools Development and Adaptive Controls, Technische Universität, Dresden, Saxony, Germany
e-mail: dario_sebastian.fernandez_riquelme@tu-dresden.de

H. Rudolph
e-mail: holger.rudolph@tu-dresden.de

L. Penter
e-mail: lars.penter@tu-dresden.de

S. Ihlenfeldt
e-mail: steffen.ihlenfeldt@tu-dresden.de

533

K. Wegener and M. Bambach (eds.), *4th International Conference on Thermal Issues in Machine Tools (ICTIMT2025)*, Lecture Notes in Production Engineering,
https://doi.org/10.1007/978-3-032-01194-7_34

models, accounting for non-linear thermal behaviors and enabling efficient, accurate simulations across diverse operational scenarios. This framework combines MOR in Matlab with Ansys-based subsystem discretization, focusing on active interfaces and matrix reduction of conductivity and capacity. Mobile components are discretized at critical zones to accurately model dynamic boundary conditions. Two thermal models are developed: a non-reduced model for high fidelity and a reduced model for faster simulations that maintain accuracy at key interfaces. This approach is particularly suited for mobile systems with dynamic thermal conditions, offering scalable, real-time analysis for electromechanical systems. Experimental validation was performed on a grease lubricated compact axle, at varying travel speeds to evaluate the performance in different operating regimes. The strong correlation between model predictions and experimental results underscores the importance of fast and high-accuracy thermal simulations enabled by MOR techniques. This approach enhances the reliability and operational efficiency of electro-mechanical axles, supporting their potential for broader industrial adoption.

Keywords Electromechanical compact axes · Thermal management · Model order reduction (MOR) · Finite element method (FEM) · Heat dissipation

1 Introduction

Electric drives perform a central function in industrial applications, with approximately 67% of Germany's electrical energy in 2020 being converted into mechanical work. Recent expert surveys underscore the growing demand for efficient, precise, and thermally optimized drive systems, reflecting trends toward compact and high-performance designs While direct linear drives offer specific benefits, rotational motor systems are more economically viable [1]. This has led to the development of electromechanical compact axes, which integrate an electric motor with a screw mechanism to convert rotary motion into precise linear movement. These systems are crucial for tasks like force and machining applications across various industries. Despite their advantages, electromechanical cylinders face thermal management challenges due to their compact design. High power density and limited heat dissipation space lead to performance degrada-tion, including reduced power output and thermal displacement that can compromise precision. Großmann et al. [2] highlight the need for a thermo-energetic approach to resolve conflicts between efficiency, accuracy, and productivity in machine tool design, a principle this work adapts for electromechanical systems. Moreover, another work by Großßmann et al. [3] emphasizes correcting thermally induced errors to enhance precision, a goal supported by adaptive thermal compensation strategies as explored in recent studies on machining centers and structural models [4, 5]. Creighton et al. [6] categorize thermal error countermeasures into three primary strategies- elimination or avoidance, resistance and control, and compensation-noting that elimination tactics, which prevent dimensional changes due to thermal fluctuations, are most effective during the design phase, aligning with the proactive optimization goals of this study. Thermal management in machine tools has driven innovative modeling approaches. For instance, lever-aging power consumption data for thermal error compensation and analyzing process-dependent thermal loads have proven effective in enhancing prediction accuracy for machine tool components [7, 8]. This work proposes a methodological framework utilizing the Finite Element (FEA) Method for thermal analysis, aiming to predict and optimize thermal behavior in electromechanical cylinders for enhanced design and operational efficiency. FEA, as detailed by Klein et al. [9] is fundamental in engineering for analyzing both mechanical and thermal behaviors, particularly under complex nonlinear conditions introduced by temperature dependencies and structural variability. However, Shi et al. [10] mentioned traditional FEA for its computational inefficiency in certain contexts, advocating instead for a Thermal Characteristic Analysis (TCA) model a lumped analytical approach offering a faster alternative for specific applications. In contrast, this study leverages FEA's strengths in multi-domain integration, combining structural mechanics with thermal dynamics to evaluate the temperature distribution and its influence on structural deformation, as demonstrated in high resolution simulations by Galant et al. [11]:

$$\mathbf{C_{TT}} \cdot \dot{\mathbf{T}} + \mathbf{L_{TT}} \cdot \mathbf{T} = \dot{\mathbf{Q}} \tag{1}$$

Here, $\mathbf{C_{TT}}$ represents the heat capacity matrix, $\mathbf{L_{TT}}$ the conductance matrix, $\mathbf{T}$ is the temperature state vector and $\dot{\mathbf{Q}}$ the heat flow vector. For constant thermal properties, static deformation effects are derived via:

$$\mathbf{K_{XX}} \cdot \mathbf{x} = \mathbf{F} - \mathbf{K_{XT}} \cdot \mathbf{T} \tag{2}$$

where $\mathbf{K_{XX}}$ is the stiffness matrix, $\mathbf{F}$ represents external forces, and $\mathbf{K_{XT}}$ the thermal expansion matrix [2]. The computational complexity of these simulations can be excessive, necessitating Model Order Reduction (MOR). MOR simplifies high-dimensional models into manageable forms, reducing computational time while preserving accuracy, as demonstrated in strategies tailored for compact machine tool models [12] and further refined by Vettermann et al. [13] for fast simulations with nonlinear component behavior during the design phase. Galant et al. [11] also propose a tailored MOR method enhancing FEA efficiency, which this work builds upon. For thermal applications, this involves transforming the system into a state-space representation.

$$\mathbf{E} \cdot \dot{\mathbf{T}} + \mathbf{A} \cdot \mathbf{T} = \mathbf{B} \cdot \mathbf{u} \tag{3}$$

$$\mathbf{y} = \mathbf{C} \cdot \mathbf{T} \tag{4}$$

where $\mathbf{u}$ denotes thermal loads, $\mathbf{E}$ is the capacity matrix, $\mathbf{A}$ is the conductivity matrix, $\mathbf{B}$ is the control matrix derived from system matrices, $\mathbf{C}$ is the observer matrix, and $\mathbf{T}$ is the temperature state vector [2]. The equivalence of $\mathbf{E} = \mathbf{C_{TT}}$ holds because both matrices scale the time derivative $\dot{\mathbf{T}}$ in their respective equations, representing thermal capacitance; the MOR uses the pre-normalized form where $\mathbf{E} = \mathbf{C_{TT}}$, before normalization sets $\mathbf{E} = \mathbf{I}$. Equations (3) and (5) apply to systems without variable structures, where no relative motion occurs between subsystems. However, the described approach must be adapted since continuous structural variability is typical in machine tools. The system is divided into structurally invariant subsystems, as their relative motion affects both thermal interactions and overall system stiffness. After decomposition, the heat flows resulting from the thermal interactions of subsystems are defined as part of the thermal loads $\dot{\mathbf{Q}}$ for the surrounding subsystems. This allows the differential Eq. (1) to be solved for each subsystem at a specific time and structural state. Addressing nonlinearities from temperature-dependent material properties requires iterative solutions, as noted by [14], who provide a comprehensive comparison of numerical methods for thermo-elastic deformation computation, validated through simulation studies. In thermal analysis of electromechanical systems, heat transfer dynamics are critical. The Eq. (5) models heat transfer with variables like $\mathbf{C_{TT}}$, $\mathbf{L_{TT}}$, and heat transfer coefficients, where thermal contact conductance, influenced by interstitial fluids as explored by [15], ensures accurate boundary conditions.

$$\mathbf{C_{TT}} \cdot \dot{\mathbf{T}} + \left(\mathbf{L_{TT}} + \sum_{i=1}^{n} h_0 \cdot \mathbf{A_i} \right) \cdot \mathbf{T} = \dot{\mathbf{Q}}(t) + \sum_{i=1}^{n} h_0 \cdot \mathbf{A_i} \cdot T_u - \sum_{i=1}^{n} \Delta h_i(t) \cdot \mathbf{A_i} \cdot (\mathbf{T} - T_u) \quad (5)$$

The Eq.(5) describes the heat transfer dynamics and accounts for environmental interactions, crucial in vacuum or controlled conditions, as investigated by Li et al. [16] through fluid-thermal coupling models, where T_u is the ambient temperature, Δh_i represent the heat transfer coefficients for each boundary surface, which are split into a constant component h_0 and n represents the number of outer surfaces $\mathbf{A_i}$. The left-hand side represents the system's thermal response, with $C_{TT} \cdot \dot{T}$ modeling heat storage and $\left(L_{TT} + \sum_{i=1}^{n} h_0 \cdot A_i \right) \cdot T$ capturing internal conduction and steady-state environmental losses. The right-hand side accounts for external heat inputs ($\dot{Q}(t)$) and environmental interactions, including steady-state ($\sum_{i=1}^{n} h_0 \cdot A_i \cdot T_u$) and time-varying ($-\sum_{i=1}^{n} \Delta h_i(t) \cdot A_i \cdot (T - T_u)$) heat exchange with the environment at temperature T_u.

Techniques like splitting heat transfer coefficients into constant and variable components ensure a regular conductance matrix, facilitating MOR processes [3]. This approach enhances model reduction and accurately represents dynamic thermal and structural interactions, as seen in simulations of thermoelastic behavior in systems with moving parts [17, 18].Electromechanical cylinders exhibit poor thermal performance due to inherent design disadvantages, necessitating optimized designs. This demands rapid simulation setups and precise models, as minor geometric changes significantly impact performance. Existing high-fidelity FEA models are time consuming and struggle with small-scale geometric effects, a challenge addressed by Schroeder et al. [19] through efficient structural component modeling for early design phases. Su et al. [20] further enhance FEA by integrating fractal analysis for thermal contact interfaces, though this may exceed the scope of this study's focus on broader system behavior. To overcome these limitations, this work presents several transferable functions, which are adaptable in term of physical and dimensional characteristics of the components for different components that are usable for different kind of interactions, also repeatable on different machine-tools (e.g. rod bar moving axially and interacting with the environmental air), directed for calculating heat conduction, power losses, and convective and radiative effects, adaptable to various machine tool designs, building on systemic thermal management insights [2] and validated through practical modeling approaches [21].

2 Modelling Workflow

This section presents in Fig. 1 the workflow for constructing a multi-domain model tailored to electromechanical cylinders , addressing their thermal and mechanical challenges. The methodology integrates insights from both thermal and mechanical domains to tackle interactions. The necessity for this approach arises from the distinct time constants of thermal and mechanical processes, prompting a dual-system strategy for model development. A CAD model design is provided, requiring defeaturing, segmentation, and optimization to prepare the geometric model for finite element analysis (FEA). Within the FEA phase, ANSYS software is utilized for geometry meshing and defining interactions between static and mobile components. This stage encompasses discretizations of the geometry, creation of subsystem models (fixed, rotary, and translational), and integration of interfaces. The system matrices for conductivity and heat capacity are then formulated for the dynamic aspects, particularly the parametrization of thermal interactions in scenarios. Therefore, it allows the integration of thermal dynamics into the system matrices, ensuring the

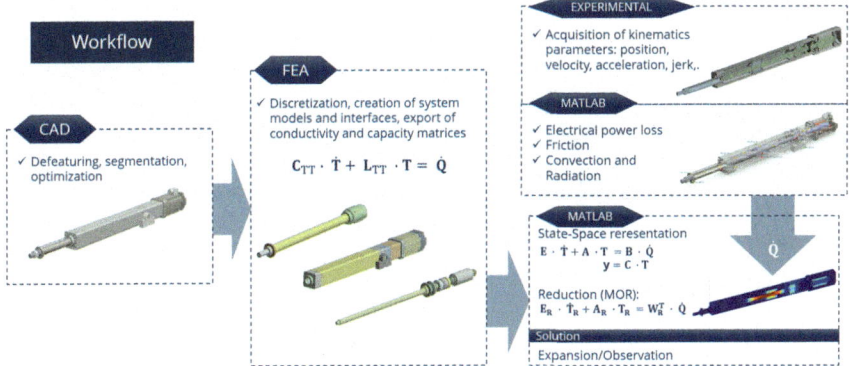

Fig. 1 Modeling workflow and software linkage

model captures both static and dynamic thermal-mechanical interactions comprehensively. The MATLAB section includes defining kinematic parameters, the respective functions for each interaction, and constants related referred to design and physical characteristics of components and boundary conditions. The final step involves the state-space representation in MATLAB, where equations for energy balance and thermal equilibrium are set up, alongside MOR for efficient simulation. The workflow concludes with solution steps for observation, providing a comprehensive framework for understanding and predicting the system behavior.

3   Experimental Setup

An experimental approach was adopted to validate the multi-domain model as detailed in the modeling workflow. The experimental setup shown in Fig. 2, includes an electromechanical compact axis (EMC-160-HP) provided by Bosch Rexroth and a load cylinder that applied the necessary load in order simulate a load on the compact axis, reaching its initial position. The control system was setup to represent a constant load applied on the piston rod for one or more strokes and study its thermal behavior.

The kinematic parameters available for tuning are travel speed, acceleration, jerk, initial and end positions for a stroke, and proportional factors for speed and force controls. In addition, temperature sensors were strategically located near to the components with higher heat generation due friction. Each of the mentioned locations has a pair of sensors on both sides of the system in order to detect potential non-symmetrical thermal behavior. Table 1 summarizes the enumeration and their respective location presented in the Fig. 3.

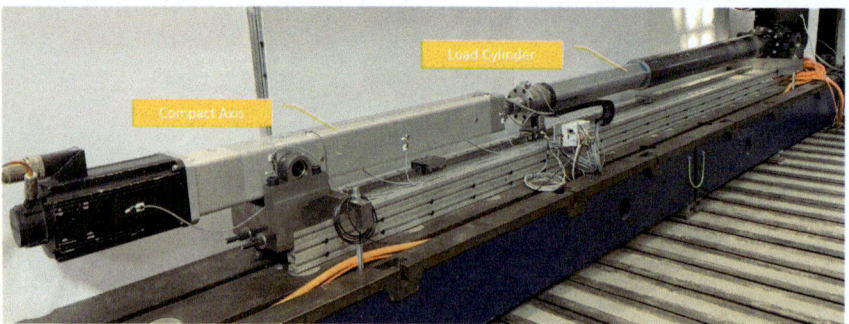

Fig. 2 Experimental setup

Table 1 Temperature sensor locations

N° sensor	Location
1, TM1, 2	Environmental air, inside the motor and motor housing respectivelly
3, 4	Near coupling and spindle bearings
5, 6	Sensor at housing midpoint
7, 8	Right end position of housing
9, 10	Outer of the cylinder head
11, 12	Inner sensor through piston rod

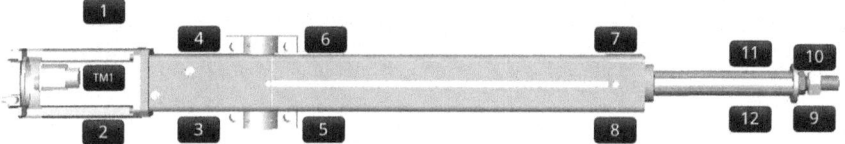

Fig. 3 Top view of symmetrical location of temperature sensors on a compact axis

4 Model Design

4.1 ANSYS Model Structure

The ANSYS model structure for the electromagnetic cylinder EMC-160-HP was meticulously developed to facilitate thermal simulations. Geometry preparation started from a STEP file provided by Bosch Rexroth AG, simplified and segmented using SOLIDWORKS 2022. This simplification was crucial to manage meshing complexity while maintaining thermal relevance, particularly for components like bearings, screw drives, and the synchronous motor. The meshing aimed for a structured hexahedral mesh with linear shape functions for better convergence and reduced element count compared to tetrahedral meshes [9]. Preparation for meshing included volume partitioning in ANSYS DesignModeler to ensure prismatic or rotationally symmetric bodies, essential for hexahedral mesh algorithms. Regarding FE model exporting, the model was segmented into three sub-models: FIX, TRL, ROT (Fixed, Translational, Rotational respectivelly), each handling different types of component motion. This segmentation allowed for detailed thermal interaction analysis. The export was facilitated through APDL macros, resulting in binary files containing system matrices for further MATLAB model creation and its thermal analysis. This methodology ensures a robust framework for further thermal analysis in MATLAB, balancing detailed geometry representation with computational efficiency, thereby setting a foundation for advanced thermal and mechanical simulations in subsequent stages.

4.2 MATLAB Model Structure

The main structure integrates a main function involving the creating of the model itself, including model construction, modular structure with nested functions and configurability according to the approach of interest and data import and processing. Efficient import of the load system matrices from binary files into MATLAB using a class library for structured organization is the first step into the MATLAB model structure. Further step is the parametrization, which includes detailing the thermal dynamics with attention to heat input from sources like friction in bearings, guides, seals, screw drive, and motor losses. This leads into heat transfer, where interactions between components are explored, either through direct contact or the air. Heat dissipation is addressed, modeling how heat is lost to the environment through convective-radiative transfer and at connection surfaces, ensuring that the system's thermal equilibrium is maintained.

4.3 Parametrization

The parametrization of thermal interactions in the axis model involves specific functions that define the thermal dynamics between subsystems and at system boundaries. These functions are invoked during model construction during simulation for temperature-dependent scenarios. Surface elements, also identified as interfaces, encompass in thermal interactions are pre-assigned in ANSYS. Key parameters include heat transfer coefficients, H_0 for constant and H_t for variable, heat flux density $H_t T$, power loss Q_t, heat transfer L_t and combined heat flows $Q_t L_t T$. The parametrization accounts for both constant and time-variable heat transfer coefficients, ensuring accurate modeling of heat distribution across components. Figure 4 summarizes the model's interactions, displaying the compact axis with function names and their thermal effects, including conduction and convection. Colored dots are a rough representation of the type and location of the mentioned thermal effects.

4.3.1 Heat Input Sources

Table 2 provides a detailed overview of the various thermal loads considered in the model. It lists the sources of heat input, the associated friction forces or torques, and the corresponding power losses or thermal loads, which are crucial for understanding and simulating the thermal dynamics within the system. Therefore, Fig. 5 shows the compact axis with all considered loads that contribute to the thermal state of the system.

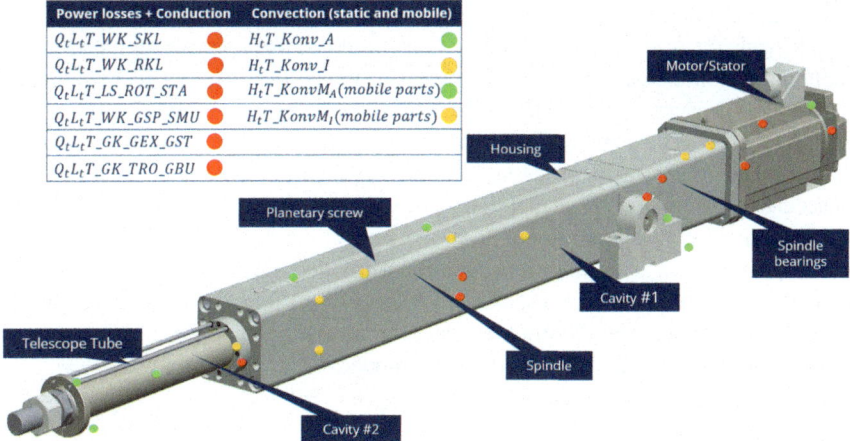

Power losses + Conduction		Convection (static and mobile)	
$Q_t L_t T\_WK\_SKL$	●	$H_t T\_Konv\_A$	●
$Q_t L_t T\_WK\_RKL$	●	$H_t T\_Konv\_I$	●
$Q_t L_t T\_LS\_ROT\_STA$	●	$H_t T\_KonvM_A(mobile\ parts)$	●
$Q_t L_t T\_WK\_GSP\_SMU$	●	$H_t T\_KonvM_I(mobile\ parts)$	●
$Q_t L_t T\_GK\_GEX\_GST$	●		
$Q_t L_t T\_GK\_TRO\_GBU$	●		

Fig. 4 EMC-180 Bosch compact axis

Table 2 Considered thermal loads

Thermal source	Friction force/torque	Power loss/thermal load
Axial seal at piston rod	$F_{R,\text{ADI},1}$, $F_{R,\text{ADI},2}$	$P_{V,\text{ADI},1}$, $P_{V,\text{ADI},2}$
Sliding bushing and piston rod	$F_{R,\text{GBU}}$	$P_{V,\text{GBU}}$
Axial guide slider and housing	$F_{R,\text{GST}}$	$P_{V,\text{GST}}$
Screw drive	$M_{R,\text{GTR}}$	$P_{V,\text{GTR}}$
Radial shaft seal	$M_{R,\text{RDI}}$	$P_{V,\text{RDI}}$
Angular contact ball bearing	$M_{R,\text{SKL}}$	$P_{V,\text{SKL}}$
Deep groove ball bearing	$M_{R,\text{RKL}}$	$P_{V,\text{RKL}}$
Synchronous drive	–	$P_{V,\text{PSM}}$

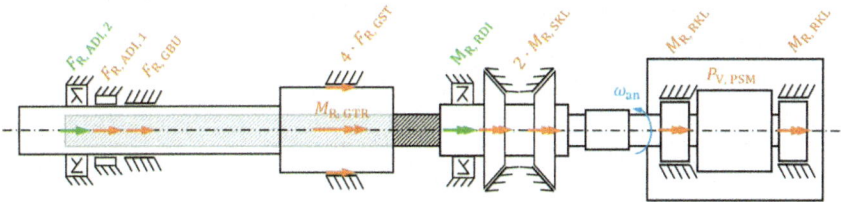

Fig. 5 Schematic representation of the compact axis with considered friction force/torque locations

4.3.2 Heat Transfer Conduction

Heat transfer by conduction is pivotal in managing thermal interactions within an electromechanical system, notably between components like the rotor and stator in the synchronous drive, where the greatest heat flux from power losses occurs, and across different components through direct contact or mediums such as the air. The transfer is modeled with a focus on component-to-component interactions, including conduction across the rotor-stator gap and within bearings-such as angular contact ball bearings-where both conduction and convection are considered, alongside surface-to-surface interactions governed by contact area and material properties. Many of these heat transfer processes depend on velocity, while others are influenced by acceleration, angular speed, applied load, and position, reflecting the dynamic and spatial variations in operating conditions. Thermal conductivity coefficients, with values derived from semi-empirical equations provided in [22], are parameterized in dedicated functions to ensure precise modeling of heat flow. These interactions are critical for accurately predicting the system's thermal behavior, ensuring a correct representation of heat flows throughout the model.

4.3.3   Heat Dissipation

The heat transfer from stationary external surfaces at temperature T_B to the environment at T_E combines convective and radiative mechanisms. The combined heat transfer coefficient h_{BE} is parameterized as

$$h_{BE} = \left(4.5 \cdot \epsilon_{BE} + 2.2 \cdot \left|\frac{T_B - T_E}{K}\right|^{\frac{1}{3}}\right) \cdot \frac{W}{m^2\,K} \tag{6}$$

according to [22]. The emissivity coefficient ϵ_{BE} is calculated from

$$\epsilon_{BE} = \epsilon_E \cdot \epsilon_B = \epsilon_E \cdot (0.4 \cdot \epsilon_{B1} + 0.6 \cdot \epsilon_{B2}), \tag{7}$$

where ϵ_{B1} and ϵ_{B2} are for black paint and aluminum oxide respectively [22]. For environmental heat transfer, $h_{BE}(t) = h_{BE,0} + \Delta h_{BE}(t)$ is used, with $h_{BE,0}$ and $h_{BE}(t)$ parameterized in the respective convective functions, where $h_{BE,0}$ is the steady-state combined heat transfer coefficient and $h_{BE}(t)$ is the time-varying version. For moving parts like the telescopic tube, a higher h_{BE} weighting factors were applied according to the respective position of the moving part. The effective area for heat transfer on the telescopic tube varies with stroke position.

5   Validation and Results

In the current section, experimental and simulation results are shown and compared. In order to achieve the steady-state, the test duration was executed for two hours, nevertheless, more time is needed to reach the steady-state completely. On the other hand, a design limitation was also suggested by the manufacturer, which is that the housing surface must not exceed 60Â°C. Figure 6shows the experimental results labeling the one sensor per location in different cylinder length positions (Fig. 3).

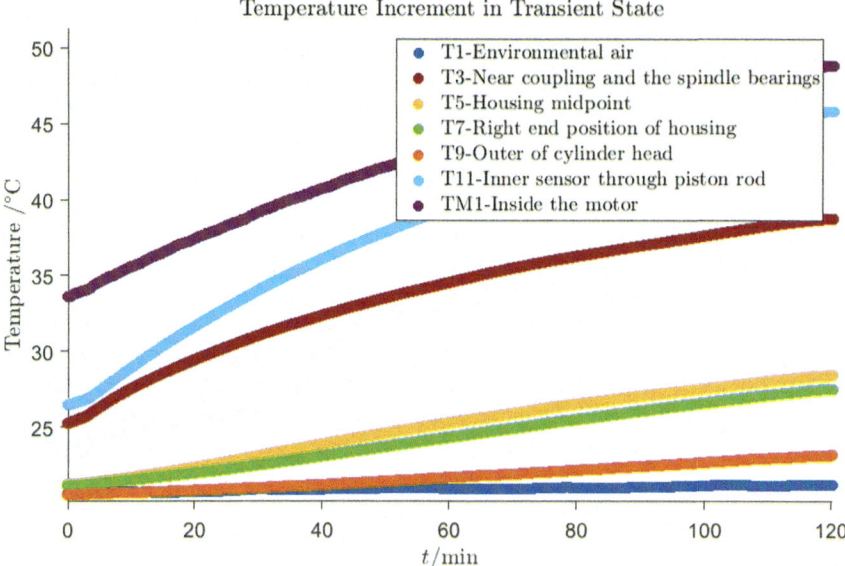

Fig. 6 Temperature measured during 2 hours at a customer application regime: velocity = 40 mm/s, load = 20 kN

The boundary conditions following a customer application regime were: velocity = 0.04 m/s; acceleration = 0.04 m/s; force = 20 kN; time pause = 13 s. From Fig. 6, it is observable that the highest increment in temperature was obtained by inner sensor located in the synchronous motor (TM1), followed by the sensor located through the piston rod (T11) and the near coupling and spindle bearings sensor (T3). The lowest temperature increment was obtained at the right end position of the housing (T7 and T9).

Regarding the simulation, Fig. 7 illustrates the thermal behavior of a compact axis under a customer application regime at the same kinematic conditions as the experimental approach. It captures the combined effects of conductivity, convection, and radiative heat transfer, revealing significant heat generation at the spindle bearings and planetary screw, consistent with their roles as critical heat sources due to friction. Besides, Fig. 8 shows a very short simulation, considering only convection and radiation. This reveal the thermal distribution across the housing, with a pronounced temperature increment at the top, where a slot for positional sensors likely amplifies the heat accumulation by limiting the local dissipation capacity and the enhancing exposure to convective currents. This observation highlights the impact of design features on thermal performance, suggesting that the housing's upper section is more susceptible to heating, potentially compromising long-term reliability and necessitating targeted cooling strategies.

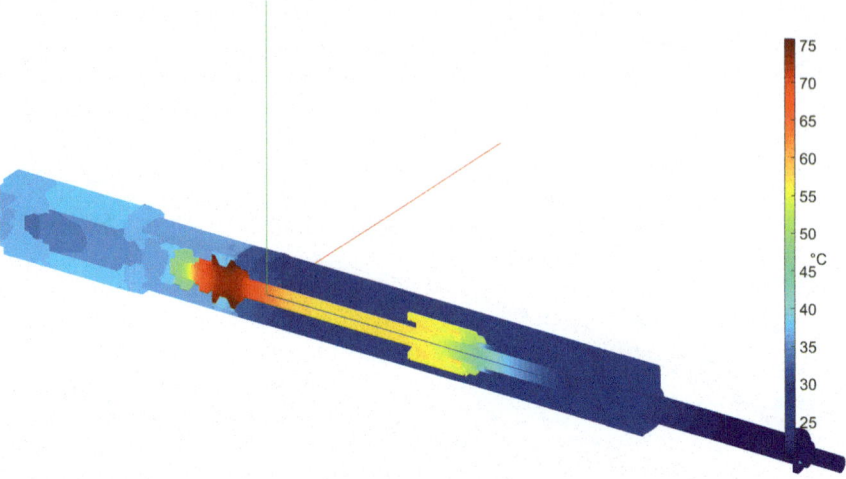

Fig. 7 Simulation for the same as for the experiments

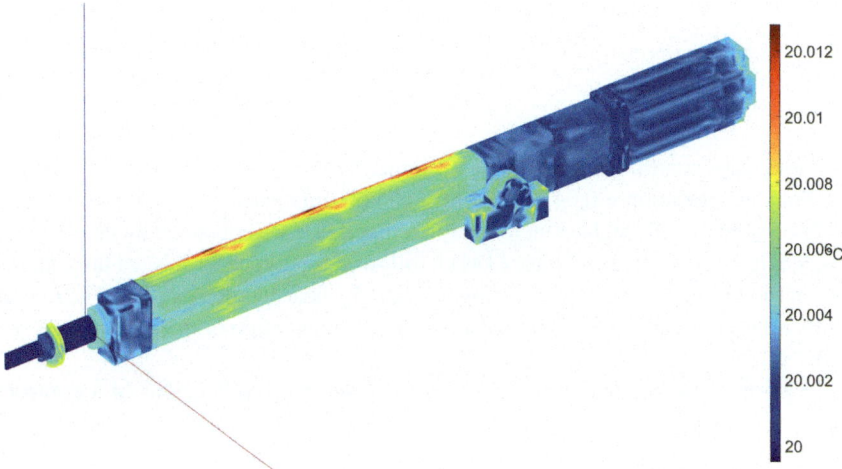

Fig. 8 Convective and radiative effects on compact axis: external housing view

Experimental and simulation results both identify the spindle bearings as the primary heat source, but the measured temperatures differ significantly. The experimental data likely reflect cumulative heating effects, including transient, non-linear phenomena like temperature-dependent lubricant viscosity in the bearings, which increases friction and heat over time. In contrast, the simulation accounts for some non-linear effects, such as temperature-dependent thermal conductivity, but may underestimate heating due to simplified semi-empirical equations for convection

Table 3 Comparison ΔT: experimental/simulation

Axis location	ΔT experimental, K	ΔT simulation (K)
Motor housing	14.7	17.5
Near coupling and spindle bearings	14.2	16
Housing midpoint	7.3	7.5
Right end position of housing	6.5	7
Outer of the cylinder head	2.6	2
Inner sensor through piston rod	19.7	17

heat transfer coefficients. These simplifications might not fully capture external factors, such as variable ambient air currents or unmodeled radiation losses. Additionally, thermal contact resistance between components (e.g., spindle and housing) may impede heat flow in experiments, increasing the temperature difference (ΔT), while the simulation assumes ideal contact. Figures 7 and 8 under comparison with results obtained experimentally shown in Fig. 6 highlights two issues: (1) The synchronous motor and surrounding components (bearings) do not heat up sufficiently, and (2) the convective heat transfer function for moving components to air inside cavities underperforms. Additionally, heat transfer through the air in both, the spindle cavity and synchronous motor, also underperforms. These ongoing research challenges are critical for improving temperature predictions in the motor zone and enhancing heat transfer from the spindle bearings, spindle, and piston rod to the housing (Table 3).

6   Conclusions and Outlook

Experimental and simulation results confirm that the spindle bearings are the primary heat source in the compact axis, with the highest temperature increases observed near the motor ($\Delta T = 14.7$ K) and spindle bearings ($\Delta T = 14.2$ K), as shown in Fig. 6 for the experimental approach. However, simulations underpredict temperatures compared to experiments, likely due to simplified convection models. In terms of modeling, the weighting factors applied in the functions for the respective calculus between component interactions should be corrected. Temperature limitations by design are also an aspect to consider in future experimental runs, due potential damage on the components may be provoked, affecting an optimal performance. Future work will refine the simulation model to better capture convective heat transfer in moving components and motor bearing heating, as identified in Fig. 8. Enhanced models for transient effects and thermal contact resistance will improve temperature predictions.

Acknowledgements This research was funded by a German Research Foundation (DFG), SFB/TR96 transfer project T09, which is gratefully acknowledged.

References

1. Haberhauer, H.,Kaczmarek, M.: Taschenbuch der Antriebstechnik. Mit 47 Tabellen. Fachbuchverl. Leipzig im Carl Hanser Verl., Munchen (2014)
2. Großmann, K. (ed.):Thermo-energetic design of machine tools: a systemic approach to solve the conflict between power efficiency, accuracy and productivity demonstrated at the example of machining production. Springer eBook collection engineering. Springer, Cham (2015)
3. Großmann, K., Muhl, A., Thiem, X.: Korrektur thermisch bedingter fehler an werkzeugmaschinen.Zeitschrift fur wirtschaftlichen Fabrikbetrieb **109**(5), 318–323 (2014)https://doi.org/10.3139/104.111148
4. Thiem, X., Rudolph, H., Krahn, R., Ihlenfeldt, S., Fetzer, C., Müller, J.: Adaptive thermal model for structure model based correction. In: International Conference on Thermal Issues in Machine Tools, pp. 67–82. Springer (2023)
5. Horejš, O., Mareš, M., Straka, M., Švéda, J., Kozlok, T.: Adaptive thermal error compensation model of a horizontal machining centre. In: International Conference on Thermal Issues in Machine Tools, pp. 83–98. Springer (2023)
6. Creighton, E., Honegger, A., Tulsian, A., Mukhopadhyay, D.: Analysis of thermal errors in a high-speed micro-milling spindle. Int. J. Mach Tools Manuf **50**(4), 386–393 (2010)
7. Lang, S., Zimmermann, N., Mayr, J., Wegener, K., Bambach, M.: Thermal error compensation models utilizing the power consumption of machine tools. In: International Conference on Thermal Issues in Machine Tools, pp. 41–53. Springer (2023)
8. Wenkler, E., Steiert, C., Boos, E., Ihlenfeldt, S.: Analysing the impact of process dependent thermal loads on the prediction accuracy of thermal effects in machine tool components. In: International Conference on Thermal Issues in Machine Tools, pp. 99–115. Springer (2023)
9. Klein, B.: FEM: Grundlagen und Anwendungen der Finite-Element-Methode Im Maschinen-und Fahrzeugbau,10., verb. aufl. edn. SpringerLink Bucher. Springer, Wiesbaden (2015)
10. Shi, K.,Yang, R., Gao, W., Zheng, Y., Hu, G., Weng, L., Fu, Y., Wang, P.: Modelling the thermal characteristics of a machine tool with external mechanical spindle motor. In: International Conference on Mechanical Design and Simulation, pp. 581–589. Springer (2024)
11. Galant, A., Beitelschmidt, M., Großmann, K.: Fast high-resolution FE-based simulation of thermo-elastic behaviour of machine tool structures. Procedia CIRP **46**, 627–630 (2016)
12. Aumann, Q., Benner, P., Saak, J., Vettermann, J.: Model order reduction strategies for the computation of compact machine tool models. In: International Conference on Thermal Issues in Machine Tools, pp. 132–145. Springer (2023)
13. Vettermann, J., Steinert, A., Brecher, C., Benner, P., Saak, J.: Compact thermo-mechanical models for the fast simulation of machine tools with nonlinear component behavior. at-Automatisierungstechnik **70**(8), 692–704 (2022)
14. Naumann, A., Lang, N., Partzsch, M., Beitelschmidt, M., Benner, P., Voigt, A., Wensch, J.: Computation of thermo-elastic deformations on machine tools a study of numerical methods. Prod. Eng. **10**, 253–263 (2016)
15. Frekers, Y.H., Kneer, R.: Numerical investigation on the influence of interstitial fluids on thermal contact conductance (2018). https://doi.org/10.1615/IHTC16.cip.023645
16. Li, T., Xi, G., Wang, H., Tang, W., Shao, Z., Sun, X.: Thermal properties prediction of large-scale machine tool in vacuum environment based on the parameter identification of fluid–thermal coupling model. Machines **10**(12), 1237 (2022)
17. Stefan Sauerzapf, M.B.: Simulation of thermoelastic behavior of technical systems with relatively moving parts–Modeling process, part coupling approaches and application to machine tools. Appl. Thermal. Eng. **224**, 119987 (2023)
18. Mares, M., Horejs, O., Nykodym, P.: An indicative model considering part of the thermo-mechanical behaviour of a large grinding machine. Int. J. Precis. Eng. **42**(6), 789–802 (2021)
19. Schroeder, S., Ihlenfeldt, S., Penter, L.: Efficient and robust creation of structural component models for thermo-elastic analysis of machine tools. MM Sci. J
20. Su, H., Lu, L., Liang, Y., Zhang, Q., Sun, Y., Liu, H.: Finite element fractal method for thermal comprehensive analysis of machine tools. Int. J. Adv. Manuf. Technol. **75**, 1517–1526 (2014)

21. Hernández-Becerro, P., Mayr, J., Wegener, K.: Efficient thermo-mechanical model of a precision 5-axis machine tool. In: Conference Proceedings on Thermal Issues 2020, p. 20116. Euspen (2020)
22. Jungnickel, G.: Simulation des Thermischen Verhaltens Von Werkzeugmaschinen. Modellierung und Parametrierung. Lehre, Forschung, Praxis. Inst. fur Werkzeugmaschinen und Steuerungstechnik, Lehrstuhl fur Werkzeugmaschinen, Dresden (2010)

AI and Data-Driven Thermal Analysis

No Warm-Up Required: Initializing Time-Dependent Thermal Error Compensation Models for Machine Tools

Sebastian Lang⑩, Mario Zorzini⑩, Markus Sprecher, Stephan Scholze⑩, and Markus Bambach⑩

Abstract Thermal errors in machine tools significantly impact the precision of manufactured workpieces, particularly in high-accuracy processes such as grinding, where even minor deviations can lead to defects. Traditional mitigation strategies, such as active cooling and warm-up cycles, increase energy consumption and reduce productivity. Data-driven compensation models offer a more sustainable alternative by predicting and compensating for thermal deformations based on sensor data. This study explores different time-dependent models such as autoregressive models with exogenous inputs (ARX), long short-term memory (LSTM) networks, and temporal convolutional networks (TCN). These effectively capture thermal hysteresis effects but require initialization strategies to function immediately after machine start-up. To address this challenge, we evaluate various initialization, or "padding," techniques, including zero initialization, leveraging prior temperature measurements, virtual predictions, and a dual-model approach that integrates a non-time-dependent (NTD) model for robust start-up compensation. Furthermore, to improve data efficiency in non-uniformly sampled scenarios, we assess different upsampling strategies such as linear interpolation, piecewise cubic Hermite interpolating polynomials (PCHIP), cubic splines, and Gaussian process regression (GPR) model-based interpolation. The proposed methods are validated on a five-axis grinding machine equipped with 13 temperature sensors and a 3D touch trigger probe. Results demonstrate that model-based interpolation significantly enhances compensation accuracy by incorporating temperature fluctuations between measurement cycles, while hybrid initialization strategies ensure accurate error compensation during critical start-up phases. The combination of optimized initialization and resampling strategies achieves a $\sim 75\%$ reduction in average thermal error to below 3.6 μm, providing a robust and energy-efficient solution for precision manufacturing.

S. Lang · M. Zorzini · M. Bambach
Advanced Manufacturing Laboratory, ETH Zurich, Zurich, Switzerland

S. Lang (✉)
inspire AG, Zurich, Switzerland
e-mail: selang@ethz.ch

M. Sprecher · S. Scholze
Agathon AG, Bellach, Switzerland

© The Author(s) 2026
K. Wegener and M. Bambach (eds.), *4th International Conference on Thermal Issues in Machine Tools (ICTIMT2025)*, Lecture Notes in Production Engineering,
https://doi.org/10.1007/978-3-032-01194-7_35

Keywords Thermal error · Machine tool · Padding · Time dependent modeling · Compensation

1   Introduction

Grinding machine tools (MT) typically have very high requirements regarding accuracy as well as productivity [26] and increasingly sustainability [24]. As grinding can achieve exceptional surface finish, dimensional accuracy and machine even very hard and brittle materials it is often a crucial manufacturing step finalizing demanding workpieces. In order to achieve this accuracy great effort has to be taken, especially thermal errors can effect the geometric errors of produced workpieces by up to 75% according to Mayr et al. [23]. Typical remediation measures include design optimizations or increased or improved cooling strategies. For example Weng et al. [28] proposed an analytical modeling method which can aid in the design of thermally balanced MT components. Kondo et al. [14] reduced the axial displacement of a spindle to one third by designing a carbon fiber reinforced plastic spindle shaft. Maurya et al. [21] used genetic algorithms, a form of artificial intelligence, to optimize cooling parameters such as the flow rate or temperature depending on the spindle speed to minimize thermally induced deformations. Denkena et al. [6] summarized the work in motor spindle cooling and concluded the growing need for efficient and intense cooling efforts to achieve high process stability.

The disadvantage of all of these approaches is that they typically require resource based efforts and lead to a higher environmental footprint during manufacturing and operation of the MT. An alternative to this is compensating the error using the MT controls. For this the error has to be known either using measurements, physical or data driven models. While measurements are the most accurate they are time consuming leading to a loss of productivity and measurement equipment may not be always available or restricted in its use through accessibility [8]. Kaftan et al. [12] developed a method of measuring one dimensional thermal errors without the use of additional measurement systems beyond the machine axis, which however still impacts productivity. Physical models allow for an understanding of the underlying mechanism, but are often limited due to their computational demands and unknown state of the system [11]. Recent work has focused on estimating the state the system is in using temperature measurements directly [13] or with the use of Kalman filters [16].

1.1   Data Driven Thermal Compensation Models

Data driven models allow a fast, often much faster than real time, prediction of the thermal errors using various kinds of model inputs often in the form of temperature sensors [20]. Any signal that contains information about the underlying thermal error can be used however such as the power consumption of the MT components as they relate directly to the heat input and therefore subsequent thermal expansion [3, 17]. Selecting a suitable model input is crucial and can even be done adaptively [29]. Updating the compensation model can improve long term stability as demonstrated by using an autoregressive model with exogenous input (ARX) [4]. Mayr et al. [22] extended this to an arbitrary sampling rate using a weighted least squares approach. Besides ARX models a wide range of regression models can be used for thermal error compensation. Fujishima et al. [7] demonstrated the use of ridge regression, a feed forward neural network (FFNN) as well as a convolutional neural network (CNN) for thermal error compensation. Liu et al. [19] used a long short term memory (LSTM) neural network, which contains a time dependency within the model, to predict the thermal errors of a spindle. Chen et al. [5] compared a number of time dependent models for a MT spindle of which the temporal convolutional network (TCN) performed well. Hsieh et al. [9] compared different time dependent neural network architectures on the X- and Y-error of a spindle such as the LSTM, TCN and the bi-directional LSTM which all showed sufficient performance. Wei et al. [27] applied Gaussian process regression (GPR) to a spindle error and showed a slightly higher but similar performance compared to ARX and LSTM models. Applying these models to more than just a spindle and a long term manufacturing process is another challenge however. Zimmermann et al. [31] manufactured impellers with and without thermal compensation and demonstrated a error reduction on the machined workpiece of 73%. Zimmermann et al. [30] introduced a new thermal test piece for 5-axis MTs that allows the separation of the different axis specific thermal errors on the machined workpieces. Besides temperature sensors also thermal imaging cameras can be used as inputs for compensation models successfully as demonstrated by Abdulshahed et al. [1].

These advantages of data-driven models are clear. However applying them in real production settings is still challenged by a lack of available and representative data. Another challenge is the applicability of time-dependent models right after a machine start when insufficient historical data is available. The initial phase after the machine start is however one with the most significant thermal effects and therefore in dire need of accurate compensation. This paper proposes a new resampling and data augmentation strategy as well as different model initialization strategies that allow the instantaneous compensation of thermal errors in MTs. To demonstrate this achievement the used compensation models are first introduced as well as different sampling strategies for data augmentation. Then the different model initialization strategies are introduced that allow compensation even with insufficient historical data available. Subsequently the measurement setup on a 5-axis grinding MT is introduced which measurements were used to validate the proposed approach.

2 Thermal Compensation Model Architecture

This section introduced the used models to allow a real time data driven compensation. Beside the time dependent ARX model, LSTM model and TCN model, a feed forward neural network (FFNN) without any time dependency is used. All these data driven models allow the use of various inputs and real time prediction of the thermal error. All model hyperparameters, such as layer count, number of units or learning rate, were chosen on a separate dataset using Optuna [2] until convergence was reached to allow a comparison between the model architectures. All neural network based models follow a multiple input multiple output (MIMO) architecture allowing one model to take all temperature sensors as inputs. Besides these models also multiple linear regression (MLR), which is similar to a ARX model without the time dependency and autoregressive component, and GPR as a nonparametric non time dependent model are used.

2.1 Feed Forward Neural Network Model

A FFNN is a simple neural network architecture that due to the used activation functions allows non linear relationships between the input data and output. In this case the input data are the temperature measurements at time t and the output is the corresponding thermal error of that timestep. A rectified linear unit (relu) activation function is used and all neurons are fully connected to the ones on the previous and subsequent layer with a trainable weight. The inputs correspond to all measured temperatures and the output corresponds to the measured thermal errors at the same time.

Figure 1 shows the used model architecture with one hidden layers of 12 units which leads to a total of 519 trainable parameters.

Fig. 1 FFNN model architecture

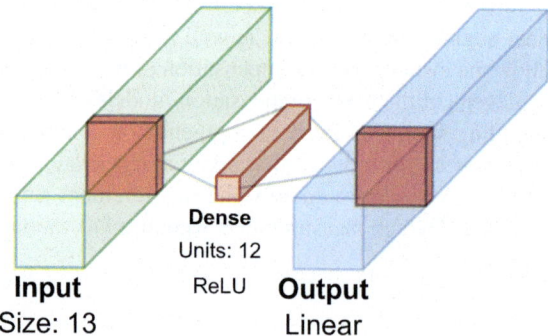

2.2 ARX Model

The ARX model is a well-established statistical approach for modeling and predicting dynamic systems with time-dependent behavior and represents a linear time-invariant system (LTI). It achieves this by leveraging historical observations of system outputs and incorporating exogenous input variables, such as temperature measurements or power consumption data, to estimate thermal errors.

The ARX model is defined as:

$$y[t] + \sum_{i=1}^{n_a} \beta_i \cdot y[t-i] = \sum_{m=1}^{M} \sum_{n=0}^{n_b(m)} \beta_{m,n} \cdot u_m[t-n] \tag{1}$$

where $y(t)$ represents the thermal error at time t, $x(t-j)$ denotes the exogenous input variables, β_i and $\beta_{m,n}$ are the model coefficients for the autoregressive and exogenous terms, respectively. The model considers n_a and $n_b(m)$ historical timesteps of the respective model predictions and input measurements.

In the context of thermal error compensation, the ARX model predicts geometric inaccuracies induced by thermal effects using a regression framework that links input signals, such as MT temperature or spindle power, to thermal deviations. The model's parameters β_i and $\beta_{m,n}$ are estimated during the training phase using historical data, often through techniques like least squares estimation. Mayr et al. [22] have extended the model's application to systems with irregular sampling rates via weighted least squares methods. Alternatively a sampling to a common frequency can be implemented as proposed in this paper and described in Sect. 3.

2.3 LSTM Model Architecture

Long Short-Term Memory (LSTM) networks, a variant of Recurrent Neural Networks (RNNs), are widely used for modeling sequential data, particularly when capturing temporal dependencies over varying time scales is critical. Unlike traditional RNNs, LSTMs are specifically designed to overcome the vanishing gradient problem by introducing memory cells and gating mechanisms that regulate the flow of information. Figure 2 shows the architecture of the LSTM based model which is based on two LSTM hidden layers with a subsequent dropout layer respectively. The outputs have to be flattened for the subsequent three dense hidden layers leading to the output.

The architecture of an LSTM cell is structured around three primary gates: the forget gate, the input gate, and the output gate. The forget gate, defined as

$$f_t = \sigma \left(W_f \begin{bmatrix} h_{t-1} \\ x_t \end{bmatrix} + b_f \right),$$

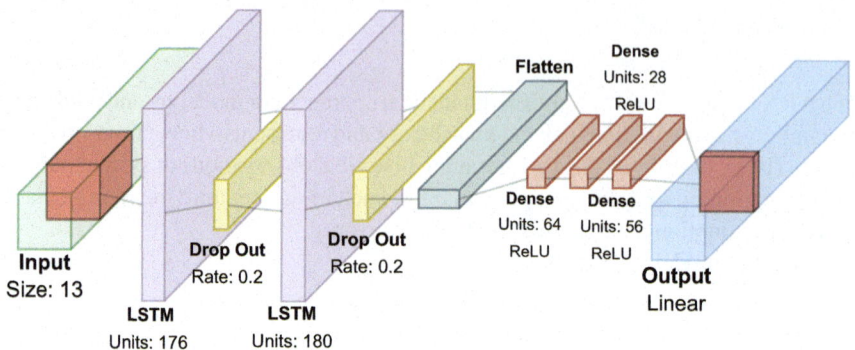

Fig. 2 LSTM model architecture

determines which components of the previous cell state C_{t-1} should be discarded. The input gate governs the addition of new information to the cell state using

$$i_t = \sigma \left(W_i \begin{bmatrix} h_{t-1} \\ x_t \end{bmatrix} + b_i \right), \quad \tilde{C}_t = \tanh \left(W_C \begin{bmatrix} h_{t-1} \\ x_t \end{bmatrix} + b_C \right),$$

where i_t denotes the gate activation and $\tilde{C}_t$ represents candidate values for the updated cell state. The overall cell state update combines the retained information from the forget gate and the new contributions from the input gate as

$$C_t = f_t \odot C_{t-1} + i_t \odot \tilde{C}_t,$$

where $\odot$ denotes element-wise multiplication. The final output of the LSTM at each time step is determined by the output gate:

$$o_t = \sigma \left(W_o \begin{bmatrix} h_{t-1} \\ x_t \end{bmatrix} + b_o \right), \quad h_t = o_t \odot \tanh(C_t).$$

This gating architecture allows LSTM networks to selectively retain or discard information over long temporal horizons, making them highly effective for capturing both transient and steady-state thermal behaviors. Their ability to model thermal hysteresis and startup conditions makes LSTMs particularly suitable for thermal error compensation tasks.

2.4 TCN Model Architecture

Temporal Convolutional Networks (TCNs) represent a modern alternative to recurrent architectures for sequential data modeling, relying on convolutional rather than recurrent operations. TCNs address the limitations of RNNs by enabling parallel computation, handling longer input sequences efficiently, and avoiding the vanishing gradient problem through their architectural design (Fig. 3).

The core of a TCN lies in its use of causal and dilated convolutions. Causal convolutions enforce temporal causality by ensuring that the output at time t depends only on inputs from time steps $\leq t$. This is mathematically expressed as

$$y_t^{(l)} = \sum_{k=0}^{K-1} w_k^{(l)} \cdot x_{t-k},$$

where $w_k^{(l)}$ are the convolutional weights, x_{t-k} represents the input sequence, and K is the kernel size. To capture long-term dependencies efficiently, TCNs employ dilated convolutions, which introduce gaps between sampled inputs. The dilation factor d modifies the convolution as

$$y_t^{(l)} = \sum_{k=0}^{K-1} w_k^{(l)} \cdot x_{t-d \cdot k}.$$

To ensure stability during training and to improve gradient flow, TCNs integrate residual connections, expressed as

$$y^{(l)} = \text{ReLU}\left(x^{(l)} + F(x^{(l)})\right),$$

where $F(x^{(l)})$ is the output of the dilated convolutional layer. These residual connections allow deeper networks to be trained effectively without degradation in performance.

Unlike LSTMs, TCNs do not rely on internal memory states and instead operate solely on exogenous inputs, such as temperature sensor readings. This simplifies

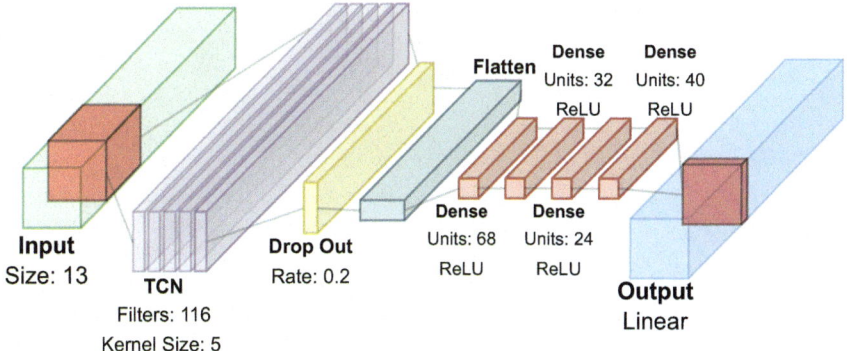

Fig. 3 TCN model architecture

the model architecture and reduces computational overhead while preserving the ability to capture both local and global temporal patterns. TCNs' hierarchical feature extraction and parallelization make them an efficient and accurate choice for thermal error compensation, particularly in scenarios requiring real-time performance.

3 Sampling Strategies for Thermal Compensation Models

As machine tasks between a thermal error measurement may not be of uniform time a resampling for model training is beneficial. Most time dependent models operate under the implicit assumption of constant time step duration. Resampling the data for model training is therefore required. At the same time up-sampling can increase the data availability and therefore model performance without any additional measurement effort and therefore cost [15].

Figure 4 illustrates a non-uniform measurement of an exemplary thermal error, represented by red crosses. Each data point is treated as a discrete measurement in space, despite the finite time required to acquire it, which inherently contributes to measurement uncertainty. Between these points, temperature measurements are typically available at much higher frequencies, such as 1 Hz, as they can be recorded concurrently with the production process. The simplest method for aligning these measurements with the thermal error data is to down-sample the temperature readings to the lower frequency of the thermal error measurements. However, this approach sacrifices temporal resolution and critical information, leading to non-uniform timesteps, which are incompatible with many time-dependent compensation model architectures. Linear interpolation, shown in blue, connects consecutive measurements with straight-line segments. This technique improves continuity in the data and maintains

Fig. 4 Sampling approaches for up-sampling uneven thermal error measurements

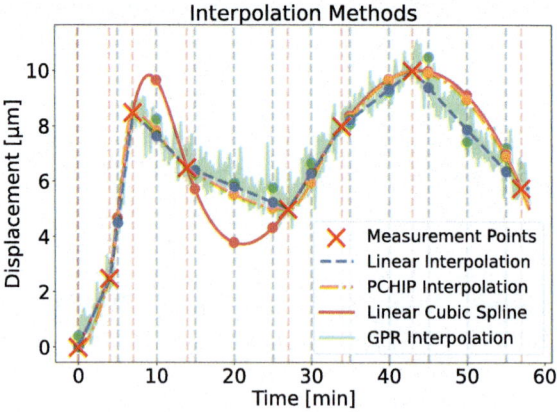

computational simplicity. However, the linear interpolation method results in discontinuities in the first derivative, visible as sharp corners, which may affect the performance of compensation models sensitive to smoothness. For higher-order continuity, methods like the Piecewise Cubic Hermite Interpolating Polynomial (PCHIP) and cubic splines with linear initialization to prevent bad conditioning at edges, shown in yellow and red, respectively, are employed. These methods generate smoother curves by ensuring continuity in the first derivative and, in the case of cubic splines, higher derivatives as well. These derivatives are obtained by the difference between the different measurement points. In the case of the cubic spline a linear interpolation is used until a sufficient number of previous measurements are available. While PCHIP prioritizes monotonicity and avoids overshooting, cubic splines excel at capturing global smoothness at the cost of potential overshooting moving further from the measurements. The GPR interpolation approach first trains a non time dependent model on the measurement points with the corresponding temperature measurements as model inputs. Subsequently all temperature measurements, or temperature measurements downsampled to an arbitrary sampling frequency, are used to predict a corresponding thermal displacement. These resulting displacements follow the temperature behavior even in between measurement points which is not possible for any other interpolation approach. At the same time the complexity of this approach is the highest and it is not possible to validate the behavior between measurement positions. In the results section all interpolation strategies as well as no resampling are compared on the same data set.

4   Model Initialization

Six compensation models are examined, namely ARX, FFNN, LSTM, GPR, MLR and TCN which are described in Sect. 2. Among these, all except FFNN, GPR and MLR belong to the class of time-dependent (TD) models. TD models incorporate past inputs and/or previously predicted outputs when making current predictions, making them particularly well-suited for thermal error compensation. This is due to thermal errors typically exhibiting strong dependencies on thermal hysteresis [25] allowing a smaller number of sensors to effectively identify the thermal state and therefore accurately compensate it. However, a key limitation of TD models arises during machine start-up, when no prior data is available. This phase is particularly challenging as the system has not yet reached a stable thermal state and therefore is strongly impacted by hysteresis effects. Accurate compensation at this stage is crucial as machine starts often show the highest rates of thermal change. To address this challenge, five initialization or padding methods are developed and analyzed. These methods aim to determine the most effective strategy for initializing the three TD compensation models, thereby improving their performance during the critical start-up phase.

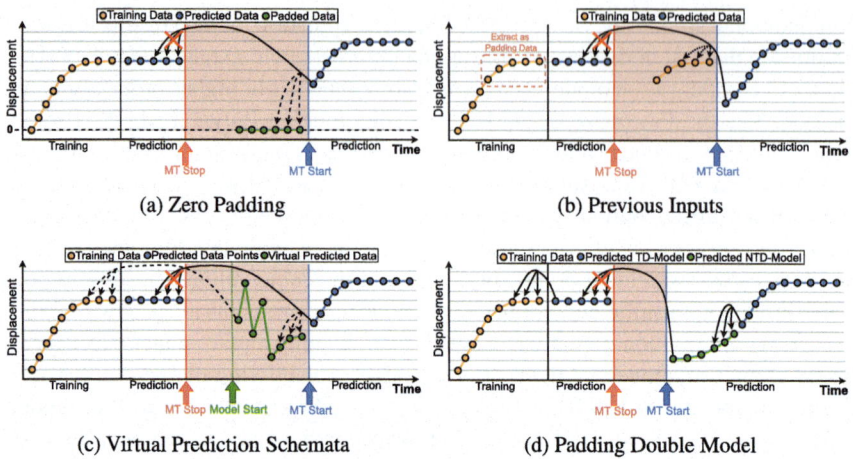

(a) Zero Padding

(b) Previous Inputs

(c) Virtual Prediction Schemata

(d) Padding Double Model

Fig. 5 Padding strategies for initializing time-dependent models

The initialization approaches developed are illustrated in Fig. 5. The first method, zero-padding, is shown in Fig. 5a. In this approach, the compensation model is initialized with zero values, assuming no thermal error at the start. The model input is also set to zero, representing no initial temperature difference. However, since a zero-value initialization may not accurately reflect all startup conditions, an alternative strategy modifies this approach by keeping zero padding for the thermal error data while using actual temperature deviations as inputs, as these values are typically continuously measured and may be available even if the machine is shut down. As zero padding may not always be suitable, another method, non-zero padding, uses previously trained thermal error values for initialization, as illustrated in Fig. 5b. These errors typically represent steady-state conditions. The model inputs remain as current temperature measurements, ensuring that the initialization aligns better with real operating conditions. Instead of explicitly defining padding values, another strategy, virtual prediction, allows the model to generate its own initial values by starting the compensation model before the MT begins operation. As shown in Fig. 5c, while the model's first predictions may be inaccurate due to improper initialization, it can self-correct after a few iterations. However, this method requires running the compensation model in advance and waiting for it to stabilize before starting the MT. The final approach eliminates the need for explicit padding or pre-initialization by using a non-time-dependent (NTD) model to initialize the final model in case it is a TD model. The FFNN, GPR and MLR models serve as the NTD models, trained on the same dataset to generate initial predictions until enough data is available for the TD compensation models to function effectively. This approach ensures a smoother transition during startup without requiring artificial initialization techniques.

Fig. 6 Agathon Evo Quinto
5-axis grinding MT. The
movement axes are
highlighted by the respective
A-, B- and C-rotational
arrow and the X- and Y-axis
linear arrow

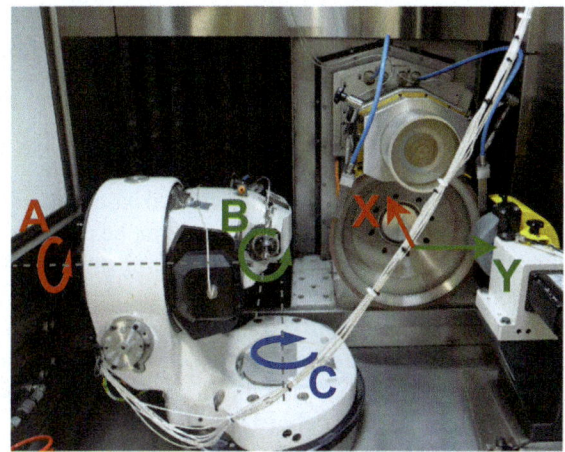

5   Measurement Setup

The investigated MT is a five-axis Agathon Evo Quinto grinding machine.
Figure 6 illustrates the workspace of the MT, highlighting the linear and rotary axes.
Its kinematic chain, according to ISO 10791-1 [10], is:

$$V\left[w - B' - A' - C' - b - Y - X - (A) - t\right] \tag{2}$$

To monitor the MTs thermal behavior, 13 temperature sensors are installed at critical points, including the spindle, linear and rotary axes, environment, and machine
bed. These sensors allow monitoring of the heat distribution and potential thermal
deformation that may affect machining accuracy and serve as the inputs for the compensation model and were positioned following the recommendations from Lang
et al. [18] which ensures a representative input set is available. A 3D touch trigger
probe is used in repeated measurement cycles to quantify the thermal displacement
of the workpiece holder.

For compensation, each measurement position is modeled and considered as one
model output. The volumetric error can be calculated as the Euclidean distance of
the average position drift, allowing one accuracy metric over all model outputs to
be used to asses the overall accuracy. Two key error metrics are used to evaluate
compensation accuracy: the root mean square error (RMSE) and peak-to-peak (PtP)
error. RMSE quantifies the average magnitude of error by computing the square root
of the mean of the squared differences between predicted and actual values. It is
particularly useful for assessing overall compensation performance by penalizing
larger errors more than smaller ones. RMSE is calculated as:

$$RMSE = \sqrt{\frac{1}{N} \sum_{i=1}^{N} (y_i - \hat{y}_i)^2}$$

where N is the number of samples, y_i represents the actual measured values of the volumetric error, and $\hat{y}_i$ represents the residual volumetric error. It has to be noted that the residual volumetric error is calculated based on each measurement positions error contribution as a volumetric error can not be used for compensation or comparison directly as it loses its directionality information it only asses the magnitude of the respective error.

The peak-to-peak (PtP) error represents the absolute difference between the maximum and minimum recorded errors, providing a measure of the worst-case scenario in thermal displacement, so the maximum thermally induced deviation within a production series. It is defined as:

$$PtP = \max(y_i) - \min(y_i)$$

A lower PtP error indicates reduced extreme deviations, which is crucial for precision machining. These metrics facilitate the evaluation of the effectiveness of different compensation and initialization strategies.

6   Results

This section introduces the results of both the different interpolation approaches as well as the different model initialization strategies for all different thermal error compensation model architectures.

6.1   Effects of Sampling Strategies on Thermal Compensation Models

The interpolation results in Fig. 7 illustrate the effect of sampling time and interpolation methods on the RMSE and PtP values. The red dashed lines represent the minimum and maximum values from the original sampling time as listed in Table 2, with a maximum RMSE of 63.5% and a maximum PtP of 58.1%. The solid red line represents the mean of the original sampling data. The boxplot shows the mean value of the respective metric as well as the highest and lowest values of all compensation models, so ARX, FFNN, LSTM and TCN, with this sampling strategy. The results in Table 1 show that the RMSE reaches its highest value with the TCN model, while the lowest RMSE is observed with the ARX model. Similarly, the PtP values are highest

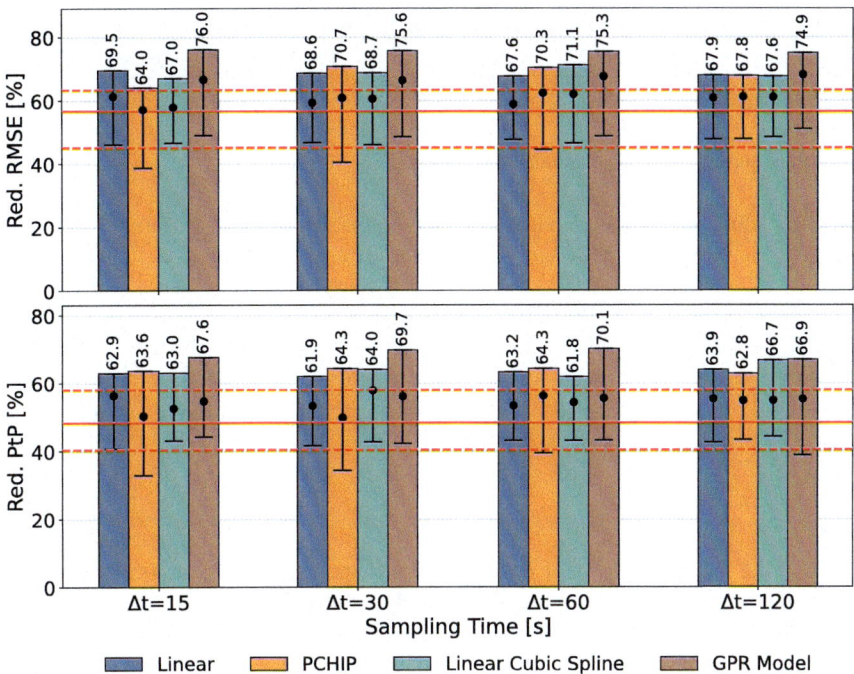

Fig. 7 Results of all sampling strategies for all models which are represented in the boxplot. The bar always shows the best performing model. The original sampling of only considering measurements is highlighted by the horizontal red lines

Table 1 Max RMSE and PTP values with the respective sampling time for each mode

Model	Sampling method	RMSE Value [%]	ΔT [s]	Sampling method	PtP value [%]	ΔT [s]
ARX	GPR model	50.9	120	GPR model	50.8	120
FFNN	GPR model	73.9	60	Linear cubic spline	64.0	30
LSTM	GPR model	75.6	30	Linear cubic spline	66.7	120
TCN	GPR model	76.0	15	GPR model	70.1	60

For the reduction metric only the resampled value closest in time to the actual measurement is considered

with the TCN model, while the lowest PtP value is recorded with the ARX model. In addition, the Linear Cubic Spline method was preferred for the LSTM and FFNN for resulting in the best PtP performance. In general, the GPR model is preferred as an interpolation method for all models, but the sampling time does not have such a generalizable impact on the overall compensation behavior when only considering the best model combinations. Figure 7 shows 30 and 60 s sampling times to slightly outperform 15 and 120 s sampling times indicating that the former is slightly too

Table 2 Original sampling intervals of only measurements

Original	ARX	FFNN	LSTM	TCN
RMSE [%]	45.5	57.9	60.3	63.5
PtP [%]	53.6	43.1	40.5	58.1

The results correspond to the horizontal red lines in Fig. 7

Table 3 Performance of compensation models for the selected interpolation strategy

Model	Comp RMSE [μm]	Uncomp RMSE [μm]	Total Red. RMSE [%]	Comp PtP [μm]	Uncomp PtP [μm]	Total Red. PtP [%]
FFNN	3.34	12.78	73.9	15.88	27.93	43.1
LSTM	3.55	12.78	72.2	10.90	27.93	61.0
TCN	3.15	12.78	75.3	8.36	27.93	70.1
ARX	6.57	12.78	48.6	14.70	27.93	47.4

fast and the latter too slow for optimal performance albeit only marginally. When compared to the original sampling time which is shown in the horizontal red lines in Fig. 7 and Table 2. Most of the RMSE values from using resampling strategies exceed the original maximum of 63.5% and most of the PtP values exceed the original maximum of 58.1%, with the exception of the ARX model in some cases. This confirms the almost universal benefit of resampling the measurement data on the overall compensation model performance. Especially the use of GPR model based resampling increased the performance and unlocked an even more fine-grained compensation of the thermal behavior even including thermal effects in between measurements that go beyond interpolation approaches (Table 3).

The outcomes of the TCN model, utilizing a 60-s sampling interval and GPR model-based resampling, are illustrated in Fig. 8. The compensation model was

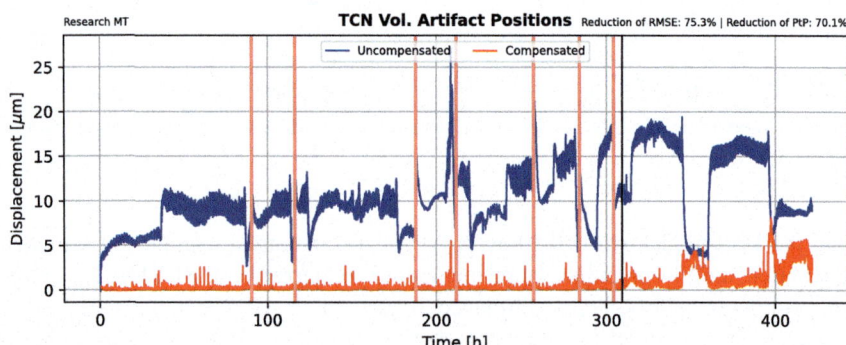

Fig. 8 Volumetric compensation results of the TCN model evaluated with GPR model-based resampling and 60 s sampling time. The red shaded are represent not to scale interruptions of the machine operation

trained on approximately 310 h of data, with the test dataset comprising a continuous, uninterrupted measurement spanning over 100 h. The vertical black line indicates the end of the training set and the subsequent start of the test set. The evaluation indicates a substantial reduction in volumetric error, achieving a 75.3% decrease in RMSE and a 70.1% reduction in PtP during the test set considering only the predictions closest in time to original measurements. However, when compared to the interpolated data fit at every resampled timestep, the RMSE increases to 81.8%, while the PtP remains nearly unchanged at 70%. However judging this accuracy is not possible as no measurements exist in between the original measurements.

One individual probing point direction, so a direct model output as compared to the aggregated volumetric error, as exemplified by the X_1 Inverse point, is shown in Fig. 9. For this specific position the reduction is even greater, with RMSE and PtP metrics decreasing by 86.4% and 80.6%, respectively considering only the points in time closest to the original measurements. The interpolated fits at this specific probing point, the RMSE and PtP increase to 90% and 88.9%, respectively.

In conclusion, when compared to the original sampling time, the majority of RMSE values surpass the initial maximum of 63.5%, and most PtP values exceed the original peak of 58.1%. The application of GPR model interpolation consistently enhances the performance across all evaluated compensation models. Moreover, an optimal sampling interval between 30 and 60 s is identified, leading to the most effective overall compensation model performance. This indicates the benefit of a physics inspired resampling process. It allows for much faster compensation compared to models being trained at their original measurement frequency as well as for uneven measurement times in the training data. At the same time the model performance increases due to the additional data availability significantly up to 16.5% percentage

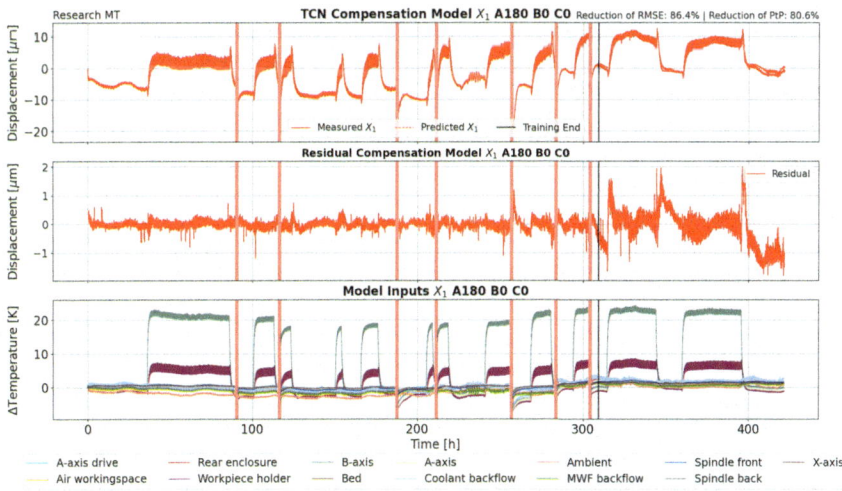

Fig. 9 Probing point of TCN model evaluated with GPR model interpolation and 60 s sampling time. The red shaded are represent not to scale interruptions of the machine operation

points increase in reduction for the RMSE and 12% percentage points increase in PtP reduction. This reduction holds generally for both only considering the original measurement data as well as the resampled data as comparison metric which generally tends to be slightly higher.

6.2   Initialization Effects on Compensation Accuracy

The different approaches for initializing TD compensation models are introduced in Sect. 4 and applied to the MTs measurement data. In order to have a comparable basis the best resampling strategy of GPR model-based interpolation at a 60 s interval is chosen as described in Sect. 6.1.

A different data set is used as the validation set should have many machine interruptions which require a re-initialization of the model which the previous data set had purposefully not to not mix the two effects of resampling and model initialization. So 130 h are used for model training while 50 h with 5 interruptions are used for validation.

Figure 10 shows the measured temperatures during the data set used to benchmark the different model initialization approaches. A very wide temperature range of more than 30 °C can be observed. Interruptions range from only around one hour to more than a month and are represented as a not to scale one hour interruption. They represent times the machine is not continuously in use such as during maintenance, set-up or during utilization below 100%. Figure 11 shows the volumetric error during the training phase and the corresponding residual. As the model initialization is not relevant during the model training but only the application there is no variation

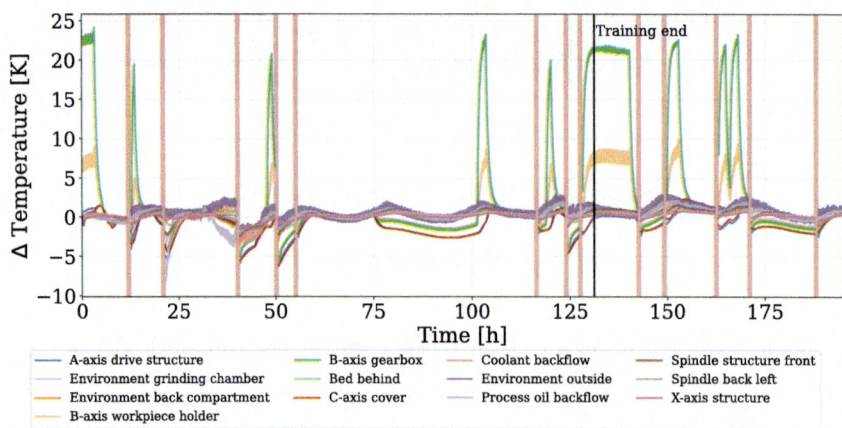

Fig. 10 Temperature measurements of the dataset used to train and evaluate models with varying initialization strategies. The red shaded are represent not to scale interruptions of the machine leading to a different thermal state after the next start

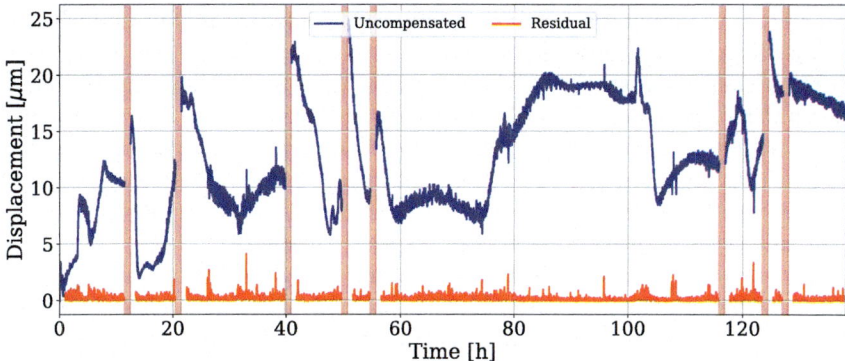

Fig. 11 Volumetric compensation of an ARX model during the training phase. A sampling time of 60 s and GPR model based interpolation to resample the training data is used. The red shaded are represent not to scale interruptions of the machine operation which do not effect the model training

between the different approaches here. A total range of up to 25 μm is observed that is reduced significantly by compensation. The residual is noise like with a couple of instantaneous outliers that are in the order of 2–3 μm which are likely due to irreproducible effects affecting the measurement.

In Fig. 12 the results for all TD compensation models with all initialization strategies is shown. The most effective initializing strategy resulted from starting with a NTD model, the padding model, before transitioning to a TD model after a sufficient number of predictions so approach (e), (f) and (g). All tested NTD models demonstrate comparable reductions in RMSE and PtP error. However, the best overall performance is achieved by the MLR model, primarily due to its minimal optimization requirements. In contrast, models such as feedforward neural networks (FFNN)

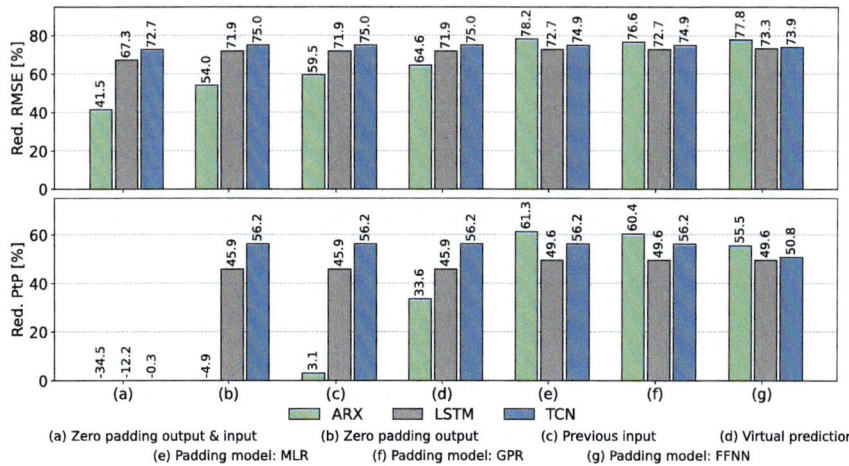

Fig. 12 Results for all model initialization strategies for the TD models. The y-axis does not show values below zero which are the case for some PtP metrics using approach (**a**) and (**b**)

and GPR have greater optimization potential and could surpass MLR with further tuning which was not carried out so far.

The initialization strategy using zero padding (a) does not lead to improvements across all models in the PtP metric compared to the uncompensated error. However, incorporating previous temperature sensors measurements as input data enhances performance, particularly for TD artificial neural networks (ANNs) in approach (b). The ARX model, on the other hand, continues to exhibit inferior performance due to its autoregressive component performing poorly when set to zero.

In strategy (c), which combines the last trained output with previous inputs, the ARX model shows slight improvement. This suggests that the initial state differs from the true starting condition, indicating that a nonzero initialization value is preferable. However, explicitly defining initialization values may not be a robust solution across different conditions. Instead, allowing the compensation model to begin earlier for calibration yields the second-best overall performance in approach (d) using virtual predictions.

As expected, LSTM and TCN initialization strategies (b), (c), and (d) yield identical results. This consistency arises because these models rely solely on past temperature data as TD model inputs: TCN strictly utilizes historical temperature values, while LSTM encodes past information implicitly in its hidden states, unlike ARX, which explicitly requires past output data as a model input.

The ARX model is particularly sensitive to the initialization strategy since it explicitly requires previous output and input data. Figure 13 shows the compensation results from the ARX model for all initialization strategies only on the validation

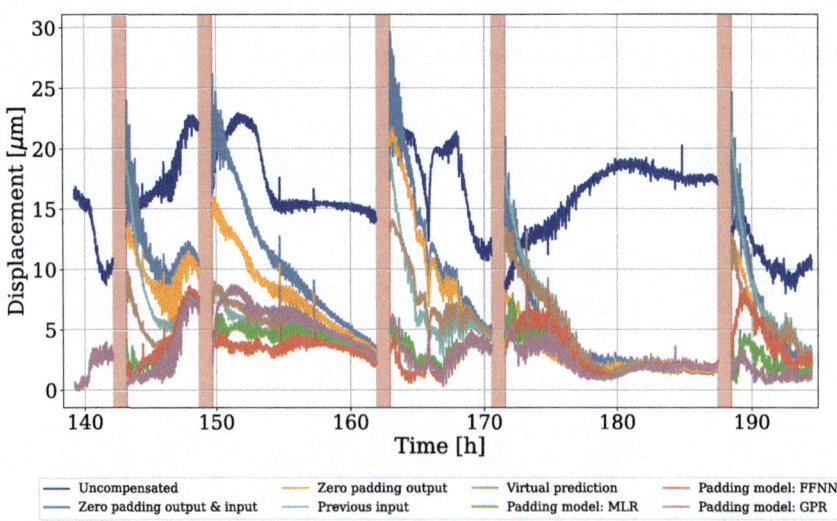

Fig. 13 Volumetric compensation of an ARX model using all initialization strategies applied on the validation data. A sampling time of 60 s and GPR model based interpolation to resample the training data is used. The red shaded are represent not to scale interruptions of the machine operation requiring a new initialization of the model

Table 4 Performance of NTD models which do not require initialization on the validation data

NTD models	Uncompensated	MLR	GPR	FFNN
RMSE [%]	16.45 μm	71.4	73.1	71.8
PtP [%]	22.02 μm	50.0	46.5	48.9

data. It can be observed that the initial behavior after a machine start is highly influenced by the initialization method, with the lowest peaks occurring when initialized with the NTD model. However, beyond the initial phase, the impact of the initialization strategy diminishes, and most approaches converge to similar performance levels indicating that the model is generally robust and converges to a stable state regardless of the initialization strategy. However the model error can be drastically improved during the critical warm up when a suitable initialization such as using virtual predictions or a second model is used. The residual model error during the training is currently the lower limit for model performance as it represents errors or slight time delays in the measurements itself that can not be explained by the model. The noise like structure indicates that this originates from measurement noise from the displacement and temperature measurements themselves and not the model architecture. Table 4 shows the comparison to the three NTD models applied to all of the experiment which therefore do not require any initialization and rather can be used as initialization for the TD models. The performance is generally good and better than some of the worse initialization strategy. However well initialized TD models generally outperform NTD models in both RMSE and P2P reduction by up to 10% points motivating the use of a suitable initialization strategy. For example the ARX model (e) reduced the volumetric RMSE error from 16.45 to 3.59 μm a 78.2% reduction and the PtP error from 22.02 to 8.53 μm by 61.3%.

7 Conclusion and Outlook

This work introduces a new approach for data-efficient thermal error compensation. Initialization strategies play a crucial role in ensuring the effectiveness of the typically beneficial time-dependent thermal compensation models, particularly in the critical start-up phase. The results indicate that the two model approach of using NTD models and subsequently the TD model offers an effective initialization strategy, as the startup is also less affected by thermal hysteresis. The RMSE and PtP errors can be reduced significantly compared to zero-padding approaches reducing the machine error reliably around 75% from 16.45 to 3.59 μm. Among the various initialization techniques, the virtual prediction method and the combination of TD models with NTD models provide the best compensation accuracy at no impact to machine operation and start-up.

Furthermore, model-based interpolation methods, using GPR, enhance compensation performance by leveraging additional temperature measurements in between displacement measurement cycles, leading to more precise models. Also linear cubic spline based resampling performed well. For the investigated MT the optimized sampling interval lies between 30 and 60 s improving model efficiency without excessive computational overhead and is also applicable to non uniformly sampled training data.

Future research should investigate the applicability of these strategies across different MT types under varying operating conditions. Fleet learning could enable models to be trained on data from multiple machines, improving generalizability and adaptability. Another promising research direction is enhancing the measurement process by both optimizing the measurement cycle as well as potentially incorporating continuous measurements. This approach could provide a more comprehensive understanding of thermal errors between standard measurements while minimizing the thermal effects introduced by the measurement process itself, ultimately improving accuracy.

Acknowledgements This work was co-financed by Innosuisse—Swiss Innovation Agency, Project no. 104.774 IP-ENG. The authors wish to thank Michel Roettig and Christian Zenger for their assistance with MT operations as well as Frank Egger and Michael Flück for technical support.

CRediT authorship contribution statement Sebastian Lang: Conceptualization, Data curation, Investigation, Software, Writing—original draft. Mario Zorzini: Data curation, Software, Writing—review & editing. Markus Sprecher: Validation, Writing—review & editing. Stephan Scholze: Funding acquisition, Project administration, Writing—review & editing. Markus Bambach: Supervision, Writing—review & editing.

Declaration of competing interest The authors declare that they have no known competing financial interests or personal relationships that could have appeared to influence the work reported in this paper.

References

1. Abdulshahed, A.M., Longstaff, A.P., Fletcher, S., Myers, A.: Thermal error modelling of machine tools based on ANFIS with fuzzy c-means clustering using a thermal imaging camera. Appl. Math. Model. **39**(7), 1837–1852 (2015). https://doi.org/10.1016/j.apm.2014.10.016
2. Akiba, T., Sano, S., Yanase, T., Ohta, T., Koyama, M.: Optuna: A Next-generation Hyperparameter Optimization Framework. In: Proceedings of the ACM SIGKDD International Conference on Knowledge Discovery and Data Mining, pp. 2623–2631 (2019). https://doi.org/10.1145/3292500.3330701
3. Bitar-Nehme, E., Mayer, J.R.: Modelling and compensation of dominant thermally induced geometric errors using rotary axes' power consumption. CIRP Ann. **67**(1), 547–550 (2018). https://doi.org/10.1016/j.cirp.2018.04.080
4. Blaser, P., Pavlíček, F., Mori, K., Mayr, J., Weikert, S., Wegener, K.: Adaptive learning control for thermal error compensation of 5-axis machine tools. J. Manuf. Syst. **44**, 302–309 (2017). https://doi.org/10.1016/j.jmsy.2017.04.011

5. Chen, Q., Mei, X., He, J., Yang, J., Liu, K., Zhou, Y., Ma, C., Liu, J., Zeng, S., Zhang, L., Gui, H., Zhou, J., Weng, S.: Modeling and compensation of small-sample thermal error in precision machine tool spindles using spatial–temporal feature interaction fusion network. Adv. Eng. Inform. **62**(PB), 102741 (2024). https://doi.org/10.1016/j.aei.2024.102741

6. Denkena, B., Bergmann, B., Klemme, H.: Cooling of motor spindles—a review. Int. J. Adv. Manuf. Technol. **110**(11–12), 3273–3294 (2020). https://doi.org/10.1007/s00170-020-06069-0

7. Fujishima, M., Narimatsu, K., Irino, N., Ido, Y.: Thermal displacement reduction and compensation of a turning center. CIRP J. Manuf. Sci. Technol. **22**, 111–115 (2018). https://doi.org/10.1016/j.cirpj.2018.04.003

8. Gao, W., Ibaraki, S., Donmez, M.A., Kono, D., Mayer, J.R., Chen, Y.L., Szipka, K., Archenti, A., Linares, J.M., Suzuki, N.: Machine tool calibration: measurement, modeling, and compensation of machine tool errors. Int. J. Mach. Tools Manuf. **187** (2023). https://doi.org/10.1016/j.ijmachtools.2023.104017

9. Hsieh, T.H., Lai, H.Y., Chou, Y.H., Wu, T.H.: Differences in applications of various neural network algorithms for thermal error compensation. Sens. Mater. **36**(8), 3573–3594 (2024). https://doi.org/10.18494/SAM4747

10. International Organization for Standardization (ISO): Test conditions for machining centers—Part 1: geometric tests for machines with horizontal spindle (horizontal Z-axis). ISO 10791-1 (148902), 8–30 (2015)

11. Irino, N., Kobayashi, A., Shinba, Y., Kawai, K., Spescha, D., Wegener, K.: Digital twin based accuracy compensation. CIRP Ann. **72**(1), 345–348 (2023). https://doi.org/10.1016/j.cirp.2023.04.088

12. Kaftan, P., Porquez, F., Mayr, J., Pomodoro, K., Keel, M., Trombert, D., Wegener, K.: Thermal error measurement and compensation with torque limit skip in Swiss-type lathe manufacturing. Precis. Eng. **88**, 315–323 (2024). https://doi.org/10.1016/j.precisioneng.2024.01.024

13. Kizaki, T., Tsujimura, S., Marukawa, Y., Morimoto, S., Kobayashi, H.: Robust and accurate prediction of thermal error of machining centers under operations with cutting fluid supply. CIRP Ann. **70**(1), 1–4 (2021). https://doi.org/10.1016/j.cirp.2021.04.074

14. Kondo, R., Kono, D., Matsubara, A.: Evaluation of machine tool spindle using carbon fiber composite. Int. J. Autom. Technol. **14**(2), 294–303 (2020). https://doi.org/10.20965/ijat.2020.p0294

15. Lang, S., Lampert, N., Mayr, J., Wegener, K., Bambach, M.: Training efficient and compensating fast : data augmentation for thermal error compensation models of machine tools . In: euspen Special Interest Group meeting on Thermal Issues, March. cuspen, Eindhoven (2024). https://doi.org/10.3929/ethz-b-000669528

16. Lang, S., Talleri, S., Mayr, J., Wegener, K., Bambach, M.: Kalman filter-driven state observer for thermal error compensation in machine tool digital twins. Manuf. Lett. **41**, 208–218 (2024). https://doi.org/10.1016/j.mfglet.2024.09.025

17. Lang, S., Zimmermann, N., Mayr, J., Wegener, K., Bambach, M.: Thermal error compensation models utilizing the power consumption of machine tools. In: International Conference on Thermal Issues in Machine Tools, vol. 3, pp. 41–53. Springer, Berlin (2023). https://doi.org/10.1007/978-3-031-34486-2_4

18. Lang, S., Zorzini, M., Scholze, S., Mayr, J., Bambach, M.: Sensor placement utilizing a digital twin for thermal error compensation of machine tools. J. Manuf. Syst. **80**, 243–257 (2025). https://doi.org/10.1016/j.jmsy.2025.03.003

19. Liu, J., Ma, C., Gui, H., Wang, S.: Thermally-induced error compensation of spindle system based on long short term memory neural networks. Appl. Soft Comput. **102**, 107094 (2021). https://doi.org/10.1016/j.asoc.2021.107094

20. Mareš, M., Horejš, O., Havlík, L.: Thermal error compensation of a 5-axis machine tool using indigenous temperature sensors and CNC integrated Python code validated with a machined test piece. Precis. Eng. **66**, 21–30 (2020). https://doi.org/10.1016/j.precisioneng.2020.06.010

21. Maurya, S.N., Luo, W.J., Panigrahi, B., Negi, P., Wang, P.T.: Input attribute optimization for thermal deformation of machine-tool spindles using artificial intelligence. J. Intell. Manuf. (2024). https://doi.org/10.1007/s10845-024-02350-1

22. Mayr, J., Blaser, P., Ryser, A., Hernandez-Becerro, P.: An adaptive self-learning compensation approach for thermal errors on 5-axis machine tools handling an arbitrary set of sample rates. CIRP Ann. **67**(1), 551–554 (2018). https://doi.org/10.1016/j.cirp.2018.04.001

23. Mayr, J., Jedrzejewski, J., Uhlmann, E., Alkan Donmez, M., Knapp, W., Härtig, F., Wendt, K., Moriwaki, T., Shore, P., Schmitt, R., Brecher, C., Würz, T., Wegener, K.: Thermal issues in machine tools. CIRP Ann. **61**(2), 771–791 (2012). https://doi.org/10.1016/j.cirp.2012.05.008

24. Singh, A.K., Kumar, A., Sharma, V., Kala, P.: Sustainable techniques in grinding: state of the art review. J. Clean. Prod. **269**, 121876 (2020). https://doi.org/10.1016/j.jclepro.2020.121876

25. Tan, B., Mao, X., Liu, H., Li, B., He, S., Peng, F., Yin, L.: A thermal error model for large machine tools that considers environmental thermal hysteresis effects. Int. J. Mach. Tools Manuf **82–83**, 11–20 (2014). https://doi.org/10.1016/j.ijmachtools.2014.03.002

26. Wegener, K., Bleicher, F., Krajnik, P., Hoffmeister, H.W., Brecher, C.: Recent developments in grinding machines. CIRP Ann. Manuf. Technol. **66**(2), 779–802 (2017). https://doi.org/10.1016/j.cirp.2017.05.006

27. Wei, X., Ye, H., Miao, E., Pan, Q.: Thermal error modeling and compensation based on Gaussian process regression for CNC machine tools. Precis. Eng. **77**, 65–76 (2022). https://doi.org/10.1016/j.precisioneng.2022.05.008

28. Weng, L., Gao, W., Zhang, D., Huang, T., Liu, T., Li, W., Zheng, Y., Shi, K., Chang, W.: Analytical modelling method for thermal balancing design of machine tool structural components. Int. J. Mach. Tools Manuf **164**, 103715 (2021). https://doi.org/10.1016/j.ijmachtools.2021.103715

29. Zimmermann, N., Lang, S., Blaser, P., Mayr, J.: Adaptive input selection for thermal error compensation models. CIRP Ann. **69**(1), 485–488 (2020). https://doi.org/10.1016/j.cirp.2020.03.017

30. Zimmermann, N., Lang, S., Mayr, J., Wegener, K.: Validating real time compensation: a thermal test piece for 5-axis machine tools to separate thermal errors in Z-direction. Precis. Eng. **91**, 263–277 (2024). https://doi.org/10.1016/j.precisioneng.2024.08.014

31. Zimmermann, N., Müller, E., Lang, S., Mayr, J., Wegener, K.: Thermally compensated 5-axis machine tools evaluated with impeller machining tests. CIRP J. Manuf. Sci. Technol. **46**, 19–35 (2023). https://doi.org/10.1016/j.cirpj.2023.07.005

Temperature Field Prediction and Convection Coefficient Estimation from Temperature Data Using PINNs

Sergio Garcia-Ferreira and Gorka Aguirre

Abstract Thermal deformation is a critical factor affecting the precision of machine tools, requiring accurate thermal modeling to predict temperature fields and thermal parameters. Traditional approaches, such as the Finite Element Method (FEM), require well-defined boundary conditions, which are often unknown or difficult to measure in real machining environments. This paper explores the use of Physics-informed neural networks (PINNs) as an alternative method for solving steady-state heat conduction problems in two dimensions. PINNs integrate sparse sensor data with physical laws, enabling temperature field prediction and convection coefficient estimation without the need for fully specified boundary conditions. We evaluate five different PINN models, varying the balance between data-driven and physics-informed constraints. Results show that enforcing the heat equation alone yields high accuracy in temperature prediction, but accurate convection coefficient estimation requires explicit enforcement of convection conditions. While PINNs successfully infer missing parameters, their sensitivity to temperature gradients can impact accuracy. Additionally, the need for retraining PINNs when conditions change limits real-time applicability, making them more suitable for offline thermal analysis rather than adaptive modeling in dynamic environments.

Keywords Physics-informed neural networks · Inverse problem · Field prediction · Parameter estimation · Heat equation · Thermal error

1 Introduction

Thermal deformation is a major source of precision errors in machine tools, significantly affecting their accuracy and performance. Temperature variations within the machine structure lead to uneven thermal expansion, which causes inconsistent

S. Garcia-Ferreira (✉) · G. Aguirre
IDEKO Member of Basque Research and Technology Alliance, Elgoibar, Spain
e-mail: sgarciaferreira@ideko.es

© The Author(s) 2026
K. Wegener and M. Bambach (eds.), *4th International Conference on Thermal Issues in Machine Tools (ICTIMT2025)*, Lecture Notes in Production Engineering,
https://doi.org/10.1007/978-3-032-01194-7_36

machining results. As modern manufacturing demands increasingly tighter tolerances, accurate thermal modeling has become essential for ensuring high precision
and reliability in machine tool operations.

Accurately modeling the thermal state of a machine tool is challenging, especially
when relying on limited sensor data. While temperature sensors provide valuable
pointwise measurements, they offer only partial information about the overall temperature distribution within the machine structure. Because heat transfer mechanisms
involve conduction through solid components and convection to the surrounding
air, determining the full thermal field from sparse sensor data requires methods that
account for these interactions beyond simple interpolation or extrapolation.

A common approach to estimating the full thermal field is through Finite Element
Method (FEM) simulations, which numerically solve the heat conduction equation
across the machine structure. However, FEM relies on a fully defined set of boundary
conditions, including heat fluxes, convection coefficients, and ambient temperatures.
In real-world machining environments, these parameters are often partially known
or difficult to measure with high accuracy, which can limit the reliability of the
simulation results.

An alternative approach to thermal modeling is Physics-informed neural networks
(PINNs), which combine data-driven learning with physical laws to approximate
the temperature field without requiring a fully defined set of boundary conditions.
Unlike traditional neural networks, PINNs incorporate the governing heat equation
and boundary conditions directly into the loss function, ensuring that the learned
solution remains consistent with the underlying physics. This makes them particularly well-suited for scenarios where only partial or sparse sensor data is available,
as they can infer the missing information by leveraging prior physical knowledge.

PINNs were first introduced as a neural network-based approach for solving differential equations in the 1990s [1, 2]. However, their modern form was popularized
by Raissi et al. [3], demonstrating their capability to effectively solve PDEs in both
forward and inverse settings. PINNs have since been widely adopted in various scientific and engineering applications, including fluid dynamics, solid mechanics, power
systems, and heat transfer [4–7].

The objective of our research is to assess the capability of PINNs in predicting
the temperature field across the domain under realistic conditions, where data is
scarce, and the underlying physics is only partially known. Additionally, we aim to
evaluate the performance of PINNs in estimating the heat convection coefficient to
the surrounding air. A key aspect of our study is to analyze how sensitive PINNs are
to the balance between physical knowledge and available data. This approach allows
us to quantify the strengths and limitations of PINNs in thermal modeling and to
determine the conditions under which they provide the most reliable results.

Our research is structured as follows: In Sect. 2, we define PINNs for both
forward and inverse problems. In Sect. 3, we introduce the heat transfer problem
and detail the methodology used in this study. The results, including temperature
prediction and convection coefficient estimation, are then presented and discussed
in Sect. 4. Finally, Sect. 5 concludes the paper by summarizing key insights and
offering recommendations for future research.

2 Physics-Informed Neural Networks

Physics-informed neural networks (PINNs) are a class of deep learning models that integrate physical laws into the training process to approximate the solution of partial differential equations (PDEs). Unlike traditional data-driven neural networks that require big amount of data, PINNs leverage governing equations—such as conservation laws or heat transfer equations—to learn solutions in a physics-consistent manner. This makes them particularly valuable for solving inverse and forward problems where experimental data may be sparse or expensive to obtain.

Forward problem PINNs are used as PDE solvers for well-posed problems where initial and boundary conditions are fully defined. Generally, PDEs can be written as

$$\begin{cases} \mathcal{N}[u(x; \alpha)] = 0, \, x \in \Omega \\ B[u(x; \alpha)] = 0, \quad x \in \partial\Omega \end{cases} \tag{1}$$

where $\mathcal{N}[\cdot]$ is a linear or nonlinear differential operator, and $B[\cdot]$ is a boundary operator such as Dirichlet, Neumann, Robin or periodic conditions. In forward problems, α parameters are known and no data is needed to find the solution. The time dimension is not explicitly written, but in transient problems x can include spatial and temporal coordinates. The unknown solution $u(x)$ is approximated by a fully connected neural network with L layers, defined as:

$$u_\theta(x) := W_L \cdot \sigma(\cdots(W_1 \cdot \sigma(W_0 x + b_0) + b_1)\cdots) + b_L \tag{2}$$

where W_i and b_i are the weight matrices and bias vectors of the i-th hidden layer, respectively. The activation function σ introduces non-linearity to the network and is typically chosen as a smooth function, such as the hyperbolic tangent or sine function. The collection of all trainable parameters, including weights and biases, is denoted as θ.

Solving the PDE is then equivalent to finding the values of θ that make $u_\theta(x)$ approximate $u(x)$ with a certain accuracy. To achieve this, a loss function is defined as

$$Loss = \frac{1}{N_r} \sum_{i=1}^{N_r} (\mathcal{N}[u_\theta(x_i)])^2 + \lambda \frac{1}{N_b} \sum_{i=1}^{N_b} (B[u_\theta(x_i)])^2 \tag{3}$$

where N_r and N_b are the number of residual points sampled within the domain Ω and on its boundary $\partial\Omega$, respectively. The first term enforces the PDE, while the second term enforces the boundary conditions. The hyperparameter λ balances the contribution of each term. The training process consists of minimizing this loss function by adjusting the neural network parameters θ, ensuring that $u_\theta(x)$ satisfies the given PDE and boundary conditions.

Inverse problem In addition to solving PDEs with known boundary conditions, PINNs can also address inverse problems where some boundary conditions or parameters are unknown. Instead of specifying these values explicitly, data collected from observations is incorporated into the loss function to infer them. A general inverse problem involves estimating unknown inputs or parameters that appear in the PDE or in the initial or boundary conditions.

To estimate these α unknown parameters, PINNs incorporate measurement data into the loss function. Let $\mathcal{D} = \{(x_i, u_i)\}_{i=1}^{N_d}$ be a set of observed data points where the solution is known. The modified loss function is:

$$
\begin{aligned}
Loss = \frac{1}{N_r} \sum_{i=1}^{N_r} (\mathcal{N}[u_\theta(x_i; \alpha)])^2 \\
+ \lambda_b \frac{1}{N_b} \sum_{i=1}^{N_b} (B[u_\theta(x_i; \alpha)])^2 \\
+ \lambda_d \frac{1}{N_d} \sum_{i=1}^{N_d} (u_\theta(x_i; \alpha) - u_i)^2
\end{aligned}
\tag{4}
$$

Same as before, the first term enforces the PDE, while the second term enforces the boundary conditions. The last term ensures that the predicted solution $u_\theta(x_i; \alpha)$ matches the observed data u_i at the measurement points, allowing the model to infer unknown parameters based on available observations. The unknown parameters α are treated as additional trainable variables alongside the neural network weights and biases. The weights λ_b and λ_d control the influence of the boundary conditions and data during training.

3 Problem Formulation and Methodology

For an initial approach to PINNs, we consider a two-dimensional steady-state heat conduction problem with a combination of convection, fixed temperature, and adiabatic boundary conditions, as defined in Eq. (5). No internal heat generation is considered. The problem dimensions follow the NAFEMS benchmark T4 [8], where the domain Ω is a 0.6×1 m rectangle. The boundary conditions are as follows: convection to an ambient temperature of 0 °C is applied on the right (BC) and top (CD) boundaries, a fixed temperature of 100 °C is imposed on the bottom boundary (AB), and a zero heat flux (adiabatic condition) is enforced on the left boundary (DA) (Fig. 1, left). The material has a thermal conductivity of 52 W/(m°C), and the convective heat transfer coefficient on the right and top boundaries is 750 W/(m^2°C). Finite Element Method (FEM) is used to obtain a reference numerical solution (Fig. 1, right).

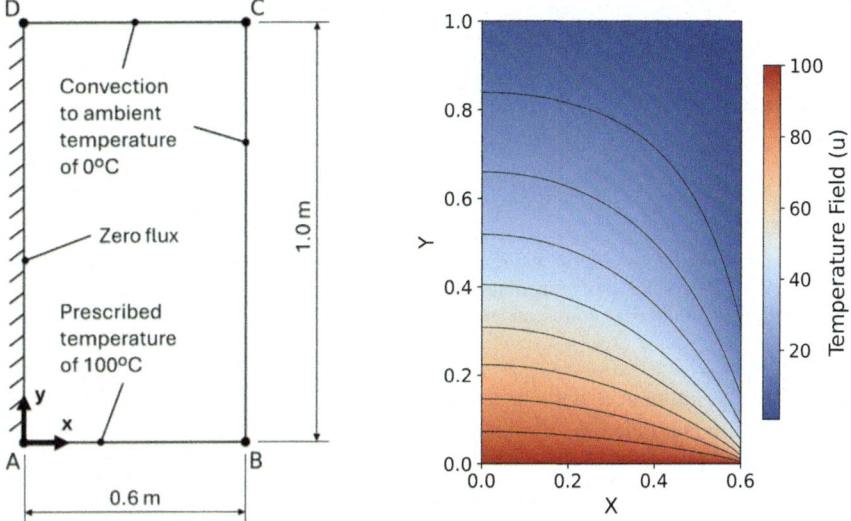

Fig. 1 Left: Problem definition based on the NAFEMS benchmark T4 [8]. Right: FEM reference solution for the temperature field, with isothermal lines displayed

$$\begin{cases} \frac{\partial^2 u}{\partial x^2} + \frac{\partial^2 u}{\partial y^2} = 0, & \forall (x, y) \in \Omega \\ k\frac{\partial u}{\partial n} = -h(u - u_\infty), & \forall (x, y) \in \Gamma_{BCD} \\ u = u_{cte}, & \forall (x, y) \in \Gamma_{AB} \\ \frac{\partial u}{\partial n} = 0, & \forall (x, y) \in \Gamma_{DA} \end{cases} \tag{5}$$

3.1 Methodology

Consider a solid body, such as a machine column or a spindle, where we aim to measure its temperature. Sensors are placed on its surface to collect temperature data at specific locations. However, the temperature distribution within the rest of the body remains unknown. In this study, we use this surface temperature data to predict the temperature field throughout the entire body and to estimate the convection coefficient to air.

The data used in this study is obtained from simulations. Equation (5) is solved using an in-house finite element software, and the solution is validated against the NAFEMS benchmark [8]. FEM data is used to simulate sensor measurements (training data) and to evaluate the models after training (test data). We choose six equidistant data points per side to emulate sensors (Fig. 2).

We use PINNs to approximate the solution as u_θ, leveraging a hybrid approach that integrates prior physical knowledge of the system with data. We analyze how

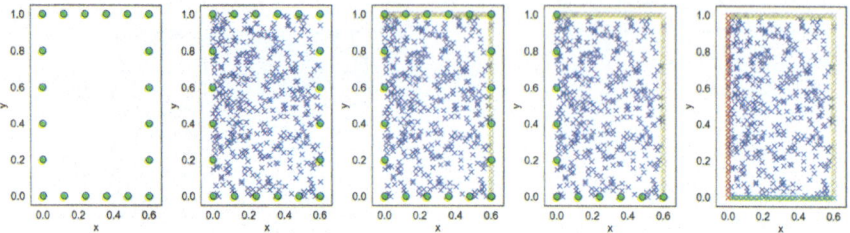

Fig. 2 Data points (green circles) and residual points (colored × markers) distribution: six data points per side, 400 residual points distributed inside the domain and 40 residual points per boundary. The images from left to right correspond to models 1 through 5

this knowledge influences both the temperature field prediction and the estimation of the convection coefficient h. To this end, we propose five different models based on the amount of physics and data incorporated during the neural network training. In the first three models, h is unknown, whereas in the last two models, h is known. The models range from minimal to maximal physics integration: (1) temperature data is available only at boundary data points, with no prior physics knowledge incorporated; (2) same as the previous model, but now the heat PDE is enforced over the domain; (3) same as the previous model, with the addition of convection at the right and top boundaries, though h remains uncertain; (4) same as the previous model, but h is now known; (5) the PDE and all boundary conditions are fully defined. Table 1 provides an overview of the different models and their corresponding physics and data integration levels.

The PINN architecture remains the same for every model: a four-hidden-layer neural network with 20 neurons per layer and a hyperbolic tangent activation function. As there is no clear consensus on the ideal architecture for PINNs, we followed the recommendations in [9], where networks with depths of 3–6 layers and widths of 128–512 neurons are suggested. However, through empirical tuning, we observed that wider architectures did not lead to significant improvements in accuracy or convergence for our problem. Training is performed for up to 10^5 iterations, with a convergence tolerance of 10^{-6}, which serves as the stopping criterion. To evaluate the PDE, 400 residual points are randomly distributed within the domain, while 40 equidistant residual points per side are used to enforce boundary conditions (Fig. 2).

Table 1 Overview of models acc. to physics and data integration

Model No	Physics integration	Data integration
Model 1	No physics	Boundary data
Model 2	Heat PDE	Boundary data
Model 3	Heat PDE + convection BC (h unknown)	Boundary data
Model 4	Heat PDE + convection BC (h known)	Left and bottom data
Model 5	Heat PDE + all BC	No data

Each layer applies a linear transformation, followed by the continuously differentiable $\tanh$ activation function. The choice of $\tanh$ is motivated by its prevalence in the PINN literature. Other popular activation functions include sinusoidal functions and GeLU, while ReLU is generally avoided due to its zero second-order derivative, which saturates the computation of PDE residuals [9]. We leverage automatic differentiation to compute the equation residuals, access the derivatives of u_θ, evaluate the heat flux, and estimate the convection coefficient h in models 1 and 2.

Temperature Field Prediction We use simulation data to evaluate the models at the finite element nodes, totaling 7177 test samples. Model accuracy is defined as the proportion of test samples where the prediction error is below 1% of the temperature range. Since the temperature varies from 0 to 100 °C, the accuracy threshold is 1 °C.

Convection Coefficient Estimation In models 1 and 2, the convection coefficient h is estimated after training the neural network. To do this, we compute the heat flux at the convection boundary points using automatic differentiation:

$$q_i = -k \frac{\partial u_\theta(x_i, y_i)}{\partial n}, \quad \forall (x_i, y_i) \in \Gamma_{BCD} \tag{6}$$

where q_i [W/m^2] represents the outward heat flux at each boundary point i. Recalling the convection condition from Eq. (5), we can express it in the form of a linear system $\Delta u_\theta h = q$, where $\Delta u_\theta \equiv [(u_\theta - u_\infty)_i]$ and $q \equiv [q_i]$ are one-dimensional vectors containing the temperature differences between the boundary and the surrounding air, as well as the heat fluxes, respectively, for every boundary point i. Once Δu_θ and q are computed, the convection coefficient h is obtained by solving a least-squares problem.

In model 3, h is not estimated post-training but is instead treated as a learnable parameter, optimized alongside the neural network's trainable parameters θ (i.e., weights and biases). Finally, in models 4 and 5, h is already known and does not need to be estimated.

Sensitivity Analysis on Temperature Data We have also investigated how the number of temperature data points—interpreted as simulated sensor data—affects the accuracy of both the predicted temperature field and the estimated convection coefficient h. To conduct this study, we trained each model with varying amounts of temperature data. Specifically, we considered four cases with different numbers of data points per side: $\{10, 8, 6, 4\}$. Additionally, we evaluated the effect of using $\{400, 100, 40\}$ residual points to enforce the PDE. This resulted in a total of 12 different configurations for each model, except for model 1, which was tested with 4 configurations due to the absence of PDE residuals in its formulation (Table 2). To enforce the boundary conditions, the number of residual points assigned per boundary side was set to one-tenth of the total residual points used in the domain.

Table 2 Overview of cases acc. to residual and data points

Model #	Model description	Residual points	Data points
Model 1	Data	–	$\{10, 8, 6, 4\} \times 4$ sides
Model 2	PDE + data	$\{400, 100, 40\}$	$\{10, 8, 6, 4\} \times 4$ sides
Model 3	PDE + conv(h) + data	$\{400, 100, 40\}$	$\{10, 8, 6, 4\} \times 4$ sides
Model 4	PDE + conv + data	$\{400, 100, 40\}$	$\{10, 8, 6, 4\} \times 2$ sides

4   Results and Discussion

In this section, we evaluate the performance of PINNs in predicting the temperature field and estimating the convection coefficient h. First, we assess the accuracy of temperature predictions across five different models, examining how incorporating physical constraints influences the solution quality. We then analyze the estimation of the convection coefficient, comparing different approaches to determining h and identifying factors that affect its accuracy.

4.1   Temperature Field Prediction

We assess the temperature prediction performance of the five models outlined in the previous section. Since only boundary data is available for training in model 1, it functions as a traditional neural network rather than a PINN. The model achieves an accuracy of 23.98%, meaning that only 1 in 4 nodes within the domain is predicted with an error below 1 °C. As expected, the most accurate predictions are concentrated near the boundaries, where the training data is provided (Fig. 3, first row). However, the model fails in the interpolation throughout the domain. Introducing prior physics in model 2, by enforcing the heat PDE, significantly improves accuracy to 98.69% (Table 3). Predictions strongly match the reference solution and both temperature fields are barely differentiable. There is a tiny region, though, near point B (lower right corner), exhibiting a high error up to 13.18 °C. This is most likely due to steep temperature gradients and is a challenge in PINNs that is already being addressed in literature through residual point resampling [10] and weight adaptation [11, 12].

Further refining the model, in model 3 the convection boundary conditions are incorporated while treating h as a learnable parameter. This results in an accuracy of 98.65%, the same as in the previous model. When h is known and explicitly enforced in model 4, accuracy drops to 95.92%. Finally, in model 5, where all boundary conditions are fully defined and no training data is used, the PINN operates as a pure forward solver, achieving an accuracy of 96.06% (Fig. 4).

With this numbers in hand, it seems that adding more physics does not necessarily improve accuracy. However, even though the accuracy in model 5 is slightly lower than in models 2 and 3, the high-error region near point B has almost disappeared

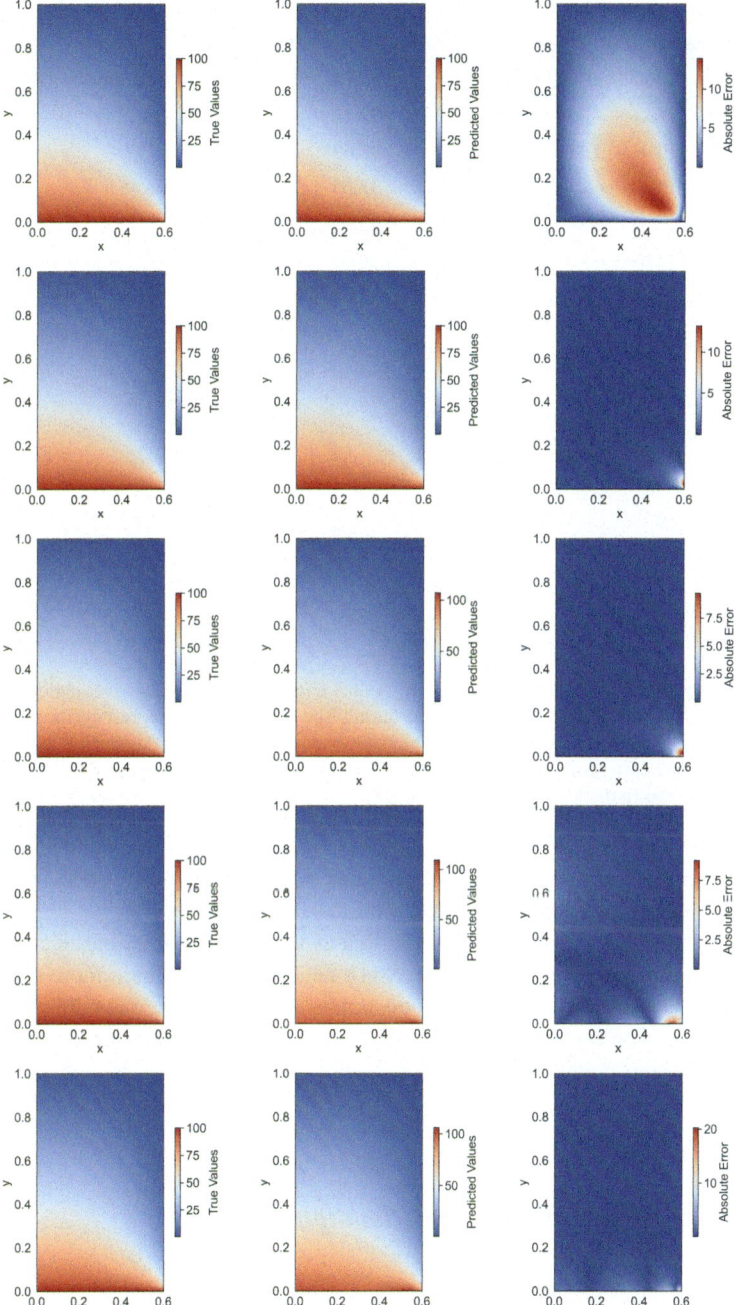

Fig. 3 Temperature field predictions for each model. The first row corresponds to model 1, and the last row to model 5. The left column presents the true temperature field from the FEM simulation, the middle column shows the PINN-predicted temperature field, and the right column displays the absolute error

Table 3 Accuracy in u temperature prediction and h estimation. The middle column percentages indicate what portion of the test dataset fell within a certain accuracy

Model #	Model description	u pred. RE < 1 (%)	h estim. accuracy (%)
Model 1	Data	23.98	18.92
Model 2	PDE + data	**98.69**	51.00
Model 3	PDE + conv(h) + data	98.65	**98.45**
Model 4	PDE + conv + data	95.91	–
Model 5	PDE + conv + tcte + adiab	96.06	–

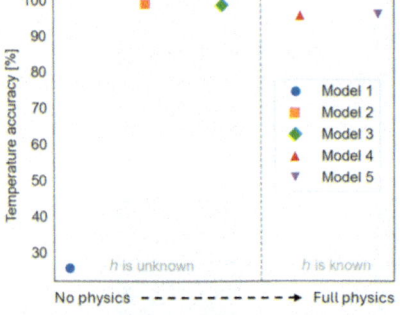
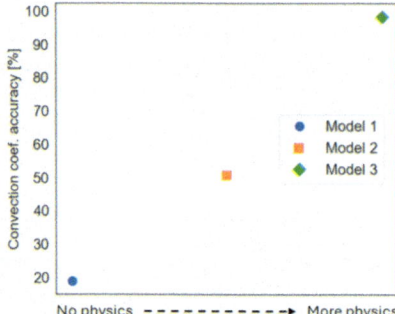

Fig. 4 Temperature prediction accuracy (left) and convection coefficient estimation accuracy (right) for different physics-informed configurations. The horizontal axis represents the inclusion of physics-based constraints, ranging from no physics to more/full physics. The vertical axis indicates temperature and convection coefficient prediction accuracy. Different markers represent different models, as shown in the legend

(Fig. 3, last row), indicating that enforcing all boundary conditions leads to a more physically consistent solution.

On the other hand, we must consider that as we added physics-based constraints to the models, we also removed data. Since neural networks learn more efficiently from data than from equation-based constraints, we expect that extending the PINN training time would eventually improve accuracy. Additionally, further tuning the loss term weights could help balance the contribution of the physics- and data-driven terms.

4.2   Convection Coefficient Estimation

We evaluate the performance of convection coefficient estimation for the first three models, where h is treated as an unknown parameter. As described in Sect. 3.1, in models 1 and 2, h is determined *post-training*, whereas in model 3, h is learned *during* training. The accuracy metrics indicate that model 3 is the only one that correctly predicts h, achieving an estimation accuracy of 98.45%, while models 1 and 2 fail to provide accurate estimations (Table 3).

To further analyze this behavior, we compute the estimated heat flux through the convective boundary and plot it against the temperature difference between the boundary and the surrounding air. According to convection condition in Eq. 5, these two variables should exhibit a linear relationship defined by h. Figure 5 presents the expected (blue-solid line) and estimated (red-dashed line) relationships for each model. Although model 2 provides the most accurate temperature field prediction, it fails to correctly estimate the convection coefficient h. In model 3, the convection boundary condition is explicitly enforced, and the convection coefficient h is learned during model training. As shown in Fig. 5, all data points align with the expected convection line, ensuring that the condition is satisfied across the entire domain.

The estimation of h depends of two key factors: the temperature prediction u_θ, which is the output of the PINN, and the temperature gradient normal to the boundary, $\frac{\partial u_\theta}{\partial n}$, computed using automatic differentiation. To analyze the source of error in h estimation—whether it arises from the predicted temperature, its gradient, or both— we compare the computed heat flux using both sides of the convection equation against the reference solution. The right-hand side (RHS) heat flux is calculated based on the predicted temperature field, $q_{rhs} = -h(u_\theta - u_\infty)$, while the left-hand side (LHS) heat flux is computed using the predicted temperature gradient, $q_{lhs} = k\frac{\partial u_\theta}{\partial n}$.

Figure 6 presents q_{rhs} and q_{lhs} along the top and right boundaries, alongside the reference heat flux. For all models and boundaries, q_{rhs} closely matches the reference solution, indicating that the PINN temperature prediction is good enough and is not the cause for the convection error. Even for the high temperature gradient region $y \in [0.0, 0.2]$ in the right boundary, where temperature predictions are the worst, the deviation from the reference heat flux curve is small. However, q_{lhs} aligns with the reference solution only in model 3 and on the top boundary in model 2, suggesting that the temperature gradients are not correctly captured elsewhere.

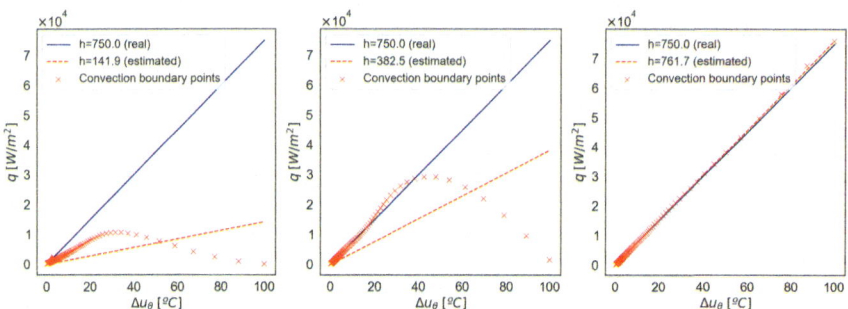

Fig. 5 Heat flux through the convective boundaries as a function of the temperature difference between the boundary and ambient (markers). The dashed red line represents the least-squares regression line, whose slope corresponds to the estimated convection coefficient h. The solid blue line indicates the real relationship between heat flux and temperature difference based on the true h value. Left: Model 1; Middle: Model 2; Right: Model 3

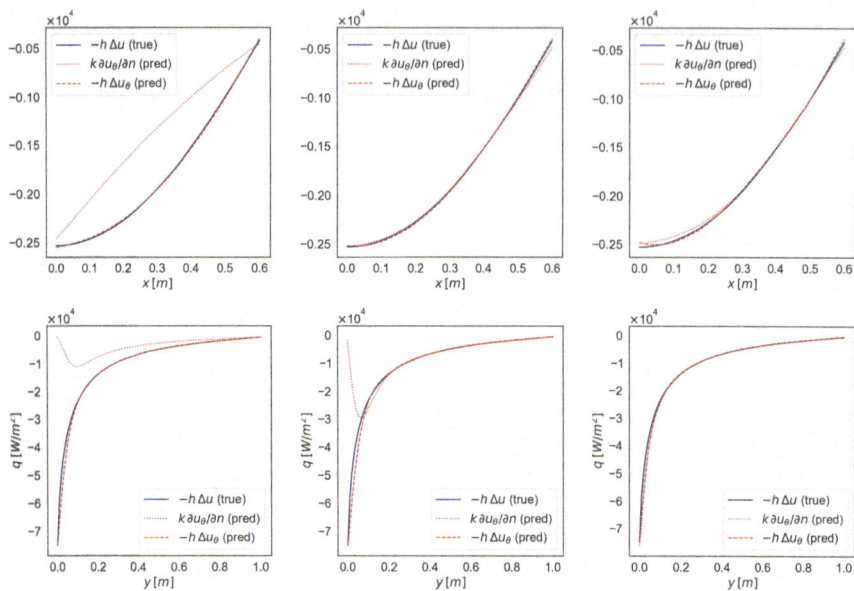

Fig. 6 Heat flux across convective boundaries: top boundary (top row) and right boundary (bottom row). The dashed red line represents the right-hand side of the convection condition, while the dotted red line represents the left-hand side. Both red lines are computed from PINN predicted temperature fields and their derivatives. The solid blue line shows the true heat flux obtained from FEM temperature data. Left: Model 1; Middle: Model 2; Right: Model 3

Model 1 incorrectly predicts the gradients on both the top and right boundaries, as expected since no prior physics is enforced. Model 2 strugles in the high-gradient region near $x = 0$ on the right boundary, where the derivative prediction deteriorates. This highlights that errors in h estimation result from inaccuracies in the temperature gradients rather than the predicted temperature field itself. In contrast, model 3 successfully enforces the convection boundary condition, ensuring consistency between temperature predictions and their gradients.

4.3 Sensitivity Analysis on Temperature Data

We now analyze the effect of the number of data points on the performance of the PINN model. Data points correspond to locations where the temperature u is known, typically obtained from sensor measurements. In Machine Learning terminology, these are referred to as training or labeled data. We focus on well-performing models, as poor-performing ones do not provide meaningful insights into the effects of data and residual points. Nevertheless, they are included in the plots (in gray) to offer additional context for the reader.

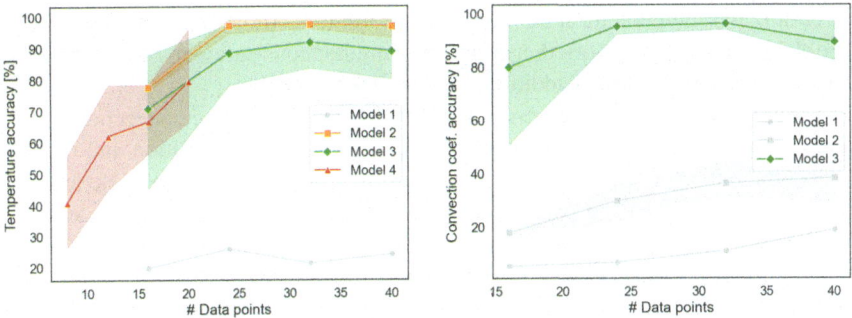

Fig. 7 Accuracy of the predicted temperature field (left) and convection coefficient (right) as a function of the number of data points. The shaded regions represent the range (min–max), while markers indicate the mean values for different numbers of residual points. Well-performing models are shown in color, while poor-performing models are displayed in gray

In general, increasing the number of data points improves the model's accuracy, but only up to a certain threshold beyond which additional data points no longer contribute significantly to accuracy (Fig. 7). In terms of both temperature and convection coefficient accuracy, this plateau is reached at approximately 24 data points (6 data points per side). Beyond this value, the models show minimal improvement, and in some cases, performance even degrades. Despite having access to the true convection coefficient, model 4 does not consistently outperform other models. Data points on convection boundaries were not included, as the convection condition is already known there. As a result, fewer data points were available during training compared to model 3, which may explain its lower performance.

For a fixed number of data points, three different models were trained, each with a different number of residual points (400, 100, or 40). The sensitivity to residual points varies across cases:

- Model 3 and model 4 exhibit greater sensitivity to the number of residual points. For instance, in model 4 with 16 data points, temperature accuracy ranges from 56.36 to 78.36%, depending on the number of residual points.
- Model 2 is more robust to changes in the number of residual points, with accuracy varying only slightly from 76.40 to 79.06% under the same conditions.

5 Conclusions

The analysis presented in this paper highlights several important findings. First, PINNs demonstrate strong capabilities in temperature field prediction when data is scarce. Results show that enforcing the heat equation alone—without requiring prior knowledge of boundary conditions—yields high accuracy in estimating the thermal state of a 2D solid surface. This suggests that PINNs are well-suited for solving inverse problems where the differential equation is known, boundary conditions are uncertain, and temperature measurements are limited. Additionally, PINNs offer a flexible modeling framework that allows for simultaneous parameter estimation and field reconstruction with no additional computational cost.

A key outcome of this study is that accurately estimating the convection coefficient requires explicitly enforcing the convection boundary condition and treating the coefficient as a learnable parameter during training. This approach leads to precise estimations of the parameter, demonstrating that PINNs can successfully infer missing thermal properties when provided with suitable constraints. In contrast, post-training estimation of the convection coefficient from temperature predictions alone is not reliable, as errors in temperature gradients lead to incorrect heat flux calculations. These findings reinforce the importance of selecting an appropriate balance between physics-based constraints and available data when applying PINNs to thermal modeling problems.

While PINNs show strong potential in inverse thermal analysis, certain limitations remain. The models are sensitive to high-gradient regions, which can impact prediction accuracy near steep temperature variations. Additionally, PINNs require retraining when conditions change, limiting their direct application in real-time, such as digital twins for machine tools. Future research should explore strategies to improve adaptability, such as integrating physics-informed neural operators or leveraging transfer learning to reduce the computational cost of retraining under varying conditions. Extending this methodology to three-dimensional and transient thermal problems would further clarify the practical advantages and challenges of using PINNs in complex machine tool geometries.

References

1. Dissanayake, M.W.M.G., Phan-Thien, N.: Neural-network-based approximations for solving partial differential equations. Commun. Numer. Methods Eng. **10**, 195–201 (1994)
2. Lagaris, I.E., Likas, A., Fotiadis, D.I.: Artificial neural networks for solving ordinary and partial differential equations. IEEE Trans. Neural Netw. **9**, 987–1000 (1998)
3. Raissi, M., Perdikaris, P., Karniadakis, G.E.: Physics-informed neural networks: a deep learning framework for solving forward and inverse problems involving nonlinear partial differential equations. J. Comput. Phys. **378**, 686–707 (2019)
4. Haghighat, E., Raissi, M., Moure, A., Gomez, H., Juanes, R.: A physics-informed deep learning framework for inversion and surrogate modeling in solid mechanics. Comput. Methods Appl. Mech. Eng. **379**, 113741 (2021)
5. Cai, S., Wang, Z., Wang, S., Perdikaris, P., Karniadakis, G.E.: Physics-informed neural networks for heat transfer problems. J. Heat Transf. **143**, 5985 (2021)
6. Wang, S., Teng, Y., Perdikaris, P.: Understanding and mitigating gradient flow pathologies in physics-informed neural networks. SIAM J. Sci. Comput. **43**, A3055–A3081 (2021)
7. Ramirez, I., Pino, J., Pardo, D., Sanz, M., del Rio, L., Ortiz, A., Morozovska, K., Aizpurua, J.I.: Residual-based attention physics-informed neural networks for spatio-temporal ageing assessment of transformers operated in renewable power plants. Eng. Appl. Artif. Intell. **139**, 109556 (2025)
8. NAFEMS: The Standard NAFEMS Benchmarks. NAFEMS Ltd. (1990)
9. Wang, S., Sankaran, S., Wang, H., Perdikaris, P.: An Expert's Guide to Training Physics-Informed Neural Networks (2023)

10. Lu, L., Meng, X., Mao, Z., Karniadakis, G.E.: DeepXDE: a deep learning library for solving differential equations. SIAM Rev. **63**, 208–228 (2021)
11. McClenny, L.D., Braga-Neto, U.M.: Self-adaptive physics-informed neural networks. J. Comput. Phys. **474**, 111722 (2023)
12. Anagnostopoulos, S.J., Toscano, J.D., Stergiopulos, N., Karniadakis, G.E.: Residual-based attention in physics-informed neural networks. Comput. Methods Appl. Mech. Eng. **421**, 116805 (2024)

Thermal Error Modeling in CNC Machine Tool Spindles Using Transfer Learning with GRU

Yue Zheng, Guoqiang Fu, J. R. R. Mayer, Sen Mu, and Sipei Zhu

Abstract Thermal error prediction forms the cornerstone of CNC machine tool accuracy compensation. Addressing the nonlinear and time-varying nature of spindle thermal deformation, this study proposes a neural network-based thermal error prediction model enhanced with transfer learning techniques. The model integrates data from multiple sensors to feed into a GRU network, effectively capturing the temporal correlations between temperature fluctuations and thermal errors. Experimental results demonstrate a prediction accuracy of ± 2.1 μm. The streamlined structure of GRU, featuring a gating mechanism and fewer parameters, enables robust learning of thermal error with reduced dependency on extensive training samples. This study validates the potential of deep time modeling for thermal error compensation in smart manufacturing systems.

Keywords CNC machine tool spindle · Thermal error prediction · Gated recurrent unit (GRU) · Temporal dependency analysis · Temperature field distribution

1 Introduction

With the rise of intelligent manufacturing, ensuring high machining accuracy through thermal error compensation is particularly important in precision manufacturing. Studies have shown that up to 40–70% of machine tool errors originate from thermally induced deformations [1]. During operation, the spindle generates significant heat due to bearing friction, causing its temperature to rise. This results in thermal deformation, which negatively impacts the machining accuracy and efficiency of the machine tool [2].

Y. Zheng (✉) · G. Fu · S. Mu · S. Zhu
Southwest Jiaotong University, Chengdu, China
e-mail: zhengyue@my.swjtu.edu.cn

Y. Zheng · J. R. R. Mayer
Polytechnique Montreal, Montreal, QC, Canada

© The Author(s) 2026
K. Wegener and M. Bambach (eds.), *4th International Conference on Thermal Issues in Machine Tools (ICTIMT2025)*, Lecture Notes in Production Engineering,
https://doi.org/10.1007/978-3-032-01194-7_37

Current thermal error control strategies include prevention, suppression and compensation. Compensation techniques based on predictive modeling offer significant economic advantages over preventive measures that require changes to the machine structure [3]. However, the development of predictive models that accurately capture the temporal correlation between temperature changes and displacements still faces the dual challenges of nonlinear fitting and computational efficiency. Recent advances in deep learning, especially recurrent neural networks (RNN), offer new possibilities for thermal error modeling by extracting features from time series data [4]. However, RNNs suffer from the gradient vanishing problem, which limits their application in modeling long-term dependencies. Long Short-Term Memory Networks (LSTMs) partially solve this problem by introducing a gating mechanism, but their complex network structure still faces challenges in industrial applications [5]. The gate recurrent unit (GRU), as an optimized variant of LSTM, not only improves the computational efficiency by simplifying the gating structure and integrating the memory unit, but also effectively captures the characteristics of dynamic systems such as machine tool spindles [6].

In order to cope with complex operating conditions, this paper proposes a dynamic spindle thermal error modeling method based on GRU neural networks and enhanced with transfer learning techniques. By drawing on the knowledge of pre-trained models in similar fields, migration learning enables the GRU network to better adapt to new operating environments and significantly improves prediction accuracy.

2   Experiments on Thermal Error Measurement

In this study, the MGKF1800 high-precision vertical composite grinder is used as an experimental object to carry out spindle dynamic error modeling tests. As shown in Fig. 1, the thermal error measurement system consisted of a spindle thermal deflection collector, a Dam-pt16 temperature acquisition board, 13 PT100 temperature sensors, a spindle check rod, a Lion Precision capacitive displacement sensor, and a sensor holder. Ambient temperature is acquired by sensor T13. In order to accurately characterize the thermal coupling effect of the spindle, the experiments are designed to synchronize the following measurement schemes: firstly, the thermal elongation of the spindle is acquired in real time by a non-contact laser displacement sensor, and at the same time, the distributed temperature sensing system is used to collect the temperature field data of the 13 key temperature measurement points of the spindle system.

The whole experiment starts from the cold start of the machine and continues until the thermal deformation of the spindle reaches the thermal equilibrium state. During the experiment, the spindle is only in idle state, i.e., it keeps rotating at constant or variable speed. The experimental results show that the thermal deformation of the spindle stabilizes and reaches thermal equilibrium in about 80 min after the start of the experiment. The sampling interval of the thermal deformation data was set to be recorded every 5 s. Figure 2 shows the data of 13 temperature sensor data and thermal

Fig. 1 Measurement of spindle axial thermal error. **a** Temperature sensor arrangement. **b** Measurement of spindle axial thermal error

error of the spindle at 1800 rpm. Several sets of data were collected, including 1500, 1800 rpm, and two types of variable speeds as shown in Fig. 3.

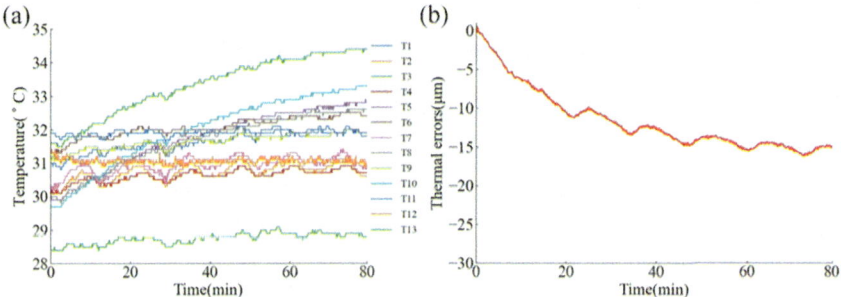

Fig. 2 The measured data of spindle at 1800 rpm

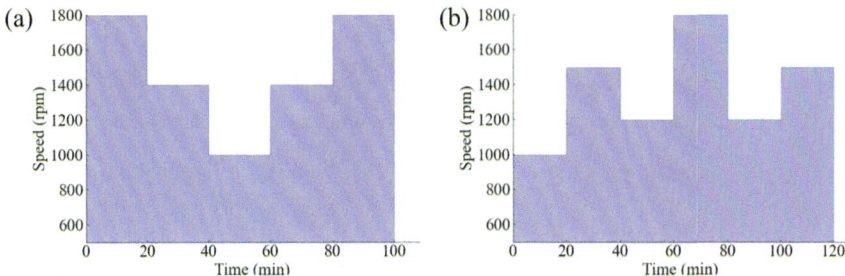

Fig. 3 Spindle speed. **a** Variable speed 1. **b** Variable speed 2

3   Thermal Error Modeling Based on GRU Network with Transfer Learning

Aiming at the dynamic characteristics of the axial thermal error of CNC machine tool spindle, this study constructs an axial thermal displacement prediction model based on GRU, which is shown in Fig. 4. The input of the model is the time series temperature field data $T(t) = [T1(t), T2(t), ..., Tn(t)]$, and the output is the axial thermal displacement prediction. The network is trained using a time series cross-validation strategy, and the loss function is defined as the axial weighted mean square error:

$$\ell = \frac{1}{N} \sum_{i=1}^{N} w_i \cdot (y_i - \hat{y}_i)^2 \tag{1}$$

where N is the number of samples; y_i is the true value of the i-th observation; $\hat{y}_i$ is the predicted value of the i-th observation; w_i is the weight of the i-th observation.

Based on the GRU model, a fine-tuning approach of transfer learning is introduced. Transfer learning is performed by fine-tuning the model based on the pre-trained model in the source domain using a small amount of data from the target domain to adapt to a specific task in the target domain [7]. In this study, the fine-tuning approach was used to optimize the modeling accuracy of the GRU model under different spindle speed conditions.

This GRU-based thermal error prediction model combines transfer learning with the ability to effectively extract dynamic features from time series data, achieving accurate prediction with limited target domain data. This provides an efficient and reliable solution for modeling the axial thermal error of CNC machine tools.

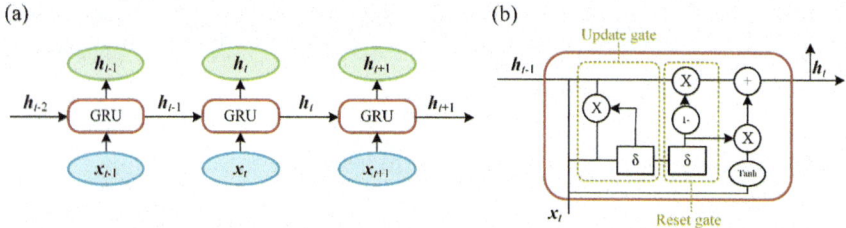

Fig. 4 GRU structure diagram. **a** GRU neural network structure. **b** GRU unit structure

4   Results and Discussion

The training set consists of variable speed 1 data, while the remaining three experimental datasets serve as 3 testing sets for 3 experiments with 20% of the data was
used for fine-tuning. GRU is used for transfer learning prediction in multi-conditional
scenarios. Three experiments were conducted using a collected dataset, with separate training and test data to evaluate generalization performance. Figures 5, 6, and
7 show the thermal error transfer predictions of both the GRU model and the GRU-
TL (Gated Recurrent Unit for Transfer Learning) model under different conditions.
The results demonstrate that while GRU effectively captures temperature trends, the
introduction of transfer learning further enhances the model's generalization ability.
This highlights the robustness of the GRU-TL model in complex scenarios and its
potential for improving prediction accuracy across different operating conditions
(Fig. 8).

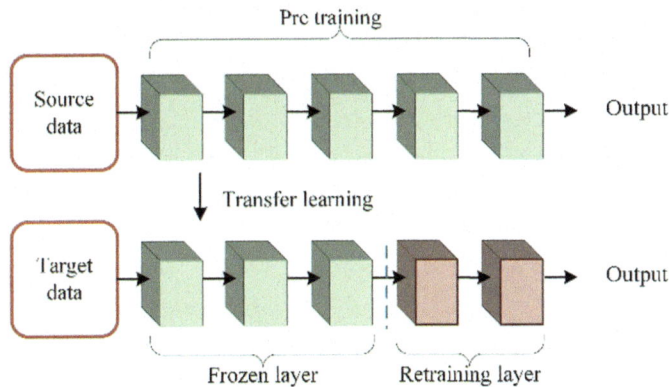

Fig. 5 Transfer learning structure diagram

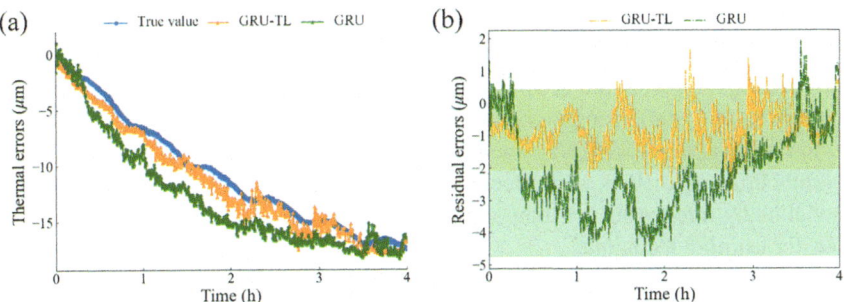

Fig. 6 Prediction results in 1500 rpm

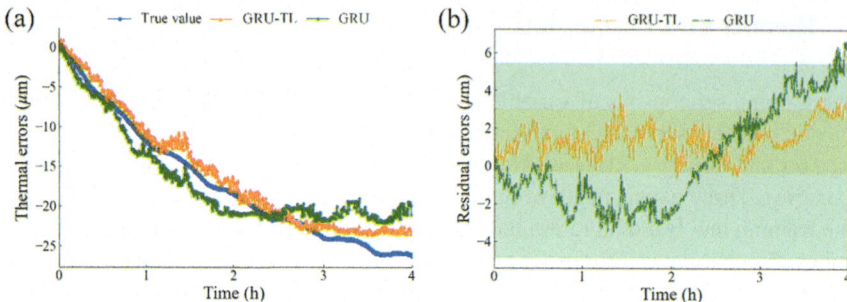

Fig. 7 Prediction results in 1800 rpm

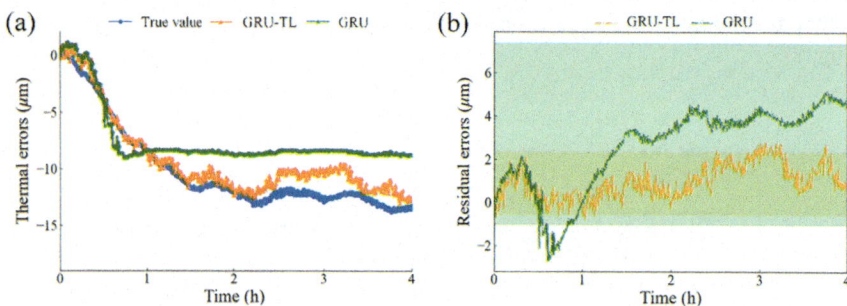

Fig. 8 Prediction results in variable speed 2

5 Conclusions

This study explores the thermal error prediction problem of grinding machine spindles and proposes a model based on GRU and transfer learning. The model effectively addresses the nonlinear and time-varying nature of thermal deformation of grinding machine spindles. By utilizing the ability of GRU network to capture the time-dependence, the transfer model has the ability to achieve multi-case prediction with a small amount of data. Experimental results show that the GRU-TL model performs well in terms of prediction accuracy and maintains strong performance despite data scarcity, highlighting its potential in spindle thermal error prediction.

In conclusion, the GRU-TL model provides an efficient framework for spindle thermal error prediction, and effectively addresses the challenges of thermal error modeling by integrating the temporal feature extraction capability of GRU and the transfer learning technique.

References

1. Ariaga, N., Longstaff, A., Fletcher, S.: A method for temperature sensor and model selection for machine tool thermal error modelling using ANFIS and ANN. Int. J. Adv. Manufact. Technol. **135**(1), 863–881 (2024)
2. Wang, J., Deng, X.: Detection of spindle thermal state accuracy of vertical CNC machine tool. Int. J. Eng. Syst. Model. Simul. **11**(2), 43–49 (2019)
3. Weng, L., Gao, W., Zhang, D., et al.: Analytical modelling of transient thermal characteristics of precision machine tools and real-time active thermal control method. Int. J. Mach. Tools Manufact. **186**, 104003 (2023)
4. Chen, Y., Chen, J., Xu, G.: A data-driven model for thermal error prediction considering thermoelasticity with gated recurrent unit attention. Measurement **184**, 109891 (2021)
5. Liu, Z., Wei, K., Chen, C., et al.: Thermal error recognition and prediction method for economical motorized spindles based on thermal characteristics knowledge Graph. Pattern Recognit. Lett. **15**, 5425 (2025)
6. Yuan, Q., Ma, C., Liu, J., et al.: Correlation analysis-based thermal error control with ITSA-GRU-A model and cloud-edge-physical collaboration framework. Adv. Eng. Inform. **54**, 101759 (2022)
7. Shi, H.L., Jiang, P.Y.: Quality control in multistage machining processes based on a machining error propagation event-knowledge graph. Adv. Manufact. **12**(4), 679–697 (2024)

Dynamic Active Temperature Control Method Based on Structure-Specific Heat Dissipation-To-Generation Ratio Matching

Weiguo Gao, Junyan Xing, Lingtao Weng, Dawei Zhang, Zheng Guo, Yingjie Zheng, and Kai Shi

Abstract This study proposes a dynamic active temperature control method for precision machine tools based on structure specific heat dissipation and power generation ratio matching. A thermal control strategy combining thermal accumulation effect with optimized cooling parameters was developed using a precision vertical coordinate machine tool as the research platform. This method determines the optimal heat dissipation rate of different structural components of precision vertical coordinate machine tools (spindle bearings: 60–80%, motors: 30–40%, feed systems: 40–60%) under various working conditions, and achieves dynamic cooling power adjustment. Experimental verification shows that the temperature of the spindle motor decreased by 7.8 °C, and the perpendicularity errors of the XY and ZX planes decreased by 80 and 22%, respectively, verifying the effectiveness of the proposed method.

Keywords Precision machine tools · Thermal error · Active temperature control · Heat dissipation ratio · Thermal accumulation

W. Gao · J. Xing · L. Weng (✉) · D. Zhang · Z. Guo · Y. Zheng · K. Shi
Tianjin University, Tianjin, People's Republic of China
e-mail: wenglingtao@tju.edu.cn

J. Xing
Beijing Precision Machinery and Engineering Research Co., Ltd., Beijing, People's Republic of China

Present Address:
W. Gao
Key Laboratory of High-end e machine tools of GT, Beijing, China

K. Wegener and M. Bambach (eds.), *4th International Conference on Thermal Issues in Machine Tools (ICTIMT2025)*, Lecture Notes in Production Engineering,
https://doi.org/10.1007/978-3-032-01194-7_38

1 Introduction

As a core equipment in modern manufacturing, the machining accuracy of CNC machine tools directly determines the quality grade of mechanical products. Among the factors affecting the accuracy of machine tools, thermal-induced errors account for 40–70%, which has become the main bottleneck restricting high-precision machining [1]. The thermal deformation caused by the change of thermal characteristics of machine tools is a complex multi-physics field coupling process. The internal heat source and environmental thermal disturbance act together, resulting in a non-uniform temperature field in the machine tool structure, and then converting into spatial geometric errors through the thermal expansion effect of materials [2, 3]. Therefore, improve the thermal characteristics of machine tools is an important way to enhance machining accuracy and reduce thermal errors of machine tools.

Faced with this demand, currently, optimization methods for the thermal characteristics of machine tools in the current academic community can be divided into two categories: one is structural optimization design for thermal characteristics, and improve the predictability of thermal behavior by constructing high-precision temperature deformation mapping models. Mori et al. [4] proposed a novel approach to design a headstock for NC lathe immune to thermal deformation caused by random temperature deviation, combined Taguchi method with CAE analysis, determined an NC lathe spindle structure and manufactured a headstock based on the design result. Oyanguren et al. [5] developed a transient thermo-mechanical finite element model for a double-nut preloaded ball screw assembly. The model accurately predicts transient temperature and preload variations, while also providing temperature and ball load distributions.

The other is thermal deformation modeling and compensation. Early error compensation mainly relied on mechanical fixed compensation, followed by error compensation technology based on numerical control instructions in the 1970s [6]. Donmez et al. [7] decomposed the errors of machine tools into geometric errors and thermal errors, and used an empirical model for error prediction based on the least squares curve fitting technique. Through a modular and structured software system, accurate thermal error prediction and compensation were achieved. Yang et al. [8] integrated modeling and compensation of thermal errors in machine tool feed systems using a CNN-GRU neural network (SABO-CNN-GRU) optimized by a subtractive average optimizer. Cheng et al. [9] studied five axis CNC milling heads, analyzed their thermal boundary loads, conducted thermal deformation simulations and experiments, and established a thermal error compensation model based on homogeneous transformation, providing an effective method for optimizing milling head design and thermal error compensation.

Although the above studies can improve the thermal characteristics of machine tools under certain operating conditions, they generally have limitations such as insufficient adaptability and delayed response. With the development of manufacturing towards high efficiency and universality, active control technology for thermal characteristics is gradually emerging. This technology actively adjusts the temperature distribution of key parts of the machine tool through a closed-loop temperature control system, achieving a paradigm shift from "passive adaptation to thermal deformation" to "active shaping of temperature field". By dynamically and actively adjusting the temperature field of the machine tool structure, can regulate the structural thermal deformation field of the machine tool and correct its structural thermal deformation. Japanese scholar Mitsuishi was the first to propose the relevant concepts and technical methods for thermal active control of machine tool structures [10, 11]. He installed fan fins and electric heaters on the machine column to change the temperature distribution of the column and achieve active control of structural thermal deformation. Liu et al. [12] proposed an active cooling control strategy to resist thermal interference from precision ball screw units. Through experiments and simulations, it has been verified that this strategy can stabilize the temperature of the screw shaft, correct thermal elongation, and has more advantages in controlling thermal errors compared to traditional strategies. Lei et al. [13] proposed an active cooling strategy based on feedback control to stabilize the thermal error of precision mechanical spindles. Through experimental verification, this strategy can effectively stabilize thermal errors, reach thermal equilibrium state in advance, and the proposed model has high reliability.

In summary, the active control technology for thermal characteristics of machine tools can better suppress the influence of heat on the structural temperature and deformation field, improve the ability of functional and structural components to resist thermal effects, and enhance the thermal accuracy and stability of the entire machine tool. This article proposes a dynamic active temperature control method based on structure-specific heat dissipation matching. Using a precision vertical boring machines as a research platform, a thermal control strategy is developed that integrates heat accumulation effects and optimizes cooling parameters. The optimal heat dissipation ratio for different structural components is determined and the cooling power is dynamically adjusted. Experimental verification shows that this method can reduce the temperature of the spindle motor and minimize errors in plane perpendicularity.

2 Active Thermal Control Strategies and Transient Simulation Analysis

2.1 Control Strategy Methodology

This section presents an integrated active temperature control strategy for precision vertical coordinate boring machines that combines dynamic heat dissipation-to-generation ratio matching with thermal accumulation-based control. The strategy optimizes cooling parameters based on structural thermal deformation patterns while accounting for temperature gradient evolution due to thermal accumulation effects.

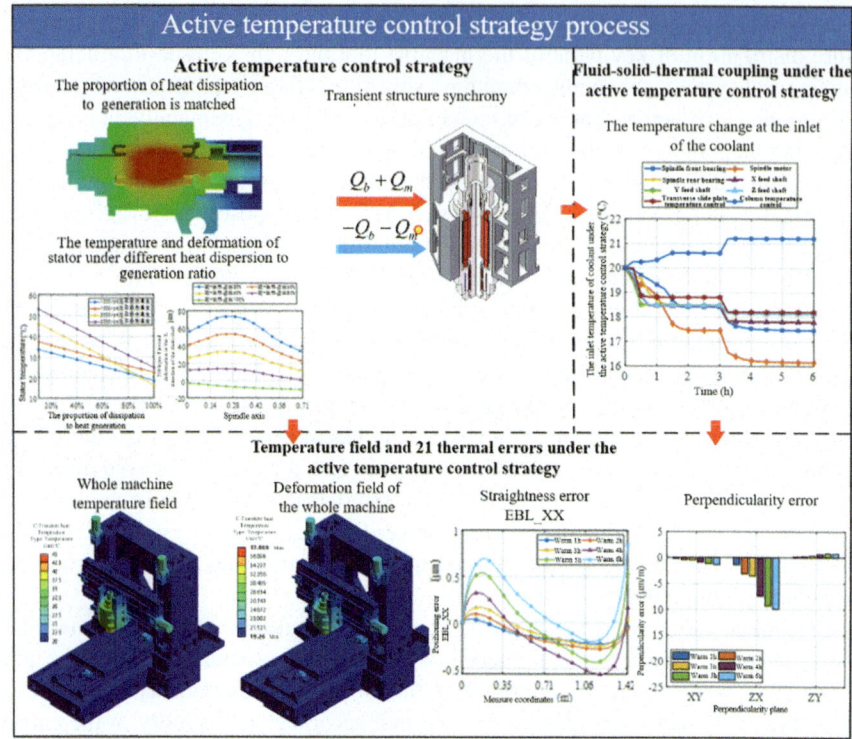

Fig. 1 Precision vertical coordinate boring machine thermal characteristics active temperature control strategy

The control framework, illustrated in Fig. 1, uses structural temperature variations as feedback to maintain optimal thermal stability.

Analysis of internal heat sources in the column reveals a unique thermal distribution pattern. The heat sources are predominantly located on the front side of the column, with the maximum thermal load concentrated at the upper section. During machining operations, this arrangement creates a significant temperature differential between the front and rear surfaces of the column. Given that the column is fixed at its base, the asymmetric thermal loading causes the entire structure to exhibit a backward-leaning tendency.

To address this thermal distortion, active temperature control devices were strategically positioned on the rear side of the column. The column's structural design features two parallel pillars on the rear side, which provide suitable mounting locations for cooling devices. Following the principle of thermal symmetry, identical temperature control sources were installed at corresponding positions on both rear pillars, mirroring the front-side heat source distribution. The specific layout of these cooling pathways is illustrated in Fig. 2.

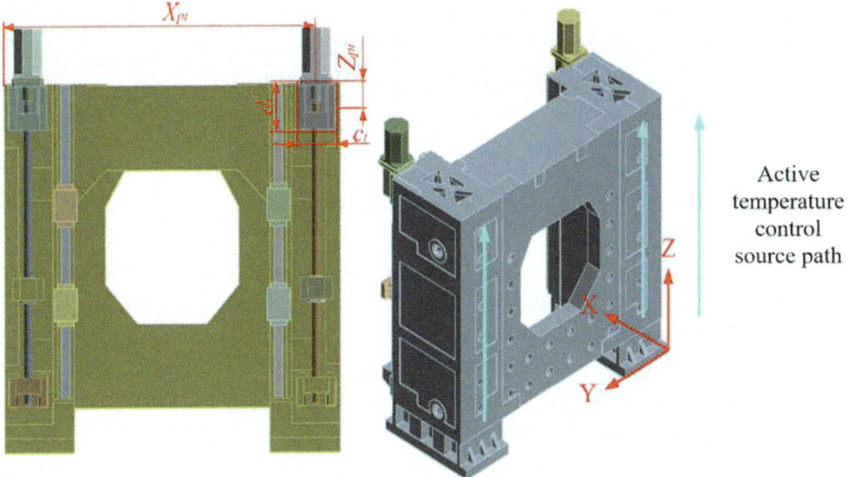

Fig. 2 Column active temperature control device layout path

This arrangement aims to establish a more balanced thermal distribution across the column structure, effectively counteracting the thermal-induced deformation caused by the front-side heat sources.

The heat dissipation-to-generation ratio (λ), defined as the ratio between the cooling power of temperature control system and the heat generation rate of internal heat sources, is a critical parameter for thermal management. This ratio can be expressed as:

$$\lambda = \frac{Q_c}{Q_h} \tag{1}$$

where Q_c represents the heat dissipation rate through active cooling and Q_h denotes the heat generation rate of machine tool components. The optimal matching of this ratio is analyzed separately for the feed system and spindle system components to determine effective cooling parameters. This analysis enables precise control of thermal balance across different machine structures.

2.2 Active Thermal Control Strategy Considering Thermal Accumulation Effects

The heat accumulation effect in machine tool structures is an inevitable phenomenon under constant operating conditions. This occurs when instantaneous heat dissipation through air convection and temperature control systems cannot fully offset continuous heat generation. In this continuous heat exchange process, the heat generated at each moment has not been completely dissipated, and at the same time there is a new heat generation, so that part of the heat generated at each time will be gradually accumulated, i.e., the heat accumulation effect occurs. This section presents a novel structural control strategy that addresses these thermal accumulation effects.

2.2.1   Selection of Key Temperature Control Points

The effectiveness of thermal control largely depends on the strategic placement of monitoring points throughout the machine structure. This selection process requires careful consideration of multiple factors including heat source characteristics, structural constraints, and practical implementation requirements. For the precision boring machine under study, a comprehensive analysis of thermal characteristics led to the development of a hierarchical monitoring strategy.

In the feed system, while ball screw-nut pairs represent significant heat sources, their rotating nature precludes direct sensor placement. Instead, monitoring points were established on the motor mount and bearing housing surfaces, where substantial heat generation occurs and sensor installation is feasible. This approach enables indirect monitoring of ball screw thermal behavior while providing reliable temperature data for control purposes, as shown in Fig. 3.

Furthermore, the spindle system presents unique challenges due to its complex internal structure and high-precision requirements; direct sensor placement on the spindle surface was deemed impractical. The spindle unit is the largest heat source. Due to its axisymmetric structure, its deformation in the X and Y directions is small and uniform, and the deformation is mainly concentrated in the thermal expansion in the Z direction. After temperature control, the heat is transferred to the spindle box and the slider. Therefore, a sensor is arranged on the spindle box to represent the heat transferred from the spindle unit to the cross slide.

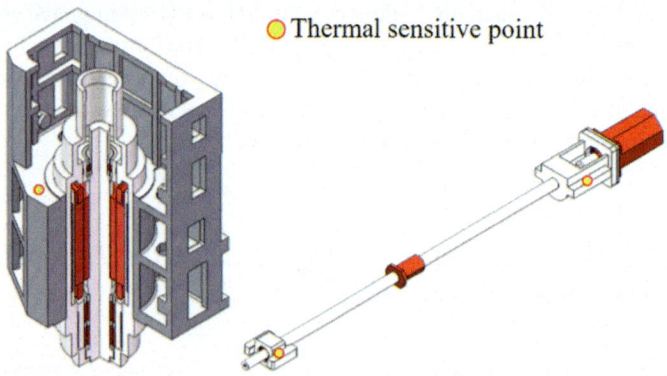

Fig. 3   Selection of thermally sensitive points

2.2.2 Temperature-Deformation Mapping Construction

The structural parts of machine tools are usually elastic materials, and according to Hooke's law, the temperature change caused by multiple heat sources can be regarded as the sum of the responses caused by a single heat source, i.e., the combined thermal effects of multiple heat sources on a machine tool are the linear superposition of the respective thermal effects of multiple single heat sources. Therefore, the thermal characteristics of each circuit can be investigated by means of the separated variable method and the superposition principle. To investigate the relationship between guideway thermal deformation and internal heat sources, a systematic simulation approach was implemented. Heat sources were analyzed using unit power settings through transient simulations, tracking both temperature evolution and guideway deformation from ambient temperature to steady state. For the feed system analysis, the X-axis motor bearing heat generation at 5 m/min was defined as the unit heat generation power (Q_x). Simulations were conducted under three scenarios: (1) Heat generation: Q_x, $2Q_x$, and $3Q_x$; (2) Cooling power: $-Q_x$, $-2Q_x$, and $-3Q_x$ (Fig. 4).

For the spindle system, due to the serial configuration of front and rear bearing cooling systems, their combined heat generation at 500 rpm was treated as a single unit (Q_b), while the spindle motor heat generation at 500 rpm was designated as Q_m. The analysis included: (1) Combined heat generation: $(Q_b + Q_m)$, $2(Q_b + Q_m)$, and $3(Q_b + Q_m)$; (2) Bearing cooling circuit: $-Q_b$, $-2Q_b$, and $-3Q_b$; (3) Motor cooling circuit: $-Q_m$, $-2Q_m$, and $-3Q_m$. For a stationary surface, the natural convection heat transfer coefficient between the surface and the surrounding air is 9.7 W/(m^2 °C).

According to Fig. 5, it is assumed that a single heat source presents a polynomial function relationship to the temperature control point at any time under unit power Q. According to the simulation results, after applying a proportionally changing heat source power, the temperature of the temperature control point also presents a proportional change. On this basis, combined with the simulation data, a comprehensive

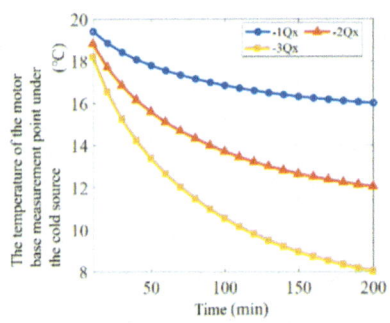

(a) Temperature changes under the action of heat sink

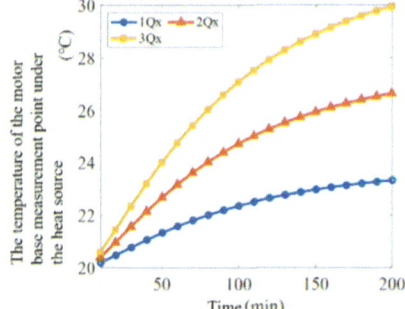

(a) Temperature changes under the action of heat sink

Fig. 4 Temperature variations under hot and cold sources in the feed system

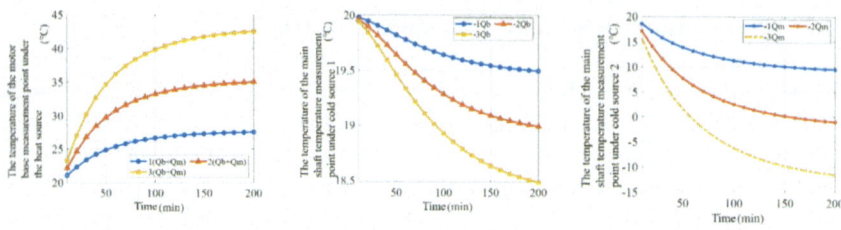

(a) Temperature changes under the action of heat sink

(b) Temperature change under the action of heat sink 1

(c) Temperature change under the action of heat sink 2

Fig. 5 Temperature changes under cold and hot sources in the spindle system

mathematical model including heating and cooling effects can be obtained:

$$T_{empX}(t) = 1.48e^{-9}t^3 - 5e^{-7}t^2 + 4.4e^{-5}t - 1e^{-4} + 20$$

$$+ \frac{Q_h}{94.09} \cdot \left(1.4981e^{-7}t^3 - 1.2346e^{-4}t^2 + 0.036159t - 0.1715\right)$$

$$- \frac{Q_c}{94.09} \cdot \left(-5.8562e^{-7}t^3 + 2.8413e^{-4}t^2 - 0.052638t - 0.1682\right) \quad (2)$$

The spindle system model incorporates multiple cooling circuits, reflecting its more complex thermal behavior:

$$T_{empm}(t) = -7.66e^{-8}t^3 + 9.5e^{-5}t^2 - 3.32e^{-2}t - 2e^{-4} + 20$$

$$+ \frac{Q_h}{597.4491} \cdot \left(1.87e^{-6}t^3 - 8.49e^{-4}t^2 + 0.133t - 0.0153\right)$$

$$- \frac{Q_{c1}}{49.365} \cdot \left(7.126e^{-9}t^3 + 1.02e^{-4}t^2 - 0.0503t - 0.0418\right)$$

$$- \frac{Q_{c2}}{549.738} \cdot \left(-1.95e^{-6}t^3 + 9.33e^{-4}t^2 - 0.16133t - 0.0574\right) \quad (3)$$

2.2.3 Active Control Strategy Implementation

In actual acquisition and control, it is difficult to implement active temperature control synchronously at the moment of measuring temperature. Therefore, combined with the actual working conditions, temperature measurements are conducted at 180-s intervals, providing a balance between control responsiveness and system stability. During each measurement cycle, both initial (T_{emp0}) and final (T_{empt}) temperatures are recorded, enabling calculation of temperature change rates and thermal power parameters.

The above formula can be used to predict the temperature of the temperature measuring point in the next time step, recorded as T_{emp2t}, so the temperature change and temperature change rate of the temperature measuring point in two time steps can be calculated respectively:

$$
\begin{cases}
\Delta T_{emp1} = T_{empt} - T_{emp0} \\
\Delta T_{emp} = T_{emp2t} - T_{empt} \\
\Delta T_{emp\prime} = \frac{\Delta T_{emp2} - \Delta T_{emp1}}{\Delta T_{emp1}}
\end{cases}
\tag{4}
$$

Through extensive thermal analysis and simulation, a critical temperature change rate threshold of -25% was established. When the calculated rate exceeds this threshold, indicating rapid thermal accumulation, the control system increases cooling power to compensate. Conversely, when the rate falls below this threshold, suggesting approach to steady state, the current cooling parameters are maintained to prevent overcooling.

2.3 Control Strategy Implementation

The developed control strategy implements a dual-mechanism approach combining heat dissipation ratio control and temperature change rate monitoring. This comprehensive strategy addresses both steady-state and transient thermal behaviors of the machine tool.

1. Heat Dissipation-to-Generation Ratio Control

The control system maintains optimal heat dissipation ratios for different cooling circuits based on their thermal characteristics and operational requirements:

a. Feed System (40–60%): Balanced cooling to maintain positioning accuracy while preventing overcooling
b. Spindle Motor (30–40%): Moderate cooling to manage high heat generation while ensuring operational stability
c. Spindle Bearings (60–80%): Intensive cooling to protect precision components and maintain thermal stability
d. Column Cooling (5–15%): Light cooling to manage structural deformation
e. Cross Slide (20–40%): Intermediate cooling to maintain geometric accuracy

2. Temperature Change Rate Monitoring

The system continuously monitors temperature change rates to optimize cooling control. When the rate exceeds the calculated threshold in Eq. (3), indicating rapid thermal accumulation, the system increases cooling power. Conversely, when the rate falls below this threshold, suggesting approach to steady state, current cooling parameters are maintained to prevent overcooling.

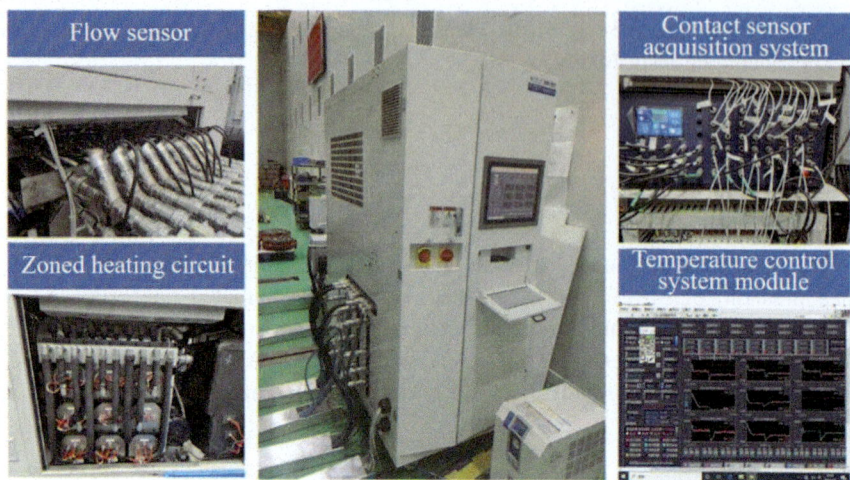

Fig. 6 Multi-source information acquisition system

3 Experimental Validation of Multi-source Information Fusion Based Active Temperature Control Strategy

3.1 Experimental Setup and Data Acquisition System

To comprehensively monitor the machine tool's thermal behavior, a multi-source information acquisition platform was developed (see Fig. 6), integrating both internal CNC system data and external sensor measurements. The system architecture consists of CNC system parameter acquisition and external sensor network monitoring module.

The sensor network monitoring module was deployed to monitor critical thermal points throughout the machine structure. The model of the sensor is the PT1000 temperature sensor, and the temperature measurement accuracy is ± 0.15 °C.

3.2 Experimental Results and Analysis

3.2.1 Temperature Field Analysis

The experimental temperature measurements revealed distinct thermal behavior patterns across different machine components and operating conditions. Under the first operating condition (0–3 h, 8 m/min feed rate, 1000 rpm spindle speed), the active control strategy demonstrated significant temperature reduction compared to constant temperature cooling, as presented in Fig. 7. The X and Y-axis motor mounts showed 28% temperature reduction, with peak temperatures decreasing from 3.5 to 2.5 °C and 2.8 to 2.0 °C respectively. The spindle housing exhibited even more substantial improvement, with a 44% reduction from 2.5 to 1.4 °C. These improvements became more pronounced during the second operating condition (3–6 h, 13 m/min feed rate, 2000 rpm spindle speed), particularly in the cooling circuit performance. The spindle bearing circuit's temperature differential increased from 1.1 to 3.4 °C, while the motor cooling circuit showed an increase from 1.6 to 2.2 °C, indicating enhanced heat dissipation capability.

Further comparing the results in Fig. 8, the active temperature control strategy maintained optimal heat dissipation ratios across different systems, with the spindle bearing circuit operating at 60–80%, motor cooling circuits at 30–40%, and feed system circuits at 40–50%. This balanced approach to thermal management resulted in significantly improved temperature stability throughout the machine structure. The cooling circuit inlet temperatures adapted dynamically to changing thermal loads, with the spindle motor cooling circuit maintaining 18.1 °C, bearing cooling circuits at 19.0 °C, and feed motor cooling circuits stabilizing around 19.2 °C during high-speed operation. The cross-slide and column cooling circuits demonstrated more moderate adjustments, maintaining 19.4 and 21.1 °C respectively, effectively managing structural thermal deformation.

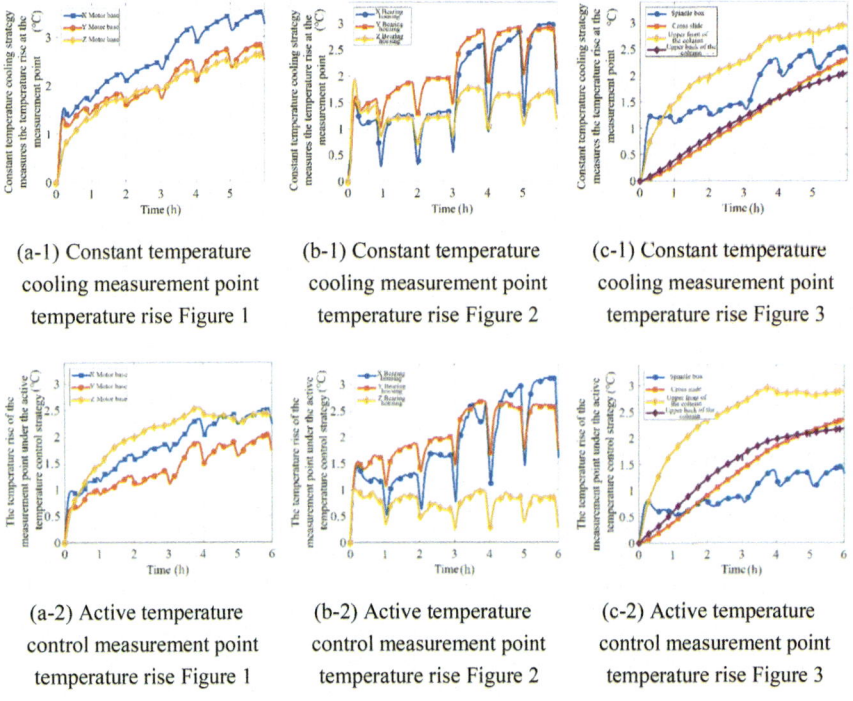

(a-1) Constant temperature cooling measurement point temperature rise Figure 1

(b-1) Constant temperature cooling measurement point temperature rise Figure 2

(c-1) Constant temperature cooling measurement point temperature rise Figure 3

(a-2) Active temperature control measurement point temperature rise Figure 1

(b-2) Active temperature control measurement point temperature rise Figure 2

(c-2) Active temperature control measurement point temperature rise Figure 3

Fig. 7 Temperature rise at each measuring point under different temperature control strategies

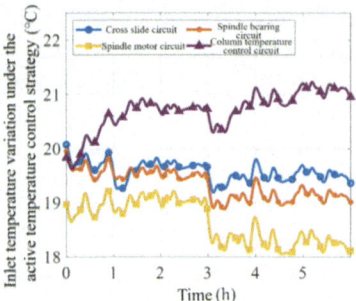 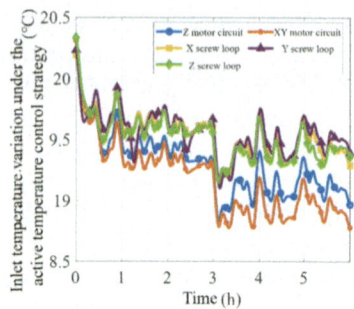

(a) Figure 1 of Inlet temperature variation under   (b) Figure 2 of Inlet temperature variation
        active temperature control strategy   under active temperature control strategy

Fig. 8 Inlet temperature change under active temperature control strategy

3.2.2 Squareness Error Analysis

The squareness error measurements, conducted using QC20 ballbar systems, revealed comprehensive improvements in machine tool accuracy under active temperature control. According to the product instruction manual of the QC20 ballbar instrument, the accuracy of its sensor is ± 0.5 μm, and the measurement accuracy of the ballbar instrument system is ± 1.25 μm (Fig. 9).

The squareness error measurements provided validation of the control strategy's effectiveness, as given in Fig. 10. The ZX plane showed steady improvement, with error reduction from -10.2 to -7.9 μm/m at 3 h, representing a 22% improvement. This improvement became more pronounced in the XY plane, where an 80% reduction from -9.2 to -1.8 μm/m was achieved after 6 h of operation. Perhaps most impressively, the ZY plane verticality error was maintained within ± 1 μm/m under active control, demonstrating exceptional stability in this critical squareness parameter. The slight variations between simulation predictions and experimental results could be attributed to factors such as sensor contact thermal resistance, placement

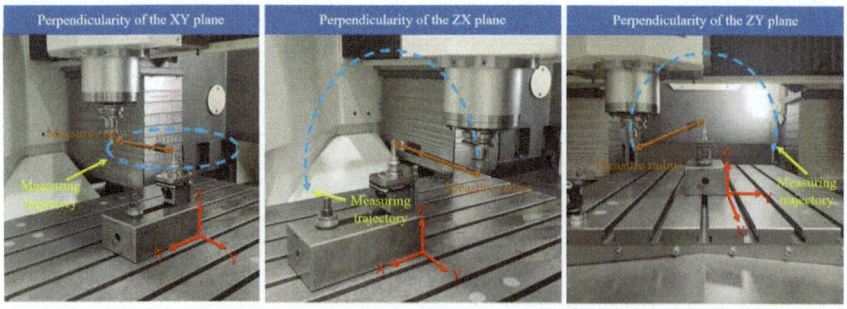

Fig. 9 QC20 the experiment flow chart

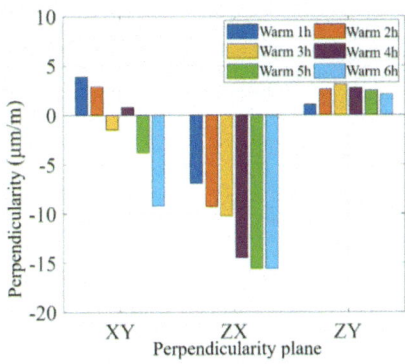 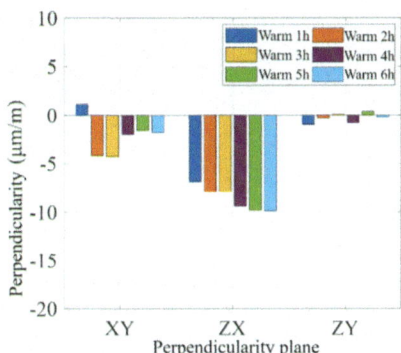

(a) The variation of thermally induced perpendicularity error under constant temperature cooling

(b) The variation of thermally induced verticality error under active temperature control

Fig. 10 Experimental verticality error under different temperature strategies

limitations, and environmental temperature effects, but these discrepancies did not significantly impact the overall effectiveness of the control strategy in maintaining geometric accuracy under thermal loads.

4 Conclusions

This study focuses on improving thermal stability and reducing thermal deformation of a precision vertical coordinate boring machine through active temperature control strategies. The main achievements are summarized as follows:

1. A novel control strategy considering both heat dissipation-to-generation ratio and thermal accumulation effects was developed. Flow-thermal-structural coupling simulation demonstrated that spindle motor temperature reduced by 7.8 °C (17%) at steady state, while ZX plane verticality error reduced by 4.81 μm/m (57%).
2. Experimental validation through a multi-source information acquisition platform confirmed the effectiveness of the control strategy. The results showed 28% temperature reduction for both X and Y-axis motors. Geometric accuracy improvements were significant: XY plane verticality error reduced by 7.4 μm/m (80%), ZX plane verticality error reduced by 2.3 μm/m (22%), and ZY plane verticality error maintained within ±1 μm/m.

Acknowledgements The authors gratefully acknowledge the funding support of the National Natural Science Foundation of China under Grant No. 52205535 and U23B20102, Shandong Province Key R&D Programme, under Grant No. 2024CXGC010201, the Postdoctoral Fellowship

Program (Grade C) of China Postdoctoral Science Foundation under Grant Number GZC20241206 and 2024M762345, the Open Research Funding of key laboratory for High-end CNC machine tools of China General Technology Group (Grant No. KLHCMT202402).

References

1. Bryan, J.: International status of thermal error research. CIRP Ann. **39**(2), 645–656 (1990)
2. Nouanegue, H., Muftuoglu, A., Bilgen, E.: Conjugate heat transfer by natural convection, conduction and radiation in open cavities. Int. J. Heat Mass Transf. **51**(25), 6054–6062 (2008)
3. Weng, L.T., Gao, W.G., Lv, Z., et al.: Influence of external heat sources on volumetric thermal errors of precision machine tools. Int. J. Adv. Manufact. Technol. **99**(1–4), 475–495 (2018)
4. Mori, M., Mizuguchi, H., Fujishima, M., et al.: Design optimization and development of CNC lathe headstock to minimize thermal deformation. CIRP Ann. Manufact. Technol. **58**(1), 331–334 (2009)
5. Oyanguren, A., Larranaga, J., Ulacia, I.: Thermo-mechanical modelling of ball screw preload force variation in different working conditions. Int. J. Adv. Manufact. Technol. **97**(1–4), 723–739 (2018)
6. Koliskor, A.S.: Compensating for automatic-cycle machining errors. Mach. Tool. **41**(5), 11–44 (1971)
7. Donmez, M.A., Blomquist, D.S., Hocken, R.J., et al.: A general methodology for machine tool accuracy enhancement by error compensation. Precis. Eng. **8**(4), 187–196 (1986)
8. Yang, T., Sun, X., Yang, H., et al.: Integrated thermal error modeling and compensation of machine tool feed system using subtraction-average-based optimizer-based CNN-GRU neural network. Int. J. Adv. Manufact. Technol. **131**(12), 6075–6089 (2024)
9. Cheng, Y., Zhang, X., Zhang, G., et al.: Thermal deformation analysis and compensation of the direct-drive five-axis CNC milling head. J. Mech. Sci. Technol. **36**(9), 4681–4694 (2022)
10. Hatamura, Y., Nagao, T., Mitsuishi, M., et al.: Development of an intelligent machining center incorporating active compensation for thermal distortion. CIRP Ann. Manufact. Technol. **42**(1), 549–552 (1993)
11. Mitsuishi, M., Warisawa, S., Hanayama, R.: Development of an intelligent highspeed machining center. CIRP Ann. Manufact. Technol. **50**(1), 275–280 (2001)
12. Liu, T., Li, C., Zhang, Y., et al.: Active coolant control onto thermal behaviors of precision ball screw unit. Int. J. Adv. Manufact. Technol. **15**, 1–16 (2021)
13. Lei, M., Gao, F., Li, Y., et al.: Feedback control–based active cooling with pre-estimated reliability for stabilizing the thermal error of a precision mechanical spindle. Int. J. Adv. Manufact. Technol. **121**(3), 2023–2040 (2022)

A Permutation Test-Based Method for Identifying Thermal Hysteresis Characteristics and Modeling Thermal Errors in Machine Tools

Guangyan Ge⑩, Xuetao Wang⑩, Zhilin Zeng⑩, and Zhengchun Du⑩

Abstract Thermal errors have a critical impact on the accuracy of machine tools, where surface-mounted temperature sensors and delayed internal temperature responses induce hysteresis that degrades model accuracy and robustness. To address this issue, this paper proposes a thermal lag characteristics identification and thermal error modeling method based on permutation tests. First, sensor-specific thermal lag time is identified through permutation-based importance evaluation, where multiple temperature sequences with lagged characteristics are processed by random forest models. Then, temperature variables are optimized through dual-criteria selection, retaining only the sensors that complies with physical constraints, followed by statistically insignificant variables being eliminated via Pearson correlation analysis. Finally, a Gated Recurrent Units (GRU) and Random Forest (RF) hybrid thermal error modeling is implemented through forward-looking temperature sequence augmentation, followed by random forest training and five-fold cross-validation. The robustness and efficiency of the proposed method has been validated through a compensation experiment of the thermal error of a spindle of a machining center. The experiment results show that under the random speed conditions of spindle, the thermal error was reduced by 55% after compensation with only 3 temperature sensors. The proposed method holds significant potential for application in thermal error modeling and compensation of other machine tool components.

1 Introduction

Thermal error is one of the key factors affecting the accuracy of CNC machine tools. It is influenced by the coupling of multiple heat sources, resulting in dynamic and time-varying characteristics [1]. Additionally, thermal transfer delays and the limitations of temperature sensor placement lead to a hysteresis effect between temperature values and thermal error. As shown in Fig. 1, there is a varying degree of lag between

G. Ge · X. Wang · Z. Zeng · Z. Du (✉)
School of Mechanical Engineering, Shanghai Jiao Tong University, Shanghai, China
e-mail: zcdu@sjtu.edu.cn

© The Author(s) 2026
K. Wegener and M. Bambach (eds.), *4th International Conference on Thermal Issues in Machine Tools (ICTIMT2025)*, Lecture Notes in Production Engineering,
https://doi.org/10.1007/978-3-032-01194-7_39

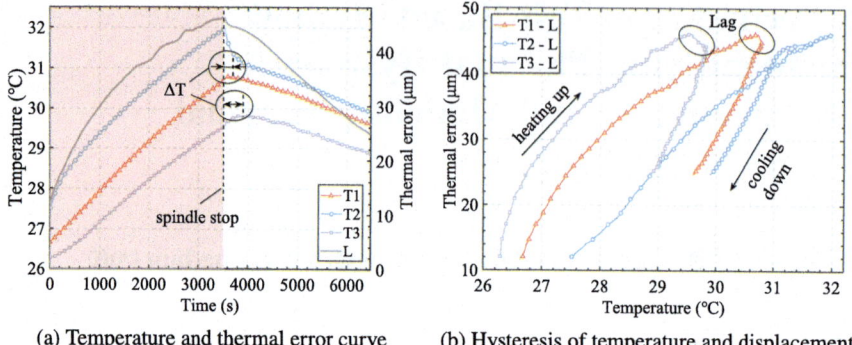

(a) Temperature and thermal error curve　　(b) Hysteresis of temperature and displacement

Fig. 1 Hysteresis effect between temperature and thermal error

the changes in temperature values at key measurement points and the thermal error, which poses significant challenges for the accurate modeling and suppression of thermal error.

As early as 2000s, Ni et al. [2] investigated the hysteresis between machine tool temperature values and thermal errors from the perspective of dynamic systems, and applied system identification theory to construct a dynamic thermal error model. Yang et al. [3] discovered early on that the lag time at different temperature points is caused by varying heat generation and conduction rates of different components. Tan et al. [4] addressed the nonlinear characteristics of hysteresis by constructing a time-varying analysis model between ambient temperature and thermal errors. In recent years, Li et al. [5] theoretically demonstrated the thermal hysteresis effect by analyzing the thermal characteristics of the spindle. Weng et al. [6] proposed a dual-layer analytical modeling method to characterize the thermal properties of structural components and analyzed the mapping relationship between different thermal condition parameters and thermal errors. Li et al. [7] established an analytical model to analyzed the hysteresis effect between spindle temperature and thermal errors.

In the field of data-driven thermal error modeling, time series models represented by long-short term memory networks (LSTM) have been widely applied in recent years due to their ability to mitigate the hysteresis effect to some extent [8]. Abdul-shahed et al. [9] combined a grey system model with an artificial neural network to develop a thermal error model for gantry five-axis machine tools, aiming to minimize the impact of the hysteresis effect. Li et al. [10] integrated static thermal errors with historical temperature sequence differences to establish a thermal error model that accounts for the hysteresis effect. Nico et al. [11] applied a self-learning method to automatically update the thermal error model based on the current thermal error and the historical thermal error sequence. Zeng et al. [12] designed a thermal error model based on LSTM and attention mechanisms to reflect the dependency of current thermal errors on historical thermal errors. Qin and Li et al. [13, 14] introduced GRU models to address the thermal hysteresis effect. Fu et al. [15] employed the mayfly algorithm to establish a spindle thermal error model under complex working conditions, achieving better prediction accuracy.

The purpose of thermal error modeling is to mitigate it, and in this regard, there are primarily two methods: active control and error compensation [16]. The active control method focuses on maintaining a constant temperature in key areas of machine tool components through heating or cooling techniques to minimize thermal expansion [17], while the error compensation method introduces a dynamic displacement equal in magnitude but opposite in direction to the original error to counteract thermal error [18]. Due to its cost-effectiveness and convenience, error compensation has been widely applied in compensating for geometric errors [19], thermal errors [20], force-induced deformation errors [21], and machining process errors [22]. Therefore, this paper also adopts the error compensation approach to suppress thermal errors.

In summary, there are mainly two methods for addressing the hysteresis characteristics between temperature values and thermal errors: mechanism modeling and data-driven analysis. However, mechanism models are complex and generally lack precision, while machine learning models suffer from poor interpretability. Therefore, this paper proposes an explicit identification method for thermal hysteresis characteristics based on permutation test analysis, selects temperature points based on physical constraints and correlation analysis, and finally employs a GRU-RF hybrid modeling approach to enhance the robustness and accuracy of thermal error modeling.

The remainder of this paper is structured as follows. Section 2 outlines the procedure of the proposed method. Section 3 describes the identification method for thermal hysteresis characteristics. Section 4 discusses the selection of temperature sensors and the thermal error modeling approach. Section 5 details the experimental setup and presents the results of the proposed method. Finally, Sect. 6 summarizes the findings and conclusions of this study.

2   Thermal Hysteresis Analysis and Method Procedure

This section first analyzes the primary influence mechanism of thermal hysteresis in machine tool, and subsequently outlines the overall procedure of the proposed method.

2.1   Mechanism Analysis of Thermal Hysteresis in Machine Tools

In the process of thermal error modeling for machine tools, the mechanism of thermal hysteresis can be attributed to two primary factors:

Thermal inertia The transfer of heat from the source to the temperature sensor is inherently time-dependent, particularly in systems with complex structures or multiple heat sources. The intricacy of the heat conduction path and the thermal resistance of the materials significantly contribute to this delay. Furthermore, the thermal inertia of the machine tool structural materials causes temperature changes to occur gradually rather than instantaneously, resulting in a lag between thermal displacement and temperature variations. The physical distance between the temperature sensor and the heat source also plays a critical role, as greater distances lead to longer heat transfer times.

Sensor and model limitations The limited number and suboptimal placement of temperature sensors may fail to adequately capture all heat sources and critical positions. Additionally, the response speed and accuracy of the sensors influence their ability to detect temperature variations effectively. In systems with multiple heat sources and complex structures, the nonlinearity and lag between temperature and thermal error are further amplified, making it difficult for simplistic models to accurately predict thermal behavior.

Overall, the lag of the temperature and thermal error in machine tool is mainly affected by the time delay of heat transfer, the placement of temperature sensors, and the complexity of the thermal error model. Among them, the time delay of heat transfer and the placement of temperature sensors are difficult to be further optimized, so it is necessary to focus on the optimization of the thermal error model, such as introducing time delay parameters or using dynamic models to improve the accuracy of thermal error prediction.

2.2 Procedure of the Proposed Method

This paper proposes a data-driven methodology for identifying temperature hysteresis characteristics and thermal error modeling, as illustrated in Fig. 2, which comprises three steps:

Step 1: Thermal hysteresis characteristic identification based on permutation-based importance analysis. First, an RF model is used with multi-sensor temperature sequences as input and thermal displacement as the label. Then, time series data

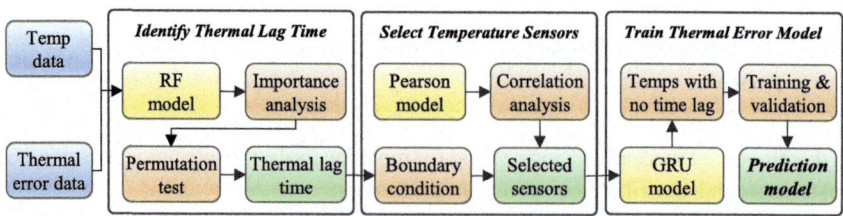

Fig. 2 Procedure of the proposed method

with lagged characteristics are constructed for each temperature sensor and used as input features to build the RF model. Subsequently, a permutation test is employed to evaluate the importance of constructed sequences for each temperature series. Finally, for each temperature sensor, the lagged time that maximizes its importance score is determined as the optimal feature lag parameter.

Step 2: Temperature sensor filtering based on physical constraint and correlation analysis. After determining the optimal lag time for each sensor, correlation analysis and physical constraint are jointly used to filter and optimize the temperature dataset. First, by combining prior knowledge of machine tool thermal characteristics, temperature sensors with optimal lag times exceeding physical limits are excluded to ensure the retained sensors meet engineering feasibility. Then, Pearson correlation analysis is performed on the optimized temperature dataset and thermal displacement data to remove weakly correlated sensors, thereby reducing sensor count and enhancing model robustness.

Step 3: GRU-RF hybrid thermal error modeling. First, temperature and thermal error datasets are generated based on the identified optimal lag times for training the random forest model. Then, to eliminate lag effects, GRU is used for forward-looking sequence prediction of the screened temperature variables, and the predicted sequence is concatenated with real-time temperature data to generate augmented temperature data. Finally, the model accuracy is validated using five-fold cross-validation.

The proposed approach explicitly characterizes time lag relationships between temperature and thermal error via permutation testing and sensor selection, enabling high-precision thermal error modeling with a reduced sensor configuration. The theoretical and technical details of each key step will be introduced subsequently.

3 Thermal Lag Time Identification Based on Permutation Test

3.1 Construction of Temperature Sequences with Lagged Characteristics

To accurately characterize the lag time between temperature values and thermal errors, it is necessary to construct temperature value matrices with lagged characteristics. Assuming the temperature value vector collected by the jth sensor is T_j, and the corresponding thermal error vectors in the three directions are $\delta_{j,x}$, $\delta_{j,y}$, and $\delta_{j,z}$, respectively. The raw data collected by all temperature sensors can be represented as $T \in \mathbb{R}^{n \times m}$, and the thermal deformation displacement labels in the three directions can be expressed as $\delta \in \mathbb{R}^{n \times 3}$. Here, n represents the number of time sampling points, and m represents the number of temperature sensors. Based on the thermal response characteristics of the machine tool, the maximum hysteresis time $\tau_{\max}$ and the step size $\Delta\tau$ are defined to generate a candidate dataset of lag time:

$$\tau_{candidates} = \{\tau_k | \tau_k = k \cdot \Delta\tau, k = 1, 2, ..., K\}, \quad K = \tau_{max}/\Delta\tau \tag{1}$$

where K is the maximum lag time determined by the physical characteristics of the machine tool. For the jth sensor, the feature vector shifted by τ is defined as:

$$T_j^{(\tau)} = \left[T_{\tau+1,j}, \ T_{\tau+2,j}, \ ..., \ T_{\tau+n,j}\right]^\mathsf{T} \in \mathbb{R}^{n-\tau} \tag{2}$$

Here, the corresponding thermal error vector is truncated as:

$$\delta^{(\tau)} = [\delta_1, \ \delta_2, \ ..., \ \delta_{n-\tau}]^\mathsf{T} \in \mathbb{R}^{(n-\tau)\times 3} \tag{3}$$

When evaluating the lag time of the jth sensor, the lag time of other sensors are set to zero. Subsequently, a composite feature matrix is constructed as:

$$\tilde{T}^{(j,\tau)} = [T_1^{(0)} \mid ... \mid T_j^{(\tau)} \mid ... \mid T_m^{(0)}] \in \mathbb{R}^{(n-\tau)\times m} \tag{4}$$

where $T_j^{(0)}$ is the original temperature sequence of the jth sensor.

3.2 Importance Evaluation of Temperature Variables

This study employs the random forest algorithm to construct a thermal error model, primarily considering its two advantages: first, it enhances the generalization capability for small sample data by integrating multiple decision trees, making it suitable for thermal error modeling under small sample conditions; second, the algorithm simultaneously achieves error prediction and feature importance evaluation based on out-of-bag (OOB) estimation, quantifying the contribution of temperature measurement points to the error and providing a basis for subsequent identification of lagging features. The algorithm and applications of random forests are very extensive, and their technical details can be found in the Ref. [23].

The proposed method adopts a feature importance calculation approach based on the OOB sample permutation test to assess the significance of each temperature sequence. For the bth decision tree in the random forest, its OOB sample set $\mathcal{O}_b$ consists of approximately one-third of the original dataset that was excluded from the training of that specific tree. These samples naturally serve as a validation set, thereby eliminating the need for additional data partitioning. During the feature importance calculation, the baseline error of the feature dataset e_b^{orig} is first determined by computing the original error of the OOB samples for each decision tree:

$$e_b^{orig} = \frac{1}{|\mathcal{O}_b|} \sum_{i\in\mathcal{O}_b} \left(y_i - \widehat{y_i^{(b)}}\right)^2 \tag{5}$$

where $\widehat{y_i^{(b)}}$ is the predicted value of the ith sample by the lth tree.

The values of the feature are randomly permuted to disrupt its association with the labels, thereby generating a perturbed out-of-bag sample dataset and recalculating the prediction error $e_b^{(j)}$. The importance score of the feature is defined as the average error increment across all trees:

$$I_j = \frac{1}{B} \sum_{b=1}^{B} \left(e_b^{(j)} - e_b^{\text{orig}} \right) \tag{6}$$

where B denotes the total number of trees in the random forest model.

Additionally, for the temperature data $\boldsymbol{T}_j^{(\tau)}$ from each temperature sensor, the importance value $I_j^{(\tau)}$ is iteratively calculated for the dataset corresponding to each candidate time $\tau \in \tau_{\text{candidates}}$. For each temperature sensor, the lag time that maximizes its importance is determined as:

$$\tau_j^* = \arg \max_{\tau \in \tau_{\text{candidates}}} I_j^{(\tau)} \tag{7}$$

where arg max represents the maximum value of the function.

Using the optimal lag time for each temperature feature, the optimized composite temperature feature matrix $\tilde{\boldsymbol{T}}^{(\tau^*)}$ is constructed. Figure 3 presents a heatmap illustrating the variations in feature importance and mean squared error (MSE) for each temperature feature under different lag times during a typical permutation test. The results indicate that the trends of feature importance and MSE are generally consistent, showing the fact that this method can effectively identify the optimal lag time for each temperature feature.

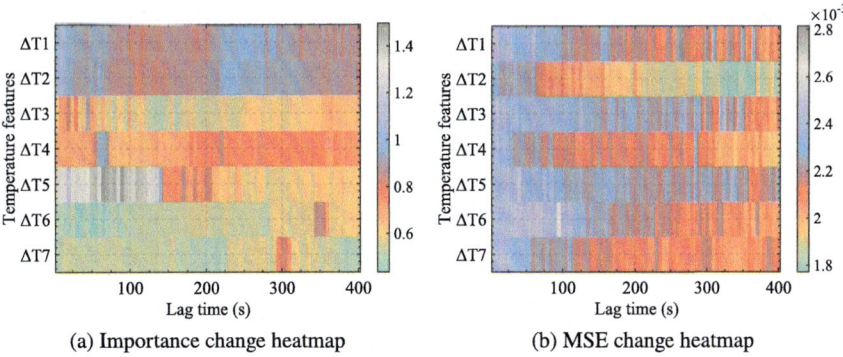

(a) Importance change heatmap   (b) MSE change heatmap

Fig. 3 Heatmaps of feature importance and MSE under different lag times

4   Thermal Error Modeling Using a GRU-RF Hybrid Approach

4.1   Selection of Temperature Sensors

For each temperature point, the feasible domain of sensor lag time is defined based on the physical characteristics of the machine tool. Assuming the thermal conduction time constant from the core heat source to each temperature point is $\tau_{j,\max}$, the condition for filtering the physically feasible sensor set is:

$$\tau_j^* \leq \tau_{j,\max} \tag{8}$$

This constraint eliminates non-physical lag parameters caused by measurement noise or abnormal operating conditions, retaining only the sensor set that complies with physical constraints. Figure 4 illustrates the relationship between the identified optimal lag time and the physical constraints, where the lag times for $\Delta T4$, $\Delta T6$, and $\Delta T7$ significantly exceed the physical constraints and thus need to be excluded.

In addition, for the temperature features filtered through physical constraints, further analysis is conducted to calculate the Pearson correlation coefficient between each temperature feature $T_j^{(\tau_j^*)}$ and the thermal error δ, as well as the multicollinearity between each temperature feature [24]. This process is used to filter out redundant temperature variables and obtain the final composite temperature feature matrix $\tilde{T}_{\text{final}}^{(\tau^*)}$, which will be used for the subsequent training of the thermal error model.

Fig. 4 Physical constraints of temperature sensors

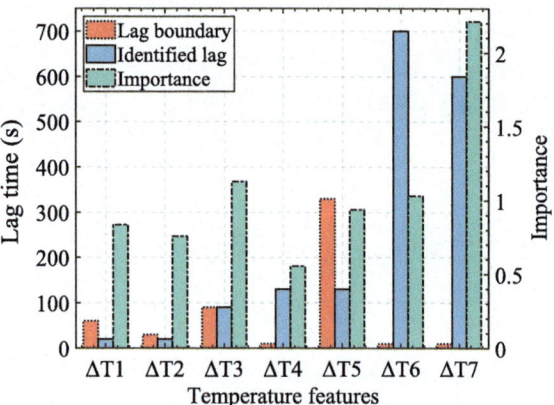

4.2 GRU-RF Hybrid Modeling of Thermal Error

Since the random forest model can simultaneously perform importance analysis and error prediction, the final composite temperature feature matrix $\widetilde{T}_{\text{final}}^{(\tau^*)}$ obtained can be used as input, and the thermal error vector $\delta^{(\tau)}$ can be used as the label to train the model for thermal error prediction and compensation. However, this process implicitly assumes that, during actual thermal error prediction and compensation, temperature data at each temperature point leading by τ_j^* time can be obtained, while it is impossible to collect future data. Therefore, a GRU model is first used to extend and forward looking the collected temperature data. The extended temperature data is then input into the trained random forest model, enabling real-time high-precision prediction and compensation of thermal errors. During the training process of the GRU model, for any time step $\tau \in [n, N_j - \tau_j^*]$ of the jth group of temperature data, the input feature matrix is composed of multiple datasets from the previous n time steps, along with the spindle speed datasets.

$$
X_j^{(\tau)} = \begin{bmatrix} T_j^{(\tau-n+1)} & S^{(\tau-n+1)} \\ T_j^{(\tau-n+2)} & S^{(\tau-n+2)} \\ \vdots & \vdots \\ T_j^{(\tau)} & S^{(\tau)} \end{bmatrix} \in \mathbb{R}^{n \times 2} \tag{9}
$$

where $X_j^{(\tau)}$ represents the temperature vector of sensor j at time τ, and $S^{(\tau)}$ represents the spindle speed vector at time τ. The label feature matrix $y_j^{(\tau)}$ corresponds to the temperature value after a future time interval of τ_j^*:

$$
y_j^{(\tau)} = T_j^{(\tau+\tau_j^*/\Delta\tau)} \in \mathbb{R} \tag{10}
$$

Considering the differences in sensor lag times, $\tau_j^*/\Delta\tau$ in the equation needs to be rounded to the nearest integer. Finally, during the actual error prediction and compensation process, the predicted temperature values of each sensor, which are the temperature values with optimal lag characteristics, can be input into the random forest model to predict the thermal error value $\delta^{(\tau)}$ in real time. Figure 5 shows the extended prediction results of a typical GRU model for three sets of temperature values. It can be observed that, compared to Fig. 1, the variation trends of the three sets of $\Delta T1$ and $\Delta T3$ in Fig. 5 are generally consistent with the start and stop of spindle, effectively eliminating their hysteresis effect.

As for $\Delta T2$, there is almost no time delay between the start and stop of spindle, and the temperature value is predicted to be the same as the original value. Furthermore, $\Delta T2$ exhibits a faster cooling rate after spindle shutdown, suggesting distinct thermal responses during heating and cooling phases. However, since this study primarily investigates thermal hysteresis, this difference in response is not considered.

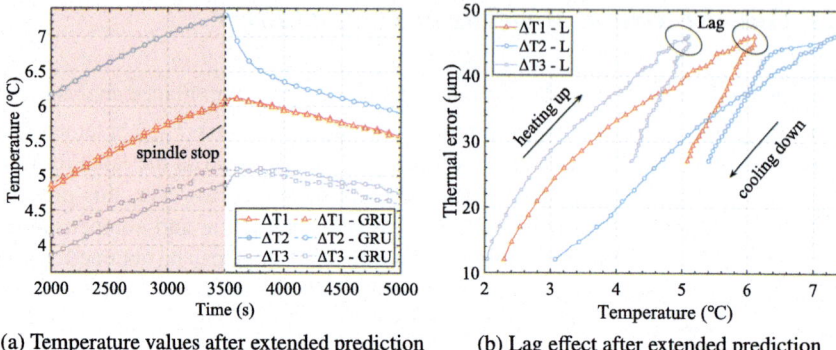

(a) Temperature values after extended prediction   (b) Lag effect after extended prediction

Fig. 5 GRU-based temperature data extension

Table 1 Temperature sensor locations in the experiment

No.	Location	No.	Location
T0	Ambient temperature	T4	Front part of spindle box
T1	Front bearing housing (lower)	T5	Middle part of spindle box
T2	Front bearing housing (upper)	T6	Bottom of spindle box
T3	Rear bearing housing (end)	T7	Rear part of spindle box

5 Experiment Validation

5.1 Experiment Setup

To validate the effectiveness and robustness of the proposed method, experiments are conducted using the external-driven spindle of an SMTCL VMC 850E vertical machining center. The experimental setup is shown in Fig. 6. The machining center is equipped with an external-driven spindle capable of reaching a maximum speed of 8000 rpm and a FANUC 0i NC system. Temperature data are collected using eight PT100 sensors, each with a measurement uncertainty of 0.05 °C, with their placement points detailed in Table 1. Spindle thermal error measurements are performed in accordance with the ISO 230-3 standard [25], utilizing five Omron ZX-ED02T eddy current displacement sensors for a five-point test, each with a measurement uncertainty of 0.8 μm. Temperature and displacement data were acquired using an NI USB-6421 data acquisition card. Real-time thermal error compensation was implemented through a previously developed compensation system, the core principles of which are outlined in our prior work [18]. The compensation system communicates with the FANUC NC system in real-time via Ethernet using the (FANUC Open

Fig. 6 Experiment setup

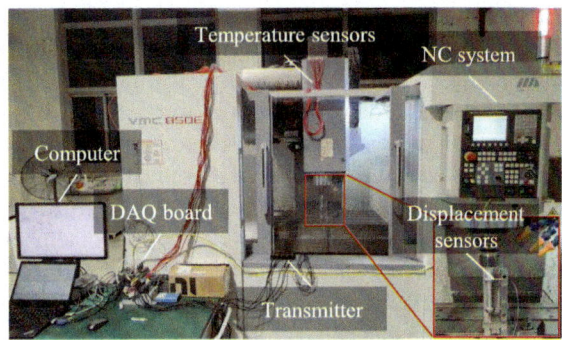

Communication Application Specification) FOCAS protocol, continuously updating compensation values with a communication cycle of about 40 ms to ensure real-time thermal error compensation.

The verification experiment mainly consists of three steps:

First, beginning from a fully cooled state, the spindle is operated under random speed conditions while synchronously collecting temperature data from eight sensors, spindle speed data, and thermal error data in the XYZ directions. These data serve as training and validation sets for subsequent temperature thermal hysteresis feature identification and thermal error modeling. The random speed condition is maintained for approximately 65,000 s, and the collected spindle speed, temperature values from eight sensors, and spindle thermal error are illustrated in Fig. 7. Notably, under this condition, the temperature trends (excluding ambient temperature) exhibit similarities and demonstrate varying degrees of lag. Among the thermal errors, the Z-direction thermal elongation is the most pronounced, followed by the Y-direction, while the X-direction is negligible.

Consequently, only the Z-direction thermal error is considered for temperature hysteresis feature identification, thermal error modeling, and compensation. Second,

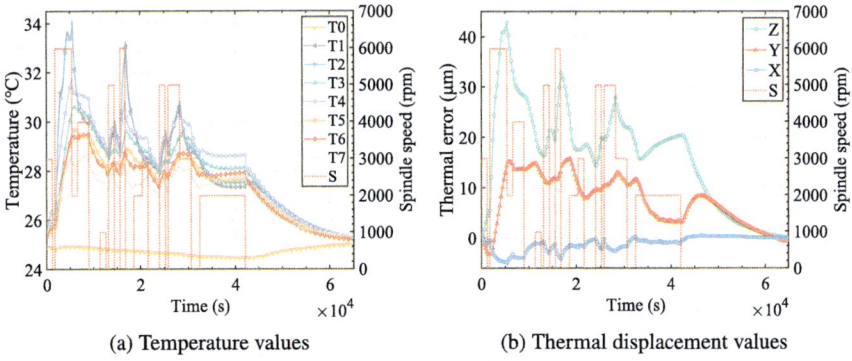

(a) Temperature values   (b) Thermal displacement values

Fig. 7 Temperature and thermal error values in training dataset

the collected training data is input into the proposed method, following the procedures detailed in Sect. 3 for identifying temperature hysteresis features, selecting temperature sensors, and modeling thermal error. Given the small sample size of the experimental data, the random forest model is configured with 60 trees and a minimum of 10 leaves per tree. Feature importance calculation and model training are conducted using five-fold cross-validation. To account for the influence of room temperature variations on the machine tool thermal characteristics, the gradient values $\Delta T1 \sim \Delta T7$ between $T1 \sim T7$ and room temperature are used as inputs, with their lag feature physical constraints set to [60, 30, 90, 10, 330, 10, 10] seconds.

Third, the established GRU-RF hybrid thermal error model is employed to perform real-time prediction and compensation of spindle thermal error. The correctness and practicality of the proposed method are validated by comparing thermal error values before and after compensation.

5.2   Results and Discussion

Using a time step of $\tau = 10s$, the seven optimal time lags identified for T1 to T7 are [20, 20, 90, 130, 130, 700, 600] s. Through physical constraints and correlation analysis, two temperature points, T1 and T3, were optimized and selected, in addition to the ambient temperature T0. Figure 8 illustrates the training and validation data for the three selected temperature sensors. It can be observed that the predicted trends of T1 and T3 using GRU for time-ahead prediction are consistent with the spindle speed, effectively eliminating the lag. Furthermore, the trends of T1 and T3 do not converge, indicating a low correlation between them.

The training datasets of three sets of temperatures T0, T1, and T3 (with T1 and T3 predicted in advance by GRU), along with the corresponding thermal error training dataset shown in Fig. 9a, were input into the RF model. The model was then validated using the validation dataset illustrated in Fig. 9b. During the model training process,

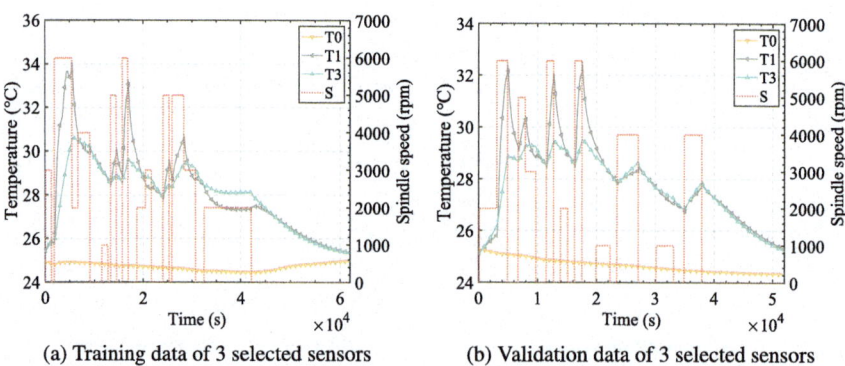

(a) Training data of 3 selected sensors   (b) Validation data of 3 selected sensors

Fig. 8 Temperature data of 3 selected sensors

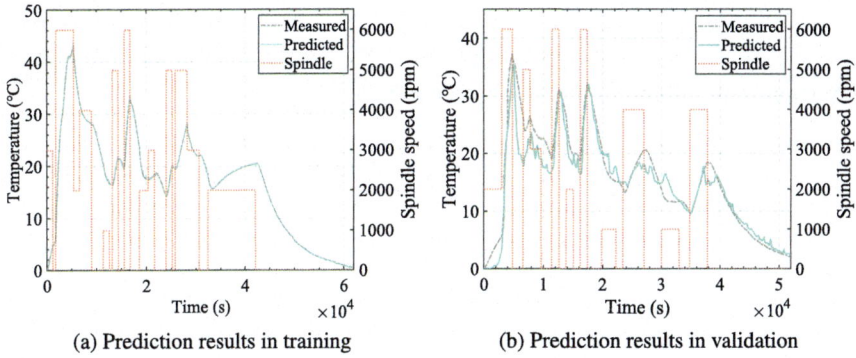

Fig. 9 Prediction results of thermal error in training and validation process

the RMSE and R-Squared values were 0.1204 and 0.9801, respectively. While on the validation process, the RMSE and R-Squared values were 3.0214 and 0.8784, respectively.

The established GRU-RF hybrid prediction model was applied to predict and compensate for spindle thermal errors in real time. The compensation curve and the compensated thermal errors are shown in Fig. 10. It can be observed that the entire random speed condition lasted approximately 55,000 s, and the random speed condition used in the validation experiment was entirely different from that in the training data. Under this condition, the spindle thermal error range without compensation was [0, 34.1] μm, while the compensated spindle thermal error range was [−4.15, 11.2] μm, representing a reduction of approximately 55.0%. The above result fully demonstrates the effectiveness and robustness of the proposed method in this study.

Fig. 10 Compensation results of the spindle thermal error

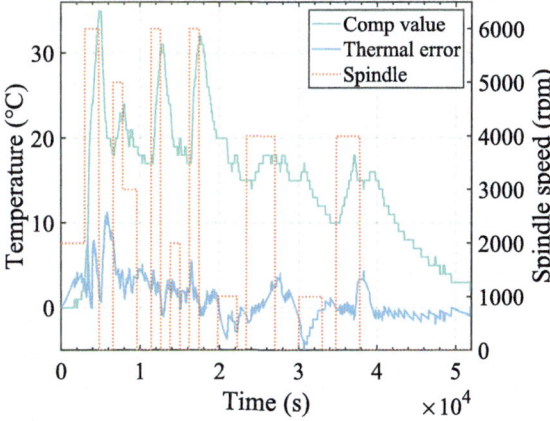

6 Conclusions

The main contributions of this study are concluded as follows:

(1) This study proposed a novel thermal error modeling method that effectively addresses hysteresis effects in machine tools using permutation-based lag time identification and hybrid GRU-RF modeling. The approach enhances model accuracy and robustness by mitigating hysteresis effects and optimizing temperature sensors selection.
(2) Compared with conventional temperature selection methods, the proposed approach explicitly identifies and compensates for hysteresis effects across all temperature sensors, resulting in more stable and reliable selections.
(3) Experimental results demonstrated that the proposed method reduced spindle thermal error by 55% under random speed conditions, using only three temperature sensors. This highlights the efficiency and practicality of the method in achieving significant error compensation with minimal sensor requirements.

The integration of statistical and machine learning techniques, such as permutation tests and GRU-RF models, provides a robust framework for thermal error modeling. The proposed method can be extended to other machine tool components, offering a scalable solution for improving machining accuracy.

Acknowledgements This research was supported by the National Natural Science Foundation of China (Grant No. 52375504, 52405556), the China Postdoctoral Science Foundation (Grant No. 2024M751963, GZC20241006) and the Open Research Funding of Key Laboratory for High-end CNC Machine Tools of China General Technology Group (Grant No. KLHCMT202408).

References

1. Mayr, J., Jedrzejewski, J., Uhlmann, E., Alkan Donmez, M., Knapp, W., Härtig, F., Wendt, K., Moriwaki, T., Shore, P., Schmitt, R., Brecher, C., Würz, T, Wegener, K.: CIRP Ann. **61**(2), 771 (2012). 10.1016/j.cirp.2012.05.008
2. Yang, H., Ni, J.: J. Manuf. Sci. Eng. **125**(2), 245 (2003). https://doi.org/10.1115/1.1557296
3. Haitao, Z., Jianguo, Y., Jinhua, S.: Int. J. Mach. Tools Manuf **47**(6), 1003 (2007). https://doi.org/10.1016/j.ijmachtools.2006.06.018
4. Tan, B., Mao, X., Liu, H., Li, B., He, S., Peng, F., Yin, L.: Int. J. Mach. Tools Manuf **82–83**, 11 (2014). https://doi.org/10.1016/j.ijmachtools.2014.03.002
5. Li, G., Wang, Z., Li, Z., Xu, K., Chen, X.: Int. J. Adv. Manuf. Technol. **126**(11–12), 5107 (2023). https://doi.org/10.1007/s00170-023-11471-5
6. Weng, L., Gao, W., Zhang, D., Huang, T.: Int. J. Heat Mass Transf. **221**, 125083 (2024). https://doi.org/10.1016/j.ijheatmasstransfer.2023.125083
7. Li, Y., Wu, K., Wang, N., Wang, Z., Li, W., Lei, M.: Heat Mass Transf. **60**(10), 1755 (2024). https://doi.org/10.1007/s00231-024-03519-3
8. Wang, Y., Cao, Y., Qu, X., Wang, M., Wang, Y., Zhang, C.: Measurement **243**, 116341 (2025). https://doi.org/10.1016/j.measurement.2024.116341
9. Abdulshahed, A.M., Longstaff, A.P., Fletcher, S., Potdar, A.: J. Manuf. Syst. **41**, 130 (2016). https://doi.org/10.1016/j.jmsy.2016.08.006

10. Li, F., Li, T., Jiang, Y., Wang, H., Ehmann, K.F.: J. Manuf. Process. **48**, 320 (2019). https://doi.org/10.1016/j.jmapro.2019.10.018
11. Zimmermann, N., Breu, M., Mayr, J., Wegener, K.: CIRP Ann. **70**(1), 431 (2021). https://doi.org/10.1016/j.cirp.2021.04.029
12. Zeng, S., Ma, C., Liu, J., Li, M., Gui, H.: Appl. Soft Comput. **138**, 110221 (2023). https://doi.org/10.1016/j.asoc.2023.110221
13. Qin, Q., Li, L., Zhao, G., Li, Z.: Precis. Eng. **83**, 159 (2023). https://doi.org/10.1016/j.precisioneng.2023.06.002
14. Li, Y., Bai, Y., Hou, Z., Nie, Z., Zhang, H.: Int. J. Adv. Manuf. Technol. **128**(11), 5519 (2023). https://doi.org/10.1007/s00170-023-12276-2
15. Fu, G., Mu, S., Zheng, Y., Lu, C., Wang, X., Wang, T.: Measurement **226**, 114183 (2024). https://doi.org/10.1016/j.measurement.2024.114183
16. Fujishima, M., Narimatsu, K., Irino, N., Ido, Y.: CIRP J. Manuf. Sci. Technol. **22**, 111 (2018). https://doi.org/10.1016/j.cirpj.2018.04.003
17. Zheng, Y., Weng, L., Gao, W., Fu, Z., Zhao, Z., Qi, J., Shi, K., Zhang, D., Huang, T.: Appl. Therm. Eng. **268**, 125911 (2025). https://doi.org/10.1016/j.applthermaleng.2025.125911
18. Ge, G., Xiao, Y., Feng, X., Du, Z.: Comput. Aided Des. **152**, 103401 (2022). https://doi.org/10.1016/j.cad.2022.103401
19. Xiao, Y., Ge, G., Deng, M., Lv, J., Du, Z.: Manuf. Lett. **41**, 31 (2024). https://doi.org/10.1016/j.mfglet.2024.09.007
20. Liu, P., Du, Z., Li, H., Deng, M., Feng, X., Yang, J.: Adv. Manuf. **9**(2), 235 (2021). https://doi.org/10.1007/s40436-020-00342-x
21. Ge, G., Xiao, Y., Du, Z.: Int. J. Adv. Manuf. Technol. **127**(5–6), 2465 (2023). https://doi.org/10.1007/s00170-023-11631-7
22. Zhu, M., Ge, G., Feng, X., Du, Z., Yang, J.: J. Manuf. Sci. Eng., Trans. ASME **143**(6), 1 (2021). https://doi.org/10.1115/1.4049036
23. Breiman, L.: Mach. Learn. **45**(1), 5 (2001). https://doi.org/10.1023/A:1010933404324
24. Winter, J.C.F., Gosling, S.D., Potter, J.: Psychol. Methods **21**(3), 273 (2016). https://doi.org/10.1037/met0000079
25. ISO Central Secretary. Test code for machine tools part 3: Determination of thermal effects (2020)

Author Index

© The Editor(s) (if applicable) and The Author(s) 2026
K. Wegener and M. Bambach (eds.), *4th International Conference on Thermal Issues in Machine Tools (ICTIMT2025)*, Lecture Notes in Production Engineering,
https://doi.org/10.1007/978-3-032-01194-7

The manufacturer's authorised representative in the EU is Springer
Nature Customer Service Centre GmbH, Europaplatz 3, 69115 Heidelberg,
Germany. If you have any concerns regarding our products, please
contact ProductSafety@springernature.com

Printed and bound by CPI Group (UK) Ltd, Croydon, CR0 4YY

23/04/2026

02095586-0005